Mycology in a Changing Planet: Applications and Perspectives from Southeast Asia

Jonathan Jaime G. Guerrero
Marian P. De Leon
Rohani B. Cena-Navarro
Kin Israel R. Notarte
Mark Angelo O. Balendres
Thomas Edison E. dela Cruz
Editors

Mycology in a Changing Planet: Applications and Perspectives from Southeast Asia

Springer

Editors
Jonathan Jaime G. Guerrero
College of Medicine
University of the Philippines Manila
Manila, Philippines

Marian P. De Leon
Museum of Natural History
University of the Philippines Los Baños
Los Baños, Laguna, Philippines

Rohani B. Cena-Navarro
National Institute of Molecular Biology
and Biotechnology
National Institutes of Health
University of the Philippines Manila
Manila, Philippines

Kin Israel R. Notarte
School of Medicine
Johns Hopkins University
Baltimore, Maryland, USA

Mark Angelo O. Balendres
Department of Biology
De La Salle University
Manila, Philippines

Thomas Edison E. dela Cruz
Department of Biological Sciences
College of Science
University of Santo Tomas
Manila, Philippines

ISBN 978-981-95-2639-0 ISBN 978-981-95-2640-6 (eBook)
https://doi.org/10.1007/978-981-95-2640-6

This Springer imprint is published by the registered company Springer Nature Singapore Pte Ltd.
The registered company address is: 152 Beach Road, #21-01/04 Gateway East, Singapore 189721, Singapore

Preface

Fungi play an essential role in our search for solutions to the complex challenges of the modern world. These remarkable organisms have given rise to natural products that address both long-standing and emerging issues in medicine, agriculture, and environmental sustainability. At the same time, fungi themselves present new challenges—spurring scientific inquiry, policy development, and economic innovation to create adaptive strategies and lasting solutions. This dynamic relationship is especially evident in Southeast Asia.

As one of the most socioculturally diverse and economically vibrant regions in the world, Southeast Asia is at the forefront of fungal research and application in the region. For centuries, fungi have been embedded in local cuisine, traditional medicine, and cultural practices. Today, they are increasingly harnessed to produce new medicines, tackle agricultural pests and diseases, and support sustainable development. Moreover, a deeper understanding of fungi is proving vital for ecosystem conservation, cultural heritage preservation, education, public health, and regional economic growth. The immense fungal diversity in Southeast Asia holds vast potential for future discoveries and innovations.

This book addresses two fundamental questions: *How are fungi helping to improve our world?* And *what challenges do fungi present in the face of a rapidly changing planet?* These questions frame the dual role of fungi—both as agents of progress and as sources of concern—in communities across Southeast Asia, and, by extension, the global stage.

Unlike traditional texts that focus on the basic biology of fungi, this book provides an up-to-date reference on their practical applications in an era of environmental and societal transformation. It brings together current literature on the diverse uses of fungi in medicine, biotechnology, nanotechnology, and food science. In addition, it explores emerging topics in conservation—such as the mycological significance of unexplored cave and marine ecosystems—as well as the preservation of cultural heritage sites and artifacts throughout Southeast Asia. The book also examines new developments in agricultural mycology, a particularly relevant focus in a region where food security remains a central concern. Finally, it addresses critical issues of biosafety and biosecurity, both in laboratory settings and in broader societal contexts.

By offering this comprehensive perspective, the book aims to serve as a modern and practical reference for scientists, policymakers, educators, and anyone with an interest in the evolving relationship between fungi and our changing world.

Manila, Philippines — Jonathan Jaime G. Guerrero
Los Baños, Laguna, Philippines — Marian P. De Leon
Manila, Philippines — Rohani B. Cena-Navarro
Baltimore, Maryland, USA — Kin Israel R. Notarte
Manila, Philippines — Mark Angelo O. Balendres
Manila, Philippines — Thomas Edison E. dela Cruz

Contents

About the Editors

Jonathan Jaime G. Guerrero is currently a medical and graduate student at the University of the Philippines Manila College of Medicine and College of Public Health. He has a bachelor's degree in Biology and a master's degree in Plant Pathology with Soil Science cognate, both of which provided focused training in Mycology. He is also a certified Biorisk Officer since 2015. He serves as editor and reviewer of various journals, and has authored multiple journal articles and book chapters. He previously worked as a full time instructor and researcher, handling biology, microbiology, and environmental science subjects, and research projects on fungal diversity, natural products, and medical and agricultural applications.

Marian P. De Leon is the Director of the Museum of Natural History (MNH) and concurrent Assistant to the Vice Chancellor for Research and Extension of the University of the Philippines Los Baños. She holds the position of University Researcher and appointed as Curator of the UPLB MNH Microbial Culture Collection. She pioneered research on Cave Microbiology in the Philippines, where she and her team explored Philippine Caves for pathogenic, novel, and potential microorganisms. She served as the Program Leader of the Niche Center for Research and Development in the Region (NICER) Center for Cave Ecosystems Research (CAVES) (2022–2025) and published several scientific papers on bat gut and guano microbiome and functional microorganisms from Philippine Caves. She is one of the Certified Biosafety Officers (CBO) in the Philippines given by the National Training Center for Biosafety and Biosecurity and the US Department of State Biosecurity Engagement Program. As CBO, she was given an appointment to serve as Biosafety Officer and Scientist Member of the UPLB Institutional Biosafety Committee. Her contributions in Microbiology have been recognized by the Philippine Academy of Microbiology, Inc. and certified her as a Registered Microbiologist. She is an active member and served in different capacities in several scientific and professional organizations, including the National Research Council of the Philippines, Philippine Society for Microbiology, Inc., Philippine Network of Microbial Culture Collection, Inc., and Mycological Society of the Philippines. Currently, she is the Treasurer of the Biology Teachers Association, Inc. (BIOTA) and the Philippine Probiotics and Functional Foods, Inc. (PhiProFF). She is happily married to Dr. Mario P. De Leon and blessed with two kids, Matthew Dominic and Madeline Dorothy.

Rohani B. Cena-Navarro is a veterinarian and Assistant Professor at the National Institute of Molecular Biology and Biotechnology, National Institutes of Health-University of the Philippines Manila, and Director at the National Center for Biosafety and Biosecurity (NTCBB), NIH-UP Manila. She earned her Doctor of Veterinary Medicine from the University of the Philippines, Los Banos, in 2010. After graduation, she pursued her MS graduate studies in Veterinary Medicine, studying pathology of viral diseases at the Laboratory of Animal Diseases, Chonnam National University, South Korea. After joining UP-NIH in 2014, she became the Chair of the Institutional Animal Care and Use Committee (IACUC). She also served as the Chair of the UP Manila Institutional Biosafety and Biosecuriy Committee (IBBC) from 2019 to 2022. She received her certification as a Biosafety Officer under the Philippine Advanced Biorisk Officers Training in 2015 and is also a Certified Professional by the International Federation of Biosafety Associations. As an advocate of biosafety, she led several capacity-building projects in promoting biosafety and biosecurity awareness in the Philippines through various local and international partners. In 2019, she received a scholarship to represent the Philippines in the Biosecurity Diplomacy Fellowship for Young Scientists from the Global South funded by the United Nations Office for Disarmament Affairs (UNODA), Geneva. She is the Course Director of the DOH-Accredited Online Biosafety Education and Awareness Training (BEAT), a virtual training program dedicated to laboratorians handling SARS-CoV as part of NCBB's response to COVID-19.

Kin Israel R. Notarte holds a Doctor of Medicine and a Master of Science in Microbiology, both from the University of Santo Tomas in the Philippines. He is currently a PhD candidate in Pathobiology (Microbiology and Infectious Disease) at the Johns Hopkins University School of Medicine in the United States. With over 60 publications, his research focuses on infectious diseases, immunology, and vaccinology. He also serves as a reviewer for more than 50 journals, including *The Lancet* and *The New England Journal of Medicine*, and contributes as an editor for *Frontiers in Medicine* and *Frontiers in Immunology*. An accomplished scientist, he was selected as a Bayer Fellow for the 72nd Lindau Nobel Laureate Meeting in Medicine and Physiology in Germany, a fellow of the American Society for Microbiology Future Leaders Mentorship Program, and the top prize awardee for Excellence in Clinical Research from the Department of Pathology at the Johns Hopkins School of Medicine. In recognition of his impactful work, his first-author paper was listed among the hot papers of 2024 in the *Journal of Clinical Medicine*, a Q1 journal in the field of medicine. He also received training in chemical biology at Waseda University in Japan and in molecular mycology at Seoul National University in Korea. In his spare time, he enjoys traveling, watching anime, and reading fiction.

Mark Angelo O. Balendres is a plant pathologist and a Professor at the Department of Biology of De La Salle University. He earned his PhD in Agricultural Science (Plant Pathology) from the University of Tasmania, Australia. His research focuses on plant disease diagnostics, host-pathogen interactions, and the sustainable management of plant diseases, especially those caused by fungi. A committed mentor, he leads the DLSU Plant and Soil Health Research Unit, trains the next generation of scientists, leads capacity-building efforts, and serves as a speaker in plant pathology, scientific writing, and research leadership. His work bridges fundamental science with practical applications to improve crop protection and support smallholder farmers in the Philippines and Southeast Asia. He is an active member of the Global Young Academy, the Australasian Plant Pathology Society, and the Philippine Phytopathological Society, Inc.

Thomas Edison E. dela Cruz is Professor of Microbiology at the Department of Biological Sciences, College of Science and the Graduate School of the University of Santo Tomas in Manila, Philippines. He serves as group leader of the Fungal Biodiversity, Ecogenomics and Systematics-Metabolomics (FBeS) group at the Research Center for the Natural and Applied Sciences in UST. His research focuses on the biodiscovery of fungi, particularly endophytic and endolichenic fungi, and myxomycetes in underexplored microhabitats in tropical ecosystems through a polyphasic approach of morphocultural, physiological, and genomic methods and the understanding of their biodiversity, ecological patterns, and interactions in response to changing landscapes. He completed his Doctor of Natural Sciences (Dr. rer. nat.) at the Technical University Braunschweig in Germany and his BSc Microbiology and MSc Biological Sciences degrees at the University of Santo Tomas. He holds a Diploma in Science Teaching and Diploma in Research and Development Management from the University of the Philippines—Open University. He serves as Philippine representative to the Asian Mycological Association and is a fellow of the Philippine Academy of Microbiology, the Indian Mycological Society, and the Global Young Academy. He served as a member of the Executive Committee of the TWAS Young Affiliate Network and the Global Young Academy. He enjoys visiting places with his wife, Jeane, and his three daughters, Tiffany Jillian, Tauriel Jasmine, and Tiana Jordyn.

Part I
Changing Mycological Perspectives

Chapter 1
Fungi and Indigenous Knowledge: Reports from Southeast Asia

Christian Lloyd R. Evangelista, Angeles M. De Leon, and Thomas Edison E. dela Cruz

Abstract Ethnomycology studies how people, particularly indigenous cultures, engage with fungi, including their use for food, medicine, crafts, and cultural rituals and beliefs. This field investigates the traditional knowledge that indigenous cultures have accumulated and passed down over generations, including their distinct methods of identifying, using, and appreciating various fungi. Southeast Asia has over 123 million indigenous people and, therefore, is a rich source of traditional knowledge on fungi, with abundant macrofungi used in culinary, medicine, and cultural rituals including ancient records of using mushrooms as psychedelics. This chapter examines the major ethnomycological studies undertaken in Southeast Asia, emphasizing how indigenous knowledge adds to our understanding of fungi and delving into its culinary, cultural, traditional, and medicinal importance. We propose practical ways to encourage greater global research on ethnomycology.

Keywords Fungal use · Indigenous people · Mushroom · Traditional knowledge

---

C. L. R. Evangelista
Fungal Biodiversity, Ecogenomics and Systematics-Metabolomics (FBeS) Group, Research Center for the Natural and Applied Sciences, University of Santo Tomas, Manila, Philippines

A. M. De Leon
Department of Biological Sciences, College of Science, Central Luzon State University, Science City of Muñoz, Nueva Ecija, Philippines

T. E. E. dela Cruz (✉)
Fungal Biodiversity, Ecogenomics and Systematics-Metabolomics (FBeS) Group, Research Center for the Natural and Applied Sciences, University of Santo Tomas, Manila, Philippines

Department of Biological Sciences, College of Science, University of Santo Tomas, Manila, Philippines
e-mail: tedelacruz@ust.edu.ph

J. J. G. Guerrero et al. (eds.), *Mycology in a Changing Planet: Applications and Perspectives from Southeast Asia*, https://doi.org/10.1007/978-981-95-2640-6_1

Introduction

Through direct interaction with their surroundings over thousands of years, indigenous people (IP) have built a deep and extensive body of knowledge. We refer to this as Indigenous Knowledge. This knowledge is rooted in ecological, spiritual, and cultural settings and includes observations, practices, innovations, and beliefs. Beyond its cultural relevance, indigenous knowledge is important as it offers solutions to modern world issues like climate change, habitat loss, and sustainable development. Through its comprehensive methods, this knowledge system has distinctive viewpoints on environmental stewardship, offers tried-and-true resource management solutions, protects cultural identity, and provides insightful information that enhances Western scientific methods (Bell et al., 2025).

Indigenous knowledge, also known as traditional knowledge (TK) or traditional ecological knowledge (TEK), is a broad category of knowledge accumulated over many generations of direct environmental interaction. Furthermore, indigenous knowledge is defined as observations, oral and written knowledge, innovations, practices, and beliefs that, through human-landscape relationships, encourage sustainability and responsible stewardship of cultural and natural resources (Jones & Choy, 2021). This information is deeply ingrained in the people who hold it, making it inseparable from their indigenous identity rather than being detached or abstract. The transmission of Indigenous Knowledge across generations is critical to its persistence and continued relevance. This is done through multiple mechanisms, including oral traditions like stories, songs, and ceremonies that encode ecological and cultural information; hands-on learning and mentorship from elders and knowledge holders; cultural practices and subsistence activities that provide opportunities for experiential learning; and indigenous languages that contain specialized terminology and concepts for environmental phenomena (Fre, 2024; U.S. National Park Service, 2024).

In the modern world, there are several obstacles to this traditional knowledge transmission process. As stated during the UN Permanent Forum on Indigenous Issues, "knowledge accumulated over thousands of years on medicine, meteorology, agriculture, and other areas is at risk of forever disappearing" (United Nations, 2019), making the preservation of indigenous languages essential. This knowledge systems' continued vitality is threatened by the disruption of conventional knowledge transmission paths caused by globalization, assimilation policies, colonization, and relocation. Thus, indigenous knowledge is extremely valuable to humanity as a whole and to the indigenous communities. Cultural identity and indigenous knowledge are inextricably linked. According to the UN Permanent Forum, "traditional knowledge is at the core of indigenous identity, culture, languages, heritage, and livelihoods" (United Nations, 2019). Therefore, preserving and reviving this knowledge is crucial to upholding cultural continuity and advancing the right to self-determination of indigenous peoples.

Ethnomycology: Fungi and Indigenous Knowledge

Indigenous knowledge significantly advances scientific knowledge and deepens scientific understanding. To improve knowledge of biological systems and create more successful conservation strategies, contemporary scientific disciplines such as ethnobotany, ethnomycology, ethnoecology, and biocultural conservation specifically reference indigenous knowledge. Humans and fungi have a lengthy history dating back to prehistoric times, long before ethnomycology became a recognized academic discipline. According to archaeological evidence, fungi have been an important part of human society for thousands of years. The Tassili Plateau in Algeria has prehistoric rock art from around 6000 years ago that shows what scientists think are mushrooms, demonstrating the cultural importance of these objects in early human communities (Samorini, 2001a, b). This graphic proof emphasizes how closely humans and fungi were related long before official ethnomycological research or scientific classification schemes were developed. Fungi exist in many cultural contexts throughout ancient civilizations, including food supplies, healing agents, spiritual aids, and folklore topics. As humans observed, classified, and passed down knowledge about fungi through years of oral and subsequently recorded traditions, these early interactions laid the groundwork for the eventual development of ethnomycology.

Ethnomycology is a scientific discipline that explores humans' cultural and traditional knowledge, beliefs, and practices regarding the use of beneficial macrofungi (mushrooms) that are found in the community surroundings. European naturalists like Christian Hendrik Persoon and Carl Linnaeus created taxonomy frameworks for fungi in the eighteenth and nineteenth centuries, which served as the scientific basis for subsequent ethnomycological research. These early classification schemes made it easier to identify different species and established a standard vocabulary for discussing mushrooms in various settings and cultures. The seminal work of Robert Gordon Wasson helped establish the formal field of ethnomycology (Wasson & Wasson, 1957). The biologist Richard Evans Schultes, an important pioneer in ethnomycology with Wasson, laid the foundation for the early development of the field. Because ethnomycology is considered a subfield of ethnobotany, Schultes' work helped establish ethnomycology's scientific legitimacy and link it to the larger subject. The focus on mushrooms shifted from solely taxonomic to a heightened interest in their cultural value and usage in various societies. Although hallucinogenic mushrooms, including psilocybin mushrooms, *Amanita muscaria*, and ergot fungus, were frequently the focus of early ethnomycological research, the topic would later broaden to include all cultural interactions with fungi (Gartz, 2021).

Since its inception in the middle of the twentieth century, ethnomycology has substantially developed its scope and methods. Initially focusing primarily on hallucinogenic or "entheogenic" mushrooms, the study eventually expanded to encompass all fungi utilized for food, medicine, crafts, and spiritual purposes in human societies. This growth resulted from a greater understanding of the diverse functions that fungi play in human civilizations, outside of their psychotropic qualities

(Dugan, 2011). To investigate the intricate links between people and fungi throughout history and civilizations, ethnomycology has developed into an interdisciplinary field incorporating ethnobotany, anthropology, and mycology (Comandini & Rinaldi, 2020). To compare indigenous knowledge and practices regarding wild mushrooms across various populations, contemporary ethnomycological research uses a variety of methodologies, such as "semi-structured interviews, focus group discussions, participant observations, and walk-in-the-woods methods," in addition to quantitative analyses (Dugan, 2011). Beyond its original focus on psychoactive fungi, ethnomycology has broadened its scope in recent years to include food and nutrition studies, folk medicine, the documentation and conservation of fungal germplasm, ethical and legal concerns related to traditional knowledge, and more extensive biotechnological applications. This growth reflects the increasing understanding of the value of fungi in tackling contemporary issues with food security, medicine, and sustainability.

Ethnomycological Studies in Southeast Asia

Outlined below are ethnomycological studies conducted in Southeast Asia as reported in many published literature. While the list may not be extensive as one hopes, it provides glaring evidence on the growth and development of ethnomycology in the region and identifies knowledge gaps for future scientific research.

The Philippines

Only few ethnomycological studies were ever conducted in the Philippines, primarily beginning with the pioneering works of De Leon and dela Cruz, two of the authors in this paper. De Leon et al. (2012) documented the use of mushrooms by the indigenous Aeta communities in Luzon Island, Northern Philippines. These mushrooms included *Auricularia auricula*, *Auricularia polytricha*, *Calvatia* sp., *Lentinus tigrinus*, *Lentinus sajor caju*, *Pleurotus* sp., *Schizophyllum commune*, *Termitomyces clypeatus*, *Termitomyces robustus*, *Termitomyces* sp., and *Volvariella volvacea*, all as food, *Mycena* sp. for medicine, and *G. lucidum* as home décor. The study also reported rituals performed by the Aetas before mushroom collection, such as tribal dancing, praying, and kissing the ground. A later study conducted by Tantengco and Ragragio (2018), also with the Aeta (Ayta) communities, revealed 15 mushroom species used as food and as alternative remedies for ailments like cough, weakness, common colds, and poor eyesight. The Gaddang communities in Nueva Vizcaya identified ten species of mushrooms as food, although only seven were collected during the study (Lazo & De Leon, 2015). These included *Auricularia auricula*, *Auricularia fuscosuccinea*, *S. commune*, *V. volvacea*, and unidentified *Lentinus*, *Pleurotus*, and *Polyporus* species. The Gaddang communities shared beliefs regarding mushrooms with the Aeta communities, that is, mushroom growth is stimulated by spontaneous lightning, and there is a need to seek permission from spirits before collecting mushrooms (Lazo & De Leon, 2015). Furthermore, the Kalanguya tribe in Carranglan, Nueva Ecija, mentioned 36 mushroom species they consumed as food, with one species used as an insect repellent (De Leon et al., 2016). However, only ten species were gathered during the collection. These included mushrooms with the local names of bang-ugan (*Meripilus giganteus*), bugatan, buo (*Scleroderma citrinum*), buttaytay (*Microporus* sp.), gum-gumot (*Leucoagaricus cepaestipes*), kuyupan (*Podocypha brasiliensis*), lingtan, uong-usa, upot (*Russula virscens*), and wek-wek (De Leon et al., 2016). Additionally, 13 beneficial mushrooms, specifically *Agaricus* sp., *A. auricula*, *Coprinellus disseminatus*, *L. sajor-caju*, *Lenzites elegans*, *Mycena* sp., *Oudemansiella canarii*, *Phellinus* sp., *Pleurotus ostreatus*, *S. commune*, *Trametes elegans*, *Vascellum pratense*, and *V. volvacea,* were collected in that study. The use of *Trametes* as a remedy for stomach pain and headaches and as body cleansing by the indigenous communities in three municipalities of Ifugao Province was also documented by De Leon et al. (2018). The Bugkalot indigenous community in Alfonso Castaneda, Nueva Vizcaya, recognized 17 mushroom species as food (*A. auricularia-judae*, *A. polytricha*, *Boletus* sp., *Clitocybe* sp., *Coprinopsis atramentaria*, *Coprinopsis lagopus*, *Coprinus cinereus*, *L. tigrinus*, *Lentinus* sp., *Mycena* sp., *Panaeolus* sp., *Pleurotus dryinus*, *Polyporus* sp., *S. commune*, and *Stereum lobatum*) and five mushroom species as medicinal (*Fomitopsis* sp., *Ganoderma applanatum*, *G. lucidum*, *Polyporus picipes*, and *Polyporus* sp.) (Torres et al., 2020).

These ethnomycological studies provide compelling evidence of the vital role that wild mushrooms play in the lives of Filipino indigenous communities and ethnic groups in the Philippines. Consequently, wild mushrooms are typically regarded as food, alternative medicine, or both. The reported medicinal benefits of the Philippine wild mushrooms also spark considerable interest in assessing their biological properties and identifying their bioactive compounds. Hence, with reports mainly concentrated from Luzon Island and from a few indigenous communities, more ethnomycological studies among the other indigenous communities in the country, particularly in the Visayas and Mindanao, are strongly suggested.

Vietnam

Vietnam boasts an abundant diversity of macrofungi. Throughout the last century, various mycologists have investigated macrofungi in Vietnam. According to the Vietnam Plant Data Center (https://www.botanyvn.com/), approximately 1400 species of macrofungi and other related taxa have been documented in Vietnam, and these are categorized as Basidiomycota (90%), Ascomycota (8%), Myxomycota (1.5%), and Glomeromycota (0.5%). The most diverse orders of macrofungi in Vietnam belong to Agaricomycetes with the following reported species: Aphyllophorales (around 350 species), Agaricales (close to 300 species), Boletales (approximately 60 species), and Russulales (over 40 species). In contrast, the orders of Tremellomycetes include only Tremellales (17 species), Dacrymycetes—Dacrymycetales (5 species), Pucciniomycetes—Septobasidiales (12 species), and Ustilagomycetes—Ustilagilales (only two species, *Ustilago maydis, Ustilago esculenta*).

The macrofungi in Vietnam also possess significant resource value. These include edible mushrooms (about 250 species), pharmaceutically active mushrooms (230 species), poisonous mushrooms (35 species), and many species that can be utilized for biotechnological and environmental conservation (Bao & Kiet, 2011). Furthermore, various species of mushrooms like straw mushrooms (*Volvariella volvacea*), button mushrooms (*Agaricus bisporus*), and shiitake mushrooms (*Lentinula edodes*) have been cultivated commercially in the country. Other cultivated mushrooms are oyster mushroom (*Pleurotus ostreatus*), wood ear mushroom (*Auricularia polytricha*), and lingzi or reishi (*Ganoderma lucidum*), and in addition *Volvariella volvacea* and *Lentinula edodes* are among the five most popular mushrooms commercially sold in Vietnam. Their nutritive values, total phenolic compounds, and antioxidant capacity were also investigated by Hung and Nhi (2012). Hoa et al. (2018) also studied seven wild edible mushroom strains for cultivation, and these were identified as species under the genera *Phlebia, Ganoderma, Tricholoma, Polyporus, Panus, Pleurotus*, and *Phlebopus*. Three edible mushrooms, *Phlebia* sp., *Ganoderma* sp., and *Pleurotus* sp., formed fruiting bodies on sawdust media supplemented with corn bran or earthworm fertilizer. Trang et al. (2023) also optimized the culture conditions for the mycelial growth and fruiting body development of two shiitake strains of *Lentinula edodes* (Berk.).

Regarding ethnomycological studies, Thao (2017) identified 20 ethnic groups in Northeast Vietnam, with eight of these groups participating in collecting *Russula griseocarnosa,* and four ethnicities are engaged in mushroom trading. Most collectors (85%) and traders (57%) were women. Harvesting was concentrated in two peak fruiting periods, at the start and end of the wet season, as rainfall affects mushroom quality and the price of the dried product. *Russula griseocarnosa* is widely collected by rural people in northeast Vietnam, where the low-quality fresh mushrooms were cooked and consumed by the gathering families, and the high-quality mushrooms were dried for local and export markets (Anh et al., 2024a, b). Anh et al. (2024a, b) also reported that *Russula griseocarnosa* has a rich nutritional composition and high mineral content. The nutritional composition of this mushroom largely mirrors values for this species.

Malaysia

Malaysia's great cultural variety is reflected in its indigenous people's deep ethnomycological knowledge, especially in Sabah, Borneo, where over 30 ethnic groups preserve traditions that are connected to wild mushroom collection, preparation, and therapeutic usage. With women central in knowledge transfer and resource gathering, recent studies show that indigenous people of Malaysia, including the Kadazan-Dusun, Bisaya, Orang Sungai, and Temuan communities have developed a sophisticated awareness of wild fungi for culinary and therapeutic uses (Fui et al., 2018).

Mushrooms are also occasionally utilized in rituals or as charms. Chang and Lee (2004) indicate that indigenous Temuan hunters wear the dried stipe of *Amauroderma* around a child's neck to fight off "fits" (Chang & Lee, 2004). The scarlet cup fungus *Aleuria aurantia* has been placed on graves in various Buddhist contexts. In Malay folk beliefs, some highly colored mushrooms were considered taboo. For example, an intensely yellow *Cantharellus cerinoalbus* is considered harmful solely because of its hue (Fui et al., 2018).

Mushrooms in Malaysia are regarded for their nutritional benefits as they are rich in fiber, proteins, and vitamins and low in fats and cholesterol (Samsudin & Abdullah, 2019). Wild mushrooms have traditionally been foraged by indigenous populations, including the Peninsular Orang Asli and the Dayak of East Malaysia. The Temuan and other Orang Asli recognize dozens of mushroom species in Peninsular Malaysia. At least nine wild mushrooms, including *Auricularia auricula-judae*, *Clavulina* spp., *Lentinus* spp., *Schizophyllum commune*, *Termitomyces heimii*, *T. microcarpus*, *Panus giganteus*, *Hygrocybe conica*, and *Cookeina* sp., were identified as food (Hamzah & Mohammad, 2021). *Auricularia* species (Kulat Korong), *Marasmiellus* species, various *Pleurotus* species (Cendawan Tiram), *Schizophyllum commune* (locally known as Kulat Kodop), and *Volvariella volvacea* (Kulat Sawit) are some of the most consumed species. To cook these, indigenous people employs four main culinary preparation techniques for wild edible mushrooms: boiling, steaming, soup preparation, and stir-frying. An example of a classic Dusun dish is linopot, which is made with hill rice wrapped in banana leaves and frequently accompanied by mushrooms, vegetables, sambal, deer meat, and wild ginger flowers. A traditional floss known as "serunding" is made by the Dusun community utilizing *Schizophyllum commune* in conjunction with the local indigenous ginger, *Etlingera* species (Tuhau). Meanwhile, *Termitomyces eurrhizus* (Kulat Temburung) is made in soup with smoked squirrel meat, while *Volvariella volvacea* is usually served with young fern shoots, garlic, and soy sauce (Abdullah & Rusea, 2009; Chang & Lee, 2004; Fui et al., 2018; Samsudin & Abdullah, 2019).

Important both culturally and commercially, medicinal mushrooms such as *Pleurotus tuber-regium* ("tumbuk-tuber" or "tumbuk-tuber raja" in Malay language or oyster mushroom in English), *Lignosus rhinocerotis* ("cendawan susu harimau" or "susu rimau" in Malay language or Tiger's Milk Mushroom in English), and *Ganoderma lucidum* ("Cendawan Lingzhi" or "Cendawan Reishi" in Malay language or reishi mushroom in English) are used to cure a variety of illnesses, such as diabetes, fevers, and skin conditions (Afolabi et al., 2024; Lau et al., 2015; Samsudin & Abdullah, 2019). In Malaysia, *Lignosus rhinocerotis* in particular is well known for its extensive traditional medicinal use (Fung & Tan, 2019), with communities drinking its infusion to increase alertness when hunting in the bush (Seow et al., 2015). Additionally, bioactive components in mushrooms, e.g., in *Lentinus squarrosulus*, have antibacterial and antioxidant qualities, supporting ethnomycological usage (Reena et al., 2020). Other mushroom species used in medicine and other folk uses are *Pycnoporus sanguineus* (bering), used by Orang Asli to treat skin sores; *Amauroderma* sp. (budak sawan), used by Temuan Orang Asli to prevent epileptic fits; *Xylaria polymorpha*, used in traditional ritual for children to stop bed-wetting; *Microporus xanthopus* (pengering), used in Temuan practices to wean infants; *Auricularia* spp., used as a general health tonic; and *Schizophyllum commune*, used by some rural folks as immune support (Chang & Lee, 2004; Seow et al., 2015).

Indonesia

Indonesia boasts a diverse range of fungal species, but estimates of 2273 species have been recorded as of 2017 despite the presence of about 2.2–3.8 million species globally (Khastini et al., 2018; Putra et al., 2022). Usually passed down orally, firmly anchored in daily activities and ceremonies, traditional fungi knowledge has constantly evolved throughout the centuries in Indonesia (Dewi et al., 2022; Khastini et al., 2018).

Fungi are important in Indonesian traditional society. Knowledge of bracket fungi (Basidiomycota) goes beyond mere practical use among the Baduy people of Banten Province. It includes classification by local names, identification by morphology, and incorporation into spiritual beliefs (Khastini et al., 2018). Sometimes employed symbolically in healing ceremonies, bracket fungi are utilized to promote local customs and resistance against diseases by their alleged powers (Khastini et al., 2018). The spiritual position of mushrooms is also reflected in Indonesian literature and oral traditions, where they are portrayed as mystical or divine in certain traditions, underscoring their significance in both everyday life and cultural lore (Ahmadi & Yulianto, 2019). Additionally, gathering and eating mushrooms in rural locations can serve as a practical survival strategy as well as a spiritually meaningful pastime, particularly when done by customary rules, e.g., only collecting mature or specific fungi at appropriate seasons (Dewi et al., 2022; Khastini et al., 2018).

Medicinal uses of fungi are deeply ingrained in Indonesia's ethnomedicinal legacy (Khastini et al., 2018). The Baduy tribe uses at least six bracket fungi species as essential resources for traditional medicine, with *Ganoderma lucidum* (locally referred to as "supa" variants), *Ganoderma applanatum*, *Fomes fomentarius*, *Rigidoporus stereum*, *Phellinus linteus*, and *Amauroderma* sp. being the most common. To prepare medicinal teas, sporocarp slices are cleaned, sliced, sun-dried, and decocted; occasionally, they are combined with honey or egg to improve flavor and effectiveness (Khastini et al., 2018). A popular conventional technique, hot water extraction, targets water-soluble polysaccharides with anti-inflammatory, antitumor, antioxidant, and anti-metabolic qualities like β-glucans (Khastini et al., 2018; Rizal et al., 2020). The propensity for strong informant consensus among communities highlights the perceived effectiveness and cultural faith in these treatments, particularly when it comes to treating particular conditions like fever, cancer, blood circulation issues, headaches, and immune deficiencies (Khastini et al., 2018).

Particularly valued in Indonesian ethnomedicine and other Asian traditions, *Ganoderma lucidum* improves general health, reduces chronic illnesses, and strengthens immunity (Zengin & Şanlier, 2020).

In Indonesian cuisine, fungi—particularly edible molds and mushrooms—play a significant role (Dewi et al., 2022; Faturrahman & Yeni, 2024; Oktavia et al., 2024). *Termitomyces* species, *Auricularia* species (wood ear mushroom), *Marasmiellus* sp., *Oudemansiella* sp., *Dacrymyces* sp., *Cookeina* sp., *Schizophyllum commune*, *Volvariella volvacea* (straw mushroom), and *Pleurotus ostreatus* (oyster mushroom) are among the diverse range of edible fungi which are frequently collected and eaten by Indonesians (Dewi et al., 2022; Faturrahman & Yeni, 2024). Mushrooms are prepared in a variety of ways in various Indonesian regions, including stir-fries, soups, bakwan (fritters), bubur pedas (spicy porridge), pepes (steamed fish or mushroom in banana leaf), and more, adding protein to regular meals (Faturrahman & Yeni, 2024; Sihotang et al., 2025). For the production of traditional Indonesian dishes like tauco (fermented soybean paste), oncom, tape (fermented cassava or rice), and tempeh (fermented soybeans), fermentation using molds like *Rhizopus oligosporus* and *Aspergillus oryzae* is essential. By increasing digestibility, lowering anti-nutritional factors, and raising levels of vitamins, amino acids, antioxidants, and probiotics, the fermentation process not only improves flavors and prolongs shelf life but also maximizes nutritional and health benefits (Oktavia et al., 2024).

Indonesian ethnomycology is a tapestry made from a wealth of traditional knowledge, cultural symbolism, and the usefulness of fungi in ritual, food, and medicine (Khastini et al., 2018; Oktavia et al., 2024). Fungi are more than just natural resources; they represent symbols of traditional knowledge and are essential to resilience, food, and communal health.

Cambodia

Natural remedies are still a big part of traditional medicine in Cambodia, where plants, animals, and mushrooms are used to treat everyday ailments due to their pharmacological activities. For example, the wild mushroom *Phellinus rimosus* is used by the Bunong community to traditionally treat backache, post-partum ailments, stomachache, tonic, diarrhea, intestinal pain, and wounds. The mushroom is prepared as a decoction with water or alcohol and then drunk. While some are chewing this mushroom to extract the "essence," the chewed mushrooms are also spat on the open wound. Additionally, *P. rimosus* is used with *Hydnophytum formicarum* to treat intestinal pain by boiling the mushroom and drinking its extract (Chassagne et al., 2016).

Psilocybe cubensis is commonly observed growing naturally in moist areas, such as rice fields and grassy areas, especially during the rainy season (Kurzbaum et al., 2025). Backpackers and other travelers seeking psychedelic experiences can openly buy magic mushrooms in some tourist destinations, such as Sihanoukville and Phnom Penh in Cambodia (Laurenson, 2015). During a visit to Angkor Wat, Allen et al. (2024) discovered a bas-relief depicting Shiva holding what appears to be a mushroom in his hand. This image, located on the second inner wall of the temple, raises intriguing possibilities about the use of neurotropic fungi in ancient Cambodia, particularly among the Mons, Khmer, or Buddhists during the eighth to twelfth centuries. While it is unclear whether Shiva is holding a lotus or a mushroom, the resemblance to a large-capped mushroom is striking. The blue color of Shiva's throat, linked to his mythological story of drinking poison, may further allude to the bluish staining of psilocybin mushrooms when bruised, adding a symbolic layer to this discovery. *Psilocybe antioquensis* stains green-blue when bruised, as described by Guzmán et al. (2006). Shiva's birth, linked to a thunderbolt, also mirrors global myths connecting mushrooms to divine or mystical origins. This fascinating find invites further speculation on the potential use of psychedelic mushrooms in ancient spiritual practices (Samorini, 2001a, b; Wasson, 1957).

Laos

Fungal species have been present in Laotian food, medicine, and cultural practices. In the markets of Luang Prabang, 54 fungal species are sold, including russuloid fungi, the most common fungi sold in open-air markets in Luang Prabang. *Russula* sp. is used in traditional Laotian cuisine, while some unidentified *Russula* on sale in Luang Prabang may be a new species (Luczaj et al., 2021). On the other hand, blackening-dull colored brittlegills or *Russula nigricans* are also collected, eaten, and sold in the market in Laos. Likewise, *Pleurotus giganteus* and *Lyophyllum* aff. *Fumosum* are also part of the locals' source of income, frequently reported as being sold on the roadside in Laos (Læssøe et al., 2018).

Since commercial medicines are not always affordable, *Auricularia polytricha* (Mont.) Sacc. or mouse ear mushroom is collected primarily in Xiengkhuang, northeast of Laos, and is used to supplement vitamin deficiency. *Dictyophora* sp. or red mushroom collected in Kie Pi Set near Vientiane is commonly powdered and put on burned skin (Deland, 2007). *Ganoderma lucidum* (Curtis) P. Karsten, or reishi mushroom, commonly found in decaying woods, is one of the sought-after ingredients for serious liver problems, making this fungus widely used in alternative medicine. This fungus is collected in large quantities in Laos from the wild (Dubost et al., 2019).

Cultural practices of harvesting mushrooms in Laos require extensive effort and deeply rooted local knowledge of the forest. In the Maikor forest, locals spend the night during the red mushroom season. For marketability, the locals harvest only the good ones, while for conservation, any damaged mushrooms are scattered and spread in other parts of the Maikor forest to propagate them (Carter & Senesombath, 2022). This shows that the locals' livelihood objectives are intertwined with efforts to conserve natural resources, as the strong market demand facilitates.

Myanmar

Thaung (2007) provided a list of mushroom species in Myanmar and listed 180 edible mushrooms. However, the ethnomycological studies in Myanmar are still lacking, and traditional knowledge on mushroom use is not well documented despite a strong link of Myanmar culture to Chinese culture, which is rich in ethnomycological knowledge (Fang et al., 2023).

Thailand

As a result of the cultural influence of mushrooms on Thailand, "magic mushrooms" are consumed for their hallucinogenic effects, and this greatly influenced the country's exotic food selections and even clothing merchandise. Between 1998 and 2000, fresh specimens of *Psilocybe cubensis* and/or *Psilocybe subcubensis* and some fresh specimens of *Copelandia cyanescens* were collected from Thailand's Koh Samui and Phuket Islands for botanical study (Allen & Merlin, 1992; Gartz et al., 1994). Allen et al. (2024) noted a rise in restaurants on Koh Pha-Ngan and Koh Tao serving mushroom-based dishes, including "hed kee kwai" omelettes upon request. Furthermore, Thai batiks for tourist trade used mushroom themes, particularly in Koh Samui Island in Thailand, but the popularity of mushroom-themed souvenirs had declined over the years (Allen et al., 2024). Nevertheless, ethnomycological studies have been recorded in Thailand, with uses that include food, food modifiers, and medicine, and with reports of poisonous and hallucinogenic fungi (Hillton & Dhitaphichit, 1993).

Concluding Points

Southeast Asia has a rich fungal diversity and cultural legacy, but ethnomycology research in the region is still comparatively understudied, and for several countries, it is lacking. Of the 312 ethnobiological studies conducted in Southeast Asia from 1960 to 2014, the majority of these studies were from Indonesia (93), Thailand (68) and Malaysia (58) (Hidayati et al., 2015). The number of indigenous groups and endangered languages in each nation correlates highly with spreading traditional knowledge. This indicates that ethnomycological studies have not yet adequately covered the region's range of ethnic variety.

In Southeast Asia, wild edible fungi are deeply culturally significant, serving as essential food sources and as components of traditional medicine and cultural customs. The rural communities in the region have a wealth of traditional knowledge related to fungal identification, ecology, and use. However, the unsustainable harvesting of wild edible fungi, especially ectomycorrhizal species associated with forest trees, is one of the most urgent issues facing ethnomycology in Southeast Asia. According to Wang et al. (2022), disorderly digging and hunting, habitat loss, and vegetation deterioration have caused many species' overexploitation and threaten the survival of fungal populations and the forests that support them. Broader environmental issues, including deforestation, shifting land uses, and the effects of climate change, further worsen the situation by reducing the acceptable habitat for numerous fungi.

Due to its limited transmission to younger generations and continued possession by older community members, traditional ethnomycological knowledge is also eroding at an alarming rate. Communities risk losing important knowledge about fungal ecology, safe consumption habits, and traditional uses that could support conservation efforts and promote community well-being if intentional efforts are not made to record and disseminate this knowledge. This degradation process is accelerated by younger generations' waning interest in traditional ecological knowledge, shifting lifestyles, and shifting economic priorities, making ethnomycological documentation projects urgent. In Southeast Asia, maintaining fungal biodiversity and related traditional knowledge will require the creation of sustainable management techniques. Future projects might concentrate on creating fungus-protected regions, enforcing harvesting laws derived from ecological understanding, and creating certification programs for wild fungi harvested responsibly. Promising models for balanced conservation and exploitation are provided by community-based conservation initiatives that enable local stakeholders to manage their fungal resources following conventional sustainable practices while reaping financial benefits from these resources.

There is a chance to lessen the impact of harvesting on natural populations while giving local communities steady sources of income by creating cultivation methods for commercially valuable wild fungi. Studies indicate that "documenting the fungal biodiversity which has a local use would allow the design and implementation of strategies to cultivate the most important wild edible fungi in specific areas and at the same time to integrate this cultivation into production systems which contribute to the recycling of local agricultural wastes" (Wang et al., 2022). While preserving cultural ties to these significant mushrooms, expanding cultivation efforts to include additional traditionally harvested species could significantly lessen the strain on wild populations. Further encouraging the switch from wild gathering to sustainable cultivation would be the development of value-added goods and processing methods that increase the financial benefits from cultivated fungi.

Traditional ethnomycological knowledge must be systematically documented to protect cultural heritage and avoid harmful misidentifications. Subsequent projects might create extensive ethnomycological databases that include ecological data, cultural applications in Southeast Asian cultures, and both traditional and scientific taxonomies. To ensure that ancient wisdom is continuously passed down, educational initiatives aimed at younger generations through schools, community workshops, and contemporary media platforms could aid in closing the knowledge gap between generations. By lowering the danger of misidentification, developing easily navigable identification guides that combine traditional knowledge with scientific explanations could benefit public health and conservation.

Ethnomycology in Southeast Asia is in a pivotal period, with enormous promise for sustainable development and cultural preservation, but also major obstacles due to overexploitation, habitat degradation, and knowledge erosion. Understanding fungal variety, ecological linkages, and sustainable management techniques can be significantly aided by the wealth of traditional knowledge that various ethnic populations in the region possess. Southeast Asian nations may create more efficient conservation plans, sustainable harvesting methods, and farming techniques that maintain biological and cultural variety by fusing this ancient knowledge with contemporary scientific methodologies. Priorities for research should be on thoroughly documenting traditional knowledge, especially among older knowledge holders, and community-based projects that enable local stakeholders to manage their fungal resources sustainably. Ultimately, the future of ethnomycology in Southeast Asian rests on appreciating and acknowledging the deep links between ecological health and cultural practices, and advocating for strategies that support the diversity of fungi that support human communities and the human communities themselves.

References

Abdullah, F., & Rusea, G. (2009). Documentation of inherited knowledge on wild edible fungi from Malaysia. *Blumea - Biodiversity, Evolution and Biogeography of Plants, 54*(1), 35–38. https://doi.org/10.3767/000651909X475996

Afolabi, F. J., Babaniyi, R. B., Obagunwa, M., Onile, F. O., Ajagun, E. J., Oyedele, O. J., & Fasiku, S. A. (2024). Ethnomycology of *Pleurotus tuber-regium* and its use in food, medicine and bioremediation. *Botanica*, 183–192. https://doi.org/10.35513/Botlit.2024.4.7

Ahmadi, A., & Yulianto, B. (2019). Indonesian literature and traditional knowledge: Interdisciplinary perspective. *International Journal of Multicultural and Multireligious Understanding, 6*(4), 102. https://doi.org/10.18415/ijmmu.v6i4.950

Allen, J. W., & Merlin, M. D. (1992). Psychoactive mushroom use in Koh Samui and Koh Pha-Ngan, Thailand. *Journal of Ethnopharmacology, 35*(3), 205–228. https://doi.org/10.1016/0378-8741(92)90020-r

Allen, J., Sihanonth, P., Gartz, J., & Toro, G. (2024). An ethnopharmacological and ethnomycological update on the occurrence, use, cultivation of known species, chemical analysis, and SEM photography of neurotropic fungi from Thailand, Cambodia, and other regions of south, Southeast Asia, Malaysia, Indonesia, and Bali. *Ethnomycological Journals: Sacred Mushroom Studies, 9*, 1–201.

Anh, C. N., Chi, N. M., & Dell, B. (2024a). Nutritional value of edible *Russula griseocarnosa* in Vietnam. *Asian Journal of Agriculture and Rural Development, 14*(3), 87–94. https://doi.org/10.55493/5005.v14i3.5162

Anh, C. N., Chi, N. M., Kiet, T. T., Ha, N. T. N., & Dell, B. (2024b). Harvest and trade of wild edible *Russula griseocarnosa* in North Vietnam. *Asian Journal of Agriculture and Rural Development, 14*(4), 128–139. https://doi.org/10.55493/5005.v14i4.5205

Bao, T. T., & Kiet, T. T. (2011). Diversity of macro fungi in Vietnam and their resource value. *1st National Scientific Conference of Vietnam Natural Museum System*, 97–104.

Bell, E., Tremblay, C., Carodenuto, S., Downie, B., Dearden, P., Ole Kileli, E., & McDougall, S. (2025). Indigenous knowledge-bridging to support ecological stewardship in Canada and Tanzania. *People and Nature, 7*, 1139–1150. https://doi.org/10.1002/pan3.70034

Carter, J., & Senesombath, K. (2022, December 5). *Cardamom, red mushrooms, broom grass and more: Biodiversity conservation through NTFP sales in Northern Laos?* https://www.helvetas.org/en/switzerland/how-you-can-help/follow-us/blog/environment/biodiversity-conservation-laos

Chang, Y. S., & Lee, S. S. (2004). Utilisation of macrofungi species in Malaysia. *Fungal Diversity, 15*, 15–22.

Chassagne, F., Hul, S., Deharo, E., & Bourdy, G. (2016). Natural remedies used by Bunong people in Mondulkiri province (Northeast Cambodia) with special reference to the treatment of 11 most common ailments. *Journal of Ethnopharmacology, 191*, 41–70. https://doi.org/10.1016/j.jep.2016.06.003

Comandini, O., & Rinaldi, A. C. (2020). Ethnomycology in Europe: The past, the present, and the future. In *Mushrooms, humans and nature in a changing world* (pp. 341–364). Springer International Publishing. https://doi.org/10.1007/978-3-030-37378-8_13

De Leon, A., Reyes, R. G., & Cruz, T. E. E. (2012). An ethnomycological survey of macrofungi utilized by Aeta communities in Central Luzon, Philippines. *Mycosphere, 3*(2), 251–259. https://doi.org/10.5943/mycosphere/3/2/9

De Leon, A. M., Kalaw, S. P., Dular, R. M., Undan, J. R., Alfonzo, D. O., Undan, J. Q., & Reyes, R. G. (2016). Ethnomycological survey of the Kalanguya indigenous community in Caranglan, Nueva Ecija, Philippines. *Current Research in Environmental & Applied Mycology, 6*(2), 61–66. https://doi.org/10.5943/cream/6/2/1

De Leon, A. M., Fermin, S. M. C., Rigor, R. P. T., Kalaw, S. P., dela Cruz, T. E. E., & Stephenson, S. L. (2018). Ethnomycological report on the macrofungi utilized by the indigenous community in Ifugao Province, Philippines. *The Philippine Agricultural Scientist, 101*(1), 84–92.

Deland, C. O. (2007). The role of medicinal plants in the provision of healthcare in Lao PDR. *Journal of Medicinal Plant Research, 1*(3), 50–59. https://academicjournals.org/article/article1380370127_Delang.pdf

Dewi, R. F., Ummah, N. R., & Hidayat, N. (2022). Ethnomycology studies of edible and medicinal macroscopic fungi in Jember, Indonesia. *BIOTIK: Jurnal Ilmiah Biologi Teknologi Dan Kependidikan, 10*(2), 218. https://doi.org/10.22373/biotik.v10i2.12513

Dubost, J. M., Phakeovilay, C., Her, C., Bochaton, A., Elliott, E., Deharo, E., Xayvue, M., Bouamanivong, S., & Bourdy, G. (2019). Hmong herbal medicine and herbalists in Lao PDR: Pharmacopeia and knowledge transmission. *Journal of Ethnobiology and Ethnomedicine, 15*(1), 27. https://doi.org/10.1186/s13002-019-0307-2

Dugan, F. M. (2011). *Conspectus of world ethnomycology: Fungi in ceremonies, crafts, diets, medicines, and myths*. American Phytopathological Society.

Fang, W., Devkota, S., Arunachalam, K., Phyo, K. M. M., & Shakya, B. (2023). Systematic review of fungi, their diversity and role in ecosystem services from the Far Eastern Himalayan Landscape (FHL). *Heliyon, 9*(1), e12756. https://doi.org/10.1016/j.heliyon.2022.e12756

Faturrahman, M. A., & Yeni, L. F. (2024). Ethnomycology study: Mushrooms traditionally utilized by the Malay Tribe in Lorong Village, Sambas Regency, West Kalimantan. *Jurnal Pijar Mipa, 19*(6), 979–986. https://doi.org/10.29303/jpm.v19i6.7723

Fre, Z. (2024, September 9). *The importance of indigenous knowledge*. UCL Press.

Fui, F. S., Saikim, F. H., Kulip, J., & Sathiya Seelan, J. S. (2018). Distribution and ethnomycological knowledge of wild edible mushrooms in Sabah (Northern Borneo), Malaysia. *Journal of Tropical Biology & Conservation (JTBC), 15*. https://doi.org/10.51200/jtbc.v15i0.1494

Fung, S.-Y., & Tan, C.-S. (2019). Tiger Milk mushroom (the Lignosus trinity) in Malaysia: A medicinal treasure trove. In *Medicinal mushrooms* (pp. 349–369). Springer Singapore. https://doi.org/10.1007/978-981-13-6382-5_14

Gartz, J. (2021). A brief overview on ethnomycology. *Journal of Bioengineering and Bioelectronics, 4*(1), 1–2.

Gartz, J., Allen, J. W., & Merlin, M. D. (1994). Ethnomycology, biochemistry, and cultivation of Psilocybe samuiensis Guzmán, Bandala and Allen, a new psychoactive fungus from Koh Samui, Thailand. *Journal of Ethnopharmacology, 43*(2), 73–80. https://doi.org/10.1016/0378-8741(94)90006-X

Guzmán, G., Allen, J., & Sihanonth, P. (2006). Distribution of the hallucinogenic mushroom Psilocybe antioquensis Guzmán et al. (Agaricomycetideae) in Colombia, Mexico, and Cambodia. *International Journal of Medicinal Mushrooms - Int J Med Mushrooms, 8*, 85–89. https://doi.org/10.1615/IntJMedMushr.v8.i1.110

Hamzah, H. N., & Mohammad, A. (2021). A study of wild mushrooms utilised for food and medicine by the Orang Asli Communities of Sungai Berua, Terengganu. In *Resource use and sustainability of Orang Asli* (pp. 75–84). Springer International Publishing. https://doi.org/10.1007/978-3-030-64961-6_6

Hidayati, S., Franco, F. M., & Bussmann, R. W. (2015). Ready for phase 5 – Current status of ethnobiology in Southeast Asia. *Journal of Ethnobiology and Ethnomedicine, 11*(1), 17. https://doi.org/10.1186/s13002-015-0005-7

Hillton, R. N., & Dhitaphichit, P. (1993). Procedures in Thai ethnomycology. *Natural History Bulletin of the Siam Society, 41*(3), 75–92. https://thesiamsociety.org/nhbss_publications/vol-41-no-2-1993/

Hoa, N. X., Huynh, L. T., Tran, T. H., Le, T. T., Luong, T. M., & Pham, T. (2018). Isolation and study on pure culture of wild edible mushrooms collected from provinces in the southeast region of Viet Nam. *Vietnam Journal of Science and Technology, 55*(1A), 124. https://doi.org/10.15625/2525-2518/55/1A/12389

Hung, P. V., & Nhi, N. N. Y. (2012). Nutritional composition and antioxidant capacity of several edible mushrooms grown in the Southern Vietnam. *International Food Research Journal, 19*(2), 611–615.

Jones, D., & Choy, D. L. (2021). *Indigenous Knowledge System and Yurlendj-nganjin*. Cambridge Scholars Publishing.

Khastini, R. O., Wahyuni, I., & Saraswati, I. (2018). Ethnomycology of bracket fungi in Baduy tribe Indonesia. *Biosaintifika: Journal of Biology & Biology Education, 10*(2), 424–432. https://doi.org/10.15294/biosaintifika.v10i2.14082

Kurzbaum, E., Páleníček, T., Shrchaton, A., Azerrad, S., & Dekel, Y. (2025). Exploring Psilocybe cubensis strains: Cultivation techniques, psychoactive compounds, genetics and research gaps. *Journal of Fungi, 11*(2), 99. https://www.mdpi.com/2309-608X/11/2/99

Læssøe, T., Pedersen, O. S. & Sysouphanthong, P. (2018). *An introduction to the edible, poisonous and medicinal fungi of Northern Laos.*

Lau, B. F., Abdullah, N., Aminudin, N., Lee, H. B., & Tan, P. J. (2015). Ethnomedicinal uses, pharmacological activities, and cultivation of *Lignosus* spp. (tiger's milk mushrooms) in Malaysia—A review. *Journal of Ethnopharmacology, 169*, 441–458. https://doi.org/10.1016/j.jep.2015.04.042

Laurenson, J. (2015). Drug dealers laughing at 'crackdown'. *Khmer Times*. https://www.khmertimeskh.com/31969/drug-dealers-laughing-at-crackdown/

Lazo, C. R. M., & De Leon, A. M. (2015). Ethnomycological survey of macrofungi utilized by Gaddang communities in Nueva Vizcaya, Philippines. *Current Research in Environmental & Applied Mycology, 5*(3), 256–262. https://doi.org/10.5943/cream/5/3/8

Luczaj, L., Lamxay, V., Tongchan, K., Xayphakatsa, K., Phimmakong, K., Radavanh, S., Kanyasone, V., Pietras, M., & Karbarz, M. (2021). Wild food plants and fungi sold in the markets of Luang Prabang, Lao PDR. *Journal of Ethnobiology and Ethnomedicine, 17*(1), 6. https://doi.org/10.1186/s13002-020-00423-y

Oktavia, A., Zainal, Djalal, M., Hidayat, S. H., & Azkiyah, M. (2024). Exploring the utilization of fungi in Indonesian traditional foods: A review. *BIO Web of Conferences, 96*, 01025. https://doi.org/10.1051/bioconf/20249601025

Putra, I. P., Nurdebyandaru, N., Amelya, M. P., & Hermawan, R. (2022). Review: Current checklist of local names and utilization information of Indonesian wild mushrooms. *Journal of Tropical Biodiversity and Biotechnology, 7*(3), 71407. https://doi.org/10.22146/jtbb.71407

Reena, R. D., Kandagalla, S., & Krishnappa, M. (2020). Exploring the ethnomycological potential of *Lentinus squarrosulus* Mont. Through GC–MS and chemoinformatics tools. *Mycology, 11*(1), 78–89. https://doi.org/10.1080/21501203.2019.1707724

Rizal, A., Sandra, F., Fadlan, M. R., & Sargowo, D. (2020). *Ganoderma lucidum* polysaccharide peptide reduce inflammation and oxidative stress in patient with atrial fibrillation. *The Indonesian Biomedical Journal, 12*(4), 384–389. https://doi.org/10.18585/inabj.v12i4.1244

Samorini, G. (2001a). New data from the ethnomycology of psychoactive mushrooms. *International Journal of Medicinal Mushrooms, 3*(2–3), 22. https://doi.org/10.1615/IntJMedMushr.v3.i2-3.1670

Samorini, G. (2001b). Funghi Allucinogeni: studi etnomicologici. *Telesterion*. https://www.samorini.it/doc1/sam/samorini-funghi-allucinogeni-studi-etnomicologici.pdf

Samsudin, N. I. P., & Abdullah, N. (2019). Edible mushrooms from Malaysia; a literature review on their nutritional and medicinal properties. *International Food Research Journal, 26*(1), 11–31.

Seow, S. L.-S., Eik, L.-F., Naidu, M., David, P., Wong, K.-H., & Sabaratnam, V. (2015). *Lignosus rhinocerotis* (Cooke) Ryvarden mimics the neuritogenic activity of nerve growth factor via MEK/ERK1/2 signaling pathway in PC-12 cells. *Scientific Reports, 5*(1), 16349. https://doi.org/10.1038/srep16349

Sihotang, S., Hermawan, R., & Putra, I. P. (2025). Characteristics of wild edible mushroom knowledge of mushroom foragers in Indonesia. *Jurnal Ilmu Pertanian Indonesia, 30*(2), 368–380. https://doi.org/10.18343/jipi.30.2.368

Tantengco, O. A. G., & Ragragio, E. M. (2018). Ethnomycological survey of macrofungi utilized by Ayta communities in Bataan, Philippines. *Current Research in Environmental & Applied Mycology, 8*(1), 104–108. https://doi.org/10.5943/cream/8/1/9

Thao, N. C. (2017). A human ecology overview of the Northeastern region. In *Human ecology and sustainable development* (pp. 38–53). Agricultural Publisher.

Thaung, M. M. (2007). A preliminary survey of Macromycetes in Burma. *Australasian Mycologist, 26*(1), 16–36. https://www.australasianmycologicalsociety.com/volume-261

Torres, M. L. S., Ontengco, D. C., Tadiosa, E. R., & Reyes, R. G. (2020). Ethnomycological studies on the Bugkalot indigenous Community in Alfonso Castañeda, Nueva Vizcaya, Philippines. *International Journal of Pharmaceutical Research and Allied Sciences, 9*(4), 43–54. https://ijpras.com/article/ethnomycological-studies-on-the-bugkalot-indigenous-community-in-alfonso-castaeda-nueva-vizcaya-philippines

Trang, N. T. H., Thuy, N. T. B., Mo, N. T., Luyen, N. T., & Nghien, N. X. (2023). Optimal culture conditions for the enhanced mycelial growth and cultivation of shiitake mushroom (Lentinula edodes). *Vietnam Journal of Agricultural Sciences, 6*(4), 1958–1968. https://doi.org/10.31817/vjas.2023.6.4.06

U.S. National Park Service. (2024, June 26). *Overview – Indigenous knowledge and traditional ecological knowledge*. National Park Service.

United Nations. (2019, April 22). *Indigenous people's traditional knowledge must be preserved, valued globally, speakers stress as permanent forum opens annual session*. Permanent Forum on Indigenous Issues.

Wang, R., Herrera, M., Xu, W., Zhang, P., Moreno, J. P., Colinas, C., & Yu, F. (2022). Ethnomycological study on wild mushrooms in Pu'er prefecture, Southwest Yunnan, China. *Journal of Ethnobiology and Ethnomedicine, 18*(1), 55. https://doi.org/10.1186/s13002-022-00551-7

Wasson, R. G. (1957, May 13). Seeking the magic mushroom. *Life Magazine, 100–102*, 109–120. https://www.cuttersguide.com/pdf/Periodical-Publications/life-by-time-inc-published-may-13-1957.pdf

Wasson, V. P., & Wasson, R. G. (1957). *Mushrooms, Russia, and history* (Vol. 1). Pantheon Books.

Zengin, F. H., & Şanlier, N. (2020). Connection of Ganoderma Lucidum with health. *Journal of Traditional Medical Complementary Therapies, 3*(1), 84–98. https://doi.org/10.5336/jtracom.2019-71988

Chapter 2
Fungi in Education: Reports from the Philippines

Jayzon G. Bitacura, Jaycee Augusto G. Paguirigan, Jayson G. Chavez, Lourdes V. Alvarez, and Thomas Edison E. dela Cruz

Abstract Fungi remain one of the most diverse organisms on the planet. Thus, the study of fungi, the science of mycology, is as essential as the other fields of biology. In the Philippines, teaching various aspects of fungal biology follows the teacher-centric lecture approach coupled with a practical and technical skills-based laboratory component. In this chapter, we present three higher education institutions in the country as case scenarios, each of which has its own unique implementation of teaching fungi. These case scenarios provide a context on how mycology is taught in the country and offer insights into innovative and inclusive learning activities. We also provided an example of technology-based learning activity in teaching fungi and called for more innovative and inclusive learning activities.

Keywords Learning activity · Mycology · PUP · Tertiary education · UST · VSU

Introduction

Higher education in the Philippines, as with most countries, is undergoing significant transformation, driven by globalization, digital innovation, and the increasing demand for industry-relevant skills. Universities adopt flexible, technology-driven learning models, incorporating online education, blended learning, and digital

---

J. G. Bitacura
Department of Biological Sciences, Faculty of Natural and Mathematical Sciences, Visayas State University, Baybay City, Leyte, Philippines

J. A. G. Paguirigan · J. G. Chavez · T. E. E. dela Cruz (✉)
Department of Biological Sciences, College of Science, University of Santo Tomas, Manila, Philippines
e-mail: tedelacruz@ust.edu.ph

L. V. Alvarez
Department of Biology, College of Science, Polytechnic University of the Philippines, Manila, Philippines

J. J. G. Guerrero et al. (eds.), *Mycology in a Changing Planet: Applications and Perspectives from Southeast Asia*, https://doi.org/10.1007/978-981-95-2640-6_2

credentials (Ignacio, 2021; Romli et al., 2022). Governments and education institutions emphasize international collaboration, research excellence, and skills-based training to enhance global competitiveness (UNESCO, 2023). However, challenges such as funding disparities, digital inequality, and the need for curriculum modernization persist, requiring coordinated efforts from policymakers, educators, and industry leaders (Khan & Anwar, 2021; Manalo et al., 2022; Chee, 2024).

A notable trend in teaching undergraduate courses is the increasing emphasis on digital learning and technology integration, even after the COVID-19 pandemic, which forced higher education institutions to innovate their delivery of programs (Joaquin et al., 2020; Ignacio, 2021). Several countries have invested in digitizing their higher education systems to expand access to quality education. For instance, Vietnam's higher education institutions are exploring the potential of digitalization to enhance learning experiences and educational outcomes (Vu & Nguyen, 2023). In the Philippines, several universities implemented synchronous and asynchronous online learning modalities, as well as hybrid learning models that combine in-person and virtual classrooms, ensuring greater accessibility and flexibility (Joaquin et al., 2020). Similarly, Singapore and Malaysia have introduced smart campuses and AI-driven learning platforms to personalize student experiences and improve digital literacy among students (UNESCO, 2023).

Transnational education and international partnerships are also on the rise. Many Southeast Asian universities collaborate with global institutions to offer dual-degree programs, student exchange opportunities, and joint research initiatives. These collaborations help universities in the region to gain international recognition and provide students with broader academic and career prospects. For example, transnational higher education rapidly reshapes Southeast Asia's educational and economic landscape, driving the region toward a future where global connections and local opportunities intersect seamlessly (Chee, 2024). Indonesia, for instance, has partnered with Australian universities to enhance research in sustainable development and digital innovation, while Vietnam has strengthened ties with institutions in the United States and Europe (UNESCO, 2023).

Despite these advancements, challenges remain in ensuring equitable access to higher education and maintaining quality standards. Urban centers often benefit from cutting-edge educational resources, while rural and underserved communities face barriers such as inadequate infrastructure, limited internet connectivity, and high tuition costs. This was primarily observed in the Philippines during the COVID-19 pandemic (Bustillo & Aguilos, 2022). Government agencies and higher education institutions are addressing these issues through scholarship programs, digital literacy initiatives, and infrastructure development to bridge these gaps (UNESCO, 2023). As Southeast Asia continues to evolve as a hub for higher education, sustained efforts in policy innovation, investment in digital infrastructure, and regional cooperation will be essential in shaping a more inclusive and globally competitive academic landscape.

Teaching Mycology in Philippine Undergraduate Courses

Responding to the need for an inclusive, digital-friendly and technology-driven higher education landscape, private and public universities in the Philippines has initiated teaching and learning activities that incorporate technology-based methods to traditional hands-on laboratory activities, particularly in the teaching of microbiology (dela Cruz, 2020; dela Cruz et al., 2023). Various efforts were also published as suggested learning activities to teach fungi (dela Cruz et al., 2012; Macabago & dela Cruz, 2012; dela Cruz et al., 2021; dela Cruz & Olayta, 2022; Estampador et al., 2022; Winsett et al., 2022). Here, we present the efforts by the Philippine higher education institutions, exemplified by one private, catholic university and two state universities—one in Metro Manila representing an urban learning setting and one in the Visayas to represent rural-based learning, in teaching fungi or mycology within their science undergraduate courses. Mycology is a branch of biology that studies fungi, such as molds, yeasts, and mushrooms, and may include other fungal-like taxonomic groups. These taxonomic groups like myxomycetes (slime molds), dictyostelids, and oomycetes are not phylogenetically related to fungi but share morphological features and, hence, are traditionally studied by mycologists. While bacteriology, the study of bacteria, including their structure, function, classification, and role in disease, dominates the traditional microbiology curricula, mycology is often underrepresented despite its importance in ecology, agriculture, biotechnology, and emerging diseases. We also outlined how research has become an integral part of the students' learning experiences.

Polytechnic University of the Philippines

The Polytechnic University of the Philippines (PUP), founded in 1904, offers a Bachelor of Science (BSc) in Biology with three specialized tracks: Animal Biology, Plant Biology, and Microbiology. While a dedicated course in Mycology is not currently offered, the study of fungi is deeply integrated into undergraduate Microbiology courses, e.g., General Microbiology, Microbial Physiology, and Environmental Microbiology. Students receive training in basic practical mycology, including techniques for isolating fungi from various environmental samples of their choice, preserving fungal cultures in agar slants and liquid broths, and preparing slides for macroscopic and microscopic examination. Students also conduct fermentation experiments, such as wine production and other fermented food products, to explore the metabolic roles of fungi in food microbiology and its industrial applications. These experiments connect basic mycological techniques with technology use, such as bioreactors.

Teaching of fungi is also augmented through fungal research as part of the undergraduate requirements for the biology program. Mycological research at PUP begins with the cultivation of edible and medicinal mushrooms, particularly different

Pleurotus species and *Calocybe indica*. These mushrooms are grown on various urban and agricultural waste substrates—including used paper, leaf litter, banana peels, corn cobs, and rice straw—promoting sustainability through waste valorization. Advanced research follows, emphasizing isolation and identification of plant pathogenic fungi, such as *Coniella, Colletotrichum, Fusarium,* and *Lasiodiplodia* from various economically important plant crops. These studies are complemented by the development of Loop-Mediated Isothermal Amplification (LAMP) protocols for the rapid detection of these pathogens, specifically *Colletotrichum asianum*, the causal agent of anthracnose, and *Lasiodiplodia theobromae*, the causal agent of stem-end rot in mangoes.

Driven by the increasing global attention on fungal research, ranging from fungal biodiversity and taxonomy to mycoremediation, endophytic symbioses, biocontrol of plant pathogens, and mycorrhizal applications in sustainable agriculture, PUP students are increasingly drawn to the field of mycology. Under the mentorship of Lourdes V. Alvarez (the fourth author), students undertake projects that involve isolating and identifying fungal pathogens from high-value crops such as papaya, tomato, jackfruit, cacao, and coffee. These projects include pure culture isolation, macroscopic and microscopic morphological characterization, DNA-based identification using internal transcribed spacer (ITS) sequencing, and phylogenetic analysis to determine evolutionary relationships. Fungal pathogens isolated in these studies include *Phytophthora palmivora, Colletotrichum* spp., *Lasiodiplodia* spp., *Pestalotiopsis* spp., *Fusarium* spp., and *Diaporthe* spp. PUP's fungal research also extends into biotechnology and environmental innovation, particularly through projects supported by the DOST Youth Innovation Program. Notable initiatives include the development of chitosan microbeads from *Rhizopus oryzae* which was isolated from rotting bread for the biosorption of heavy metals, demonstrating potential for environmental cleanup; microbial fuel cells using mushroom spent substrate as a source of bioenergy; and aerogels derived from coconut coir and mushroom spent for possible use in insulation, packaging, or filtration.

Furthermore, chitosan extracted from *R. oryzae, Aspergillus niger, Pleurotus florida*, and *Auricularia* sp. has been tested for antimicrobial activity. The results demonstrated inhibition of Gram-positive (*Staphylococcus aureus, Bacillus subtilis*) and Gram-negative (*Escherichia coli, Pseudomonas aeruginosa*) bacteria, indicating promising applications in biomedicine and food safety. Growing research interest in fungi led to the establishment of the PUP Mycology Laboratory, which now serves as the central hub for all mycological studies in the University.

Students' contributions to mycological research are actively disseminated through presentations at national and international scientific conferences, such as the Asian Mycological Congress (AMC), and the annual scientific meetings and symposia of the Mycological Society of the Philippines (MSP) and the Philippine Society for Microbiology (PSM). In addition to these presentations, students' research outputs are published in reputable scientific journals. Notable studys include the PUPMCR: an R package for image-based identification of color based on Rayner's (1970) terminology and known fungal pigments (Zapanta et al., 2025).

These efforts demonstrate the active role of students in advancing fungal research and contributing to the global body of scientific knowledge.

Despite the absence of a stand-alone mycology course, integrating fungal science into the curriculum and research agenda at PUP has cultivated a vibrant learning environment. Through a combination of practical training, research mentorship, and innovation-focused projects, PUP continues to inspire and equip students to contribute to mycology, biotechnology, and sustainable development in the Philippines and beyond.

University of Santo Tomas

The University of Santo Tomas, a catholic university founded in 1611 and hence, the oldest university in Asia, currently offers specialized programs in biology—Bachelor of Science in Environmental Biology, Bachelor of Science in Industrial Biology, and Bachelor of Science in Medical Biology—and is the first to offer a Bachelor of Science in Microbiology in the country. Under the different bachelor programs in biology, mycology is traditionally discussed within the undergraduate courses Botany and General Microbiology but includes only a handful of topics related to fungi, mainly on its taxonomy or phylogeny and a limited practical experience. For the laboratory activities, this involves only the isolation and culture of filamentous fungi and yeasts. However, for the BSc Microbiology program, Mycology is offered as one of the foundation courses that is taken during the students' first year of studies. It is taken as a 5-unit credit course, with independent lecture (equivalent to 3 hours of contact time) and laboratory (equivalent to 6 hours of contact time) components. Lecture topics covered in Mycology include diversity and phylogeny of fungi and other fungus-like groups, fungal structure, genetics and reproduction, and fungal nutrition and ecology. The laboratory component offers practical experiences in isolation, culture, identification, and preservation of fungi as well as methods to study fungal reproduction and growth. Applications of fungi in food and industry are often included in higher courses.

Fungal studies are also given a significant part in the teaching of other core courses such as Microbial Systematics & Evolution and Microbial Ecology where fungi are treated distinct from techniques used in studying bacteria. Fungi are also given attention in the teaching of specialization courses such as Industrial Microbiology for the concepts of fermentation technology and Food Microbiology on food safety as it relates to mycotoxins. Occasionally, fungi are mentioned or compared to other microorganisms in some lessons discussed within other microbiology courses, e.g., Microbial Physiology, Medical Microbiology, and Pharmaceutical Microbiology.

The Commission on Higher Education (CHED) of the Philippines, the country's tertiary education agency, issued memorandum order CMO No. 49 series of 2017 that details the policies, standards, and guidelines for BSc Biology programs. The teaching of fungi within CMO No. 49 is treated only as a lesson component for foundation courses such as General Botany, Systematics and Microbiology. Nevertheless, under the Mycology course of the BSc Microbiology program, the teaching of fungi is given much attention with the lecture component focusing on taxonomy and diversity of fungi and other fungal-related groups (myxomycetes, straminopilans) with comparative and detailed discussion on fungal reproduction, nutrition, genetics, and ecology. The learning of mycological techniques under the laboratory component in the undergraduate Mycology course has shifted from taxonomy-based approaches to technical skill-building activities and understanding key concepts such as baiting techniques for isolation and culture, and species identification through polyphasic morphological and molecular approaches. The laboratory exercises also include techniques to evaluate fungal growth, reproduction, and interactions with other organisms.

Fungal research is also integral in the teaching of fungi at UST. Annually, at least 10 undergraduate student groups, with 3 to 4 members per group, conduct research on various topics related to fungi, lichens, and other fungal-like taxa. The undergraduate theses ranged from basic to applied research and involved field- and laboratory-based activities, occasionally also *in silico* studies with ecological modeling and machine learning. Their research is mainly conducted at the Research Center for Applied and Natural Sciences (RCNAS), the science and technology research arm at UST. At RCNAS, two research groups have focused their research themes on fungi and lichens, with another group focusing on fungal-like oomycetes. The Fungal Biodiversity, Ecogenomics and Systematics—Metabolomics (FBeS) group under Thomas Edison E. dela Cruz (the last author) certainly leads mycological research in the university. The research program of the FBeS group mainly focuses on biodiversity, taxonomy, and systematics of fungi, with specific emphasis on fungal endophytes associated with mangroves and marine and terrestrial plants, endolichenic fungi, macrofungi, and myxomycetes. In addition, the FBeS group also focuses in applied fungal research, particularly on fungal natural products with emphasis on bioactivities, bioremediation, and biocontrol. The group pioneered ethnomycology in the country, with the documentation of fungal use by several Filipino indigenous communities. For lichen studies, the Philippine Lichen Systematics and Metabolomics (PLSMe) group under Jaycee Augusto G. Paguirigan (the second author) leads research on lichen biodiversity and taxonomy, mainly from species collected in Luzon Island, Northern Philippines. The group also tests lichens for bioactivities against bacteria, cancer cells, and other pathogens. Other research groups also conduct and/or use fungi and other fungal-like groups within their research group. For instance, Fr. Nicanor Pier Giorgio Austriaco's laboratory uses the yeast *Saccharomyces boulardii* as a vector in several of their genetic transformation and cell biology studies. The laboratory of Reuel M. Bennett works on zoosporic fungi and oomycetes, with strong interests on taxonomy and phylogeny leading to the discovery of novel species while the laboratory of Gina R. Dedeles uses thraustochytrids and yeasts for aquaculture research. These research endeavors highlight the efforts done by the university on fungal research.

Visayas State University

The Visayas State University (VSU), established in 1924 in the province of Leyte, is one of the leading universities in the Visayas and is one of the universities in the country that has strong teaching and research in agriculture. The study and teaching of fungi are integral components of various academic programs, particularly within undergraduate courses in Plant Pathology and Botany. The university also offers specialized courses in Advanced Mycology and Plant Pathogenic Fungi under its graduate program. These courses are designed to provide students with a comprehensive understanding of fungal biology, ecology, and their roles in plant diseases. These courses delve into the taxonomy, physiology, and management of fungi affecting agricultural crops, equipping students with both theoretical knowledge and practical skills essential for careers in agriculture and environmental science.

VSU is also home to the Plant Disease Diagnostic Laboratory, which offers services such as fungal identification and field diagnosis. This facility serves as a practical training ground for students, allowing them to apply classroom knowledge to real-world scenarios. Through hands-on experiences, students learn to identify fungal pathogens and develop management strategies, thereby enhancing their diagnostic and problem-solving skills. In addition, VSU faculty actively participate in international collaborations to advance their expertise in mycology. For instance, faculty members from the Faculty of Veterinary Medicine have attended workshops focused on emerging fungal infections, antifungal resistance, and advanced diagnostic methods. These engagements not only enhance the faculty's knowledge but also enrich the learning experiences of their students by integrating cutting-edge research and global perspectives into the curriculum.

For the undergraduate programs at VSU, only the Bachelor of Science in Agriculture major in Plant Protection with specialization in Plant Pathology offers a course on Introductory Mycology, while other aspects of fungal biology are integrated in Microbiology courses. For example, General Microbiology is offered by the Department of Biological Sciences and is taken by students enrolled in Bachelor of Science in Food Technology, Bachelor of Science in Biology, and Bachelor of Science in Biotechnology programs. In the course General Microbiology, students learn the classification of fungi as part of microbial eukaryotes. Under the Bachelor of Science in Marine Biology program, the course Marine Microbiology and Fungi covers fungal biology in marine environments, with emphasis on classification and ecology. In addition, fungal diseases and fungal endosymbionts of marine plants and animals are also discussed as well as the role of fungi in marine bioremediation.

For all the mentioned courses, practical skills in the laboratory component are designed to allow the students to learn how to isolate fungi from samples of their choice. For example, students in the food technology program are encouraged to isolate fungi from food products, whereas students in the marine biology program isolate fungi from marine organisms or samples. Students are taught how to produce axenic cultures of their isolated fungi. The isolated fungi are then characterized by the students morphologically based on their colonial (agar growth) and microscopic characteristics, and thus, students learned slide culture preparations and photomicroscopy (Fig. 2.1). Students are also taught how to screen for the production of enzymes of biotechnological importance. After characterization, students also conduct antibacterial assays to determine the bacteriostatic activities of the fungal endophytes against known pathogens as well as antagonistic assays for determining potential antifungal capacities of their fungal isolates (Fig. 2.1). Lastly, students are taught molecular phylogenetic analysis necessary for identifying their fungal isolates. Finally, aside from Microbiology courses, fungi are also used as model organisms for some teaching activities in the Cell and Molecular Biology course. In the laboratory, students conduct cellular proliferation and cytotoxicity assays using *Saccharomyces cerevisiae* as a test organism, an activity developed by the first author (Bitacura, 2018). Students are introduced to some bioinformatic tools for *in silico* molecular characterization of proteins using sequences of fungi-derived enzymes.

Through a combination of rigorous academic programs, practical laboratory experiences, and active research initiatives, Visayas State University remains at the forefront of fungal studies in the region. Some notable research accomplishments of VSU are on the discovery of fungal strains with oil- and plastic-degrading capabilities, heavy-metal tolerance, and heavy-metal biosorption capabilities under the mentorship of Jayzon G. Bitacura (first author). In agriculture, plant pathogenic and entomopathogenic fungi were identified and characterized from the region. The university's commitment to excellence in teaching and research ensures that students are well prepared to contribute to advancements in agriculture, environmental sustainability, and beyond.

It is clear from these case reports that efforts are made to improve the teaching of fungi in undergraduate courses in the Philippines. Integral to the effort is the use of

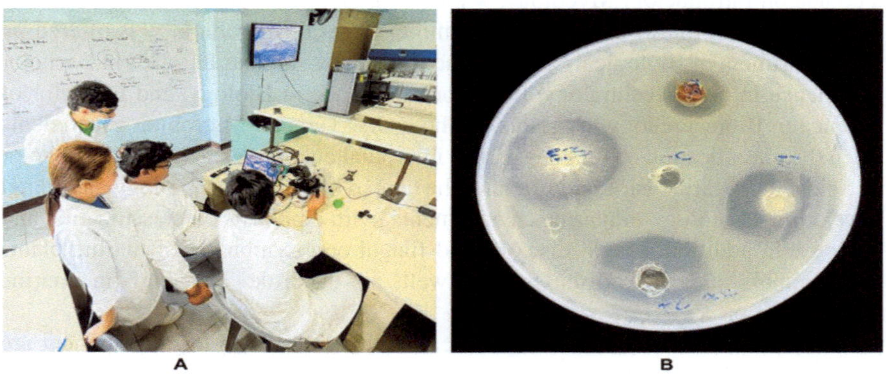

Fig. 2.1 Some activities related to the teaching of mycology at Visayas State University. (**A**) Students are taught photomicroscopy and proper image processing and fungal slide culture preparations which is also simultaneously shown on television for the rest of the class to see. (**B**) Sample result of agar plug diffusion assay which tested the bacteriostatic activity of the fungal endophytes the students isolated from a seagrass against *Escherichia coli*

research activities to augment the learning experiences of students. There are also efforts to provide more inclusive, technology-driven learning activities, particularly in the teaching of fungi.

Technology-Driven Teaching of Fungi

The opportunity for students to engage in practical work is an essential component in biology education; hence, it is pivotal that supplementary materials are available to make the connections between scientific content, technologies, and real-world interactions (Goodman et al., 2006). Learning experiences, whether it be in the laboratory, *in silico*, or field (1) reinforces understanding of content, (2) provides opportunities to develop competence, (3) engages students with the scientific method, (4) provides an arena for critical evaluation of data and information, and (5) enhances the materialization of learning outcomes and twenty-first century skills (Lewis, 2014). Hands-on learning, a way of teaching students using resources for testing what they learned, helps modify, handle, and test the learning; facilitates the development of perceptual skills (Haury & Rillero, 1994), making students the focus of learning; and ignites their thinking.

Color in taxonomy is empirical in species identification. Nevertheless, the interpretation of color is considered subjective, as individual differences, and cultural and personal experiences lead to variations on how we perceive this important source of taxonomic character. Attempts to mitigate subjectivity are increasingly being developed by using quantitative methods and color measurement tools to describe colors in more objective ways, e.g., Rayner, 1970; Beentje, 2016. In present time, Machine Learning (ML), and Artificial Intelligence (AI) are increasingly used in taxonomy to improve the efficiency and accuracy of species characterization, identification, and classification. AI algorithms can analyze images and help identify subtle differences between species, even in cases where traditional methods are challenging. ML models and AI can automate tasks that would otherwise require significant human effort that rely on the quality and representativeness of the data, e.g., color identification. The Polytechnic University of the Philippines (PUP) has recently addressed issues related to color identification in fungal taxonomy. Researchers from the university (including the fourth author) have developed the PUPMCR (Zapanta et al., 2025). It is a programming language created to approximate the color name based on the pixel coordinates of an uploaded image using the R interface. It uses Rayner's colors for fungal colonies (Fig. 2.2). R is a programming language and environment used primarily for statistical computing, data analysis, and visualization. It uses default R packages, but also a few specialized ones that need to be installed by the user, e.g., PUPMCR.

The following command in the R interface can be used to install the PUPMCR package. It will work, provided that the computer is connected to the Internet.

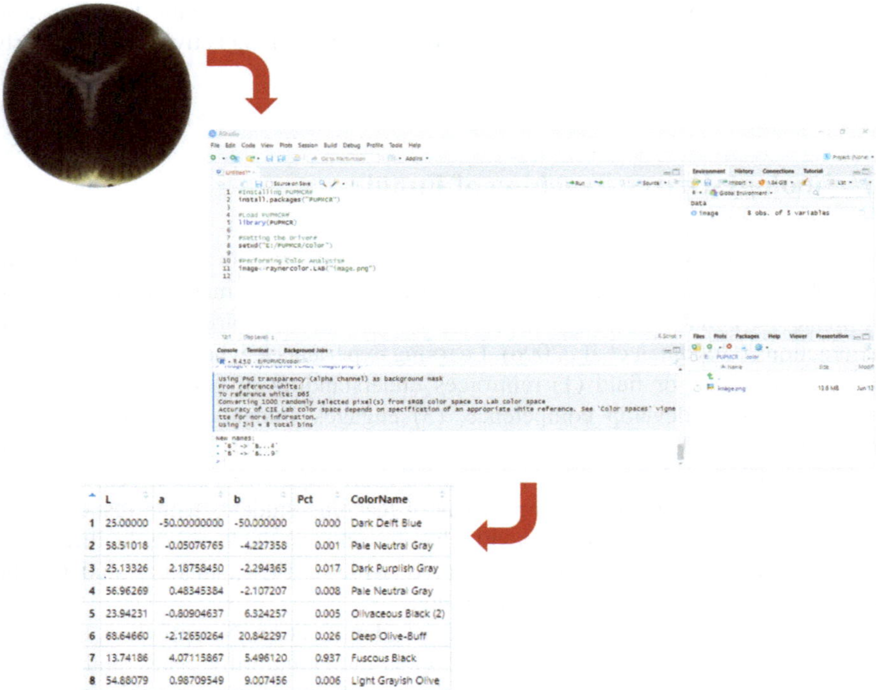

	L	a	b	Pct	ColorName
1	25.00000	-50.00000000	-50.000000	0.000	Dark Delft Blue
2	58.51018	-0.05076765	-4.227358	0.001	Pale Neutral Gray
3	25.13326	2.18758450	-2.294365	0.017	Dark Purplish Gray
4	56.96269	0.48345384	-2.107207	0.008	Pale Neutral Gray
5	23.94231	-0.80904637	6.324257	0.005	Olivaceous Black (2)
6	68.64660	-2.12650264	20.842297	0.026	Deep Olive-Buff
7	13.74186	4.07115867	5.496120	0.937	Fuscous Black
8	54.88079	0.98709549	9.007456	0.006	Light Grayish Olive

Fig. 2.2 The PUPMCR package of the R environment

```
>install.packages("PUPMCR")
```

Once the package is installed, it is already available for use after being loaded in memory by the user, which is usually done by entering:

```
>library(PUPMCR)
```

When R is started, a working directory that contains the necessary files can be modified and set.

```
>setwd("E:/PUPMCR/color")
```

The main pipeline on how PUPMCR works consists of the following steps: (1) image preparation—JPG or PNG files of colored images must be prepared with backgrounds being masked using a graphic editor software, (2) color matching—images are imported in R as 3D arrays, and non-background pixels are grouped into color categories, wherein pixel color coordinates are compared from the reference lookup table, and (3) results generation—the top eight results of the analysis are shown, i.e., color names and hues (Zapanta et al., 2025). To do this, the raynercolor.

LAB function is used, which defines the Rayner's nomenclature for colors in the image using the CIELAB color space for matching.

```
>raynercolor.LAB("image.png")
```

The imported image will result in a data frame showing qualitative (color names) and quantitative values (Fig. 2.2).

Students using this software can objectively identify colony colors, providing important characters for species identification. While more relevant for use in research, it can be incorporated into a learning activity for the laboratory component of microbiology, botany or mycology courses as exemplified in the class activities prepared by Jayson G. Chavez (third author).

Concluding Points

The teaching of fungi in the Philippines, while ingrained within general biology or general microbiology courses, remains limited to the taxonomic aspects of fungi. Technology-driven learning activities like the PUPMCR R program can make the lessons more engaging and student-centered. This tool can augment the traditional culture-based approach in laboratory activities related to fungal taxonomy. Other aspects of fungal biology are given emphasis on core undergraduate courses like Mycology and specialization courses like Plant Pathology. In these courses, lecture and laboratory components are aligned to provide a meaningful learning experience for Filipino students. However, for the three universities presented here, which may also reflect the current scenario for other universities, whether private or public, the teaching of fungi remains dependent on traditional modes of learning. This paper calls for further improvement and innovation of our learning activities and teaching strategies that incorporate better use of technologies, particularly with the advent of artificial intelligence. Keeping track with the updated technologies and innovations and diversifying mycology education, as exemplified by the Philippine experience, may be replicated by neighboring Southeast Asian countries, aiming to enrich biology education and fungal research capability in the region.

References

Beentje, H. (2016). *The Kew plant glossary – An illustrated dictionary of plant terms* (p. 184). Kew.

Bitacura, J. G. (2018). The use of baker's yeast in the Resazurin reduction test: A simple, low-cost method for determining cell viability in proliferation and cytotoxicity assays. *Journal of Microbiology & Biology Education, 19*(2), 10–1128.

Bustillo, E., & Aguilos, M. (2022). The challenges of modular learning in the wake of COVID-19: A digital divide in the Philippine countryside revealed. *Education Sciences, 12*(7), 449.

Chee, D. (2024). Empowering Southeast Asia: The impact of transnational higher education on regional development. *Fulcrum*. Retrieved from https://fulcrum.sg/aseanfocus/empowering-southeast-asia-the-impact-of-transnational-higher-education-on-regional-development/.

dela Cruz, T. E. E. (2020). Learning is fun! Developing an outcome-based, engaging learning activities in microbiology. In A. K. Raymundo & G. T. Pawilen (Eds.), *Outcomes-based education (OBE) in microbiology* (Monograph Series 23). National Academy of Science and Technology (NAST). ISSN: 1655-4299.

dela Cruz, T. E. E., Paguirigan, J. A. G., Aril-dela Cruz, J. V. (2021). Easy-to-do assessment tasks to create infoposters and infographics for communicating hot-button science issues. American Biology Teacher 83(7):482–485.

dela Cruz, T. E. E., & Olayta, C. O. M. (2022). Citizen taxonomy in social media: The use of Facebook for mapping species distribution of myxomycetes. *American Biology Teacher, 84*(4), 189–194.

dela Cruz, T. E. E., Pangilinan, M. V. B., & Litao, R. A. (2012). Printed identification key or web-based identification guide: An effective tool for species identification? *Journal of Microbiology & Biology Education, 13*(2), 180–182.

dela Cruz, T. E. E., Bennett, R. M., Balolong, M. P., Leaño, B. T. D., De Leon, A. M., Jacob, J. K. S., Magday, J. C., Jr., Valencia, A. D. L. C., Eloreta, M. B. F. M., Serrano, J. E., Bitacura, J. G., Apurillo, C. C. S., Nogodula, J. N., Pecundo, M. H., Santiago, K. A. A., & Aril-dela Cruz, J. V. (2023). Chapter 14 – Innovative learning activities for teaching mycology in the Philippines. In J. J. Guerrero, T. Dalisay, M. De Leon, M. A. Balendres, K. I. Notarte, & T. E. dela Cruz (Eds.), *Mycology in the tropics: Updates on Philippines fungi* (pp. 343–365). Elsevier Academic Press.

Estampador, F. R. F., Esteva, M. S. D. S., Garcia, N. F. G., Baylosis, D. C. G., Dagamac, N. H. A., & dela Cruz, T. E. E. (2022). A home-based experiment with myxomycetes for teaching basic ecological concepts. *SlimeMolds, 2*, V2A13.

Goodman, B. E., Freeburg, E. M., Rasmussen, K., & Meng, D. (2006). Elementary education majors experience hands-on learning in introductory biology. *Advances in Physiology Education, 30*, 195–203.

Haury, D. K., & Rillero, P. (1994). *Perspectives of hands-on science teaching* (p. 151). ERIC Clearinghouse for Science, Mathematics, and Environmental Education.

Ignacio, A. E. (2021). Online classes and learning in the Philippines during the COVID-19 pandemic. *International Journal on Integrated Education, 4*(3), 1–6.

Joaquin, J. J. B., Biana, H. T., & Dacela, M. A. (2020). The Philippine higher education sector in the time of COVID-19. In *Frontiers in education* (Vol. 5, p. 576371). Frontiers Media SA.

Khan, A., & Anwar, M. (2021). Higher education in peril: Challenges to Southeast Asian academics. In *Higher education challenges in South-East Asia* (pp. 238–249). IGI Global.

Lewis, D. I. (2014). The pedagogical benefits and pitfalls of virtual tools for teaching and learning laboratory practices in the biological sciences. *The Higher Education Academy: STEM*, 1–30.

Macabago, S. A. B., & dela Cruz, T. E. E. (2012). Development of a myxomycete photoguide as a teaching tool for microbial taxonomy. *Journal of Microbiology & Biology Education, 13*(1), 67–69.

Manalo, F. K. B., Reyes, V. P., & Bundalian, A. M. B. (2022). Challenges and opportunities in online distance learning modality in one public secondary school in the Philippines. *IOER International Multidisciplinary Research Journal, 4*(1), 89–99.

Rayner, R. W. (1970). *Mycological colour chart* (p. 34). Kew.

Romli, M. H., Wan Yunus, F., Cheema, M. S., Abdul Hamid, H., Mehat, M. Z., Md Hashim, N. F., & Jaafar, M. H. (2022). A meta-synthesis on technology-based learning among healthcare students in Southeast Asia. *Medical Science Educator, 32*(3), 657–677.

UNESCO. (2023). *Southeast Asia report – Technology in education*. Retrieved from https://www.unesco.org/gem-report/en/2023-southeast-asia

Vu, D. T., & Nguyen, T. T. (2023). Digitalization of higher education in Vietnam. *Journal of Comparative & International Higher Education, 15*(5), 92–108. Retrieved from https://ojed.org/jcihe/article/view/5710

Winsett, K. E., dela Cruz, T. E. E., & Wrigley de Basanta, D. (2022). Myxomycetes in education. The use of slime molds in promoting active and engaged learning. In S. L. Stephenson & C. A. Rojas (Eds.), *Myxomycetes: Biology, systematics, biogeography and ecology* (pp. 533–563). Elsevier Academic Press.

Zapanta, N. R., Santos, R. H., Pineda, J. I., Pedrosa, J. S., Rabelas, K. J., Samontan, C., Alvarez, L., & Deocaris, C. (2025). PUPMCR: An R package for image-based identification of color based on Rayner's (1970). Terminology and known fungal pigments. *Biology Methods & Protocols, 10*(1), bpaf004.

Chapter 3
Biosafety and Biosecurity in the Context of Mycology

Rohani B. Cena-Navarro, Jonathan Jaime G. Guerrero, Rodel Jonathan Vitor II, and Angelo dela Tonga

Abstract Biorisk management consideration is an essential component when working with fungi in a laboratory setting. This chapter highlights the unique challenges when handling fungi specimens including collection, transport, and testing. Risk assessment as a key component in the evaluation and management of risks are emphasized. Various mitigation strategies to minimize the risks are described. Moreover, a list of fungal pathogens of concern in Southeast Asia will be described in this chapter. Lastly, this chapter will also feature the steps in performance evaluation after risk mitigation.

Keywords Biorisk management · Laboratory · Diagnostics · Laboratory-acquired infection

Introduction

Fungi are increasingly recognized as critical threats to public health, agriculture, and environmental stability. Once overlooked in the global biosecurity agenda, fungal pathogens are now emerging with greater frequency and severity, particularly in

---

R. B. Cena-Navarro (✉) · A. dela Tonga
National Institute of Molecular Biology and Biotechnology, National Institutes of Health, University of the Philippines, Manila, Philippines

National Center for Biosafety and Biosecurity, National Institutes of Health, University of the Philippines Manila, Manila, Philippines
e-mail: rbcena@up.edu.ph

J. J. G. Guerrero
College of Medicine, University of the Philippines Manila, Manila, Philippines

R. J. Vitor II
Department of Biology, College of Arts and Sciences, University of the Philippines Manila, Manila, Philippines

J. J. G. Guerrero et al. (eds.), *Mycology in a Changing Planet: Applications and Perspectives from Southeast Asia*, https://doi.org/10.1007/978-981-95-2640-6_3

regions like Southeast Asia, where high biodiversity (Sodhi et al., 2010), climate sensitivity (Raktham et al., 2015), and rapid human development converge (Arfanuzzaman & Dahiya, 2019). Unlike bacteria and viruses, fungal pathogens are relatively difficult to diagnose and treat, exhibit growing antifungal resistance (Perlin et al., 2017), and affect both human and plant hosts. As global temperatures continue to rise and regional mobility increases (Lian et al., 2016), fungal outbreaks in hospitals, farms, and natural ecosystems are becoming more frequent and severe—necessitating urgent investment in biosafety and biosecurity systems.

Southeast Asia is uniquely vulnerable to these emerging fungal threats. Home to over 680 million people, the region has one of the world's highest levels of biodiversity, dense urban populations, widespread informal agriculture, and high rates of transboundary migration. Yet despite these risk factors, fungal pathogens remain underrepresented in regional health surveillance, agricultural biodiversity, and national biosafety frameworks. The result is a dangerous blind spot: limited detection capacity, delayed response to outbreaks, and unchecked environmental dissemination of high-risk fungal species.

From a public health perspective, the burden of fungal disease in Southeast Asia is substantial and growing. Invasive fungal infections are especially lethal among immunocompromised individuals such as people living with HIV, tuberculosis, or diabetes, all of which remain prevalent in the region. For example, Cryptococcal meningitis, primarily caused by *Cryptococcus neoformans* and *Cryptococcus gattii*, is responsible for 15% AIDS-related death (Rajasingham et al., 2017), with the highest mortality in sub-Saharan Africa and Southeast Asia. Similarly, invasive aspergillosis, often caused by *Aspergillus fumigatus*, has an estimated case fatality rate of 58% (Lin et al., 2001), especially among patients with pulmonary tuberculosis, still highly prevalent in the Philippines, Indonesia, and Myanmar (Zheng et al., 2017).

Another growing concern is *Candida auris*, an emerging multidrug-resistant yeast that causes bloodstream infections and healthcare-associated outbreaks. Since its global emergence in 2009, *C. auris* has been reported in Singapore and India (Khan & Ahmad, 2017; Suphavilai et al., 2024), which may proceed to a silent spread across other Southeast Asian countries. Its resistance to first-line antifungals (Chaabane et al., 2019) and ability to persist on hospital surfaces (Piedrahita et al., 2017; Welsh et al., 2017) make it a formidable nosocomial threat in health systems already burdened by antimicrobial resistance. However, fungal infections are rarely included in AMR national action plans, leaving a critical gap in regional health preparedness.

Agriculture and food systems are also heavily impacted by fungal pathogens. Southeast Asia's reliance on staple crops like rice, banana, and maize places it at significant risk of fungal plant pathogens. *Fusarium oxysporum* f.sp. *cubense* Tropical Race 4, responsible for Panama disease in bananas, has caused severe losses in the Philippines and spreading to Vietnam, Laos, and Malaysia (Chittarath et al., 2022; Hung et al., 2018; Ordonez et al., 2015; Salvacion et al., 2019).

Magnaporthe oryzae, the causative agent of rice blast, can reduce yields by up to 30% during outbreaks, threatening the food security of millions (Lim et al., 2023; Saleh et al., 2014; Zhang & Islam, 2022). Aflatoxin contamination, primarily by *Aspergillus flavus*, also affects maize, peanuts, and cassava, particularly under poor post-harvest storage conditions. In some Southeast Asian countries, studies have found aflatoxin levels in food that far exceed safe thresholds, posing serious risks for liver cancer and growth impairment in children (Benkerroum, 2020; Huong et al., 2019; Phan et al., 2023).

Climate change is intensifying these fungal threats. Rising temperatures and changes in rainfall patterns are altering fungal ecology, increasing the viability and spread of thermotolerant and spore-producing species. Pathogens like *A. flavus* thrive under hot, humid conditions, which are becoming more frequent in tropical lowlands. Flooding events linked to climate change also contribute to soil disruption and the release of dormant fungal spores, potentially causing fungal skin infections (Dayrit et al., 2018; Dayrit et al., 2022; Gadre et al., 2022). Meanwhile, warming trends are facilitating the northward expansion of fungal diseases into previously non-endemic zones, blurring traditional boundaries of fungal risks (Bonifaz et al., 2024; Salazar-Hamm & Torres-Cruz, 2024).

A One Health approach is essential to understanding and mitigating these risks. This integrated framework acknowledges the interconnectedness of human, animal, plant, and environmental health. Fungi exemplify these intersections. For example, azole-resistant *A. fumigatus* has been linked to the overuse of azole fungicides in agriculture (Berger et al., 2017; Cao et al., 2021; Chowdhary et al., 2013). Resistant strains can develop in the environment and affect immunocompromised individuals, demonstrating how agricultural practices can directly affect human health. Moreover, fungal infections in animals—such as chytridiomycosis in amphibians or dermatophytes in livestocks—can destabilize ecosystems or reduce productivity, with cascading effects on human livelihoods.

Migration and trade further complicate the fungal landscape. Southeast Asia has over 11 million intra-regional migrants, many of whom work in informal sectors like agriculture, construction, and caregiving—environments conducive to fungal exposure. Migrants often lack access to health services, making fungal infections harder to diagnose and treat. Meanwhile, informal cross-border trade of plant materials, especially banana and rice seedlings, enables the spread of fungal plant pathogens across national borders. This is particularly concerning given the limited capacity of phytosanitary agencies in low-resourced areas to inspect or quarantine high-risk imports.

With all these, this chapter describes the considerations in biosafety and biosecurity as well as mitigation strategies to minimize the risk when working with fungal pathogens in the laboratory. The importance of risk assessment will be discussed emphasizing the need for the oversight of the institutional biorisk committee.

Biosafety Considerations in Mycological Specimen Handling

Working with mycological specimens presents biosafety challenges in the laboratory. Whether dealing with fungi for research, teaching, or diagnostics, laboratory safety is critical to prevent personnel exposure and environmental release. Fungal infection can pose a health risk to humans, animals, and the environment, and therefore, safety considerations must be in place. Southeast Asia (SEA) is a "hotspot" of fungal pathogens, many of which are opportunistic and may cause infection especially in immunocompromised individuals or animals. Common fungal diseases in SEA are candidiasis, aspergillosis, mucormycosis, pythiosis, scedosporiosis, fusariosis, trichosporonosis, and cryptococcosis (Wang et al., 2016; Slavin & Chakrabarti, 2012). There are also endemic fungal diseases in specific regions in SEA which cause infections such as histoplasmosis, penicilliosis, sporotrichosis, and blastomycosis (Chakrabarti & Slavin, 2011). Various genera and species of microsporidia are now classified as fungi which may pose occupational risk of infection (BMBL, 2020).

Biosafety poses significant challenges for humans, animals, and the environment. Key challenges include high prevalence of fungal infections in SEA. Another challenge is the limited diagnostic capability. Many Asian countries have limited access to diagnostic tools which hampers early detection, diagnosis, and reporting. A survey in SEA laboratories showed that only 32 out of 219 or 16.9% of the laboratories perform sequencing and 27 out of 219 or 12.3% utilize matrix-assisted laser desorption/ionization time-of-flight mass spectroscopy (MALDI-TOF MS) for isolate identification (Chindamporn et al., 2018). The emergence of antifungal resistance also has a serious impact on human and animal health. Additionally, many laboratories in SEA lack high containment facilities to handle highly pathogenic fungi, which may lead to laboratory-acquired infection (LAI) or unwanted accidental release. LAIs have been reported in Asia where 7% of the LAIs are due to fungi (Siengsanan-Lamont & Blacksell, 2018). Several studies have also documented fungal contamination in healthcare settings underscoring the need to implement safety practices to prevent exposure. A prior study has documented fungal contamination in medical masks in healthcare workers (Merad et al., 2023). A similar finding has shown the presence of fungal pathogens in the personal protective equipment of medical workers (Lin et al., 2022). This underscores the need for training and awareness on safety precautions in handling fungal pathogens.

In Southeast Asia, several assessment biosafety regulations are available, such as in Singapore, Malaysia, the Philippines, and Thailand (Tin & Preiser, 2018; Kalidasan & Theva Das, 2021; Destura et al., 2021; Soisangwan, 2021). These regulations are critical in the management of biological risks in the region.

Risk Assessment

Risk assessment is important in identifying and evaluating the risks associated with handling fungi. When performing risk assessment, the personnel should evaluate the pathogenicity, virulence, and transmissibility of the fungal pathogens. Based on the risk, appropriate mitigation measures are implemented. Among the control measures are containment (engineering) controls, standard operating procedures, implementing policies and procedures, and the use of personal protective equipment.

Laboratory Safety Considerations

Engineering Considerations

Depending on the risk assessment, fungi may be handled in biosafety level (BSL) 1 to BSL4 laboratories. In general, non-pathogenic organisms may be handled in BSL-1 or BSL-2, while highly pathogenic fungi will require a high containment facility such as BSL-3 or BSL-4. High-risk fungi are organisms that cause severe infections particularly in people with weak immune system, while low-risk fungi causes milder infections. BSL-2 practices are recommended for identifying cultures that may contain *Histoplasma* sp. in diagnostic laboratories (BMBL, 2020). While international guidelines are available, laboratories must comply with their national regulations to determine appropriate biosafety level containment. Low- and middle-income countries face challenges in biocontainment (Fellizar et al., 2024), but implementation of biorisk management not only relies on engineering control but also on a combination of practices and procedures.

The use of a biological safety cabinet (BSC) serves as primary containment and barrier between the personnel and the pathogen. Commonly used BSCs are class II cabinets while class III cabinets offer the highest protection when handling highly pathogenic fungi. BSCs must be calibrated and certified regularly to ensure that they provide protection. Any culture identifying dimorphic fungi should be handled in a class II BSC (BMBL, 2020).

Personal Protective Equipment

Selection of personal protective equipment (PPE) should be based on the risk assessment. The type of containment level, fungal pathogen, and nature of activity should be considered when selecting PPE. Appropriate respiratory protection such as respirators or gas masks is used to prevent inhalation of fungal spores. Laboratory gowns are typically used when handling fungi in typical diagnostic laboratories. Additional layers of protection such as protective suits are worn in BSL-3 while a positive pressure protective suit is worn in BSL-4 laboratories. Face shields and goggles for eye protection are worn to prevent exposure from fungal spores.

Specimen Handling and Transport

Fungi are hardy organisms and transport media is not generally required. However, sterile containers must be used to prevent bacterial contamination. Specimens for culture are usually from anatomic sites such as respiratory specimens, tissue biopsies, cutaneous scrapings from skin, blood, bodily fluids including cerebrospinal fluid, and bone marrow. Therefore, safety considerations must be in place to prevent potential infection during collection, handling, and transport. Considerations in safe handling must also be in place when handling fungal specimens from the environment.

Standard Operating Procedures

Developing standard operating procedures (SOPs) is one of the most effective and cost-efficient mitigation controls to minimize the risk of exposure to fungal pathogens. Adherence to biosafety protocols will lead to successful containment of pathogenic fungi. Developing SOPs should be based on risk assessment and tailored to the activities in the laboratory. Examples of SOPs include the use of PPE, specimen transport, waste management, and others. These SOPs should be validated by the end users and revised as necessary.

Competency and Training

Training and competency of laboratory personnel are essential in ensuring safety and security of fungal containment. Understanding the risks associated with handling fungal pathogens are important to identify the mitigation strategies to prevent laboratory-acquired infection. Competency assessments are conducted to evaluate the ability of the personnel to perform tasks safely and effectively. In Southeast Asia, mycology training is available. However, inadequacies in medical mycology training are also one of the identified challenges in managing fungal diseases (Tan et al., 2020). Training and competency programs allow continuous improvement of personnel to ensure compliance with regulatory requirements, new diagnostics, and procedures. Several mycology training programs are available in Southeast Asian countries such as in the Philippines and Malaysia to strengthen laboratory capacity in diagnostics, surveillance, and safety.

Biosecurity Considerations in Mycological Specimen Handling

Biosecurity is critical to scientific research, especially when dealing with biological agents such as fungi, which can pose significant threats to public health and the environment. Mycological specimens, including fungal cultures, are studied for various purposes, including medical research, biotechnology, and agricultural applications. However, as with all biological agents, these specimens must be handled carefully to prevent unintended release or misuse (Paterson, 2006). Further, fungal pathogens are responsible for a range of diseases that can have devastating effects on humans, animals (Gnat et al., 2021), and plants (Fisher et al., 2020). Given the potential of fungi to cause infections, their pathogenic nature, and their ability to spread rapidly in suitable environments, robust biosecurity frameworks are necessary to ensure their safe management and research.

Fungal pathogens such as *Aspergillus*, *Candida*, and *Cryptococcus* can cause serious diseases in immunocompromised individuals, and the emergence of resistant strains further complicates treatment options (Garnacho-Montero et al., 2024). Fungi may also play a role in bioterrorism due to their ability to be weaponized and their resilience in various environmental conditions (Bokan & Orahovec, 2004). The rapid growth of biotechnological methods has increased the ease of access to genetic manipulation tools, which makes the regulation and oversight of fungal research more complex. This underscores the importance of a comprehensive biosecurity approach to mitigate these risks (Vitor & Estrella, 2024).

Biosecurity is important to various research institutions and laboratories, including universities, government facilities, and private research entities. To ensure that fungal specimens are handled safely, biosecurity measures should focus on material control, information security, personnel reliability, physical security, transport control, emergency response, and the prevention of dual-use research that could potentially lead to malicious activities (World Health Organization, 2024b). A multifaceted, proactive approach is essential to protecting public health, maintaining scientific integrity, and preventing the misuse of fungal research.

This section outlines the key pillars of biosecurity for handling fungi and mycological specimens, focusing on Southeast Asia, where certain fungal pathogens may present unique challenges. This also looks into the possibility of fungal pathogens being used as bioterrorism agents and the vital role institutional oversight, such as the Institutional Biosafety and Biosecurity Committee (IBBC), plays in ensuring safe and responsible research.

Pillars of Biosecurity

The pillars of biosecurity are essential components that help mitigate risks associated with biological agents, ensuring safety in research, agriculture, and public health settings. These pillars provide a comprehensive framework for managing and controlling biological threats, including pathogens and other hazardous biological materials (Vitor & Estrella, 2024; World Health Organization, 2024b).

Material Control and Accountability

Material control is a crucial component of biosecurity, ensuring that all mycological specimens are properly stored, tracked, and accounted for throughout their lifecycle. The risks associated with misplacement, theft, or improper handling of fungal specimens necessitate a well-defined system for inventory management (Baldwin et al., 2009). This includes the proper labeling of samples, a secure method for storing fungal cultures, and a procedure for tracking the movement of these materials within the laboratory and across institutions. Institutions must maintain a comprehensive record of all fungal specimens, including their origin, handling procedures, and relevant research findings to ensure transparency and accountability (World Health Organization, 2024b). Furthermore, institutions must implement robust systems for disposing of fungal materials. This involves ensuring that cultures are destroyed in a manner that eliminates any viable spores or pathogens, thus preventing environmental contamination or the unintentional spread of dangerous fungi. Safe disposal practices, such as autoclaving or chemical decontamination, must also be followed to maintain biosecurity integrity (World Health Organization, 2020a).

Information Control, Data Security, and Cyberbiosecurity

As mycological research increasingly depends on digital technologies for data storage, analysis, and sharing, the significance of information control and data security has grown markedly. Research on fungal pathogens involves gathering sensitive data, including genetic sequences, infection pathways, and therapeutic strategies. Unauthorized access to this information could result in the misuse of research, theft of intellectual property, or, in extreme cases, the weaponization of fungal pathogens (Crawford et al., 2023). Institutions must establish rigorous cybersecurity measures to reduce these risks and safeguard research data. These measures encompass encrypting sensitive information, implementing multi-factor authentication systems, and utilizing firewalls along with other digital protections. Furthermore, institutions should have policies in place to manage access to proprietary research and ensure that only authorized personnel can access the most sensitive data (Patel et al., 2023). The emergence of "cyberbiosecurity" as a specialized field underscores the growing need for biosecurity measures that address both biological and digital threats. This interdisciplinary approach ensures that research infrastructure, data repositories, and communication networks are protected against cyberattacks, which could hinder scientific progress or facilitate the misuse of dangerous biological agents (Murch, 2023).

Personnel Control and Reliability

Personnel control ensures that only authorized and trained individuals handle fungal specimens. Laboratories and research institutions must have strict hiring protocols, including background checks to assess the reliability and trustworthiness of potential staff members. Personnel should also undergo thorough training on biosecurity practices, including properly handling fungal cultures, identifying potential hazards, and taking steps to prevent accidental exposure or contamination. Personnel must remain vigilant and compliant with established biosecurity guidelines. This requires monitoring staff adherence to protocols and regular refresher training to keep them up to date with the latest developments in biosecurity. Security clearance levels should be established in high-security laboratories, particularly those dealing with highly pathogenic fungi, and personnel should be subject to continuous assessment (Higgins et al., 2013).

Physical Security Control

Physical security is a cornerstone of biosecurity, ensuring that fungal specimens and research materials are protected from unauthorized access, theft, or sabotage. Laboratories that house fungal cultures must be equipped with physical barriers, such as locked doors, secure storage areas, and surveillance systems. Additional measures, such as biometric access controls and reinforced security systems, may be required in facilities dealing with high-risk pathogens (Fellizar et al., 2024). BSL-3 and BSL-4 laboratories handle the most hazardous fungal pathogens and must adhere to strict physical security standards. These include isolated environments, air filtration systems, and negative-pressure rooms to prevent the accidental release of fungal spores into the environment (World Health Organization, 2024b).

Transport Control

Transporting fungal specimens, particularly pathogenic ones, requires strict adherence to national and international regulations governing the safe shipment of biological materials. Packaging and labeling must meet established guidelines and transport routes should be carefully planned to minimize the risk of exposure or contamination during transit. Institutions must ensure that fungal specimens are securely contained within specialized containers to prevent leakage or spore dispersal. Additionally, institutions should maintain a chain of custody for all transported fungal samples, ensuring that every transfer is tracked and monitored to prevent mishandling or misplacement (World Health Organization, 2024a).

Emergency and Incident Response

Biosecurity protocols must include well-established emergency and incident response plans to address accidents or security breaches. In the event of a spill, exposure, or release of fungal pathogens, laboratories must have immediate response strategies in place to contain the threat. This includes readily available appropriate personal protective equipment (PPE), quarantine protocols for exposed personnel, and access to emergency decontamination facilities. Effective communication and coordination are essential during an emergency. Institutions must have a defined chain of command and established lines of communication to ensure that all relevant personnel and external authorities, such as public health agencies, are promptly informed and engaged in the response effort (World Health Organization, 2024b).

Emerging Biological Risks

Emerging biological risks refer to newly identified pathogens or those undergoing rapid evolution that may threaten public health, agriculture, or ecosystems. In the realm of mycology, emerging fungal pathogens, such as *C. auris* and *A. flavus*, are a growing concern. These fungi have developed resistance to antifungal treatments, making them harder to control and treat. Emerging risks also include the increasing occurrence of fungal diseases in regions where they were previously uncommon, as well as the spread of antibiotic-resistant strains. To detect and respond to these emerging threats before they can cause widespread harm, surveillance systems and early warning mechanisms must be established. This requires cooperation between public health authorities, research institutions, and international organizations to monitor and track fungal disease outbreaks (Fisher et al., 2012).

Dual-Use Research of Concern

Dual-use research can be used for both beneficial and harmful purposes. In the context of fungal pathogens, this includes research that could lead to developing more effective antifungal therapies. However, it also has potential misuse, such as creating bioweapons. The dual-use nature of mycological research raises significant biosecurity concerns, as research intended for the public good could be diverted for malicious purposes. To mitigate these risks, institutions must implement oversight mechanisms to assess the potential dual-use implications of fungal research. This includes ensuring that research involving genetically modified organisms (GMOs) or pathogenic strains of fungi is conducted under strict regulations and oversight. In some cases, research that could lead to the creation of bioweapons should be subject to additional scrutiny to prevent misuse (Yoshizawa et al., 2024).

Fungal Pathogens of Concern in Southeast Asia

Southeast Asia is home to a rich diversity of fungi, many of which are responsible for significant public health and agricultural concerns. Tropical climates provide an ideal environment for the growth and spread of fungal pathogens, some of which have high pathogenic potential. The region is particularly vulnerable to fungal infections due to its large, dense population and agricultural practices, involving widespread use of crops susceptible to fungal contamination. Fungal pathogens are a significant biosecurity concern due to their ability to form durable spores that can persist in the environment for extended periods (Chakrabarti & Slavin, 2011). These spores are highly resistant to heat, desiccation, and disinfectants, allowing them to survive routine cleaning procedures and spread through air, water, or fomites. In research, agricultural, and clinical settings, spore-forming fungi like *Aspergillus*, *Fusarium*, and *Coccidioides* can cause severe infections in animals and humans, especially in immunocompromised individuals (Fisher et al., 2012, 2020). Their airborne transmission makes containment difficult, requiring specialized ventilation systems and personal protective equipment (Guerrero et al., 2024). Additionally, spores can contaminate laboratory surfaces, instruments, and biological samples, potentially leading to cross-contamination or accidental release. Therefore, rigorous biosecurity measures—including environmental monitoring, decontamination protocols, and secure handling procedures—are essential to prevent fungal outbreaks and protect public and animal health (Chakrabarti & Slavin, 2011; Fisher et al., 2012, 2020; Gnat et al., 2021; Casalini et al., 2024).

1. *Aspergillus Species: Aspergillus* is a fungal genus with pathogenic and non-pathogenic strains. Some species, such as *Aspergillus flavus*, produce aflatoxins, which are potent carcinogens and pose a significant risk to food safety. Aflatoxin contamination in crops such as rice, peanuts, and maize is a significant concern in Southeast Asia, where the climate is conducive to fungal growth. Additionally, *A. fumigatus* is an opportunistic pathogen that can cause respiratory infections, particularly in immunocompromised individuals (Fisher et al., 2012, 2020).
2. *Candida albicans: Candida* species, particularly *Candida albicans*, are responsible for many human infections, including thrush, vaginitis, and systemic infections. The rising incidence of *Candida* infections and increasing resistance to antifungal agents make this genus a significant concern in Southeast Asia. The high prevalence of human immunodeficiency virus/acquired immunodeficiency syndrome (HIV-AIDS) in the region further exacerbates the risk of opportunistic fungal infections (Fisher et al., 2020).
3. *Cryptococcus neoformans: Cryptococcus neoformans* is a major cause of fungal meningitis, particularly in individuals with weakened immune systems, such as those living with HIV/AIDS. The prevalence of *C. neoformans* in Southeast Asia is significant, and the rising incidence of cryptococcal infections underscores the need for stringent biosecurity measures when working with this pathogen (Fisher et al., 2020; Gnat et al., 2021).

4. *Fusarium solani:* This soil-borne fungus causes keratitis, a serious eye infection. It is also a plant pathogen, causing diseases in crops such as bananas and tomatoes. The ability of *Fusarium* species to infect humans and plants makes them a dual threat to public health and agriculture in Southeast Asia (Fisher et al., 2012; Gnat et al., 2021).
5. *Coccidioides immitis:* This dimorphic fungus is responsible for coccidioidomycosis, or valley fever, primarily affecting the lungs. *C. immitis* thrives in arid environments, making it a concern in Southeast Asia, where environmental changes can exacerbate fungal dust dispersion. It can be transmitted by inhaling airborne spores. Laboratories must follow strict containment protocols, including air filtration systems and proper handling procedures, to prevent airborne exposure to researchers. Personal protective equipment (PPE), including respiratory protection, is essential (Casalini et al., 2024).
6. *Histoplasma capsulatum: Histoplasma capsulatum* causes histoplasmosis, primarily affecting the lungs. It is found in soils enriched with bird or bat guano. In Southeast Asia, where agriculture and bat roosts are prevalent, this fungus poses a significant risk. Fungal spores are inhaled during soil disturbance, leading to infection. Biosecurity practices must include containment and appropriate respiratory protection to avoid inhalation during field research or lab handling. Institutions must ensure that fungal cultures are securely stored and properly disposed of to prevent unintended environmental release (Casalini et al., 2024).

The following fungi are primarily plant pathogens that pose only threats to plants. While the discussion of this section primarily looks into the biosecurity aspect as it applies primarily to human and animal health, the role of biosecurity in preventing the introduction of plant pathogens in regions where it is uncommon or unknown is of interest primarily to mycologists (Doehlemann et al., 2017; Jain et al., 2019; Limbo-Dizon et al., 2023; Balendres et al., 2024; Guerrero et al., 2024).

1. *Colletotrichum coffeanum var. virulans:* This fungal pathogen is responsible for anthracnose in coffee plants, a significant concern in Southeast Asia's coffee-growing regions. *C. coffeanum* var. *virulans* can cause significant agricultural damage. Biosecurity measures for agricultural pathogens like this include the use of containment and quarantine procedures for field samples, as well as decontamination practices for tools and equipment. Preventing the spread of infected plant material to unaffected regions is essential for controlling outbreaks (Balendres et al., 2024).
2. *Puccinia graminis:* A notorious wheat pathogen, *Puccinia graminis* causes stem rust, which significantly impacts crop yields. In Southeast Asia, where wheat production is expanding, this pathogen poses a serious threat. The fungal spores can travel long distances, making transport control vital to prevent international spread. Research facilities must ensure proper containment during work with wheat and rust samples, and transport protocols should follow strict international guidelines (Jain et al., 2019).
3. *Tilletia indica:* This fungal pathogen affects wheat and barley, causing Karnal bunt, a serious disease in wheat production. In Southeast Asia, where wheat

cultivation is expanding, biosecurity measures must include strict surveillance, control of infected plant material, and quarantining procedures for imported grains. Proper handling of infected plant samples in laboratories is critical to preventing further spread (Jain et al., 2019).

4. *Sclerotinia sclerotiorum:* Responsible for stem rot in a wide variety of crops, *Sclerotinia sclerotiorum* is a significant agricultural pathogen in Southeast Asia, where it affects crops like soybeans, sunflowers, and beans. Containment in laboratories and field studies is necessary to prevent contamination of healthy plants. Additionally, laboratory equipment and tools must be thoroughly decontaminated after use with infected material (Doehlemann et al., 2017).
5. *Dothistroma pini (Scirrhia pini):* This pathogen affects pine trees, causing needle cast and blight. *D. pini* thrives in humid tropical climates and can be a major threat to forestry industries in Southeast Asia. Biosecurity measures must include quarantining infected plant material and implementing hygiene protocols in forestry research to prevent accidental spread (Doehlemann et al., 2017).
6. *Puccinia striiformis: Puccinia striiformis* causes stripe rust in wheat, a major threat to food security. The pathogen's spores are easily spread by wind and can travel long distances, making it critical to implement containment strategies. International transport of wheat and wheat samples must adhere to quarantine protocols to prevent the spread of this pathogen between countries (Doehlemann et al., 2017).
7. *Pyricularia oryzae:* As one of the most devastating rice pathogens in Southeast Asia, *Pyricularia oryzae* causes rice blast, severely impacting rice yields. Biosecurity measures for handling *P. oryzae* in laboratories include using BSL-2 containment, preventing cross-contamination of rice samples, and applying strict protocols for the disposal of infected plant material. Surveillance systems should be in place to detect outbreaks early (Jain et al., 2019).
8. *Ustilago maydis:* A serious pathogen of maize, *Ustilago maydis* causes corn smut, leading to significant crop losses. In Southeast Asia, where corn is a staple food crop, biosecurity measures should include quarantine procedures for contaminated corn, as well as strict hygiene protocols for research environments to prevent the spread of fungal spores (Doehlemann et al., 2017).
9. *Claviceps purpurea:* This fungal pathogen infects cereal crops such as rye, producing toxic alkaloids known as ergot, which can cause ergotism in humans and animals. In Southeast Asia, ergot can affect rice and other grain crops, posing both health and economic risks. Biosecurity measures for handling *C. purpurea* include maintaining strict containment protocols to prevent accidental exposure and controlling the transport of infected grain samples to avoid contamination of unaffected regions (Doehlemann et al., 2017).
10. *Peronospora hyoscyami f.sp. tabacina:* This pathogen affects tobacco plants, causing downy mildew, which can devastate tobacco crops. In Southeast Asia, where tobacco cultivation is significant, biosecurity measures must include quarantine for infected tobacco plants and rigorous surveillance of crops for early detection. Proper containment and disposal of infected plant material are necessary to prevent the spread of the pathogen (Jain et al., 2019).

11. *Peronosclerospora philippinensis*: *Peronosclerospora philippinensis*, the causal agent of Philippine downy mildew, is a significant biosecurity concern due to its potential to devastate maize crops. This pathogen can cause severe yield losses, particularly in tropical and subtropical regions, threatening food security and agricultural economies. Its ease of spread through infected plant material and favorable environmental conditions make early detection and containment critical. Strict quarantine measures are essential to prevent its introduction into maize-growing regions currently free from the disease (Limbo-Dizon et al., 2023).

Fungal Pathogens as Possible Bioterrorism Agents

Bioterrorism is a growing concern in global security, with non-traditional warfare agents such as biological agents gaining attention due to their potential for mass disruption and public health impact. Among the various categories of biological agents, fungal pathogens represent a unique and significant threat as bioterrorism agents. Although less commonly discussed than viral or bacterial agents, certain fungal species have characteristics that make them possible candidates for biological warfare (Klassen-Fischer, 2006).

The paper "Biological Warfare Agents, Toxins, Vectors, and Pests as Biological Terrorism Agents" by Bokan and Orahovec (2004) explores the potential of various biological organisms, including fungi, to be used in bioterrorism. They discussed the traits that define an effective biological warfare agent, such as ease of dissemination, infectivity, pathogenicity, and environmental resilience, all of which are critical in evaluating the potential for fungi to serve as bioterrorism agents. Fungal pathogens, in particular, present an unconventional threat because they possess characteristics that could make them highly effective in causing widespread disease, economic disruption, and public panic.

While fungal pathogens have not been commonly used in bioterrorism to date, their potential for causing widespread disease and economic harm makes them worthy of consideration as biological warfare agents. Fungal pathogens that could be used as bioterrorism agents generally share several key characteristics. First and foremost, many fungi are highly resilient in adverse environmental conditions, allowing them to survive and thrive in diverse environments. This resilience makes them an ideal candidate for deployment in bioterrorism, as they can remain viable even in challenging conditions, such as in extreme temperatures or soil and plant matter. Additionally, many fungal pathogens can be disseminated in an aerosolized form, which increases their potential for widespread exposure and infection. Fungal spores are airborne and can be inhaled by humans or animals, making them particularly dangerous in densely populated areas or agricultural zones. The ability of fungi to survive in spores or conidia form allows them to be easily aerosolized, making them feasible for use in biological warfare (Paterson, 2006).

Another important characteristic is the ability of certain fungi to cause severe infections in humans, animals, and plants. Fungal infections can lead to chronic and debilitating diseases, sometimes with high mortality rates, especially in immunocompromised individuals. Furthermore, fungal diseases can cause substantial damage to crops and agriculture, leading to food insecurity and significant economic losses. The ability of fungi to target both human health and agriculture gives them the potential to cause multidimensional harm as part of a bioterrorist attack (Fisher et al., 2012, 2020).

Several fungal pathogens, which have been discussed above, stand out due to their potential to be used in bioterrorism. These include both human pathogens and plant pathogens, which could be deployed to inflict harm on both individuals and entire food systems. Given the potential for fungal pathogens to be used in bioterrorism, it is essential for both public health and agricultural systems to implement stringent biosecurity measures. These measures should include surveillance programs to detect early signs of fungal outbreaks, rapid response teams for containment, and research into the development of effective countermeasures, such as antifungal treatments or resistant crop strains (World Health Organization, 2024b).

Laboratories studying fungal pathogens must adhere to biosafety and biosecurity protocols to prevent accidental release or misuse. This includes ensuring proper containment of fungal cultures, as well as maintaining strict personnel security and monitoring. Research institutions should implement access control measures to prevent unauthorized personnel from gaining access to potentially dangerous pathogens. Additionally, the international community must establish and enforce guidelines for the safe transport of fungal pathogens, especially in the case of dual-use research, where research that is intended for beneficial purposes could also be exploited for harmful purposes. Coordinated efforts among governments, international organizations, and research institutions are crucial for preventing the misuse of fungal pathogens as bioterrorism agents (Vitor & Estrella, 2024).

Role of the Institutional Biorisk Management Committee

The Institutional Biorisk and Biosecurity Committee (IBBC) is crucial in ensuring that fungal research and laboratory practices adhere to biosecurity standards. The IBBC oversees biosecurity policies, conducts risk assessments, and provides guidance on safe research practices (Johnson & Dobos, 2019; Canoy & Vitor, 2024).

Key responsibilities of the IBBC include:

- *Risk Assessments*: Conducting thorough risk assessments to identify potential hazards associated with mycological research and implement risk mitigation strategies.
- *Policy Development*: Developing and enforcing biosecurity policies to protect researchers, the environment, and the public from fungal pathogens.
- *Training and Education*: Ensuring laboratory personnel are adequately trained in biosecurity protocols and regularly updated on new risks and mitigation strategies.

- *Monitoring and Compliance*: Monitoring the implementation of biosecurity measures and ensuring compliance with institutional and regulatory guidelines.

By maintaining a strong focus on safety and security, the IBBC ensures that mycological research is conducted responsibly and controlled, minimizing the risks associated with fungal pathogens.

Biosecurity in fungal research and handling is crucial to protecting human health, the environment, and the integrity of scientific progress. The complexity of biosecurity considerations in mycological research requires a multifaceted approach that incorporates material control, data security, personnel management, and emergency response protocols. As fungal pathogens evolve and present new challenges, research institutions must maintain rigorous oversight and adhere to biosecurity best practices to mitigate potential risks. The growing threat of bioterrorism and emerging fungal diseases highlights the importance of international cooperation and continuous vigilance in ensuring the safe management of fungal specimens.

Risk Management Cycle

Risk Assessment

Risk assessment is a very common theme across different industries such as finance and engineering. Similarly with biological safety, the aim of risk assessment is to identify the hazard in a particular environment. Once a hazard has been identified, accompanying risks can be listed to determine the likelihood and severity. Listing of hazards and the corresponding risks needs to be thorough since this would be the basis for the succeeding steps in reducing risks in the laboratory (Fig. 3.1).

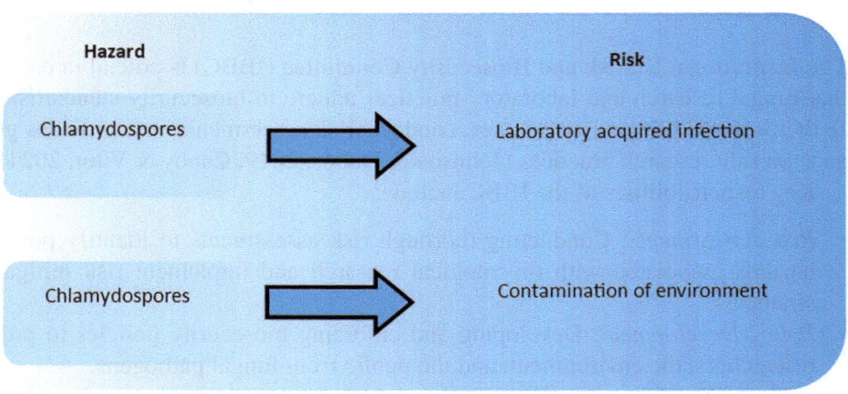

Fig. 3.1 Example of hazard and corresponding risks

Risk Assessment Process

The following are the usual steps in performing risk assessments (Canadian Centre for Occupational Health and Safety, 2025). Models such as Plan-Do-Check-Act (PDCA) and Assessment, Mitigation and Performance (Gribble et al., 2015) provide a framework on the conduct of the whole process starting with risk assessment.

1. Create a risk assessment team—Assessment should be done as a group in a collaborative manner that would produce a management approach that caters to the risk acceptability of all the individuals in the facility as much as possible. The team should be composed of individuals who have knowledge and experience of working with the procedures being assessed. In addition, external safety and security experts, legal consultants, and engineers can be included to determine other factors that will have an effect in the assessment.
2. Select the project/procedure to assess—In an ideal setting, risk assessment should be performed for all the procedures in the laboratory. Risk assessment is done for new processes that are being introduced in the laboratory or when changes in the current procedures are done. Development of new information regarding the biological agent and legislation requirements are also scenarios where risk assessment is needed. In cases where resources are not adequate, prioritization is usually performed.
3. Break down the processes into tasks—Dividing the process into small steps provides a level of granularity where hazards and risks are easily identified.
4. Identify the hazard in each task—Hazards are identified in each of the tasks listed. Existing records on incidents and accidents together with discussion with workers are valuable resources for this step.
5. List the risks accompanying each task—List each risk identified taking into consideration the likelihood of the risk happening and the severity once it occurred. Describe possible scenarios such as emergencies, extreme weather, and maintenance shutdowns that may impact the risk. Several risk assessment tools are available for use such as the WHO risk assessment template (World Health Organization, 2020b).
6. Risk evaluation—Once the risk assessments have been completed, all of the risks are reviewed and evaluated. Risk acceptability of each is discussed by the team and the next steps are formulated to reduce the likelihood and severity.

Mitigation Strategies

Once risk assessment has been completed, the recommended mitigation controls are listed to lower down the likelihood and severity of the risk (Fig. 3.2). There are several mitigation controls that can be adopted by the institution depending on the need to lower down the risk and the resources available.

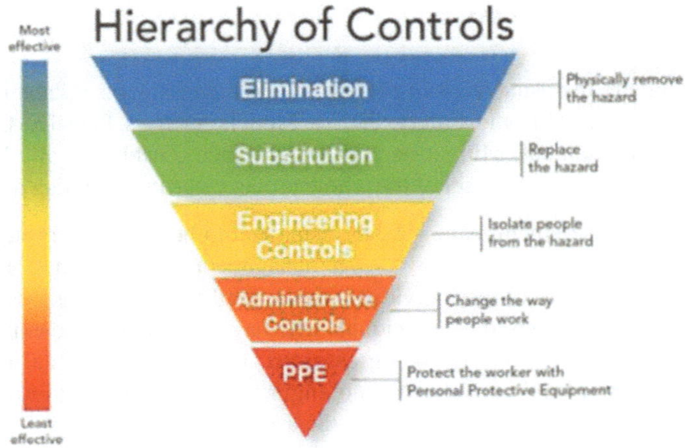

Fig. 3.2 Hierarchy of mitigation controls used in Biorisk management. (Source: National Institute for Occupational Safety and Health (NIOSH))

The Swiss Cheese model used in risk management illustrates the efficient use of different risk mitigation to reduce risk. The model likens slices of Swiss cheese placed in layers as mitigation controls that have different strengths and weaknesses or "holes." The goal is to prevent risk (red arrow) from going from one end to the other. Risk assessment-based mitigation strategy is the most sound and efficient way to reduce the risk in a facility. The current approach provided by the WHO Laboratory Biosafety Manual (LBM4) emphasizes evidence-based localized risk assessment as a foundation for mitigation strategies (Fig. 3.3).

Performance Evaluation

Performance evaluation is an integral part of risk management that validates the effectiveness of the whole program. The AMP and PDCA models rely on effective checks for continuous improvement. Performance measurement is already incorporated in the early stages of planning when doing risk assessment to include objectives, targets, and appropriate data collection plan for future evaluation. Data collection materials that are used in performance evaluation are important since it provides historical information about the institution. Forms like incident reports, non-compliance reports, and record books on worker competence are important in gauging performance. It should be kept in mind that performance evaluation is done to promote positive change and progress in an institution and not to create a blame culture.

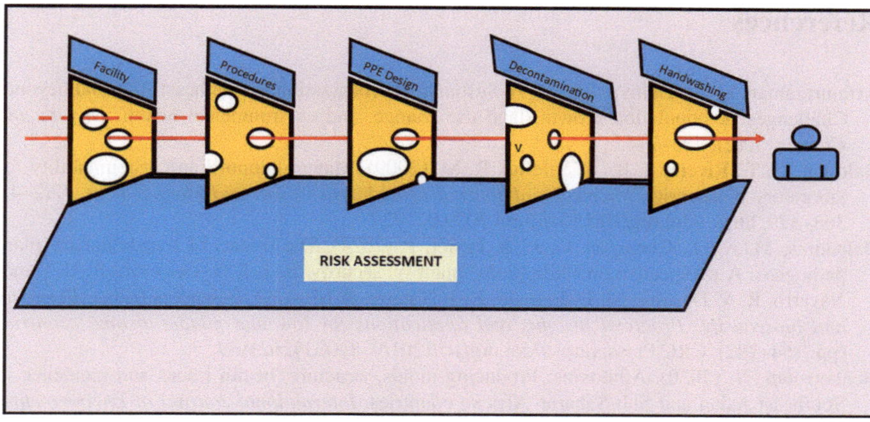

Fig. 3.3 Swiss cheese model showing five layers of mitigation controls to reduce the risk. The red broken arrow is a representation of a risk not fully prevented by the mitigation controls (Gmuender, 2023)

Risk Communication

Awareness is a key component in risk management. Risk communication is done to increase protection of workers by making them understand the risks in the institution and what are the necessary controls being made to minimize the risk. By doing risk communication, the institution promotes trust within its workforce and maintains its credibility. Having risk communication also ensures compliance to national regulations and guidelines. On the other hand, with the changes in technology and spread of information, organizations must first plan the proper way to communicate risk to avoid misinformation and the creation of panic.

Conclusion

Working with fungi requires complex biosafety and biosecurity management. Risk assessment is essential to determine the risks involved with handling specific agents using particular procedures. It provides effective foresight on the issues in the laboratory that might arise and could compromise worker and environment safety. Mitigation strategies are measures needed to reduce the risk to an acceptable level. There are numerous mitigation strategies that can be placed in a facility; mitigation controls rely on robust risk assessment to be successful. To complete a biorisk management cycle, performance measurement is needed to provide the institution an overview of the current mitigation strategies and its effect on lowering the risks. The support of the management is vital to ensure implementation of biorisk management in the laboratories.

References

Arfanuzzaman, M., & Dahiya, B. (2019). Sustainable urbanization in Southeast Asia and beyond: Challenges of population growth, land use change, and environmental health. *Growth and Change, 50*(2), 725–744.

Baldwin, G. T., Rivera, S. B., & Salerno, R. M. (2009). Material control and accountability for laboratory biosecurity. *International Journal of Risk Assessment and Management, 12*(2–4), 368–379. https://doi.org/10.1504/ijram.2009.025927

Balendres, M. A. O., Cumagun, C. J., & Taylor, P. (2024). The threats of transboundary plant pathogens: A perspective on Philippines' plant biosecurity. In J. J. G. Guerrero, R. B. Cena-Navarro, R. V. Destura, M. P. de Leon, K. I. Notarte, & M. A. O. Balendres (Eds.), *Biosafety and biosecurity: Practical insights and applications for low and middle-income countries* (pp. 154–182). CRC Press. https://doi.org/10.1201/9781003426219-9

Benkerroum, N. (2020). Aflatoxins: Producing-molds, structure, health issues and incidence in Southeast Asian and Sub-Saharan African countries. *International Journal of Environmental Research and Public Health, 17*(4), 1215.

Berger, S., El Chazli, Y., Babu, A. F., & Coste, A. T. (2017). Azole resistance in Aspergillus fumigatus: A consequence of antifungal use in agriculture? *Frontiers in Microbiology, 8*, 1024.

Bokan, S., & Orahovec, Z. (2004). Biological warfare agents, toxins, vectors and pests as biological terrorism agents. In P. J. Stopa & Z. Orahovec (Eds.), *Technology for combating WMD terrorism* (pp. 19–28). Springer Science+Business Media.

Bonifaz, A., García-Salazar, E., & Frías-De-León, M. G. (2024). Climate change exacerbating fungal disease disparities. *Current Fungal Infection Reports, 18*(1), 1–12.

Canadian Centre for Occupational Health and Safety. (2025). *How is risk assessment done? Hazard identification, risk assessment, hazard control, and evaluation*. https://www.ccohs.ca/oshanswers/hsprograms/hazard/risk_assessment.html

Canoy, R. J. C., & Vitor, R. J. S. (2024). Performing risk assessments in biological laboratories. In J. J. G. Guerrero, R. B. Cena-Navarro, R. V. Destura, M. P. de Leon, K. I. Notarte, & M. A. O. Balendres (Eds.), *Biosafety and biosecurity: Practical insights and applications for low and middle-income countries* (pp. 76–92). CRC Press. https://doi.org/10.1201/9781003426219-5

Cao, D., Wang, F., Yu, S., Dong, S., Wu, R., Cui, N., et al. (2021). Prevalence of azole-resistant Aspergillus fumigatus is highly associated with azole fungicide residues in the fields. *Environmental Science & Technology, 55*(5), 3041–3049.

Casalini, G., Giacomelli, A., & Antinori, S. (2024). The WHO fungal priority pathogens list: A crucial reappraisal to review the prioritisation. *The Lancet Microbe, 5*(7), 717–724. https://doi.org/10.1016/S2666-5247(24)00042-9

Chaabane, F., Graf, A., Jequier, L., & Coste, A. T. (2019). Review on antifungal resistance mechanisms in the emerging pathogen Candida auris. *Frontiers in Microbiology, 10*, 2788.

Chakrabarti, A., & Slavin, M. A. (2011). Endemic fungal infections in the Asia-Pacific region. *Medical Mycology, 49*(4), 337–344. https://doi.org/10.3109/13693786.2010.551426

Chindamporn, A., Chakrabarti, A., Li, R., Sun, P. L., Tan, B. H., Chua, M., Wahyuningsih, R., Patel, A., Liu, Z., Chen, Y. C., & Chayakulkeeree, M. (2018). Survey of laboratory practices for diagnosis of fungal infection in seven Asian countries: An Asia Fungal Working Group (AFWG) initiative. *Medical Mycology, 56*(4), 416–425. https://doi.org/10.1093/mmy/myx066. PMID: 29036605.

Chittarath, K., Nguyen, C. H., Bailey, W. C., Zheng, S. J., Mostert, D., Viljoen, A., et al. (2022). Geographical distribution and genetic diversity of the banana Fusarium wilt fungus in Laos and Vietnam. *Journal of Fungi, 8*(1), 46.

Chowdhary, A., Kathuria, S., Xu, J., & Meis, J. F. (2013). Emergence of azole-resistant Aspergillus fumigatus strains due to agricultural azole use creates an increasing threat to human health. *PLoS Pathogens, 9*(10), e1003633.

Crawford, E., Bobrow, A., Sun, L., Joshi, S., Vijayan, V., Blacksell, S., Venugopalan, G., & Tensmeyer, N. (2023). Cyberbiosecurity in high-containment laboratories [Original Research]. *Frontiers in Bioengineering and Biotechnology, 11*. https://doi.org/10.3389/fbioe.2023.1240281

Dayrit, J. F., Bintanjoyo, L., Andersen, L. K., & Davis, M. D. P. (2018). Impact of climate change on dermatological conditions related to flooding: Update from the International Society of Dermatology Climate Change Committee. *International Journal of Dermatology, 57*(8), 901–910.

Dayrit, J. F., Sugiharto, A., Coates, S. J., Lucero-Prisno, D. E., III, Davis, M. D. D., & Andersen, L. K. (2022). Climate change, human migration, and skin disease: Is there a link? *International Journal of Dermatology, 61*(2), 127–138.

Destura, R. V., Lam, H. Y., Navarro, R. C., Lopez, J. C. F., Sales, R. K. P., Gomez, M. I. F. A., Dela Tonga, A., & Ulanday, G. E. (2021). Assessment of the biosafety and biosecurity landscape in the philippines and the development of the national biorisk management framework. *Applied Biosafety, 26*(4), 232–244. https://doi.org/10.1089/apb.20.0070. Epub 2021 Nov 24. PMID: 36034094; PMCID: PMC9134340.

Doehlemann, G., Ökmen, B., Zhu, W., & Sharon, A. (2017). Plant pathogenic fungi. *Microbiology. Spectrum, 5*(1). https://doi.org/10.1128/microbiolspec.funk-0023-2016

Fellizar, A. L., Adalem, B. M. L., Vitor, R. J. S., Gibson, A., & Navarro, R. C. (2024). Engineering designs for biological laboratories. In J. J. G. Guerrero, R. B. Cena-Navarro, R. V. Destura, M. P. de Leon, K. I. Notarte, & M. A. O. Balendres (Eds.), *Biosafety and biosecurity: Practical insights and applications for low and middle-income countries* (pp. 320–346). CRC Press. https://doi.org/10.1201/9781003426219-15

Fisher, M. C., Henk, D. A., Briggs, C. J., Brownstein, J. S., Madoff, L. C., McCraw, S. L., & Gurr, S. J. (2012). Emerging fungal threats to animal, plant and ecosystem health. *Nature, 484*(7393), 186–194. https://doi.org/10.1038/nature10947

Fisher, M. C., Gurr, S. J., Cuomo, C. A., Blehert, D. S., Jin, H., Stukenbrock, E. H., Stajich, J. E., Kahmann, R., Boone, C., Denning, D. W., Gow, N. A. R., Klein, B. S., Kronstad, J. W., Sheppard, D. C., Taylor, J. W., Wright, G. D., Heitman, J., Casadevall, A., & Cowen, L. E. (2020). Threats posed by the fungal kingdom to humans, wildlife, and agriculture. *MBio, 11*(3). https://doi.org/10.1128/mBio.00449-20

Gadre, A., Enbiale, W., Andersen, L. K., & Coates, S. J. (2022). The effects of climate change on fungal diseases with cutaneous manifestations: A report from the International Society of Dermatology Climate Change Committee. *The Journal of Climate Change and Health, 6*, 100156.

Garnacho-Montero, J., Barrero-García, I., & León-Moya, C. (2024). Fungal infections in immunocompromised critically ill patients. *Journal of Intensive Medicine, 4*(3), 299–306. https://doi.org/10.1016/j.jointm.2024.01.005

Gmuender, F.. (2023). *Containment talk 2: The Swiss cheese risk management model for BSL-2 laboratories*. Basler and Hofmann. https://doi.org/10.13140/RG.2.2.20799.89769.

Gnat, S., Łagowski, D., Nowakiewicz, A., & Dyląg, M. (2021). A global view on fungal infections in humans and animals: Opportunistic infections and microsporidioses. *Journal of Applied Microbiology, 131*(5), 2095–2113. https://doi.org/10.1111/jam.15032

Gribble, L. A., Tria, E. S., & Wallis, L. (2015). The AMP model. In R. M. Salerno & J. Gaudioso (Eds.), *Laboratory biorisk management: Biosafety and biosecurity* (pp. 31–43). CRC Press, Taylor & Francis Group.

Guerrero, J. J. G., Regalado, R. R. H., Penceilla, G. L. M., Malonzo, C. A., General, M. A., & Junsay, A. T. (2024). Mycological biosafety and biosecurity in The Philippines. In J. J. G. Guerrero, R. B. Cena-Navarro, R. V. Destura, M. P. de Leon, K. I. Notarte, & M. A. O. Balendres (Eds.), *Biosafety and biosecurity: Practical insights and applications for low and middle-income countries* (pp. 206–234). CRC Press. https://doi.org/10.1201/9781003426219-11

Higgins, J. J., Weaver, P., Fitch, J. P., Johnson, B., & Pearl, R. M. (2013). Implementation of a personnel reliability program as a facilitator of biosafety and biosecurity culture in BSL-3 and BSL-4 laboratories. *Biosecurity and Bioterrorism, 11*(2), 130–137. https://doi.org/10.1089/bsp.2013.0024

Hung, T. N., Hung, N. Q., Mostert, D., Viljoen, A., Chao, C. P., & Molina, A. B. (2018). First report of Fusarium wilt on Cavendish bananas, caused by Fusarium oxysporum f. sp. cubense tropical race 4 (VCG 01213/16), in Vietnam. *Plant Disease, 102*(2), 448–448.

Huong, B. T. M., Tuyen, L. D., Madsen, H., Brimer, L., Friis, H., & Dalsgaard, A. (2019). Total dietary intake and health risks associated with exposure to aflatoxin B1, ochratoxin A and fumonisins of children in Lao Cai Province, Vietnam. *Toxins, 11*(11), 638.

Jain, A., Sarsaiya, S., Wu, Q., Lu, Y., & Shi, J. (2019). A review of plant leaf fungal diseases and its environment speciation. *Bioengineered, 10*(1), 409–424. https://doi.org/10.1080/21655979.2019.1649520

Johnson, C. M., & Dobos, K. M. (2019). The evolving landscape of institutional biosafety committees and biosafety programs: Results from a national survey on organizational structure, resources, and practices. *Applied Biosafety, 24*(4), 213–219. https://doi.org/10.1177/1535676019886175

Kalidasan, V., & Theva Das, K. (2021). Is Malaysia ready for human gene editing: A regulatory, biosafety and biosecurity perspective. *Frontiers in Bioengineering and Biotechnology, 9*, 649203. https://doi.org/10.3389/fbioe.2021.649203. PMID: 33777918; PMCID: PMC7992004.

Khan, Z., & Ahmad, S. (2017). Candida auris: An emerging multidrug-resistant pathogen of global significance. *Current Medicine Research and Practice, 7*(6), 240–248.

Klassen-Fischer, M. K. (2006). Fungi as bioweapons. *Clinics in Laboratory Medicine, 26*(2), 387–395, ix. https://doi.org/10.1016/j.cll.2006.03.008

Lian, K. F., Rahman, M. M., & Alas, Y. B. (2016). Making sense of inter and intraregional mobility in Southeast Asia. *International Migration in Southeast Asia: Continuities and Discontinuities*, 1–12.

Lim, J. A., Yaacob, J. S., Mohd Rasli, S. R. A., Eyahmalay, J. E., El Enshasy, H. A., & Zakaria, M. R. S. (2023). Mitigating the repercussions of climate change on diseases affecting important crop commodities in Southeast Asia, for food security and environmental sustainability—A review. *Frontiers in Sustainable Food Systems, 6*, 1030540.

Limbo-Dizon, J. E., Aldover, G. C. C., Dagamac, N. H. A., & Bennett, R. M. (2023). *Peronosclerospora philippinensis* (Philippine corn downy mildew): Predicting plant disease emergence and distribution. *Philippine Journal of Systematic Biology, 17*(1), 3–8. https://doi.org/10.26757/pjsb202317002

Lin, S. J., Schranz, J., & Teutsch, S. M. (2001). Aspergillosis case-fatality rate: Systematic review of the literature. *Clinical Infectious Diseases, 32*(3), 358–366.

Lin, X., Li, Y. Z., Chen, T., Min, S. H., Wang, D. F., Ding, M. M., & Jiang, G. (2022). Effects of wearing personal protective equipment during COVID-19 pandemic on composition and diversity of skin bacteria and fungi of medical workers. *Journal of the European Academy of Dermatology and Venereology, 36*(9), 1612–1622. https://doi.org/10.1111/jdv.18216. Epub 2022 May 18. PMID: 35538594; PMCID: PMC9348071.

Merad, Y., Belmokhtar, Z., Hadjazi, O., Belkacemi, M., Matmour, D., Merad, Z., Bassaid, A., & Megherbi, O. (2023). Fungal contamination of medical masks among forensic healthcare workers in the COVID19 era. *New Microbes New Infections, 53*, 101134. https://doi.org/10.1016/j.nmni.2023.101134. PMID: 37206638; PMCID: PMC10132829.

Murch, R. (2023). Introduction: Origin and intent for the new field of cyberbiosecurity. In D. Greenbaum (Ed.), *Cyberbiosecurity: A new field to deal with emerging threats* (pp. 7–15). Springer International Publishing. https://doi.org/10.1007/978-3-031-26034-6_2

Ordonez, N., Seidl, M. F., Waalwijk, C., Drenth, A., Kilian, A., Thomma, B. P., et al. (2015). Worse comes to worst: Bananas and Panama disease—When plant and pathogen clones meet. *PLoS Pathogens, 11*(11), e1005197.

Patel, A. U., Williams, C. L., Hart, S. N., Garcia, C. A., Durant, T. J. S., Cornish, T. C., & McClintock, D. S. (2023). Cybersecurity and information assurance for the clinical laboratory. *The Journal of Applied Laboratory Medicine, 8*(1), 145–161. https://doi.org/10.1093/jalm/jfac119

Paterson, R. R. M. (2006). Fungi and fungal toxins as weapons. *Mycological Research, 110*(9), 1003–1010. https://doi.org/10.1016/j.mycres.2006.04.004

Perlin, D. S., Rautemaa-Richardson, R., & Alastruey-Izquierdo, A. (2017). The global problem of antifungal resistance: Prevalence, mechanisms, and management. *The Lancet Infectious Diseases, 17*(12), e383–e392.

Phan, L. T. K., De Saeger, S., Eeckhout, M., & Jacxsens, L. (2023). Public health risk due to aflatoxin and fumonisin contamination in rice in the Mekong Delta, Vietnam. *Food Safety and Risk, 10*(1), 4.

Piedrahita, C. T., Cadnum, J. L., Jencson, A. L., Shaikh, A. A., Ghannoum, M. A., & Donskey, C. J. (2017). Environmental surfaces in healthcare facilities are a potential source for transmission of Candida auris and other Candida species. *Infection Control & Hospital Epidemiology, 38*(9), 1107–1109.

Rajasingham, R., Smith, R. M., Park, B. J., Jarvis, J. N., Govender, N. P., Chiller, T. M., et al. (2017). Global burden of disease of HIV-associated cryptococcal meningitis: An updated analysis. *The Lancet Infectious Diseases, 17*(8), 873–881.

Raktham, C., Bruyère, C., Kreasuwun, J., Done, J., Thongbai, C., & Promnopas, W. (2015). Simulation sensitivities of the major weather regimes of the Southeast Asia region. *Climate Dynamics, 44*, 1403–1417.

Salazar-Hamm, P., & Torres-Cruz, T. J. (2024). The impact of climate change on human fungal pathogen distribution and disease incidence. *Current Clinical Microbiology Reports, 11*(3), 140–152.

Saleh, D., Milazzo, J., Adreit, H., Fournier, E., & Tharreau, D. (2014). South-East Asia is the center of origin, diversity and dispersion of the rice blast fungus, Magnaporthe oryzae. *New Phytologist, 201*(4), 1440–1456.

Salvacion, A. R., Cumagun, C. J. R., Pangga, I. B., Magcale-Macandog, D. B., Cruz, P. C. S., Saludes, R. B., et al. (2019). Banana suitability and Fusarium wilt distribution in The Philippines under climate change. *Spatial Information Research, 27*, 339–349.

Siengsanan-Lamont, J., & Blacksell, S. D. (2018). A review of laboratory-acquired infections in the Asia-Pacific: Understanding risk and the need for improved biosafety for veterinary and zoonotic diseases. *TropIcal Medicine and Infectious Disease., 3*(2), 36. https://doi.org/10.3390/tropicalmed3020036. PMID: 30274433; PMCID: PMC6073996.

Slavin, M. A., & Chakrabarti, A. (2012). Opportunistic fungal infections in the Asia-Pacific region. *Medical Mycology, 50*(1), 18–25. https://doi.org/10.3109/13693786.2011.602989. Epub 2011 Sep 12.

Sodhi, N. S., Posa, M. R. C., Lee, T. M., Bickford, D., Koh, L. P., & Brook, B. W. (2010). The state and conservation of Southeast Asian biodiversity. *Biodiversity and Conservation, 19*, 317–328.

Soisangwan, P. (2021). Biosafety and biosecurity law in Thailand: From legislation to practice. *Journal of Biosafety and Biosecurity, 3*(2), 91–98.

Suphavilai, C., Ko, K. K. K., Lim, K. M., Tan, M. G., Boonsimma, P., Chu, J. J. K., et al. (2024). Detection and characterisation of a sixth Candida auris clade in Singapore: A genomic and phenotypic study. *The Lancet Microbe, 5*(9), 100878.

Tan, B. H., Chakrabarti, A., Patel, A., Chua, M. M. M., Sun, P. L., Liu, Z., Rotjanapan, P., Li, R., Wahyuningsih, R., Chayakulkeeree, M., & Chen, Y. C. (2020). Asia Fungal Working Group (AFWG). Clinicians' challenges in managing patients with invasive fungal diseases in seven Asian countries: An Asia Fungal Working Group (AFWG) Survey. *International Journal of Infectious Diseases, 95*, 471–480. https://doi.org/10.1016/j.ijid.2020.01.007. Epub 2020 Jan 13. PMID: 31945491.

Tin, T., & Preiser, P. R. (2018). Biosafety and biosecurity programme: Its implementation, requirements and continuous development at the Nanyang Technological University in Singapore. *Korean Journal of ClinicalLaboratory Science, 50*, 77–84. https://doi.org/10.15324/kjcls.2018.50.2.77

Vitor, R. J. S., & Estrella, F. P. (2024). Principles and concepts of biosecurity. In J. J. G. Guerrero, R. B. Cena-Navarro, R. V. Destura, M. P. de Leon, K. I. Notarte, & M. A. O. Balendres (Eds.), *Biosafety and biosecurity: Practical insights and applications for low and middle-income countries* (pp. 65–75). CRC Press. https://doi.org/10.1201/9781003426219-4

Wang, H., Xu, Y. C., & Hsueh, P. R. (2016). Epidemiology of candidemia and antifungal susceptibility in invasive Candida species in the Asia-Pacific region. *Future Microbiology, 11*, 1461–1477. https://doi.org/10.2217/fmb-2016-0099. Epub 2016 Oct 18. PMID: 27750452.

Welsh, R. M., Bentz, M. L., Shams, A., Houston, H., Lyons, A., Rose, L. J., & Litvintseva, A. P. (2017). Survival, persistence, and isolation of the emerging multidrug-resistant pathogenic yeast *Candida auris* on a plastic health care surface. *Journal of Clinical Microbiology, 55*(10), 2996–3005.

World Health Organization. (2020a). *Laboratory biosafety manual* (6th ed.). World Health Organization.

World Health Organization. (2020b). *Risk assessment*. World Health Organization. isbn:978-92-4-001131-1.

World Health Organization. (2024a). *Guidance on regulations for the transport of infectious substances, 2023–2024*. World Health Organization.

World Health Organization. (2024b). *Laboratory biosecurity guidance* (2nd ed.). World Health Organization.

Yoshizawa, G., Shinomiya, N., Kawamoto, S., Kawahara, N., Kiga, D., Hanaki, K.-I., & Minari, J. (2024). Limiting open science? Three approaches to bottom-up governance of dual-use research of concern. *Pathogens and Global Health, 118*(4), 285–294. https://doi.org/10.1080/20477724.2023.2265626

Zhang, H. F., & Islam, T. (2022). Integrated pest management programme for cereal blast fungus Magnaporthe oryzae. *Journal of Integrative Agriculture, 21*(12), 3420–3433.

Zheng, C., Hu, M., & Gao, F. (2017). Diabetes and pulmonary tuberculosis: A global overview with special focus on the situation in Asian countries with high TB-DM burden. *Global Health Action, 10*(1), 1264702.

Part II
Mycology and Its Evolving Medication Applications

Chapter 4
Unearthing Medicines: Soil Fungi and the Potential Pharmacopeia of Southeast Asia

Jonathan Jaime G. Guerrero, Ric Ryan H. Regalado, and Roselle E. Madayag-Villanueva

Abstract Soil fungi represent an immense, largely untapped source of bioactive secondary metabolites with significant pharmaceutical potential. In southeast Asia, where tropical climates, diverse ecosystems, and complex soil systems converge, fungal communities are particularly rich yet remain critically understudied. This chapter synthesizes current research on soil-derived fungi across the region, highlighting notable genera such as *Aspergillus*, *Penicillium*, *Trichoderma*, and *Ganoderma*, which have produced metabolites with antimicrobial, anticancer, antioxidant, and immunomodulatory properties. Over 130 novel compounds have been identified in the past decade alone, predominantly in Thailand, Indonesia, and Malaysia. However, the pace of fungal discovery lags behind studies on bacteria, endophytes, and mycorrhizae, reflecting longstanding biases in microbial research. Methodological challenges include taxonomic complexity, culturing difficulty, and limited access to genomic and metabolomic tools. These limitations are further compounded by soil degradation, particularly in dominant soil types such as Acrisols and Ferralsols, which are low in fertility and prone to erosion, acidification, and nutrient depletion. Anthropogenic pressures—from deforestation and monoculture farming to peatland drainage—threaten fungal biodiversity across the region. Recognizing soil fungi as both vulnerable ecological assets and promising biomedical resources is key to transforming the region's microbial wealth into sustainable health innovations.

J. J. G. Guerrero (✉)
College of Medicine, University of the Philippines Manila, Manila, Philippines
e-mail: jgguerrero1@up.edu.ph

R. R. H. Regalado
University of the Philippines Diliman, Quezon City, Philippines

R. E. Madayag-Villanueva
Agricultural Systems Institute, College of Agriculture and Food Science, University of the Philippines Los Baños, College, Los Baños, Laguna, Philippines

J. J. G. Guerrero et al. (eds.), *Mycology in a Changing Planet: Applications and Perspectives from Southeast Asia*, https://doi.org/10.1007/978-981-95-2640-6_4

Keywords Bioprospecting · Metabolites · Soil fungi

Introduction

The soil is a complex, competitive ecosystem where microorganisms engage in chemical warfare to survive. Reliable estimates posit that a single gram of soil can contain approximately 10^{10} bacteria, 10^{6} fungi, 10^{4} protozoa, and 10^{2} nematodes, including annelids and arthropods (Yang et al., 2022; Trevors, 2010; Young & Crawford, 2004). Remarkably, the majority of this biomass is microbial. Thus, with such a substantial microbial presence, it is reasonable to expect that the biosynthetic potential and genetic diversity present in soil can well be correspondingly vast.

Among these, soil fungi are master chemists, producing an array of bioactive compounds that have quietly shaped the course of modern medicine. From the discovery of penicillin, which revolutionized infection treatment (Gaynes, 2017), to the development of statins that transformed cardiovascular care (Sadowska et al., 2023), to immunosuppressants like cyclosporine (Survase et al., 2011) that made organ transplantation possible, many of the world's most important drugs have origins in fungal metabolites sourced from the soil. Thus far, the overwhelming majority of natural products discovered come from organisms that inhabit the soil (Traxler and Kolter, 2015; Li and Vederas, 2009). Members of the fungal kingdom, particularly filamentous fungi, are ubiquitous in soil microbial communities and renowned for their versatile and highly efficient metabolic machinery. The production of secondary metabolites stands out as one of the most prominent biochemical attributes of this taxon (Zhgun, 2023). These secondary metabolites, evolved over millennia to defend, communicate, and adapt, offer structures and bioactivities far beyond the reach of synthetic chemistry. And yet, despite this legacy, the vast majority of soil fungi remain unstudied—locked away in unexplored ecosystems, their medicinal potential still unearthed.

Southeast Asia stands at the forefront of this untapped microbial wealth. As one of the world's most biologically diverse regions (Morand & Lajaunie, 2024; Von Rintelen et al., 2017), its ecosystems—from dense tropical rainforests and mangrove swamps to fertile volcanic soils and agroecosystems—harbor complex microbial communities shaped by rich ecological interactions and varied environmental pressures. The soils of this region, often enriched with decaying biomass and high microbial turnover, provide an ideal habitat for diverse fungal taxa. Many of these fungi, including both well-known and endemic genera, possess biosynthetic pathways capable of producing bioactive compounds with pharmacological properties yet to be explored.

However, research into Southeast Asia's soil fungi has lagged behind that of temperate regions, leaving a significant gap in the global mapping of microbial diversity. This is a missed opportunity, as the region's fungal species represent a potential pharmacopeia—a living, natural repository of therapeutic compounds waiting to be discovered. In light of growing global health challenges, unlocking this pharmacological treasure trove could contribute not only to drug discovery but also to more equitable and sustainable healthcare innovations.

Moreover, this scientific imperative is accompanied by a conservation urgency (Hughes, 2017; Sodhi et al., 2010; Strang & Rusli, 2021). Deforestation, land degradation, urbanization, and climate change are rapidly altering or destroying fungal habitats, threatening to erase species before their medical potential can be understood. At the same time, advances in molecular biology, genome mining, and metabolomic profiling now allow researchers to investigate these organisms with unprecedented depth and speed.

This chapter aims to spotlight southeast Asia as a critical frontier for fungal bioprospecting. It explores the diversity of soil fungi in the region, summarizes known medicinal metabolites, and highlights both methodological approaches and pressing challenges. By framing Southeast Asia's fungal biodiversity as a potential pharmacopeia, this chapter advocates for stronger research, collaboration, and conservation efforts to transform ecological wealth into biomedical opportunity.

Soil Systems in Southeast Asia

Climate and Agroecology

Southeast Asia is a vast region, covering approximately 4.5 million square kilometers, around 10% of Asia's total land area (JRC & FAO, 2023). It stretches southwards from China, extending toward Australia and includes many island chains or island arcs, marked by a mountainous landscape with volcanic activities (FAO & ITPS, 2015). The region is characterized by a hot and humid climate with high rainfall levels. Most countries also have distinct wet and dry seasons due to the influence of monsoon winds. Numerous islands and island groups also significantly affect local conditions in many areas. The mainland portion features a mountainous northern region where multiple mountain ranges extend southward, separated by expansive plains and plateaus. Major river systems often drain these areas, forming extensive deltaic systems as they flow into the sea (JRC & FAO, 2023). Geologically, Southeast Asia is highly diverse. It is anchored by a stable cratonic core surrounded by a seismically and volcanically active outer arc system. This region features steep volcanic slopes, rolling granitic hills, steep mountains, and scarp-and-vale topography developed over sedimentary rocks (Gupta and Wong, 1993). This environmental diversity is reflected in the region's wide-ranging climatic conditions, vegetation types, physiographic features, and lithological conditions—all of which contribute to significant variation in soil characteristics across the region.

Distribution of Major Soils of Southeast Asia

Southeast Asia has been divided into 19 soil regions and 10 soil subregions (Fig. 4.1). According to the most recent *Soil Atlas of Asia* (JRC & FAO, 2023), the predominant soil types in the region include Acrisols, which correspond to Ultisols in the USDA classification and constitute 38% of the area. Cambisols (Inceptisols) account for 21%, while ferralsol (Oxisols) represent 10%. Gleysols make up 8% of the region, followed by Histosols (also Histosols in the USDA system) and Luvisols (Alfisols), each covering 4%. Fluvisols and Leptosols, both listed under Entisols—with Leptosols belonging to lithic subgroup—each represents 3% of the total area. Andosols (Andisols) comprise 2%, and the remaining 9% accounts for other soil types.

The following discussion will focus exclusively on the first three soil groups, which are the most predominant across Southeast Asia according to the report of the FAO-UNESCO soil map of the world.

"Acrisol-Alisol group" (USDA equivalent: *Ultisol*) is the dominant soil type in this region (FAO-UNESCO 1976, 1990, 2004; Driessen et al., 2000; Fig. 4.2). However, in 1990, the Alisol group was newly introduced, resulting in the reclassification of certain soils previously categorized under the Acrisol group, specifically those that are more chemically unstable and low in base saturation. Alisols are now distinguished from the post-1990 Acrisols group. Nonetheless, the Acrisol group

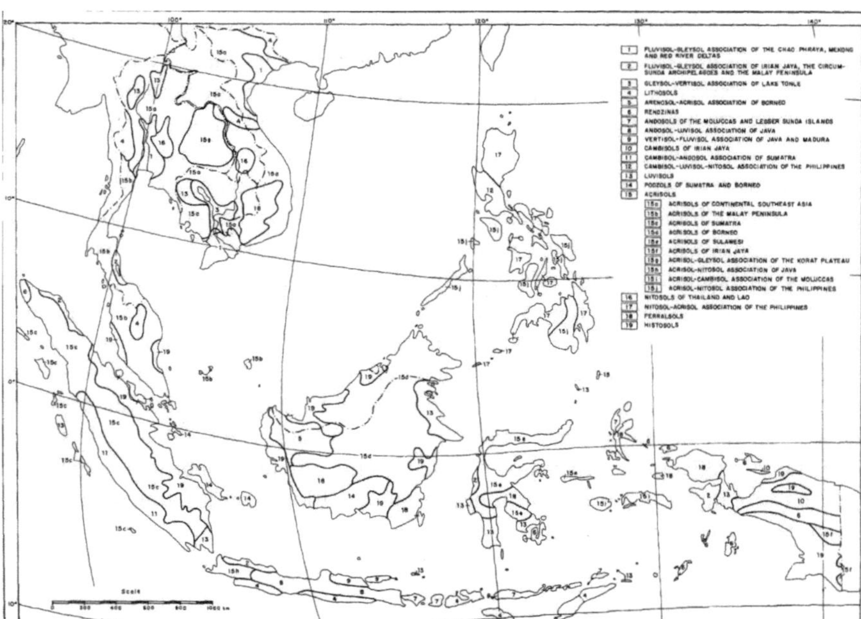

Fig. 4.1 Distribution of different soil types across Southeast Asia (FAO-Unesco, 1979)

has not yet been remapped into Alisol soils and post-1990 Acrisol soils. Therefore, the pre-1990 Acrisol soils shown in Fig. 4.2 are referred to as part of the combined Acrisol-Alisol group.

As illustrated in Figs. 4.1 and 4.2, a prominent feature of the soils of Southeast Asia is the predominance of Acrisols. The greater part of the steeply, lithic, stony, and petric members of Acrisols have serious physical limitations due to continuous cultivation. This, together with their inherent low fertility, leads to very low agricultural potential. On exposed slopes, the deep, gently sloping Acrisols without physical limitations within rooting depth have a low intrinsic fertility and are vulnerable to erosion. Expanding agricultural production without necessary management practices poses a significant risk and could lead to widespread soil degradation.

Cambisols, along with Nitisols and Luvisols, generally offer moderate to high potential for agriculture, except for members with steep slopes or poor physical characteristics. While their natural fertility ranges from low to moderate, they respond positively to the nitrogen and phosphorus fertilizers. These soils can support good yields of various annual and perennial food crops and tree plantations.

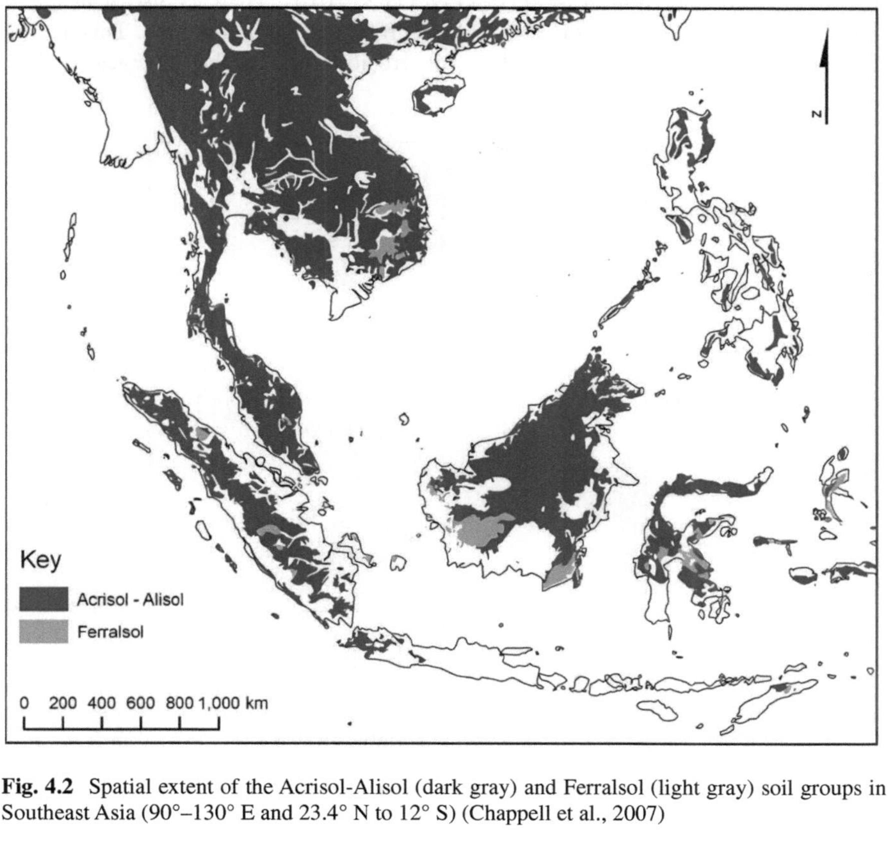

Fig. 4.2 Spatial extent of the Acrisol-Alisol (dark gray) and Ferralsol (light gray) soil groups in Southeast Asia (90°–130° E and 23.4° N to 12° S) (Chappell et al., 2007)

Ferralsols' petric and petroferric types have serious physical constraints in the root zone and low natural fertility, making them unsuitable for long-term farming. However, deeper and gently sloping members have favorable physical terrains that help compensate for their low productivity and also offer moderate potential for agriculture when managed properly.

Major Soil-Related Issues in Southeast Asia

Southeast Asia faces a range of critical soil-related challenges, including soil acidification and salinization, waterlogging, erosion, nutrient depletion, low organic matter content, poor water retention capacity, declining soil fertility and productivity, and the ecological impacts of peatland drainage. Figure 4.3 highlights region-specific soil-related constraints in the 11 Southeast Asian countries.

Soil erosion is a major concern in almost all regions. In the Philippines, erosion by water is one of the major causes of land degradation, with an estimated soil loss for agricultural land for the whole country of about 61.8 tonnes ha^{-1} $yr.^{-1}$ based on the 2003 Land Use System Map (FAO & ITPS, 2015). There is also widespread nutrient deficiency and low organic matter across nearly all countries. This decline in soil fertility was aggravated by soil erosion due to the changes in land

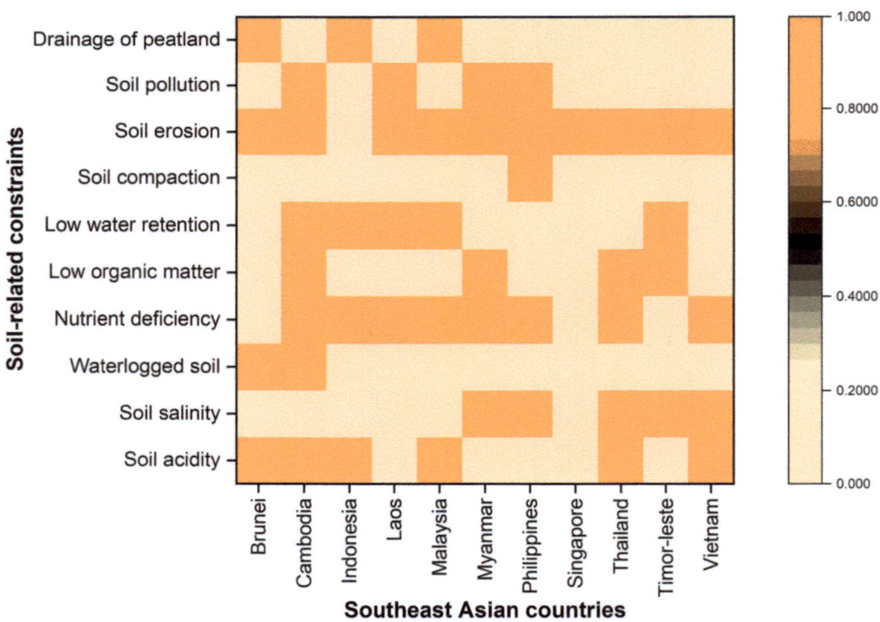

Fig. 4.3 Distribution of soil-related issues across Asia (JRC & FAO, 2023). *Note: The heatmap was generated based solely on data reported in the Soil Atlas of Asia regarding soil-related issues*

management practices (Kazuyuki, 2015). Meanwhile, soil acidity is prevalent in Brunei, Cambodia, Indonesia, Malaysia, Thailand, and Vietnam. In Southeast Asia, 7.5 million ha are acid sulfate soils (Shamshuddin et al., 2014). The ferralitic, basaltic, and gray degraded soil in Vietnam, which constitutes about one-third of the country, has a huge potential for acidification and degradation because of its nature. However, soil acidification is also primarily caused by unbalanced and unsuitable applications of chemical fertilizers (FAO & ITPS, 2015). On the other hand, the threat of soil salinization is caused mainly by saltwater intrusion (FAO & ITPS, 2015). Drainage of peatland is observed in Brunei, Indonesia, and Malaysia. The increasing demands for food and fiber due to rapid human population growth have resulted in the drainage of natural peatland ecosystems for conversion to agricultural use (Koh et al., 2011; Miettinen et al., 2011; Nurulita et al., 2014; Page et al., 2002).

These soil-related issues contribute to the broader and more urgent concern of soil degradation which involves the deterioration of the soil's organic matter, reduced fertility, nutrient imbalances, weakened aggregate stability, structure deterioration, acidification, salinization, severe soil erosion, and a loss of regional biodiversity (Chaudhary et al., 2009; Mann, 2009). The pace of agricultural transition from subsistence to intensive farming to meet the demands of rapid demographic growth has accelerated human-induced soil degradation in this region (Tassilo & Douxchamps, 2023). The most damaging activities contributing as early as 1991 to the degradation of 31% of the soils in inhabited areas in Asia are deforestation, agriculture, and overgrazing (Oldeman, 1992).

Land use change is likely the most significant driver of soil biodiversity loss in Asia (FAO & ITPS, 2015). However, alongside human-induced pressure, natural events such as tsunamis can also impact soil biodiversity. In tsunami-affected areas in the Phang Nga province of Thailand, prokaryotes (archaea and bacteria) made up a significantly higher proportion of the microbial community (83.25%) compared to non-affected areas (72.5%). In contrast, non-affected areas supported a greater presence of eukaryotes, including animals, plants, fungi, and protists (Somboonna et al., 2014). Nonetheless, as of now, there has been no comprehensive assessment of the threats to soil biodiversity even in Asia. Conducting large-scale scientific evaluations has proven challenging due to the microscopic size of soil organisms, their vast abundance and diversity, and significant research gaps and data limitations (FAO & ITPS, 2015).

Soil Fungal and Metabolite Diversity

Given the soil context of the region, soil fungi face critical challenges. The widespread degradation of soil quality due to intensive agriculture, mining, and pollution alters key physical and chemical parameters such as pH, organic matter content, and moisture levels—factors essential to fungal survival and reproduction. In monoculture plantations, the reduction in plant diversity limits symbiotic fungal networks, especially mycorrhizal associations, while excessive pesticide and fertilizer use can directly inhibit fungal growth or shift community composition toward more disturbance-tolerant, less functionally diverse species (McGuire et al., 2015; Monkai et al., 2017). In karst and ultramafic soils—already characterized by nutrient imbalances—further anthropogenic stressors may disproportionately affect endemic and specialized fungal taxa, compounding their vulnerability (Suetrong et al., 2023).

Despite these pressures, soil fungi in Southeast Asia continue to exhibit considerable diversity and resilience, particularly in remnant natural habitats and traditionally managed agroecosystems. In forested areas, complex plant-microbe interaction continues to support diverse fungal communities, including endophytes and mycorrhizae that contribute to forest regeneration and nutrient cycling (Corrales et al., 2022; Hyde et al., 2018; Peay, 2011; Robinson et al., 2020). Even in degraded and marginal lands, certain fungal taxa have adapted to stress conditions, potentially harboring stress-tolerant genes or secondary metabolites of biotechnological interest. This persistence underscores the ecological importance and adaptive capacity of soil fungi, but also highlights the urgency of documenting and conserving this diversity before it is eroded beyond recovery.

Preserving this soil fungal diversity not only supports ecosystem functioning and resilience but also sustains the potential for novel discoveries in the region's soil pharmacopeia. As pressures on tropical soils intensify, understanding the ecological roles, functional traits, and biogeographic patterns of soil fungi becomes a key priority for biodiversity conservation, sustainable land management, and bioresource innovation in southeast Asia.

In parallel with their ecological significance, soil fungi also serve as prolific chemical innovators, capable of producing a vast array of bioactive molecules with pharmacological relevance. The very traits that enable fungi to thrive in competitive and often resource-limited soil environments—such as the ability to synthesize diverse secondary metabolites—also underpin their value to drug discovery. These metabolites, evolved as survival strategies in complex microbial communities, represent a molecular expression of the fungi's ecological roles (Karlovsky, 2008; Macheleidt et al., 2016). Thus, preserving fungal diversity is not only essential for sustaining soil ecosystems but also for maintaining the biochemical potential embedded within these organisms. Understanding the environmental conditions that shape fungal metabolic expression is therefore critical to unlocking their medicinal value.

Secondary metabolites are chemical compounds (molecular weight of <1500 Da) resulting from specific biosynthetic pathways. Their production is not necessary for the normal growth and development of the organism; however, their persistence throughout evolution suggests a fitness cost and confers a competitive advantage against surrounding organisms via within- and between-species interactions. The biosynthesis of these molecules is influenced by environmental and physiological triggers. Much of the published data indicate that the most important environmental signal triggering secondary metabolism is nutrient starvation, particularly that of iron and nitrogen (Gressler et al., 2015; Tudzynski, 2014; Wiemann et al., 2014).

The signaling networks behind the regulation of this fungal secondary metabolism have also been reviewed in the literature (Haas, 2015; Brakhage, 2013).

Fungi stand out as the most prolific producers of bioactive compounds, more than 15,000 active compounds identified from microorganisms are currently known to be produced by fungi. This represents approximately 50% of all known biologically active metabolites from microorganisms (Zhgun, 2023; Chandra et al., 2020; Demain, 2014). To date, the majority of these fungal bioactive secondary metabolites are known to be from soil sources and show an enormous chemical diversity. Apart from antibiotics, which constitute the largest group of bioactive secondary metabolites, fungal compounds exhibit a range of other biological activities. These can be categorized broadly into four classes based on their bioactivity: (1) antagonistic agents, encompassing antibacterials, antifungals, antivirals, and antiprotozoals including antimalarials; (2) pharmacological agents, including antitumorals, immunomodulators, neurological agents, and enzyme inhibitors; (3) agrobiologicals such as insecticides and herbicides; and (4) regulatory agents, such as growth factors, hormones, siderophores, quorum sensing molecules, and morphogenic agents.

Southeast Asia represents a prominent hotspot for natural product drug discovery and development, boasting a taxonomically diverse assemblage of fungal taxa from which a remarkable number of chemical entities with unprecedented chemical backbones are continuously being reported (Kuephadungphan et al., 2021). This huge potential in the region arguably remains underexplored. Aside from various taxonomic novelties, numerous unique and biologically active secondary metabolites were also reported. Since 2011, a large number of compounds have been obtained from soil-derived fungi, a total of 130 new compounds were isolated and characterized during the past decade, mainly including alkaloids, terpenes, steroids, ketones, quinones, esters, and lactones. Our meta-analysis of the data collected from across the Southeast Asian region revealed that alkaloids (26.2%), terpenoids, and steroids (24.8%), followed by ketones (24.1%), constitute the main chemical classes (Fig. 4.4). Among the total of 130 new compounds reported, 45 compounds (approximately 35%) exhibit considerable biological activities, ranging from anticancer, antimicrobial, antioxidant, antimalarial to cholesterol-lowering activities. Of the 35%, anticancer (41.4%) and antimicrobial (32.8%) activities accounted for the highest proportion.

As of writing, Thailand leads in publication records, followed by Indonesia and Malaysia, while there are currently no records from other Southeast Asian neighbors. Some examples highlighting the rich biochemical resources of the region include the discovery of three new anticancer reverse prenylated indole derivatives, sartoryglabrins A–C, identified during a chemical investigation of the fungus *Neosartorya glabra* from Thailand (Kijjoa et al., 2011). Additionally, two cytotoxic polyketides of the gymnopalyne type (Thongbai et al., 2013) and the mildly antibiotic deconins (Surup et al., 2015) were obtained from soil-derived *Gymnopus* and *Deconica*, respectively, likely representing new fungal species. Two novel terpene antibiotics, named pyristriatins, were obtained from a novel species of *Cyathus* (Richter et al., 2016). A new aszonalenin analogue was isolated from the culture of

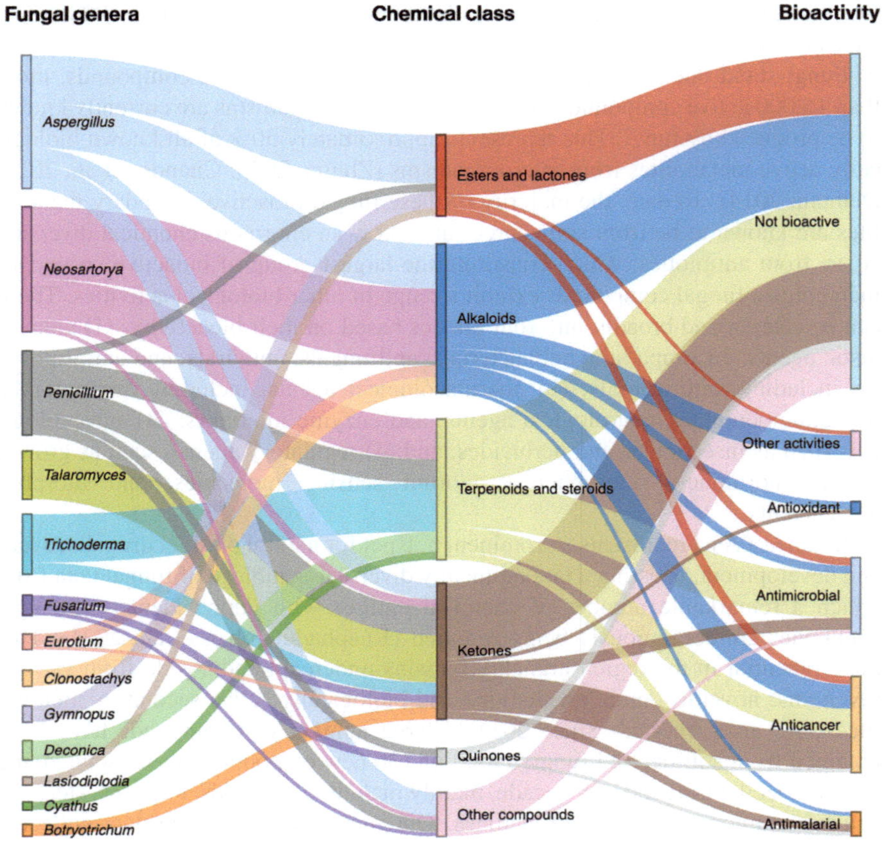

Fig. 4.4 Sankey plot showing the distribution of secondary metabolite classes from their fungal sources with their associated biological activities. Data are compiled from the 40 available studies conducted in Southeast Asia (mainly Thailand) between 2011 and 2022, highlighting trends and patterns in the region's natural product research

the soil fungus *Neosartorya fischeri* (Eamvijarn et al., 2013). Furthermore, two new cyclic carbonate derivatives, aspergillusols A and B, featuring an unusual cyclic carbonate functionality, along with a new eutypinic acid derivative, aspergillusic acid, were discovered in *Aspergillus* sp. (Rukachaisirikul et al., 2014). Similarly, three new lovastatin analogues from the soil-derived fungus *Aspergillus sclerotiorum* showed interesting cholesterol-lowering activities via HMG-CoA reductase inhibition (Phainuphong et al., 2016). Meanwhile, fermentation of the soil-derived fungus *Clonostachys rosea* in solid culture afforded three new cyclic heptapeptides as well as a new γ-lactam named clonostalactam (Abdel-Wahab et al., 2019). Additionally, two new meroterpenoid pyrones, chevalone G and aszonapyrone C, along with a novel indole alkaloid, 7-chlorofischerindoline, and a newly discovered bicyclic brasiliamide, brasiliamide H, have been identified from an isolate of

Neosartorya hiratsukae (Paluka et al., 2020). More recently, two new sesquiterpenes were discovered in the soil-derived fungus *Trichoderma citrinoviride*, including trichocitrinovirene B, which possesses a rare bicyclic sesquiterpene skeleton (Yaosanit et al., 2022).

A number of natural products from fungi associated with the plant rhizosphere have also been isolated and characterized. The cytotoxic rubrumazines A–C as well as the antioxidant 2-benzofuran-1(3*H*)-one derivative, (±)-europhenol A were isolated and identified from a culture extract of *Eurotium rubrum*, a fungus obtained from mangrove-derived rhizospheric soil (Meng et al., 2015, 2016). Similarly, *Penicillium citrinum*, associated with the roots of the mangrove *Bruguiera gymnorrhiza*, produced pravastatin (Emine et al., 2015), a known statin medication used to prevent cardiovascular disease. These few examples suggest that the rhizosphere is equally a rich source of natural products. Surprisingly, while endophyte-derived secondary metabolites have garnered considerable attention, those originating from rhizosphere microorganisms have been comparatively neglected (Gunatilaka, 2006). Microorganisms that inhabit the rhizosphere have apparently learned to deal with the ecology of the niches present and developed key genetic strategies allowing survival by chemical warfare accordingly. Overall, the majority of the compounds in Southeast Asia originated from *Aspergillus, Penicillium,* and *Neosartorya* strains collectively accounting for approximately 60%, which is typical for species belonging to Eurotiomycetes, known for their robust metabolic productivity (Robey et al., 2021). Figure 4.4 provides a detailed classification and analysis of the compounds derived from these various strains.

Challenges and Moving Forward

Despite the immense pharmacological potential of southeast Asia's soil fungi, the field remains significantly underexplored due to a host of interconnected challenges. Chief among these is the chronic underfunding and lack of dedicated infrastructure for fungal research in the region. While selected institutions in countries like Thailand and Indonesia have made progress, high-throughput platforms such as genome mining, advanced metabolomics, and compound isolation are largely inaccessible to most laboratories. This technological gap hampers the pace and scale of fungal bioprospecting and contributes to a fragmented research landscape with limited collaboration between mycologists, chemists, ecologists, and drug development specialists.

An additional concern is the current imbalance in microbial research focus. Studies on soil bacteria—especially those related to nitrogen fixation, plant growth promotion, and antibiotic production—far outnumber those of soil fungi. This reflects both a historical bias in microbiology and the practical ease of culturing and manipulating bacterial species. Similarly, within fungal research, endophytes and mycorrhizal fungi dominate the literature due to their clear plant associations and ecological roles. This uneven research attention may narrow the scope of discovery and may overlook potentially novel fungal pathways.

Compounding the scientific and technological limitations are pressing environmental issues. Much of Southeast Asia is dominated by Acrisols and Ferralsols—soils that are acidic, nutrient-poor, and vulnerable to erosion. Intensive monoculture agriculture, mining, deforestation, and peatland conversion further degrade these fragile soils, altering key properties. Such shifts severely impact fungal communities, especially those adapted to specific microhabitats or sensitive to chemical disturbance.

Moving forward, there is a critical need to elevate soil fungal research to the same priority level currently given to bacterial studies and plant-associated fungi. This entails targeted investments in research infrastructure, capacity-building for fungal taxonomy, and metabolite analysis and the development of regional, open access databases cataloging southeast Asia's fungal biodiversity and its bioactive potential. Broader soil microbial research should also shift toward ecological balance, ensuring that saprophytic and free-living fungi receive focused attention. Conservation efforts must integrate fungal diversity into land management and restoration programs, while policies on access, benefit-sharing, and ethical bioprospecting should be strengthened to ensure local ownership and sustainability.

References

Abdel-Wahab, N. M., Harwoko, H., Müller, W. E., Hamacher, A., Kassack, M. U., Fouad, M. A., Kamel, M. S., Lin, W., Ebrahim, W., Liu, Z., & Proksch, P. (2019). Cyclic heptapeptides from the soil-derived fungus *Clonostachys rosea. Bioorganic & Medicinal Chemistry, 27*(17), 3954–3959.

Brakhage, A. A. (2013). Regulation of fungal secondary metabolism. *Nature Reviews Microbiology, 11*(1), 21–32.

Chandra, P., Sharma, R. K., Arora, D. S. (2020). Antioxidant compounds from microbial sources: A review. *Food Research International, 129*, 108849.

Chappell et al. (2007). Runoff processes in Southeast Asia: Role of soil, regolith, and rock type. In Sawada H., Araki M. , Chappell N., LaFrankie J & Shimizu, Forest environments in the Mekong River Basin (pp. 3–23). Springer.

Chaudhary, V. B., Bowker, M. A., O'Dell, T. E., Grace, J. B., Redman, A. E., Rillig, M. C., & Johnson, N. C. (2009). Untangling the biological contributions to soil stability in semiarid shrublands. *Ecological Applications, 19*(1), 110–122. https://doi.org/10.1890/07-2076

Corrales, A., Koch, R. A., Vasco-Palacios, A. M., Smith, M. E., Ge, Z. W., & Henkel, T. W. (2022). Diversity and distribution of tropical ectomycorrhizal fungi. *Mycologia, 114*(6), 919–933.

Driessen, P., Deckers, J., Spaargaren, O., & Nachtergaele, F. (2000). Lecture notes on the major soils of the world. *World Soil Resources Report, 94*, 94. http://edepot.wur.nl/82729

Demain, A. L. (2014). Importance of microbial natural products and the need to revitalize their discovery. *Journal of Industrial Microbiology and Biotechnology, 41*(2), 185–201.

Eamvijarn, A., Gomes, N. M., Dethoup, T., Buaruang, J., Manoch, L., Silva, A., Pedro, M., Marini, I., Roussis, V., & Kijjoa, A. (2013). Bioactive meroditerpenes and indole alkaloids from the soil fungus *Neosartorya fischeri* (KUFC 6344), and the marine-derived fungi *Neosartorya laciniosa* (KUFC 7896) and *Neosartorya tsunodae* (KUFC 9213). *Tetrahedron, 69*(40), 8583–8591.

Emine, S., Zainol, N., Salihon, J., & Convey, P. (2015). Mangrove rhizosphere soils: A unique natural source of pravastatin-producing *Penicillium* microfungi. *International Journal of Extensive Research, 5*, 79–87.

FAO and ITPS. (2015). *Status of the world's soil resources (SWSR)—Technical summary*. Food and Agriculture Organization of the United Nations and Intergovernmental Technical Panel on Soils, Rome, Italy.

FAO-UNESCO (1990). Soil map of the world. Revised legend. Soils Bulletin 66. FAO, Rome

FAO-UNESCO (2004). Digital soil map of the world and derived soil properties. FAO, Rome

FAO-UNESCO (1979). Soil map of the world 1:5 000 000 Volume IX Southeast Asia. FAO, Rome

FAO-UNESCO (1976). Soil map of the world, sheet IX. FAO, Rome

Gaynes, R. (2017). The discovery of penicillin—New insights after more than 75 years of clinical use. *Emerging Infectious Diseases, 23*(5), 849.

Gressler, M., Meyer, F., Heine, D., Hortschansky, P., Hertweck, C., Brock, M. (2015). Phytotoxin production in Aspergillus terreus is regulated by independent environmental signals. *eLife, 4*, e07861.

Gunatilaka, A. L. (2006). Natural products from plant-associated microorganisms: Distribution, structural diversity, bioactivity, and implications of their occurrence. *Journal of Natural Products, 69*(3), 509–526.

Gupta, A., Wong, P. (1993). Physical Geography in Southeast Asia: Past themes, current status and future trends. Singapore Journal of Tropical Geography, 14(2), 252–264.https://doi.org/10.1111/j.1467-9493.1993.tb00051.x

Haas, H. (2015). Microbial ecology: How to trigger a fungal weapon. *eLife, 4*, e10504.

Hughes, A. C. (2017). Mapping priorities for conservation in Southeast Asia. *Biological Conservation, 209*, 395–405.

Hyde, K. D., Norphanphoun, C., Chen, J., Dissanayake, A. J., Doilom, M., Hongsanan, S., et al. (2018). Thailand's amazing diversity: Up to 96% of fungi in northern Thailand may be novel. *Fungal Diversity, 93*, 215–239.

Karlovsky, P. (2008). Secondary metabolites in soil ecology. In *Secondary metabolites in soil ecology* (pp. 1–19). Springer Berlin Heidelberg.

Kazuyuki Yagi (2015) Chapter 10, Regional Assessment of Soil Changes in Asia, in: Status of the World's Soil Resources- Main Report. Food and Agriculture Organization of the United Nations, Intergovernmental Technical Panel on Soils, Rome, 287–329

Kijjoa, A., Santos, S., Dethoup, T., Manoch, L., Almeida, A. P., Vasconcelos, M. H., Silva, A., Gales, L., & Herz, W. (2011). Sartoryglabrins, analogs of ardeemins, from *Neosartorya glabra*. *Natural Product Communications, 6*(6), 1934578X1100600615.

Koh, L. P., Miettinen, J., Liew, S. C., & Ghazoul, J. (2011). Remotely sensed evidence of tropical peatland conversion to oil palm. *Proceedings of the National Academy of Sciences, 108*(12), 5127–5132. https://doi.org/10.1073/pnas.1018776108

Kuephadungphan, W., Macabeo, A. P. G., Luangsa-Ard, J. J., & Stadler, M. (2021). Discovery of novel biologically active secondary metabolites from Thai mycodiversity with anti-infective potential. *Current Research in Biotechnology, 3*, 160–172.

Li, J. W. H., & Vederas, J. C. (2009). Drug discovery and natural products: end of an era or an endless frontier?. *Science, 325*(5937), 161–165.

Macheleidt, J., Mattern, D. J., Fischer, J., Netzker, T., Weber, J., Schroeckh, V., et al. (2016). Regulation and role of fungal secondary metabolites. *Annual Review of Genetics, 50*(1), 371–392.

Mann, C. C. (2009). Addicted to rubber. *Science, 325*(5940), 564–566. https://doi.org/10.1126/science.325_564

McGuire, K. L., D'angelo, H., Brearley, F. Q., Gedallovich, S. M., Babar, N., Yang, N., Gillikin, C. M., Gradoville, R., Bateman, C., Turner, B. L., & Mansor, P. (2015). Responses of soil fungi to logging and oil palm agriculture in Southeast Asian tropical forests. *Microbial Ecology, 69*(4), 733–747.

Meng, L. H., Du, F. Y., Li, X. M., Pedpradab, P., Xu, G. M., & Wang, B. G. (2015). Rubrumazines A–C, indolediketopiperazines of the isoechinulin class from *Eurotium rubrum* MA-150, a fungus obtained from marine mangrove-derived rhizospheric soil. *Journal of Natural Products, 78*(4), 909–913.

Meng, L. H., Mándi, A., Li, X. M., Liu, Y., Kurtán, T., & Wang, B. G. (2016). Isolation, stereochemical study, and antioxidant activity of benzofuranone derivatives from a mangrove-derived fungus *Eurotium rubrum* MA-150. *Chirality, 28*(8), 581–584.

Miettinen, J., Shi, C., & Liew, S. C. (2011). Two decades of destruction in Southeast Asia's peat swamp forests. *Frontiers in Ecology and the Environment, 10*, 124–128. https://doi.org/10.1890/100236

Monkai, J., Hyde, K. D., Xu, J., & Mortimer, P. E. (2017). Diversity and ecology of soil fungal communities in rubber plantations. *Fungal Biology Reviews, 31*(1), 1–11.

Morand, S., & Lajaunie, C. (2024). Biodiversity, health, and societies in Southeast Asia. *On the Edge of the Sixth Mass Extinction, 281.*

Nurulita, Y., Adetutu, E. M., Kadali, K. K., Zul, D., Mansur, A. A., & Ball, A. S. (2014). The assessment of the impact of oil palm and rubber plantations on the biotic and abiotic properties of tropical peat swamp soil in Indonesia. *International Journal of Agricultural Sustainability, 13*(2), 150–166. https://doi.org/10.1080/14735903.2014.98632

Oldeman, L. (1992). *Global extent of soil degradation* (pp. 19–36). ISRIC Bi-Annual Report 1991-1992. http://edepot.wur.nl/299739

Page, S., Siegert, F., Rieley, J. O., Boehm, H.-D. V., Jaya, A., & Limin, S. (2002). The amount of carbon released from peat and forest fires in Indonesia during 1997. *Nature, 420*, 61. https://doi.org/10.1038/nature01131

Paluka, J., Kanokmedhakul, K., Soytong, M., Soytong, K., Yahuafai, J., Siripong, P., & Kanokmedhakul, S. (2020). Meroterpenoid pyrones, alkaloid and bicyclic brasiliamide from the fungus *Neosartorya hiratsukae*. *Fitoterapia, 142*, 104485.

Peay, K. G. (2011). Building a molecular foundation for tropical mycorrhizal biology: Sporocarp surveys of ectomycorrhizal fungal diversity of southeast Asian dipterocarp forests. *NSF Award Number 1119795. Directorate for Biological Sciences, 11*(1119795), 19795.

Phainuphong, P., Rukachaisirikul, V., Saithong, S., Phongpaichit, S., Bowornwiriyapan, K., Muanprasat, C., Srimaroeng, C., Duangjai, A., & Sakayaroj, J. (2016). Lovastatin analogues from the soil-derived fungus *Aspergillus sclerotiorum* PSU-RSPG178. *Journal of Natural Products, 79*(6), 1500–1507.

Richter, C., Helaly, S. E., Thongbai, B., Hyde, K. D., & Stadler, M. (2016). Pyristriatins A and B: Pyridino-cyathane antibiotics from the basidiomycete *Cyathus cf. striatus*. *Journal of Natural Products, 79*(6), 1684–1688.

Robey, M. T., Caesar, L. K., Drott, M. T., Keller, N. P., & Kelleher, N. L. (2021). An interpreted atlas of biosynthetic gene clusters from 1,000 fungal genomes. *Proceedings of the National Academy of Sciences, 118*(19), e2020230118.

Robinson, S. J., Elias, D., Johnson, D., Both, S., Riutta, T., Goodall, T., et al. (2020). Soil fungal community characteristics and mycelial production across a disturbance gradient in lowland dipterocarp rainforest in Borneo. *Frontiers in Forests and Global Change, 3*, 64.

Rukachaisirikul, V., Rungsaiwattana, N., Klaiklay, S., Phongpaichit, S., Borwornwiriyapan, K., & Sakayaroj, J. (2014). γ-Butyrolactone, cytochalasin, cyclic carbonate, eutypinic acid, and phenalenone derivatives from the soil fungus *Aspergillus* sp. PSU-RSPG185. *Journal of Natural Products, 77*(11), 2375–2382.

Sadowska, A., Osiński, P., Roztocka, A., Kaczmarz-Chojnacka, K., Zapora, E., Sawicka, D., & Car, H. (2023). Statins—From fungi to pharmacy. *International Journal of Molecular Sciences, 25*(1), 466.

Shamshuddin, J., Azura, A. E., Shazana, M. A. R. S., Fauziah, C. I., Panhwar, Q. A., Naher, U. A. (2014). Properties and measurement of acid sulfate soils in Southeast Asia for sustainable cultivation of rice, oil palm, and cocoa. *Advances in Agronomy, 124*, 91–142.

Sodhi, N. S., Posa, M. R. C., Lee, T. M., Bickford, D., Koh, L. P., & Brook, B. W. (2010). The state and conservation of Southeast Asian biodiversity. *Biodiversity and Conservation, 19*, 317–328.

Strang, K., & Rusli, N. (2021). The challenges of conserving biodiversity: A spotlight on Southeast Asia. In *Wildlife biodiversity conservation: Multidisciplinary and forensic approaches* (pp. 47–66). Springer International Publishing.

Somboonna, N., Wilantho, A., Jankaew, K., Assawamakin, A., Sangsrakru, D., Tangphatsornruang, S. Tongsima, S. (2014) Microbial ecology of Thailand tsunami and non-tsunami affected terrestrials. *PLoS ONE*, 9.

Suetrong, S., Preedanon, S., Kobmoo, N., Srihom, C., Somrithipol, S., Saengkaewsuk, S., et al. (2023). Unravelling the hidden diversity of cave mycobiota in Thailand's Satun Geopark. *Scientific Reports, 13*(1), 19162.

Surup, F., Thongbai, B., Kuhnert, E., Sudarman, E., Hyde, K. D., & Stadler, M. (2015). Deconins A–E: cuparenic and mevalonic or propionic acid conjugates from the basidiomycete *Deconica* sp. 471. *Journal of Natural Products, 78*(4), 934–938.

Survase, S. A., Kagliwal, L. D., Annapure, U. S., & Singhal, R. S. (2011). Cyclosporin A—A review on fermentative production, downstream processing and pharmacological applications. *Biotechnology Advances, 29*(4), 418–435.

Thongbai, B., Surup, F., Mohr, K., Kuhnert, E., Hyde, K. D., & Stadler, M. (2013). Gymnopalynes A and B, chloropropynyl-isocoumarin antibiotics from cultures of the basidiomycete *Gymnopus* sp. *Journal of Natural Products, 76*(11), 2141–2144.

Tiemann, T., Douxchamps, S. (2023) Opportunities and challenges for integrated smallholder farming systems to improve soil nutrient management in Southeast Asia. *World Development Sustainability, 3*, 100080. https://doi.org/10.1016/j.wds.2023.100080

Trevors, J. T. (2010). One gram of soil: a microbial biochemical gene library. *Antonie Van Leeuwenhoek, 97*, 99–106.

Traxler, M. F., Kolter, R. (2015). Natural products in soil microbe interactions and evolution. *Natural Product Reports, 32*(7), 956–970.

Tudzynski, B. (2014). Nitrogen regulation of fungal secondary metabolism in fungi. *Frontiers in Microbiology, 5*, 656.

Von Rintelen, K., Arida, E., & Häuser, C. (2017). A review of biodiversity-related issues and challenges in megadiverse Indonesia and other Southeast Asian countries. *Research Ideas & Outcomes, 3*, e20860.

Wiemann, P., Lechner, B. E., Baccile, J. A., Velk, T. A., Yin, W. B., Bok, J. W., Pakala, S., Losada, L., Nierman, W. C., Schroeder, F. C., Haas, H., Keller, N. P. (2014). Perturbations in small molecule synthesis uncovers an ironresponsive secondary metabolite network in Aspergillus fumigatus. *Frontiers in Microbiology, 5*, 530.

Yang, G., Ryo, M., Roy, J., Lammel, D. R., Ballhausen, M. B., Jing, X., ... & Rillig, M. C. (2022). Multiple anthropogenic pressures eliminate the effects of soil microbial diversity on ecosystem functions in experimental microcosms. *Nature communications, 13*(1), 4260.

Yaosanit, W., Rukachaisirikul, V., Phongpaichit, S., Preedanon, S., & Sakayaroj, J. (2022). Sesquiterpenes from the soil-derived fungus *Trichoderma citrinoviride* PSU-SPSF346. *Beilstein Journal of Organic Chemistry, 18*(1), 479–485.

Young, I. M., Crawford, J. W. (2004). Interactions and self-organization in the soil-microbe complex. *Science, 304*(5677), 1634–1637.

Zhgun, A. A. (2023). Fungal BGCs for production of secondary metabolites: main types, central roles in strain improvement, and regulation according to the piano principle. *International Journal of Molecular Sciences, 24*(13), 11184.

Chapter 5
Endemic Mycoses of Southeast Asia: Epidemiology, Diagnostics, Immunologic Insights and Vaccine Development

Kin Israel R. Notarte, Connie Cañete-Gibas, Timothy Hudson David Culasino Carandang, Adriel Pastrana, Gerald Manalo, Jeremy Ace Ng, and Po-Ying Chia

Abstract Endemic mycoses, including cryptococcosis, aspergillosis, mucormycosis, talaromycosis, histoplasmosis, sporotrichosis, *Pneumocystis jirovecii*, and tinea capitis, present significant health challenges in Southeast Asia. We summarize the available evidence of the burden of these diseases in Southeast Asia, as well as the current available diagnostic tests, and their associated disease manifestations. Recent advancements in understanding the pathogenesis underlying these infections are summarized in this chapter as well in terms of innate and adaptive immunity. Current state-of-the-art vaccine development efforts for each of these mycoses

K. I. R. Notarte (✉)
Department of Pathology, Johns Hopkins University School of Medicine, Baltimore, MD, USA

C. Cañete-Gibas
Department of Pathology and Laboratory Medicine, University of Texas Health Science Center at San Antonio, San Antonio, TX, USA

T. H. D. C. Carandang
Research Development Unit, Dr. Paulino J. Garcia Memorial Research and Medical Center, Cabanatuan, Nueva Ecija, Philippines

Department of Clinical Epidemiology, College of Medicine, University of the Philippines Manila, Manila, Philippines

A. Pastrana · G. Manalo
University of Santo Tomas Faculty of Medicine & Surgery, Manila, Philippines

J. A. Ng
University of the Philippines Manila, College of Medicine, Manila, Philippines

P.-Y. Chia
Department of Infectious Diseases, Tan Tock Seng Hospital, Singapore, Singapore

Lee Kong Chian School of Medicine, Nanyang Technological University, Singapore, Singapore

National Center for Infectious Diseases, Singapore, Singapore

J. J. G. Guerrero et al. (eds.), *Mycology in a Changing Planet: Applications and Perspectives from Southeast Asia*, https://doi.org/10.1007/978-981-95-2640-6_5

exploring innovative approaches to induce effective immune responses in a safe manner are also presented in detail.

Keywords Fungal infections · Diagnostics · Immune response · Prevalence · Vaccines

Overview of Endemic Mycoses

Asia bears one of the highest burdens of fungal diseases globally, primarily due to its large population and tropical to subtropical climate, which create an optimal environment for fungal growth (Tan et al., 2020). In Southeast Asia, Indonesia, for instance, has a warm and humid climate, a large population, and a high prevalence of HIV infection and tuberculosis—both major risk factors for invasive fungal diseases (IFDs) (Wahyuningsih et al., 2021). Similarly, in the Philippines, tuberculosis remains a leading cause of morbidity, while the prevalence of AIDS, malignancies, and autoimmune diseases requiring chronic steroid use continues to rise (Batac & Denning, 2017), further increasing the risk of fungal infections. In Thailand, IFDs were historically linked to HIV infections; however, with the widespread availability of antiretroviral therapy (ART), fungal infections are now more frequently reported among patients undergoing organ and stem cell transplantation, those receiving immunosuppressive therapy, and critically ill individuals (Chayakulkeeree & Denning, 2017). As these trends persist, fungal diseases remain a significant public health concern across Southeast Asia.

This chapter explores mycoses in Southeast Asia, including histoplasmosis, cryptococcosis, sporotrichosis, talaromycosis, mucormycosis, aspergillosis, tinea capitis, and *Pneumocystis* pneumonia, with a focus on their immunological aspects. Given the ubiquity of fungi in the environment and the numerous opportunities they present for infections, the effectiveness of human defense mechanisms against fungal infections suggests a high degree of innate immunity to mycoses. Both specific and nonspecific defense mechanisms are believed to play a critical role in resisting the invasion of fungal pathogens into tissues. An in-depth understanding of the host's immune response to these pathogens is crucial for early presumptive diagnosis and timely initiation of antifungal therapy, particularly when pathogen isolation for definitive diagnosis is not feasible. The impact of individual fungal diseases on human health is profound and is elaborated further in this chapter. These mycoses continue to rise in incidence and remain a leading cause of morbidity and mortality worldwide, especially in immunocompromised hosts. Despite their significant health implications, these diseases remain largely neglected.

The incidence of fungal infections is challenging to determine due to several factors, including but not limited to the following: (1) Most fungal pathogens are opportunistic, primarily affecting immunosuppressed, neutropenic, or critically ill individuals with underlying conditions, and may not be recognized as part of a syndrome of systemic mycoses. (2) While some fungal infections, such as coccidioidomycosis and histoplasmosis, are considered notifiable in select states in the United States, they are not notifiable in many other countries, leading to underreporting and an unknown burden of these diseases. (3) There is a lack of rapid identification tests and affordable diagnostic kits for fungal pathogens. (4) Many clinical laboratories and healthcare institutions lack expertise in classical laboratory methods in mycology.

This chapter aims to provide an overview of endemic mycoses in Southeast Asia, the currently available diagnostic tests, and the specific fungal species and their associated disease manifestations. Given the inadequacy of current diagnostic techniques, physicians, and medical mycologists must engage in research on the epidemiology, risk factors, prevention, and clinical management of these infections. Furthermore, there is a pressing need for the development of new diagnostic methods to enhance infection control and improve patient prognosis. This chapter also summarizes recent advancements in understanding the pathogenesis underlying these infections, with a focus on the roles of innate and adaptive immunity. In addition, this chapter highlights innovative vaccine development efforts for each mycosis, focusing on strategies to induce effective and safe immune responses.

Cryptococcosis

Overview

Cryptococcus genus was traditionally classified into three varieties, five serotypes, and eight molecular subtypes, and the known species to cause disease in humans are *Cryptococcus neoformans* (Fig. 5.1) and *Cryptococcus gattii,* with the former being the most common (Maziarz & Perfect, 2016). Due to recent molecular identification techniques, there are now better-defined molecular and genetic divisions (Maziarz & Perfect, 2016). There are two forms involved in the life cycle of *Cryptococcus*: the asexual form usually found in clinical and environmental samples, and the sexual form observed only under specific laboratory conditions. Cryptococcosis is mainly caused by inhalation of infectious poorly encapsulated yeast cells or basidiospores from environmental sources like droppings from birds such as pigeons and chickens (Emmons, 1955) and hollows of certain tree species (Maziarz & Perfect, 2016), with its human pathogenicity exacerbated by environmental scavengers such as ameba (Steenbergen et al., 2001). Risk factors for cryptococcosis include HIV/AIDS (human immunodeficiency virus/acquired immunodeficiency syndrome), malignancy, organ transplant, and certain immunosuppressive treatments.

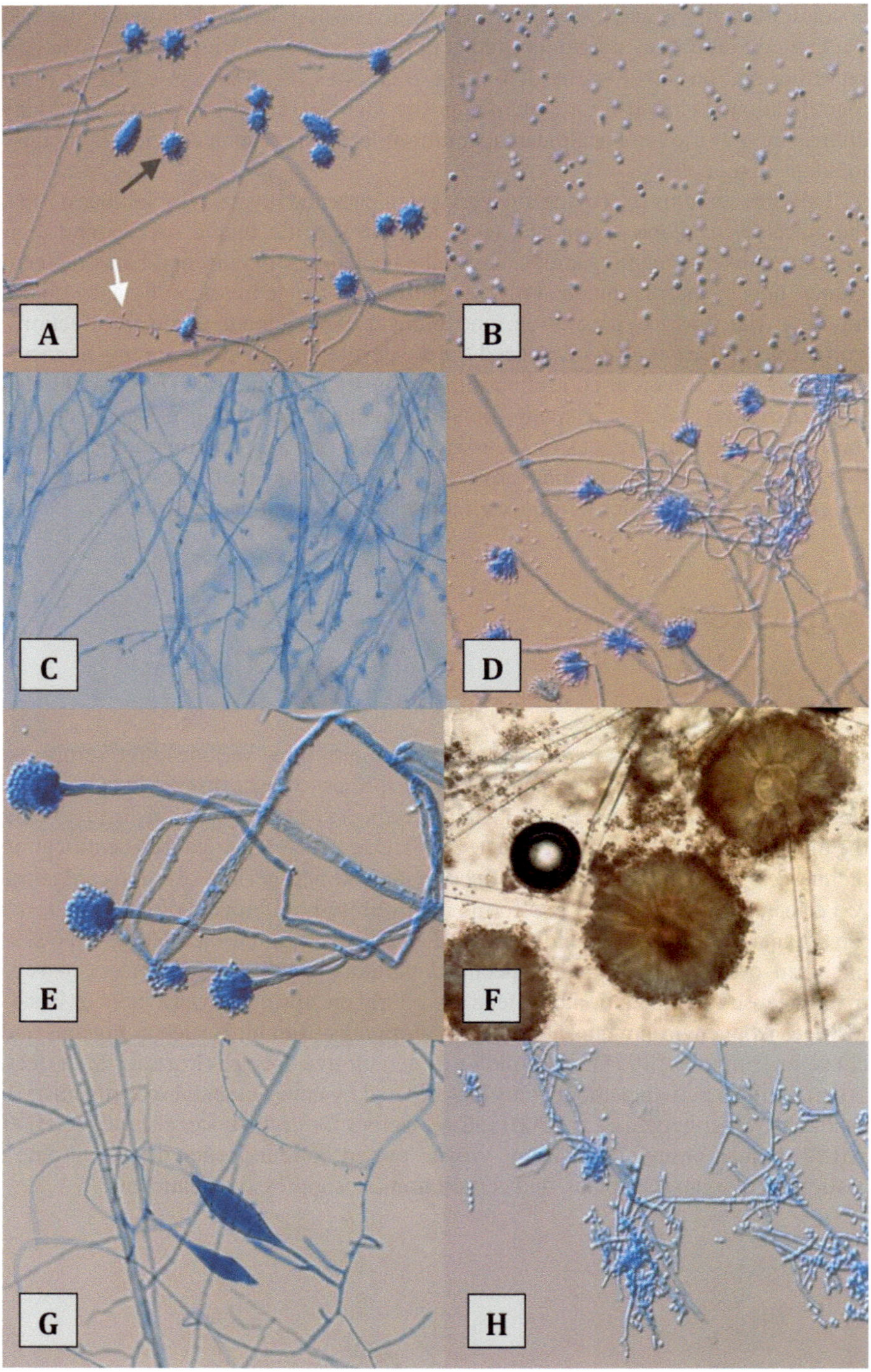
A
B
C
D
E
F
G
H

Epidemiology

In terms of epidemiology, cryptococcosis was considered uncommon before the AIDS pandemic, and its incidence increased in the 1980s with 80% of the cases being associated with AIDS (Hajjeh et al., 1999; Perfect & Casadevall, 2002; Perfect, 2015). With the broad adaptation of effective antiretroviral therapy (ART), the occurrence of HIV-associated cryptococcosis has markedly declined in most developed countries, but its incidence among other at-risk populations has remained unchanged (Bratton et al., 2012). In regions where access to ART and essential healthcare resources is limited, particularly in sub-Saharan Africa and parts of Asia, the prevalence, morbidity, and mortality rates of cryptococcal meningitis remain unacceptably high (Maziarz & Perfect, 2016). One study approximated globally that there were 223,100 incident cases, and 181,000 deaths due to cryptococcal meningitis, with the highest burden in sub-Saharan Africa (Rajasingham et al., 2017). In South and Southeast Asia, a yearly incidence of 120,000 cases and 66,000 deaths is estimated (Park et al., 2009). Multiple studies in Southeast Asian countries estimated the incidence of cryptococcosis as presented in Table 5.1.

Diagnostics

The definitive diagnosis of cryptococcosis includes isolating *Cryptococcus* from a clinical specimen or by directly detecting the fungus through India ink staining of body fluids (Maziarz & Perfect, 2016). The fastest way to diagnose cryptococcal meningitis is through direct microscopic examination of cerebrospinal fluid (CSF) using an India ink preparation to identify encapsulated yeasts. The yeasts appear as a globular, encapsulated cell, with or without budding, and measures between 5 and 20 μm in diameter (Maziarz & Perfect, 2016). *Cryptococcus* can be easily cultured from biological samples like CSF, sputum, and skin biopsies using standard fungal and bacterial culture media. They typically appear on solid agar plates after 48–72 hours of incubation at 30–35 °C under aerobic conditions (Maziarz & Perfect, 2016). Initially, they present as opaque, white-to-cream colonies, then turn to orange-tan or brown with extended incubation (Maziarz & Perfect, 2016).

Fig. 5.1 (continued) Photomicrographs of selected fungi endemic in Southeast Asia: (**a**) *Cryptococcus neoformans*—round and budding yeast cells. (**b**) *Aspergillus fumigatus*—monoseriate vesicles, phialides, and conidia at the tip of conidiophores. (**c**) *Aspergillus niger*—biseriate Fig. 5.1 (continued) vesicles with phialides and conidia covering the whole vesicle. (**d**) *Talaromyces marneffei*—biverticillate, smooth conidiophores bearing metulae, phialides, and conidia. (**e**) *Histoplasma capsulatum var. capsulatum* (black arrow: tuberculate macroconidia; white arrow: microconidia). (**f**) *Sporothrix* sp.—narrow conidiophores bearing tear-shaped conidia on delicate denticles. (**g**) *Microsporum canis*—rough spindle-shaped macroconidia tapering to a beaklike end and a few tiny club shaped microconidia forming along the hyphae. (**h**) *Trichophyton* sp.—macro and microconidia

Table 5.1 Prevalence, host immune response, and diagnosis of endemic mycotic infections from Southeast Asia

Endemic mycoses	Southeast Asian country reported / prevalence; annual incidence	Diagnostic and supportive tests	Host immune response
Cryptococcosis	**Philippines**: 0.09/100,000 (Batac & Denning, 2017) **Malaysia**: 2.8/100,000 (Velayuthan et al., 2018) **Indonesia**: 8.7/100,000 (Wahyuningsih et al., 2021) **Vietnam**: 0.5/100,000 (Duong et al., 2023) **Thailand**: 4.2/100,000 (Chayakulkeeree & Denning, 2017)	Direct microscopy: encapsulated yeast on India ink (Maziarz & Perfect, 2016) Culture Histopathology: narrow-based budding yeast (Maziarz & Perfect, 2016) Serology: cryptococcal polysaccharide capsular antigen (CrAg) (Maziarz & Perfect, 2016) Molecular (Zhou et al., 2022): PCR-fingerprinting, randomly amplified polymorphic DNA (RAPD) PCR-restriction fragment length polymorphism (PCR-RFLP) Amplified fragment length polymorphism (AFLP) Microsatellite typing Multilocus microsatellite typing (MLMT) Multilocus sequence typing (MLST) Whole-genome sequencing (WGS)	Innate: Inflammatory monocytes and macrophages inhibit the germination of fungal conidia (Hardison et al., 2012) Polymorphonuclear leukocytes (PMLs) combat *Cryptococcus* through various strategies to contain it or restrict its growth (Hardison et al., 2012) Adaptive: Th1 cells generate IFN-γ which is important for combatting intracellular fungal pathogens (Pathakumari et al., 2020) . B cells undergo class switching and generate various classes of antibodies with direct and indirect fungicidal effects (Dunne et al., 2021)

Endemic mycoses	Southeast Asian country reported / prevalence; annual incidence	Diagnostic and supportive tests	Host immune response
Aspergillosis	**Malaysia** (Velayuthan et al., 2018) Chronic pulmonary aspergillosis prevalence: 24.9/100,000 Invasive aspergillosis prevalence: 3.3/100,000 Acute bronchopulmonary aspergillosis prevalence: 98/100,000 **Philippines** (Batac & Denning, 2017) Chronic pulmonary aspergillosis prevalence: 78/100,000 Invasive aspergillosis annual incidence: 3/100,000 Allergic bronchopulmonary aspergillosis prevalence: 123/100,000 **Indonesia** (Wahyuningsih et al., 2021) Chronic pulmonary aspergillosis prevalence: 142/100,000 Invasive aspergillosis annual incidence: 18.6/100,000 Allergic bronchopulmonary aspergillosis prevalence: 126/100,000 **Vietnam** (Duong et al., 2023) Chronic pulmonary aspergillosis (5-year period) prevalence: 120.2/100,000 Invasive aspergillosis incidence: 24.4/100,000 Allergic bronchopulmonary aspergillosis incidence: 77.6/100,000	Direct microscopy (Hope et al., 2005) Wet mounts Potassium hydroxide Periodic acid–Schiff stain Gomori methenamine silver stain Calcofluor white Uvitex 2B Blankophor Immunohistochemistry FISH Culture (Hope et al., 2005) Serology (Hope et al., 2005; Cadena et al., 2021; Agarwal et al., 2024; Sojinrin et al., 2017) Galactomannan ELISA Fungus-specific IgE assays (1,3)-β-D glucan assays Gold nanoparticles Molecular (Hope et al., 2005; Vidal-Acuña et al., 2018) PCR MALDI-TOF MS	Innate: NLRP3 gets activated in response to infections by *Aspergillus* hyphae (Mancuso et al., 2022) Classical monocytes inhibit the conidial germination of the fungi (Pathakumari et al., 2020) Human plasmacytoid dendritic cells (pDCs) inhibit hyphal growth of *A. fumigatus* (Pathakumari et al., 2020) Natural killer (NK) cells target hyphal form of *A. fumigatus* (Pathakumari et al., 2020) Adaptive: *Aspergillus* triggers Th1 immune response contributing to the control of infection (Dunne et al., 2021) Th2 immune response induce the alternatively activated macrophages (Pathakumari et al., 2020) B cells undergo class switching and generate various classes of antibodies with direct and indirect fungicidal effects (Dunne et al., 2021)

Endemic mycoses	Southeast Asian country reported / prevalence; annual incidence	Diagnostic and supportive tests	Host immune response
	Thailand (Chayakulkeeree & Denning, 2017) Chronic pulmonary aspergillosis prevalence: 29.2/100,000 Invasive aspergillosis prevalence: 1.4/100,000 Acute bronchopulmonary aspergillosis prevalence: 58.4/100,000		
Mucormycosis	**Thailand** Annual prevalence: 0.2/100,000 (Chayakulkeeree & Denning, 2017) **Philippines** Annual incidence: 0.02/100,000 (Batac & Denning, 2017) **Indonesia** Annual incidence: 0.20/100,000 (Wahyuningsih et al., 2021) **Vietnam** Annual incidence: 0.20/100,000 (Duong et al., 2023)	Direct microscopy (Lackner et al., 2014) Wet mounts Potassium hydroxide (KOH) Calcofluor white Periodic acid–Schiff stain Gomori methenamine silver stain Immunohistochemistry FISH Culture (Lackner et al., 2014) API ID50C, API ID32C Serology (Lackner et al., 2014; Sojinrin et al., 2017) ELISpot Gold nanoparticles Molecular (Lackner et al., 2014; Someshwaran & Appalaraju, 2023) PCR RFLP MALDI-TOF MS RT-PCR Sequencing	Innate: Neutrophils produce cytokines and form neutrophil extracellular traps (NETs) (Hassan & Voigt, 2019 Resting (dormant) spores and germ tubes stimulate maturation of dendritic cells, initiating antifungal response (Hassan & Voigt, 2019) Adaptive: T cells produce cytokines damaging fungal hyphae (Hassan & Voigt, 2019)

Endemic mycoses	Southeast Asian country reported / prevalence; annual incidence	Diagnostic and supportive tests	Host immune response
Pneumocystis jirovecii pneumonia	**Philippines** Annual incidence 0.40/100,000 (Batac & Denning, 2017) **Indonesia** Annual incidence: 11.5/100,000 (Wahyuningsih et al., 2021) **Malaysia** Annual incidence: 4.2/100,000 (Velayuthan et al., 2018) **Vietnam** Annual incidence: 1.1/100,000 (Duong et al., 2023) **Thailand** Annual incidence: 2.6/100,000 (Chayakulkeeree & Denning, 2017)	Direct microscopy: Histochemical or immunofluorescent stains (Franconi et al., 2022) Serology: Serum (1,3)-β-D glucan (Burzio et al., 2023) Molecular: PCR using BALF or induced sputum (Burzio et al., 2023) Sputum or lavage fluid: Antibody detection through immunofluorescence (Burzio et al., 2023)	Innate: Macrophages function as antigen-presenting cells and/or as effector cells by phagocytosis (Charpentier et al., 2021) Adaptive: CD8 T cells play a protective role and maintain memory function (Dunne et al., 2021) B cells undergo class switching and generate various classes of antibodies with direct and indirect fungicidal effects (Dunne et al., 2021)
Talaromycosis (Penicilliosis)	**Malaysia** Prevalence: 2.1% (Qin et al., 2020) Annual prevalence: 1.1/100,000 (Velayuthan et al., 2018) **Indonesia** Annual incidence: 0.4/100,000 (Wahyuningsih et al., 2021) **Vietnam** Prevalence: 6.4% (Qin et al., 2020) Incidence: 1.7/100,000 (Duong et al., 2023) Incidence among hospitalized patients with AIDS: 4.4% (Le et al., 2011) **Thailand** Prevalence: 3.9% (Qin et al., 2020) Annual prevalence: 0.3/100,000 (Chayakulkeeree & Denning, 2017)	Direct microscopy (Supparatpinyo et al., 1994; Wright stain Periodic acid–Schiff stain Gomori methenamine silver stain Immunohistochemistry Culture (Narayanasamy et al., 2021) Bone marrow and lymph node biopsies Skin lesions Blood culture Serology (Ly et al., 2020; Thu et al., 2020; Kinnamon et al., 2023; Sojinrin et al., 2017) ELISA (Mp1p) Microfluidic D4 Gold nanoparticle Molecular (Narayanasamy et al., 2021) PCR	Innate: Macrophages and neutrophils facilitate the phagocytosis of the fungi (Wang et al., 2023) Adaptive: $CD4^+$ Th1 responses help control *Talaromyces marneffei* (Dunne et al., 2021) *T. marneffei* yeast induce NLRP3 inflammasome-triggered production of the cytokines IL-1β and TNF-α, which may help boost adaptive Th1 and Th17 immune responses (Wang et al., 2023)

Endemic mycoses	Southeast Asian country reported / prevalence; annual incidence	Diagnostic and supportive tests	Host immune response
Histoplasmosis	**Thailand:** 0/100,000 (Chayakulkeeree & Denning, 2017) **Indonesia:** 0.4/100,000 (Wahyuningsih et al., 2021) **Malaysia:** 0.6/100,000 (Velayuthan et al., 2018) **Vietnam:** 0.2/100,000 (Duong et al., 2023)	Culture (Rippon, 1988a, 1988b) Histopathology: intracellular ovoid yeast cells (Rippon, 1988a, 1988b) Serology: *Histoplasma* antibodies and polysaccharide antigen (Hage et al., 2011, 2015; Guimarães et al., 2004; Nacher et al., 2018; Scheel & Gómez, 2014) Molecular: PCR (Vasconcellos et al., 2019)	Innate: Dectin-1 recognition of β-(1,3)-glucan initiates proinflammatory cytokine production (Horwath et al., 2015) TNF-α promotes macrophage activation and granuloma formation, enabling early control of fungal infection (Merkhofer & Klein, 2020) Neutrophils, macrophages, and dendritic cells phagocytose pathogenic fungi (Horwath et al., 2015) Adaptive: Th1 cells generate IFN-γ which activates intracellular killing and controls primary infection (Horwath et al., 2015) B cells undergo class switching and generate various classes of antibodies with direct and indirect fungicidal effects (Dunne et al., 2021) TNF-α supports the establishment of adaptive immune memory and the control of fungal infection (Merkhofer & Klein2020)

Endemic mycoses	Southeast Asian country reported / prevalence; annual incidence	Diagnostic and supportive tests	Host immune response
Sporotrichosis	Limited data	Direct microscopy: oval to round, frequently elongated, resembling "cigar-shaped" forms on KOH stain (de Hoog & Guarro, 1995) Culture (Morris-Jones, 2002; Oyarce et al., 2016) Histopathology: inflammatory infiltrate and fungal structures (Morris-Jones, 2002) Serology: *Sporothrix* antibodies (Orofino-Costa et al., 2017) Molecular (Rodrigues et al., 2016)	Innate: *S. schenckii* cells are mainly eliminated by phagocytosis and production of reactive oxygen species (García Carnero et al., 2018) Adaptive: NLRP3 mediates the Th1 response (García Carnero et al., 2018) B cells undergo class switching and generate various classes of antibodies with direct and indirect fungicidal effects (García Carnero et al., 2018)
Tinea capitis	**Philippines** Annual incidence: 0.86/100,000 (Batac & Denning, 2017) **Indonesia** Prevalence: 270/100,000 (Wahyuningsih et al., 2021) **Vietnam** Annual incidence: 0.21/100,000 Duong et al., 2023) **Thailand** Annual incidence: 0.1/100,000 (Chayakulkeeree & Denning, 2017) **Cambodia** 30% of cases of superficial fungal infection (Uhrlaß et al., 2024)	Direct visualization Microscopy using 10% KOH (Afshar et al., 2018) Trichoscopy (Dhaille et al., 2019) Wood light examination (Mayser et al., 2020) Fungal culture (Gupta et al., 2022)	Innate: Neutrophils, macrophages, keratinocytes, and dendritic cells recognize the pathogen and initiate phagocytosis and fungicidal activity (Celestrino et al., 2021) IFN-γ and IL-12 recruit and activate neutrophils and macrophages to combat pathogen (Celestrino et al., 2021) Adaptive: Th1-associated cytokines IFN-γ and IL-12 activate and recruit macrophages and neutrophils, enhancing pathogen clearance (Celestrino et al., 2021)

FISH fluorescence in situ hybridization, *ELISA* enzyme-linked immunosorbent assay, *PCR* polymerase chain reaction, *MALDI-TOF MS* matrix-assisted laser ionization-time of flight mass spectrometry, *RFLP* restriction fragment length polymorphism, *RT-PCR* reverse transcriptase-polymerase chain reaction, *BALF* bronchoalveolar lavage fluid

Cryptococcus can also be identified through histologic staining of tissues from the lungs, skin, bone marrow, brain, and other organs (Shibuya et al., 2001). The yeast reproduces by narrow-based budding (Maziarz & Perfect, 2016). The development of serologic tests for the cryptococcal polysaccharide capsular antigen (CrAg) has significantly improved the diagnosis of cryptococcosis (Maziarz & Perfect, 2016), and latex agglutination and enzyme immunoassay techniques have been widely available for use with both serum and CSF, with the former being the most commonly used method until recently (Tanner et al., 1994; Wu & Koo, 1983). A lateral flow assay has also been approved for use in serum and CSF (Jarvis et al., 2011). This is a semiquantitative test that provides advantages over other serological methods with a rapid turnaround time of 15 minutes, minimal laboratory infrastructure requirements, stability at room temperature, low cost, and a broader detection range for cryptococcal polysaccharides (Jarvis et al., 2011). Finally, although molecular methods such as PCR-fingerprinting, randomly amplified polymorphic DNA (RAPD), PCR-restriction fragment length polymorphism (PCR-RFLP), amplified fragment length polymorphism (AFLP), microsatellite typing, multilocus microsatellite typing (MLMT), multilocus sequence typing (MLST), and whole-genome sequencing (WGS) are available and widely used for research, they are not currently employed in routine clinical practice (Maziarz & Perfect, 2016; Zhou et al., 2022).

Aspergillosis

Overview

Aspergillosis is a mycosis caused by the various, ubiquitous species of the *Aspergillus* genus (*Aspergillus fumigatus* [Fig. 5.1], *Aspergillus flavus*, *Aspergillus niger* [Fig. 5.1], and *Aspergillus terreus*); the most prevalent etiologic agent is *Aspergillus fumigatus* (Agarwal et al., 2024). Simple exposure rarely progresses to an overt manifestation of disease except in immunocompromised individuals. Pulmonary aspergillosis presents as invasive aspergillosis (IA), chronic pulmonary aspergillosis (CPA), saprobic aspergillosis, and allergic aspergillosis. Bronchopulmonary aspergillosis characterized by bronchitis, eosinophilia, bronchiectasis and/or mucus plugs, and the presence of *Aspergillus* spp. in the lungs has been designated as allergic bronchopulmonary aspergillosis (ABPA), whereas related mycoses caused by filamentous fungi, such as *Penicillium* and *Schizophyllum commune,* other than *Aspergillus* spp. are classified as allergic mycosis (ABPM) (Agarwal et al., 2024).

Epidemiology

Aspergillus colonization and sensitization are significantly associated with increased frequency of exacerbations in chronic obstructive pulmonary disease (COPD) patients as observed in a previous study involving a Southeast Asian cohort (Singapore, Malaysia, and Hong Kong; $n = 446$) (Tiew et al., 2020). Nevertheless, rates of prevalence vary greatly between countries and specific types of aspergillosis, as summarized in Table 5.1. ABPA is the most prevalent among the aspergilloses, with prevalence rates of approximately 125/100,000 in Indonesia and the Philippines (Batac & Denning, 2017; Wahyuningsih et al., 2021; Reyes et al., 2024). The prevalence of CPA is highest in Indonesia with a rate of 142/100,000 and lowest in Malaysia, with a rate of 24.9/100,000. Underlying diseases accompanying IA as bases for incidence estimation include cancer (particularly acute myeloid leukemia), organ transplantation, HIV/AIDS, ICU admission, and COPD, with the highest total annual incidence of 24.4/100,000 in Vietnam (Velayuthan et al., 2018; Batac & Denning, 2017; Wahyuningsih et al., 2021; Duong et al., 2023; Chayakulkeeree & Denning, 2017).

Diagnostics

Slender septate hyphae with angular, dichotomous branching are characteristic of *Aspergillus* spp. on direct microscopy visualized using various methods listed in Table 5.1 (Hope et al., 2005). The most commonly used diagnostic test for *Aspergillus* sensitization is the fungus-specific IgE immunoassay (cut-off 0.35 $kUA \cdot L^{-1}$, fluorescent enzyme immunoassay) (Agarwal et al., 2024). *A. fumigatus*-specific IgE (99–100%) is more sensitive than the *Aspergillus* skin test (88–94%) and can detect sensitization even against other *Aspergillus* spp., hence is the recommended screening tool for *Aspergillus* sensitization and ABPA (Agarwal et al., 2024). Detection of galactomannan, an integral component of the *Aspergillus* cell wall, has been approved for invasive aspergillosis but recommended against ABPA testing owing to its poor accuracy per the Delphi expert consensus group (Agarwal et al., 2024). (1,3)-β-D glucan (BDG) is a fungal cell wall component and less specific serologic marker; BDG detection assays employ its ability to activate a coagulation cascade in amoebocyte lysate preparations (Cadena et al., 2021; Hope et al., 2005). Thoracoscopy and subsequent histologic examination may be auxiliary and are considered to be pivotal in establishing invasive aspergillosis diagnoses in challenging cases and immunocompetent infections (Huang et al., 2024; Aljutaily & Al-Shamrani, 2022).

Mucormycosis

Overview

Mucormycosis is an opportunistic mycosis caused by the ubiquitous, saprobic, filamentous fungi of the subphylum *Mucoromycotina* whose spores enter humans via inhalation, ingestion, or direct inoculation. Rhino-orbital-cerebral mucormycosis is the most common manifestation of the disease, comprising 69.2% of cases, followed by disseminated (5.8%), cutaneous (4.1%), gastrointestinal (1.4%), and renal mucormycosis (0.7%) (Sigera & Denning, 2023).

Epidemiology

Mucorales is the second-leading invasive mycotic agent in transplant recipients and patients with malignancies (Skiada et al., 2020). Reports of mucormycosis are more common in Asia than in Europe (Sigera & Denning, 2023). The annual incidence/prevalence of mucormycosis in Thailand, Indonesia, the Philippines, and Vietnam is summarized in Table 5.1. With a prevalence of 0.14/1000, mucormycosis in India is around 80 times higher than that of developed countries (Chander et al., 2016). Up to 60% of mucormycosis infections are accompanied by diabetes as the most frequent underlying disease, followed by COVID-19 (17%) (Sigera & Denning, 2023). Since the advent of the COVID-19 pandemic, mucormycosis cases have risen, and several studies corroborate COVID-19 infection as a potential risk factor for mucormycosis (Sundaram et al., 2021; Dave et al., 2021; Srivastava et al., 2022; Notarte et al., 2023). Other underlying diseases include hematologic diseases (acute myeloid leukemia, acute lymphoblastic leukemia, aplastic anemia, non-Hodgkin lymphoma, myelodysplastic syndrome), neutropenia, and transplant recipients. Fifty-three percent of patients were on immunosuppressive treatment, while 2.1% did not have obvious predisposing factors (Sigera & Denning, 2023). Throughout the year, seasonal variations in the prevalence of mucormycosis infection are discernible with over half (53.4%) of the cases presenting in summer and autumn between August and October (Aksoy & Ulas, 2024).

Diagnostics

The most prevalent diagnostic method is direct microscopy using potassium hydroxide (KOH) or Calcofluor white, followed by histopathology and culture methods. Molecular techniques and immunofluorescence are employed to a much lesser extent (Sigera & Denning, 2023). Biopsy of tissues including subcutaneous fat with concomitant molecular diagnostic testing is recommended; in particular, species-specific 18S rRNA-targeted fluorescence in situ hybridization (FISH) with polymerase chain reaction (PCR) assays are recommended (Rickerts et al., 2006, 2013). Most PCR, restriction fragment length polymorphism (RFLP), and sequencing molecular assays are based on the internal transcribed spacer (ITS) region as the pan-fungal barcode, while some reverse-transcriptase PCR (RT-PCR) detection tools are based on 28S RNA followed by melt curve analysis (Lackner et al., 2014; Kasai et al., 2008). In contrast to KOH or culture, matrix-assisted laser ionization-time of flight mass spectrometry (MALDI-TOF MS) has also shown to more specifically and rapidly detect mucormycosis agents at the genus or species level (Someshwaran & Appalaraju, 2023). A plasmonic gold nanoparticle-based system has been described to detect cutaneous *Mucor hiemalis*, *Aspergillus oryzae*, *Penicillium chrysogenum*, and *Aspergillus niger* infection without prior sample preparation or extraneous instrumentation (Sojinrin et al., 2017). Imaging modalities such as the fluorine-18 fluorodeoxyglucose positron emission tomography/computed tomography (18F-FDG PET/CT) have also been documented to be helpful in the localization of an intestinal mucormycosis fungal ball and mucormycosis-associated head and neck lesions (Gallo et al., 2019; Liu et al., 2013).

Pneumocystis Pneumonia

Overview

Pneumocystis jirovecii, previously named as *Pneumocystis carinii*, is an opportunistic yeast-like fungi which normally resides within mammalian lung in trophic and cystic forms (Kibbler et al., 2018). It causes pneumonia when the host is immunocompromised, typically presenting as exertional dyspnea, malaise, dry cough, low-grade fever, poor inspiratory effort and may include pleuritic chest pain, hemoptysis, and pneumothorax (Kibbler et al., 2018). Persons living with HIV (PLHIV) with CD4 T-cell counts of less than 200 cells per microliter or CD4 T-cell percentage of all lymphocytes less than 14% account for almost 90% of *Pneumocystis jirovecii* pneumonia (PJP) cases (Kibbler et al., 2018). Recently, incidence of PJP in patients without HIV infection was noted to be increasing and is linked to more aggressive use of immunosuppressive agents, increased implementation of organ and stem cell transplantation, application of more sensitive diagnostic methods, and improved awareness of PJP by physicians (Xue et al., 2023).

Data inadequacy suggests that active surveillance for serious fungal infections is not done in any of the Southeast Asian countries (Batac & Denning, 2017). National surveillance programs for fungal infections are lacking in Southeast Asia, though these statistics are considered the gold standard for estimating disease prevalence and incidence in developing countries (Batac & Denning, 2017; Velayuthan et al., 2018). The complexity and cost of these programs hinder their implementation. Therefore, all the studies estimating national epidemiology of PJP in Southeast Asia cited here have opted for actuarial estimates; national HIV incidence partnered with assumptions usually derived from small studies investigating PJP were used to estimate national incidence or prevalence of PJP (Rodriguez-Tudela et al., 2015). Data on PJP in non-HIV cases are therefore lacking in Southeast Asia.

Epidemiology

The annual incidence of PJP in the Philippines is estimated to be 0.40/100,000, the lowest in Southeast Asia. This is based on an assumption that 31% of new HIV cases have PJP, the rationale of which was not explicitly stated (Batac & Denning, 2017). On the other hand, the annual incidence of PJP in Malaysia is estimated to be 4.2/100,000, derived from the assumption that 15% of all newly presenting HIV/AIDS patients have PJP (Velayuthan et al., 2018). This proportion was calculated from Thai retrospective study (Ratanasuwan et al., 2005). Indonesia used the same assumption, leading to an estimated annual incidence of 11.5/100,000, the highest in Southeast Asia (Wahyuningsih et al., 2021). The estimated annual incidence for Vietnam is 1.1/100,000, based on the assumption that 6.2% of cases of newly diagnosed AIDS and HIV drug resistance involve PJP (Duong et al., 2023). This assumption was based on a regional retrospective study of selected hospitals in Japan, Northern Thailand, Northern Vietnam, and the Philippines (Gangcuangco et al., 2017). Thailand was estimated to have an annual incidence of 2.6/100,000, based on the assumption that 21% of new HIV cases present with PJP (Chayakulkeeree & Denning, 2017). This proportion was calculated from a local study which used Flinders Technology Associates (FTA) filter paper together with PCR in noninvasive induced sputum specimens (Jaijakul et al., 2005). The epidemiology of PJP in Southeast Asia is summarized in Table 5.1.

Diagnostics

The gold standard for diagnosing PJP is direct visualization of respiratory samples stained histochemically or via immunofluorescence (Franconi et al., 2022). However, the sensitivity of microscopic visualization is low since it is highly dependent on the skill of the observer, quality of sample, and fungal burden, which is low in non-HIV patients (Burzio et al., 2023). Serum BDG has been proposed as first-line test in several algorithms due to its high sensitivity [91%; 95%CI (87–94%)], though it has a lower specificity [84%; 95% CI (83–86%)] (Burzio et al., 2023; Apostolopoulou & Fishman, 2022; Onishi et al., 2012). PCR using bronchoalveolar lavage fluid (BALF) has higher specificity [91.0%; 95% CI (82.7–95.5%)] and can therefore be used to confirm results from serum BDG. Alternatively, PCR using induced sputum can be used if fiber optic bronchoscopy cannot be performed, such as in cases of high-grade tracheal obstruction, life-threatening arrhythmias, recent myocardial infarction, uncorrectable coagulopathy, or even the inability or unwillingness to hold still for the procedure (Burzio et al., 2023; Dezube, 2023). The excellent specificity of antibody detection through immunofluorescence [100%; 95% CI (91–100%)] is also notable, making it particularly useful for further confirmation, in case of negative PCR result (Burzio et al., 2023; Senécal et al., 2022). Details about the diagnosis of PJP are summarized in Table 5.1.

Talaromycosis

Overview

Talaromycosis, formerly known as penicilliosis, is an invasive mycosis caused by the thermally dimorphic fungus *Talaromyces marneffei* (previously *Penicillium marneffei*; Fig. 5.1) endemic to Southeast Asia, Southern China, and North-Eastern India (Narayanasamy et al., 2021). Airborne spores released by the environmental mycelium form are inhaled and settle in the lungs. Alveolar macrophages engulf the spores, and body temperature (37 °C) triggers a mold-to-yeast morphologic transformation. Uninucleate yeast cells divide by binary fission, forming fission arthroconidium cells, and multiply until it ruptures the macrophage, thereby spreading within the lungs and to other organs (Vanittanakom et al., 2006; Wang et al., 2023).

Epidemiology

Talaromycosis has been primarily described in immunocompromised individuals, with a pooled prevalence of approximately 3.6% (0.13–19.63%) in persons living with HIV/AIDS (PLHIV) in the endemic region (Qin et al., 2020). Talaromycosis has been increasingly reported in individuals without HIV, representing 12.3% (82 out of 668) of cases (Hu et al., 2012). The prevalence of talaromycosis in PLHIV varies from 0.1% to 26.7% in countries such as China (including Hong Kong), Vietnam, Thailand, and India. Furthermore, the mortality rates associated with this condition in these regions vary between 8% and 40% (Qin et al., 2020); WHO (2022) estimates mortality in adult PLHIV to be approximately 12–21%. AIDS-associated talaromycosis of the central nervous system has a mortality rate of 81% in a Vietnamese study, emphasizing the imperative of prompt diagnosis and treatment (Le et al., 2010). Reported incidence rates detailed in Table 5.1 range from 0.3/1000,000 in Thailand to 1.7/100,000 in Vietnam (Chayakulkeeree & Denning, 2017; Duong et al., 2023). Humidity is the most important environmental predictor of talaromycosis admission, with increasing incidence during the rainy months; no significant associations were found with wind and temperature (Bulterys et al.,

2013). In persons without HIV, tuberculosis is a common misdiagnosis in 80.7% of cases, and bacterial pneumonia in 20.5% of cases (Guo et al., 2019).

Diagnostics

At present, the gold standard for diagnosis of talaromycosis is still biopsy and culture, which can take up to 28 days and result in treatment delays; however, antigen and PCR-based assays are reported to be superior to culture-based diagnostics (Narayanasamy et al., 2021). The Mp1p enzyme immunoassay based on a *T. marneffei*-specific cell wall mannoprotein is reported to be highly specific (98.1%), more rapid, and more sensitive compared to standard BACTEC blood cultures (86.3% vs. 72.8%) and real-time PCR assays (86.3% vs. 70–80%) (Ly et al., 2020; Thu et al., 2020). Development of a microfluidic D4 point-of-care test for talaromycosis detection in advanced HIV disease reported a sensitivity of 92.3% (CI: 75.9–98.6%), a specificity of 100% (CI: 67.6–100.0%), and a turnaround time of 30–45 minutes (Kinnamon et al., 2023). Pleural effusions with lower protein content and lymphocyte count and higher neutrophil count relative to tuberculous pleural effusion is a differential characteristic of talaromycosis among other opportunistic mycoses such as mucormycosis and aspergillosis (Narayanasamy et al., 2021).

Histoplasmosis

Overview

Histoplasmosis is caused by *Histoplasma capsulatum* (Fig. 5.1). Three varieties were originally used for classification: *H. capsulatum var. capsulatum*, *H. capsulatum var. duboisii*, and *H. capsulatum var. farciminosum* (Chakrabarti & Slavin, 2011). In humans, the first two variants cause histoplasmosis (Araúz & Papineni, 2021), and the most common variant is *H. capsulatum var. capsulatum* which is found worldwide (Myint et al., 2020). *H. capsulatum* is a dimorphic fungus with two different morphologies based on environmental conditions: in the soil, it grows as a mold or mycelium, and in mammalian hosts it develops as a budding yeast (Maresca & Kobayashi, 1989). In the mycelial phase, it exists as either macroconidia or microconidia (Kauffman, 2007). Being a soil saprophyte, it is mainly found in microfoci like damp soil with large amounts of bird excreta or bat guano and can cause infection when the aerosolized microconidia are inhaled (Araúz & Papineni, 2021). As such, demolitions, constructions, farming, raising chickens, archaeological digging, and cave explorations increase the risk of infection (McKinsey et al., 1997). Meanwhile, AIDS (Marukutira et al., 2014), primary immunodeficiencies

(Zerbe & Holland, 2005), other immunosuppressive disorders (Wheat et al., 1982; Smith & Kauffman, 2009), immunosuppressive drugs (Cuellar-rodriguez et al., 2009), and extreme ages (Wheat et al., 1982; Odio et al., 1999) are risk factors for dissemination.

Epidemiology

Epidemiological data regarding histoplasmosis are mostly described in the United States (Baker et al., 2019), and older studies show variable distribution of histoplasmosis in different areas worldwide (Cano & Hajjeh, 2001; Bahr et al., 2015). In Southeast Asia, its epidemiology has historically been neglected (Baker et al., 2019), except in Thailand where 1253 disseminated histoplasmosis in patients with HIV were reported to the Ministry of Public Health from 1984 to 2010 (Norkaew et al., 2013). Recently, Baker et al. (2019) conducted a systematic literature review of cases of histoplasmosis in Southeast Asia from 1932 to 2018. Of the 407 cases obtained, the number of cases varied per country, and the most abundant cases were seen in Thailand (233), Malaysia (76), Indonesia (48), and Singapore (21) with minimal or no cases in other countries (Baker et al., 2019). Recent studies in Indonesia, Malaysia, and Vietnam estimated the incidence and prevalence of serious fungal infections including histoplasmosis in their respective regions as seen in Table 5.1.

Diagnostics

In terms of diagnosis, the gold standard is still isolation of the mold from cultures of clinical specimens (Azar & Hage, 2017). *H. capsulatum* typically grows within 2–4 weeks, although it can occasionally take up to 8 weeks (Rippon, 1988a, 1988b). The resulting colonies are white or buff-brown, exhibiting a suede-like to cottony texture, with a pale yellow-brown reverse when grown at 25 °C on Sabouraud's dextrose agar (Rippon, 1988a, 1988b). Histopathological evidence showing the presence of yeast cells compatible with *H. capsulatum* in tissue is part of the diagnostic criteria in international guidelines for proven histoplasmosis (Donnelly et al., 2020). Cytology (i.e., peripheral blood cell smear, bone marrow smear, bronchoalveolar lavage fluid, or fine needle aspiration) can also help in diagnosis, but is usually only useful if the disease is disseminated (Gupta et al., 2010). Serology is primarily used in diagnosing subacute and chronic forms of histoplasmosis since antibodies become detectable 4–8 weeks following the initial infection (Hage et al., 2011, 2015). The serological tests mainly used for identifying *H. capsulatum* antibodies include immunodiffusion, complement fixation, and enzyme immunoassay (Guimarães et al., 2004). Antigen testing has improved the diagnosis of histoplasmosis, offering a quick, noninvasive method that is both sensitive and specific, particularly in individuals with HIV (Nacher et al., 2018; Scheel & Gómez, 2014). Monitoring treatment response can also be done by observing the clearance of the *Histoplasma* antigen (Hage et al., 2011). Finally, molecular tests such as PCR offer an alternative for obtaining rapid diagnostic results, although they still need standardization (Vasconcellos et al., 2019).

Sporotrichosis

Overview

Sporotrichosis is the most common subcutaneous mycosis caused by the species of the *Sporothrix schenckii* complex, which consists of over six species of closely related dimorphic fungi (Vásquez-del-Mercado et al., 2012). They can be found in soil or decaying matter, and some animals can be vectors for the disease such as cats, dogs, horses, camels, donkeys, or rats (Crothers et al., 2009). Sporotrichosis was historically referred to as "rosebush mycosis" or "gardener's mycosis" because the infection typically occurs when the skin or mucous membrane is inoculated through trauma with contaminated plant material (Orofino-Costa et al., 2017). However, there have also been reports of zoonotic transmission and, less commonly, cases where inhaled fungal propagules lead to systemic mycosis (Rippon, 1988a, 1988b; Morris-Jones, 2002). Risk factors for sporotrichosis include working with wood, plants, flowers, or infected animals, alcoholism, diabetes, immunosuppression regardless of the cause, and laboratory manipulation of the fungus (Vásquez-del-Mercado et al., 2012).

Epidemiology

Sporotrichosis is found globally but occurs more frequently in tropical and subtropical regions that have warm temperatures (15–25 °C) and high humidity (90%) (Vásquez-del-Mercado et al., 2012). Sporotrichosis is frequently observed in various parts of the Asia-Pacific region with a broad range of climatic and geographic conditions, e.g., India, Japan, Australia (Rajendran et al., 1990; Houh et al., 1995; Ghosh et al., 1999; Haldar et al., 2007; Feeney et al., 2007; Ishizaki et al., 2009; Kikuchi et al., 2006). However, in some Southeast Asian countries with similar climatic conditions like Thailand, Vietnam, or Indonesia, it is rarely reported (Chakrabarti & Slavin, 2011). Liewchanpattana et al. (2020) reviewed the existing literature and found a total of 19 microbiological confirmed cases of sporotrichosis in Southeast Asia. Malaysia has the highest reported percentage (73.7%), followed by Thailand (21.1%), and Laos PDR (5.3%) (Liewchanpattana et al., 2020). Moreover, the same study discussed only two proven cases of sporotrichosis seen in

Thailand since 1990, but the authors noted possible underreporting because of under-recognition and limited laboratory availability (Liewchanpattana et al., 2020). In Malaysia, six case reports of ocular sporotrichosis were noted (Ahmad-Fauzi et al., 2021).

Diagnostics

The gold standard for diagnosing sporotrichosis involves isolating and identifying the *Sporothrix* species (Fig. 5.1) from various clinical samples including skin lesions, biopsies, aspirates from floating abscesses, sputum, pus, synovial fluid, blood, and cerebrospinal fluid (Morris-Jones, 2002; Kauffman et al., 2007). Direct microscopy using potassium hydroxide preparations, with or without dimethyl sulfoxide, seldom shows fungal elements (Orofino-Costa et al., 2017), and in tissue, *Sporothrix* spp. display various yeast-like structures, ranging from oval to round, and frequently elongated, resembling "cigar-shaped" forms, typically measuring around 5–8 μm (de Hoog & Guarro, 1995). *Sporothrix* spp. can be cultivated on standard culture media at room temperature (25–30 °C) (Morris-Jones, 2002; Oyarce et al., 2016), and isolation of *Sporothrix* spp. occurs within 4–6 days for samples collected from skin lesions and within 10–19 days for extracutaneous lesions (Oyarce et al., 2016; Hessler et al., 2017; Pappas et al., 2000; Zhao et al., 2015). Colonies are phenotypically identified by their membranous appearance on the surface, which typically appears off-white to cream with a black halo (Orofino-Costa et al., 2017). They may also be dark from the outset, varying according to the species and environmental factors (Orofino-Costa et al., 2017). The sensitivity of histopathology tests in humans is limited because fungal elements in the tissue are sparse (Orofino-Costa et al., 2017). The inflammatory infiltration is better seen with hematoxylin-eosin stains, while PAS or methenamine silver stains are used to detect fungal structures (Morris-Jones, 2002). In terms of serology, ELISA (enzyme-linked immunosorbent assay) and Western blot are currently used which are quick and non-invasive diagnostic methods for sporotrichosis. These tests serve as supplementary tools for diagnosing cutaneous forms and identifying systemic or atypical presentations of sporotrichosis, and are valuable for screening, monitoring treatment progress, and guiding medication adjustments in challenging clinical scenarios (Orofino-Costa et al., 2017). Finally, distinct molecular boundaries now exist among *Sporothrix* spp., facilitating the creation of multiple genetic markers for the recognition and identification of clinical specimens (Rodrigues et al., 2016). The advancement of rapid and cost-effective genotyping techniques is crucial for both diagnosis and epidemiological research (Orofino-Costa et al., 2017).

Tinea Capitis

Overview

Tinea capitis is a superficial mycosis of the scalp and hair caused by dermatophyte molds; *Microsporum canis* (Fig. 5.1) is the most prevalent zoophilic agent, while *Trichophyton violaceum* and *Trichophyton tonsurans* are the most common anthropophilic agents (Rodríguez-Cerdeira et al., 2021). Recent studies in Thailand and Cambodia show that the most common etiology of tinea capitis is *T. rubrum*, followed by *T. mentagrophytes* and *M. canis* (Udomphan et al., 2019; Uhrlaß et al., 2024). A study in the Philippines shows a different picture; the most common causes of tinea capitis were *M. canis, Trichophyton tonsurans, T. mentagrophytes, and M. gypseum* (Simuangco et al., 1963). Philippine data are more similar to global epidemiologic data, though more studies are needed to update these estimates. It mainly affects children, aged 3–7 years, though traditional estimates show that between 3% and 11% of all cases occur in adults (Leung et al., 2020; Hill et al., 2024). Other risk factors include crowded living conditions, close proximity to animals, and immunosuppression (Hill et al., 2024). Tinea capitis can be transmitted through direct contact, fomites, and asymptomatic carriers (Hill et al., 2024). It has non-inflammatory subtypes, ectothrix and endothrix infections, and inflammatory subtypes, which are favus and kerion. Ectothrix infections may present as single or multiple scaly, erythematous patches with gray, circular areas of hair loss (gray-patch ringworm) or white scales which resemble dandruff, while endothrix infections may manifest as black dots on a patch of bald scalp (black dot ringworm) due to hair breakage (Hill et al., 2024). Favus and kerion, on the other hand, can cause greater damage, causing alopecia, permanent scarring, and even disfigurement of the scalp (Hill et al., 2024).

Epidemiology

National community-based surveillance programs for superficial fungal infections are also lacking in Southeast Asia because of their complexity and cost (Batac & Denning, 2017; Velayuthan et al., 2018). Therefore, all studies estimating the national epidemiology of tinea capitis in Southeast Asia cited here have opted for actuarial estimates; data from small studies were used to estimate national incidence or prevalence of tinea capitis (Rodriguez-Tudela et al., 2015).

Based on cases seen at ten dermatological training hospitals in 2015, the annual incidence of tinea capitis in the Philippines was estimated to be 0.86/100,000 (Batac & Denning, 2017). This statistic was used in estimating the annual incidence in Vietnam, which is 0.21/100,000, though this estimate is limited to children <15 years old (Duong et al., 2023). Thailand recorded the lowest annual incidence of 0.1/100,000 based on the number of cases per year at its national pediatric hospital

(Chayakulkeeree & Denning, 2017). The prevalence of tinea capitis in Indonesian schoolchildren was calculated to be 270/100,000 based on a weighted average of estimates from six prevalence studies dated 2005–2017 (Wahyuningsih et al., 2021). Almost all the estimates were limited to school-aged children; only the estimates from the Philippines were intended for the general population. Furthermore, underestimates are likely due to stigma of getting medical care and limited generalizability of data from small studies (Duong et al., 2023). No national estimate was computed in Cambodia; however, a pioneering epidemiological study cited that 30% of patients with superficial fungal infections have tinea capitis (Uhrlaß et al., 2024). The epidemiology of tinea capitis in Southeast Asia is summarized in Table 5.1.

Diagnostics

Fungal culture remains to be the gold standard for the diagnosis of tinea capitis (Gupta et al., 2022). However, culture typically takes 4 weeks or more for results. Trichoscopy and direct microscopy are more rapid and commonly used tests in clinical practice. Trichoscopy, or the direct examination of the scalp through a dermatoscope, is a rapid and accurate test, with a sensitivity of 94% (95% CI 88–100) and specificity of 83% (95% CI 72–94) (Dhaille et al., 2019). *T. tonsurans* infection manifests as multiple comma-shaped hairs, while *M. canis* infection presents as dystrophic and elbow-shaped hairs, with different height levels of broken hair (Hay, 2017). Likewise, direct microscopy using 10% potassium hydroxide provides a rapid result, with low sensitivity (66%) but high specificity (98%) (Afshar et al., 2018). Better results were observed with microscopy using Chicago sky blue 6B, recording sensitivity and specificity of 97% and 100%, respectively (Afshar et al., 2018). Another commonly used test is the Wood light examination, which may be useful in epidemics and identification of affected animals (Mayser et al., 2020). Yellow-green fluorescence confirms *M. canis* infection, though absence of a yellow-green fluorescence is not definitive to rule out *M. canis* infection (Mayser et al., 2020). Details about the diagnosis of tinea capitis are summarized in Table 5.1.

Immune Responses to Mycotic Infection (Fig. 5.2)

Innate Immunity

The innate immune system is considered the initial line of defense against infections, using a variety of cellular and molecular processes to identify and remove fungal pathogens. Despite its lack of specificity, it successfully differentiates self from non-self and activates adaptive immune mechanisms by providing unique

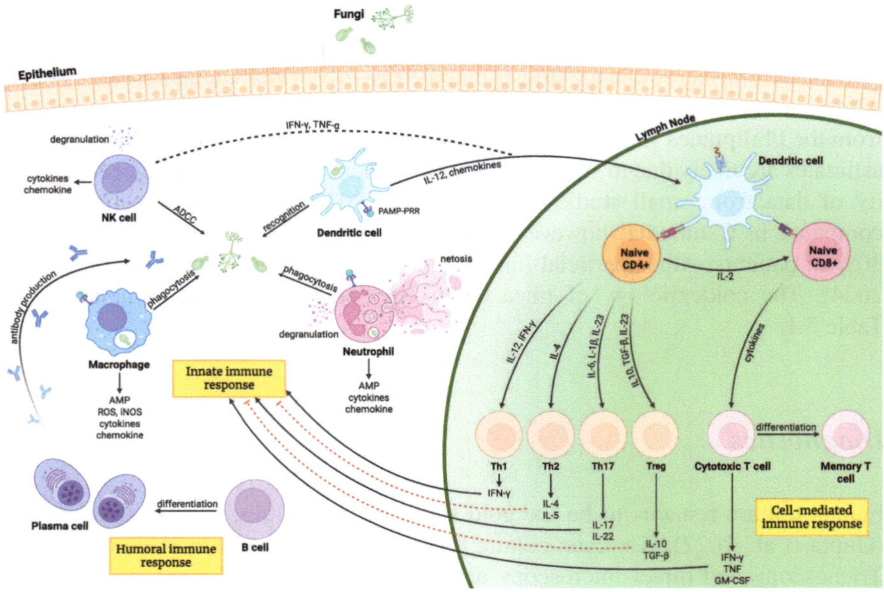

Fig. 5.2 Innate and adaptive immunity against fungal infection: The innate immune response is the initial line of defense, and it includes immune cells like neutrophils, macrophages, natural killer (NK) cells, and dendritic cells (DCs). These cells recognize fungal infections using pattern recognition receptors (PRRs), which detect pathogen-associated molecular patterns (PAMPs). Activated DCs and macrophages produce pro-inflammatory cytokines, which attract more immune cells to the infection site. NK cells also play a role by directly killing fungi and producing cytokines; they also process fungal antigens and present them to naive T cells in lymph nodes. Antibody-dependent cellular cytotoxicity (ADCC), in which antibodies linked to fungal cells are identified by NK cells, also contributes to the targeted killing of fungi. The adaptive immune response includes the activation and proliferation of particular T and B cells. CD4+ T helper cells generate cytokines including IFN-γ and IL-17, which boost the antifungal activity of macrophages and neutrophils. CD8+ cytotoxic T lymphocytes (CTLs) destroy infected cells directly and may develop into memory cells, allowing them to respond more quickly when re-exposed to the same antigen. B cells create antifungal antibodies, which destroy infections and aid in their elimination via opsonization. Together, these responses ensure effective fungal pathogen elimination and long-term immunity. The figure was created using Biorender

signals. It provides fast identification of infection through a limited collection of germline-encoded receptors that recognize a group of conserved molecular patterns that is shared by large groups of microbial species (Romani, 2004).

The innate immune response begins when immune cells recognize fungal pathogens via pattern recognition receptors (PRRs) on immune cells. These receptors recognize pathogen-associated molecular patterns (PAMPs) specific to the fungi, including β-glucans, mannans, and chitin found in the fungal cell wall. The key PRRs involved in antifungal immunity include Toll-like receptors (TLRs), C-type lectin receptors (CLRs), and NOD-like receptors (NODs) (Ward & Vyas, 2020). TLR-2 and TLR-4 detect mannan and lipomannan derivatives on fungal cell walls, and then the adaptor molecule MyD88 acts as a key component in the downstream signaling of TLRs, leading to the production of inflammatory cytokines that initiate the naive T cells to differentiate into Th1 and Treg cells. TLR-9, located in the endosomes of myeloid and plasmacytoid dendritic cells, recognizes fungal DNA (CpG) and triggers the production of type-I interferons (Kawasaki & Kawai, 2014; Pathakumari et al., 2020). CLRs, such as Dectin-1, Dectin-2, Dectin-3, and macrophage inducible C-type lectin (Mincle), play an important role in the recognition of fungal carbohydrates. TLR2 and Dectin-1 work together to identify β-glucans, which are a key element of fungal cell walls. Dectin-2 recognizes α-mannans, and therefore the detection of fungal hyphae of different fungal pathogens, including *Candida albicans*, *Cryptococcus neoformans*, *Aspergillus fumigatus*, *Saccharomyces cerevisiae*, *Paracoccidioides brasiliensis*, *Histoplasma capsulatum*, *Microsporum audouinii*, and *Trichophyton rubrum*. The detection of β-glucans and α-mannans by Dectin-1 and Dectin-2 leads to the activation of NLRP3 and NF-kB, which are vital for producing inflammatory cytokines and promoting Th1 and Th17 differentiation. Dectin-2 and Dectin-3 can create heterodimers that identify the hyphal forms of *C. albicans* to stimulate pro-inflammatory production. Mincle is expressed on macrophages, neutrophils, myeloid dendritic cells, and B cells, where it primarily detects fungal mannose ligands, triggering the production of TNF-α and IL-10. Dectin-1, Dectin-2, and Mincle operate through the CARD9-BCL10-MALT1 intracellular signaling pathway. This signaling cascade is critical for initiating immune responses, leading to the production of various cytokines, including TNF-α, IL-6, IL-10, and IL-23, which play essential roles in inflammation and the activation and differentiation of T-helper cells necessary for effective antifungal immunity (del Fresno et al., 2018; Tang et al., 2018; Pathakumari et al., 2020). Among the various NLRs (NOD-like receptors), NLRP3 is a particularly well-studied inflammasome component. It plays a critical role in controlling the spread of *C. albicans* infections throughout the body. Additionally, NLRP3 has been shown to activate in response to infections by *Aspergillus* hyphae, indicating its broad significance in the immune response against different fungal pathogens. This activation of NLRP3 triggers the production of pro-inflammatory cytokines, such as IL-1β and IL-18, which are vital for initiating and sustaining effective antifungal defenses (Mancuso et al., 2022).

NK cells are another crucial component of the innate immune system that provides early defense against fungal pathogens. NK cells can directly kill infected cells by releasing cytotoxic granules containing perforin and granzymes, a process known as degranulation. They also produce a variety of cytokines, such as IFN-γ and TNF-α, and chemokines, which enhance the activation and recruitment of other immune cells to the site of infection. Human NK cells are known to engage with dendritic cells and secrete IFNγ, which promotes the differentiation of IFNγ-producing $CD4^{+}$Th1 cells, while some studies show that the maturation of dendritic cells induced by resting NK cells was observed to be dependent on the production of endogenous TNF-α. Additionally, NK cells can mediate antibody-dependent cellular cytotoxicity (ADCC), where they recognize and kill antibody-coated fungal cells. The role of NK cells in fungal infections is particularly significant in the early stages of infection, where their rapid response can contain the spread of the pathogen before the adaptive immune system is fully activated (Piccioli et al., 2002; Schmidt et al., 2017; Perera Molligoda Arachchige, 2021).

Phagocytic cells including macrophages, dendritic cells, and polymorphonuclear lymphocytes (PMLs) play a role in the ingestion and destruction of fungal pathogens. Macrophages, as crucial cells of the innate immune system, serve as a bridge between innate and adaptive immunity. In the early stages of inflammation, macrophages focus on eradicating pathogens and preventing their spread. They achieve this by engulfing fungal cells through phagocytosis and initiating an acute inflammatory response through the production of cytokines and chemokines, which then attract and activate other immune cells to the infection site. Additionally, inflammatory monocytes and macrophages are vital in defending against aspergillosis and candidiasis by inhibiting the germination of fungal conidia. Recent studies indicate that monocytes may have a dual role in *Cryptococcus* infections: they benefit the host's innate response against *C. neoformans* but can also harm the host by protecting the fungus, promoting its spread and brain invasion (Hardison et al., 2012; Pathakumari et al., 2020). Dendritic cells (DCs) are crucial antigen-presenting cells that can migrate to lymph nodes in the presence of IL-12 and chemokines, and activate naive T cells, leading to their differentiation into T-helper (Th) subsets such as Th1, Th2, and Th17 cells. These Th cells are essential for protective immune responses against fungal infections. In humans, there are two primary subsets of DCs: conventional DCs (cDCs) and plasmacytoid DCs (pDCs). Human pDCs can inhibit the hyphal growth of *A. fumigatus*, and their depletion reduces the survival rate in mice infected with *Aspergillus* (Ramirez-Ortiz et al., 2011; Ward & Vyas, 2020; Pathakumari et al., 2020). Among the PMLs, neutrophils are the most abundant phagocytes and are among the first responders to fungal infections like *Candida* and *Aspergillus*. They play a secondary role in infections caused by other fungi such as *C. neoformans*, *P. jirovecii*, *H. capsulatum,* and *S. schenckii*, and they use various strategies to combat fungal pathogens, including the production of chemokines, enzymes (such as myeloperoxidase, NADPH oxidase, and SOD), and antimicrobial peptides. Additionally, neutrophils generate neutrophil extracellular traps (NETs), which help restrict fungal growth and contain hyphal forms of fungal pathogens (Urban et al., 2006; García Carnero et al., 2018; Mancuso et al., 2022).

Understanding the multifaceted processes involved in the innate immune response to mycotic infections, including pathogen recognition by PRRs, activation and recruitment of phagocytic cells, and cytokine and chemokine production, is vital for devising strategies to address fungal infections, particularly in immunocompromised patients.

Adaptive Immunity

While the innate immune system acts as the initial barrier against fungal infections, the adaptive immune response is essential for long-term protection and immune memory. Effective adaptive immune responses are initiated by the presentation of fungal antigens to T cells. Dendritic cells (DCs), as primary antigen-presenting cells (APCs), play a critical role in capturing, processing, and presenting fungal antigens to T cells via MHC (major histocompatibility complex) molecules. DCs recognize fungal pathogens through PRRs and internalize them via phagocytosis. After processing the fungal antigens, DCs migrate to lymph nodes, where they present the antigens to naïve T cells, thereby initiating their activation and differentiation. This process is pivotal for developing a specific and effective adaptive immune response. MHC class I molecules present endogenous fungal antigens to CD8+ T cells, while MHC class II molecules present exogenous fungal antigens to CD4+ T cells. This antigen presentation is essential for the activation and proliferation of T cells, leading to a robust adaptive immune response (Pathakumari et al., 2020).

T lymphocytes, or T cells, are essential for adaptive protection against fungal infections. They are widely classified as CD4+ helper T cells (Th cells) and CD8+ cytotoxic T cells (CTLs), with each serving a specific role in antifungal defense. CD4+ T cells are essential for coordinating the immune response by releasing cytokines, which then stimulate the development of naïve T cells into diverse CD4 T helper (Th) subsets. These cells develop into several subsets, including Th1, Th2, Th17, and Treg cells, each of which contributes independently to antifungal immunity (Pathakumari et al., 2020). Th1 cells generate IFN-γ, activating macrophages and increasing their fungicidal activity. This reaction is especially crucial for battling intracellular fungal diseases like *H. capsulatum* and *C. neoformans* (Herring et al., 2002). IFN-γ also increases macrophages' capacity to destroy ingested fungus by promoting the generation of nitric oxide (NO) and reactive oxygen species (ROS). The Th2 immune response is generally associated with an increased fungal burden and exacerbation of disease (Pathakumari et al., 2020; Shankar et al., 2024). It has been found that overexpression of GATA-3, a Th2 cell transcription factor, increases susceptibility to systemic *Candida* infection by reducing IFN-γ levels. Th2 cells mainly produce anti-inflammatory cytokines like IL-4, IL-5, IL-10, TGF-β, and IL-13, which promote the differentiation of naive T cells into Th2 cells and inhibit Th1 differentiation, leading to reduced pro-inflammatory cytokine production. Therapeutic inhibition of IL-4 signaling and IL-10 knockout mice have shown increased resistance to candidiasis. However, the role of Th2 cytokines is complex; while IL-4 is necessary for protective immunity against *Candida*, IL-4R knockout mice displayed a higher pathogen burden after pulmonary infection with *C. neoformans*, indicating that IL-4 is protective in the early stages of infection. Additionally, Th2 responses can induce alternatively activated macrophages (M2), which may protect against *A. fumigatus*. Th2-dependent antibodies might also provide protection by enhancing Th1 immunity. Therefore, the outcome of fungal infections depends on the balance between Th1 and Th2 responses (Pathakumari et al., 2020). Interleukin-17 (IL-17) and IL-22, which are secreted by Th17 cells, are essential for drawing neutrophils and enhancing their antifungal properties. Th17 responses play a special role in mucosal immunity, supporting the body's defense against infections caused by *C. albicans*. IL-17 improves barrier integrity and stimulates the synthesis of antimicrobial peptides, providing an additional line of defense (Pathakumari et al., 2020; Shankar et al., 2024). In addition, regulatory T cells (Treg cells) are essential for maintaining immune balance and preventing excessive immune reactions during fungal infections. In this context, Treg cells play a key role in fine-tuning the immune response to ensure its effectiveness without causing undue tissue damage. They achieve this by releasing anti-inflammatory cytokines like IL-10 and TGF-β, which dampen the activation and proliferation of other immune cells, including effector T cells. This regulatory function is critical for averting hyperactive immune responses that might result in chronic inflammation or tissue injury. Moreover, Treg cells help minimize damage to host tissues while still enabling a robust response to clear the fungal infection. By orchestrating this delicate balance in the immune response, Treg cells contribute to resolving inflammation and promoting tissue healing and equilibrium once the pathogen is eliminated (Lee & Lau, 2017; Pathakumari et al., 2020).

CD8+ T cells play a direct role in eliminating infected cells by detecting fungal antigens displayed by MHC class I molecules. This cytotoxic function is essential for eradicating cells containing intracellular fungi, thus halting the dissemination of infection. Furthermore, CD8+ T cells have the capacity to generate cytokines like IFN-γ, TNF, GM-CSF, and generation of protective memory responses to fungi like *Aspergillus fumigatus* and *Blastomyces* spp., which bolster the broader immune response. These cells have also demonstrated protective capabilities in mouse models lacking CD4 cells when infected with *C. albicans*, *C. neoformans*, *H. capsulatum*, and *P. jirovecii*. Furthermore, they can sustain memory function even without CD4 T cells or prior exposure to fungal antigens (Merkhofer & Klein, 2020; Dunne et al., 2021). Emerging evidence indicates that the reliance of CD8+ T cells on assistance from CD4+ T cells is largely dependent on their unique interactions with dendritic cells during activation. This suggests that CD8+ T cells, when adequately primed and possessing inherent memory capabilities, can autonomously mount protective responses and recall immunity. Nevertheless, further investigation is essential to comprehend the specific fungal antigens involved and the precise contribution of dendritic cells in the priming process of CD8+ T cells (Sharma et al., 2022).

B cells and the antibodies they produce are crucial elements of the adaptive immune response against fungal infections. Upon encountering fungal antigens, B cells undergo differentiation into plasma cells, which secrete specific antibodies. These antibodies play multiple roles in combating fungal pathogens. They bind to fungal antigens, neutralizing the pathogen and marking it for destruction by other immune cells—a process known as opsonization. This enhances the efficiency of phagocytosis by macrophages and neutrophils. Additionally, antibodies can neutralize fungal toxins and hinder fungal adherence to host cells, thereby reducing the risk of infection spread (Pathakumari et al., 2020).

In response to fungal antigens, B cells undergo class switching and generate various classes of antibodies with direct and indirect fungicidal effects. Direct effects include antibody binding to fungi, inhibiting fungal growth and metabolism, while indirect methods involve antibody-mediated activation of immune effector pathways, such as complement and opsonization. Protective antibodies against major fungal species, including *A. fumigatus*, *C. albicans*, *C. neoformans*, *H. capsulatum*, *P. brasiliensis*, and *Pneumocystis* spp., have been identified. For example, *A.*

fumigatus proteins trigger the production of *Aspergillus*-specific IgG antibodies in invasive aspergillosis (Pathakumari et al., 2020; Dunne et al., 2021).

Furthermore, B cells have been shown to support anti-fungal T cells independently of antibodies. In models of *C. albicans* infection, B cells respond to the fungus, producing cytokines like IL-6 and GM-CSF, and provide potent costimulatory signals to naive T cells through CD80/CD86 in an MHC-dependent manner. The observation that patients with agammaglobulinemia maintain normal antifungal immunity while those undergoing B cell-depleting therapies exhibit increased fungal infection incidence underscores the importance of B cell-mediated support for effector T cells in the antifungal immune response (Dunne et al., 2021).

The adaptive immune response plays a pivotal role in providing long-term protection and immune memory against fungal infections. Comprehensive understanding of the intricate interactions among these immune cells and molecules is essential for devising effective therapeutic approaches against fungal infections.

Fungal Vaccines

Fungal infections are often overshadowed by bacterial and viral infections in clinical research and public health initiatives. In this age of climate change, fungal stress and thermo-adaptation responses induce enhanced morphogenesis, biofilm formation, and toxin production that enable pathogenic fungi to adapt to environments of higher temperatures, including the human body (Seidel et al., 2024). Besides the consequent enhanced virulence and immune evasion, rising temperatures have also been theorized to contribute to the emergence of new pathogens such as *Candida auris* and emerging outbreaks of *Batrachochytrium dendrobatidis* and *Cryptococcus deuterogattii* (Nnadi & Carter, 2021). Displacement and dissemination of disaster-associated fungi as well as anthropogenic factors including freshwater body contamination with agricultural and industrial pollutants significantly increase the risk for fungal disease acquisition (Nnadi & Carter, 2021; Seidel et al., 2024). Moreover, the global rise in the agricultural use of azole fungicides, which have broad-spectrum activity, has implications not only for crop pathogens but also in promoting tolerance and resistance among human opportunistic, pathogenic fungi (Fisher et al., 2022). Antibiotic overuse, which creates permissive environments for *Candida* overgrowth, and the development of multidrug resistance, especially in *C. auris* and *Trichophyton indotineae*, warrant the development of novel approaches in treatment, surveillance and diagnostic options, and education at large (Lockhart et al., 2017; McCormick & Ghannoum, 2024).

Current antifungal treatments are often inadequate for several reasons. The limited repertoire of antifungal agents includes the three primary classes of antifungal drugs—polyenes, azoles, and echinocandins—each having significant limitations including host toxicity, suboptimal pharmacokinetics, limited spectrum of activity, and resistance development (Lee et al., 2023). In addition, many antifungal drugs, such as itraconazole, voriconazole, posaconazole, and flucytosine, have narrow therapeutic indices or unpredictable pharmacokinetic profiles, thereby imposing constraints and requiring special considerations especially for the critically ill and pediatric populations (John et al., 2019). Lastly, access to antifungal medications can be limited in low- and middle-income countries due to prohibitive cost, availability, and the need for specialized healthcare infrastructure.

Given the limitations of current antifungal therapies and the increasing incidence of fungal infections, the development of fungal vaccines represents a critical unmet need. The development of fungal vaccines is a promising approach to address the aforementioned growing threats, offering a preventive strategy that could reduce both the incidence of infections and the emergence of antifungal resistance (Rivera et al., 2022; Casadevall & Pirofski, 2012).

While there are currently no approved fungal vaccines for human use, current vaccine development platforms include live attenuated, killed whole-cell, recombinant, conjugate, pan-fungal, DNA-, and nanoparticle-based fungal vaccines (Alapan et al., 2024). Pan-fungal vaccines, based on highly conserved sequences of NXT-2 and calnexin, show potential as broad-spectrum vaccine candidates against an array of pathogens that is particularly valuable for immunocompromised patients, who are at risk of multiple opportunistic infections (Rayens et al., 2022; Alapan et al., 2024). While immunocompromised patients may have suboptimal immunogenicity to fungal vaccines, an improvement in host response to pathogenic mycoses is still beneficial. In vivo murine experiments show that fatty acyl analogs of muramyl dipeptide integrated into liposomal nanoparticles show promising potential as fungal vaccine adjuvants by stimulating a Th1-biased immune response and promoting cellular proliferation (Effenberg et al., 2017). Previous attempts at developing DNA-based vaccines for *Paracoccidioides brasiliensis* and *Pneumocystis* using plasmids that encode particular target antigens show promise in eliciting cellular and humoral immune responses; however, unresolved challenges posed by the lack of clinical data and potential unintended genomic integration remain (De Amorim et al., 2013; Fan et al., 2016; Alapan et al., 2024). No human clinical trials have been reported in the past 5 years against the fungal pathogens previously discussed; several in vivo vaccine studies have been employed in animal models. No in vivo vaccine studies have been reported against *Mucor* sp., *Talaromyces marneffei*, and *Histoplasma capsulatum* in the last 5 years, although in vitro and in silico studies on their immunogenicity are augmenting our understanding toward vaccine development.

Cryptococcus neoformans

Several vaccine strategies have been developed and tested against *C. neoformans* in various mouse models, each employing a different vaccine platform and targeting different aspects of the host immune response to enhance survival and reduce fungal burden. Specht et al. (2024) explored a live attenuated vaccine approach using the cda1Δ2Δ3Δ strain of *C. neoformans*, which resulted in high survival rates among vaccinated mice across different strains (C57BL/6, BALB/c, CBA/J, NSG), alongside a significant reduction in colony-forming units (CFUs) in the lungs. Wang et al. (2023) investigated the efficacy of the GP-Cda1 and GP-Cda2 subunit vaccines in wild-type and various genetically modified mouse models. Both vaccines achieved over 50% survival at 70 days post-infection (DPI) and elicited strong Th1 and Th17 immune responses. Another subunit vaccine approach by Hester et al. (2020) demonstrated that the GP-Cda1, GP-Cda2, GP-Blp4, GP-Glo1, and GP-Cpd1 subunits provided varying degrees of protection in BALB/c and C57BL/6 mice. GP-Cda1 and GP-Blp4 showed 100% survival in BALB/c mice, with significant protection in C57BL/6 mice. GP-Glo1 and GP-Cpd1 also improved survival and reduced lung CFUs, suggesting that these subunits are effective in modulating the host immune response to mitigate fungal burden. Rizzo et al. (2021) explored an extracellular vesicle (EV)-based subunit vaccine, which led to significantly prolonged survival in BALB/c mice, with 100% of the EV-immunized mice producing antibodies against EV proteins. Normile and del Poeta (2022) investigated the efficacy of a heat-killed *C. neoformans* Δsgl1 strain. Their findings showed that administering two doses of this vaccine led to 100% survival in CBA/J mice, including models deficient in CD4+ T cells. Crawford et al. (2024) examined two synthetic decasaccharide vaccines, DECA-CRM197 and DECA-PA63, in different mouse strains (A/J, BALB/c). DECA-CRM197 showed median survival times of 18 days for the H99 strain and 27 days for the KN99-α strain, with a significant reduction in brain CFUs for the KN99-α strain. However, DECA-PA63 elicited only a modest immune response, with low binding affinity to glucuronoxylomannan (GXM) and an inconsistent antibody isotype response.

Aspergillus *species*

Escherichia coli Nissle 1917 is a well-studied probiotic strain with known immunomodulatory properties, particularly due to its high antigenic α-Gal content (Hodžić et al., 2020). In the context of vaccine efficacy against *Aspergillus* infection, *E. coli* Nissle administration as an oral probiotic-based vaccine exhibited no significant difference in *A. fumigatus*-specific IgM and IgY levels versus a placebo, both pre- and post-vaccination (Thorel et al., 2022).

Sterylglucosidase (SGL) is a fungus-specific enzyme involved in the pathogenesis of *A. fumigatus* in murine models; *sgl* gene knockouts induced the accumulation of the sterylglucoside ergosterol 3β-d-glucoside, conferring sterilizing immunity and virulence attenuation (Normile et al., 2020). When immunized with *A. fumigatus* ΔsglA conidia, CBA/J mice showed 100% survival until the experiment endpoint at 75 days post-challenge (Fernandes et al., 2022). *A. fumigatus* ΔsglA conidia have shown to be avirulent with reduced adhesion and were fully eliminated from the lungs with no extrapulmonary dissemination in murine models.

In Swiss-Webster mice, immunization with the *A. fumigatus* surface antigen 3 and 9 (Af3/9) subunit demonstrated significantly higher survival rates following infection with azole-resistant *A. fumigatus* strain V45 (Slarve et al., 2022). The immunized mice also showed higher levels of anti-spore antibodies, particularly IgG1 and IgG2a isotypes. Additionally, there was increased production of IFN-γ and IL-4, with IL-4-secreting cells being significantly more numerous than those producing IFN-γ (Slarve et al., 2022). These findings suggest that the Af3/9 subunit elicits a robust and balanced Th1/Th2 immune response, enhancing both humoral and cellular immunity against *A. fumigatus*.

NXT-2, a recombinant pan-fungal vaccine developed from a consensus sequence in the KEX1 regions of several fungal pathogens, tested in BALB/c mice demonstrated a reduction in *Aspergillus*-associated morbidity and mortality (Rayens et al., 2022). Following immunosuppression and *Aspergillus* challenge, NXT-2-vaccinated mice showed significantly lower mortality and reduced lung fungal burden compared to sham-immunized controls. The vaccine elicited high anti-NXT-2 IgG reciprocal endpoint titers ($4.64 \times 10^5 \pm 2.92 \times 10^5$), generating a strong humoral response and a protection level comparable to that achieved with the species-specific AF.KEX1 vaccine (Rayens et al., 2022).

Blastomyces dermatitidis

Most vaccine development toward *B. dermatitidis* has revolved around testing various adjuvant formulations combined with *Blastomyces* endoglucanase 2 (Bl-Eng2) and its derivatives for their immunogenic effects in murine (C57BL/6 mice) models. One such adjuvant of potential, ΔN rBl-Eng2, was synthesized through the deglycosylation of the N- and O-glycans from Bl-Eng2, yielding a protein backbone with a reduced molecular weight and is rendered unrecognized by Dectin-2 reporter cells (Dias et al., 2021).

The combination of Bl-Eng2 with Advax3™, which includes a TLR9 agonist, resulted in a notable induction of CD4+ T cells producing single and double cytokines (IFN-γ and IL-17). This formulation achieved a significant reduction in lung CFU by 18-fold and exhibited 100% survival at day 14, in contrast to naive controls and Bl-Eng2 antigen-only immunized mice who all succumbed to infection by then (Wüthrich et al., 2021). The combination of Bl-Eng2 with Advax8™ likewise led to an increase in CD4+ T-cell populations; however, the increase in IL-17-producing CD4+ T cells did not reach statistical significance (Wüthrich et al., 2021).

Mice treated with the combination of either Bl-Eng2 peptide or calnexin with ΔN rBl-Eng2 displayed significant T cell expansion and differentiation into Th1 and Th17 subsets (Dias et al., 2021). The pairing of Calnexin with ΔN rBl-Eng2 also increased recruitment of Th17 cells to the lung, while the Bl-Eng2 + ΔN rBl-Eng2 formulation induced notable migration of primed CD4+ T cells into the lung parenchyma, reduction of lung CFU, and weight loss (Dias et al., 2021).

Dermatophytes

The efficacy of various dermatophyte vaccines and extracts has been evaluated in different animal models, exhibiting diverse protective and immunological responses. Both live and inactivated vaccines demonstrated satisfactory protection rates, particularly against *T. verrucosum* and *T. mentagrophytes*, inasmuch as conferring complete and lasting resistance. Most in vivo studies reported no significant adverse effects, although mild local reactions and infection in some cases with *M. canis* have been associated with live non-adjuvanted polyvalent dermatophytes (Abo-Elyazeed et al., 2023).

In adult female guinea pig models, the inactivated non-adjuvanted polyvalent dermatophytes vaccine demonstrated varying levels of protection against different dermatophyte species (Abo-Elyazeed et al., 2023). The protection rates indicate strong efficacy against *T. verrucosum* and *T. mentagrophytes* and lower efficacy against *M. canis* and *T. rubrum* (Abo-Elyazeed et al., 2023). On the other hand, the live non-adjuvanted polyvalent dermatophytes vaccine conferred superior efficacy across all dermatophyte species tested, particularly achieving complete protection against *T. verrucosum* and *T. mentagrophytes* (Abo-Elyazeed et al., 2023). Lower protection rates against *M. canis* and *T. rubrum* highlight the need for further optimization to enhance the polyvalent vaccine's broad-spectrum efficacy.

The inactivated *T. mentagrophytes* vaccine administered to rabbits demonstrated complete prophylactic efficacy, with the subjects showing resistance to infection for over 6 months. Additionally, the vaccine exhibited therapeutic efficacy within 8 days. This suggests a robust and long-lasting immune response, making it a promising candidate for both prophylactic and therapeutic applications against *T. mentagrophytes* infections (Al-Janabi & Al-Khikani, 2021).

The cytoplasmic crude extract of *T. verrucosum* was tested by Salahi et al. (2020) in BALB/c mice to evaluate its immunological effects. The extract significantly increased IFN-γ production, indicating an enhanced Th1 immune response. However, it had no significant effect on IL-4 production, suggesting that it did not influence the Th2 response. Moreover, the MTT assay revealed that the extract stimulated lymphocyte metabolic activity, indicating a potential role in enhancing cellular immunity. These findings support its potential use as an immunostimulatory agent in vaccine formulations or as an adjunctive treatment to enhance the immune response against dermatophyte infections.

Pneumocystis jirovecii

The NXT-2 recombinant pan-fungal vaccine elicited a significant reduction in *P. jirovecii* infection in non-human primates. The NXT-2 vaccine, designed as a broad-spectrum antifungal vaccine, was highly effective in reducing the burden of infection, as evidenced by the significant titers of anti-NXT-2 IgG

($1.08 \times 10^5 \pm 1.04 \times 10^5$) (Rayens et al., 2022). Rabacal et al. (2022) explored the use of an iNKT cell-activating adjuvant, α-galactosylceramide (α-GC), in combination with the KEX1 antigen in rhesus macaques. The study demonstrated a significant increase in KEX1-specific IgG titers, correlated with a marked reduction in the incidence of *Pneumocystis* co-infection, with only 14.3% of vaccinated animals in Cohort 1 and 0% in Cohort 2 developing the infection compared to 80% and 71.4% in the respective sham groups. The use of α-GC as an adjuvant appears to enhance the immunogenicity of the KEX1 antigen significantly, offering a promising strategy to boost vaccine efficacy and protect against *Pneumocystis* infections in high-risk populations. In a murine model, Tong et al. (2021) assessed the immunogenicity of the recombinant A121–85 antigen against *P. jirovecii*. Following four immunizations, serum IgG titers reached 1:32000, indicating a strong humoral response. Additionally, there was a significant increase in the levels of pro-inflammatory cytokines IFN-γ, IL-12, and IL-17 in lung tissues.

Sporothrix schenckii

A multi-epitope vaccine consisting of six peptides derived from *Sporothrix* spp. enolase formulated with Freund's adjuvant (FA) demonstrated a robust humoral response characterized by high titers of specific IgG, IgG1, and IgG2b antibodies in mice (Portuondo et al., 2022). In a previous study, recombinant enolase subunit (rSsEno) elicited significantly higher IgG-specific antibody production compared to controls and elevated production of cytokines IL-2, IFN-γ, IL-4, and IL-6, further supporting the activation of both Th1 and Th2 responses (Portuondo et al., 2019).

Challenges to Fungal Vaccines

The development of fungal vaccines is filled with challenges, such as identifying suitable antigens, overcoming fungal evasion strategies, and navigating regulatory barriers. The complex cell walls of fungi make it difficult to find conserved antigens that can trigger protective immune responses across various species. Additionally, fungi have evolved strategies to evade the host's immune system, including altering their cell wall composition and secreting immunosuppressive molecules (Hopke et al., 2018; Oliveira et al., 2021).

To address these challenges, researchers are exploring the use of novel adjuvants, which can enhance the immunogenicity of vaccines by promoting stronger and more sustained immune responses, and the use of advanced delivery systems. Some studies incorporated synthetic decasaccharide and α-galactosylceramide into their vaccine formulation to boost specific antibody production and reduce fungal burden in vaccinated animals (Crawford et al., 2024; Rabacal et al., 2022; Oliveira et al., 2021). Researchers are also investigating advanced delivery systems, such as nanoparticles, to improve the stability and targeted delivery of fungal antigens (Rizzo et al., 2021; Gao & Guo, 2023).

Despite these advances, potential adverse effects remain a concern. Some vaccines may trigger unwanted inflammatory responses or fail to provide adequate protection. In order to address these issues, extensive preclinical studies are being conducted to evaluate the safety and efficacy of these vaccine candidates. Animal models, including mice and non-human primates, are being used to identify and mitigate potential adverse effects before proceeding to human trials. Additionally, efforts are being made to design vaccines that target multiple antigens or combine antigens to enhance protective efficacy and reduce the risk of immune evasion (Rayens et al., 2022).

The successful development and deployment of fungal vaccines could significantly impact global health, particularly for vulnerable groups such as immunocompromised individuals and those living in areas with high rates of fungal infections. Vaccination could reduce the incidence of severe fungal infections, lower healthcare costs associated with long-term antifungal treatments, and improve the quality of life for patients at risk of recurrent infections. Continued research and investment in fungal vaccine development are critical to realizing these benefits and addressing the growing threat posed by fungal diseases (Oliveira et al., 2021).

Fungal vaccines hold substantial potential in addressing the increasing burden of fungal infections. Advances in vaccine technology, including the development of recombinant proteins and mRNA vaccines, have shown promising results in preclinical studies. These vaccines stimulate strong immune responses, including the development of long-lasting immune memory, which is essential for preventing recurrent infections (Oliveira et al., 2021). With ongoing research and innovation, fungal vaccines could become a crucial tool in the global fight against fungal diseases, providing a new means of prevention and control.

Conclusion

Endemic mycoses present a significant challenge in Southeast Asia but also offer opportunities for progress in the knowledge of their epidemiology and diagnostics, as well as immunologic responses to endemic mycoses and development of vaccines against them. As climate change alters temperature and humidity patterns, the epidemiology of these diseases is likely to shift, exacerbating the situation. Currently, baseline epidemiologic surveillance for endemic mycoses in the region is lacking, and the limited understanding of their evolving epidemiology may lead to insufficient awareness among healthcare providers. This challenge is further complicated by the fact that endemic mycoses often present with non-specific symptoms. When coupled with the growing use of immunosuppressants and biologics—both of which increase the at-risk population—these factors can lead to delays in diagnosis and treatment, ultimately leading to poorer outcomes. However, advances in molecular biology and a deeper understanding of immune responses are enabling the development of more sensitive, specific, and rapid diagnostics. These advancements will strengthen surveillance and monitoring, enabling the early detection of emerging outbreaks. Scientific progress in these fields also allows more targeted development of fungal vaccine candidates. While there are vaccine platforms being investigated, fungal vaccine development is complex and multifaceted, especially given the diversity of fungal strains and unclear epidemiologic patterns in Southeast Asia. To effectively address these challenges, coordinated efforts among healthcare professionals, researchers, industry, and policymakers are crucial to capitalize on the opportunities these innovations present.

References

Abo-Elyazeed, H., Soliman, R., Hassan, H., El-Seedy, F. R., & Aboul-Ella, H. (2023). Development, preparation, and evaluation of a novel non-adjuvanted polyvalent dermatophytes vaccine. *Scientific Reports, 13*(1), 157. https://doi.org/10.1038/s41598-022-27675-7

Afshar, P., Larijani, L. V., & Rouhanizadeh, H. (2018). A comparison of conventional rapid methods in diagnosis of superficial and cutaneous mycoses based on KOH, Chicago sky blue 6B and calcofluor white stains. *Iranian Journal of microbiology, 10*(6), 433–440.

Agarwal, R., Sehgal, I. S., Muthu, V., Denning, D. W., Chakrabarti, A., Soundappan, K., Garg, M., Rudramurthy, S. M., Dhooria, S., Armstrong-James, D., Asano, K., Gangneux, J., Chotirmall, S. H., Salzer, H. J., Chalmers, J. D., Godet, C., Joest, M., Page, I., Nair, P., & Aggarwal, A. N. (2024). Revised clinical practice guidelines for diagnosing, classifying, and treating allergic bronchopulmonary aspergillosis/mycoses: a Delphi statement from the ISHAM-ABPA working group. *European Respiratory Journal, 2400061.* https://doi.org/10.1183/13993003.00061-2024

Ahmad-Fauzi, S., Abd-Manan, N., Yusof, N. S., Ibrahim, M., Mohamad, S. A., & Muhammed, J. (2021). Ocular sporotrichosis from a tertiary referral center in Malaysia and review of literature in Southeast Asia. *Taiwan Journal of Ophthalmology, 12*(2), 237–241. https://doi.org/10.4103/tjo.tjo_93_20

Aksoy, M., & Ulas, B. (2024). Prognostic factors and clinical features of rhino-orbital-mucormycosis cases: an update for patient and visual survivals. *International Journal of Ophthalmology, 17*(5), 916–923. https://doi.org/10.18240/ijo.2024.05.17

Alapan, D., Bisweswar, O., Prasenjit, S., Prasanjit, D., & Arkapal, B. (2024). Recent advances in the clinical development of antifungal vaccines: A narrative review. *Frontiers in Tropical Diseases, 5.* https://doi.org/10.3389/fitd.2024.1446477

Al-Janabi, A. A., & Al-Khikani, F. H. (2021). Prophylaxis and therapeutic ability of inactivated dermatophytic vaccine against dermatophytosis in the rabbits as an animal model. *Turkish Journal of Pharmaceutical Sciences, 18*(3), 326–332. https://doi.org/10.4274/tjps.galenos.2020.64187

Aljutaily, H. I., & Al-Shamrani, A. (2022). Invasive Pulmonary aspergillosis in Children: A case report and literature review. *The American Journal of Case Reports, 23.* https://doi.org/10.12659/ajcr.935971

Apostolopoulou, A., & Fishman, J. A. (2022). The pathogenesis and diagnosis of Pneumocystis jirovecii pneumonia. *Journal of Fungi (Basel, Switzerland), 8*(11), 1167. https://doi.org/10.3390/jof8111167

Araúz, A. B., & Papineni, P. (2021). Histoplasmosis. *Infectious Disease Clinics of North America, 35*(2), 471–491. https://doi.org/10.1016/j.idc.2021.03.011

Azar, M. M., & Hage, C. A. (2017). Laboratory Diagnostics for Histoplasmosis. *Journal of Clinical Microbiology, 55*(6), 1612–1620. https://doi.org/10.1128/JCM.02430-16

Bahr, N. C., Antinori, S., Wheat, L. J., & Sarosi, G. A. (2015). Histoplasmosis infections worldwide: thinking outside of the Ohio River valley. *Current Tropical Medicine Reports, 2*(2), 70–80. https://doi.org/10.1007/s40475-015-0044-0

Baker, J., Setianingrum, F., Wahyuningsih, R., & Denning, D. W. (2019). Mapping histoplasmosis in South East Asia – Implications for diagnosis in AIDS. *Emerging Microbes & Infections, 8*(1), 1139–1145. https://doi.org/10.1080/22221751.2019.1644539

Batac, M. C. R., & Denning, D. (2017). Serious fungal infections in The Philippines. *European journal of clinical microbiology & infectious diseases : official publication of the European Society of Clinical Microbiology, 36*(6), 937–941. https://doi.org/10.1007/s10096-017-2918-7

Bratton, E. W., El Husseini, N., Chastain, C. A., Lee, M. S., Poole, C., Stürmer, T., Juliano, J. J., Weber, D. J., & Perfect, J. R. (2012). Comparison and temporal trends of three groups with cryptococcosis: HIV-infected, solid organ transplant, and HIV-negative/non-transplant. *PLoS One, 7*(8), e43582. https://doi.org/10.1371/journal.pone.0043582

Bulterys, P. L., Le, T., Quang, V. M., Nelson, K. E., & Lloyd-Smith, J. O. (2013). Environmental predictors and incubation period of AIDS-associated penicillium marneffei infection in Ho Chi Minh City, Vietnam. *Clinical Infectious Diseases : An Official Publication of the Infectious Diseases Society of America, 56*(9), 1273–1279. https://doi.org/10.1093/cid/cit058

Burzio, C., Balzani, E., Corcione, S., Montrucchio, G., Trompeo, A. C., & Brazzi, L. (2023). Pneumocystis jirovecii pneumonia after heart transplantation: Two case reports and a review of the literature. *Pathogens (Basel, Switzerland), 12*(10), 1265. https://doi.org/10.3390/pathogens12101265

Cadena, J., Thompson, G. R., & Patterson, T. F. (2021). Aspergillosis. *Infectious Disease Clinics of North America, 35*(2), 415–434. https://doi.org/10.1016/j.idc.2021.03.008

Cano, M. V., & Hajjeh, R. A. (2001). The epidemiology of histoplasmosis: a review. *Seminars in Respiratory Infections, 16*(2), 109–118. https://doi.org/10.1053/srin.2001.24241

Casadevall, A., & Pirofski, L. A. (2012). Immunoglobulins in defense, pathogenesis, and therapy of fungal diseases. *Cell Host & Microbe, 11*(5), 447–456. https://doi.org/10.1016/j.chom.2012.04.004

Celestrino, G. A., Verrinder Veasey, J., Benard, G., & Sousa, M. G. T. (2021). Host immune responses in dermatophytes infection. *Mycoses, 64*(5), 477–483. https://doi.org/10.1111/myc.13246

Chakrabarti, A., & Slavin, M. A. (2011). Endemic fungal infections in the Asia-Pacific region. *Medical Mycology, 49*(4), 337–344. https://doi.org/10.3109/13693786.2010.551426

Chander, J., Singla, N., Kaur, M., Punia, R. S., Attri, A., Alastruey-Izquierdo, A., Cano-Lira, J. F., Stchigel, A. M., & Guarro, J. (2016). Saksenaea erythrospora, an emerging mucoralean fungus causing severe necrotizing skin and soft tissue infections—A study from a tertiary care hospital in North India. *Infectious Diseases, 49*(3), 170–177. https://doi.org/10.1080/23744235.2016.1239027

Charpentier, E., Ménard, S., Marques, C., Berry, A., & Iriart, X. (2021). Immune response in *Pneumocystis* infections according to the host immune system status. *Journal of Fungi (Basel, Switzerland), 7*(8), 625. https://doi.org/10.3390/jof7080625

Chayakulkeeree, M., & Denning, D. W. (2017). Serious fungal infections in Thailand. *European journal of clinical microbiology & infectious diseases : official publication of the European Society of Clinical Microbiology, 36*(6), 931–935. https://doi.org/10.1007/s10096-017-2927-6

Crawford, C. J., Liporagi-Lopes, L., Coelho, C., Santos Junior, S. R., Moraes Nicola, A., Wear, M. P., Vij, R., Oscarson, S., & Casadevall, A. (2024). Semisynthetic Glycoconjugate Vaccine Candidates against *Cryptococcus neoformans*. *ACS Infectious Diseases, 10*(6), 2089–2100. https://doi.org/10.1021/acsinfecdis.4c00094

Crothers, S. L., White, S. D., Ihrke, P. J., & Affolter, V. K. (2009). Sporotrichosis: a retrospective evaluation of 23 cases seen in northern California (1987-2007). *Veterinary Dermatology, 20*(4), 249–259. https://doi.org/10.1111/j.1365-3164.2009.00763.x

Cuellar-Rodriguez, J., Avery, R. K., Lard, M., Budev, M., Gordon, S. M., Shrestha, N. K., van Duin, D., Oethinger, M., & Mawhorter, S. D. (2009). Histoplasmosis in solid organ transplant recipients: 10 years of experience at a large transplant center in an endemic area. *Clinical Infectious Diseases : An Official Publication of the Infectious Diseases Society of America, 49*(5), 710–716. https://doi.org/10.1086/604712

Dave, T. V., Gopinathan Nair, A., Hegde, R., Vithalani, N., Desai, S., Adulkar, N., Kamal, S., Mittal, R., & Bradoo, R. A. (2021). Clinical Presentations, Management and Outcomes of Rhino-Orbital-Cerebral Mucormycosis (ROCM) Following COVID-19: A Multi-Centric Study. *Ophthalmic Plastic and Reconstructive Surgery, 37*(5), 488–495. https://doi.org/10.1097/IOP.0000000000002030

De Amorim, J., Magalhães, A., Muñoz, J. E., Rittner, G. M. G., Nosanchuk, J. D., Travassos, L. R., et al. (2013). DNA vaccine encoding peptide P10 against experimental paracoccidioidomycosis induces long-term protection in presence of regulatory T cells. *Microbes and Infection, 15*(3), 181–191. https://doi.org/10.1016/j.micinf.2012.11.007

de Hoog, G. S., & Guarro, J. (1995). Atlas of clinical fungi. Baarn: Centralbureau voor Schimmelcultures. *Mycoses, 39*(1–2), 71–71. https://doi.org/10.1111/j.1439-0507.1996.tb00088.x

Del Fresno, C., Iborra, S., Saz-Leal, P., Martínez-López, M., & Sancho, D. (2018). Flexible signaling of myeloid C-type lectin receptors in immunity and inflammation. *Frontiers in Immunology, 9*, 804. https://doi.org/10.3389/fimmu.2018.00804

Dezube, R. (2023). *Bronchoscopy – Bronchoscopy*. MSD Manual Professional Edition. https://www.msdmanuals.com/professional/pulmonary-disorders/diagnostic-and-therapeutic-pulmonary-procedures/bronchoscopy

Dhaille, F., Dillies, A. S., Dessirier, F., Reygagne, P., Diouf, M., Baltazard, T., Lombart, F., Hébert, V., Chopinaud, M., Verneuil, L., Becquart, C., Delaporte, E., Lok, C., & Chaby, G. (2019). A single typical trichoscopic feature is predictive of tinea capitis: a prospective multicentre study. *The British Journal of Dermatology, 181*(5), 1046–1051. https://doi.org/10.1111/bjd.17866

Dias, L. D. S., Dobson, H. E., Bakke, B. K., Kujoth, G. C., Huang, J., Kohn, E. M., Taira, C. L., Wang, H., Supekar, N. T., Fites, J. S., Gates, D., Gomez, C. L., Specht, C. A., Levitz, S. M., Azadi, P., Li, L., Suresh, M., Klein, B. S., & Wüthrich, M. (2021). Structural basis of Blastomyces Endoglucanase-2 adjuvancy in anti-fungal and -viral immunity. *PLoS Pathogens, 17*(3), e1009324. https://doi.org/10.1371/journal.ppat.1009324

Donnelly, J. P., Chen, S. C., Kauffman, C. A., Steinbach, W. J., Baddley, J. W., Verweij, P. E., Clancy, C. J., Wingard, J. R., Lockhart, S. R., Groll, A. H., Sorrell, T. C., Bassetti, M., Akan, H., Alexander, B. D., Andes, D., Azoulay, E., Bialek, R., Bradsher, R. W., Bretagne, S., Calandra, T., et al. (2020). Revision and update of the consensus definitions of invasive fungal disease from the European Organization for Research and Treatment of Cancer and the Mycoses Study Group Education and Research Consortium. *Clinical Infectious Diseases : An Official Publication of the Infectious Diseases Society of America, 71*(6), 1367–1376. https://doi.org/10.1093/cid/ciz1008

Dunne, M. R., Wagener, J., Loeffler, J., Doherty, D. G., & Rogers, T. R. (2021). Unconventional T cells – New players in antifungal immunity. *Clinical immunology (Orlando, Fla.), 227*, 108734. https://doi.org/10.1016/j.clim.2021.108734

Duong, T. N., Le, M. H., Beardsley, J., Denning, D. W., Le, N. H., & Nguyen, B. T. (2023). Updated estimation of the burden of fungal disease in Vietnam. *Mycoses, 66*(4), 346–353. https://doi.org/10.1111/myc.13559

Effenberg, R., Knötigová, P. T., Zyka, D., Čelechovská, H., Mašek, J., Bartheldyová, E., Hubatka, F., Koudelka, Š., Lukáč, R., Kovalová, A., Šaman, D., Křupka, M., Barkocziova, L., Kosztyu, P., Šebela, M., Drož, L., Hučko, M., Kanásová, M., Miller, A. D., et al. (2017). Nonpyrogenic molecular adjuvants based on NorAbu-Muramyldipeptide and norAbu-Glucosaminyl muramyldipeptide: Synthesis, molecular mechanisms of action, and biological activities in vitro and in vivo. *Journal of Medicinal Chemistry, 60*(18), 7745–7763. https://doi.org/10.1021/acs.jmedchem.7b00593

Emmons, C. W. (1955). Saprophytic sources of Cryptococcus neoformans associated with the pigeon (Columba livia). *American Journal of Hygiene, 62*(3), 227–232. https://doi.org/10.1093/oxfordjournals.aje.a119775

Fan, H., Guo, J. Y., Ma, S. L., Zhang, N., & An, C. L. (2016). Synthetic p55 tandem DNA vaccine against *Pneumocystis carinii* in rats. *Microbiology and Immunology, 60*(6), 397–406. https://doi.org/10.1111/1348-0421.12386

Feeney, K. T., Arthur, I. H., Whittle, A. J., Altman, S. A., & Speers, D. J. (2007). Outbreak of sporotrichosis, Western Australia. *Emerging Infectious Diseases, 13*(8), 1228–1231. https://doi.org/10.3201/eid1308.061462

Fernandes, C. M., Normile, T. G., Fabri, J. H. T. M., Brauer, V. S., de S Araújo, G. R., Frases, S., Nimrichter, L., Malavazi, I., & Del Poeta, M. (2022). Vaccination with live or heat-killed Aspergillus fumigatus Δ*sglA* conidia fully protects immunocompromised mice from invasive aspergillosis. *MBio, 13*(5), e0232822. https://doi.org/10.1128/mbio.02328-22

Fisher, M. C., Alastruey-Izquierdo, A., Berman, J., et al. (2022). Tackling the emerging threat of antifungal resistance to human health. *Nature Reviews Microbiology, 20*(9), 557–571. https://doi.org/10.1038/s41579-022-00720-1

Franconi, I., Leonildi, A., Erra, G., Fais, R., Falcone, M., Ghelardi, E., & Lupetti, A. (2022). Comparison of different microbiological procedures for the diagnosis of Pneumocystis jirovecii pneumonia on bronchoalveolar-lavage fluid. *BMC Microbiology, 22*(1), 143. https://doi.org/10.1186/s12866-022-02559-1

Gallo, F., Vija, L., Grand, S. L., Moukarbel, N., Mortele, K., Gabiache, E., Courbon, F., Tavitian, S., & Dierickx, L. O. (2019). Diagnosis of an intestinal mucormycosis 'fungus ball' located with PET/CT with [18F] FDG-PET/CT. *European Journal of Hybrid Imaging, 3*(1). https://doi.org/10.1186/s41824-019-0068-0

Gangcuangco, L. M. A., Sawada, I., Tsuchiya, N., Do, C. D., Pham, T. T. T., Rojanawiwat, A., Alejandria, M., Leyritana, K., Yokomaku, Y., Pathipvanich, P., & Ariyoshi, K. (2017). Regional Differences in the Prevalence of Major Opportunistic Infections among Antiretroviral-Naïve Human Immunodeficiency Virus Patients in Japan, Northern Thailand, Northern Vietnam, and The Philippines. *The American Journal of Tropical Medicine and Hygiene, 97*(1), 49–56. https://doi.org/10.4269/ajtmh.16-0783

Gao, Y., & Guo, Y. (2023). Research progress in the development of natural-product-based mucosal vaccine adjuvants. *Frontiers in Immunology, 14*, 1152855. https://doi.org/10.3389/fimmu.2023.1152855

García Carnero, L. C., Lozoya Pérez, N. E., González Hernández, S. E., & Martínez Álvarez, J. A. (2018). Immunity and treatment of sporotrichosis. *Journal of Fungi (Basel, Switzerland), 4*(3), 100. https://doi.org/10.3390/jof4030100

Ghosh, A., Chakrabarti, A., Sharma, V. K., Singh, K., & Singh, A. (1999). Sporotrichosis in Himachal Pradesh (north India). *Transactions of the Royal Society of Tropical Medicine and Hygiene, 93*(1), 41–45. https://doi.org/10.1016/s0035-9203(99)90173-6

Guimarães, A. J., Pizzini, C. V., de Matos Guedes, H. L., Albuquerque, P. C., Peralta, J. M., Hamilton, A. J., & Zancopé-Oliveira, R. M. (2004). ELISA for early diagnosis of histoplasmosis. *Journal of Medical Microbiology, 53*(Pt 6), 509–514. https://doi.org/10.1099/jmm.0.05469-0

Guo, J., Li, B., Li, T., Wei, F., Fu, Y., Zheng, Y., Pan, K., Huang, C., & Cao, C. (2019). Characteristics and Prognosis of Talaromyces marneffei Infection in Non-HIV-Infected Children in Southern China. *Mycopathologia, 184*(6), 735–745. https://doi.org/10.1007/s11046-019-00373-4

Gupta, N., Arora, S. K., Rajwanshi, A., Nijhawan, R., & Srinivasan, R. (2010). Histoplasmosis: Cytodiagnosis and review of literature with special emphasis on differential diagnosis on cytomorphology. *Cytopathology : Official Journal of the British Society for Clinical Cytology, 21*(4), 240–244. https://doi.org/10.1111/j.1365-2303.2009.00693.x

Gupta, A. K., Friedlander, S. F., & Simkovich, A. J. (2022). Tinea capitis: An update. *Pediatric Dermatology, 39*(2), 167–172. https://doi.org/10.1111/pde.14925

Hage, C. A., Azar, M. M., Bahr, N., Loyd, J., & Wheat, L. J. (2015). Histoplasmosis: Up-to-date evidence-based approach to diagnosis and management. *Seminars in Respiratory and Critical Care Medicine, 36*(5), 729–745. https://doi.org/10.1055/s-0035-1562899

Hage, C. A., Kirsch, E. J., Stump, T. E., Kauffman, C. A., Goldman, M., Connolly, P., Johnson, P. C., Wheat, L. J., & Baddley, J. W. (2011). Histoplasma antigen clearance during treatment of histoplasmosis in patients with AIDS determined by a quantitative antigen enzyme immunoassay. *Clinical and Vaccine Immunology: CVI, 18*(4), 661–666. https://doi.org/10.1128/CVI.00389-10

Hajjeh, R. A., Conn, L. A., Stephens, D. S., Baughman, W., Hamill, R., Graviss, E., Pappas, P. G., Thomas, C., Reingold, A., Rothrock, G., Hutwagner, L. C., Schuchat, A., Brandt, M. E., & Pinner, R. W. (1999). Cryptococcosis: Population-based multistate active surveillance and risk factors in human immunodeficiency virus-infected persons. Cryptococcal Active Surveillance Group. *The Journal of Infectious Diseases, 179*(2), 449–454. https://doi.org/10.1086/314606

Haldar, N., Sharma, M. K., & Gugnani, H. C. (2007). Sporotrichosis in north-east India. *Mycoses, 50*(3), 201–204. https://doi.org/10.1111/j.1439-0507.2006.01344.x

Hardison, S. E., Herrera, G., Young, M. L., Hole, C. R., Wozniak, K. L., & Wormley, F. L., Jr. (2012). Protective immunity against pulmonary cryptococcosis is associated with STAT1-mediated classical macrophage activation. *Journal of Immunology (Baltimore, Md. : 1950), 189*(8), 4060–4068. https://doi.org/10.4049/jimmunol.1103455

Hassan, M. I. A., & Voigt, K. (2019). Pathogenicity patterns of mucormycosis: epidemiology, interaction with immune cells and virulence factors. *Medical Mycology, 57*(Supplement_2), S245–S256. https://doi.org/10.1093/mmy/myz011

Hay, R. J. (2017). Tinea Capitis: Current Status. *Mycopathologia, 182*(1–2), 87–93. https://doi.org/10.1007/s11046-016-0058-8

Herring, A. C., Lee, J., McDonald, R. A., Toews, G. B., & Huffnagle, G. B. (2002). Induction of interleukin-12 and gamma interferon requires tumor necrosis factor alpha for protective T1-cell-mediated immunity to pulmonary Cryptococcus neoformans infection. *Infection and Immunity, 70*(6), 2959–2964. https://doi.org/10.1128/IAI.70.6.2959-2964.2002

Hessler, C., Kauffman, C. A., & Chow, F. C. (2017). The Upside of Bias: A Case of Chronic Meningitis Due to Sporothrix Schenckii in an Immunocompetent Host. *The Neurohospitalist, 7*(1), 30–34. https://doi.org/10.1177/1941874416641468

Hester, M. M., Lee, C. K., Abraham, A., Khoshkenar, P., Ostroff, G. R., Levitz, S. M., & Specht, C. A. (2020). Protection of mice against experimental cryptococcosis using glucan particle-based vaccines containing novel recombinant antigens. *Vaccine, 38*(3), 620–626. https://doi.org/10.1016/j.vaccine.2019.10.051

Hill, R. C., Gold, J. A. W., & Lipner, S. R. (2024). Comprehensive review of tinea capitis in adults: Epidemiology, risk factors, clinical presentations, and management. *Journal of Fungi (Basel, Switzerland), 10*(5), 357. https://doi.org/10.3390/jof10050357

Hodžić, A., Hernandez, L. M., De La Fuente, J., & Cruz, A. C. (2020). α-GaL-Based vaccines: Advances, opportunities, and perspectives. *Trends in Parasitology, 36*(12), 992–1001. https://doi.org/10.1016/j.pt.2020.08.001

Hope, W., Walsh, T., & Denning, D. (2005). Laboratory diagnosis of invasive aspergillosis. *The Lancet Infectious Diseases, 5*(10), 609–622. https://doi.org/10.1016/s1473-3099(05)70238-3

Hopke, A., Brown, A. J. P., Hall, R. A., & Wheeler, R. T. (2018). Dynamic Fungal Cell Wall Architecture in Stress Adaptation and Immune Evasion. *Trends in Microbiology, 26*(4), 284–295. https://doi.org/10.1016/j.tim.2018.01.007

Horwath, M. C., Fecher, R. A., & Deepe, G. S., Jr. (2015). Histoplasma capsulatum, lung infection and immunity. *Future Microbiology, 10*(6), 967–975. https://doi.org/10.2217/fmb.15.25

Houh, W., Chung, S. R., & Ro, B. I. (1995). Itraconazole in the treatment of sporotrichosis: a Korean experience. *Tropical Doctor, 25*(3), 107–109. https://doi.org/10.1177/004947559502500306

Hu, Y., Zhang, J., Li, X., Yang, Y., Zhang, Y., Ma, J., & Xi, L. (2012). Penicillium marneffei infection: An emerging disease in mainland China. *Mycopathologia, 175*(1–2), 57–67. https://doi.org/10.1007/s11046-012-9577-0

Huang, X., Kuang, Y., Liu, Q., & Liao, H. (2024). Pleural aspergillosis in an otherwise healthy patient diagnosed by medical thoracoscopy. *Clinical Microbiology and Infection, 30*(4), 489–491. https://doi.org/10.1016/j.cmi.2023.12.031

Ishizaki, H., Kawasaki, M., Anzawa, K., Mochizuki, T., Chakrabarti, A., Ungpakorn, R., Torres Guererro, H., Toriello, C., & Arenas, R. (2009). Mitochondrial DNA analysis of Sporothrix schenckii in India, Thailand, Brazil, Colombia, Guatemala and Mexico. *Nihon Ishinkin Gakkai zasshi = Japanese journal of medical mycology, 50*(1), 19–26. https://doi.org/10.3314/jjmm.50.19

Jaijakul, S., Saksirisampant, W., Prownebon, J., Yenthakam, S., Mungthin, M., Leelayoova, S., & Nuchprayoon, S. (2005). Pneumocystis jiroveci in HIV/AIDS patients: Detection by FTA filter paper together with PCR in noninvasive induced sputum specimens. *Journal of the Medical Association of Thailand = Chotmaihet thangphaet, 88 Suppl 4*, S294–S299.

Jarvis, J. N., Percival, A., Bauman, S., Pelfrey, J., Meintjes, G., Williams, G. N., Longley, N., Harrison, T. S., & Kozel, T. R. (2011). Evaluation of a novel point-of-care cryptococcal antigen test on serum, plasma, and urine from patients with HIV-associated cryptococcal meningitis. *Clinical Infectious Diseases : An Official Publication of the Infectious Diseases Society of America, 53*(10), 1019–1023. https://doi.org/10.1093/cid/cir613

John, J., Loo, A., Mazur, S., & Walsh, T. J. (2019). Therapeutic drug monitoring of systemic antifungal agents: A pragmatic approach for adult and pediatric patients. *Expert Opinion on Drug Metabolism & Toxicology, 15*(11), 881–895. https://doi.org/10.1080/17425255.2019.1671971

Kasai, M., Harrington, S. M., Francesconi, A., Petraitis, V., Petraitiene, R., Beveridge, M. G., Knudsen, T., Milanovich, J., Cotton, M. P., Hughes, J., Schaufele, R. L., Sein, T., Bacher, J., Murray, P. R., Kontoyiannis, D. P., & Walsh, T. J. (2008). Detection of a molecular biomarker for zygomycetes by quantitative PCR assays of plasma, bronchoalveolar lavage, and lung tissue in a rabbit model of experimental pulmonary zygomycosis. *Journal of Clinical Microbiology, 46*(11), 3690–3702. https://doi.org/10.1128/JCM.00917-08

Kauffman, C. A. (2007). Histoplasmosis: a clinical and laboratory update. *Clinical Microbiology Reviews, 20*(1), 115–132. https://doi.org/10.1128/CMR.00027-06

Kauffman, C. A., Bustamante, B., Chapman, S. W., Pappas, P. G., & Infectious Diseases Society of America. (2007). Clinical practice guidelines for the management of sporotrichosis: 2007 update by the Infectious Diseases Society of America. *Clinical Infectious Diseases, 45*, 1255–1265.

Kawasaki, T., & Kawai, T. (2014). Toll-like receptor signaling pathways. *Frontiers in Immunology, 5*, 461. https://doi.org/10.3389/fimmu.2014.00461

Kibbler, C. C., Barton, R., Gow, N., Howell, S., MacCallum, D., & Manuel, R. (Eds.). (2018). *Oxford textbook of medical mycology* (Oxford textbooks in infectious disease and microbiology). Oxford University Press.

Kikuchi, I., Morimoto, K., Kawana, S., & Tanuma, H. (2006). Usefulness of itraconazole for sporotrichosis in Japan: study of three cases and literature comparison of therapeutic effects before and after release on the market. *European Journal of Dermatology, 16*(1), 42–47.

Kinnamon, D. S., Heggestad, J. T., Liu, J., Nguyen, T., Ly, V., Hucknall, A. M., Fontes, C. M., Britton, R. J., Cai, J., Chan, J. F., Yuen, K., Le, T., & Chilkoti, A. (2023). Environmentally resilient microfluidic Point-of-Care immunoassay enables rapid diagnosis of talaromycosis. *ACS Sensors, 8*(6), 2228–2236. https://doi.org/10.1021/acssensors.3c00209

Lackner, M., Caramalho, R., & Lass-Flörl, C. (2014). Laboratory diagnosis of mucormycosis: current status and future perspectives. *Future Microbiology, 9*(5), 683–695. https://doi.org/10.2217/fmb.14.23

Le, T., Chi, N. H., Cuc, N. T. K., Sieu, T. P. M., Shikuma, C. M., Farrar, J., & Day, J. N. (2010). AIDS-Associated Penicillium Marneffei Infection of the central nervous system. *Clinical Infectious Diseases (Online. University of Chicago. Press), 51*(12), 1458–1462. https://doi.org/10.1086/657400

Le, T., Wolbers, M., Chi, N. H., Quang, V. M., Chinh, N. T., Lan, N. P. H., Lam, P. S., Kozal, M. J., Shikuma, C. M., Day, J. N., & Farrar, J. (2011). Epidemiology, Seasonality, and Predictors of Outcome of AIDS-Associated Penicillium marneffei Infection in Ho Chi Minh City, Viet Nam. *Clinical Infectious Diseases/Clinical Infectious Diseases (Online. University of Chicago. Press), 52*(7), 945–952. https://doi.org/10.1093/cid/cir028

Lee, P. P., & Lau, Y. L. (2017). Cellular and molecular defects underlying invasive fungal infections-revelations from endemic mycoses. *Frontiers in Immunology, 8*, 735. https://doi.org/10.3389/fimmu.2017.00735

Lee, Y., Robbins, N., & Cowen, L. E. (2023). Molecular mechanisms governing antifungal drug resistance. *npj Antimicrobials and Resistance, 1*(1), 7. https://doi.org/10.1038/s44259-023-00007-2

Leung, A. K. C., Hon, K. L., Leong, K. F., Barankin, B., & Lam, J. M. (2020). Tinea Capitis: An Updated Review. *Recent Patents on Inflammation & Allergy Drug Discovery, 14*(1), 58–68. https://doi.org/10.2174/1872213X14666200106145624

Liewchanpattana, W., Arayawichanont, A., & Suwannawong, P. (2020). Sporotrichosis: The case series in Thailand and literature review in Southeast Asia. *Medical Mycology Case Reports, 27*, 13–17. https://doi.org/10.1016/j.mmcr.2020.02.001

Liu, Y., Wu, H., Huang, F., Fan, Z., & Xu, B. (2013). Utility of 18F-FDG PET/CT in diagnosis and management of mucormycosis. *Clinical Nuclear Medicine, 38*(9), e370–e371. https://doi.org/10.1097/rlu.0b013e3182867d13

Lockhart, S. R., Etienne, K. A., Vallabhaneni, S., Farooqi, J., Chowdhary, A., Govender, N. P., & Chiller, T. (2017). Simultaneous emergence of multidrug-resistant *Candida auris* on three continents confirmed by whole-genome sequencing and epidemiological analyses. *Clinical Infectious Diseases, 64*(2), 134–140. https://doi.org/10.1093/cid/ciw691

Ly, V. T., Thanh, N. T., Thu, N. T. M., Chan, J., Day, J. N., Perfect, J., Nga, C. N., Van Vinh Chau, N., & Le, T. (2020). Occult talaromyces Marneffei infection unveiled by the novel MP1P Antigen Detection assay. *Open Forum Infectious Diseases, 7*(11). https://doi.org/10.1093/ofid/ofaa502

Mancuso, G., Midiri, A., Gerace, E., & Biondo, C. (2022). Role of the innate immune system in host defence against fungal infections. *European Review for Medical and Pharmacological Sciences, 26*(4), 1138–1147. https://doi.org/10.26355/eurrev_202202_28105

Maresca, B., & Kobayashi, G. S. (1989). Dimorphism in Histoplasma capsulatum: a model for the study of cell differentiation in pathogenic fungi. *Microbiological Reviews, 53*(2), 186–209. https://doi.org/10.1128/mr.53.2.186-209.1989

Marukutira, T., Huprikar, S., Azie, N., Quan, S. P., Meier-Kriesche, H. U., & Horn, D. L. (2014). Clinical characteristics and outcomes in 303 HIV-infected patients with invasive fungal infections: data from the Prospective Antifungal Therapy Alliance registry, a multicenter, observational study. *HIV/AIDS (Auckland, N.Z.), 6*, 39–47. https://doi.org/10.2147/HIV.S53910

Mayser, P., Nenoff, P., Reinel, D., Abeck, D., Brasch, J., Daeschlein, G., Effendy, I., Ginter-Hanselmayer, G., Gräser, Y., Hipler, U. C., Höger, P., Kolb-Mäurer, A., Ott, H., Schaller, M., & Zidane, M. (2020). S1 guidelines: Tinea capitis. *Journal der Deutschen Dermatologischen Gesellschaft = Journal of the German Society of Dermatology : JDDG, 18*(2), 161–179. https://doi.org/10.1111/ddg.14026

Maziarz, E. K., & Perfect, J. R. (2016). Cryptococcosis. *Infectious Disease Clinics of North America, 30*(1), 179–206. https://doi.org/10.1016/j.idc.2015.10.006

McCormick, T., & Ghannoum, M. (2024). Time to think antifungal resistance. *Pathogens and Immunity, 8*(2), 158–176. https://doi.org/10.20411/pai.v8i2.656

McKinsey, D. S., Spiegel, R. A., Hutwagner, L., Stanford, J., Driks, M. R., Brewer, J., Gupta, M. R., Smith, D. L., O'Connor, M. C., & Dall, L. (1997). Prospective study of histoplasmosis in patients infected with human immunodeficiency virus: incidence, risk factors, and pathophysiology. *Clinical Infectious Diseases : An Official Publication of the Infectious Diseases Society of America, 24*(6), 1195–1203. https://doi.org/10.1086/513653

Merkhofer, R. M., & Klein, B. S. (2020). Advances in understanding human genetic variations that influence innate immunity to fungi. *Frontiers in Cellular and Infection Microbiology, 10*, 69. https://doi.org/10.3389/fcimb.2020.00069

Morris-Jones, R. (2002). Sporotrichosis. *Clinical and Experimental Dermatology, 27*(6), 427–431. https://doi.org/10.1046/j.1365-2230.2002.01087.x

Myint, T., Leedy, N., Villacorta Cari, E., & Wheat, L. J. (2020). HIV-associated histoplasmosis: Current perspectives. *HIV/AIDS (Auckland, N.Z.), 12*, 113–125. https://doi.org/10.2147/HIV.S185631

Nacher, M., Blanchet, D., Bongomin, F., Chakrabarti, A., Couppié, P., Demar, M., Denning, D. W., Djossou, F., Epelboin, L., Govender, N., Leitão, T., Mac Donald, S., Mandengue, C., Marques da Silva, S. H., Oladele, R., Panizo, M. M., Pasqualotto, A., Ramos, R., Swaminathan, S., et al. (2018). Histoplasma capsulatum antigen detection tests as an essential diagnostic tool for patients with advanced HIV disease in low and middle income countries: A systematic review of Diagnostic Accuracy Studies. *PLOS Neglected Tropical Diseases, 12*(10). https://doi.org/10.1371/journal.pntd.0006802

Narayanasamy, S., Dougherty, J., Van Doorn, H. R., & Le, T. (2021). Pulmonary Talaromycosis: A Window into the Immunopathogenesis of an Endemic Mycosis. *Mycopathologia, 186*(5), 707–715. https://doi.org/10.1007/s11046-021-00570-0

Nnadi, N. E., & Carter, D. A. (2021). Climate change and the emergence of fungal pathogens. *PLoS Pathogens, 17*(4), Article e1009503. https://doi.org/10.1371/journal.ppat.1009503

Norkaew, T., Ohno, H., Sriburee, P., Tanabe, K., Tharavichitkul, P., Takarn, P., Puengchan, T., Bumrungsri, S., & Miyazaki, Y. (2013). Detection of environmental sources of Histoplasma capsulatum in Chiang Mai, Thailand, by nested PCR. *Mycopathologia, 176*(5–6), 395–402. https://doi.org/10.1007/s11046-013-9701-9

Normile, T. G., & Del Poeta, M. (2022). Three Models of Vaccination Strategies Against Cryptococcosis in Immunocompromised Hosts Using Heat-Killed *Cryptococcus neoformans* Δ*sgl1*. *Frontiers in Immunology, 13*, 868523. https://doi.org/10.3389/fimmu.2022.868523

Normile, T. G., McEvoy, K., & Del Poeta, M. (2020). Steryl glycosides in fungal pathogenesis: An understudied immunomodulatory adjuvant. *Journal of Fungi (Basel, Switzerland), 6*(1), 25. https://doi.org/10.3390/jof6010025

Notarte, K. I. R., Pastrana, A. M., Ver, A. T. M., Velasco, J. V. L., Gellaco, M. M. L. D., & Pecundo, M. H. (2023). Mycosis in The Philippines: Epidemiology, clinical presentation, diagnostics and interventions. In *Mycology in the tropics* (pp. 213–233). https://doi.org/10.1016/b978-0-323-99489-7.00005-6

Odio, C. M., Navarrete, M., Carrillo, J. M., Mora, L., & Carranza, A. (1999). Disseminated histoplasmosis in infants. *The Pediatric Infectious Disease Journal, 18*(12), 1065–1068. https://doi.org/10.1097/00006454-199912000-00007

Oliveira, L. V. N., Wang, R., Specht, C. A., & Levitz, S. M. (2021). Vaccines for human fungal diseases: close but still a long way to go. *npj Vaccines, 6*(1), 33. https://doi.org/10.1038/s41541-021-00294-8

Onishi, A., Sugiyama, D., Kogata, Y., Saegusa, J., Sugimoto, T., Kawano, S., Morinobu, A., Nishimura, K., & Kumagai, S. (2012). Diagnostic accuracy of serum 1,3-β-D-glucan for pneumocystis jiroveci pneumonia, invasive candidiasis, and invasive aspergillosis: systematic review and meta-analysis. *Journal of Clinical Microbiology, 50*(1), 7–15. https://doi.org/10.1128/JCM.05267-11

Orofino-Costa, R., Macedo, P. M., Rodrigues, A. M., & Bernardes-Engemann, A. R. (2017). Sporotrichosis: an update on epidemiology, etiopathogenesis, laboratory and clinical therapeutics. *Anais Brasileiros de Dermatologia, 92*(5), 606–620. https://doi.org/10.1590/abd1806-4841.2017279

Oyarce, J. A., García, C., Alave, J., & Bustamante, B. (2016). Epidemiological clinical and laboratory characterization of sporotrichosis in patients of a tertiary care hospital in Lima, Peru, from 1991 to 2014. *Revista Chilena de Infectología, 33*(3), 315–321.

Pappas, P. G., Tellez, I., Deep, A. E., Nolasco, D., Holgado, W., & Bustamante, B. (2000). Sporotrichosis in Peru: description of an area of hyperendemicity. *Clinical Infectious Diseases : An Official Publication of the Infectious Diseases Society of America, 30*(1), 65–70. https://doi.org/10.1086/313607

Park, B. J., Wannemuehler, K. A., Marston, B. J., Govender, N., Pappas, P. G., & Chiller, T. M. (2009). Estimation of the current global burden of cryptococcal meningitis among persons living with HIV/AIDS. *AIDS (London, England), 23*(4), 525–530. https://doi.org/10.1097/QAD.0b013e328322ffac

Pathakumari, B., Liang, G., & Liu, W. (2020). Immune defence to invasive fungal infections: A comprehensive review. *Biomedicine & Pharmacotherapy = Biomedecine & Pharmacotherapie, 130*, 110550. https://doi.org/10.1016/j.biopha.2020.110550

Perera Molligoda Arachchige, A. S. (2021). Human NK cells: From development to effector functions. *Innate Immunity, 27*(3), 212–229. https://doi.org/10.1177/17534259211001512

Perfect, J. R. (2015). Cryptococcus neoformans and Cryptococcus gattii. In J. E. Bennett, R. Dolin, & M. J. Blaser (Eds.), *Mandell, Douglas, and Bennett's principles and practice of infectious diseases* (8th ed., pp. 2934–2948). Elsevier Saunders.

Perfect, J. R., & Casadevall, A. (2002). Cryptococcosis. *Infectious Disease Clinics of North America, 16*(4), 837–8vi. https://doi.org/10.1016/s0891-5520(02)00036-3

Piccioli, D., Sbrana, S., Melandri, E., & Valiante, N. M. (2002). Contact-dependent stimulation and inhibition of dendritic cells by natural killer cells. *The Journal of Experimental Medicine, 195*(3), 335–341. https://doi.org/10.1084/jem.20010934

Portuondo, D. L., Batista-Duharte, A., Cardenas, C., de Oliveira, C. S., Borges, J. C., Téllez-Martínez, D., Santana, P. A., Gauna, A., Mercado, L., de Castilho, B. M., & Costa, P. (2022). A *Sporothrix* spp. enolase-derived multi-epitope vaccine confers protective response in BALB/c mice challenged with *Sporothrix brasiliensis*. *Microbial Pathogenesis, 166*, 105539. https://doi.org/10.1016/j.micpath.2022.105539

Portuondo, D. L., Dores-Silva, P. R., Ferreira, L. S., de Oliveira, C. S., Téllez-Martínez, D., Marcos, C. M., de Aguiar Loesch, M. L., Guzmán, F., Gava, L. M., Borges, J. C., & Pereira, S. A. (2019). Immunization with recombinant enolase of *Sporothrix* spp. (rSsEno) confers effective protection against sporotrichosis in mice. *Scientific Reports, 9*(1), 17179. https://doi.org/10.1038/s41598-019-53742-1

Qin, Y., Huang, X., Chen, H., Liu, X., Li, Y., Hou, J., Li, A., Yan, X., & Chen, Y. (2020). Burden of Talaromyces marneffei infection in people living with HIV/AIDS in Asia during ART era: a systematic review and meta-analysis. *BMC Infectious Diseases, 20*(1). https://doi.org/10.1186/s12879-020-05260-8

Rabacal, W., Schweitzer, F., Kling, H. M., Buzzelli, L., Rayens, E., & Norris, K. A. (2022). A therapeutic vaccine strategy to prevent Pneumocystis pneumonia in an immunocompromised host in a non-human primate model of HIV and Pneumocystis co-infection. *Frontiers in Immunology, 13*, 1036658. https://doi.org/10.3389/fimmu.2022.1036658

Rajasingham, R., Smith, R. M., Park, B. J., Jarvis, J. N., Govender, N. P., Chiller, T. M., Denning, D. W., Loyse, A., & Boulware, D. R. (2017). Global burden of disease of HIV-associated cryptococcal meningitis: an updated analysis. *The Lancet. Infectious Diseases, 17*(8), 873–881. https://doi.org/10.1016/S1473-3099(17)30243-8

Rajendran, C., Ramesh, V., Misra, R. S., & Kandhari, S. (1990). Sporotrichosis in Nepal. *International Journal of Dermatology, 29*(10), 716–718. https://doi.org/10.1111/j.1365-4362.1990.tb03776.x

Ramirez-Ortiz, Z. G., Lee, C. K., Wang, J. P., Boon, L., Specht, C. A., & Levitz, S. M. (2011). A nonredundant role for plasmacytoid dendritic cells in host defense against the human fungal pathogen Aspergillus fumigatus. *Cell Host & Microbe, 9*(5), 415–424. https://doi.org/10.1016/j.chom.2011.04.007

Ratanasuwan, W., Anekthananon, T., Techasathit, W., Rongrungruang, Y., Sonjai, A., & Suwanagool, S. (2005). Estimated economic losses of hospitalized AIDS patients at Siriraj Hospital from January 2003 to December 2003: Time for aggressive voluntary counseling and HIV testing. *Journal of the Medical Association of Thailand = Chotmaihet thangphaet, 88*(3), 335–339.

Rayens, E., Rabacal, W., Willems, H. M. E., Kirton, G. M., Barber, J. P., Mousa, J. J., Celia-Sanchez, B. N., Momany, M., & Norris, K. A. (2022). Immunogenicity and protective efficacy of a pan-fungal vaccine in preclinical models of aspergillosis, candidiasis, and pneumocystosis. *PNAS Nexus, 1*(5), pgac248. https://doi.org/10.1093/pnasnexus/pgac248

Reyes, A., Lucero-Prisno, D. E., Catahay, J. A., Gibas, C., Uy-Lumandas, M., De Paz-Silava, S. L., Flores, M. J., Carandang, T. H. D. C., Tantengco, O. A., Climacosa, F. M., Moron-Espiritu, L., Pastrana, A., De Guia, J. J., Ilagan-Tagarda, D., & Notarte, K. I. (2024). Reinvigorating vigilance: Updates on emerging and re-emerging infectious diseases in The Philippines. In *Biosafety and Biosecurity* (pp. 259–307). CRC Press. https://doi.org/10.1201/9781003426219-13

Rickerts, V., Just-Nübling, G., Konrad, F., Kern, J., Lambrecht, E., Böhme, A., Jacobi, V., & Bialek, R. (2006). Diagnosis of invasive aspergillosis and mucormycosis in immunocompromised patients by seminested PCR assay of tissue samples. *European Journal of Clinical Microbiology & Infectious Diseases, 25*(1), 8–13. https://doi.org/10.1007/s10096-005-0078-7

Rickerts, V., McCormick Smith, I., Mousset, S., Kommedal, O., & Fredricks, D. N. (2013). Deciphering the aetiology of a mixed fungal infection by broad-range PCR with sequencing and fluorescence in situ hybridisation. *Mycoses, 56*(6), 681–686. https://doi.org/10.1111/myc.12083

Rippon, J. W. (1988a). *Medical mycology: The pathogenic fungi and the pathogenic actinomycetes* (3rd ed.). Saunders.

Rippon, J. W. (1988b). Sporotrichosis. In J. W. Rippon (Ed.), *Medical mycology: The pathogenic fungi and the pathogenic actinomycetes* (pp. 325–352). WB Saunders.

Rivera, A., Lodge, J., & Xue, C. (2022). Harnessing the immune response to fungal pathogens for vaccine development. *Annual Review of Microbiology, 76*(1), 703–726. https://doi.org/10.1146/annurev-micro-041020-111511

Rizzo, J., Wong, S. S. W., Gazi, A. D., Moyrand, F., Chaze, T., Commere, P. H., Novault, S., Matondo, M., Péhau-Arnaudet, G., Reis, F. C. G., Vos, M., Alves, L. R., May, R. C., Nimrichter, L., Rodrigues, M. L., Aimanianda, V., & Janbon, G. (2021). *Cryptococcus* extracellular vesicles properties and their use as vaccine platforms. *Journal of Extracellular Vesicles, 10*(10), e12129. https://doi.org/10.1002/jev2.12129

Rodrigues, A. M., de Hoog, G. S., & de Camargo, Z. P. (2016). Sporothrix Species Causing Outbreaks in Animals and Humans Driven by Animal-Animal Transmission. *PLoS Pathogens, 12*(7), e1005638. https://doi.org/10.1371/journal.ppat.1005638

Rodríguez-Cerdeira, C., Martínez-Herrera, E., Szepietowski, J. C., Pinto-Almazán, R., Frías-De-León, M. G., Espinosa-Hernández, V. M., Chávez-Gutiérrez, E., García-Salazar, E., Vega-Sánchez, D. C., Arenas, R., Hay, R., & Saunte, D. M. (2021). A systematic review of worldwide data on tinea capitis: analysis of the last 20 years. *Journal of the European Academy of Dermatology and Venereology : JEADV, 35*(4), 844–883. https://doi.org/10.1111/jdv.16951

Rodriguez-Tudela, J. L., Alastruey-Izquierdo, A., Gago, S., Cuenca-Estrella, M., León, C., Miro, J. M., Nuñez Boluda, A., Ruiz Camps, I., Sole, A., Denning, D. W., & University of Manchester in association with the LIFE program. (2015). Burden of serious fungal infections in Spain. *Clinical Microbiology and Infection : The Official Publication of the European Society of Clinical Microbiology and Infectious Diseases, 21*(2), 183–189. https://doi.org/10.1016/j.cmi.2014.07.013

Romani, L. (2004). Immunity to fungal infections. *Nature Reviews Immunology, 4*(1), 11–24. https://doi.org/10.1038/nri1255

Salahi, Z., Yadegari, M. H., Bagheri, M., Khani, M. H., Zahmatkesh, A., & Khaki, P. (2020). The effect of cytoplasmic crude extracts of *Trichophyton verrucosum* on cell mediated immunity. *Journal de Mycologie Médicale, 30*(3), 100969. https://doi.org/10.1016/j.mycmed.2020.100969

Scheel, C. M., & Gómez, B. L. (2014). Diagnostic methods for histoplasmosis: Focus on endemic countries with variable infrastructure levels. *Current Tropical Medicine Reports, 1*(2), 129–137. https://doi.org/10.1007/s40475-014-0020-0

Schmidt, S., Tramsen, L., & Lehrnbecher, T. (2017). Natural killer cells in antifungal immunity. *Frontiers in Immunology, 8*, 1623. https://doi.org/10.3389/fimmu.2017.01623

Seidel, D., Wurster, S., Jenks, J. D., Sati, H., Gangneux, J. P., Egger, M., Alastruey-Izquierdo, A., Ford, N. P., Chowdhary, A., Sprute, R., Cornely, O., Thompson, G. R., 3rd, Hoenigl, M., &

Kontoyiannis, D. P. (2024). Impact of climate change and natural disasters on fungal infections. *The Lancet Microbe, 5*(6), e594–e605. https://doi.org/10.1016/S2666-5247(24)00039-9

Senécal, J., Smyth, E., Del Corpo, O., Hsu, J. M., Amar-Zifkin, A., Bergeron, A., Cheng, M. P., Butler-Laporte, G., McDonald, E. G., & Lee, T. C. (2022). Non-invasive diagnosis of Pneumocystis jirovecii pneumonia: a systematic review and meta-analysis. *Clinical Microbiology and Infection : The Official Publication of the European Society of Clinical Microbiology and Infectious Diseases, 28*(1), 23–30. https://doi.org/10.1016/j.cmi.2021.08.017

Shankar, J., Thakur, R., Clemons, K. V., & Stevens, D. A. (2024). Interplay of cytokines and chemokines in aspergillosis. *Journal of Fungi (Basel, Switzerland), 10*(4), 251. https://doi.org/10.3390/jof10040251

Sharma, J., Mudalagiriyappa, S., & Nanjappa, S. G. (2022). T cell responses to control fungal infection in an immunological memory lens. *Frontiers in Immunology, 13*, 905867. https://doi.org/10.3389/fimmu.2022.905867

Shibuya, K., Coulson, W. F., Wollman, J. S., Wakayama, M., Ando, T., Oharaseki, T., Takahashi, K., & Naoe, S. (2001). Histopathology of cryptococcosis and other fungal infections in patients with acquired immunodeficiency syndrome. *International Journal of Infectious Diseases : IJID : Official Publication of the International Society for Infectious Diseases, 5*(2), 78–85. https://doi.org/10.1016/s1201-9712(01)90030-x

Sigera, L. S. M., & Denning, D. W. (2023). A systematic review of the therapeutic outcome of mucormycosis. *Open Forum Infectious Diseases, 11*(1). https://doi.org/10.1093/ofid/ofad704

Simuangco, S. A., Bocobo, F. C., & Lacuna, L. (1963). Review of literature on medical mycology in The Philippines, 1955–1962. *Mycopathologia et Mycologia Applicata, 20*, 145–156. https://doi.org/10.1007/BF02054886

Skiada, A., Pavleas, I., & Drogari-Apiranthitou, M. (2020). Epidemiology and diagnosis of mucormycosis: An update. *Journal of Fungi (Basel, Switzerland), 6*(4), 265. https://doi.org/10.3390/jof6040265

Slarve, M., Holznecht, N., Reza, H., Gilkes, A., Slarve, I., Olson, J., Ernst, W., Ho, S. O., Adler-Moore, J., & Fujii, G. (2022). Recombinant Aspergillus fumigatus antigens Asp f 3 and Asp f 9 in liposomal vaccine protect mice against invasive pulmonary aspergillosis. *Vaccine, 40*(31), 4160–4168. https://doi.org/10.1016/j.vaccine.2022.05.057

Smith, J. A., & Kauffman, C. A. (2009). Endemic fungal infections in patients receiving tumour necrosis factor-alpha inhibitor therapy. *Drugs, 69*(11), 1403–1415. https://doi.org/10.2165/00003495-200969110-00002

Sojinrin, T., Conde, J., Liu, K., Curtin, J., Byrne, H. J., Cui, D., & Tian, F. (2017). Plasmonic gold nanoparticles for detection of fungi and human cutaneous fungal infections. *Analytical and Bioanalytical Chemistry/Analytical & Bioanalytical Chemistry, 409*(19), 4647–4658. https://doi.org/10.1007/s00216-017-0414-7

Someshwaran, R., & Appalaraju, B. (2023). MALDI-TOF-MS for Rapid Detection of Fungi causing Mucormycosis during COVID-19 Pandemic from a Tertiary Care Hospital in Southern India: A Retrospective Cohort Study. *Journal of Clinical and Diagnostic Research*. https://doi.org/10.7860/jcdr/2023/64875.18409

Specht, C. A., Wang, R., Oliveira, L. V. N., Hester, M. M., Gomez, C., Mou, Z., Carlson, D., Lee, C. K., Hole, C. R., Lam, W. C., Upadhya, R., Lodge, J. K., & Levitz, S. M. (2024). Immunological correlates of protection mediated by a whole organism, *Cryptococcus neoformans*, vaccine deficient in chitosan. *MBio, 15*(8), e0174624. https://doi.org/10.1128/mbio.01746-24

Srivastava, S. R., Ganguly, P., Barman, D., Das, S., Bandyopadhyay, M., Ghosh, A. K., Sarkar, S., Sengupta, A., Swaika, S., Chatterjee, P., Gupta, A. K., Mondal, A. R., Guha, S., Dutta, S., Adhikari, S., Kaushik, A., Biswas, P. S., & Ayub, A. (2022). Clinical outcome of orbital apex syndrome in COVID associated mucormycosis patients in a tertiary care hospital. *International Journal of Ophthalmology, 15*(4), 527–532. https://doi.org/10.18240/ijo.2022.04.01

Steenbergen, J. N., Shuman, H. A., & Casadevall, A. (2001). Cryptococcus neoformans interactions with amoebae suggest an explanation for its virulence and intracellular pathogenic strategy in macrophages. *Proceedings of the National Academy of Sciences of the United States of America, 98*(26), 15245–15250. https://doi.org/10.1073/pnas.261418798

Sundaram, N., Bhende, T., Yashwant, R., Jadhav, S., & Jain, A. (2021). Mucormycosis in COVID-19 patients. *Indian Journal of Ophthalmology/Indian Journal of Ophthalmology, 69*(12), 3728. https://doi.org/10.4103/ijo.ijo_1316_21

Supparatpinyo, K., Khamwan, C., Baosoung, V., Nelson, K. E., & Sirisanthana, T. (1994). Disseminated Penicillium marneffei infection in southeast Asia. *Lancet (London, England), 344*(8915), 110–113. https://doi.org/10.1016/s0140-6736(94)91287-4

Tan, B. H., et al. (2020). Clinicians' challenges in managing patients with invasive fungal diseases in seven Asian countries: An Asia Fungal Working Group (AFWG) survey. *International Journal of Infectious Diseases, 95*, 471–480. https://doi.org/10.1016/j.ijid.2020.01.007

Tang, J., Lin, G., Langdon, W. Y., Tao, L., & Zhang, J. (2018). Regulation of C-type lectin receptor-mediated antifungal immunity. *Frontiers in Immunology, 9*, 123. https://doi.org/10.3389/fimmu.2018.00123

Tanner, D. C., Weinstein, M. P., Fedorciw, B., Joho, K. L., Thorpe, J. J., & Reller, L. (1994). Comparison of commercial kits for detection of cryptococcal antigen. *Journal of Clinical Microbiology, 32*(7), 1680–1684. https://doi.org/10.1128/jcm.32.7.1680-1684.1994

Thorel, M., Mateos-Hernandez, L., Mulot, B., Azzouni, M. N., Hodžić, A., Gaillot, H., Ruel, Y., Desoubeaux, G., Delaye, J. B., Obregon, D., Wu-Chuang, A., de la Fuente, J., Bermúdez-Humarán, L. G., Risco-Castillo, V., Leclerc, A., & Cabezas-Cruz, A. (2022). Assessment of the Safety and Efficacy of an Oral Probiotic-Based Vaccine Against Aspergillus Infection in Captive-Bred Humboldt Penguins (*Spheniscus humboldti*). *Frontiers in Immunology, 13*, 897223. https://doi.org/10.3389/fimmu.2022.897223

Thu, N. T. M., Chan, J. F. W., Ly, V. T., Ngo, H. T., Hien, H. T. A., Lan, N. P. H., Chau, N. V. V., Cai, J., Woo, P. C. Y., Day, J. N., Van Doorn, R., Thwaites, G., Perfect, J., Yuen, K., & Le, T. (2020). Superiority of a novel MP1P antigen detection enzyme immunoassay compared to standard BACTEC blood culture in the diagnosis of talaromycosis. *Clinical Infectious Diseases/Clinical Infectious Diseases (Online. University of Chicago. Press), 73*(2), e330–e336. https://doi.org/10.1093/cid/ciaa826

Tiew, P. Y., Ko, F. W. S., Pang, S. L., Matta, S. A., Sio, Y. Y., Poh, M. E., Lau, K. J., Mac Aogáin, M., Jaggi, T. K., Ivan, F. X., Gaultier, N. E., Uchida, A., Drautz-Moses, D. I., Xu, H., Koh, M. S., Hui, D. S. C., Tee, A., Abisheganaden, J. A., Schuster, S. C., et al. (2020). Environmental fungal sensitisation associates with poorer clinical outcomes in COPD. *European Respiratory Journal, 56*(2), 2000418. https://doi.org/10.1183/13993003.00418-2020

Tong, T., Wang, Z., Xu, Y., & Shen, J. (2021). Immunization with *Pneumocystis carinii* A12 1–85 antigen activates immune function against *P. carinii*. *BMC Immunology, 22*, 1–8. https://doi.org/10.1186/s12865-021-00590-2

Udomphan, P., Bunyaratavej, S., Leeyaphan, C., Matthapan, L., Lertrujiwanit, K., & Pattanaprichakul, P. (2019). Review of adult tinea capitis cases presenting to Siriraj hospital, Bangkok, Thailand. *The Southeast Asian Journal of Tropical Medicine and Public Health, 50*(5), 905–911.

Uhrlaß, S., Mey, S., Koch, D., Mütze, H., Krüger, C., Monod, M., & Nenoff, P. (2024). Dermatophytes and skin dermatophytoses in Southeast Asia-First epidemiological survey from Cambodia. *Mycoses, 67*(4), e13718. https://doi.org/10.1111/myc.13718

Urban, C. F., Reichard, U., Brinkmann, V., & Zychlinsky, A. (2006). Neutrophil extracellular traps capture and kill Candida albicans yeast and hyphal forms. *Cellular Microbiology, 8*(4), 668–676. https://doi.org/10.1111/j.1462-5822.2005.00659.x

Vanittanakom, N., Cooper, C. R., Jr., Fisher, M. C., & Sirisanthana, T. (2006). Penicillium marneffei infection and recent advances in the epidemiology and molecular biology aspects. *Clinical Microbiology Reviews, 19*(1), 95–110. https://doi.org/10.1128/CMR.19.1.95-110.2006

Vasconcellos, I. C. D. S., Dalla Lana, D. F., & Pasqualotto, A. C. (2019). The role of molecular tests in the diagnosis of disseminated histoplasmosis. *Journal of Fungi (Basel, Switzerland), 6*(1), 1. https://doi.org/10.3390/jof6010001

Vásquez-del-Mercado, E., Arenas, R., & Padilla-Desgarenes, C. (2012). Sporotrichosis. *Clinics in Dermatology, 30*(4), 437–443. https://doi.org/10.1016/j.clindermatol.2011.09.017

Velayuthan, R. D., Samudi, C., Lakhbeer Singh, H. K., Ng, K. P., Shankar, E. M., & Denning, D. W. (2018). Estimation of the burden of serious human fungal infections in Malaysia. *Journal of Fungi (Basel, Switzerland), 4*(1), 38. https://doi.org/10.3390/jof4010038

Vidal-Acuña, M. R., Ruiz-Pérez de Pipaón, M., Torres-Sánchez, M. J., & Aznar, J. (2018). Identification of clinical isolates of Aspergillus, including cryptic species, by matrix assisted laser desorption ionization time-of-flight mass spectrometry (MALDI-TOF MS). *Medical Mycology, 56*(7), 838–846. https://doi.org/10.1093/mmy/myx115

Wahyuningsih, R., Adawiyah, R., Sjam, R., Prihartono, J., Ayu Tri Wulandari, E., Rozaliyani, A., Ronny, R., Imran, D., Tugiran, M., Siagian, F. E., & Denning, D. W. (2021). Serious fungal disease incidence and prevalence in Indonesia. *Mycoses, 64*(10), 1203–1212. https://doi.org/10.1111/myc.13304

Wang, R., Oliveira, L. V. N., Lourenco, D., Gomez, C. L., Lee, C. K., Hester, M. M., Mou, Z., Ostroff, G. R., Specht, C. A., & Levitz, S. M. (2023). Immunological correlates of protection following vaccination with glucan particles containing Cryptococcus neoformans chitin deacetylases. *npj Vaccines, 8*(1), 6. https://doi.org/10.1038/s41541-023-00606-0

Ward, R. A., & Vyas, J. M. (2020). The first line of defense: effector pathways of anti-fungal innate immunity. *Current Opinion in Microbiology, 58*, 160–165. https://doi.org/10.1016/j.mib.2020.10.003

Wheat, L. J., Slama, T. G., Norton, J. A., Kohler, R. B., Eitzen, H. E., French, M. L., & Sathapatayavongs, B. (1982). Risk factors for disseminated or fatal histoplasmosis. Analysis of a large urban outbreak. *Annals of Internal Medicine, 96*(2), 159–163. https://doi.org/10.7326/0003-4819-96-2-159

World Health Organization. (2022). WHO fungal priority pathogens list to guide research, development and public health action. https://www.who.int/publications/i/item/9789240060241.

Wu, T. C., & Koo, S. Y. (1983). Comparison of three commercial cryptococcal latex kits for detection of cryptococcal antigen. *Journal of Clinical Microbiology, 18*(5), 1127–1130. https://doi.org/10.1128/jcm.18.5.1127-1130.1983

Wüthrich, M., Dobson, H. E., Ledesma Taira, C., Okaa, U. J., dos Santos Dias, L., Isidoro-Ayza, M., Petrovsky, N., & Klein, B. S. (2021). Combination adjuvants enhance recombinant protein vaccine protection against fungal infection. *MBio, 12*(4), e010128. https://doi.org/10.1128/mBio.01012-21

Xue, T., Kong, X., & Ma, L. (2023). Trends in the epidemiology of Pneumocystis pneumonia in immunocompromised patients without HIV infection. *Journal of Fungi (Basel, Switzerland), 9*(8), 812. https://doi.org/10.3390/jof9080812

Zerbe, C. S., & Holland, S. M. (2005). Disseminated histoplasmosis in persons with interferon-gamma receptor 1 deficiency. *Clinical Infectious Diseases : An Official Publication of the Infectious Diseases Society of America, 41*(4), e38–e41. https://doi.org/10.1086/432120

Zhao, M. D., Zhou, X., Liu, T. T., & Yang, Z. B. (2015). Morphological and physiological comparison of taxa comprising the Sporothrix schenckii complex. *Journal of Zhejiang University. Science. B, 16*(11), 940–947. https://doi.org/10.1631/jzus.B1500055

Zhou, Z., Zhu, C., Ip, M., Liu, M., Zhu, Z., Liu, R., Li, X., Zeng, L., & Wu, W. (2022). Molecular epidemiology and antifungal resistance of Cryptococcus neoformans from human immunodeficiency virus-negative and human immunodeficiency virus-positive patients in Eastern China. *Frontiers in Microbiology, 13*, 942940. https://doi.org/10.3389/fmicb.2022.942940

Chapter 6
The Changing Face of *Candida* in Southeast Asia

Llewelyn M. Espiritu, Paul Rainier D. Capule, Lance Yurik Casulla, Paul K. Yu, Mark B. Carascal, Daniel Jose N. Ombao, Kin Israel R. Notarte, Jonathan Jaime G. Guerrero, and Vishnu Chaturvedi

Abstract The phenotypic and genotypic plasticity of *Candida* species has been extensively characterized in Western literature, while countries in the Eastern Hemisphere are increasingly incorporating these findings into both research and clinical strategies. In Southeast Asia, however, the landscape is more complex. Some nations have taken a more limited role, focusing primarily on clinical identification, with research and surveillance receiving less emphasis. Conversely, other countries in the region actively contribute to global efforts to monitor *Candida,* including the surveillance of multidrug-resistant strains such as *Candidozyma auris.* The historical context, distribution patterns, and regional insights into *Candida* across Southeast Asian countries have been explored in various individual publications; however, a more comprehensive and comparative analysis remains limited. This chapter addresses the gap by presenting an up-to-date, region-wide synthesis of *Candida*-related data throughout Southeast Asia.

Keywords *Candida* spp. · *Candidozyma auris* · Fungal epidemiology · Southeast Asia · Clinical mycology

---

L. M. Espiritu (✉) · P. R. D. Capule · L. Y. Casulla
Department of Biology, College of Science, De La Salle University, Manila, Philippines

P. K. Yu
Bioinformatics and Genomics, Huck Institutes of the Life Sciences, The Pennsylvania State University, University Park, PA, USA

M. B. Carascal · D. J. N. Ombao
Clinical and Translational Research Institute, The Medical City, Pasig, Philippines

K. I. R. Notarte
Department of Pathology, Johns Hopkins University School of Medicine, Baltimore, MD, USA

J. J. G. Guerrero
College of Medicine, University of the Philippines Manila, Manila, Philippines

V. Chaturvedi
Westchester Medical Center, New York Medical College, Valhalla, NY, USA

J. J. G. Guerrero et al. (eds.), *Mycology in a Changing Planet: Applications and Perspectives from Southeast Asia*, https://doi.org/10.1007/978-981-95-2640-6_6

Introduction

Candida spp. are a group of commensal fungi that coexist with the normal microbiota of the human body. However, when the immune system becomes compromised, these fungi can transition to a pathogenic state. Initially, they may colonize mucosal surfaces, causing superficial candidiasis. If left untreated, *Candida* can breach the epithelial barrier, invade deeper tissues, and enter the bloodstream, leading to *invasive candidiasis*. This serious condition occurs when *Candida* spp. (historically dominated by *Candida albicans*, though more non-*albicans* species are emerging) infiltrate the bloodstream and disrupt the body's internal environment (Lass-Flörl et al., 2024). Without proper management, this fungal invasion can be fatal. *C. albicans* and related species have long posed a significant clinical challenge, particularly in the Western hemisphere, where extensive research has focused on their epidemiology, drug resistance, and clinical impact. Recently, research efforts have expanded to Asia, where countries, including those in Southeast Asia, face challenges depending on their situations in economics, policies, and priorities in managing *Candida* infections. These challenges comprise a high prevalence of *Candida tropicalis*, emerging antifungal resistance, and limited access to advanced diagnostics like molecular testing and MALDI-TOF-MS. Economic constraints further restrict access to antifungal drugs, leading to delayed or inadequate treatment. While these issues are global, they are particularly severe in low- and middle-income countries (LMICs), where healthcare infrastructure and surveillance systems are underdeveloped. Southeast Asia's burden is shaped by regional species prevalence, resistance patterns, and resource limitations (Chakrabarti et al., 2023; Salmanton-García et al., 2023; Tan et al., 2015, 2020). Factors such as high population density, diverse healthcare infrastructure, and limited epidemiological data complicate efforts to control *Candida* in the region. This chapter explores the risk factors, diagnostic methods, antifungal treatments, epidemiology, and virulence factors associated with *Candida* spp. in Southeast Asia, with a particular focus on the emerging threat of multidrug-resistant *Candidozyma auris* and its ongoing management in the region.

Risk Factors for Invasive Candidiasis: Intrinsic and Iatrogenic Origins

Invasive candidiasis can stem from either intrinsic factor inherent to the patient or iatrogenic causes induced by medical treatments. Both categories have been identified through extensive medical observation (Pappas et al., 2018).

General intrinsic risk factors include conditions such as *Candida* spp. colonization, diabetes mellitus, gastrointestinal perforation, advanced age, pancreatitis, sepsis, severe chronic illnesses, and genetic defects that impair host immune responses (Lass-Flörl et al., 2024). Genetic mutations in crucial genes of the innate immune system, like ***DOCK8***, ***MYD88***, ***CARD9***, ***NCF1***, ***NOX*s**, ***TLR1***, ***MPO***, ***CYBA***, and ***CYBB***, can predispose individuals to conditions such as chronic mucocutaneous candidiasis (CMC), chronic granulomatous disease (CGD), candidemia, and recurrent invasive candidiasis (Abd Elaziz et al., 2020; Carvalho et al., 2010). Similarly, mutations in genes like *STAT1* and *STAT3*, which are essential for macrophage activation, Th1 cell immune response, and cytokine signaling pathways, can lead to deficiencies in protective mucocutaneous immunity against *C. albicans* (Liu et al., 2023; van de Veerdonk et al., 2011). Other genetic deficiencies, such as those in the *AIRE* and *GATA2* genes, further increase susceptibility to *C. albicans* infections by impairing immune responses and blood cell differentiation (Donadieu et al., 2018; Vedula et al., 2020). General iatrogenic risk factors arise from medical interventions and include procedures and treatments such as dialysis and the use of broad-spectrum antibiotics. Other risk factors include central venous catheters, corticosteroids, and immunosuppressants. Major gastrointestinal surgeries, left ventricular assist devices, prolonged hospital stays, intensive care unit admissions, mechanical ventilation, and total parenteral nutrition also contribute to the risk (Cortés et al., 2018).

Additional risk factors in immunocompromised patients further elevate the risk of candidemia and invasive candidiasis, with associated high mortality rates (Garnacho-Montero et al., 2024). Intrinsic factors in this population include graft-versus-host disease, mucositis, and profound neutropenia (absolute neutrophil count <100/μL), while iatrogenic factors include solid organ and stem cell transplantation (Pappas et al., 2018).

Understanding these multifaceted risk factors highlights the complex interplay between intrinsic and iatrogenic elements in predisposing individuals to invasive candidiasis. Recognizing these vulnerabilities is crucial for effective management and mitigation of the impact of candidiasis.

Diagnostics and Prevention

Direct Visualization

Many laboratories in limited-resource healthcare facilities still rely on visualizing yeast cells from fresh clinical specimens as a crude method for detecting *Candida* infections. Routine Gram staining, Swartz-Lamkins fungal stain, or saline/potassium hydroxide wet mounts of blood, urine, mucous, and systemic patient samples could initially hint at the presence of clinically significant *Candida* yeast cells and pseudohyphal structures (Knoll et al., 2023). However, a definitive diagnosis relies on correlating microscopy results with clinical evaluation and other tests, as direct visualization alone could miss up to 50% of clinical cases (Sobel & Mitchell, 2024).

Culture-Based Diagnostics

C. albicans and non-*albicans* species are traditionally identified using phenotypic characteristics, such as biotyping, morphotyping, serotyping, and chemical resistance testing. These methods rely on isolating and growing microorganisms in culture media. However, their reproducibility and ability to differentiate between species are often low, limiting their diagnostic accuracy (Arafa et al., 2023). Despite this, blood cultures remain the gold standard for diagnosing invasive candidiasis, though these are often supplemented with molecular and antigen-based assays to improve sensitivity and accuracy (Camp et al., 2020).

Several culture-based diagnostic methods, including the API 20C AUX system, CHROMagar, and Vitek YBC, are widely used in Southeast Asia. The API 20C AUX system differentiates *Candida* species through carbohydrate assimilation tests, although it struggles to identify *Pichia kudriavzevii* (formerly *Candida krusei*) (Arafa et al., 2023). CHROMagar uses chromogenic substrates to distinguish species based on colony color, and it has been used for diagnosing vulvovaginal candidiasis and invasive *Candida* infections in Southeast Asia (Tan et al., 2016). The Vitek YBC system automates biochemical testing to identify species like *C. albicans*, *Nakaseomyces glabratus* (Takashima & Sugita, 2022) (formerly *Candida glabrata*), and *C. tropicalis* (Arafa et al., 2023).

Non-culture-Based Diagnostics

Initial immune assays for diagnosing *Candida* infections detected mannan and antimannan antibodies but had low sensitivity (55%) and specificity (65%) (Barantsevich & Barantsevich, 2022). The detection of β-D-glucan, a fungal cell wall component, has significantly improved diagnostics, with 92% sensitivity and 81% specificity for invasive candidiasis. The *C. albicans* germ tube antibody test (CAGTA) helps differentiate between infection and colonization, with varying sensitivity and specificity depending on the species (Pfaller, 2015). Combining a positive β-D-glucan test with procalcitonin improves the detection of invasive candidiasis, with 66% sensitivity and 98% specificity (Giacobbe et al., 2017).

PCR-based methods offer rapid identification of *Candida* species within 2–4 hours, with 95% sensitivity and 92% specificity. The FDA-approved T2Candida panel, which uses PCR combined with T2 magnetic resonance, has 91% sensitivity and 99% specificity for detecting various species in blood (Barantsevich & Barantsevich, 2022; Zacharioudakis et al., 2018).

In Southeast Asia, non-culture-based diagnostics are limited, with tests like β-D-glucan and galactomannan available in only a few laboratories (Chindamporn et al., 2018). The VITEK® 2 Yeast Identification (YST) system is used for rapid biochemical identification of *Candida* species, while matrix-assisted laser desorption/ionization time-of-flight mass spectrometry (MALDI-TOF MS) provides quick and cost-effective species identification through protein fingerprinting. If there are inconclusive or low-confidence results from MALDI-TOF MS, ribosomal DNA (rDNA) sequencing can be employed as an alternative. Molecular methods such as multi-locus sequence typing (MLST), pulsed-field gel electrophoresis (PFGE), and random amplified polymorphic DNA (RAPD) are used to differentiate and characterize species across the region.

Surveys conducted across Southeast Asia, including Indonesia, the Philippines, Singapore, and Thailand, indicate that although advanced fungal diagnostic techniques such as MALDI-TOF MS, DNA sequencing, antifungal susceptibility testing, and PCR are utilized in both hospital and research environments, their availability is not consistent. These technologies, which are frequently used in academic laboratories for molecular characterization, have also been incorporated into well-funded hospitals for the identification and differentiation of *Candida* species. However, numerous hospitals, particularly in lower-income countries, encounter challenges in accessing these technologies due to financial constraints, limited funding, and budget priorities that emphasize emergency care and the management of high-burden infectious diseases. For instance, MALDI-TOF MS is available in only 12.3% of surveyed laboratories, predominantly in wealthier nations, while DNA sequencing, essential for identifying drug-resistant *Candida* strains, is accessible in just 16.9% of laboratories, with increased availability in hospitals located in higher-GDP countries. Antifungal susceptibility testing is conducted in 58.9% of facilities, with complete availability in Singapore (100%) but significantly lower in Indonesia (45.5%), the Philippines (46.2%), and Thailand (28.6%). Moreover, PCR, which is critical for fungal diagnostics, is utilized in 66% of laboratories. While the deployment of these technologies enhances the management of fungal infections, their implementation in Southeast Asian hospitals faces significant economic challenges (Chindamporn et al., 2018; Salmanton-García et al., 2023).

Antifungal Susceptibility Testing

Culture-confirmed clinical isolates are recommended to undergo antifungal susceptibility testing to guide therapy. Testing could be done using commercial kits such as ETest® (bioMérieux, France) and Sensititre® YeastOne (Thermo Fisher Scientific, USA), which are based on agar diffusion and microbroth dilution principles, respectively. Meanwhile, automated testing could also be done using the Vitek ® 2 system. Minimum inhibitory concentration (MIC) and zone diameter breakpoints are provided by both the Clinical and Laboratory Standards Institute (CLSI) and the European Committee on Antimicrobial Susceptibility Testing (EUCAST) for *C. albicans, Candida parapsilosis, C. tropicalis, N. glabratus*, and *P. kudriavzevii* against amphotericin B, anidulafungin, caspofungin, fluconazole, micafungin, rezafungin, and voriconazole (CLSI, 2022; EUCAST, 2023). Limited breakpoint data are available for other species such as *C. auris, C. dubliniensis, C. lusitaniae*, and *Meyerozyma guilliermondii* (formerly *Candida guilliermondii*).

Finally, species-specific breakpoints were suggested for rare but clinically significant *Candida* species, including *C. bovina, C. fermentati, C. inconspicua, C. intermedia, C. kefyr, C. lipolytica, C. magnoliae, C. metapsilosis, C. nivariensis, C. norvegensis, C. orthopsilosis, C. palmioleophila, C. pararugosa, C. pelliculosa,* and *C. utilis* (EUCAST, 2024).

Diagnosis of Oral Candidiasis (OC)

Oral candidiasis is diagnosed through clinical evaluation, medical history, and risk factors. A biopsy may be required in certain cases to confirm the diagnosis, along with empirical antifungal treatment. Cultures are performed if initial antifungal therapy fails. Mild cases are usually resolved with topical antifungal treatment and proper oral hygiene, while systemic antifungals are reserved for patients unable to use topical treatments or at high risk of systemic infections (Akpan & Morgan, 2002).

Diagnosis of Vulvovaginal Candidiasis (VVC)

Candida vaginitis is diagnosed based on symptoms like dysuria, vulvar itching, swelling, redness, and thick, curd-like discharge (CDC, 2021). Diagnosis can be confirmed with a 10% KOH wet prep revealing yeasts, hyphae, or pseudohyphae, or through a positive yeast culture, which remains the gold standard. For complicated cases or non-*albicans Candida*, vaginal cultures or PCR are recommended. *N. glabratus*, which lacks pseudohyphae, is difficult to detect microscopically. Due to rising azole resistance, culture and susceptibility testing are advised for persistent cases (CDC, 2021).

Diagnosis of Invasive Candidiasis

Guidelines for diagnosing invasive candidiasis include four key methods: (1) histopathological or microscopic identification of *Candida* hyphae from sterile sites, (2) positive cultures from normally sterile sites, (3) PCR and DNA sequencing of yeasts in tissue samples, and (4) positive blood cultures for *Candida* species (Donnelly et al., 2020).

Prevention of Invasive Candidiasis in ICU Settings

The 2016 Infectious Diseases Society of America (IDSA) guidelines recommend fluconazole or nystatin prophylaxis for neonates in high-risk nurseries, with alternative options like bovine lactoferrin. For adults in ICUs with high risk of invasive candidiasis, fluconazole or echinocandins are suggested. Daily chlorhexidine baths are also effective in reducing bloodstream infections, including candidemia (Pappas et al., 2016).

In Southeast Asia, hospitals emphasize monitoring *Candida* species distribution and resistance patterns to guide treatment. Echinocandins are recommended as first-line empirical treatment in ICUs, especially where fluconazole resistance is high. Continuous surveillance of species distribution aids in refining public health and ICU protocols (Anh et al., 2021; Boonsilp et al., 2021; Ngamchokwathana et al., 2021; Thunyaharn et al., 2021).

Antifungal Agents for Treating Invasive Candidiasis

Antifungal antibiotic drugs have been in use since the discovery and introduction of Nystatin in the 1950s (Homei & Worboys, 2013; Vanreppelen et al., 2023). Since then, both natural (polyenes, echinocandins) and synthetic (pyrimidine analogs, azoles, allylamines) antifungals have been developed. However, continuous use has led to reduced efficacy due to antifungal resistance, driving the need for new drug discoveries and updated treatment guidelines.

Polyenes were the first antifungal class, introduced with Nystatin in 1954. Polyenes disrupt fungal membranes by binding to ergosterol, causing membrane instability and eventual fungal cell death (Anderson et al., 2014). Despite their nephrotoxicity, low resistance rate, and poor water solubility (Lemke et al., 2005; Wingard et al., 1999), polyenes remain a last-resort treatment for severe fungal infections (Vanreppelen et al., 2023). A commonly used polyene antifungal agent in the clinical setting is amphotericin B.

Pyrimidine analogs, like flucytosine (5-FC), inhibit fungal DNA and protein synthesis (Waldorf & Polak, 1983). Originally used as monotherapy, widespread resistance led to its use in combination with other drugs like azoles (Louie et al., 1999).

The azoles, introduced in the 1960s, inhibit ergosterol biosynthesis by targeting the cytochrome P450 enzyme (Berman & Krysan, 2020). Triazoles (i.e., fluconazole, voriconazole, itraconazole, posaconazole, isavuconazole) replaced imidazoles due to their better safety profile and efficacy (Sheehan et al., 1999). However, resistance has emerged in both medical and agricultural sectors, and azoles are prone to drug interactions (Jørgensen & Heick, 2021).

Allylamines, like terbinafine (introduced in 1992), also inhibit ergosterol synthesis but target an earlier step in the pathway, leading to membrane destabilization (Birnbaum, 1990; Jessup et al., 2000).

The echinocandins (i.e., anidulafungin, caspofungin, micafungin), introduced in 2001, have become the first-line treatment for invasive candidiasis. They work by inhibiting β-(1,3)-glucan synthase, essential for fungal cell wall integrity, causing cell death (Fera et al., 2009). Echinocandins are preferred for their efficacy and low toxicity when administered intravenously (Balkovec et al., 2014; Capoor & Bal, 2022). Rezafungin, a new drug under this class, was shown to have a prolonged half-life and can be used for the treatment of invasive candidiasis and candidemia, or as prophylaxis (Kriegl et al., 2025).

Clinical Management

In general, clinical management of invasive candidiasis involves antifungal treatment with approaches (could be overlapping) that depend on the patient status or available diagnostic information: (1) directed/targeted strategy—for microbiologically confirmed cases, (2) empiric therapy—for patients with signs and risk factors for invasive candidiasis, (3) diagnosis-driven—for patients with microbiologically and biomarker-confirmed invasive candidiasis, and (4) prophylactic—for critically ill and immunosuppressed patients (Chua & Rosario, 2024). Localized candidiasis (i.e., mucocutaneous, tissue-specific) are treated based on anatomic sites of infection. For instance, flucytosine is a preferred additive for *Candida* meningitis because of its excellent penetration in the cerebrospinal fluid (Kauffman & Vazquez, 2024). Meanwhile, for the majority of candidemia, an initial fluconazole therapy is recommended. In non-neutropenic cases, echinocandins (including the newer antifungal, rezafungin) are preferred (Vazquez, 2023). For targeted therapy in adults, species-specific approach is recommended: amphotericin B for fluconazole- and echinocandin-resistant *C. albicans*; voriconazole for *C. parapsilosis*; and echinocandins for *C. auris, M. guilliermondii, N. glabratus*, and *P. kudriavzevii* (Vazquez, 2023). For pediatric infections, amphotericin B is preferred for most *Candida* species due to its broad antifungal activity (Campbell & Palazzi, 2023).

Virulence Factors of *Candida* spp.

Candida species possess a variety of genetic factors that enable them to infect susceptible hosts, contributing to their pathogenicity. These genetic factors contribute to virulence, especially in opportunistic strains that can lead to invasive candidiasis. Among the various species, *C. albicans* has been the most studied due to its clinical relevance, and much of the understanding of virulence mechanisms is derived from this species.

Key virulence traits in *C. albicans* include the ability to undergo morphological transitions, form biofilms, and secrete hydrolytic enzymes such as proteases, phospholipases, and lipases. These enzymes facilitate tissue invasion and immune evasion by degrading host cell membranes and proteins (Naglik et al., 2003). Moreover, *C. albicans* can adapt to different environmental conditions within the host, such as pH changes, which further enhance its survival and pathogenicity (Gow et al., 2012).

Phenotypic Switching

One of the critical virulence traits of *Candida* species, especially *C. albicans*, is the ability to switch between different forms: a white, yeast-like form and an opaque, multicellular filamentous form. This switch is crucial for mating, pathogenesis, and immune evasion. Phenotypic switching is essential to *Candida* species, as it is needed for mating competence, as only opaque cells can engage in mating. Opaque cells release pheromones that prompt white cells to become adhesive and form biofilms, which protect the fragile pheromone gradients necessary for chemotropism and fusion during mating. Additionally, phenotypic switching plays a critical role in pathogenesis and adaptation. It provides *Candida* species with phenotypic plasticity, enabling it to thrive in various host niches and adapt to environmental challenges. The opaque cells facilitate biofilm formation by signaling white cells, thereby enhancing mating by preserving pheromone gradients. This biofilm formation is also considered a virulence factor, contributing to successful colonization and persistence within the host. Furthermore, the white and opaque phenotypes display distinct virulence characteristics: white cells tend to be more virulent in systemic infections, whereas opaque cells demonstrate increased virulence in cutaneous infections (Soll, 2009). While opaque cells are less effectively phagocytosed by macrophages, they remain more susceptible to neutrophil-mediated killing once recognized. In contrast, dendritic cells can effectively phagocytose both white and opaque cells (Pande et al., 2013).

Beyond the well-known white-opaque switch, *C. albicans* exhibits an even greater degree of morphological plasticity with the recent identification of additional cell types that further enhance its adaptability. For example, GUT cells (gastrointestinally induced transition cells) are a specialized commensal cell type that arises when *C. albicans* passes through the gastrointestinal system. These cells exhibit a unique gut-optimized transcriptional program, repressing genes related to glucose catabolism and iron uptake while activating those involved in fatty acid and N-acetylglucosamine catabolism. This adaptation allows *C. albicans* to thrive in the nutrient-rich gut environment, highlighting its ability to fine-tune its metabolism and behavior to specific host niches (Pande et al., 2013). Importantly, the discovery of GUT cells underscores that *C. albicans*-mediated infections are not solely dependent on impaired host immunity but also on the pathogen's ability to control its cell identity during commensal-pathogen transitions.

In addition to GUT cells, *C. albicans* can transition into a gray phenotype, an intermediate state between white and opaque. Gray cells exhibit unique virulence characteristics, with a high ability to cause cutaneous infections but a reduced capacity to colonize internal organs like the kidney, lung, and brain. This tri-stable white-gray-opaque switching system, regulated by transcription factors Wor1 and Efg1, further enhances *C. albicans*' ability to adapt to diverse host environments and evade immune responses (Tao et al., 2014). A similar gray state has also been observed in *C. tropicalis* (Xie et al., 2012).

Another intriguing morphological adaptation is the formation of Trimeras, a three-cell structure composed of a mother, daughter, and granddaughter bud. Trimeras arise in response to antifungal stress, such as exposure to fluconazole, and are associated with aneuploidy (Harrison et al., 2014). This altered ploidy may serve as a resistance mechanism, allowing *C. albicans* to survive and proliferate under drug pressure (Selmecki et al., 2006).

The ability to switch between these forms is regulated primarily by the *WOR1* gene; the main trigger for phenotypic switching is the master regulator of the white-opaque transition, along with other regulators like *EFG1*, *CZF1*, *AHR1*, and *WOR2*. The expression of *WOR1* is essential for the shift from white to opaque phenotype. Other triggers include environmental factors, evidenced by the opaque phenotype exhibiting instability at physiological temperatures (37 °C). Increasing the temperature of opaque cell cultures above 35 °C triggers a mass conversion to the white phenotype. Elevated CO_2 levels also factor in activation, which can be found in specific host niches like the gastrointestinal tract, promoting the transition from white to opaque and stabilizing the opaque phenotype at 37 °C. Even oxygen levels have been observed to stabilize the opaque phenotype, specifically hypoxia (low oxygen), indicating that oxygen levels within the host environment can affect the switching process. Pheromone signaling is another important trigger. Pheromones released by opaque cells can prompt white cells to form biofilms, which facilitate mating. This signaling is a crucial aspect of the switching process, directly linking it to mating competence. Lastly, stochastic events can also lead to spontaneous switching, where white cells can randomly produce opaque cells and vice versa at a low frequency. This stochasticity is influenced by the regulatory network involving the *WOR1* gene and other transcription factors (Soll, 2009).

The *EFG1* gene is crucial for regulating morphogenesis, virulence, and biofilm formation in *C. albicans*. It both promotes and inhibits hyphal formation depending on environmental cues (Doedt et al., 2004; Mulhern et al., 2006). As a core transcription factor in biofilm development, *EFG1* regulates key adhesion genes like *ALS1* and *HWP1* (Braun & Johnson, 2000). Mutations in *EFG1* can impair biofilm formation and hyphal growth, reducing the organism's virulence (Lo et al., 1997; Pande et al., 2013).

Adhesion Factors

The ability of *Candida* species to adhere to epithelial and endothelial cells is a fundamental step in establishing infection. Several genes contribute to this ability, including the Agglutinin-Like Sequence (*ALS*) gene family, the Hyphal Wall Protein

1 (*HWP1*) family, the Integrin-Like Protein 1 (*INT1*) gene, and Mannosyl transferase Type I (*MNT1*). Adhesins also facilitate the attachment of *Candida* species to host cells or host-cell ligands.

ALS genes encode glycoproteins that are essential for adhesion. There are eight ALS genes identified in *C. albicans*: *ALS1, ALS2, ALS3, ALS4, ALS5, ALS6, ALS7,* and *ALS9* (Hoyer & Cota, 2016). *ALS1* plays a critical role in adhesion to both epithelial and endothelial cells, facilitating colonization and infection (Nobile et al., 2008). *ALS3* shares functional similarities with *ALS1*; however, it is disposable for the pathogen's virulence (Cleary et al., 2011). Additionally, *ALS5* contributes to cell surface hydrophobicity (CSH), which enhances adhesion to host tissues and extracellular matrix components, particularly in systemic infections. Studies show that *C. tropicalis* has the highest CSH levels among different species (Saiprom et al., 2023).

The *HWP1* gene encodes a hypha-specific cell wall protein essential for biofilm formation and attachment to host cells. *INT1* encodes a surface protein in *C. albicans* important for adhesion and promoting the yeast-to-hypha transition, which is critical for virulence (Gale et al., 1998). Meanwhile, *MNT1* is responsible for O-glycosylation of cell walls and secreted proteins; therefore, it is important in the adhesion and dispersion mechanisms of *Candida* species (Munro et al., 2005; Zarnowski et al., 2021).

Hyphal Growth and Morphogenesis

Hyphal growth is vital for tissue invasion and virulence in *Candida* species. *C. albicans* and *C. tropicalis* form true hyphae, while *C. parapsilosis* forms pseudohyphae, and *N. glabratus* does not demonstrate filamentous growth (Saiprom et al., 2023). Several genes, such as *CRK1*, *ECE1*, and *EFG1*, are involved in regulating hyphal growth and morphogenesis.

CRK1 is a critical gene for the development of hyphae as its absence has been shown to impair hyphal development and produce avirulent strains. There is also a strong association between its presence and the health of the hyphae (Chen et al., 2000).

The *ECE1* gene encodes a protein (ECE1p) critical for damaging epithelial tissues, facilitating colonization. It also produces candidalysin, a cytolytic peptide toxin that damages epithelial cells and activates immune responses (Mogavero et al., 2022). Consequently, this toxin is released by the hyphae in low concentrations and triggers immune responses, whereas higher concentrations are required to induce significant epithelial damage (Moyes et al., 2016).

EFG1 is a key regulator of hyphal morphogenesis as it directly influences the expression of critical hyphal regulators and genes associated with hyphal formation. In addition, it functions in inhibiting cell separation. Numerous environmental signals that trigger hyphal initiation operate through *EFG1*, thereby modulating the expression of hyphal-specific genes (Lassak et al., 2011; Lu et al., 2008; Stoldt et al., 1997).

PRA1 is another virulence factor that helps *Candida* evade the immune system by binding to complement regulators like Factor H, protecting the fungus from immune attack (Luo et al., 2009).

Biofilm Formation

Biofilm formation on surfaces is another important virulence factor for *Candida* species. This process begins with the adhesion of fungal cells to a surface, followed by growth and the development of hyphal structures. These biofilms are encased in a protective extracellular matrix, which not only shields the fungal cells from the host's immune responses but also from antifungal agents, allowing the infection to persist (Cavalheiro & Teixeira, 2018). Notably, *C. auris* can form biofilms on the skin (Horton et al., 2020). An essential final step in biofilm formation is dissemination, where cells break away from the biofilm and spread to other parts of the host. This dissemination is a key contributor to the virulence of *Candida*, enabling the pathogen to colonize new areas and establish secondary infections in various tissues (Uppuluri et al., 2010).

The formation of biofilms in *Candida* species varies by species. In *C. albicans* and *C. tropicalis*, the biofilm structure is unique, consisting of various morphological forms. *C. albicans* can transition smoothly between yeast and hyphal forms (Priya & Pandian, 2020). Meanwhile, clinical isolates of *N. glabratus* produce compact biofilms made of multiple layers, containing proteins, carbohydrates, and ergosterol in their matrix.

Among the *Candida* species, *C. tropicalis* has been shown to be a prolific biofilm producer, more so than *C. albicans* (Saiprom et al., 2023). Recent data from Thailand indicate that *C. tropicalis* is the most prolific biofilm producer among the *Candida* species examined, followed by *C. parapsilosis*, *N. glabratus*, and *C. albicans*. These findings align with previously reported data suggesting that non-*albicans* species tend to produce more biofilm compared to *C. albicans* (Saiprom et al., 2023). The robust biofilm-forming capacity of non-*albican* species, particularly *C. tropicalis*, poses significant challenges in clinical settings. Biofilms formed on medical devices such as catheters, heart valves, and prosthetics are difficult to treat due to their increased resistance to antifungal drugs. The extracellular matrix acts as a physical barrier, preventing drug penetration and allowing fungal cells within the biofilm to survive despite treatment (Nobile & Johnson, 2015). As a result, infections caused by biofilm-forming *Candida* species often become chronic and recurrent, leading to complications, especially in immunocompromised patients or those with indwelling medical devices (Nett & Andes, 2020).

Biofilm formation is regulated by several genes, including *BCR1* (biofilm and cell wall regulator), *TEC1* (transposon enhancement control), *EFG1* (enhanced filamentous growth), *NDT80* (meiosis-specific transcription factor), *ROB1* (biofilm regulator), and *BRG1* (biofilm regulator) (Fanning & Mitchell, 2012; Nobile et al., 2012). Quorum sensing, the process of cell-to-cell communication, also plays a crucial role in biofilm development. Farnesol, an established quorum-sensing molecule, inhibits the formation of hyphae and limits biofilm growth, acting as a density-dependent repressor during biofilm maturation (Alem et al., 2006). Tyrosol, another quorum-sensing molecule identified in *C. albicans*, has an opposing effect; it reduces the lag phase of growth in low-density cultures and promotes early-stage biofilm formation by inducing germ tube development and hyphal growth (Chen et al., 2004). Although tyrosol and farnesol exhibit antagonistic properties (i.e., tyrosol drives initial biofilm formation and farnesol suppresses it later), farnesol tends to dominate during the later stages of biofilm development (Alem et al., 2006). This balance between quorum-sensing molecules is crucial for the dynamic regulation of biofilms, allowing *Candida* species to adapt to different environmental conditions and control the transition between yeast and hyphal forms.

Secreted Enzymes

Candida species secrete various enzymes to degrade host tissues and evade immune responses. These include carbohydrate-active enzymes (CAZymes), secreted aspartic proteinases (Saps), phospholipases (PLs), and lipases (Lips).

CAZymes help break down complex carbohydrates, contributing to fungal cell wall remodeling. CHT2p, CRH11p, PGA4p, and PHR1p are among the enzymes involved in this process (de Groot et al., 2004). CHT2p is a chitinase, involved in cell wall remodeling and expansion supporting its role in virulence by enhancing cell wall robustness under stress conditions (Iranzo et al., 2002). CRH11p functions in cell cycle regulation and surface localization, which implies its contribution to morphogenesis (Rodríguez-Peña et al., 2000). PGA4p is a GPI-anchored protein involved in shaping the fungal cell wall and contributes to forming a strong fungal cell wall that will resist host defenses. Thus, it plays a crucial role in the fungus survival in hostile environments. Additionally, PHR1p plays a crucial role in the filamentous growth of the fungus especially in response to alkaline pH levels (Saporito-Irwin et al., 1995). It aids in pH adaptation, alongside its counterpart PHR2p, which facilitates the pathogen's survival and growth in acidic environments (Mühlschlegel & Fonzi, 1997). Finally, SCW1p, also known as MP65, is implicated in glucan remodeling and plays a significant role in biofilm formation and contributes to the ability to evade immune detection by modifying cell wall glucans (Gomez et al., 1996).

The ability of *C. albicans* to lyse proteins was first documented in 1965 (Staib, 1966). Saps are a family of proteases that degrade host proteins and immune molecules, helping *Candida* evade the immune system. The *SAP* gene family in *C. albicans* comprises at least 10 members, with Saps 1–8 being secreted and Saps 9–10 being membrane-anchored. Saps 1–6 are particularly important in mucosal infections and exhibit differential expression across various *Candida* species and infection sites (Abad-Zapatero et al., 1996; Hube, 2000; Naglik et al., 2003). This is notable for *C. tropicalis* as the expression of proteinases is found in 52% of its isolates (Deorukhkar et al., 2014).

Phospholipases and lipases degrade host cell membranes, facilitating tissue invasion. The various types of phospholipases include Phospholipase A1 (PLA1), Phospholipase A2 (PLA2), Phospholipase B (PLB), Phospholipase C (PLC), and Phospholipase D (PLD), with each type specifically targeting different phospholipid bonds (Sarode et al., 2023). *PLA* and *PLB* are crucial phospholipases, with *PLB* playing a key role in human cell lysis (Ibrahim et al., 1995). Lipases, such as *LIP1-10*, help *Candida* utilize host lipids for growth and invasion (Fu et al., 1997).

Lipase (LIP) genes in *C. albicans* form a large gene family with up to 10 members (Fu et al., 1997; Sheridan & Ratledge, 1996). These genes encode secreted enzymes that catalyze the hydrolysis of ester bonds in insoluble triacylglycerols. *LIP1* was the first identified gene belonging to this family. Subsequently, nine additional lipase genes (*LIP2–LIP10*) were identified. Most LIP genes have homologs that are present in other *Candida* species, except for *N. glabratus* and *Saccharomyces cerevisiae* (Hube et al., 2000; Wurner, 1966). The extensive and adaptable expression of these genes in *C. albicans* may indicate their diverse lipolytic activity, which could play a role in its virulence and ability to persist in human tissues.

Data from Thailand highlight that *C. albicans* exhibits strong phospholipase activity, while non-*albicans* species show varying levels of enzyme production, with *C. tropicalis* showing the highest levels of protease production and biofilm formation, indicating high virulence (Saiprom et al., 2023).

Recently, the discovery of candidalysin, the first peptide toxin identified in any human fungal pathogen, has provided new insights into the pathogenicity of *C. albicans*. It is regarded as the first true classical virulence factor of the species, playing a critical role in its infections by directly damaging host cell membranes. It achieves this by forming pores in the cell's plasma membrane, leading to cell lysis and tissue damage. Beyond this, it also triggers protective immune responses in the host such as a MAPK-mediated danger response activation, neutrophil recruitment, and the induction of Type 17 immunity. These immune responses are essential to infection control but can contribute to inflammation and tissue damage if no regulation occurs (Sellam & Whiteway, 2016; Talapko et al., 2021).

Regulatory Circuits

C. albicans utilizes regulatory circuits that manage essential processes such as biofilm formation, stress response, and metabolic adaptation. These enable the organism to adjust to various host environments and effectively address challenges like nutrient limitations, immune responses, and antifungal therapies (Sellam & Whiteway, 2016). For example, the heat shock response, facilitated by factors such as Hsf1 and Hsp90, is vital for stress adaptation, chromatin remodeling, and metabolic regulation, enabling *C. albicans* to withstand host defenses and environmental stressors (Leach et al., 2016).

Metabolic flexibility is another defining feature of *C. albicans* virulence, allowing it to utilize various carbon sources and niches within the host. This adaptability is driven by transcriptional rewiring, wherein orthologous transcription factors govern distinct metabolic pathways in *C. albicans* compared to other fungi, such as *S. cerevisiae*. For instance, Gal4 regulates glucose metabolism in *C. albicans* while controlling the Leloir pathway in *S. cerevisiae*; similarly, Ppr1 is involved in purine catabolism in *C. albicans* but shifts to pyrimidine biosynthesis in *S. cerevisiae* (Askew et al., 2009; Tebung et al., 2016). These rewiring events illustrate evolutionary adaptations to specific environmental challenges, such as low-oxygen conditions, and demonstrate the complexity of *C. albicans* virulence mechanisms.

Immune Evasion

C. albicans employs a range of strategies to evade the host immune system, enabling it to persist and cause infections. A primary mechanism involves the masking of cell wall components, such as β-glucan, to evade detection by host pattern recognition receptors (PRRs) like Dectin-1 (Gow & Hube, 2012). This is dynamically regulated in response to environmental factors, including changes in pH, iron availability, and carbon sources. For instance, elevated iron levels decrease the visibility of mannans and chitin while increasing β-glucan, a process mediated by the lactate-induced transcription factor Crz1 (Tripathi et al., 2020). These adjustments not only facilitate immune evasion but also diminish susceptibility to antifungal medications that target the cell wall. Furthermore, *C. albicans* can modify its cell wall architecture in response to sub-therapeutic doses of antifungals, such as caspofungin, which can inadvertently increase β-glucan exposure and provoke pro-inflammatory responses (Wheeler et al., 2008; Wheeler & Fink, 2006).

Another significant strategy for immune evasion is the adaptation to the phagosomal environment within immune cells. Upon phagocytosis, *C. albicans* activates metabolic pathways like gluconeogenesis and fatty acid degradation to survive the nutrient-poor, acidic conditions (Lorenz et al., 2004). The fungus also produces ammonia to neutralize phagosomal pH, assisting the transition from yeast to hyphal forms that exhibit increased resistance to immune clearance. This morphological transformation is essential for escaping phagocytes and establishing invasive infections (Vylkova & Lorenz, 2014; Westman et al., 2018).

C. albicans further enhances its immune evasion capabilities through the inhibition of the complement system. The zinc-binding protein Pra1 not only scavenges zinc from host tissues but also disrupts the complement system by blocking C3 activation, thereby hindering fungal clearance. This dual function of Pra1 in nutrient acquisition and immune evasion emphasizes its critical role in virulence (Luo et al., 2018).

Finally, *C. albicans* exhibits tissue-specific evasion strategies during systemic infections. For example, in the kidney, the fungus produces hyphae that are less vulnerable to neutrophil attack, while in the liver and spleen, it primarily exists in the yeast form, which is more readily managed by immune cells (Lionakis et al., 2010). These tissue-specific adaptations underscore the fungus ability to tailor its evasion mechanisms to various host environments.

Epidemiology of *Candida* in Southeast Asia

A 2022 survey by the European Confederation of Medical Mycology (ECMM) and the International Society for Human and Animal Mycology (ISHAM) assessed laboratory mycology across the Asia-Pacific region. All Southeast Asian (SEA) countries, except Timor-Leste, participated. As shown in Fig. 6.1, *C. albicans* was the most predominant species reported in most countries. Although non-*albicans* species were said to be more common in Asia (Falagas et al., 2010) and have had an increasing number over the years (Taei et al., 2019), *C. albicans* remains to be the predominant species throughout the region (Wang et al., 2016). This may be attributed to *C. albicans*' colonization as part of the normal flora of the human oropharyngeal cavity, and gastrointestinal and vaginal tracts (Talapko et al., 2021).

Additionally, the antifungal susceptibility in invasive *Candida* bloodstream isolates within the Asia-Pacific region, focusing on Brunei, the Philippines, Singapore,

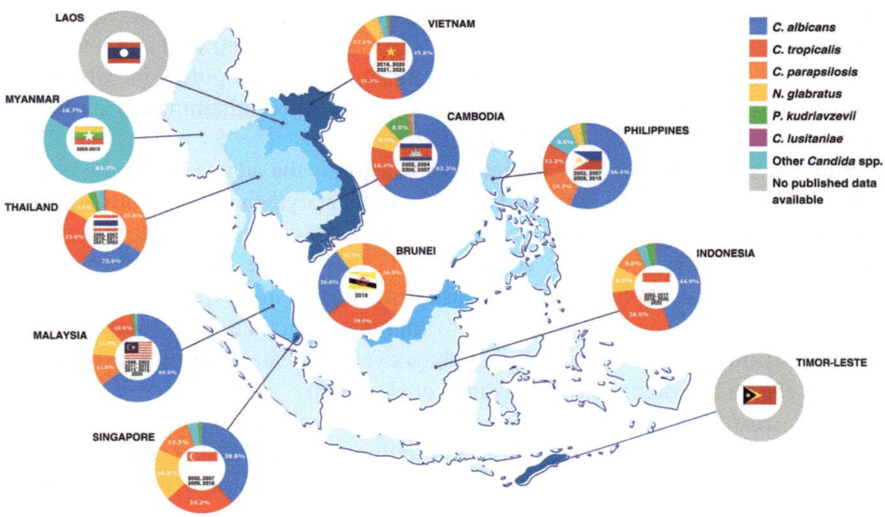

Fig. 6.1 Epidemiology of *Candida* spp. across Southeast Asia, with Brunei, Indonesia, Malaysia, Myanmar, the Philippines, Singapore, Thailand, and Vietnam contributing to various studies analyzing the distribution of these species

Thailand, and Vietnam, was studied (Tan et al., 2016). Figure 6.2 illustrates the antifungal susceptibility patterns of *Candida* species across these nations.

Antimicrobial susceptibility testing (AST) is crucial for evaluating the resistance and sensitivity of pathogens to different antimicrobial agents. The Clinical and Laboratory Standards Institute (CLSI) guidelines are widely adopted in the United States and throughout Southeast Asia. Notably, laboratories in Thailand, Laos, and Cambodia, including those affiliated with the Mahidol–Oxford Tropical Health Network (MORU), utilize CLSI disk-diffusion guidelines for both diagnostic and research purposes. While the European Committee for Antimicrobial Susceptibility Testing (EUCAST) guidelines are standard in Europe, CLSI remains the reference for AST in Southeast Asia (Cusack et al., 2019). Many laboratories in the region primarily rely on direct microscopy and culture for diagnosing fungal infections, though not all perform antifungal susceptibility testing using the standard microbroth dilution method (Chindamporn et al., 2018).

Since 2016, non-*albicans* species, particularly *C. tropicalis* and *C. parapsilosis*, have been commonly isolated from bloodstream infections in the Asia-Pacific Region. *C. tropicalis* has shown reduced susceptibility to fluconazole, raising concerns about the reliability of fluconazole susceptibility predictions for non-*albicans* species. Nevertheless, echinocandins remain the preferred antifungal choice for high-risk patients with documented candidemia. Regular AST for non-*albicans* species is recommended prior to initiating azole therapy and determining the appropriateness of fluconazole as empirical treatment (Tan et al., 2016).

Country-Specific Findings on Candida *Research*

Brunei

The first research on *Candida* in Brunei was conducted in 1998 by Sivapalan et al., who cultured samples from durian fruits. Further identification was not pursued, as these organisms were considered weak pathogens. Reports on candidiasis incidence in Brunei are limited. However, a study by Tan et al. (2016) involving 50 invasive bloodstream isolates indicated that non-*albicans* species, particularly *C. parapsilosis* (36.0%) and *C. tropicalis* (28.0%), were predominant, with *C. albicans* at 26.0%. *N. glabratus* exhibited the highest fluconazole resistance among the countries in Southeast Asia that participated in the 2016 study of Tan et al.

Indonesia

Indonesia's first documented *Candida* study dates back to 1964, when Nemec et al. (1964) used *Candida pseudotropicalis* to assess soil fungi's antibiotic activity. Research in the 1970s primarily focused on yeast isolation via streak and pour plate techniques. Candidemia currently affects approximately ten individuals per 100,000,

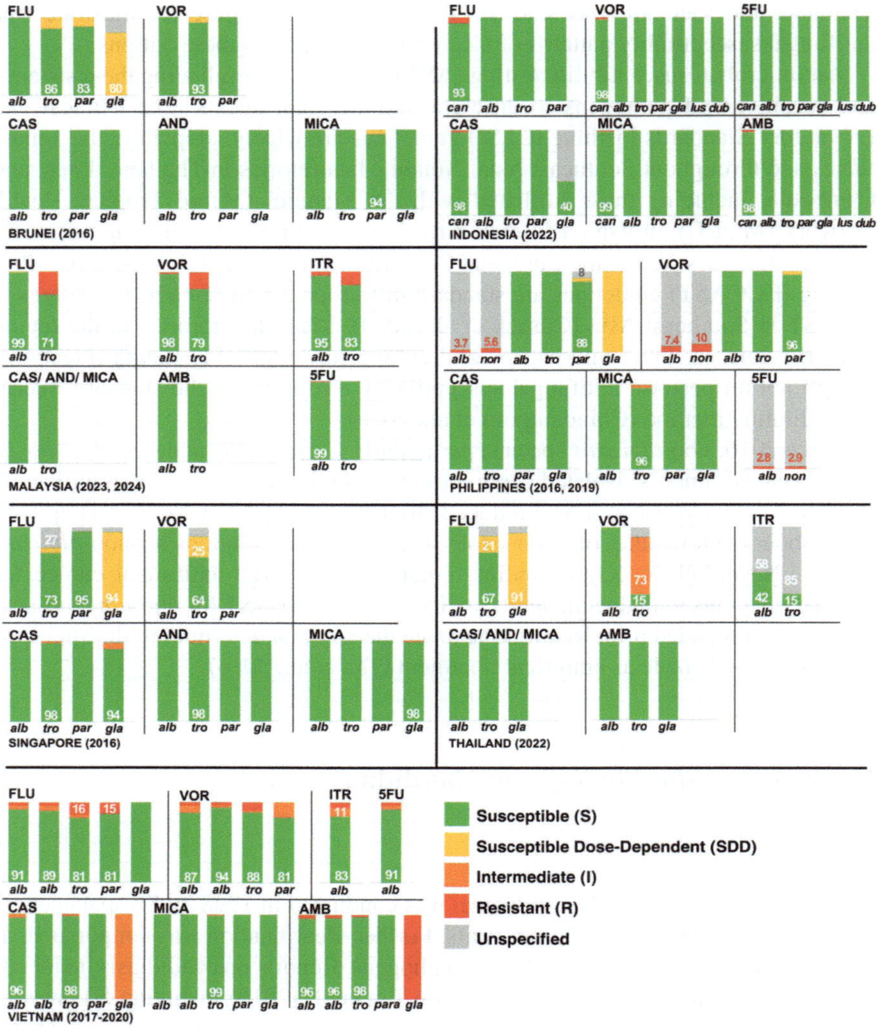

Fig. 6.2 Antifungal susceptibility of the different *Candida* species in Southeast Asian countries: *can* (*Candida* spp.), *non* (*non-Candida* spp.), *alb* (*C. albicans*), *tro* (*C. tropicalis*), *par* (*C. parapsilosis*), *gla* (*N. glabratus*), *lus* (*Candida lusitaniae*), and *dub* (*Candida dubliniensis*). The antifungals are fluconazole (FLU), voriconazole (VOR), itraconazole (ITR), 5-fluorocytosine (5FU), caspofungin (CAS), micafungin (MICA), and amphotericin B (AMB). Percentages represent the total isolate population from various study years, with colors indicating susceptibility status

totaling 26,710 cases annually, while invasive candidiasis accounts for 66,750 cases each year. Recurrent vulvovaginal candidiasis impacts about five million women each year (Wahyuningsih et al., 2021). An increase in non-*albicans Candida* has been noted, with *C. tropicalis* reported as the dominant species in several studies (Suhartono et al., 2020; Wulandari et al., 2021). Recent findings indicate *C. parapsilosis* prevalence in neurological patients and *C. albicans* dominance among HIV/AIDS patients (Rinawati et al., 2022). *Candida kefyr* has been detected in cases of cutaneous candidiasis, and *Candida haemulonii* has emerged as resistant to amphotericin B (Hanifah et al., 2024).

Thailand

Candida research in Thailand began in the 1960s, with initial studies focusing on *C. albicans* cultured from skin scrapings. The 1990s saw increased research on *Candida* prevalence among AIDS patients (Teanpaisan & Nittayananta, 1998). Current estimates suggest approximately 3310 individuals per 100,000 experience recurrent vulvovaginal candidiasis, with around 8650 annual cases of candidemia (Chayakulkeeree & Denning, 2017). Oral manifestations of *Candida* are frequently observed and remain predominant among high-risk populations, particularly in HIV patients (Reichart et al., 2003).

Malaysia

The earliest documentation of *Candida* species in Malaysia by Ngeow and Soo-Hoo (1989) analyzed vaginal swabs, identifying *C. albicans* as the most frequently detected species. *Candida* infections are becoming increasingly common in Malaysia, with recurrent vaginal candidiasis (RVC) reported as the most prevalent type which accounts for over 500,000 cases annually (Velayuthan et al., 2018). Candidemia primarily impacts hospitalized cancer patients and those in intensive care units (Siti Umairah et al., 2012). The distribution of species has also changed over time, with *C. albicans* as the most frequently detected species in clinical specimens, followed by non-*albicans* such as *C. parapsilosis, C. tropicalis*, and *N. glabratus* (Ng, 1998). More recent studies have indicated an increase in the prevalence of *C. tropicalis* in blood samples. Antifungal susceptibility studies show that while amphotericin B remains effective against most *Candida* species, fluconazole resistance is notable in *N. glabratus* and *C. parapsilosis* (Amran et al., 2011).

Singapore

The first Singaporean study on *Candida* examined *Torulopsis* (*Nakaseomyces*) *glabrata* among diabetic patients (Tan, Lim, & Ishak, 1977; Tan, Lim, & Teo, 1977). In the early 2000s, *C. tropicalis* emerged as the leading species in bloodstream infections, surpassing *C. albicans*. By 2016, *C. albicans* remained common, followed by *N. glabratus* and *C. tropicalis*, with varying antifungal susceptibilities. Interestingly, *C. auris* was first reported in the country in 2018 (Tan & Tan, 2018). Notably, the *C. auris* isolates were initially misidentified as *Candida duobushaemulonii* (Tan et al., 2019).

Philippines

In the early 1990s, a multi-centered study conducted by Sorongon et al. (1995) documented *Candida* spp. in 4.9% of blood cultures and 5.5% of urine cultures, with *C. albicans* being the most common species. Another study was by Bulmer et al. (1999), where 579 isolates across 16 medical centers were sent for identification and susceptibility testing to the University of Santo Tomas, Manila, and revealed high number of *C. albicans* (49.16%), followed by *C. parapsilosis* (17.2%) and *C. tropicalis* (14.9%). A report by Moron and Cabrera (2018, 2019) isolated azole-resistant *C. albicans,* and in Linggawaty et al., 2021, Linggawaty et al. reported the first instance of *C. auris* fungemia in the country. The isolate was identified in a patient who had traveled to Cebu and was later evacuated to Japan. Oh et al. (2023) documented a case of *Candida vulturna* fungemia in an infant, misidentified initially, but confirmed through sequencing. Annually, around 1.9% of the Philippine population is affected by serious fungal infections, with aspergillosis and candidiasis being the most prevalent. Recurrent vulvovaginal candidiasis impacts approximately 1.48 million women, while candidemia affects 1968 people (Batac & Denning, 2017). A study by Juayang et al. (2019) isolated 184 clinical specimens, predominantly *C. albicans* (61%) and *C. tropicalis* (15%), noting antifungal resistance in both *C. albicans* and non-*albicans* species. Fluconazole resistance was particularly noted in *Candida ciferri* (42.1%) and non-*albicans* species, echoing findings from Bulmer et al. (1999) regarding fluconazole susceptibility in yeast isolates, where *C. albicans* showed high susceptibility while non-*albicans* exhibited varying resistance levels.

Cambodia

Reports on *Candida* started in the early 2000s, with studies highlighting oral candidiasis among HIV and leprosy patients (Reichart et al., 2003; Schmidt-Westhausen et al., 2004). In 2004, Schmidt-Westhausen et al. identified *C. albicans* as the most common species among isolates from both exposed and non-exposed groups to antimycotics. A study involving 101 HIV-positive patients showed a high incidence of oral candidiasis, with 52.5% exhibiting pseudomembranous candidiasis (Bendick et al., 2002; Reichart et al., 2003). *C. albicans* was found to be the most common species among yeast isolates from immunocompromised individuals.

Vietnam

Vietnam's first *Candida* study focused on cutaneous infections in US military personnel (Blank et al., 1969). In the 2000s, a case report identified *C. albicans* in a cerebral abscess, revealing fungal hyphae through pathological analysis (Van Pham et al., 2008). Bac et al. (2019) identified *C. tropicalis* as the most prevalent species among bloodstream isolates. In the 2020s, the first detection of *C. auris* occurred in a septic patient transferred from Vietnam to Australia (Xie et al., 2020). Additionally, Dang-Vu et al. (2024) achieved the first local isolation of *C. auris*. Tan et al. (2016) reported species distribution in bloodstream isolates: *C. albicans* (39.9%), *C. tropicalis* (39.2%), *C. parapsilosis* (7.8%), *N. glabratus* (7.8%), and *M. guilliermondii* (0.7%). *C. tropicalis* displayed reduced antifungal susceptibility, with 61.7% being resistant to fluconazole and 46.7% to voriconazole. A similar trend was observed by Ngo et al. (2023), where *C. albicans* constituted 48% and *C. tropicalis* 16% of 196 strains tested, with fluconazole resistance noted in 12.5% and voriconazole resistance in 18.8% of *C. tropicalis* isolates. Sinh et al. (2021) found that *C. tropicalis* represented 45.7% of 186 isolates from ICU patients, with significant resistance rates observed for fluconazole and voriconazole. Estimates indicate that there were 2,389,661 serious fungal infections in Vietnam in 2020 and suggest that approximately 12 per 100,000 individuals experience candidemia annually, with recurrent vaginal candidiasis affecting 4047 women per 100,000 (Duong et al., 2023).

Rising Threat of Candidozyma auris in Southeast Asia

Candidozyma auris formerly *Candida auris* is a relatively recent addition to the well-known groups of human pathogenic fungi that has emerged as a significant global concern. This fungus was first recognized as a new species by Japanese researchers from the *C. haemulonii* species complex by sequencing the 26S rDNA D1/D2 domain, nuclear ribosomal DNA ITS region sequences in 2009 (Satoh et al., 2009). The initial isolation of the new *Candida* species came from the ear canal of a Japanese patient, hence the name *C. auris*. Soon after, investigators in South Korea described the bloodstream isolations of *C. auris* (Lee et al., 2011). The bloodstream isolates were resistant to fluconazole, an important insight into the drug-resistance potential of the new pathogen. After the first identification in East Asia, this spread outward, highlighting its potential for global spread. Many case clusters of *C. auris* were being recognized around 2013–2016 among hospitalized patients in India, South Africa, the UK, and the United States (Chowdhary et al., 2013; Magobo et al., 2014; Sarma et al., 2013; Schelenz et al., 2016; Vallabhaneni et al., 2017). The confirmation of the etiologic role of *C. auris* remained dependent on molecular methods such as PCR, real-time PCR, or ITS sequencing, as colony and microscopic features were not distinctive for identification (Dennis et al., 2021) (Fig. 6.3). As of 2024, this fungus was reclassified to *C*andidozyma auris based on genomic metrics that included Average Amino Acid Identity (AAI), Percentage of Conserved

Proteins (POCP), and the Presence-Absence Patterns of Orthologs (PAPO) (Liu et al., 2024).

This multidrug-resistant species poses a high risk of mortality. Since 2018, seven documented cases of *C. auris* have been reported in Southeast Asia. In Malaysia, *C. auris* was isolated from the blood of a 63-year-old neutropenic patient (Mohd Tap et al., 2018). A study by Tan and Tan (2018) described three cases of *C. auris* infection in Singapore. In Thailand, two *C. auris* were found resistant to both fluconazole and amphotericin B (Pharkjaksu et al., 2021). In Vietnam, the first case of bloodstream colonization of *C. auris* and *C. duobushaemulonii* was reported (Xie et al., 2020). It has since been identified in over 40 countries across six continents (Keighley et al., 2021; Sabino et al., 2020). More recently, in 2024, three cases of *C. auris* from a new clade, Clade VI, were identified in Singapore and Bangladesh. The independent detection of these isolates, especially the two isolates found in Singapore, raises the possibility that these strains may have been circulating undetected in Southeast Asia (Suphavilai et al., 2024).

As the *C. auris* epidemic spreads around the globe, distinctive physiological features of the pathogen are being discovered, such as its heat tolerance to 42 °C, high salt tolerance, and the ability to metabolize dulcitol. The scientists at the Centers used these unique physiologic features to formulate a selective isolation medium for *C. auris*, setting the stage for its widespread use for surveillance samples (Zhu et al., 2020). However, the regular workflow for yeast identifications in the clinical laboratories in the Western hemisphere depended on automated or semi-automated identification systems, which initially yielded incorrect reports for *C. auris*. Subsequent changes in the commercial systems corrected these anomalies; *C. auris* culture

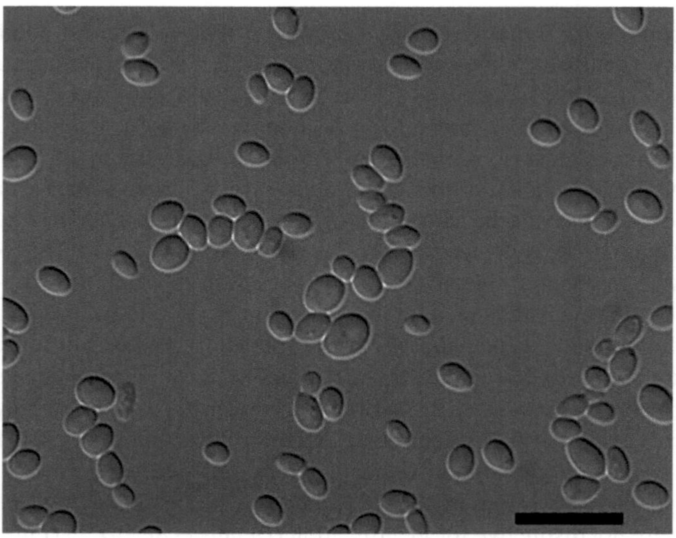

Fig. 6.3 *Candidozyma auris* yeast cells from yeast extract-peptone broth at 25 °C; bar 10 µm. (Reproduced with permission from *Microbiol Immunol* 2009; 53: 41–44)

confirmation by rapid mass spectrometry using commercial MALDI-TOF-MS instruments also became feasible in the same period (Zhu et al., 2020). It is also noteworthy to highlight the selective isolation and identification medium ChromAgar Candida plus®, which enabled the distinction of *C. auris* directly from primary clinical specimens by the color of its colonies on the culture plate (Marathe et al., 2022). Thus, 2015–2020 were crucial years for the development of various laboratory tools for the selective isolation, presumptive identification, and confirmation of *C. auris* from clinical and surveillance samples, as summarized in Fig. 6.4. Many investigative teams have now developed rapid DNA identification methods using ambient temperature isothermal PCR. This simplified technology can extend rapid molecular methods to resource-challenged settings because sophisticated machines are unnecessary for PCR setup or for reading color-change results (Yamamoto et al., 2018). The quick dissemination of *C. auris* revealed multiple significant diagnostic challenges, hindering timely identification and management, as discussed in the following.

Phenotypic and biochemical methods, such as API 20C, Vitek 2, Phoenix, and MicroScan, frequently misidentify *C. auris* isolates as various other *Candida* species, including *C. haemulonii*, *Candida famata*, *Candida sake*, as well as *Rhodotorula* and *Saccharomyces* species. Misidentifications also occasionally occur with other *Candida* species, or isolates as these methods identify isolates only at the genus level, leading to a lack of specificity (Jeffery-Smith et al., 2018). The misidentification can delay appropriate treatment, contributing to poor patient outcomes and increased healthcare costs. Fortunately, these identification systems have been updated with improved databases that enhance the accurate identification of *C. auris*; however, they still require yeast culture, which can be a time-intensive process. MALDI-TOF MS now provides reliable identification of *C. auris* when supported by up-to-date databases (Ahmad & Alfouzan, 2021). Previous versions of these databases had occasionally misidentified *C. auris* as *C. haemulonii* or other species (Kim et al., 2016), but these issues have been largely resolved with recent updates. Polymerase chain reaction (PCR) assays specifically designed for *C. auris* and its closely related species have emerged as a promising approach for rapid and precise identification, particularly during outbreak situations (Kordalewska et al., 2017). In South Korea, pulsed-field gel electrophoresis (PFGE) analysis of 15 *C. auris* isolates from patients' ear specimens across three hospitals demonstrated diverse PFGE patterns, suggesting potential clonal transmission in certain instances (Oh et al., 2011). Although sequencing genetic loci such as D1/D2, RPB1, RPB2, and ITS domains of rRNA has facilitated reliable *C. auris* identification, this method is not commonly employed for *Candida* species isolates. It is generally accessible only in specialized reference laboratories (Katsiari et al., 2023).

Globally, *C. auris* is divided into five primary clades: clade I (South Asian), clade II (East Asian), clade III (South African), clade IV (South American), and clade V (Iranian) (Narayanan et al., 2022; Spruijtenburg et al., 2022). Each clade has distinct characteristics that include variations in mating-type locus, virulence, and drug resistance. Isolates from clades I and IV possess the MTLa allele, while those from clades II and III carry the MTLα allele. Clade III isolates tend to form

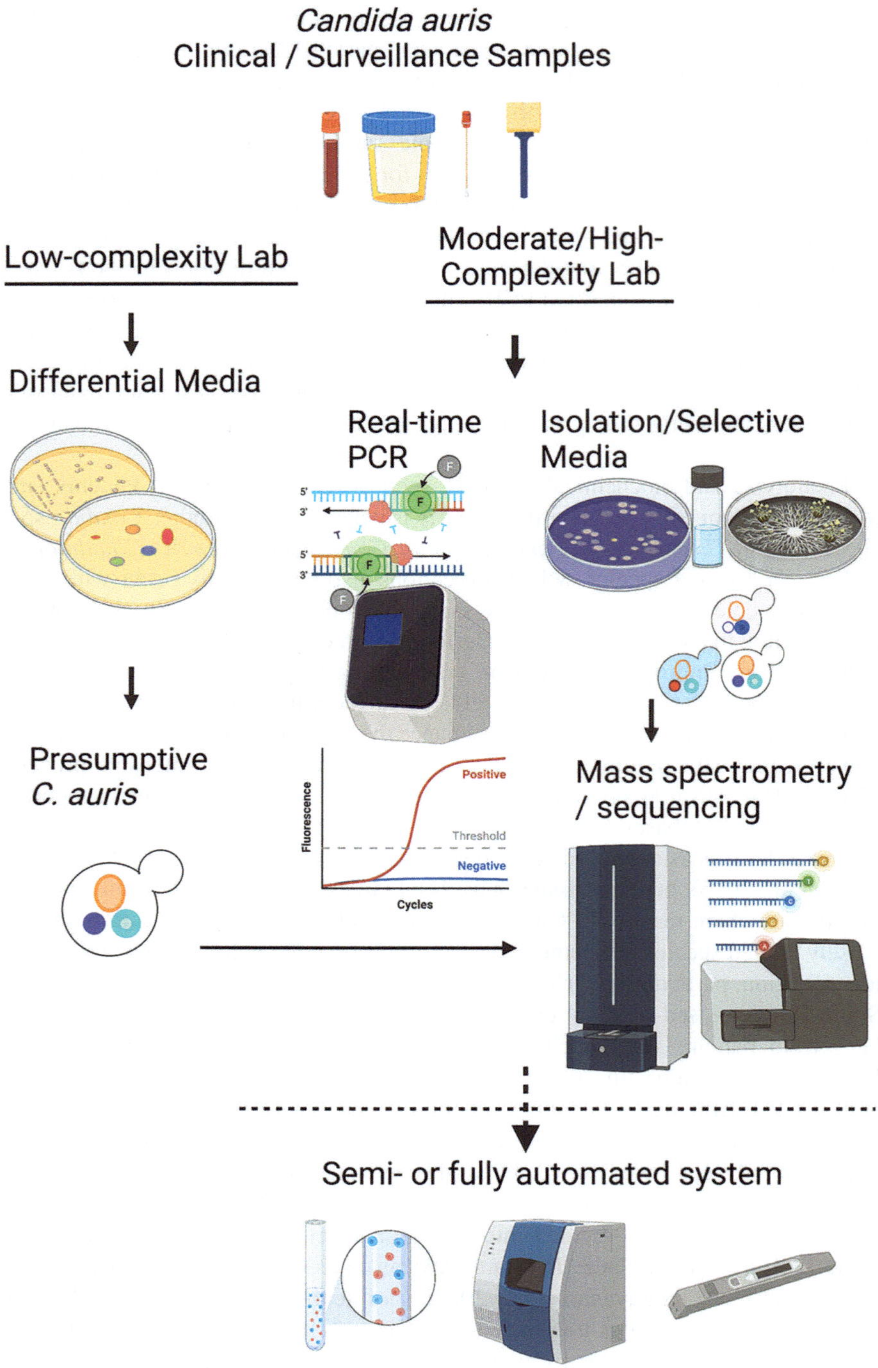

Fig. 6.4 A conceptual illustration of *Candidozyma auris (formerly Candida auris)* testing method differences between low-complexity and moderate- to high-complexity laboratories. (Reproduced from *Front Microbiol*, 12:757835, 2021)

large fungal cellular aggregates, whereas clade I is characterized by a pronounced formation of pseudohyphae. Additionally, clade II is generally susceptible to antifungal treatments, contrasting with the multidrug-resistant isolates found in the other clades (Narayanan et al., 2022). Clade V, identified in Iran, is particularly noteworthy due to its genomic divergence, containing over 200,000 SNPs compared to the other clades, and sharing L-rhamnose utilization traits with clade III (Spruijtenburg et al., 2022). Recently, Clade VI has been proposed based on genomic analyses of isolates from Singapore and Bangladesh that demonstrate considerable genetic distance from clade IV, its nearest relative (Suphavilai et al., 2024). The implementation of active surveillance programs, particularly in Singapore, is crucial for early detection and containment. However, genomic surveillance efforts remain limited, with a lack of data on the emerging Clade VI isolates (Suphavilai et al., 2024). It is a requirement to understand the genetic diversity of *C. auris*, as not only do these data aid in the identification of clades, but they also inform strategies that can help mitigate and control the infection of *C. auris* through the documentation of a clade's patterns, especially toward resistance to certain medicinal interventions.

Currently, there is no standardized protocol for environmental sanitation aimed at controlling the transmission of *C. auris* within healthcare settings. Ku et al. (2018) summarize that major health organizations have issued guidelines and recommendations for managing *C. auris*. The Centers for Disease Control and Prevention (CDC) recommends using a hospital-grade disinfectant approved by the Environmental Protection Agency (EPA) that is effective against *Clostridium difficile* spores for cleaning surfaces contaminated with *C. auris*. Public Health England (PHE) advises employing hypochlorite products at a concentration of 1000 ppm for all cleaning procedures, irrespective of the use of alternative methods such as gaseous hydrogen peroxide or UV-C light. The European Centre for Disease Prevention and Control (ECDC) advocates for terminal cleaning with disinfectants and techniques that have verified antifungal properties. Similarly, the Public Health Agency of Canada (PHAC) and the South African Centre for Opportunistic, Tropical and Hospital Infections (COTHI) have released temporary guidelines for managing *C. auris* infections, promoting routine and thorough cleaning with a chlorine-releasing agent at a concentration of 1000 ppm. Additionally, COTHI recommends incorporating hydrogen peroxide vapor into the cleaning process whenever feasible. Finally, the Pan American Health Organization/World Health Organization (PAHO/WHO) advises cleaning with soap and water followed by disinfection using 0.1% bleach. Despite these rigorous control measures, *C. auris* maintains a substantial clinical challenge, especially among vulnerable patient populations.

Clinically, *C. auris* has been reported from catheters, blood, urine, wound, respiratory tract, and bile. Common risk factors include prolonged hospital stay and underlying comorbidities. Treatment involves echinocandins with careful monitoring of resistance development. Echinocandin-resistant strains could be treated with amphotericin B (Vazquez, 2023). *C. auris* exhibits high rates of azole resistance; hence, these drugs are not typically used for therapy.

C. auris has proven to be a highly adept pathogen for the colonization of skin and its bloodstream invasion among patients receiving care at various healthcare facilities (Lyman et al., 2023; Vallabhaneni et al., 2017). The pathogen spreads readily among patients via fomites in the hospital environment (Zhu et al., 2020). An effective *C. auris* infection control and prevention effort depends upon robust screening of patients and their environment (Tsay et al., 2018). It is crucial to baseline the presence and extent of *C. auris* presence in a facility to design adequate control efforts. Both culture-based and molecular methods have proven useful for *C. auris* surveillance, although rapid DNA-based methods are preferred as they allow quicker implementation of the remedial measure (Zhu et al., 2020). DNA sequencing is equally valuable for the genotyping of *C. auris*. The genotype of a *Candida* isolate allows its placement among four major or two minor clades; the clades have origins from distinct geographic areas of the world, and they show variable resistance to various antifungal drugs (Lockhart et al., 2017). Further analysis by whole genome sequencing is the only accurate approach to find out the exact genetic relationship among isolates from a patient, within a hospital, from a region or a country (Rhodes et al., 2017).

*Advancements in Anti-*Candida *Vaccine Development*

Despite significant efforts, no anti-*Candida* vaccine has been approved for human use to date. However, researchers remain committed to developing an effective vaccine. To date, three candidates have progressed to clinical trials, while recent advancements since 2019 have focused on in silico vaccine design, pan-fungal vaccines, and Trained Immunity-Based Vaccines (TIBVs) (Rivera et al., 2022).

One of the leading candidates, *PEV7*, is a virosomal vaccine that incorporates the secreted aspartic protease 2 (SAP2) protein from *C. albicans*. It has been evaluated for its efficacy against vaginal candidiasis, showing potential protection in preclinical models, and has advanced to phase I clinical trials. The results from these trials indicated favorable safety and demonstrated the generation of specific and functional B cell memory in all 48 healthy volunteers who received the vaccine. These findings underscore its potential as a therapeutic option for recurrent vulvovaginal candidiasis (De Bernardis et al., 2012, 2018). Research has also indicated that anti-Sap2 antibodies offer protective benefits against mucosal infections and help decrease fungal burden in systemic candidiasis models. Furthermore, Sap2-specific monoclonal antibodies have shown cross-reactivity with other Sap proteins, which may enhance their protective efficacy (Vilanova et al., 2004). Recent studies have shown that adjuvants enhance the efficacy of PEV7, specifically on mucosal immunity and the potential for cross-protection against other fungal pathogens (Wüthrich et al., 2021). Another promising candidate is *NDV-3A*, a recombinant vaccine developed from the agglutinin-like sequence 3 (ALS3) protein of *C. albicans*. NDV-3A has demonstrated immunogenicity and protective effects in both preclinical and clinical trials. Importantly, it is the first *Candida* vaccine candidate to advance to human trials, with a focus on preventing both mucosal and systemic candidiasis (Edwards Jr et al., 2018; Schmidt et al., 2012). NDV-3A has been shown to induce robust Th1 and Th17 responses, while the antibodies generated against Als3 effectively inhibit *Candida* adhesion, filamentation, and biofilm formation. Moreover, passive transfer of anti-Als3 monoclonal antibodies has successfully reduced fungal burden in infected mice (Uppuluri et al., 2018). Recent research explored the combination of this vaccine with adjuvants that increase its efficacy, particularly in immunocompromised populations and to address better drug-resistant *Candida* strains (Oliveira et al., 2021). *Mycograb*, a different approach, is a human recombinant monoclonal antibody developed by NeuTec Pharma. It targets heat shock protein 90 (Hsp90), a protein essential for *Candida* survival. When used in combination with lipid-associated amphotericin B, Mycograb has shown a synergistic antifungal effect against several *Candida* species, including drug-resistant strains (Pachl et al., 2006). Mycograb enhances fungal clearance by disrupting Hsp90's role in *Candida* stress response and morphogenesis, leading to fungal cell death. It has been noted though that Mycograb faced regulatory challenges due to concerns over safety and quality, thus limiting further developments (Matthews et al., 2003). Studies have suggested revisiting Mycograb's potential in combination with newer antifungal agents or as part of a multi-epitope vaccine strategy (Louie et al., 2011; Richie et al., 2012). In addition to these candidates, recent advancements in vaccine development are now focusing on pan-fungal vaccines, aiming to provide broad-spectrum protection against multiple fungal pathogens, including *Candida*, *Aspergillus*, and *Pneumocystis*. For example, the *NXT-2* vaccine, a pan-fungal recombinant peptide with vaccine-like features, has shown promising results in pre-clinical models. It offers protection against aspergillosis, candidiasis, and pneumocystosis (Rayens et al., 2022). Furthermore, the concept of trained immunity, where innate immune cells are "trained" to respond more effectively to subsequent infections, has gained traction. This is evident with studies using exposure to avirulent *Candida* strains or β-glucan inducing trained immunity, which in turn provides non-specific protection against fungal infections (Alapan et al., 2024; Kaur et al., 2023; Netea et al., 2020; Quintin et al., 2012). Studies have also demonstrated the potential of β-glucan-based vaccines in combating fungal infections. A glyco-conjugate vaccine, created by attaching laminarin to diphtheria toxoid (CRM197), has shown protective efficacy against systemic candidiasis in murine models by inducing anti-β-glucan antibodies that directly inhibit fungal growth and adherence (Adamo et al., 2011). Recent research has also emphasized the use of immunoinformatics for vaccine design, particularly for the identification of conserved epitopes across *Candida* species. For example, Akhtar et al. (2023) used computational tools to design a multi-epitope vaccine targeting *C. dubliniensis*, which showed potential immunogenicity in silico and could be a model for future *Candida* vaccines. Additionally, the exploration of oral vaccines, such as those using *Saccharomyces cerevisiae* or *Lactobacillus casei* as delivery vehicles, has shown potential in pre-clinical models. These offer a convenient and cost-effective alternative to injectable vaccines (Shibasaki et al., 2013, 2014). And developments in synthetic glycopeptide vaccines, including those that integrate β-mannan with *Candida* cell wall proteins, such as the Fba peptide, have demonstrated potential in generating protective immune responses against candidiasis. Notably, when these vaccines are conjugated with tetanus toxoid, they can stimulate strong antibody responses without requiring additional adjuvants, underscoring their promise as viable candidates for antifungal vaccines (Shukla et al., 2021; Xin et al., 2008, 2012). While these five candidates are promising, no additional vaccines targeting *Candida* have been published recently. The field is rapidly evolving, with the ongoing research into novel adjuvants, multivalent vaccines, and integration of trained immunity concepts could pave the way for the next generation of *Candida* vaccines (Alapan et al., 2024; Inácio et al., 2023; Kaur et al., 2023; Sahu et al., 2022).

Conclusion

The situation regarding *Candida* species in Southeast Asia mirrors trends observed in the Western Hemisphere, where non-*albicans* species are progressively displacing *C. albicans* as the predominant cause of infections. This shift is compounded by rising resistance to the latest antifungal agents, raising concerns about an impending crisis characterized by the widespread emergence of highly resistant strains.

To address this growing threat, a multifaceted approach is essential. Collaborative research efforts should be prioritized to better understand the evolving resistance patterns and to develop effective treatment strategies. Educational initiatives aimed at healthcare professionals and the general public must be implemented to raise awareness about antifungal resistance and the importance of appropriate prescribing practices. Moreover, the establishment and strict adherence to infection control protocols within medical facilities are crucial in mitigating the spread of resistant *Candida* species. This requires a coordinated effort among local and national government agencies to ensure that resources are allocated effectively and that health policies are aligned. Given that vulnerable populations—such as those with weakened immune systems, the elderly, and individuals with chronic illnesses—are at heightened risk for *Candida* infections, proactive management of this issue is imperative. Without immediate action, the risks associated with *Candida* species infections will continue to escalate, ultimately affecting broader segments of the population. Therefore, addressing this ongoing crisis should be a top priority for public health stakeholders, as all individuals can eventually become part of this susceptible group.

References

Abad-Zapatero, C., Goldman, R., Muchmore, S. W., Hutchins, C., Stewart, K., Navaza, J., Payne, C. D., & Ray, T. L. (1996). Structure of a secreted aspartic protease from *C. albicans* complexed with a potent inhibitor: Implications for the design of antifungal agents. *Protein Science, 5*(4), 640–652.

Abd Elaziz, D., Abd El-Ghany, M., Meshaal, S., El Hawary, R., Lotfy, S., Galal, N., Ouf, S. A., & Elmarsafy, A. (2020). Fungal infections in primary immunodeficiency diseases. *Clinical Immunology, 219*, 108553. https://doi.org/10.1016/j.clim.2020.108553

Adamo, R., Tontini, M., Brogioni, G., Romano, M. R., Costantini, G., Danieli, E., Proietti, D., Berti, F., & Costantino, P. (2011). Synthesis of laminarin fragments and evaluation of a β-(1,3) glucan hexasaccaride-CRM197 conjugate as vaccine candidate against *Candida albicans*. *Journal of Carbohydrate Chemistry, 30*(4–6), 249–280. https://doi.org/10.1080/07328303.2011.604453

Ahmad, S., & Alfouzan, W. (2021). *Candida auris*: Epidemiology, diagnosis, pathogenesis, antifungal susceptibility, and infection control measures to combat the spread of infections in healthcare facilities. *Microorganisms, 9*(4). https://doi.org/10.3390/microorganisms9040807

Akhtar, N., Magdaleno, J. S., Ranjan, S., Wani, A. K., Grewal, R. K., Oliva, R., Shaikh, A. R., Cavallo, L., & Chawla, M. (2023). Secreted aspartyl proteinases targeted multi-epitope vaccine design for *Candida dubliniensis* using immunoinformatics. *Vaccine, 11*(2). https://doi.org/10.3390/vaccines11020364

Akpan, A., & Morgan, R. (2002). Oral candidiasis. *Postgraduate Medical Journal, 78*(922), 455–459.

Alapan, D., Bisweswar, O., Prasenjit, S., Prasanjit, D., & Arkapal, B. (2024). Recent advances in the clinical development of antifungal vaccines: A narrative review. *Frontiers in Tropical Diseases, 5*. https://www.frontiersin.org/journals/tropical-diseases/articles/10.3389/fitd.2024.1446477

Alem, M. A., Oteef, M. D., Flowers, T. H., & Douglas, L. J. (2006). Production of tyrosol by *Candida albicans* biofilms and its role in quorum sensing and biofilm development. *Eukaryotic Cell, 5*(10), 1770–1779.

Amran, F., Aziz, M. N., Ibrahim, H. M., Atiqah, N., Parameswari, S., Hafiza, M., & Ifwat, M. (2011). In vitro antifungal susceptibilities of *Candida* isolates from patients with invasive candidiasis in Kuala Lumpur Hospital, Malaysia. *Journal of Medical Microbiology, 60*(9), 1312–1316.

Anderson, T. M., Clay, M. C., Cioffi, A. G., Diaz, K. A., Hisao, G. S., Tuttle, M. D., Nieuwkoop, A. J., Comellas, G., Maryum, N., Wang, S., Uno, B. E., Wildeman, E. L., Gonen, T., Rienstra, C. M., & Burke, M. D. (2014). Amphotericin forms an extramembranous and fungicidal sterol sponge. *Nature Chemical Biology, 10*(5), 400–406. https://doi.org/10.1038/nchembio.1496

Anh, D. N., Hung, D. N., Tien, T. V., Dinh, V. N., Son, V. T., Luong, N. V., Van, N. T., Quynh, N. T. N., Van Tuan, N., & Tuan, L. Q. (2021). Prevalence, species distribution and antifungal susceptibility of *Candida albicans* causing vaginal discharge among symptomatic non-pregnant women of reproductive age at a tertiary care hospital, Vietnam. *BMC Infectious Diseases, 21*(1), 523.

Arafa, S. H., Elbanna, K., Osman, G. E. H., & Abulreesh, H. H. (2023). *Candida* diagnostic techniques: A review. *Journal of Umm Al-Qura University for Applied Sciences, 9*(3), 360–377. https://doi.org/10.1007/s43994-023-00049-2

Askew, C., Sellam, A., Epp, E., Hogues, H., Mullick, A., Nantel, A., & Whiteway, M. (2009). Transcriptional regulation of carbohydrate metabolism in the human pathogen *Candida albicans*. *PLoS Pathogens, 5*(10), e1000612. https://doi.org/10.1371/journal.ppat.1000612

Bac, N. D., Luc, N. K., Nga, T. T. T., Nagi, M., Yoshitsugu, M., Ha, H. T. T., Anh, D. D., & Quyet, D. (2019). Prevalence of *Candida* bloodstream isolates from patients in two hospitals in Vietnam. *Iranian Journal of Microbiology, 11*(2), 108.

Balkovec, J. M., Hughes, D. L., Masurekar, P. S., Sable, C. A., Schwartz, R. E., & Singh, S. B. (2014). Discovery and development of first in class antifungal caspofungin (CANCIDAS®)—A case study. *Natural Product Reports, 31*(1), 15–34. https://doi.org/10.1039/C3NP70070D

Barantsevich, N., & Barantsevich, E. (2022). Diagnosis and treatment of invasive candidiasis. *Antibiotics, 11*(6), 718.

Batac, M. C. R., & Denning, D. (2017). Serious fungal infections in The Philippines. *European Journal of Clinical Microbiology & Infectious Diseases, 36*, 937–941.

Bendick, C., Scheifele, C., & Reichart, P. (2002). Oral manifestations in 101 Cambodian patients with HIV infection and AIDS. *Journal of Oral Pathology & Medicine, 31*(1), 1–4.

Berman, J., & Krysan, D. J. (2020). Drug resistance and tolerance in fungi. *Nature Reviews Microbiology, 18*(6), 319–331. https://doi.org/10.1038/s41579-019-0322-2

Birnbaum, J. E. (1990). Pharmacology of the allylamines. *Journal of the American Academy of Dermatology, 23*(4, Part 2), 782–785. https://doi.org/10.1016/0190-9622(90)70288-S

Blank, H., Taplin, D., & Zaias, N. (1969). Cutaneous Trichophyton mentagrophytes infections in Vietnam. *Archives of Dermatology, 99*(2), 135–144.

Boonsilp, S., Homkaew, A., Phumisantiphong, U., Nutalai, D., & Wongsuk, T. (2021). Species distribution, antifungal susceptibility, and molecular epidemiology of *Candida* species causing Candidemia in a tertiary care hospital in Bangkok, Thailand. *Journal of Fungi, 7*(7). https://doi.org/10.3390/jof7070577

Braun, B. R., & Johnson, A. D. (2000). TUP1, CPH1 and EFG1 make independent contributions to filamentation in *Candida albicans*. *Genetics, 155*(1), 57–67.

Bulmer, G. S., Marquez, M. L., Co-Barcelona, L., & Fromtling, R. A. (1999). Yeasts and fluconazole susceptibility in The Philippines. *Mycopathologia, 146*, 117–120.

Camp, I., Spettel, K., & Willinger, B. (2020). Molecular methods for the diagnosis of invasive candidiasis. *Journal of Fungi, 6*(3). https://doi.org/10.3390/jof6030101

Campbell, J. & Palazzi, D. (2023). Candidemia and invasive candidiasis in children: Management. *UpToDate*. Retrieved October 7, 2024 from https://www.uptodate.com/contents/candidemia-and-invasive-candidiasis-in-children-management

Capoor, M. R., & Bal, A. M. (2022). 7.22—Echinocandins. In T. Kenakin (Ed.), *Comprehensive pharmacology* (pp. 363–371). Elsevier. https://doi.org/10.1016/B978-0-12-820472-6.00211-5

Carvalho, A., Cunha, C., Pasqualotto, A. C., Pitzurra, L., Denning, D. W., & Romani, L. (2010). Genetic variability of innate immunity impacts human susceptibility to fungal diseases. *International Journal of Infectious Diseases, 14*(6), e460–e468. https://doi.org/10.1016/j.ijid.2009.06.028

Cavalheiro, M., & Teixeira, M. C. (2018). *Candida* biofilms: Threats, challenges, and promising strategies. *Frontiers in Medicine, 5*, 28.

Centers for Disease Control and Prevention. (2021, July 22). *Vulvovaginal Candidiasis—STI treatment guidelines*. https://www.cdc.gov/std/treatment-guidelines/candidiasis.htm

Chakrabarti, A., Patel, A. K., Soman, R., & Todi, S. (2023). Overcoming clinical challenges in the management of invasive fungal infections in low- and middle-income countries (LMIC). *Expert Review of Anti-Infective Therapy, 21*(10), 1057–1070. https://doi.org/10.1080/14787210.2023.2257895

Chayakulkeeree, M., & Denning, D. (2017). Serious fungal infections in Thailand. *European Journal of Clinical Microbiology & Infectious Diseases, 36*, 931–935.

Chen, H., Fujita, M., Feng, Q., Clardy, J., & Fink, G. R. (2004). Tyrosol is a quorum-sensing molecule in *Candida albicans*. *Proceedings of the National Academy of Sciences, 101*(14), 5048–5052.

Chen, J., Zhou, S., Wang, Q., Chen, X., Pan, T., & Liu, H. (2000). Crk1, a novel Cdc2-related protein kinase, is required for hyphal development and virulence in *Candida albicans*. *Molecular and Cellular Biology, 20*(23), 8696–8708. https://doi.org/10.1128/MCB.20.23.8696-8708.2000

Chindamporn, A., Chakrabarti, A., Li, R., Sun, P.-L., Tan, B.-H., Chua, M., Wahyuningsih, R., Patel, A., Liu, Z., Chen, Y.-C., & Chayakulkeeree, M. (2018). Survey of laboratory practices for diagnosis of fungal infection in seven Asian countries: An Asia Fungal Working Group (AFWG) initiative. *Medical Mycology, 56*(4), 416–425. https://doi.org/10.1093/mmy/myx066

Chowdhary, A., Sharma, C., Duggal, S., Agarwal, K., Prakash, A., Singh, P. K., Jain, S., Kathuria, S., Randhawa, H., Hagen, F., & Meis, J. (2013). New clonal strain of *Candida auris*,

Delhi, India. *Emerging Infectious Disease Journal, 19*(10), 1670. https://doi.org/10.3201/eid1910.130393

Chua, M. M. M., & Rosario, M. C. O. (2024). Common invasive fungal infections. In *The PCP textbook of internal medicine* (Vol. 2, pp. 1051–1056). Philippine College of Physicians.

Cleary, I. A., Reinhard, S. M., Miller, C. L., Murdoch, C., Thornhill, M. H., Lazzell, A. L., Monteagudo, C., Thomas, D. P., & Saville, S. P. (2011). *Candida albicans* adhesin ALS3p is dispensable for virulence in the mouse model of disseminated candidiasis. *Microbiology, 157*(6), 1806–1815.

CLSI. (2022). *Performance standards for antifungal susceptibility testing of yeasts*, 3rd ed. CLSI M27M44S-ED3:2022. Clinical and Laboratory Standards Institute.

Cortés, J. A., Corrales, I. F., Cortés, J. A., & Corrales, I. F. (2018). Invasive candidiasis: Epidemiology and risk factors. In *Fungal infection*. IntechOpen. https://doi.org/10.5772/intechopen.81813

Cusack, T. P., Ashley, E. A., Ling, C. L., Roberts, T., Turner, P., Wangrangsimakul, T., & Dance, D. A. B. (2019). Time to switch from CLSI to EUCAST? A southeast Asian perspective. *Clinical Microbiology and Infection, 25*(7), 782–785. https://doi.org/10.1016/j.cmi.2019.03.016

Dang-Vu, T., Lam-Quoc, D., Duong-Minh, N., Vu, H., Nguyen-Ho, L., Truong-Thien, P., & Nguyen-Dang, K. (2024). The first report of *Candida auris* infection in Vietnam. *European Journal of Case Reports in Internal Medicine, 11*(3).

De Bernardis, F., Amacker, M., Arancia, S., Sandini, S., Gremion, C., Zurbriggen, R., Moser, C., & Cassone, A. (2012). A virosomal vaccine against candidal vaginitis: Immunogenicity, efficacy and safety profile in animal models. *Vaccine, 30*(30), 4490–4498.

De Bernardis, F., Graziani, S., Tirelli, F., & Antonopoulou, S. (2018). *Candida* vaginitis: Virulence, host response and vaccine prospects. *Medical Mycology, 56*(suppl_1), S26–S31. https://doi.org/10.1093/mmy/myx139

de Groot, P. W. J., de Boer Albert, D., Cunningham, J., Dekker, H. L., de Jong, L., Hellingwerf, K. J., de Koster, C., & Klis, F. M. (2004). Proteomic analysis of *Candida albicans* cell walls reveals covalently bound carbohydrate-active enzymes and adhesins. *Eukaryotic Cell, 3*(4), 955–965. https://doi.org/10.1128/ec.3.4.955-965.2004

Dennis, E. K., Chaturvedi, S., & Chaturvedi, V. (2021). So many diagnostic tests, so little time: Review and preview of Candida auris testing in clinical and public health laboratories. *Frontiers in Microbiology, 12*. https://www.frontiersin.org/journals/microbiology/articles/10.3389/fmicb.2021.757835

Deorukhkar, S. C., Saini, S., & Mathew, S. (2014). Virulence factors contributing to pathogenicity of *Candida tropicalis* and its antifungal susceptibility profile. *International Journal of Microbiology, 2014*(1), 456878. https://doi.org/10.1155/2014/456878

Doedt, T., Krishnamurthy, S., Bockmuhl, D. P., Tebarth, B., Stempel, C., Russell, C. L., Brown, A. J., & Ernst, J. F. (2004). APSES proteins regulate morphogenesis and metabolism in *Candida albicans*. *Molecular Biology of the Cell, 15*(7), 3167–3180.

Donadieu, J., Lamant, M., Fieschi, C., de Fontbrune, F. S., Caye, A., Ouachee, M., Beaupain, B., Bustamante, J., Poirel, H. A., Isidor, B., Van Den Neste, E., Neel, A., Nimubona, S., Toutain, F., Barlogis, V., Schleinitz, N., Leblanc, T., Rohrlich, P., Suarez, F., et al. (2018). Natural history of GATA2 deficiency in a survey of 79 French and Belgian patients. *Haematologica, 103*(8), 1278–1287. https://doi.org/10.3324/haematol.2017.181909

Donnelly, J. P., Chen, S. C., Kauffman, C. A., Steinbach, W. J., Baddley, J. W., Verweij, P. E., Clancy, C. J., Wingard, J. R., Lockhart, S. R., & Groll, A. H. (2020). Revision and update of the consensus definitions of invasive fungal disease from the European Organization for Research and Treatment of Cancer and the Mycoses Study Group Education and Research Consortium. *Clinical Infectious Diseases, 71*(6), 1367–1376.

Duong, T. N., Le, M., Beardsley, J., Denning, D. W., Le, N., & Nguyen, B. T. (2023). Updated estimation of the burden of fungal disease in Vietnam. *Mycoses, 66*(4), 346–353.

Edwards, J. E., Jr., Schwartz, M. M., Schmidt, C. S., Sobel, J. D., Nyirjesy, P., Schodel, F., Marchus, E., Lizakowski, M., DeMontigny, E. A., & Hoeg, J. (2018). A fungal immunotherapeutic vaccine (NDV-3A) for treatment of recurrent vulvovaginal candidiasis—A phase 2 randomized, double-blind, placebo-controlled trial. *Clinical Infectious Diseases, 66*(12), 1928–1936.

EUCAST. (2023). *Overview of antifungal ECOFFs and clinical breakpoints for yeasts, moulds and dermatophytes using the EUCAST E.Def 7.4, E.Def 9.4 and E.Def 11.0 procedures*. Version 4.0. European Committee on Antimicrobial Susceptibility Testing.

EUCAST. (2024). *EUCAST guidance on interpretation of MICs for rare yeast without breakpoints in breakpoint tables*. European Committee on Antimicrobial Susceptibility Testing.

Falagas, M. E., Roussos, N., & Vardakas, K. Z. (2010). Relative frequency of *albicans* and the various non-*albicans Candida* spp. among candidemia isolates from inpatients in various parts of the world: A systematic review. *International Journal of Infectious Diseases, 14*(11), e954–e966. https://doi.org/10.1016/j.ijid.2010.04.006

Fanning, S., & Mitchell, A. P. (2012). Fungal biofilms. *PLoS Pathogens, 8*(4), e1002585.

Fera, M. T., La Camera, E., & De Sarro, A. (2009). New triazoles and echinocandins: Mode of action, in vitro activity and mechanisms of resistance. *Expert Review of Anti-Infective Therapy, 7*(8), 981–998. https://doi.org/10.1586/eri.09.67

Fu, Y., Ibrahim, A. S., Fonzi, W., Zhou, X., Ramos, C. F., & Ghannoum, M. A. (1997). Cloning and characterization of a gene (LIP1) which encodes a lipase from the pathogenic yeast *Candida albicans*. *Microbiology, 143*(2), 331–340.

Gale, C. A., Bendel, C. M., McClellan, M., Hauser, M., Becker, J. M., Berman, J., & Hostetter, M. K. (1998). Linkage of adhesion, filamentous growth, and virulence in *Candida albicans* to a single gene, INT1. *Science, 279*(5355), 1355–1358.

Garnacho-Montero, J., Barrero-García, I., & León-Moya, C. (2024). Fungal infections in immunocompromised critically ill patients. *Journal of Intensive Medicine*. https://doi.org/10.1016/j.jointm.2024.01.005

Giacobbe, D. R., Mikulska, M., Tumbarello, M., Furfaro, E., Spadaro, M., Losito, A. R., Mesini, A., De Pascale, G., Marchese, A., Bruzzone, M., Pelosi, P., Mussap, M., Molin, A., Antonelli, M., Posteraro, B., Sanguinetti, M., Viscoli, C., Del Bono, V., & on behalf of ISGRI-SITA (Italian Study Group on Resistant Infections of the Società Italiana Terapia Antinfettiva). (2017). Combined use of serum (1,3)-β-d-glucan and procalcitonin for the early differential diagnosis between candidaemia and bacteraemia in intensive care units. *Critical Care, 21*(1), 176. https://doi.org/10.1186/s13054-017-1763-5

Gomez, M. J., Torosantucci, A., Arancia, S., Maras, B., Parisi, L., & Cassone, A. (1996). Purification and biochemical characterization of a 65-kilodalton mannoprotein (MP65), a main target of anti-*Candida* cell-mediated immune responses in humans. *Infection and Immunity, 64*(7), 2577–2584.

Gow, N. A., & Hube, B. (2012). Importance of the *Candida albicans* cell wall during commensalism and infection. *Host--Microbe Interactions: Fungi/Parasites/Viruses, 15*(4), 406–412. https://doi.org/10.1016/j.mib.2012.04.005

Gow, N. A., Van De Veerdonk, F. L., Brown, A. J., & Netea, M. G. (2012). *Candida albicans* morphogenesis and host defence: Discriminating invasion from colonization. *Nature Reviews Microbiology, 10*(2), 112–122.

Hanifah, R., Wahid, M., & Yasmon, A. (2024). First report of amphotericin B resistant *Candida haemulonii* isolated from the ICU of a referral hospital in Indonesia. *Iranian Journal of Microbiology, 16*(2), 280.

Harrison, B. D., Hashemi, J., Bibi, M., Pulver, R., Bavli, D., Nahmias, Y., Wellington, M., Sapiro, G., & Berman, J. (2014). A tetraploid intermediate precedes aneuploid formation in yeasts exposed to fluconazole. *PLoS Biology, 12*(3), e1001815. https://doi.org/10.1371/journal.pbio.1001815

Homei, A., & Worboys, M. (2013). Candida. In *Fungal disease in Britain and the United States 1850–2000: Mycoses and modernity*. Palgrave Macmillan. https://www.ncbi.nlm.nih.gov/books/NBK169215/

Horton, M., Chad, J., Kernien, J., Patel, T., Lam, B., Cheong, J. Z. A., Meudt, J., Shanmuganayagam, D., Kalan, L., & Nett, J. (2020). *Candida auris* forms high-burden biofilms in skin niche conditions and on porcine skin. *mSphere, 5*(1). https://doi.org/10.1128/msphere.00910-19

Hoyer, L. L., & Cota, E. (2016). *Candida albicans* agglutinin-like sequence (Als) family vignettes: A review of Als protein structure and function. *Frontiers in Microbiology, 7*, 280.

Hube, B. (2000). Extracellular proteinases of human pathogenic fungi. *Contributions to Microbiology, 5*, 126–137.

Hube, B., Stehr, F., Bossenz, M., Mazur, A., Kretschmar, M., & Schäfer, W. (2000). Secreted lipases of *Candida albicans*: Cloning, characterisation and expression analysis of a new gene family with at least ten members. *Archives of Microbiology, 174*, 362–374.

Ibrahim, A. S., Mirbod, F., Filler, S. G., Banno, Y., Cole, G. T., Kitajima, Y., Edwards, J. E., Jr., Nozawa, Y., & Ghannoum, M. A. (1995). Evidence implicating phospholipase as a virulence factor of *Candida albicans*. *Infection and Immunity, 63*(5), 1993–1998.

Inácio, M. M., Moreira, A. L., Cruz-Leite, V. R., Mattos, K., Silva, L. O., Venturini, J., Ruiz, O. H., Ribeiro-Dias, F., Weber, S. S., Soares, C. M., & Borges, C. L. (2023). Fungal vaccine development: State of the art and perspectives using immunoinformatics. *Journal of Fungi, 9*(6). https://doi.org/10.3390/jof9060633

Iranzo, M., Aguado, C., Pallotti, C., Cañizares, J. V., & Mormeneo, S. (2002). The use of trypsin to solubilize wall proteins from *Candida albicans* led to the identification of chitinase 2 as an enzyme covalently linked to the yeast wall structure. *Research in Microbiology, 153*(4), 227–232.

Jeffery-Smith, A., Taori, S. K., Schelenz, S., Jeffery, K., Johnson, E. M., Borman, A., Candida auris Incident Management Team, Manuel, R., & Brown, C. S. (2018). *Candida auris*: A review of the literature. *Clinical Microbiology Reviews, 31*(1), 10–1128.

Jessup, C., Ghannoum, M., & Ryder, N. (2000). An evaluation of the in vitro activity of terbinafine. *Medical Mycology, 38*(2), 155–159.

Jørgensen, L. N., & Heick, T. M. (2021). Azole use in agriculture, horticulture, and wood preservation—Is it indispensable? *Frontiers in Cellular and Infection Microbiology, 11*, 730297. https://doi.org/10.3389/fcimb.2021.730297

Juayang, A., Lim, J., de los Reyes, Z., Tuante, M., Batiles, Z., Guino-o, J., Villanueva, F., & de los Reyes, G. (2019). Antifungal resistance of *Candida* species in Bacolod City, Philippines. *Journal of Infectious Diseases and Epidemiology, 5*(76), 10–23937.

Katsiari, M., Mavroidi, A., Kesesidis, N., Palla, E., Zourla, K., Ntorlis, K., Konstantinidis, K., Laskou, M., Strigklis, K., Sakkalis, A., Nikolaou, C., Platsouka, E. D., Karakasiliotis, I., Vrioni, G., & Tsakris, A. (2023). Emergence of clonally-related South Asian clade I clinical isolates of *Candida auris* in a Greek COVID-19 intensive care unit. *Journal of Fungi, 9*(2). https://doi.org/10.3390/jof9020243

Kauffman, C. & Vazquez, J. (2024). *Candida* infections of the central nervous system. *UpToDate*. Accessed October 07, 2024 from https://www.uptodate.com/contents/candida-infections-of-the-central-nervous-system

Kaur, G., Chawla, S., Kumar, P., & Singh, R. (2023). Advancing vaccine strategies against *Candida* infections: Exploring new frontiers. *Vaccine, 11*(11). https://doi.org/10.3390/vaccines11111658

Keighley, C., Garnham, K., Harch, S. A. J., Robertson, M., Chaw, K., Teng, J. C., & Chen, S. C.-A. (2021). *Candida auris*: Diagnostic challenges and emerging opportunities for the clinical microbiology laboratory. *Current Fungal Infection Reports, 15*(3), 116–126. https://doi.org/10.1007/s12281-021-00420-y

Kim, T.-H., Kweon, O. J., Kim, H. R., & Lee, M.-K. (2016). Identification of uncommon *Candida* species using commercial identification systems. *Journal of Microbiology and Biotechnology, 26*(12), 2206–2213.

Knoll, M. A., Steixner, S., & Lass-Flörl, C. (2023). How to use direct microscopy for diagnosing fungal infections. *Clinical Microbiology and Infection, 29*(8), 1031–1038.

Kordalewska, M., Zhao, Y., Lockhart, S. R., Chowdhary, A., Berrio, I., & Perlin, D. S. (2017). Rapid and accurate molecular identification of the emerging multidrug-resistant pathogen *Candida auris*. *Journal of Clinical Microbiology, 55*(8), 2445–2452.

Kriegl, L., Egger, M., Boyer, J., Hoenigl, M., & Krause, R. (2025). New treatment options for critically important WHO fungal priority pathogens. *Clinical Microbiology and Infection, 31*(6), 922–930.

Ku, T. S., Walraven, C. J., & Lee, S. A. (2018). *Candida auris*: Disinfectants and implications for infection control. *Frontiers in Microbiology, 9*, 726.

Lassak, T., Schneider, E., Bussmann, M., Kurtz, D., Manak, J. R., Srikantha, T., Soll, D. R., & Ernst, J. F. (2011). Target specificity of the *Candida albicans* Efg1 regulator. *Molecular Microbiology, 82*(3), 602–618.

Lass-Flörl, C., Kanj, S. S., Govender, N. P., Thompson, G. R., Ostrosky-Zeichner, L., & Govrins, M. A. (2024). Invasive candidiasis. *Nature Reviews Disease Primers, 10*(1), 1–18. https://doi.org/10.1038/s41572-024-00503-3

Leach, M. D., Farrer, R. A., Tan, K., Miao, Z., Walker, L. A., Cuomo, C. A., Wheeler, R. T., Brown, A. J. P., Wong, K. H., & Cowen, L. E. (2016). Hsf1 and Hsp90 orchestrate temperature-dependent global transcriptional remodelling and chromatin architecture in *Candida albicans*. *Nature Communications, 7*(1), 11704. https://doi.org/10.1038/ncomms11704

Lee, W. G., Shin, J. H., Uh, Y., Kang, M. G., Kim, S. H., Park, K. H., & Jang, H.-C. (2011). First three reported cases of nosocomial Fungemia caused by *Candida auris*. *Journal of Clinical Microbiology, 49*(9), 3139–3142. https://doi.org/10.1128/jcm.00319-11

Lemke, A., Kiderlen, A. F., & Kayser, O. (2005). Amphotericin B. *Applied Microbiology and Biotechnology, 68*(2), 151–162. https://doi.org/10.1007/s00253-005-1955-9

Linggawaty, L., Marcellana, G., Zabat, G., & Villalobos, R. (2021). A case report on the first documented case of *Candida auris* fungemia in a tertiary hospital in Philippines. *International Journal of Antimicrobial Agents, 58*, 21003678.

Lionakis, M. S., Lim, J. K., Lee, C.-C. R., & Murphy, P. M. (2010). Organ-specific innate immune responses in a mouse model of invasive candidiasis. *Journal of Innate Immunity, 3*(2), 180–199. https://doi.org/10.1159/000321157

Liu, L., Huang, Y., Liao, Y., & Shu, S. (2023). Autosomal dominant chronic mucocutaneous candidiasis with STAT1 mutation can be associated with chronic active hepatitis: A case report. *Frontiers in Pediatrics, 10*. https://www.frontiersin.org/journals/pediatrics/articles/10.3389/fped.2022.990729

Liu, F., Hu, Z.-D., Zhao, X.-M., Zhao, W.-N., Feng, Z.-X., Yurkov, A., Alwasel, S., Boekhout, T., Bensch, K., & Hui, F.-L. (2024). Phylogenomic analysis of the *Candida auris-Candida haemuli* clade and related taxa in the *Metschnikowiaceae*, and proposal of thirteen new genera, fifty-five new combinations and nine new species. *Persoonia-Molecular Phylogeny and Evolution of Fungi, 52*(1), 22–43.

Lo, H. J., Köhler, J. R., DiDomenico, B., Loebenberg, D., Cacciapuoti, A., & Fink, G. R. (1997). Nonfilamentous *C. albicans* mutants are avirulent. *Cell, 90*(5), 939–949.

Lockhart, S. R., Etienne, K. A., Vallabhaneni, S., Farooqi, J., Chowdhary, A., Govender, N. P., Colombo, A. L., Calvo, B., Cuomo, C. A., Desjardins, C. A., Berkow, E. L., Castanheira, M., Magobo, R. E., Jabeen, K., Asghar, R. J., Meis, J. F., Jackson, B., Chiller, T., & Litvintseva, A. P. (2017). Simultaneous emergence of multidrug-resistant *Candida auris* on 3 continents confirmed by whole-genome sequencing and epidemiological analyses. *Clinical Infectious Diseases, 64*(2), 134–140. https://doi.org/10.1093/cid/ciw691

Lorenz, M. C., Bender, J. A., & Fink, G. R. (2004). Transcriptional response of *Candida albicans* upon internalization by macrophages. *Eukaryotic Cell, 3*(5), 1076–1087. https://doi.org/10.1128/ec.3.5.1076-1087.2004

Louie, A., Liu, W., Miller, D. A., Sucke, A. C., Liu, Q.-F., Drusano, G. L., Mayers, M., & Miller, M. H. (1999). Efficacies of high-dose fluconazole plus amphotericin B and high-dose fluconazole plus 5-fluorocytosine versus amphotericin B, fluconazole, and 5-fluorocytosine monotherapies in treatment of experimental endocarditis, endophthalmitis, and pyelonephritis due to *Candida albicans*. *Antimicrobial Agents and Chemotherapy, 43*(12), 2831–2840. https://doi.org/10.1128/aac.43.12.2831

Louie, A., Stein, D. S., Zack, J. Z., Liu, W., Conde, H., Fregeau, C., VanScoy, B. D., & Drusano, G. L. (2011). Dose range evaluation of Mycograb C28Y variant, a human recombinant antibody fragment to heat shock protein 90, in combination with amphotericin B-Desoxycholate for treatment of murine systemic candidiasis. *Antimicrobial Agents and Chemotherapy, 55*(7), 3295–3304. https://doi.org/10.1128/aac.01324-10

Lu, Y., Su, C., Mao, X., Raniga, P. P., Liu, H., & Chen, J. (2008). Efg1-mediated recruitment of NuA4 to promoters is required for hypha-specific Swi/Snf binding and activation in *Candida albicans*. *Molecular Biology of the Cell, 19*(10), 4260–4272.

Luo, S., Dasari, P., Reiher, N., Hartmann, A., Jacksch, S., Wende, E., Barz, D., Niemiec, M. J., Jacobsen, I., Beyersdorf, N., Hünig, T., Klos, A., Skerka, C., & Zipfel, P. F. (2018). The secreted *Candida albicans* protein Pra1 disrupts host defense by broadly targeting and blocking complement C3 and C3 activation fragments. *Molecular Immunology, 93*, 266–277. https://doi.org/10.1016/j.molimm.2017.07.010

Luo, S., Poltermann, S., Kunert, A., Rupp, S., & Zipfel, P. F. (2009). Immune evasion of the human pathogenic yeast *Candida albicans*: Pra1 is a factor H, FHL-1 and plasminogen binding surface protein. *Molecular Immunology, 47*(2), 541–550. https://doi.org/10.1016/j.molimm.2009.07.017

Lyman, M., Forsberg, K., Sexton, D. J., Chow, N. A., Lockhart, S. R., Jackson, B. R., & Chiller, T. (2023). Worsening spread of *Candida auris* in the United States, 2019 to 2021. *Annals of Internal Medicine, 176*(4), 489–495. https://doi.org/10.7326/M22-3469

Magobo, R., Corcoran, C., Seetharam, S., & Govender, N. (2014). *Candida auris*–associated Candidemia, South Africa. *Emerging Infectious Disease Journal, 20*(7), 1250. https://doi.org/10.3201/eid2007.131765

Marathe, A., Zhu, Y., Chaturvedi, V., & Chaturvedi, S. (2022). Utility of CHROMagar™ Candida plus for presumptive identification of *Candida auris* from surveillance samples. *Mycopathologia, 187*(5), 527–534. https://doi.org/10.1007/s11046-022-00656-3

Matthews, R. C., Rigg, G., Hodgetts, S., Carter, T., Chapman, C., Gregory, C., Illidge, C., & Burnie, J. (2003). Preclinical assessment of the efficacy of Mycograb, a human recombinant antibody against fungal HSP90. *Antimicrobial Agents and Chemotherapy, 47*(7), 2208–2216. https://doi.org/10.1128/aac.47.7.2208-2216.2003

Mogavero, S., Höfs, S., Lauer, A. N., Müller, R., Brunke, S., Allert, S., Gerwien, F., Groth, S., Dolk, E., Wilson, D., Gutsmann, T., & Hube, B. (2022). Candidalysin is the hemolytic factor of *Candida albicans*. *Toxins, 14*(12). https://doi.org/10.3390/toxins14120874

Mohd Tap, R., Lim, T. C., Kamarudin, N. A., Ginsapu, S. J., Abd Razak, M. F., Ahmad, N., & Amran, F. (2018). A fatal case of *Candida auris* and *Candida tropicalis* Candidemia in neutropenic patient. *Mycopathologia, 183*(3), 559–564. https://doi.org/10.1007/s11046-018-0244-y

Moron, L. S., & Cabrera, E. C. (2018). Detection of azole resistance and ERG11 point mutations in *Candida albicans* isolates from tertiary hospitals in The Philippines. *Current Research in Environmental & Applied Mycology, 8*(3), 298–305.

Moron, L. S., & Cabrera, E. C. (2019). ABC genotyping and putative virulence factors of *Candida albicans* clinical isolates. *Malaysian Journal of Microbiology, 15*(5).

Moyes, D. L., Wilson, D., Richardson, J. P., Mogavero, S., Tang, S. X., Wernecke, J., Höfs, S., Gratacap, R. L., Robbins, J., Runglall, M., Murciano, C., Blagojevic, M., Thavaraj, S., Förster, T. M., Hebecker, B., Kasper, L., Vizcay, G., Iancu, S. I., Kichik, N., et al. (2016). Candidalysin is a fungal peptide toxin critical for mucosal infection. *Nature, 532*(7597), 64–68. https://doi.org/10.1038/nature17625

Mühlschlegel, F. A., & Fonzi, W. A. (1997). PHR2 of *Candida Albicans* encodes a functional homolog of the pH-regulated gene PHR1 with an inverted pattern of pH-dependent expression. *Molecular and Cellular Biology, 17*(10), 5960–5967. https://doi.org/10.1128/MCB.17.10.5960

Mulhern, S. M., Logue, M. E., & Butler, G. (2006). *Candida albicans* transcription factor Ace2 regulates metabolism and is required for filamentation in hypoxic conditions. *Eukaryotic Cell, 5*(12), 2001–2013.

Munro, C. A., Bates, S., Buurman, E. T., Hughes, H. B., MacCallum, D. M., Bertram, G., et al. (2005). Mnt1p and Mnt2p of *Candida albicans* are partially redundant α-1, 2-mannosyltransferases that participate in O-linked mannosylation and are required for adhesion and virulence. *Journal of Biological Chemistry, 280*(2), 1051–1060.

Naglik, J. R., Rodgers, C. A., Shirlaw, P. J., Dobbie, J. L., Fernandes-Naglik, L. L., Greenspan, D., Agabian, N., & Challacombe, S. J. (2003). Differential expression of *Candida albicans* secreted aspartyl proteinase and phospholipase B genes in humans correlates with active oral and vaginal infections. *The Journal of Infectious Diseases, 188*(3), 469–479.

Narayanan, A., Selvakumar, P., Siddharthan, R., & Sanyal, K. (2022). ClaID: A rapid method of clade-level identification of the multidrug resistant human fungal pathogen *Candida auris*. *Microbiology Spectrum, 10*(2), e00634–e00622.

Nemec, P., Betina, V., Bacíkková, D., Baráth, Z., Betinova, M., Rusinko, M., & Takácova, Z. (1964). Antibacterial, antifungal, antiprotozoal and antinematodal activity of microscopic fungi isolated from soils in Indonesia. *The Journal of Antibiotics, Series A, 17*(4), 140–143.

Netea, M. G., Domínguez-Andrés, J., Barreiro, L. B., Chavakis, T., Divangahi, M., Fuchs, E., Joosten, L. A. B., van der Meer, J. W. M., Mhlanga, M. M., Mulder, W. J. M., Riksen, N. P., Schlitzer, A., Schultze, J. L., Stabell Benn, C., Sun, J. C., Xavier, R. J., & Latz, E. (2020). Defining trained immunity and its role in health and disease. *Nature Reviews Immunology, 20*(6), 375–388. https://doi.org/10.1038/s41577-020-0285-6

Nett, J. E., & Andes, D. R. (2020). Contributions of the biofilm matrix to *Candida* pathogenesis. *Journal of Fungi, 6*(1). https://doi.org/10.3390/jof6010021

Ng, K. (1998). *Candida* biotypes isolated from clinical specimens in Malaysia. *Mycopathologia, 144*(3), 135–140.

Ngamchokwathana, C., Chongtrakool, P., Waesamaae, A., & Chayakulkeeree, M. (2021). Risk factors and outcomes of non-albicans *Candida* bloodstream infection in patients with Candidemia at Siriraj Hospital—Thailand's Largest National Tertiary Referral Hospital. *Journal of Fungi, 7*(4). https://doi.org/10.3390/jof7040269

Ngeow, Y., & Soo-Hoo, T. (1989). Incidence and distribution of vaginal yeasts in Malaysian women: Häufigkeit und Verteilung vaginaler Hefen bei malaysischen Frauen. *Mycoses, 32*(11), 563–567.

Ngo, T. M. C., Santona, A., Fiamma, M., Nu, P. A. T., Do, T. B. T., Cappuccinelli, P., & Paglietti, B. (2023). Azole non-susceptible *C. tropicalis* and polyclonal spread of *C. albicans* in Central Vietnam hospitals. *The Journal of Infection in Developing Countries, 17*(04), 550–558.

Nobile, C. J., Fox, E. P., Nett, J. E., Sorrells, T. R., Mitrovich, Q. M., Hernday, A. D., Tuch, B. B., Andes, D. R., & Johnson, A. D. (2012). A recently evolved transcriptional network controls biofilm development in *Candida albicans*. *Cell, 148*(1), 126–138.

Nobile, C. J., & Johnson, A. D. (2015). *Candida albicans* biofilms and human disease. *Annual Review of Microbiology, 69*(1), 71–92.

Nobile, C. J., Schneider, H. A., Nett, J. E., Sheppard, D. C., Filler, S. G., Andes, D. R., & Mitchell, A. P. (2008). Complementary adhesin function in *C. albicans* biofilm formation. *Current Biology, 18*(14), 1017–1024.

Oh, B. J., Shin, J. H., Kim, M.-N., Sung, H., Lee, K., Joo, M. Y., Shin, M. G., Suh, S. P., & Ryang, D. W. (2011). Biofilm formation and genotyping of *Candida haemulonii*, *Candida pseudohaemulonii*, and a proposed new species (*Candida auris*) isolates from Korea. *Medical Mycology, 49*(1), 98–102.

Oh, Y. B., Lim, H. J., Byun, S. A., Choi, M. J., Choi, H.-J., Shin, M. G., Lee, S. Y., & Shin, J. H. (2023). *Candida vulturna* fungemia in an infant with congenital megacolon from The Philippines. *Annals of Laboratory Medicine, 43*(4), 395–397.

Oliveira, L. V. N., Wang, R., Specht, C. A., & Levitz, S. M. (2021). Vaccines for human fungal diseases: Close but still a long way to go. *npj Vaccines, 6*(1), 33. https://doi.org/10.1038/s41541-021-00294-8

Pachl, J., Svoboda, P., Jacobs, F., Vandewoude, K., van der Hoven, B., Spronk, P., Masterson, G., Malbrain, M., Aoun, M., Garbino, J., Takala, J., Drgona, L., Burnie, J., Matthews, R., & Mycograb Invasive Candidiasis Study Group. (2006). A randomized, blinded, multicenter trial of lipid-associated amphotericin B alone versus in combination with an antibody-based inhibitor of heat shock protein 90 in patients with invasive candidiasis. *Clinical Infectious Diseases, 42*(10), 1404–1413. https://doi.org/10.1086/503428

Pande, K., Chen, C., & Noble, S. M. (2013). Passage through the mammalian gut triggers a phenotypic switch that promotes *Candida albicans* commensalism. *Nature Genetics, 45*(9), 1088–1091.

Pappas, P. G., Kauffman, C. A., Andes, D. R., Clancy, C. J., Marr, K. A., Ostrosky-Zeichner, L., Reboli, A. C., Schuster, M. G., Vazquez, J. A., & Walsh, T. J. (2016). Clinical practice guideline for the management of candidiasis: 2016 update by the Infectious Diseases Society of America. *Clinical Infectious Diseases, 62*(4), e1–e50.

Pappas, P. G., Lionakis, M. S., Arendrup, M. C., Ostrosky-Zeichner, L., & Kullberg, B. J. (2018). Invasive candidiasis. *Nature Reviews Disease Primers, 4*(1), 1–20. https://doi.org/10.1038/nrdp.2018.26

Pfaller, M. A. (2015). Application of culture-independent rapid diagnostic tests in the management of invasive candidiasis and cryptococcosis. *Journal of Fungi, 1*(2), 217–251.

Pharkjaksu, S., Boonmee, N., Mitrpant, C., & Ngamskulrungroj, P. (2021). Immunopathogenesis of emerging *Candida auris* and *Candida haemulonii* strains. *Journal of Fungi, 7*(9). https://doi.org/10.3390/jof7090725

Priya, A., & Pandian, S. K. (2020). Piperine impedes biofilm formation and hyphal morphogenesis of *Candida albicans*. *Frontiers in Microbiology, 11*, 756.

Quintin, J., Saeed, S., Martens, J. H. A., Giamarellos-Bourboulis, E. J., Ifrim, D. C., Logie, C., Jacobs, L., Jansen, T., Kullberg, B.-J., Wijmenga, C., Joosten, L. A. B., Xavier, R. J., van der Meer, J. W. M., Stunnenberg, H. G., & Netea, M. G. (2012). *Candida albicans* infection affords protection against reinfection via functional reprogramming of monocytes. *Cell Host & Microbe, 12*(2), 223–232. https://doi.org/10.1016/j.chom.2012.06.006

Rayens, E., Rayens, M. K., & Norris, K. A. (2022). Demographic and socioeconomic factors associated with fungal infection risk, United States, 2019. *Emerging Infectious Diseases, 28*(10), 1955.

Reichart, P., Khongkhunthian, P., & Bendick, C. (2003). Oral manifestations in HIV-infected individuals from Thailand and Cambodia. *Medical Microbiology and Immunology, 192*, 157–160.

Rhodes, J., Abdolrasouli, A., Farrer, R. A., Cuomo, C. A., Aanensen, D. M., Armstrong-James, D., Fisher, M. C., & Schelenz, S. (2017). Rapid genome sequencing for outbreak analysis of the emerging human fungal pathogen *Candida auris*. *bioRxiv, 201343*. https://doi.org/10.1101/201343

Richie, D. L., Ghannoum, M. A., Isham, N., Thompson, K. V., & Ryder, N. S. (2012). Nonspecific effect of mycograb on amphotericin B MIC. *Antimicrobial Agents and Chemotherapy, 56*(7), 3963–3964. https://doi.org/10.1128/aac.00435-12

Rinawati, W., Kumalawati, J., Bardosono, S., Immanuel, S., Sukartini, N., & Indrasari, N. D. (2022). Invasive candidiasis among high prevalence neurological patients. *The Journal of Infection in Developing Countries, 16*(05), 871–880.

Rivera, A., Lodge, J., & Xue, C. (2022). Harnessing the immune response to fungal pathogens for vaccine development. *Annual Review of Microbiology, 76*, 703–726. https://doi.org/10.1146/annurev-micro-041020-111511

Rodríguez-Peña, J. M., Cid, V. J., Arroyo, J., & Nombela, C. (2000). A novel family of cell wall-related proteins regulated differently during the yeast life cycle. *Molecular and Cellular Biology, 20*(9), 3245–3255.

Sabino, R., Veríssimo, C., Pereira, Á. A., & Antunes, F. (2020). *Candida auris*, an agent of hospital-associated outbreaks: Which challenging issues do we need to have in mind? *Microorganisms, 8*(2), 181.

Sahu, S. R., Bose, S., Singh, M., Kumari, P., Dutta, A., Utkalaja, B. G., Patel, S. K., & Acharya, N. (2022). Vaccines against candidiasis: Status, challenges and emerging opportunity. *Frontiers in Cellular and Infection Microbiology, 12*. https://www.frontiersin.org/journals/cellular-and-infection-microbiology/articles/10.3389/fcimb.2022.1002406

Saiprom, N., Wongsuk, T., Oonanant, W., Sukphopetch, P., Chantratita, N., & Boonsilp, S. (2023). Characterization of virulence factors in *Candida* species causing Candidemia in a tertiary care hospital in Bangkok, Thailand. *Journal of Fungi, 9*(3). https://doi.org/10.3390/jof9030353

Salmanton-García, J., Au, W.-Y., Hoenigl, M., Chai, L. Y. A., Badali, H., Basher, A., Brockhoff, R. A., Chen, S. C.-A., Chindamporn, A., Chowdhary, A., Heath, C. H., Jabeen, K., Lee, J., Matar, M., Taj-Aldeen, S. J., Tan, B. H., Uno, K., Wahyuningsih, R., Zhu, L., et al. (2023). The current state of laboratory mycology in Asia/Pacific: A survey from the European Confederation of Medical Mycology (ECMM) and International Society for Human and Animal Mycology (ISHAM). *International Journal of Antimicrobial Agents, 61*(3), 106718. https://doi.org/10.1016/j.ijantimicag.2023.106718

Saporito-Irwin, S. M., Birse, C. E., Sypherd, P. S., & Fonzi, W. A. (1995). PHR1, a pH-regulated gene of *Candida albicans*, is required for morphogenesis. *Molecular and Cellular Biology, 15*(2), 601–613. https://doi.org/10.1128/MCB.15.2.601

Sarma, S., Kumar, N., Sharma, S., Govil, D., Ali, T., Mehta, Y., & Rattan, A. (2013). Candidemia caused by amphotericin B and fluconazole resistant *Candida auris*. *Indian Journal of Medical Microbiology, 31*(1), 90–91. https://doi.org/10.4103/0255-0857.108746

Sarode, G., Ghone, U., Sarode, S., Sengupta, N., & Anand, R. (2023). Chapter 6—Phospholipases in candidiasis. In S. Chakraborti (Ed.), *Phospholipases in physiology and pathology* (pp. 93–100). Academic. https://doi.org/10.1016/B978-0-443-21800-2.00024-5

Satoh, K., Makimura, K., Hasumi, Y., Nishiyama, Y., Uchida, K., & Yamaguchi, H. (2009). *Candida auris* sp. Nov., a novel ascomycetous yeast isolated from the external ear canal of an inpatient in a Japanese hospital. *Microbiology and Immunology, 53*(1), 41–44. https://doi.org/10.1111/j.1348-0421.2008.00083.x

Schelenz, S., Hagen, F., Rhodes, J. L., Abdolrasouli, A., Chowdhary, A., Hall, A., Ryan, L., Shackleton, J., Trimlett, R., Meis, J. F., Armstrong-James, D., & Fisher, M. C. (2016). First hospital outbreak of the globally emerging *Candida auris* in a European hospital. *Antimicrobial Resistance & Infection Control, 5*(1), 35. https://doi.org/10.1186/s13756-016-0132-5

Schmidt, C. S., White, C. J., Ibrahim, A. S., Filler, S. G., Fu, Y., Yeaman, M. R., Edwards, J. E., Jr., & Hennessey, J. P., Jr. (2012). NDV-3, a recombinant alum-adjuvanted vaccine for *Candida* and *Staphylococcus aureus*, is safe and immunogenic in healthy adults. *Vaccine, 30*(52), 7594–7600.

Schmidt-Westhausen, A., Bendick, C., Reichart, P., & Samaranayake, L. (2004). Oral candidosis and associated *Candida* species in HIV-infected Cambodians exposed to antimycotics. *Mycoses, 47*(9–10), 435–441.

Sellam, A., & Whiteway, M. (2016). Recent advances on *Candida albicans* biology and virulence. *F1000Research, 5*(2582). https://doi.org/10.12688/f1000research.9617.1

Selmecki, A., Forche, A., & Berman, J. (2006). Aneuploidy and isochromosome formation in drug-resistant *Candida albicans*. *Science, 313*(5785), 367–370. https://doi.org/10.1126/science.1128242

Sheehan, D. J., Hitchcock, C. A., & Sibley, C. M. (1999). Current and emerging azole antifungal agents. *Clinical Microbiology Reviews, 12*(1), 40–79. https://doi.org/10.1128/cmr.12.1.40

Sheridan, R., & Ratledge, C. (1996). Changes in cell morphology and carnitine acetyltransferase activity in *Candida albicans* following growth on lipids and serum and after in vivo incubation in mice. *Microbiology, 142*(11), 3171–3180.

Shibasaki, S., Aoki, W., Nomura, T., Miyoshi, A., Tafuku, S., Sewaki, T., & Ueda, M. (2013). An oral vaccine against candidiasis generated by a yeast molecular display system. *Pathogens and Disease, 69*(3), 262–268. https://doi.org/10.1111/2049-632X.12068

Shibasaki, S., Karasaki, M., Tafuku, S., Aoki, W., Sewaki, T., & Ueda, M. (2014). Oral immunization against candidiasis using *lactobacillus casei* displaying enolase 1 from *Candida albicans*. *Scientia Pharmaceutica, 82*(3), 697–708. https://doi.org/10.3797/scipharm.1404-07

Shukla, M., Chandley, P., & Rohatgi, S. (2021). The role of B-cells and antibodies against *Candida* vaccine antigens in invasive candidiasis. *Vaccine, 9*(10). https://doi.org/10.3390/vaccines9101159

Sinh, C. T., Loi, C. B., Minh, N. T. N., Lam, N. N., Quang, D. X., Quyet, D., Anh, D. N., Hien, T. T. T., Su, H. X., & Tran-Anh, L. (2021). Species distribution and antifungal susceptibility pattern of *Candida* recovered from intensive care unit patients, Vietnam national hospital of burn (2017–2019). *Mycopathologia, 186*(4), 543–551.

Siti Umairah, B., Tan, S., Siti Nurwani, B., Mohd Syazwan, C., Ramli, R. B., & Tzar Mohd, N. (2012). Antifungal susceptibility patterns among *Candida* species isolated from blood at Universiti Kebangsaan Malaysia medical Centre. *Sains Malaysiana, 41*(8), 961–967.

Sivapalan, A., Metussin, R., Harndan, F., & Zain, R. M. (1998). Fungi associated with postharvest fruit rots of *Durio graveolens* and *D. kutejensis* in Brunei Darussalam. *Australasian Plant Pathology, 27*(4), 274–277.

Sobel, J. & Mitchell, C. (2024). *Candida* vulvovaginitis: Clinical manifestations and diagnosis. *UpToDate*. Retrieved October 7, 2024, from https://www.uptodate.com/contents/candida-vulvovaginitis-clinical-manifestations-and-diagnosis

Soll, D. R. (2009). Why does *Candida albicans* switch? *FEMS Yeast Research, 9*(7), 973–989. https://doi.org/10.1111/j.1567-1364.2009.00562.x

Sorongon, E. D., Mesola, V., Mendoza, M., Pena, A., & Tupasi, T. (1995). A multicenter case-control study on *Candida* infections. *The Philippine Journal of Microbiology and Infectious Diseases, 24*(2), 43–46.

Spruijtenburg, B., Badali, H., Abastabar, M., Mirhendi, H., Khodavaisy, S., Sharifisooraki, J., Taghizadeh Armaki, M., de Groot, T., & Meis, J. F. (2022). Confirmation of fifth *Candida auris* clade by whole genome sequencing. *Emerging Microbes & Infections, 11*(1), 2405–2411. https://doi.org/10.1080/22221751.2022.2125349

Staib, F. (1966). Serum-proteins as nitrogen source for yeast-like fungi. *Sabouraudia: Journal of Medical and Veterinary Mycology, 4*(3), 187–193.

Stoldt, V. R., Sonneborn, A., Leuker, C. E., & Ernst, J. F. (1997). Efg1p, an essential regulator of morphogenesis of the human pathogen *Candida albicans*, is a member of a conserved class of bHLH proteins regulating morphogenetic processes in fungi. *The EMBO Journal., 16*, 1982.

Suhartono, S., Mahdani, W., Masthura, A., & Rusmana, I. (2020). *Candida* species distribution of clinical specimens in Banda Aceh, Indonesia. *Biosaintifika: Journal of Biology & Biology Education, 12*(2), 262–267.

Suphavilai, C., Ko, K. K. K., Lim, K. M., Tan, M. G., Boonsimma, P., Chu, J. J. K., Goh, S. S., Rajandran, P., Lee, L. C., Tan, K. Y., Shaik Ismail, B. B., Aung, M. K., Yang, Y., Sim, J. X. Y., Venkatachalam, I., Cherng, B. P. Z., Spruijtenburg, B., Chan, K. S., Oon, L. L. E., et al. (2024). Detection and characterisation of a sixth *Candida auris* clade in Singapore: A genomic and phenotypic study. *The Lancet Microbe, 5*(9). https://doi.org/10.1016/S2666-5247(24)00101-0

Taei, M., Chadeganipour, M., & Mohammadi, R. (2019). An alarming rise of non-*albicans Candida* species and uncommon yeasts in the clinical samples; a combination of various molecular techniques for identification of etiologic agents. *BMC Research Notes, 12*(1), 779. https://doi.org/10.1186/s13104-019-4811-1

Takashima, M., & Sugita, T. (2022). Taxonomy of pathogenic yeasts *Candida*, *Cryptococcus*, *Malassezia*, and *Trichosporon*. *Medical Mycology Journal, 63*(4), 119–132. https://doi.org/10.3314/mmj.22.004

Talapko, J., Juzbašić, M., Matijević, T., Pustijanac, E., Bekić, S., Kotris, I., & Škrlec, I. (2021). *Candida albicans*—The virulence factors and clinical manifestations of infection. *Journal of Fungi, 7*(2). https://doi.org/10.3390/jof7020079

Tan, B. H., Chakrabarti, A., Li, R. Y., Patel, A. K., Watcharananan, S. P., Liu, Z., Chindamporn, A., Tan, A. L., Sun, P.-L., Wu, U.-I., Chen, Y.-C., Xu, Y.-C., Wang, H., Sun, Z.-Y., Wang, L.-L., Lu, J., Yang, Q., Zhang, Q.-Q., Shao, H.-F., et al. (2015). Incidence and species distribution of candidaemia in Asia: A laboratory-based surveillance study. *Clinical Microbiology and Infection, 21*(10), 946–953. https://doi.org/10.1016/j.cmi.2015.06.010

Tan, B. H., Chakrabarti, A., Patel, A., Chua, M. M. M., Sun, P.-L., Liu, Z., Rotjanapan, P., Li, R., Wahyuningsih, R., Chayakulkeeree, M., & Chen, Y.-C. (2020). Clinicians' challenges in managing patients with invasive fungal diseases in seven Asian countries: An Asia Fungal Working Group (AFWG) Survey. *International Journal of Infectious Diseases, 95*, 471–480. https://doi.org/10.1016/j.ijid.2020.01.007

Tan, T. Y., Hsu, L. Y., Alejandria, M. M., Chaiwarith, R., Chinniah, T., Chayakulkeeree, M., Choudhury, S., Chen, Y. H., Shin, J. H., & Kiratisin, P. (2016). Antifungal susceptibility of invasive *Candida* bloodstream isolates from the Asia-Pacific region. *Sabouraudia, 54*(5), 471–477.

Tan, R., Lim, E., & Ishak, B. (1977). *Torulopsis glabrata*—Urinary tract infections in diabetic patients in Singapore. *Australian and New Zealand Journal of Medicine, 7*(1), 56–59.

Tan, R. J., Lim, E.-W., & Teo, H.-K. (1977). *Torulopsis glabrata* infections in Singapore: A 4-year study. *Epidemiology and Infection, 78*(2), 129–134.

Tan, Y. E., & Tan, A. L. (2018). Arrival of *Candida auris* fungus in Singapore: Report of the first 3 cases. *Annals of the Academy of Medicine, Singapore, 47*(7), 260–262.

Tan, Y. E., Teo, J. Q.-M., Rahman, N. B. A., Ng, O. T., Kalisvar, M., Tan, A. L., Koh, T. H., & Ong, R. T. H. (2019). *Candida auris* in Singapore: Genomic epidemiology, antifungal drug resistance, and identification using the updated 8.01 VITEK® 2 system. *International Journal of Antimicrobial Agents, 54*(6), 709–715.

Tao, L., Du, H., Guan, G., Dai, Y., Nobile, C. J., Liang, W., Cao, C., Zhang, Q., Zhong, J., & Huang, G. (2014). Discovery of a "white-gray-opaque" tristable phenotypic switching system in *Candida albicans*: Roles of non-genetic diversity in host adaptation. *PLoS Biology, 12*(4), e1001830. https://doi.org/10.1371/journal.pbio.1001830

Teanpaisan, R., & Nittayananta, W. (1998). Prevalence of *Candida* species in AIDS patients and HIV-free subjects in Thailand. *Journal of Oral Pathology & Medicine, 27*(1), 4–7.

Tebung, W. A., Choudhury, B. I., Tebbji, F., Morschhäuser, J., & Whiteway, M. (2016). Rewiring of the Ppr1 zinc cluster transcription factor from purine catabolism to pyrimidine biogenesis in the Saccharomycetaceae. *Current Biology, 26*(13), 1677–1687. https://doi.org/10.1016/j.cub.2016.04.064

Thunyaharn, S., Santimaleeworagun, W., Khoprasert, C., Kesakomol, P., Theeraapisakkun, M., & Visawapoka, U. (2021). Activity of triazoles and echinocandins against *Candida* bloodstream isolates at Phramongkutklao hospital, Thailand. *Journal of Southeast Asian Medical Research, 5*(2), 84–90.

Tripathi, A., Liverani, E., Tsygankov, A. Y., & Puri, S. (2020). Iron alters the cell wall composition and intracellular lactate to affect *Candida albicans* susceptibility to antifungals and host immune response. *Journal of Biological Chemistry, 295*(29), 10032–10044. https://doi.org/10.1074/jbc.RA120.013413

Tsay, S., Kallen, A., Jackson, B. R., Chiller, T. M., & Vallabhaneni, S. (2018). Approach to the investigation and management of patients with *Candida auris*, an emerging multidrug-resistant yeast. *Clinical Infectious Diseases, 66*(2), 306–311. https://doi.org/10.1093/cid/cix744

Uppuluri, P., Chaturvedi, A. K., Srinivasan, A., Banerjee, M., Ramasubramaniam, A. K., Köhler, J. R., Kadosh, D., & Lopez-Ribot, J. L. (2010). Dispersion as an important step in the *Candida albicans* biofilm developmental cycle. *PLoS Pathogens, 6*(3), e1000828.

Uppuluri, P., Lin, L., Alqarihi, A., Luo, G., Youssef, E. G., Alkhazraji, S., Yount, N. Y., Ibrahim, B. A., Bolaris, M. A., Edwards, J. E., Jr., Swidergall, M., Filler, S. G., Yeaman, M. R., & Ibrahim, A. S. (2018). The Hyr1 protein from the fungus *Candida albicans* is a cross kingdom immunotherapeutic target for *Acinetobacter* bacterial infection. *PLoS Pathogens, 14*(5), e1007056. https://doi.org/10.1371/journal.ppat.1007056

Vallabhaneni, S., Kallen, A., Tsay, S., Chow, N., Welsh, R., Kerins, J., Kemble, S. K., Pacilli, M., Black, S. R., Landon, E., Ridgway, J., Palmore, T. N., Zelzany, A., Adams, E. H., Quinn, M., Chaturvedi, S., Greenko, J., Fernandez, R., Southwick, K., et al. (2017). Investigation of the first seven reported cases of *Candida auris*, a globally emerging invasive, multidrug-resistant fungus—United States, May 2013–August 2016. *American Journal of Transplantation, 17*(1), 296–299. https://doi.org/10.1111/ajt.14121

van de Veerdonk, F. L., Plantinga, T. S., Hoischen, A., Smeekens, S. P., Joosten, L. A. B., Gilissen, C., Arts, P., Rosentul, D. C., Carmichael, A. J., Smits-van der Graaf, C. A. A., Kullberg, B. J., van der Meer Jos W.M., Lilic Desa, Veltman, J. A., & Netea, M. G. (2011). STAT1 mutations in autosomal dominant chronic mucocutaneous candidiasis. *New England Journal of Medicine, 365*(1), 54–61. https://doi.org/10.1056/NEJMoa1100102

Van Pham, L., Quang, A. T., Nu, P. A. T., Duc, T. T., & Thi, H. N. (2008). *Cladophialophora bantiana* and *Candida albicans* mixed infection in cerebral abscess of an HIV-negative patient. *The Journal of Infection in Developing Countries, 2*(03), 245–248.

Vanreppelen, G., Wuyts, J., Van Dijck, P., & Vandecruys, P. (2023). Sources of antifungal drugs. *Journal of Fungi, 9*(2), Article 2. https://doi.org/10.3390/jof9020171

Vazquez, J. (2023). Management of candidemia and invasive candidiasis in adults. *UpToDate*. Retrieved October 7, 2024 from https://www.uptodate.com/contents/management-of-candidemia-and-invasive-candidiasis-in-adults

Vedula, R. S., Cheng, M. P., Ronayne, C. E., Farmakiotis, D., Ho, V. T., Koo, S., Marty, F. M., Lindsley, R. C., & Bold, T. D. (2020). Somatic GATA2 mutations define a subgroup of myeloid malignancy patients at high risk for invasive fungal disease. *Blood Advances, 5*(1), 54–60. https://doi.org/10.1182/bloodadvances.2020002854

Velayuthan, R. D., Samudi, C., Lakhbeer Singh, H. K., Ng, K. P., Shankar, E. M., & Denning, D. W. (2018). Estimation of the burden of serious human fungal infections in Malaysia. *Journal of Fungi, 4*(1), 38.

Vilanova, M., Teixeira, L., Caramalho, Í., Torrado, E., Marques, A., Madureira, P., Ribeiro, A., Ferreira, P., Gama, M., & Demengeot, J. (2004). Protection against systemic candidiasis in mice immunized with secreted aspartic proteinase 2. *Immunology, 111*(3), 334–342. https://doi.org/10.1111/j.1365-2567.2004.01819.x

Vylkova, S., & Lorenz, M. C. (2014). Modulation of phagosomal pH by *Candida albicans* promotes hyphal morphogenesis and requires Stp2p, a regulator of amino acid transport. *PLoS Pathogens, 10*(3), e1003995. https://doi.org/10.1371/journal.ppat.1003995

Wahyuningsih, R., Adawiyah, R., Sjam, R., Prihartono, J., Ayu Tri Wulandari, E., Rozaliyani, A., Ronny, R., Imran, D., Tugiran, M., & Siagian, F. E. (2021). Serious fungal disease incidence and prevalence in Indonesia. *Mycoses, 64*(10), 1203–1212.

Waldorf, A. R., & Polak, A. (1983). Mechanisms of action of 5-fluorocytosine. *Antimicrobial Agents and Chemotherapy, 23*(1), 79–85. https://doi.org/10.1128/AAC.23.1.79

Wang, H., Xu, Y.-C., & Hsueh, P.-R. (2016). Epidemiology of Candidemia and antifungal susceptibility in invasive *Candida* species in the Asia-Pacific region. *Future Microbiology, 11*(11), 1461–1477. https://doi.org/10.2217/fmb-2016-0099

Westman, J., Moran, G., Mogavero, S., Hube, B., & Grinstein, S. (2018). *Candida albicans* hyphal expansion causes phagosomal membrane damage and luminal alkalinization. *MBio, 9*(5). https://doi.org/10.1128/mbio.01226-18

Wheeler, R. T., & Fink, G. R. (2006). A drug-sensitive genetic network masks fungi from the immune system. *PLoS Pathogens, 2*(4), e35. https://doi.org/10.1371/journal.ppat.0020035

Wheeler, R. T., Kombe, D., Agarwala, S. D., & Fink, G. R. (2008). Dynamic, morphotype-specific *Candida albicans* β-Glucan exposure during infection and drug treatment. *PLoS Pathogens, 4*(12), e1000227. https://doi.org/10.1371/journal.ppat.1000227

Wingard, J. R., Kubilis, P., Lee, L., Yee, G., White, M., Louise, W., Bowden, R., Anaissie, E., Hiemenz, J., & Lister, J. (1999). Clinical significance of nephrotoxicity in patients treated with amphotericin B for suspected or proven Aspergillosis. *Clinical Infectious Diseases, 29*(6), 1402–1407. https://doi.org/10.1086/313498

Wulandari, A., Hapsari, R., Sari, D., Puspitasari, I., & Pramukarso, D. T. (2021). Antifungal susceptibility profile of *Candida* spp. causing candidemia in an Indonesian tertiary hospital. *Journal of Clinical Microbiology and Infectious Diseases, 1*(2), 28–32.

Wurner, H. (1966). Studies of the lipase activity of yeasts and yeast like fungi. *Zentralbl Bakteriol Orig, 200*(1), 113–124.

Wüthrich, M., Dobson, H. E., Ledesma, T. C., Okaa, U. J., dos Santos, D. L., Isidoro-Ayza, M., Petrovsky, N., & Klein, B. S. (2021). Combination adjuvants enhance recombinant protein vaccine protection against fungal infection. *MBio, 12*(4). https://doi.org/10.1128/mbio.02018-21

Xie, J., Du, H., Guan, G., Tong, Y., Kourkoumpetis, T. K., Zhang, L., Bai, F., & Huang, G. (2012). N-Acetylglucosamine induces white-to-opaque switching and mating in *Candida tropicalis*, providing new insights into adaptation and fungal sexual evolution. *Eukaryotic Cell, 11*(6), 773–782. https://doi.org/10.1128/ec.00047-12

Xie, O., Streitberg, R., Hughes, C., Stuart, R., & Graham, M. (2020). *Candida duobushaemulonii* sepsis and *Candida auris* co-isolation following hospitalisation in Vietnam. *Pathology, 52*(5), 590–591.

Xin, H., Cartmell, J., Bailey, J. J., Dziadek, S., Bundle, D. R., & Cutler, J. E. (2012). Self-adjuvanting glycopeptide conjugate vaccine against disseminated candidiasis. *PLoS One, 7*(4), e35106. https://doi.org/10.1371/journal.pone.0035106

Xin, H., Dziadek, S., Bundle, D. R., & Cutler, J. E. (2008). Synthetic glycopeptide vaccines combining β-mannan and peptide epitopes induce protection against candidiasis. *Proceedings of the National Academy of Sciences, 105*(36), 13526–13531. https://doi.org/10.1073/pnas.0803195105

Yamamoto, M., Alshahni, M. M., Tamura, T., Satoh, K., Iguchi, S., Kikuchi, K., Mimaki, M., & Makimura, K. (2018). Rapid detection of *Candida auris* based on loop-mediated isothermal amplification (LAMP). *Journal of Clinical Microbiology, 56*(9). https://doi.org/10.1128/jcm.00591-18

Zacharioudakis, I. M., Zervou, F. N., & Mylonakis, E. (2018). T2 magnetic resonance assay: Overview of available data and clinical implications. *Journal of Fungi, 4*(2), 45.

Zarnowski, R., Noll, A., Chevrette, M. G., Sanchez, H., Jones, R., Anhalt, H., et al. (2021). Coordination of fungal biofilm development by extracellular vesicle cargo. *Nature Communications, 12*(1), 6235.

Zhu, Y., Clarke, A., Bates, M., Adams, E., Ostrowsky, B., Quinn, M., Dufort, E., Southwick, K., Erazo, R., Haley, V. B., Bucher, C., Chaturvedi, V., Limberger, R. J., Blog, D., Lutterloh, E., & Chaturvedi, S. (2020). Laboratory analysis of an outbreak of *Candida auris* in New York from 2016 to 2018: Impact and lessons learned. *Journal of Clinical Microbiology, 58*(4), e01503–e01519. https://doi.org/10.1128/jcm.01503-19

Chapter 7
Myco-pharming the Seas: Bioactive Metabolites from Marine-Derived Fungi

Carlo Chris S. Apurillo, Kustiariyah Tarman, Trung Thanh Nguyen, Joel C. Magday Jr., Lloyd Christian J. Llames, and Thomas Edison E. dela Cruz

Abstract Marine-derived fungi represent a promising resource for biotechnological and pharmaceutical applications. The Southeast Asian region, known for its rich marine biodiversity, harbors a diverse assemblage of fungal species that could be exploited to produce bioactive secondary metabolites. This chapter highlights the potential of this fungal group for drug discovery, industrial processes, and sustainable bioproducts as exemplified by reports on species identified from Southeast Asia. The immense diversity of marine fungi from various sources collected along

C. C. S. Apurillo
Department of Science and Technology - Center for Research in Science and Technology (CReST), Philippine Science High School - Eastern Visayas Campus in Leyte, Palo, Leyte, Philippines

Center of Excellence in Fungal Research, Mae Fah Luang University, Chiang Rai, Thailand

K. Tarman
Department of Aquatic Product Technology, Faculty of Fisheries and Marine Sciences, IPB University, Kampus IPB Dramaga, Bogor, Indonesia

Center for Coastal and Marine Resources Studies, International Research Institute for Maritime, Ocean and Fisheries, IPB University, Kampus IPB Baranangsiang, Bogor, Indonesia

T. T. Nguyen
Center for Pharmaceutical Biotechnology, College of Medicine and Pharmacy, Duy Tan University, Danang, Vietnam

J. C. Magday Jr.
Philippine Science High School – Cagayan Valley Campus in Nueva Vizcaya, Bayombong, Nueva Vizcaya, Philippines

L. C. J. Llames · T. E. E. dela Cruz (✉)
Fungal Biodiversity, Ecogenomics and Systematics-Metabolomics (FBeS) Group, Research Center for the Natural and Applied Sciences, University of Santo Tomas, Manila, Philippines

Department of Biological Sciences, College of Science, University of Santo Tomas, Manila, Philippines
e-mail: tedelacruz@ust.edu.ph

J. J. G. Guerrero et al. (eds.), *Mycology in a Changing Planet: Applications and Perspectives from Southeast Asia*, https://doi.org/10.1007/978-981-95-2640-6_7

the coastlines and coastal waters in selected countries in Southeast Asia is also presented. This highlights the immense potential of the region for bioprospecting of fungal natural products.

Keywords Bioactivities · Bioprospecting · Drug discovery · Fungal natural products · Metabolite production

Introduction

Marine fungi were first described as those that grow and reproduce in seawater (Barghoorn & Linder, 1944). Later, Johnson and Sparrow (1961) refined this group and defined it as those that attain reproductive maturity at some stage of their life cycle despite exposure to salinities of 3.0% or higher. They emphasized the ecological resilience of these organisms, which enables them to persist in marine environment under varying conditions, including continuous submersion and periodic tidal inundation. Tubaki and Nishihara (1969) also proposed a related definition and stated that marine fungi exhibit superior growth and reproduction on substrates across a broad salinity range compared to freshwater conditions. However, Sparrow (1937) pointed out that some marine fungi exhibited traits like those of non-marine species. This observation is further supported by Pitt and Hocking (1997) who noted that many of these fungi are not strictly salt-tolerant but rather osmotolerant. Osmotolerant fungi have the capacity to thrive in low-water-activity environments, allowing them to grow not only in elevated concentrations of sugars and inland saline habitats but also in high-salinity conditions (Jin et al., 2005; Gal-Hemed et al., 2011). Given the high adaptability of fungi (Jones & Jennings, 1964; Byrne & Jones, 1975), defining marine fungi solely based on physiological parameters presents a challenge, as many species can successfully colonize both freshwater and saline environments. As a result, the definition of marine fungi was refined, categorizing them into two ecological groups: obligate and facultative marine fungi (Kohlmeyer & Kohlmeyer, 1979). Obligate marine fungi are species that exclusively grow and sporulate in estuarine or marine environments, while facultative marine fungi, although primarily associated with freshwater or terrestrial ecosystems, can also grow and sporulate in marine conditions (Kohlmeyer & Kohlmeyer, 1979). This definition also extends to marine yeasts. However, the ambiguity necessitated a more precise and objective framework for defining marine fungi. Jones (2011) identified obligate marine fungi as those isolated from submerged substrates or sediments in marine environments, including saprobic species that are frequently isolated from intertidal substrates such as grasses and mangrove plants. Mahé et al. (2013) added three classifications to describe fungi from oceanic environment based on their functional roles: (A) strict, endemic, active marine fungi; (B) ubiquist, metabolically active marine fungi; and (C) ubiquist, passive fungi. Strict, endemic, active marine fungi produce fruiting bodies on any marine substrates, while ubiquist, passive fungi are mainly asexual fungi that exhibit limited metabolic activity in the marine environment. By integrating this three-tiered classification with the definition proposed by Kohlmeyer and Kohlmeyer (1979), a more comprehensive framework for understanding fungal diversity in marine ecosystems can be established, emphasizing ecological function and habitat specificity. This definition was further expanded by Pang et al. (2016) to encompass fungi that grow and sporulate in marine environments and form symbiotic relationships with other marine organisms, enabling them to sustain metabolic processes in these habitats and eventually adapt or evolve under marine conditions. Additionally, the term "marine-derived fungi" was also introduced. This group encompasses fungi isolated from marine or marine-associated habitats and substrates, which are subsequently explored in natural product chemistry for the discovery of bioactive compounds (Engel et al., 2002; Jones et al., 2015). However, this term remains inherently ambiguous, as it fails to reflect the ecological or physiological functions of the cells that generate the cultured strain studied in the laboratory. The presence of a fungus growing from a propagule within a substrate does not necessarily imply that the substrate serves as its natural habitat, nor does it confirm a clear ecological or physiological connection to the marine environment.

Marine fungi have been described based on their ability to colonize a wide variety of substrates. Lignicolous species, the largest and most extensively studied group, are commonly found on submerged woody materials, including driftwood and trapped wood (Garzoli et al., 2015), while manglicolous fungi grow on mangroves including mangrove leaves (Abdel-Wahab et al., 2014, 2019; Devadatha et al., 2018; Moron et al., 2018; Apurillo et al., 2019; Ramirez et al., 2020). Additionally, there are fungi that have been isolated from marine algae (Solis et al., 2010; Jones & Pang, 2012; Lavadia et al., 2017; Raghukumar & Raghukumar, 2017), corals (Marchese et al., 2021), seagrasses (Pasqualetti et al., 2020), and sponges and other invertebrates (Crous et al., 2019; Godinho et al., 2019; Shaumi et al., 2021; Chin et al., 2021). Numerous species were documented in other marine habitats such as coastal oligotrophic and upwelling waters, sandy beaches, deep-sea environments, ocean crusts, and anoxic sediment zones (Raghukumar, 2008; Gutiérrez et al., 2016; De Padua & dela Cruz, 2021). In terms of numbers, several studies showed increasing diversity from 530 species (Jones et al., 2009) to 624 species (Abdel-Wahab & Bahkali, 2012) to 1112 species (Jones et al., 2015) and 1257 species (Jones et al., 2019). Recently, Calabon et al. (2023) provided an updated documented diversity of marine fungi, reporting the existence of nine phyla, 33 classes, 107 orders, 273 families, 767 genera, and 1898 species globally. With this enormous number, it is not surprising that marine fungi are exploited for their natural products. In this chapter, we present studies on marine fungi and their natural products, with emphasis on reports made in Southeast Asia.

Natural Products from Marine Fungi

Marine fungi face significant challenges that set them apart from terrestrial fungi (Gonçalves et al., 2022). Marine environments are often harsh and dynamic, with fluctuating factors such as salinity, pressure, temperature, and light. In response to the complex and ever-changing conditions of marine environments, marine fungi have evolved unique metabolic systems and defense mechanisms (Pang et al., 2016). These adaptations contribute to their biochemical diversity, making them a promising source of novel bioactive natural compounds.

Studies reported the production of these compounds by marine fungi: alkaloids (Willems et al., 2020), lactones (Küppers et al., 2017), polyketides (El-Kashef et al., 2021), polyketones (Tan et al., 2016), peptides (Youssef et al., 2022), steroids (Dos Santos Dias et al., 2019), terpenes (Jiang et al., 2020), and polysaccharides (Sun et al., 2009). These metabolites are mainly associated with antimicrobial (Wang et al., 2021), anticancer (Deshmukh et al., 2018), antiviral (Yasuhara-Bell & Lu, 2010), antioxidant (Vitale et al., 2020), anti-inflammatory (Xu et al., 2019), and trypanocidal activities (Notarte et al., 2017, 2018).

Studies conducted between 1970 and 2002 identified 272 natural products from 80 fungal isolates, with algae (27%) and sponges (28%) being the primary marine substrata (Bugni & Ireland, 2004), which cumulated in over 1000 new metabolites as described by the middle of 2010 (Overy et al., 2014). The pace of discovery accelerated from 2005 to mid-2010, during which over 600 new metabolites were extracted from more than 200 fungal strains, mainly originating from algae (21%), sponges (19%), sediments (16%), and mangroves (15%) (Rateb & Ebel, 2011). This rapid increase in identified bioactive compounds highlights the largely untapped potential of marine fungi in biomedical research.

In recent years, approximately 22,000 bioactive microbial metabolites have been identified, with marine fungi accounting for 38% of these discoveries (Shin, 2020). Their unique adaptation to extreme marine environments has led to the production of structurally diverse and pharmacologically active compounds, many of which exhibit antimicrobial, anticancer, and anti-inflammatory properties. Moreover, the large-scale cultivation of marine fungi is relatively straightforward, making them a sustainable source for drug development. Advances in mutagenesis and metabolic engineering further enhance their potential by facilitating the production of novel compounds with improved bioactivity and higher yields (Hou et al., 2019; Pandey, 2019). Given their vast biodiversity and biochemical versatility, marine fungi continue to represent a promising frontier in the search for innovative therapeutic agents (dela Cruz et al., 2020, 2023).

Marine Fungal Natural Product Research in Southeast Asia

Indonesia

Fungi isolated from Indonesian marine habitats have been explored since the 1990s, and in 2000s, they gained more significant attention. The bioactive compounds from Indonesian marine fungi showed various biological activities, such as antibacterial (Lutfiah et al., 2021; Sibero et al., 2021; Sumilat & Lintang, 2021; Tarman et al., 2011; Trianto et al., 2021; Isti'anah et al., 2024a, b), antifungal (Fadillah et al., 2021; Tarman et al., 2011; Syam et al., 2022), anticancer (Nursid et al., 2020; Pramana et al., 2022; Setyowati et al., 2021; Tarman et al., 2011), anti-obesity (Fadillah et al., 2022), enzyme producer (Andhikawati et al., 2014; Ayuningtyas et al., 2021; Trianto et al., 2021; Julia et al., 2022; Tarman et al., 2024; Budiman et al., 2024), and immunostimulant (Wahjuningrum et al., 2020). Most Indonesian marine fungi were isolated from sponges, seaweeds, and mangroves.

Marine natural products of Indonesian origin were still dominated by bioactive compounds derived from marine bacteria and other macro-organisms such as sponges, corals, seaweeds, and mangroves. Despite the recognized potential of marine fungi as a source of bioactive compounds, research output in this area has been relatively low from 2023 to the present, with only 12 articles published. Among these studies, the primary sources of fungal isolates were mangroves (33%), followed by a notable proportion from sponges (27%), seaweeds (20%), soft coral (13%), and seagrass (7%) (Fig. 7.1).

Seaweeds as hosts of marine fungi were identified from Indonesia, and these included *Sargassum aquifolium*, *Padina australis*, *Turbinaria ornata* (Wiradana et al., 2024), *Cladophora* sp., and *Vaucheria* sp. (Budiman et al., 2024). Among mangroves, identified hosts are *Rhizophora* sp. (Kurnia et al., 2024; Isti'anah et al., 2024a, b; Nabila et al., 2024; Nurhayati et al., 2024), *Rhizophora mucronata* (Budiman et al., 2024), and *Acanthus ilicifolius* L. (Sandrawati et al., 2023). Sponges as fungal hosts comprised of *Xestospongia testudinaria* (Fauziah et al., 2023) and *Stylissa* sp. (Fadillah et al., 2024), with some unidentified sponges (Azhari et al.,

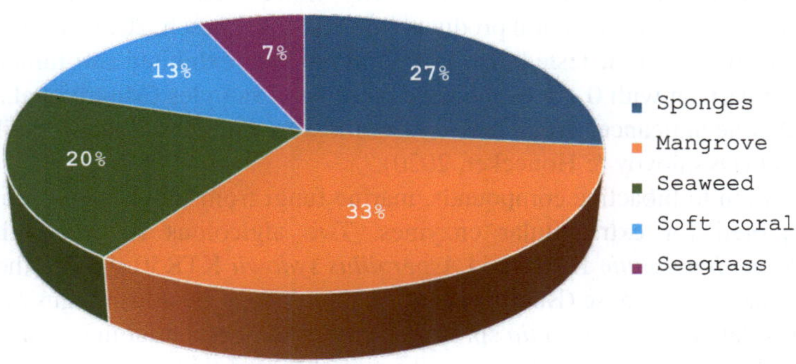

Fig. 7.1 Marine fungi from Indonesian waters based on hosts

2025). In addition, some fungi were collected from soft coral *Sarcophyton* sp. (Wulandari & Primastia, 2024) and another unidentified coral (Azhari et al., 2025).

In terms of taxonomy, species of marine fungi reported from Indonesian waters were dominated by *Aspergillus*, specifically *Aspergillus sydowii* (Janarkho et al., 2023; Budiman et al., 2024), *Aspergillus terreus* (Kurnia et al., 2024; Isti'anah et al., 2024a, b; Nabila et al., 2024; Nurhayati et al., 2024), *Aspergillus hortai* (Fadillah et al., 2024), *Aspergillus ostianus*, and *Aspergillus flavus* (Azhari et al., 2025). Other fungal species were *Penicillium citrinum* (Fadillah et al., 2022), *Trichoderma yunnanense* (Sandrawati et al., 2023), *Trichoderma harzianum* (Budiman et al., 2024), *Pestalotiopsis* sp. (Sandrawati et al., 2023), *Phlebiopsis magnicystidiat*, *Neurospora crassa*, and *Peniophora* sp. (Wiradana et al., 2024). From those reported marine fungi, the substrates or hosts were collected from various localities, including Buton Island (42%), followed by Java Island (33%), Sumatra Island (25%), and Bali (17%). Table 7.1 presents the origin of the marine fungi in Indonesia.

In terms of bioactivities, metabolites from Indonesian marine fungi exhibited various pharmacological activities. The isolated fungi exhibited antibacterial and anticancer properties. Most of the marine fungi were active against pathogenic bacteria, including *Escherichia coli (*Kurnia et al., 2024; Fadillah et al., 2024,; Wiradana et al., 2024; Wulandari & Primastia, 2024), *Salmonella typhimurium* (Nabila et al., 2024), *Staphylococcus aureus* (Wulandari & Primastia, 2024), and *Mycobacterium smegmatis* (Azhari et al., 2025). For fish pathogenic bacteria, the marine fungal extracts were tested against *Vibrio harveyi* (Isti'anah et al., 2024a, b), *Vibrio alginolytucus*, and *Vibrio parahaemolyticus* (Janarkho et al., 2023). Marine fungi also showed anti-diabetic activity (Fauziah et al., 2023). Ethyl acetate extract of *A. terreus* was also used as a feed additive for day-old chicken (Nurhayati et al., 2024). The ethyl acetate extracts of two isolates of the marine fungi *Pestalotiopsis* sp. AIA1 and *Trichoderma yunnanense* AIKB3 showed anticancer activity. Both species were active against the T47D cell line (Sandrawati et al., 2023). The cancer-drug development has been growing significantly. The interest in the exploration of marine organisms, including marine fungi, for cancer-drug discovery, is therefore increasing.

A comparative analysis of chemical novelty showed that marine natural products are superior to terrestrial natural products (Montaser & Luesch, 2011). Furthermore, approximately 1% of the tested marine natural products exhibited anti-tumor potential in comparison with 0.1% of the tested terrestrial samples (Munro et al., 1999). By 2020, nine anticancer drugs derived from marine fungi are already available in the market (Dyshlovoy & Honecker, 2020).

In addition to bioactive compounds, marine fungi from Indonesian waters were also explored for extracellular enzymes. Two algicolous fungi identified as *Trichoderma harzianum* KTR3 and *Aspergillus sydowii* KTR50 showed their ability to produce chitosanase (Budiman et al., 2024). A manglicolous fungus *Fusarium equiseti* isolated from *Avicennia* sp. showed the highest potential in producing protease, while the highest cellulase activity was observed with *Hypocrea* sp., which was isolated from mangrove sediment (Trianto et al., 2021). The potential of marine

Table 7.1 Origin of marine fungi recorded in Indonesia from 2023 to present

Host or substrate	Locality	Fungi species	Compounds/ extracts	Bioactivities	Reference
Sponges	Lampung Bay	*Aspergillus sydowii*	Ethyl acetate extract	Antibacterial (*Vibrio alginolyticus, Vibrio parahaemolyticus*)	Janarkho et al. (2023)
Sponges (*Xestospongia testudinaria*)	West Sumatera	*Penicillium citrinum* Xt6	Ethyl acetate extract	Antidiabetic (induce adipocyte differentiation)	Fauziah et al. (2023)
Mangroves *Acanthus ilicifolius* L	Pariaman, West Sumatera	*Pestalotiopsis* sp. *AIA1* *Trichoderma yunnanense AIKB3*	Ethyl acetate extract (flavonoid, triterpenoid, phenolic)	Cytotoxic against T47D cell line and antibacterial activity	Sandrawati et al. (2023)
Seaweeds (*Cladophora* sp., *Vaucheria* sp.) Mangroves (*Rhizophora mucronata*)	Sukabumi, West Java, and Buton Island	*Trichoderma harzianum* KTR3 *Aspergillus sydowii* KTR50	Extracellular extract	Chitosanase	Budiman et al. (2024)
Mangroves *Rhizophora* sp.	Buton Island	*Aspergillus terreus*	Ethyl acetate extract	Antibacterial against *Escherichia coli*	Kurnia et al. (2024)
Sponges (*Stylissa* sp.)	Pramuka Island, North Jakarta	*Aspergillus hortai*	Ethyl acetate extract	Antibacterial against *Escherichia coli*	Fadillah et al. (2024)
Mangroves *Rhizophora* sp.	Buton Island	*Aspergillus terreus*	Ethyl acetate extract	Antibacterial against *Vibrio harveyi*	Isti'anah et al. (2024a, b)
Mangroves *Rhizophora* sp.	Buton Island	*Aspergillus terreus*	Ethyl acetate extract	Antibacterial against *Salmonella* Typhimurium	Nabila et al. (2024)
Mangroves *Rhizophora* sp.	Buton Island	*Aspergillus terreus*	Ethyl acetate extract	Feed additive for day-old chicken	Nurhayati et al. (2024)
Brown algae (*Sargassum aquifolium, Padina australis, Turbinaria ornata*)	Sanur Beach, Bali	*Phlebiopsis magnicystidiat, Neurospora crassa, Peniophora* sp.	Ethyl acetate extract (alkaloids, triterpenoids, saponin, phenolic)	Antibacterial against *Escherichia coli*	Wiradana et al. (2024)

(continued)

Table 7.1 (continued)

Host or substrate	Locality	Fungi species	Compounds/ extracts	Bioactivities	Reference
Soft coral *Sarcophyton* sp.	Pramuka Island, North Jakarta	Unidentified	Ethyl acetate extract	Antibacterial against *Escherichia coli, Staphylococcus aureus*	Wulandari and Primastia (2024)
Seagrass, sponges, coral, seawater and marine sediment	Java and Bali	*Aspergillus ostianus*, *Aspergillus flavus*	Ethyl acetate extract	Antibacterial against *Mycobacterium smegmatis*	Azhari et al. (2025)

fungi in producing numerous enzymes opens opportunities for industrial purposes. Cellulolytic activity is useful in the saccharification process for bioethanol production as well as in pulp and paper industry. Singh et al. (2016) reported that cellulase contributes 10% of the worldwide industrial enzyme demand.

Vietnam

Vietnam's lengthy coastline of over 3600 kilometers, featuring diverse habitats such as mangroves, coral reefs, and estuaries, provides an ideal environment to study the biodiversity of marine fungi and the natural products they yield. These varied ecosystems, including Ha Long Bay, Lang Co Bay, Son Tra Peninsula, Nha Trang Bay, Van Phong Bay, and Ha Tien Bay, among others, harbor an extensive diversity of marine fungi. Research in Vietnam has successfully isolated fungi from seaweeds, sponges, soft corals, sediments, and seawater. Ngoc et al. (2023) conducted a study in the North Sea of Vietnam and isolated 61 fungal strains from various samples. These were classified under the genera *Penicillium*, *Aspergillus*, *Pleosporales*, *Talaromyces*, and *Hypomontagnella*. Similarly, Pham et al. (2021) explored the fungal biodiversity in Nha Trang Bay and Van Phong Bay and identified 75 strains across 3 phyla, 5 subdivisions, 7 classes, 12 orders, 17 families, 22 genera, and approximately 40 species. The phylum *Ascomycota* dominated, comprising 91.22% of the isolated strains. Notably, this study documented 12 species as new records with 28 being novel to Vietnamese marine environments. Furthermore, the study also highlighted that the biodiversity of marine fungi is influenced by both human and ecological factors. For instance, the lowest fungal density (10^2 CFU L^{-1}) occurs in samples from beaches experiencing intense tourism activity, while the highest density (10^5 CFU L^{-1}) is observed in marine protected areas with minimal human disturbance. Additionally, deep-water environments yield the most significant number of filamentous fungal isolates, despite exhibiting a relatively low fungal density.

Among the most abundant genera of marine fungi in Vietnam, *Aspergillus* stands out as a key producer of secondary metabolites. Species like *A. niveoglaucus*,

A. flocculosus, and *A. terreus* have been thoroughly investigated for their bioactive compounds (Trinh et al., 2018; Choi et al., 2019; Girich et al., 2020). Another prevalent genus, *Penicillium*, is also recognized for generating a diverse array of bioactive substances. Notably, *P. chrysogenum* and *P. chermesinum* are well-documented for their biological activities, especially in the production of bioactive compounds (Trinh et al., 2018; Ngoc et al., 2024, 2025). Additionally, a recently identified novel species, *P. vietnamese*, distinguished by its unique conidiophore structure, has been described (Nguyen & Pham, 2022).

Marine fungi from Vietnam have proven to be a rich source of diverse bioactive compounds. Many studies have focused on the antimicrobial activities of marine fungi, showing that they can produce compounds active against a range of bacteria and fungi. For example, a study of 100 fungal isolates from Nha Trang Bay found that 59 fungal strains displayed antimicrobial activity against at least two tested pathogens. Notably, fungal strains that showed the highest antimicrobial activity belonged to *Aspergillus* (Trinh et al., 2018). The compound citreohybridonol which was extracted from the marine fungus *Penicillium chrysogenum* 045-357-2 had high antibacterial activity against two pathogens, *Bacillus cereus* and *Streptococcus faecalis* (Trinh et al., 2018). Furthermore, phenol A acid isolated from *Penicillium* sp. OM07 exhibits the same antifungal activity toward *Candida albicans* with a minimum inhibitory concentration (MIC) value of 8 μg/mL (Dao et al., 2024).

The secondary metabolites, 4-hydroxyscytalone (**1**), 4-hydroxy-6-dehydroxyscytalone (**2**), and demethylcitreoviranol (**3**) (Fig. 7.2), which were isolated from the marine fungus *Penicillium* sp., showed significant neuroprotective and anti-ROS activities by increasing the viability of paraquat-and rotenone-treated Neuro-2a cell (Girich et al., 2020). The marine fungus *Aspergillus niveoglaucus*, isolated from sediment samples obtained in Nha Trang Bay, has been identified as a producer of polyketides, namely niveoglaucin A and flavoglaucin, along with several echinulin-related compounds. These metabolites have exhibited antioxidant and neuroprotective properties across various in vitro models of Parkinson's disease (Yurchenko et al., 2020; Smetanina et al., 2020). Furthermore, restricticin B (**1**) and *N*-acetyl restricticin (**2**) (Fig. 7.3), extracted from *Penicillium janthinellum* 168CLC-17.1, demonstrated anti-neuroinflammatory properties by reducing the production of pro-inflammatory mediators in lipopolysaccharide (LPS)-stimulated

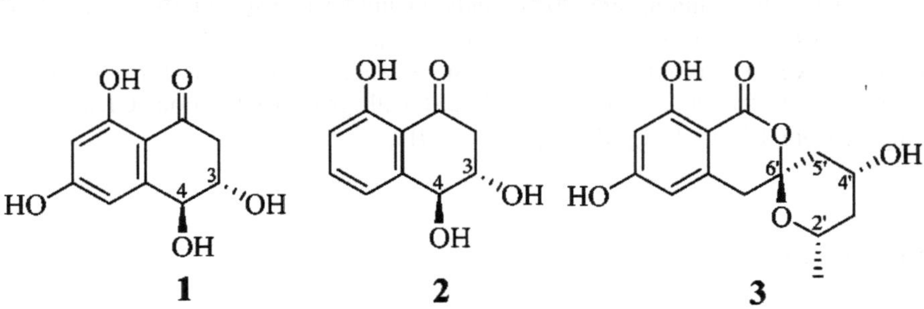

1 **2** **3**

Fig. 7.2 Structures of isolated compounds (**1-3**) from marine *Penicillium* sp. in Vietnam (Girich et al., 2020)

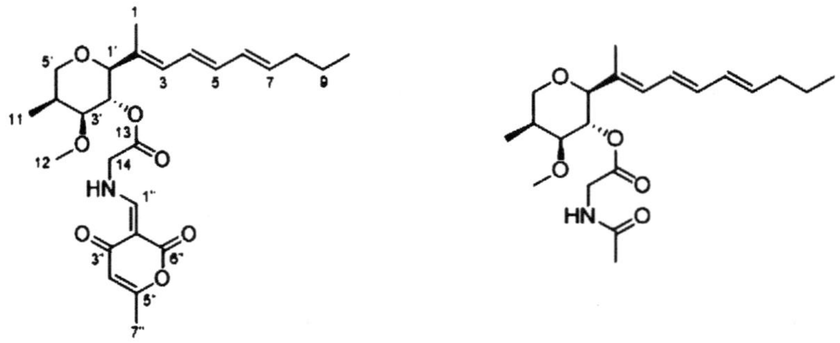

1 **2**

Fig. 7.3 Structures of isolated compounds (**1-2**) from marine-derived fungi *Penicillium janthinellum* (Choi et al., 2020)

BV-2 microglia cells (Choi et al., 2020). The antioxidant and cytoprotective activities have been found in crude extracts of marine fungi belonging to the genera *Aspergillus*, *Penicillium*, *Pleosporales*, *Talaromyces*, and *Hypomontagnella*. However, the research has not yet determined which compounds in the crude extracts are responsible for those activities (Ngoc et al., 2023).

A study of Ngoc et al. (2025) on an ergosterol derivative isolated from the marine fungus *P. chermesinum* 2104NT-1.3 shows that 3β,15β-dihydroxy-(22*E*, 24 *R*)-ergosta-5,8(14),22-trien-7-one (diHET) has cardioprotective activity. In detail, diHET has been shown to promote cell cycle progression in normal H9c2 cardiomyocytes, thereby enhancing their proliferative capacity. Additionally, diHET exhibits a protective effect against $CoCl_2$-induced chronic hypoxia in cardiomyocytes, mediated through endoplasmic reticulum (ER)-dependent signaling pathways.

The potential cytotoxic activity of Vietnamese marine fungi against cancer cells is also of great interest. In particular, 31 out of 66 fungal strains isolated from sponge samples in Nha Trang Bay show potential cytotoxic activity against two human cancer cell lines, including cervical cancer (HeLa) and breast cancer (MCF-7). Those strains are then identified as belonging to *Aspergillus* sp., *Talaromyces* sp., *Aspergillus subramanianii*, *Phoma* sp., and *Penicillium* sp. (Trinh et al., 2018). In addition, nine compounds, including five sesterterpenes and four ophiobolins, isolated from the marine fungus *Aspergillus flocculosus* derived from seaweed, exhibit cytotoxic activities against six cancer cell lines, HCT-15, NUGC-3, NCI-H23, ACHN, PC-3, and MDA-MB-231 with growth inhibition (GI_{50}) values ranging from 0.14 to 2.01 μM (Choi et al., 2019).

Philippines

The Philippines is considered a global hotspot of marine biodiversity. Despite the richness of marine species, studies on marine fungi remain limited. Among the early studies were those of Gacutan and Uyenco (1983) with the isolation of *Antennospora quadricornuta* and *Lulworthia* sp. in Aklan and Jones (1988) with 31 manglicolous fungi recorded from mangroves in Quezon. Jones et al. (1996) also reported *Tirisporella beccariana* from *Nypa fruticans*, which was collected in Pampanga and Quezon. Alias et al. (1999) documented 51 taxa from mangrove driftwoods and included the report of a novel species, *Acrocordiopsis sphaerica*. Other reports are the studies of Besitulo et al. (2010), which identified 66 fungal taxa from the mangroves *Rhizophora apiculata*, *Xylocarpus granatum*, and *Nypa fruticans*, including driftwoods, and that of Su et al. (2013), which identified seven marine fungi from water samples collected in Manila Bay. Calabon et al. (2019) recently reported 110 sponge-associated fungi from mangrove waters.

Fungal endophytes from mangroves have also been given attention in the Philippines. Moron et al. (2018) reported 40 fungal endophytes from 12 mangrove species, and these included *Arthrinium phaeospermum*, *Colletotrichum siamense*, *Fusarium oxysporum*, *Lasiodiplodia theobromae*, and *Pestalotiopsis microspora*, which also exhibited antimicrobial activities. Among these, *P. microspora* showed a strong activity against gram-positive bacteria, while *Lasiodiplodia* species were highly active against *Saccharomyces cerevisiae*. Apurillo et al. (2019) also isolated 73 mangrove fungal endophytes from *Sonneratia alba*, *Rhizophora mucronata*, *Aegiceras floridum*, and *Avicennia marina* in Leyte and Samar, with *R. mucronata* from Samar exhibiting the highest diversity. The study also tested 16 mangrove fungal endophytes and reported *Pestalotiopsis adusta* having the strongest antibacterial activity against *Pseudomonas aeruginosa* and *Xylaria cubensis* exhibited cytotoxicity against K562 leukemia cells. Tan et al. (2015) identified secondary metabolites from *Phyllosticta* sp. from a Philippine mangrove, and elucidated the bioactive compounds as tyrosol C, cytosporone B, dothiorelone A, and dothiorelone C. Tyrosol C displayed antioxidant and anticancer activity, while cytosporone B exhibited moderate antibacterial activity against *Escherichia coli*. Ramirez et al. (2020) also studied mangrove fungal endophytes from 19 mangrove hosts, particularly those belonging to the genera *Aspergillus*, *Cladosporium*, *Colletotrichum*, *Fusarium*, *Nigrospora*, *Penicillium*, *Pestalotiopsis*, *Phialophora*, and *Trichoderma*. Many of these mangrove fungal endophytes produced metabolites that were potent against the gram-positive bacteria *Staphylococcus aureus* and *Micrococcus luteus*. Other studies explored fungi associated with seaweeds and seagrasses (Notarte et al., 2017, 2018; Lavadia et al., 2017), which were also explored for their biological activities.

Fungi from marine environments were also tested for their ability to clean up environmental pollutants. Sadaba and Niego (2017) investigated the hydrocarbon-degrading potential of fungi from oil-contaminated beach and mangrove soils in Iloilo following the Napocor Power Barge 103 oil spill. *Aspergillus fumigatus* was the most prevalent isolate (43.06%), while *Penicillium* sp1 (EB331) exhibited the highest degradation of total petroleum hydrocarbons (TPH) (64.70%) and polycyclic aromatic hydrocarbons (PAH) (43.42%). De Padua and dela Cruz (2021) evaluated *Trichoderma* strains for nickel biosorption, revealing 68% uptake efficiencies. Meanwhile, Torres and dela Cruz (2012) explored mangrove fungal xylanases for recycled paper pulp processing, demonstrating enzymatic pretreatment that released

70–650 mg reducing sugars per gram of pulp, highlighting its industrial potential. Torres et al. (2011) also explored the potential of marine-derived fungi to decolorize dyes.

Seagrass-associated fungi have also been explored for bioactive compounds. Notarte et al. (2017) identified Malformin A1, a cyclic pentapeptide from *Aspergillus tubingensis* IFM 63452, isolated from the seagrass *Enhalus acoroides* (Linnaeus) Royle. Malformin A1 exhibited potent trypanocidal activity against *Trypanosoma congolense* (IC_{50}: 15.08 ng/mL) and cytotoxicity against HeLa (IC_{50}: 50.15 ng/mL) and P388 (IC_{50}: 70.38 ng/mL) cells. Immunofluorescence staining revealed epigenetic modifications in histone H3 and H4, suggesting its chemotherapeutic potential. Additionally, Notarte et al. (2018) isolated 16 fungal morphospecies from macroalgae and seagrasses, with *A. tubingensis* (grown with marine salt) exhibiting the highest cytotoxicity against cancer cells and *A. fumigatus* (grown without marine salt) demonstrating strong activity against *T. congolense*.

Malaysia

Malaysia's extensive coastline and rich marine ecosystems provide an ideal environment for diverse marine fungi. Research efforts in Malaysia have focused on isolating and characterizing fungal strains from various marine environments, including seaweeds, mangrove ecosystems, and marine sponges. Recent studies have significantly advanced the understanding of their diversity and distribution, particularly in mangrove ecosystems. Devadatha et al. (2021) recorded 850 taxa in Malaysian mangrove ecosystems, with Ascomycota being the most dominant (773 species), followed by Basidiomycota (58) and smaller groups from other phyla. Malaysia ranks among the top countries for mangrove fungal diversity, alongside India (339 taxa) and Thailand (303). The Pacific Ocean harbors the highest mangrove fungal diversity (553 taxa), followed by the Indian (408) and Atlantic Oceans (259).

A study by Alias et al. (2010) examined decaying mangrove substrates in Malaysia, identifying 139 marine fungal species (115 Ascomycota, 2 Basidiomycota, and 22 anamorphic taxa). *Lignincola laevis* (16.8%), *Verruculina enalia* (13.3%), and *Trichocladium achrasporum* (12.1%) were the most prevalent. Notably, 84% of these fungi are exclusive to tropical and subtropical regions, with their distribution influenced by substrate type, sampling frequency, vertical stratification, and geographical factors. Alias and Jones (2000) also examined fungal colonization on test blocks of *Avicennia marina* and *Bruguiera parviflora* in a Kuala Selangor mangrove stand over 90 weeks. The most common fungi on both wood types included *Halosarpheia marina*, *Lignincola laevis*, *Lulworthia grandispora*, *Savoryella lignicola*, and *Trichocladium achrasporum*. Colonization followed a clear pattern: early colonizers (6–18 weeks) included *Halosarpheia marina*, *H. retorquens*, and *Lignincola laevis*; intermediate colonizers (26–54 weeks) featured *Dictyosporium pelagicum* and *Halocyphina villosa*; and late colonizers (60–96 weeks) included

Aigialus parvus and *Leptosphaeria australiensis*. These findings align with those of Alias et al. (2010), who investigated fungal colonization on submerged mangrove wood in the same region and highlighted the dominance of species such as *Lignincola laevis* and *Trichocladium achrasporum* and provided insights into the sequential colonization patterns of marine fungi over time. These studies provide a comprehensive overview of Malaysia's diverse marine fungal communities, highlighting their ecological roles and biotechnological potential.

Extensive research has also identified numerous bioactive metabolites from Malaysian marine fungi, exhibiting antimicrobial properties. This aligns with the findings of Jones et al. (2013), who documented the diversity and ecological distribution of marine fungi in the South China Sea, where about 560 species—constituting a third of all described marine fungi—have been identified. Most of these species have been reported from Hong Kong, Malaysia, Thailand, and Taiwan, with Malaysia contributing 25 newly described species. Many fungi have been found on mangrove substrates, leading to studies on their distribution and colonization patterns. The researchers have screened marine fungi for bioactive compounds, discovering 42 novel and 35 known chemical structures. Despite the significant research on filamentous fungi, information on chytrids, mucoraceous fungi, and fungal-like organisms remains limited, highlighting the need for further study. The following works highlight key findings on marine fungal diversity and their potential applications in biotechnology and drug discovery.

Anwar bin Azlan et al. (2024) investigated *Hypoxylon monticulosum*, isolated from the marine macroalga *Ulva lactuca* in Desaru Coast, Johor, Malaysia. Bioassay-guided fractionation using solvents of increasing polarity identified 128 compounds, including dihydrocordoin, D-pantothenoyl-L-cysteine, caffeine, and tumonoic A acid, all known for their antibacterial properties. Latif et al. (2022) isolated 18 marine fungal endophytes from eight seaweed species in Teluk Kemang, Negeri Sembilan, Malaysia, and evaluated their antimicrobial potential. ITS sequencing identified the 11 strains, including six *Aspergillus* species and other taxa like *Exophiala dermatitidis*, *Diaporthe pseudomangiferae*, *Arthrinium xenocordella*, *Phanerochaete carnosa*, and *Psathyrella purpureobadia*. Antifungal assays revealed significant activity against *Candida albicans* and *Trichophyton rubrum*, with inhibition zones exceeding those of Amphotericin B. Salinity (3% sea salt) influenced fungal growth, spore production, and antifungal properties. Paliany et al. (2014) explored the antibacterial potential of marine-derived endophytic and manglicolous fungi from Peninsular Malaysia. Fungal strains were isolated from marine-associated plants, including *Vitex rotundifolia* (11 strains), *Ipomoea pes-caprae* (18 strains), and decaying mangrove wood (10 strains). Endophytes from *I. pes-caprae* exhibited stronger antibacterial potential than those from *V. rotundifolia*, with *Minimidochium* sp. and *Bipolaris* sp. showing activity against five or more bacterial strains. Further analysis identified *Minimidochium* sp. and *Dyfrolomyces rhizophorae* as promising candidates, with MIC values ≤0.5 mg/ml. Bioactivity-guided fractionation of *D. rhizophorae* led to the isolation of palmitic and linoleic acids, both of which have commercial applications as health supplements and biodiesel alternatives.

Mosadeghzad et al. (2013) investigated the antimicrobial potential of *Paecilomyces* sp., a marine fungus isolated from a sponge collected at Tinggi Island, Malaysia. Purification of the extract led to the isolation of one novel compound, *(3R,5E,7E,9R*,11E,13Z)*-1-((3a*S**,6a*R**)-2-amino-5-oxo-3a,5,6,6a-tetrahydrofuro[3,2-b]furan-3-yl)-3,7,9,11,13-tetramethylheptadeca-5,7,11,13-tetraene-1,2-dione*, along with four known metabolites: stigmasterol, cholesteryl linoleate, linoleic acid, and 2,5-furandimethanol. Zainuddin et al. (2010) systematically screened marine fungi from decaying mangrove wood, driftwood, and *Nypa fruticans* samples collected across multiple locations in Peninsular Malaysia. The majority of fungal strains (90.2%) exhibited antibacterial effects, primarily against Gram-positive bacteria, while only 9.6% demonstrated antifungal properties. *Fasciatispora nypae* displayed the broadest antimicrobial spectrum, leading to bioactivity-guided fractionation and the isolation of 2,2,7-trimethyl-2H-chromen-5-ol, a compound previously known only as a synthetic intermediate, now identified as a natural product via mass spectrometry. While Anwar bin Azlan et al. (2024) and Latif et al. (2022) focused on fungi associated with seaweeds, demonstrating antibacterial and antifungal properties, respectively, Paliany et al. (2014) extended the exploration to endophytic and manglicolous fungi from marine plants, identifying promising antibacterial strains. Similarly, Mosadeghzad et al. (2013) and Zainuddin et al. (2010) underscored the antimicrobial capabilities of fungi isolated from marine sponges and decaying mangrove wood, leading to the discovery of novel and commercially relevant compounds. Together, these findings reinforce the immense pharmaceutical potential of marine fungal metabolites and the need for continued exploration and characterization of these natural products.

Thailand

Marine fungi are a rich source of bioactive metabolites with diverse pharmacological properties, making them valuable candidates for drug discovery. In Thailand, numerous bioactive compounds have been identified from marine-derived fungi, highlighting their potential in antimicrobial, anti-inflammatory, and other therapeutic applications. These fungi have been isolated from a variety of marine hosts and environments, including sponges, sea fans, corals, seagrasses, unidentified marine organisms, and marine sediments. Beyond their structural uniqueness, many of these compounds exhibit significant bioactivity, reinforcing the importance of marine fungi as a promising yet underexplored reservoir of natural products. A closer look at these fungi and their bioactive properties, categorized by their marine sources, provides further insight into their potential and the challenges in advancing them for medical and pharmaceutical applications.

Marine fungi associated with sponges are emerging as significant sources of bioactive compounds with diverse therapeutic potentials. The sponge-associated fungus PSU-SP2/4, isolated from a marine sponge in Trang province, Thailand, yielded a new pentacyclic cytochalasin, diaporthalasin, and a new ethyl trihydroxytrideca-trienoate, diaporthacol, along with five known compounds (Khamthong et al., 2014). Diaporthalasin exhibited strong antibacterial activity against both *Staphylococcus aureus* and methicillin-resistant *S. aureus* (MRSA), with minimal inhibitory concentration (MIC) values of 2 mg/mL, highlighting its potential as an antimicrobial agent (Khamthong et al., 2014). Another sponge-associated fungus, *Aspergillus aculeatus* CRI323–04, isolated from the marine sponge *Xestospongia testudinaria*, produced a new tyrosine-derived metabolite, aspergillusol A, along-side a methyl ester of 4-hydroxyphenylpyruvic acid oxime and secalonic acid A (Ingavat et al., 2009). Aspergillusol A selectively inhibited R-glucosidase from *Saccharomyces cerevisiae* while showing inactivity against the enzyme from *Bacillus stearothermophilus*, showcasing its potential for targeted enzyme inhibition (Ingavat et al., 2009). Additionally, a marine-derived fungus of the order *Pleosporales*, isolated from unidentified sponge tissues, produced new compounds, including a diketopiperazine and a phthalide derivative, both exhibiting cytotoxic activity with IC_{50} values ranging from 25.8 to 44.0 µg/mL, suggesting potential anticancer properties (Prachyawarakorn et al., 2008). *Talaromyces minioluteus*, isolated from an unidentified marine sponge, yielded four new sesquiterpene lactones and three known compounds. While the new compounds did not inhibit caspase-3, several exhibited weak cytotoxic activity, further demonstrating the diverse biological activities associated with sponge-derived fungi (Ngokpol et al., 2015). Furthermore, *Aspergillus aculeatus* CRI322–03 from the marine sponge, *Stylissa flabelliformis*, yielded a novel diketopiperazine, pre-aurantiamine, alongside itaconic acid derivatives, with pre-aurantiamine potentially serving as a biosynthetic precursor to aurantiamine, a potent anticancer agent (Antia et al., 2011). Lastly, the isolation of *Neosartorya quadricincta* from the marine sponge *Clathria reinwardti* revealed two new pentaketides, a benzofuran-1-one derivative, and several benzoic acid derivatives. However, none of the isolated compounds showed significant antibacterial or antifungal activity against various microbial strains or substantial inhibitory activity against cancer cell lines, indicating the need for further investigations into their bioactive properties (Prompanya et al., 2016). These studies highlight the importance of sponge-associated fungi as a rich source of bioactive metabolites, revealing a variety of biological activities that warrant further exploration for potential therapeutic applications.

Marine fungi associated with sea fans have garnered attention due to their ability to produce diverse secondary metabolites with significant bioactivities. Various fungal strains have been isolated from sea fans in Thailand, particularly from the genus *Annella*, and characterized for their bioactive compounds (Preedanon et al., 2016). In one study, 163 strains of fungi were isolated from sea fans, and nearly half of the fungal isolates exhibited antimicrobial activity against at least one pathogen, with minimum inhibitory concentrations (MIC) ranging from 1 to 1280 µg/mL. Several isolates inhibited *S. aureus*, MRSA, *Escherichia coli*, and *Pseudomonas aeruginosa*. Additionally, 18% of the fungal isolates demonstrated antifungal activity against *Microsporum gypseum*, while a smaller proportion inhibited *Cryptococcus neoformans* and *Candida albicans*. Among the identified bioactive compounds, griseofulvin from *Nigrospora* sp. PSU-F13 showed the strongest antifungal activity against *M. gypseum*, and the same fungal strain also produced dechlorogriseofulvin (Preedanon et al., 2016). The chemical diversity of these fungi from sea fans was highlighted by the presence of anthraquinones, cyclohexenes, cyclopentanones, cytochalasins, diketopiperazines, epoxydons, lactones, modiolides, naphthoquinones, phenols, and pyrone derivatives (Preedanon et al., 2016). Notable metabolites include 9α-hydroxydihydrodesoxybostrycin from *Fusarium* sp. PSU-F14, nigrosporapyrone A from *Nigrospora* sp. PSU-F18, and antibacterial compounds from *Penicillium* sp. PSU-F40. Crude fungal extracts exhibited stronger bioactivity than isolated compounds, indicating potential synergistic effects among multiple constituents (Preedanon et al., 2016). This finding underscores the importance of considering the synergism of bioactive metabolites, as the combined action of various compounds may enhance therapeutic efficacy and lead to more effective natural products. *Curvularia* sp. PSU-F22, a fungus isolated from a sea fan collected near the Similan Islands in Thailand, was studied for the production of bioactive metabolites (Trisuwan et al., 2011). Three novel compounds were identified: curvulapyrone, an α-pyrone derivative; curvulalide, a modiolide macrolide; and curvulalic acid, an uncyclized metabolite of modiolide macrolides. Additionally, six known compounds were isolated, including modiolides A and B, pyrenolide A, stagonolide E, mycoepoxydiene, and deacetylmycoepoxydiene (Trisuwan et al., 2011) However, despite their structural novelty, the isolated metabolites from *Curvularia* sp. PSU-F22 displayed no significant antimicrobial activity against *S. aureus*, MRSA, or *M. gypseum (*Trisuwan et al., 2011*)*. Nonetheless, this marked the first reported isolation of an α-pyrone derivative from the genus *Curvularia* from sea fans, underscoring the importance of marine-derived fungi as reservoirs of chemically diverse compounds with potential pharmaceutical applications.

Marine fungi associated with corals in Thailand are increasingly recognized for their potential to produce bioactive compounds with significant therapeutic properties. One such compound is nodulisporacid A, a new tetronic acid isolated from the marine-derived fungus *Nodulisporium* sp., which was obtained from an unidentified soft coral (Kasettrathat et al., 2008). Nodulisporacid A has demonstrated promising antimalarial activity, with an IC_{50} value in the range of 1–10 μM (Kasettrathat et al., 2008).

Marine-derived fungi from unidentified marine organisms have emerged as significant producers of bioactive metabolites with promising pharmaceutical potential. The fungus *Aspergillus* sp. AMF2431 produced two new compounds, which exhibited antibacterial activity against *S. aureus* ATCC 25922 and MRSA (Widyasti & Rukachaisirikul, 2017). Similarly, the marine-derived *Aspergillus unguis* CRI282–03 was utilized for the production of novel depsidone derivatives through directed biosynthesis involving biohalogenation (Sureram et al., 2013). Cultivation in media containing different halide salts led to the formation of distinct metabolites, including aspergillusidones D–F, aspergillusphenols A and B, as well as brominated and chlorinated depsidones. The production of these metabolites was influenced by the halide composition of the medium, demonstrating the adaptability of fungal biosynthesis. Notably, aspergillusidones D–F and unguinol were found to inhibit aromatase, suggesting their potential therapeutic applications (Sureram et al., 2013). Another marine-derived fungus, CRI247–01, classified within the order *Pleosporales*, produced vermelhotin, a tetramic acid known for its cytotoxic, antimalarial, and calmodulin-inhibitory activities (Kasettrathat et al., 2008; Pansanit et al., 2013). Additionally, vermelhotin shows potential anti-inflammatory effects by suppressing inducible nitric oxide synthase expression and selectively inhibiting p38 phosphorylation (Leyte-Lugo et al., 2012). Similarly, a marine-derived *Trichoderma harzianum* PSU-MF79 produced two new compounds—a pyrone (trichoharzianone) and a decalin (trichoharzianin)—along with eight known metabolites, including three decalins, two δ-lactones, two carboxylic acids, and one isochroman (Nuansri et al., 2022). Among these, (−)-massoia lactone exhibited mild antifungal activity against *Cryptococcus neoformans* and *Candida albicans*, with weak cytotoxic activity against HCT-116 and MCF-7 cell lines, while remaining noncytotoxic to Vero cells (Nuansri et al., 2022).

Endophytic fungi associated with seagrasses in southern Thailand have demonstrated significant antimicrobial potential, with diverse isolates exhibiting activity against human pathogens (Supaphon et al., 2014). One hundred sixty (160) fungal strains were isolated from *Cymodocea serrulata*, *Halophila ovalis*, and *Thalassia hemprichii*, of which 69% displayed antimicrobial activity against at least one of ten tested pathogenic microorganisms, with the antifungal activity more pronounced than the antibacterial activity. These included *S. aureus* ATCC 25923, clinical isolates of MRSA, *E. coli* ATCC 25923, *Pseudomonas aeruginosa* ATCC 27853, *Candida albicans* ATCC 90028 and NCPF 3153, *Cryptococcus neoformans* ATCC 90112 and ATCC 90113, and clinical isolates of *Microsporum gypseum* and *Penicillium marneffei* (Supaphon et al., 2014). Seven fungal isolates showed strong antimicrobial properties against these human pathogens: *Hypocreales* sp. PSU-ES26, *Trichoderma* spp. PSU-ES8 and PSU-ES38, *Penicillium* sp. PSU-ES43, *Fusarium* sp. PSU-ES73, *Stephanonectria* sp. PSU-ES172, and an unidentified endophyte PSU-ES190. These isolates exhibited minimum inhibitory concentrations (MIC) below 10 μg/ml, demonstrating potent antimicrobial effects. Furthermore, extracts at concentrations four times their MIC resulted in cellular destruction (Supaphon et al., 2013). These findings highlight the potential of seagrass-derived endophytic fungi as promising sources of antimicrobial compounds.

Studies on marine sediment fungi from Thai mangrove ecosystems revealed their acid tolerance and bioactive potential. A total of 237 acidophilic or aciduric fungal isolates were identified, with *Penicillium*, *Aspergillus*, and *Talaromyces* being the dominant genera (Gao et al., 2021). Remarkably, 95% of these fungi were able to grow under extremely acidic conditions (pH 2), with six *Talaromyces* strains exhibiting strict acidophilic characteristics, marking the first report of such traits in the genus. Bioactivity screening showed that nearly 75% of fungal extracts exhibited cytotoxic activity, while a smaller fraction displayed antimicrobial or anti-H1N1 activity. One strain, OUCMDZ-5130, demonstrated broad-spectrum inhibition against multiple human and aquatic pathogens. Additionally, 63.4% of tested strains exhibited more than 50% cytotoxic inhibition, with eight strains showing over 60% inhibition against A549 and MCF-7 cancer cell lines (Gao et al., 2021). Notably, *Penicillium oxalicum* OUCMDZ-5207 displayed distinct metabolite profiles under acid stress, producing unique compounds such as tetrahydroauroglaucin, flavoglaucin, and auroglaucin at pH 3, with auroglaucin selectively inhibiting A549 cells ($IC_{50} = 5.67\ \mu M$) (Gao et al., 2021). These studies underscore the potential of marine fungi in Thailand as valuable sources of bioactive metabolites, with applications spanning antibacterial, antifungal, anti-inflammatory, anticancer, and other therapeutic uses. However, despite the wealth of bioactive compounds identified over the years, none have progressed to commercial drug development. This gap highlights the challenge of translating promising leads from basic research into clinical applications. Bridging this gap requires further biochemical and pharmacological studies and investments in scalable production, safety assessments, and regulatory approvals. Strengthening collaborations between researchers, industry, and policymakers will be key to unlocking the full potential of these marine-derived compounds and transforming them into viable therapeutic solutions.

Concluding Points

The rich biodiversity and unique bioactive compounds of marine fungi offer significant potential for various applications. Thus, marine fungal natural products can be further explored for development into novel drugs for treating infectious diseases, cancer, neurodegenerative disorders, and other ailments. The antioxidant and anti-inflammatory compounds derived from marine fungi may also find applications in the cosmetic industry. In addition, marine fungi could be utilized for various biotechnological applications, including the production of enzymes, biofuels, and other industrial products.

Despite their promising potential, several challenges hinder the full utilization of marine fungal natural products. Cultivating marine fungi under laboratory conditions remains difficult, limiting access to their secondary metabolites. Additionally, the structural complexity of marine fungal metabolites presents challenges for isolation, characterization, and synthesis. Addressing these issues requires advancements in cultivation techniques, the application of omics technologies for metabolic profiling, and the exploration of sustainable harvesting methods. Future research should integrate genome mining approaches with biotechnological tools to enhance the discovery of novel bioactive compounds. Collaborative efforts between academia and industry are essential for translating marine fungal metabolites into commercially viable pharmaceutical agents.

References

Abdel-Wahab, M. & Bahkali, A. (2012). 5 Taxonomy of filamentous anamorphic marine fungi: morphology and molecular evidence. In E. Jones & K. Pang (Ed.), Marine Fungi (pp. 65-90). Berlin, Boston: De Gruyter.

Abdel-Wahab, M. A., Hodhod, M. S., Bahkali, A. H., & Jones, E. G. (2014). Marine fungi of Saudi Arabia. *Botanica Marina, 57*(4), 323–335. https://doi.org/10.1515/bot-2014-0010

Abdel-Wahab, M. A., Jones, E. G., Bahkali, A. H., & El-Gorban, A. M. (2019). Marine fungi from Red Sea mangroves in Saudi Arabia with *Fulvocentrum rubrum* sp. nov. (*Torpedosporales, Ascomycota*). *Nova Hedwigia, 108*(3), 1–13. https://doi.org/10.1127/nova_hedwigia/2018/0511

Alias, S. A., & Jones, E. B. G. (2000). Colonization of mangrove wood by marine fungi at Kuala Selangor mangrove stand, Malaysia. *Fungal Diversity, 5*, 9–21.

Alias, S. A., Jones, E. G., & Torres, J. (1999). Intertidal fungi from the Philippines, with a description of *Acrocordiopsis sphaerica* sp. nov. (Ascomycota). *Fungal Diversity, 2*, 35–41.

Alias, S. A., Zainuddin, N., & Jones, E. G. (2010). Biodiversity of marine fungi in Malaysian mangroves. *Botanica Marina, 53*(6), 545–554. https://doi.org/10.1127/nova_hedwigia/2018/0511

Andhikawati, A., Oktavia, Y., Ibrahim, B., & Tarman, K. (2014). Isolation and screening of endophytic marine fungi for cellulase production. *Jurnal Ilmu Dan Teknologi Kelautan Tropis, 6*(1), 219–227. https://doi.org/10.29244/jitkt.v6i1.8643

Antia, B. S., Aree, T., Kasettrathat, C., Wiyakrutta, S., Ekpa, O. D., Ekpe, U. J., Mahidol, C., Ruchirawat, S., & Kittakoop, P. (2011). Itaconic acid derivatives and diketopiperazine from the marine-derived fungus *Aspergillus aculeatus* CRI322-03. *Phytochemistry, 72*(8), 816–820. https://doi.org/10.1016/j.phytochem.2011.02.013

Anwar bin Azlan, M. H., Ain Zainee, N. F., & Ibrahim, N. (2024). Antibacterial activity and bioactive compounds of a marine macroalgae endophytic fungi, *Hypoxylon monticulosum*. *Malaysian Journal of Microbiology, 20*(2). https://doi.org/10.21161/mjm.230352

Apurillo, C. C. S., Cai, L. , dela Cruz, T. E. E. (2019). Diversity and bioactivities of mangrove fungal endophytes from Leyte and Samar, Philippines. *Philippine Science Letters*, 12: 33–48. https://scienggj.org/2019-special-issue-3/

Ayuningtyas, E. P., Sibero, M. T., Hutapea, N. E. B., Frederick, E. H., Murwani, R., Zilda, D. S., Wijayanti, D. P., Sabdono, A., Pringgenies, D., & Radjasa, O. K. (2021). Screening of extracellular enzyme from *Phaeophyceae*-associated fungi. *IOP Conference Series: Earth and Environmental Science. https://doi.org/10.1088/17551315/750/1/012005*

Azhari, M., Litanjuasari, A. P., Singgih, M., Arai, M., Handayani, D., Artasasta, M. A., & Julianti, E. (2025). Activity of ethyl acetate extracts of marine-derived fungi against active and hypoxia-induced dormant *Mycobacterium. Journal of Pharmacy & Pharmacognosy Research, 13*(1), 16–26. https://doi.org/10.56499/jppres24.1953_13.1.16

Barghoorn, E. S., & Linder, D. H. (1944). Marine fungi: Their taxonomy and biology. *Farlowia, 1*, 395–467.

Besitulo, A., Moslem, M. A., & Hyde, K. D. (2010). Occurrence and distribution of fungi in a mangrove forest on Siargao Island, Philippines. *Botanica Marina, 53*(6), 535–543. https://doi.org/10.1515/bot.2010.065

Budiman, M. A., Tarman, K., Hardiningtyas, S. D., & Nurazizah, M. A. (2024). Eksplorasi aktivitas enzimatik dari fungi endofit laut serta aplikasinya untuk hidrolisis kitosan. *Jurnal Pengolahan Hasil Perikanan Indonesia, 27*(11), 1035–1049. https://doi.org/10.17844/jphpi.v27i11.58419

Bugni, T., & Ireland, C. (2004). Marine-derived fungi: A chemically and biologically diverse group of microorganisms. *Natural Product Reports, 21*(2), 143–163. https://doi.org/10.1039/B301926H

Byrne, P. J., & Jones, E. B. G. (1975). Effect of salinity on the reproduction of terrestrial and marine fungi. *Transactions of the British Mycological Society, 65*(2), 185–200. https://doi.org/10.1016/S0007-1536(75)80002-7

Calabon, M. S., Sadaba, R. B., & Campos, W. L. (2019). Fungal diversity of mangrove-associated sponges from New Washington, Aklan, Philippines. *Mycology, 10*(1), 6–21. https://doi.org/10.1080/21501203.2018.1518934

Calabon, M. S., Jones, E. G., Pang, K. L., Abdel-Wahab, M. A., Jin, J., Devadatha, B., et al. (2023). Updates on the classification and numbers of marine fungi. *Botanica Marina, 66*(4), 213–238. https://doi.org/10.1515/bot-2023-0032

Chin, J. M., Puchooa, D., Bahorun, T., & Jeewon, R. (2021). Antimicrobial properties of marine fungi from sponges and brown algae of Mauritius. *Mycology, 12*(4), 231–244. https://doi.org/10.1080/21501203.2021.1895347

Choi, B. K., Trinh, P. T. H., Lee, H. S., Choi, B. W., Kang, J. S., Ngoc, N. T. D., Van, T. T. T., & Shin, H. J. (2019). New ophiobolin derivatives from the marine fungus *Aspergillus flocculosus* and their cytotoxicities against cancer cells. *Marine Drugs, 17*, 346. https://doi.org/10.3390/md17060346

Choi, B. K., Jo, S. H., Choi, D. K., Trinh, P. T. H., Lee, H. S., Cao, V. A., Van, T. T. T., & Shin, H. J. (2020). Anti-neuroinflammatory agent, restricticin B, from the marine-derived fungus *Penicillium janthiellum* and its inhibitory activity on NO production in BV-2 microglia cells. *Marine Drugs, 18*(9), 465. https://doi.org/10.3390/md18090465

Crous, P. W., Wingfield, M. J., Lombard, L., Roets, F., Swart, W. J., Alvarado, P., et al. (2019). Fungal Planet description sheets: 951–1041. *Persoonia: Molecular Phylogeny and Evolution of Fungi, 43*, 223–425. https://doi.org/10.3767/persoonia.2019.43.06

Dao, P. T., Linh, N. T., Nam, V. V., Quyen, V. T., Anh, N. M., Huyen, V. T. T., Minh, L. T. H., Cuong, P. V., & Huong, D. T. M. (2024). Antimicrobial secondary metabolites from a marine-derived fungus *Penicillium* sp. OM07. *Vietnam. Journal of Marine Science and Technology, 24*(2), 175–184. https://doi.org/10.15625/1859-3097/19564

De Padua, J. C., & dela Cruz, T. E. E. (2021). Isolation and characterization of nickel-tolerant *Trichoderma* strains from marine and terrestrial environments. *Journal of Fungi, 7*(8), 591. https://doi.org/10.3390/jof7080591

dela Cruz, T. E. E., Notarte, K. I. R., Apurillo, C. C. S., Tarman, K., & Bungihan, M. E. (2020). Biomining fungal endophytes from tropical plants and seaweeds for drug discovery. In *Biodiversity and biomedicine* (pp. 51–62). Academic Press. https://doi.org/10.1016/B978-0-12-819541-3.00004-9

dela Cruz, T. E. E., Bennett, R. M., & Apurillo, C. C. S. (2023). Diversity and applications of fungi associated with mangrove leaves in the Philippines. In *Mycology in the Tropics* (pp. 17–32). Academic Press. https://doi.org/10.1016/B978-0-323-99489-7.00006-8

Deshmukh, S. K., Prakash, V., & Ranjan, N. (2018). Marine fungi: A source of potential anticancer compounds. *Frontiers in Microbiology, 8*, 2536. https://doi.org/10.3389/fmicb.2017.02536

Devadatha, B., Sarma, V. V., Jeewon, R., Wanasinghe, D. N., Hyde, K. D., & Jones, E. B. G. (2018). *Thyridariella*, a novel marine fungal genus from India: Morphological characterization and phylogeny inferred from multigene DNA sequence analyses. *Mycological Progress, 17*, 791–804. https://doi.org/10.1007/s11557-018-1387-4

Devadatha, B., Jones, E. B. G., Pang, K. L., Abdel-Wahab, M. A., Hyde, K. D., Sakayaroj, J., et al. (2021). Occurrence and geographical distribution of mangrove fungi. *Fungal Diversity, 106*, 137–227. https://doi.org/10.1007/s13225-020-00468-0

Dos Santos Dias, A. C., Couzinet-Mossion, A., Ruiz, N., Lakhdar, F., Etahiri, S., Bertrand, S., et al. (2019). Steroids from marine-derived fungi: Evaluation of antiproliferative and antimicrobial activities of eburicol. *Marine Drugs, 17*(6), 372. https://doi.org/10.3390/md17060372

Dyshlovoy, S. A., & Honecker, F. (2020). Marine compounds and cancer: Updates 2020. *Marine Drugs., 18*(12), 643. https://doi.org/10.3390/md18120643

El-Kashef, D. H., Youssef, F. S., Reimche, I., Teusch, N., Mueller, W. E., Lin, W., et al. (2021). Polyketides from the marine-derived fungus *Aspergillus falconensis*: In silico and in vitro cytotoxicity studies. *Bioorganic & Medicinal Chemistry, 29*, 115883. https://doi.org/10.1016/j.bmc.2020.115883

Engel, S., Jensen, P. R., & Fenical, W. (2002). Chemical ecology of marine microbial defense. *Journal of Chemical Ecology, 28*, 1971–1985. https://doi.org/10.1023/A:1020793726898

Fadillah, W. N., Sukarno, N., Iswantini, D., Rahminiwati, M., & Listiyowati, S. (2021). *New record of Gymnoascus udagawae associated with Clathria sp. sponge from Indonesia and the potency as anti-Candida*. Earth and Environmental Science. https://doi.org/10.1088/1755-1315/948/1/012072

Fadillah, W. N., Sukarno, N., Iswantini, D., Rahminiwati, M., Hanif, N., & Waite, M. (2022). *In vitro* pancreatic lipase inhibition by marine fungi *Purpureocillium lilacinum* associated with *Stylissa* sp. sponge as an anti-obesity agent. *HAYATI Journal of Biosciences, 29*, 76–86. https://doi.org/10.4308/hjb.29.1.76-86

Fadillah, W. N., Sukarno, N., Iswantini, D., Rahminiwati, M., Franco, C. M., Zhang, W., Hanif, N., & Waite, M. (2024). Exploring Indonesian sponge-associated marine *Aspergillus hortai*: Characterization of bioactive compounds with potential anti-*Escherichia coli* properties. *HAYATI Journal of Biosciences, 31*(4), 712–724. https://doi.org/10.4308/hjb.31.4.712-724

Fauziah, F., Ali, H., Ilmiawati, Bakhtra, D. D. A., Taher, M., & Handayani, D. (2023). Secondary metabolites of marine sponge-derived fungus *Penicillium citrinum* Xt6 induce adipocyte differentiation on 3T3-L1 preadipocytes. *AIP Conference Proceedings, 2730*(1), 080006. https://doi.org/10.1063/5.0127873

Gacutan, V. C., & Uyenco, F. R. (1983). Marine fungi from Batan, Aklan. *Natural and Applied Science Bulletin, 35*, 1–16.

Gal-Hemed, I., Atanasova, L., Komon-Zelazowska, M., Druzhinina, I. S., Viterba, A., & Yarde, O. (2011). Marine isolates of *Trichoderma* spp. as potential halotolerant agents of biological control for arid-zone agriculture. *Applied and Environmental Microbiology, 77*, 5100–5109. https://doi.org/10.1128/AEM.00541-11

Gao, H., Wang, Y., Luo, Q., Yang, L., He, X., Wu, J., Kachanuban, K., Wilaipun, P., Zhu, W., & Wang, Y. (2021). Bioactive metabolites from acid-tolerant fungi in a Thai mangrove sediment. *Frontiers in Microbiology, 11*, 609952. https://doi.org/10.3389/fmicb.2020.609952

Garzoli, L., Gnavi, G., Tamma, F., Tosi, S., Varese, G. C., & Picco, A. M. (2015). Sink or swim: Updated knowledge on marine fungi associated with wood substrates in the Mediterranean Sea and hints about their potential to remediate hydrocarbons. *Progress in Oceanography, 137*, 140–148. https://doi.org/10.1016/j.pocean.2015.05.028

Girich, E. V., Yurchenko, A. N., Smetanina, O. F., Trinh, P. T. H., Ngoc, N. T. D., & Yurchenko, E. A. (2020). Neuroprotective metabolites from Vietnamese marine-derived fungi of *Aspergillus* and *Penicillium* genera. *Marine Drugs, 18*, 608. https://doi.org/10.3390/md18120608

Godinho, V. M., de Paula, M. T. R., Silva, D. A. S., Paresque, K., Martins, A. P., Colepicolo, P., et al. (2019). Diversity and distribution of hidden cultivable fungi associated with marine animals of Antarctica. *Fungal Biology, 123*(7), 507–516. https://doi.org/10.3390/md18120608

Gonçalves, M. F., Esteves, A. C., & Alves, A. (2022). Marine fungi: Opportunities and challenges. *Encyclopedia, 2*(1), 559–577. https://doi.org/10.3390/encyclopedia2010037

Gutiérrez, M. H., Jara, A. M., & Pantoja, S. (2016). Fungal parasites infect marine diatoms in the upwelling ecosystem of the Humboldt current system off central Chile. *Environmental Microbiology, 18*(5), 1646–1653. https://doi.org/10.1111/1462-2920.13257

Hou, X. M., Hai, Y., Gu, Y. C., Wang, C. Y., & Shao, C. L. (2019). Chemical and bioactive marine natural products of coral-derived microorganisms (2015–2017). *Current Medicinal Chemistry, 26*(38), 6930–6941. https://doi.org/10.2174/0929867326666190626153819

Ingavat, N., Dobereiner, J., Wiyakrutta, S., Mahidol, C., Ruchirawat, S., & Kittakoop, P. (2009). Aspergillusol A, an α-glucosidase inhibitor from the marine-derived fungus *Aspergillus aculeatus*. *Journal of Natural Products, 72*(11), 2049–2052. https://doi.org/10.1021/np9003883

Isti'anah, I., Tarman, K., Suseno, S. H., Nugraha, R., & Effendi, I. (2024a). Screening of bioactive antibacterial compounds from marine endophytic fungi from Buton Island. *Indonesia. Jurnal Pengolahan Hasil Perikanan Indonesia, 27*(7), 553–563. https://doi.org/10.17844/jphpi.v27i7.50489

Isti'anah, I., Tarman, K., Suseno, S. H., Nugraha, R., Effendi, I., Lee, W. C., et al. (2024b). Diversity and antibacterial potential of marine endophytic fungi by submerged fermentation from Buton Island, Indonesia. *BIO Web of Conferences, 147*, 01007. https://doi.org/10.1051/bioconf/202414701007

Janarkho, G. F., Trianto, A., Sedjati, S., & Listari, R. P. (2023). Study on the anti-vibrio activity of marine fungi *Aspergillus sydowii* and *Rhizopus* sp. using OSMAC approach. *Jurnal Kelautan Tropis, 26*(3), 536–542. https://doi.org/10.14710/jkt.v26i3.20451

Jiang, M., Wu, Z., Guo, H., Liu, L., & Chen, S. (2020). A review of terpenes from marine-derived fungi: 2015–2019. *Marine Drugs, 18*(6), 321. https://doi.org/10.3390/md18060321

Jin, Y., Weining, S., & Nevo, E. (2005). A MAPK gene from Dead Sea fungus confers stress tolerance to lithium salt and freezing–thawing: Prospects for saline agriculture. *Proceedings of the National Academy of Sciences, 102*(51), 18992–18997. https://doi.org/10.1073/pnas.0509653102

Johnson, T. W. & Sparrow, F. K. (1961). Fungi in Oceans and Estuaries. J. Cramer, Germany

Jones, E. G. (1988). Do fungi occur in the sea? *Mycologist, 2*(4), 150–157. https://doi.org/10.1016/S0269-915X(88)80035-1

Jones, E. B. G. (2011). Are there more marine fungi to be described? *Fungal Biology Reviews, 25*(2), 152–159. https://doi.org/10.1515/bot.2011.043

Jones, E. G., & Jennings, D. H. (1964). The effect of salinity on the growth of marine fungi in comparison with non-marine species. *Transactions of the British Mycological Society, 47*(4), 619–625. https://doi.org/10.1016/S0007-1536(64)80041-3

Jones, E. G., & Pang, K. L. (2012). Tropical aquatic fungi. *Biodiversity and Conservation, 21*, 2403–2423. https://doi.org/10.1007/s10531-011-0198-6

Jones, E. B. G., Read, S. J., Moss, S. T., Alias, S. A., & Hyde, K. D. (1996). *Tirisporella* gen. nov., an ascomycete from the mangrove palm *Nypa fruticans*. *Canadian Journal of Botany, 74*(9), 1487–1495. https://doi.org/10.1139/b96-179

Jones, E. B. G., Sakayaroj, J., Suetrong, S., Somrithipol, S., & Pang, K. L. (2009). Classification of marine *Ascomycota*, anamorphic taxa and *Basidiomycota*. *Fungal Diversity, 35*, 1-187.

Jones, E. G., Alias, S. A., & Pang, K. L. (2013). Distribution of marine fungi and fungus-like organisms in the South China Sea and their potential use in industry and pharmaceutical application. *Malaysian Journal of Science, 32*, 95–106. https://doi.org/10.22452/mjs.vol32no3.9

Jones, E. G., Suetrong, S., Sakayaroj, J., Bahkali, A. H., Abdel-Wahab, M. A., Boekhout, T., & Pang, K. L. (2015). Classification of marine *Ascomycota, Basidiomycota, Blastocladiomycota,* and *Chytridiomycota*. *Fungal Diversity, 73*, 1–72. https://doi.org/10.1007/s13225-015-0339-4

Jones, E. B. G., Pang, K. L., Abdel-Wahab, M. A., Scholz, B., Hyde, K. D., Boekhout, T., et al. (2019). An online resource for marine fungi. *Fungal Diversity, 96*, 347–433. https://doi.org/10.1007/s13225-019-00426-5

Julia, Z. A. R., Sukarno, N., Ardie, S. W., Batubara, I., Tjitrosoedirdjo, S., & Waite, M. (2022). Endophytic fungi isolated from the mangrove species *Rhizophora apiculata* and their efficacy as herbicides. *HAYATI Journal of Biosciences, 29*, 605–620. https://doi.org/10.4308/hjb.29.5.605-620

Kasettrathat, C., Ngamrojanavanich, N., Wiyakrutta, S., Mahidol, C., Ruchirawat, S., & Kittakoop, P. (2008). Cytotoxic and antiplasmodial substances from marine-derived fungi, *Nodulisporium* sp. and CRI247-01. *Phytochemistry, 69*(14), 2621–2626. https://doi.org/10.1016/j.phytochem.2008.08.005

Khamthong, N., Rukachaisirikul, V., Phongpaichit, S., Preedanon, S., & Sakayaroj, J. (2014). An antibacterial cytochalasin derivative from the marine-derived fungus *Diaporthaceae* sp. PSU-SP2/4. *Phytochemistry Letters, 10*, 5–9. https://doi.org/10.1016/j.phytol.2014.06.014

Kohlmeyer, J., & Kohlmeyer, E. (1979). *Marine mycology—The higher fungi*. Academic Press, New York.

Küppers, L., Ebrahim, W., El-Neketi, M., Özkaya, F. C., Mándi, A., Kurtán, T., et al. (2017). Lactones from the sponge-derived fungus *Talaromyces rugulosus*. *Marine Drugs, 15*(11), 359. https://doi.org/10.3390/md15110359

Kurnia, A. D., Suharti, S., & Tarman, K. (2024). Antibacterial activity assay of marine fungal extract *Aspergillus terreus* (WB 1–2) against *Escherichia coli* cause of colibacillosis disease. *IOP Conference Series: Earth and Environmental Science, 1359*, 012090. https://doi.org/10.1088/1755-1315/1359/1/012090

Latif, P. S. M. A., Noor, N. M. M., Mazlan, S., Razali, N., Ramasamy, K., & Ariffin, S. A. (2022). Saline tolerant of marine endophytic fungi from Teluk Kemang Malaysia – A rich source of bioactive material. *Malaysian Applied Biology, 51*(5), 193–200. https://doi.org/10.55230/mabjournal.v51i5.2366

Lavadia, M. G. B., Dagamac, N. H. A., & dela Cruz, T. E. E. (2017). Diversity and biofilm inhibition activities of algicolous fungi collected from two remote islands of the Philippine archipelago. *Current Research in Environmental & Applied Mycology, 7*(4), 309–321.

Leyte-Lugo, M., González-Andrade, M., González, M. D. C., Glenn, A. E., Cerda-García-Rojas, C. M., & Mata, R. (2012). (+)-Ascosalitoxin and vermelhotin, a calmodulin inhibitor, from an endophytic fungus isolated from *Hintonia latiflora*. *Journal of Natural Products, 75*(9), 1571–1577. https://doi.org/10.1021/np300327y

Lutfiah, R., Juliasih, N. L. G. R., Hendri, J., & Setiawan, A. (2021). Screening extract EtOAc sponge-derived fungi against clinical *Staphylococcus aureus* to obtain sustainable natural products. *IOP Conference Series: Earth and Environmental Science*. https://doi.org/10.1088/1755-1315/940/1/012043

Mahé, S., Rédou, V., Calvez, T. L., Vandenkoornhuyse, P., & Burgaud, G. (2013). Fungi in deep-sea environments and metagenomics. *The Ecological Genomics of Fungi*, 325–354. https://doi.org/10.1002/9781118735893.ch15

Marchese, P., Garzoli, L., Young, R., Allcock, L., Barry, F., Tuohy, M., & Murphy, M. (2021). Fungi populate deep-sea coral gardens as well as marine sediments in the Irish Atlantic Ocean. *Environmental Microbiology, 23*(8), 4168–4184. https://doi.org/10.1111/1462-2920.15560

Montaser, R., & Luesch, H. (2011). Marine natural products: A new wave of drugs? *Future Medicinal Chemistry., 3*(12), 1475–1489. https://doi.org/10.4155/fmc.11.118

Moron, L. S., Lim, Y. W., & dela Cruz, T. E. E. (2018). Antimicrobial activities of crude culture extracts from mangrove fungal endophytes collected in Luzon Island, Philippines. *Philippine Science Letters, 11*, 28–36. https://scienggj.org/2018-special-issue-4/

Mosadeghzad, Z., Zuriati, Z., Asmat, A., Gires, U., Wickneswari, R., Pittayakhajonwut, P., & Farahani, G. H. N. (2013). Chemical components and bioactivity of the marine-derived fungus *Paecilomyces* sp. collected from Tinggi Island, Malaysia. *Chemistry of Natural Compounds, 49*, 621–625. https://doi.org/10.1007/s10600-013-0693-y

Munro, R. C., Wilson, J., Jefwa, J., & Mbuthia, K. W. (1999). A low-cost method of mycorrhizal inoculation improves growth of *Acacia tortilis* seedlings in the nursery. *Forest Ecology and Management, 113*(1), 51–56. https://doi.org/10.1016/S0378-1127(98)00414-9

Nabila, S., Suharti, S., & Tarman, K. (2024). Investigated of marine endophytic fungi *Aspergillus terreus* extract's antibacterial activity against *Salmonella Typhimurium*. *IOP Conference Series: Earth and Environmental Science, 1359*, 012093. https://doi.org/10.1088/1755-1315/1359/1/012093

Ngoc, N. T. D., Trinh, P. T. H., Trung, D. T., Yurchenko, A. N., Khanh, H. H. N., Hung, L. D., Trang, V. T. D., Hoa, L. T., & Thinh, P. D. (2023). Screening of antibacterial and antioxidant activities of marine fungi isolated from the North Sea of Vietnam. *Vietnam Journal of Marine Science and Technology, 23*(2), 189–201. https://doi.org/10.15625/1859-3097/17458

Ngoc, N. T. D., Trinh, P. T. H., Yurchenko, E. A., Yurchenko, A. N., Hang, C. T. T., Trang, V. T. D., Van, T. T. T., Thinh, P. D., Khanh, H. H. N., & Phuong, B. T. N. (2024). Antioxidant and cytoprotective activities of marine fungi isolated from brown seaweeds in Nha Trang Bay, Khanh Hoa Province, Central Vietnam. *Vietnam Journal of Marine Science and Technology, 24*(3), 297–306. https://doi.org/10.15625/1859-3097/18869

Ngoc, N. T. D., Yurchenko, E. A., Trinh, P. T. H., Menchinskaya, E. S., Dieu, T. V. T., Savagina, A. D., Minin, A., Thinh, P. D., Khanh, H. H. N., Van, T. T. T., & Yurchenko, A. N. (2025). Secondary metabolites of Vietnamese marine fungus *Penicillium chermesinum* 2104NT-1.3

and their cardioprotective activity. *Regional Studies in Marine. Science, 81*, 104003. https://doi.org/10.1016/j.rsma.2024.104003

Ngokpol, S., Suwakulsiri, W., Sureram, S., Lirdprapamongkol, K., Aree, T., Wiyakrutta, S., Mahidol, C., Ruchirawat, S., & Kittakoop, P. (2015). Drimane sesquiterpene-conjugated amino acids from a marine isolate of the fungus *Talaromyces minioluteus* (*Penicillium minioluteum*). *Marine Drugs, 13*(6), 3567–3580. https://doi.org/10.3390/md13063567

Nguyen, V. D., & Pham, T. T. (2022). *Penicillium vietnamense* sp. nov., the first novel marine fungi species described from Vietnam with a unique conidiophore structure and molecular phylogeny of *Penicillium* section *Charlesia*. *Mycobiology, 50*(3), 155–165. https://doi.org/10.1080/12298093.2022.2068750

Notarte, K. I., Nakao, Y., Yaguchi, T., Bungihan, M., Suganuma, K., & dela Cruz, T. E. (2017). Trypanocidal activity, cytotoxicity and histone modifications induced by malformin A1 isolated from the marine-derived fungus *Aspergillus tubingensis* IFM 63452. *Mycosphere, 8*(1), 10. https://doi.org/10.5943/mycosphere/8/1/10

Notarte, K. I., Yaguchi, T., Suganuma, K., & dela Cruz, T. E. (2018). Antibacterial, cytotoxic and trypanocidal activities of marine-derived fungi isolated from Philippine macroalgae and seagrasses. *Acta Botanica Croatica, 77*(2), 141–151. https://doi.org/10.2478/botcro-2018-0016

Nuansri, S., Rukachaisirikul, V., Rungwirain, N., Kaewin, S., Yimnual, C., Phongpaichit, S., Preedanon, S., Sakayaroj, J., & Muanprasat, C. (2022). α-Pyrone and decalin derivatives from the marine-derived fungus *Trichoderma harzianum* PSU-MF79. *Natural Product Research, 36*(21), 5462–5469. https://doi.org/10.1080/14786419.2021.2015593

Nurhayati, N., Lisna, L., Ramadan, F., & Tarman, K. (2024). Utilization of marine fungi as a feed additive on the performance of local chickens' digestive organs. *BIO Web of Conferences, 92*, 02015. https://doi.org/10.1051/bioconf/20249202015

Nursid, M., Dewi, A. S., & Maya, D. (2020). Cytotoxicity of marine-derived fungi collected from Kepulauan Seribu Marine National Park. In *IOP Conference Series: Earth and Environmental Science* (Vol. 404, No. 1, 012006). IOP Publishing. https://doi.org/10.1088/1755 1315/404/1/012006

Overy, D. P., Bayman, P., Kerr, R. G., & Bills, G. F. (2014). An assessment of natural product discovery from marine (*sensu strictu*) and marine-derived fungi. *Mycology, 5*(3), 145–167. https://doi.org/10.1080/21501203.2014.931308

Paliany, A. S., Sivasothy, Y., Awang, K., Rizman-Idid, M., & Alias, S. A. (2014). Marine-derived fungi of Peninsular Malaysia—A biochemical perspective. *Chiang Mai Journal of Science, 41*(4), 894–909.

Pandey, A. (2019). Pharmacological significance of marine microbial bioactive compounds. *Environmental Chemistry Letters, 17*(4), 1741–1751. https://doi.org/10.1007/s10311-019-00908-7

Pang, K. L., Overy, D. P., Jones, E. G., da Luz Calado, M., Burgaud, G., Walker, A. K., et al. (2016). 'Marine fungi' and 'marine-derived fungi' in natural product chemistry research: Toward a new consensual definition. *Fungal Biology Reviews, 30*(4), 163–175. https://doi.org/10.1016/j.fbr.2016.08.001

Pansanit, A., Park, E.-J., Kondratyuk, T. P., Pezzuto, J. M., Lirdprapamongkol, K., & Kittakoop, P. (2013). Vermelhotin, an anti-inflammatory agent, suppresses nitric oxide production in RAW 264.7 cells via p38 inhibition. *Journal of Natural Products, 76*(9), 1824–1827. https://doi.org/10.1021/np400565e

Pasqualetti, M., Giovannini, V., Barghini, P., Gorrasi, S., & Fenice, M. (2020). Diversity and ecology of culturable marine fungi associated with *Posidonia oceanica* leaves and their epiphytic algae *Dictyota dichotoma* and *Sphaerococcus coronopifolius*. *Fungal Ecology, 44*, 100906. https://doi.org/10.1016/j.funeco.2019.100906

Pham, T. T., Dinh, K. V., & Nguyen, V. D. (2021). Biodiversity and enzyme activity of marine fungi with 28 new records from the tropical coastal ecosystems in Vietnam. *Mycobiology, 49*(6), 559–581. https://doi.org/10.1080/12298093.2021.2008103

Pitt, J. I., & Hocking, A. D. (1997). *Fungi and food spoilage* (2nd ed., pp. 203–338). Springer. https://doi.org/10.1007/978-0-387-92207-2

Prachyawarakorn, V., Mahidol, C., Sureram, S., Sangpetsiripan, S., Wiyakrutta, S., Ruchirawat, S., & Kittakoop, P. (2008). Diketopiperazines and phthalides from a marine-derived fungus of the order *Pleosporales*. *Planta Medica, 74*(1), 69–72. https://doi.org/10.1055/s-2007-993783

Pramana, A. A. C., Ramadhani, E., Priyambada, F., Pertiwi, G. A., Setiawibawa, R. A. A., & Wijayanti, N. (2022). Sponge-associated fungi isolate from *Ancorina* sp. showed anti-cancer activity against HeLa cell lines. *Journal of Microbiology, Biotechnology and Food Sciences, 12*. https://doi.org/10.55251/jmbfs.3352

Preedanon, S., Phongpaichit, S., Sakayaroj, J., Rukachaisirikul, V., Trisuwan, K., & Plathong, S. (2016). Antimicrobial activities of fungi derived from the gorgonian sea fan *Annella* sp. and their metabolites. *Indian Journal of Marine Sciences, 45*(11), 1491-1498.

Prompanya, C., Dethoup, T., Gales, L., Lee, M., Pereira, J. A., Silva, A. M., et al. (2016). New polyketides and new benzoic acid derivatives from the marine sponge-associated fungus *Neosartorya quadricincta* KUFA 0081. *Marine Drugs, 14*(7), 134. https://doi.org/10.3390/md14070134

Raghukumar, C. (2008). Marine fungal biotechnology: An ecological perspective. *Fungal Diversity, 31*, 19–35.

Raghukumar, S., & Raghukumar, S. (2017). The marine environment and the role of fungi. In *Fungi in coastal and oceanic marine ecosystems: Marine fungi* (pp. 17–38). Springer. https://doi.org/10.1007/978-3-319-54304-8_2

Ramirez, C. S. P., Notarte, K. I. R., & dela Cruz, T. E. E. (2020). Antibacterial activities of mangrove leaf endophytic fungi from Luzon Island, Philippines. *Studies in Fungi, 5*(1), 320–331. https://doi.org/10.5943/sif/5/1/14

Rateb, M. E., & Ebel, R. (2011). Secondary metabolites of fungi from marine habitats. *Natural Product Reports, 28*(2), 290–344. doi: 10.1039/c0np00061b.

Sadaba, R. B., & Niego, A. G. T. (2017, May). Biodegradation of hydrocarbon by marine-derived fungi isolated from oil-contaminated beach and mangrove soil in Western Visayas, Philippines. *International Oil Spill Conference Proceedings, 2017*(1), 2017357. International Oil Spill Conference. https://doi.org/10.7901/2169-3358-2017.1.000357

Sandrawati, N., Ningsih, W., Layla, R., Putra, A. E., Ismed, F., Tallei, T. E., & Handayani, D. (2023). Endophytic fungi from mangrove plant *Acanthus ilicifolius* L.: Antimicrobial, anticancer, and species determination. *Trends in Sciences, 20*(7), 5089–5089. https://doi.org/10.48048/tis.2023.5089

Setyowati, E. P., Purwantiningsih, E., Erawan, F. M. Y., Rahmanti, S., Hanum, N. M. R., & Devi, N. C. M. (2021). Cytotoxic and antimicrobial activities of ethyl acetate extract from fungus *Trichoderma reesei* strain JCM 2267, *Aspergillus flavus* strain MC-10-L, *Penicillium* sp., and *Aspergillus fumigatus* associated with marine sponge *Stylissa flabelliformis*. *Research Journal of Pharmacy and Technology, 14*, 5126–5132. DOI:10.7324/JAPS.2021.1101207

Shaumi, A., Cheang, U. C., Yang, C. Y., Chang, C. W., Guo, S. Y., Yang, C. H., et al. (2021). Culturable fungi associated with the marine shallow-water hydrothermal vent crab *Xenograpsus testudinatus* at Kueishan Island, Taiwan. *Botanica Marina, 64*(4), 289–300. https://doi.org/10.1515/bot-2021-0034

Shin, H. J. (2020). Natural products from marine fungi. *Marine Drugs, 18*(5), 230. https://doi.org/10.3390/md18050230

Sibero, M. T., Pribadi, R., Larasati, S. J. H., Calabon, M. S., Sabdono, A., Subagiyo, S., & Frederick, E. H. (2021). Diversity of sponge-associated fungi from a mangrove forest in Kemujan Island, Karimunjawa National Park, Indonesia. *Biodiversitas, 22*, 5695–5705. https://doi.org/10.13057/biodiv/d221256

Singh, S., Singh, V. K., Aamir, M., Dubey, M. K., Patel, J. S., Upadhyay, R. S., & Gupta, V. K. (2016). Cellulase in pulp and paper industry. In *New and future developments in microbial biotechnology and bioengineering: Microbial cellulase system properties and applications*. Elsevier Inc. https://doi.org/10.1016/B978-0-444-63507-5.00013-7

Smetanina, O. F., Yurchenko, A. N., Ivanets, E. V. G., Trinh, P. T., Antonov, A. S., Dyshlovoy, S. A., et al. (2020). Biologically active echinulin-related indolediketopiperazines from the marine sediment-derived fungus *Aspergillus niveoglaucus*. *Molecules, 25*, 61. https://doi.org/10.3390/molecules25010061

Solis, M. J. L., Draeger, S., & dela Cruz, T. E. E. (2010). Marine-derived fungi from *Kappaphycus alvarezii* and *K. striatum* as potential causative agents of ice-ice disease in farmed seaweeds. *Botanica Marina, 53*(6), 587–594. https://doi.org/10.1515/bot.2010.071

Sparrow, F. K., Jr. (1937). The occurrence of saprophytic fungi in marine muds. *The Biological Bulletin, 73*(2), 242–248. https://doi.org/10.2307/1537586

Su, G. S., Dueñas, K., Roderno, K., Sison, M. A., Su, M. L. S., Ragragio, E., et al. (2013). Distribution and diversity of marine fungi in Manila Bay, Philippines. Annual Research & Review in Biology, 4 (24), 4166–4173. DOI: 10.9734/ARRB/2014/12338

Sumilat, D. A., & Lintang, R. A. J. (2021). Antibacterial potential of marine fungus *Aspergillus nomius* isolated from green algae *Bornetella* sp. *Jurnal Ilmiah PLATAX, 9*, 49–49. https://doi.org/10.35800/jip.9.1.2021.33781

Sun, H. H., Mao, W. J., Chen, Y., Guo, S. D., Li, H. Y., Qi, X. H., et al. (2009). Isolation, chemical characteristics, and antioxidant properties of the polysaccharides from marine fungus *Penicillium* sp. F23–2. *Carbohydrate Polymers, 78*(1), 117–124. https://doi.org/10.1016/j.carbpol.2009.04.017

Supaphon, P., Phongpaichit, S., Rukachaisirikul, V., & Sakayaroj, J. (2013). Antimicrobial potential of endophytic fungi derived from three seagrass species: *Cymodocea serrulata, Halophila ovalis*, and *Thalassia hemprichii*. *PLoS One, 8*(8), e72520. https://doi.org/10.1371/journal.pone.0072520

Supaphon, P., Phongpaichit, S., Rukachaisirikul, V., & Sakayaroj, J. (2014). Diversity and antimicrobial activity of endophytic fungi isolated from the seagrass *Enhalus acoroides*. *Indian Journal of Geo-Marine Sciences, 43*, 785–797.

Sureram, S., Kesornpun, C., Mahidol, C., Ruchirawat, S., & Kittakoop, P. (2013). Directed biosynthesis through biohalogenation of secondary metabolites of the marine-derived fungus *Aspergillus unguis*. *RSC Advances, 3*(6), 1781–1788. https://doi.org/10.1039/C2RA23021F

Syam, N. A., Tondok, E. T., Tarman, K., & Widodo, W. (2022). Screening of marine fungi as biological control agents of *Colletotrichum acutatum* on chili. *Jurnal Fitopatologi Indonesia, 18*(2), 53–65. https://doi.org/10.14692/jfi.18.2.53-65

Tan, M. A., dela Cruz, T. E., Apurillo, C. C. S., & Proksch, P. (2015). Chemical constituents from a Philippine mangrove endophytic fungi *Phyllosticta* sp. *Der Pharma Chemica, 7*(2), 43–45.

Tan, C., Liu, Z., Chen, S., Huang, X., Cui, H., Long, Y., et al. (2016). Antioxidative polyketones from the mangrove-derived fungus *Ascomycota* sp. SK2YWS-L. *Scientific Reports, 6*(1), 36609. https://doi.org/10.1038/srep36609

Tarman, K., Lindequist, U., Wende, K., Porzel, A., Arnold, N., & Wessjohann, L. A. (2011). Isolation of a new natural product and cytotoxic and antimicrobial activities of extracts from fungi of Indonesian marine habitats. *Marine Drugs, 9*, 294–306. https://doi.org/10.3390/md9030294

Tarman, K., Supinah, P., Dewanti, E. W., Santoso, J., & Nurjanah. (2024). Characteristics of carrageenan from seaweed hydrolysis using marine fungi as hard-shell capsule material. *Jurnal Pengolahan Hasil Perikanan Indonesia, 27*(8), 642–653. https://doi.org/10.17844/jphpi.v27i8.51946

Torres, J. M. O., & dela Cruz, T. E. E. (2012). Production of xylanases by mangrove fungi from the Philippines and their application in enzymatic pretreatment of recycled paper pulps. *World Journal of Microbiology and Biotechnology, 29*, 645–655. https://doi.org/10.1007/s11274-012-1220-1

Torres, J. M. O., Cardenas, C. V., Moron, L. S., Guzman, A. P. A., & dela Cruz, T. E. E. (2011). Dye decolorization activities of marine-derived fungi isolated from Manila Bay and Calatagan Bay, Philippines. *Philippine Journal of Science, 140*(2), 133–143.

Trianto, A., Radjasa, O. K., Subagiyo, S., Purnaweni, H., Bahry, M. S., Djamaludin, R., et al. (2021). Potential of fungi isolated from a mangrove ecosystem in Northern Sulawesi, Indonesia: Protease, cellulase and anti-microbial capabilities. *Biodiversitas, 22*(4), 1717–1724. https://doi.org/10.13057/biodiv/d220415

Trinh, P. T. H., Tien, P. Q., Ngoc, N. T. D., Ly, B. M., & Van, T. T. T. (2018). Isolation and screening of marine fungi with antimicrobial activity from samples collected in Nha Trang Bay, Vietnam. *Vietnam Journal of Biotechnology, 16*(1), 181–187. DOI: https://doi.org/10.15625/1811-4989/16/1/13463

Trisuwan, K., Rukachaisirikul, V., Phongpaichit, S., Preedanon, S., & Sakayaroj, J. (2011). Modiolide and pyrone derivatives from the sea fan-derived fungus *Curvularia* sp. PSU-F22. *Archives of Pharmacal Research, 34*(5), 709–714. https://doi.org/10.1007/s12272-011-0502-8

Tubaki, K., & Nishihara, N. (1969). Alternaria helianthi (Hansf.) comb. nov. *Transactions of the British Mycological Society, 53*(1), 147–IN14. https://doi.org/10.1016/S0007-1536(69)80022-7

Vitale, G. A., Coppola, D., Palma Esposito, F., Buonocore, C., Ausuri, J., Tortorella, E., & de Pascale, D. (2020). Antioxidant molecules from marine fungi: Methodologies and perspectives. *Antioxidants, 9*(12), 1183. https://doi.org/10.3390/antiox9121183

Wahjuningrum, D., Efianda, T. R., Tarman, K., Yuhana, M., Effendi, I., & Saputra, F. (2020). Supplementation of *Nodulisporium* sp. KT29 induced by *Vibrio harveyi* as an immunostimulant for controlling vibriosis in vannamei white shrimp under a marine culture system. *Jurnal Akuakultur Indonesia, 19*, 95–105. DOI: 10.19027/jai.19.2.95-105

Wang, C., Tang, S., & Cao, S. (2021). Antimicrobial compounds from marine fungi. *Phytochemistry Reviews, 20*, 85–117. https://doi.org/10.1007/s11101-020-09705-5

Widyasti, A. R., & Rukachaisirikul, V. (2017). Secondary metabolites of the marine-derived fungi: *Aspergillus* sp. AMF243. *Indonesian Journal of Chemistry and Environment, 1*(1), 1–6.

Willems, T., De Mol, M. L., De Bruycker, A., De Maeseneire, S. L., & Soetaert, W. K. (2020). Alkaloids from marine fungi: Promising antimicrobials. *Antibiotics, 9*(6), 340. https://doi.org/10.3390/antibiotics9060340

Wiradana, P. A., Permatasari, A. A. A. P., Sari, N. K. Y., Widhiantara, I. G., Sandhika, I. M. G. S., Waruwu, E., Fimaputra, J. M., Panjaitan, N. S. D., Putri, F. R., & Ansori, A. N. M. (2024). Isolation and molecular identification of endophytic fungi associated with brown algae for inhibiting *Escherichia coli* ESBL. *Journal of Applied Veterinary Science and Technology, 5*(1), 31–41. https://doi.org/10.20473/javest.V5.I1.2024.31-41

Wulandari, A. P., & Primastia, N. (2024). Skrining aktivitas antibakteri dari fungi simbion karang lunak *Sarcophyton* sp. hasil isolasi di perairan Pulau Pramuka, Taman Nasional Laut Kepulauan Seribu. *Jurnal Mikologi. Indonesia, 6*(1), 48–56.

Xu, J., Yi, M., Ding, L., & He, S. (2019). A review of anti-inflammatory compounds from marine fungi, 2000–2018. *Marine Drugs, 17*(11), 636. https://doi.org/10.3390/md17110636

Yasuhara-Bell, J., & Lu, Y. (2010). Marine compounds and their antiviral activities. *Antiviral Research, 86*(3), 231–240. https://doi.org/10.1016/j.antiviral.2010.03.009

Youssef, D. T., Shaala, L. A., & Genta-Jouve, G. (2022). Asperopiperazines A and B: Antimicrobial and cytotoxic dipeptides from a tunicate-derived fungus *Aspergillus* sp. DY001. *Marine Drugs, 20*(7), 451. https://doi.org/10.3390/md20070451

Yurchenko, A. N., Smetanina, O. F., Ivanets, E. V., Phan, T. T. H., Ngo, N. T. D., Zhuravleva, O. I., Rasin, A. B., Dyshlovoy, S. A., Menchinskaya, E. S., Pislyagin, E. A., von Amsberg, G., Afiyatullov, S. S., & Yurchenko, E. A. (2020). Auroglaucin-related neuroprotective compounds from Vietnamese marine sediment-derived fungus *Aspergillus niveoglaucus*. *Natural Product Research, 34*, 2589–2594. https://doi.org/10.1080/14786419.2018.1547293

Zainuddin, N., Alias, S. A., Lee, C. W., Ebel, R., Othman, N. A., Mukhtar, M. R., & Awang, K. (2010). Antimicrobial activities of marine fungi from Malaysia. *Botanica Marina, 53*(6), 507–513. https://doi.org/10.1515/bot.2010.067

Chapter 8
Techniques in Detecting Fungal Elements in Biopsy Samples: Perspectives from Local Practices in Southeast Asia

Marvin C. Masalunga

Abstract Opportunistic infections, including mycotic diseases, are increasingly being recognized as a major public health concern in Southeast Asia. The diagnosis of fungal infections from clinical specimens relies mainly on culture and, at select institutions, molecular methods. In certain instances, it is possible to detect fungal elements, such as hyphae and yeast forms, in tissue samples submitted to the histopathology laboratory. Routine examination with hematoxylin and eosin (H and E) staining is sufficient to detect the presence of organisms in most cases. On occasion, special stains such as periodic acid-Schiff (PAS) and Gomori methenamine silver (GMS) may be utilized to highlight fungal structures. Definite species identification of fungi detected on routine histology still relies on correlation with clinical findings and ancillary tests. Among the challenges encountered in the region are the lack of facilities, equipment, and supplies in regular diagnostic laboratories; overreliance on referral laboratories; lack of skilled laboratory personnel; and few opportunities for training and subspecialization in diagnostic mycology. The introduction of newer techniques, such as MALDI-TOF and nucleic acid-based tests, may augment the services related to fungal diagnosis on biopsy specimens. Furthermore, international linkages in the region, especially in tropical medicine, may pave the way toward enhancing mycology services in the region.

Keywords Fungi · Histopathology · Mycology · Opportunistic infections · Special stains · Southeast Asia

M. C. Masalunga (✉)
School of Medicine, Southwestern University PHINMA, Cebu City, Philippines

Chong Hua Hospital, Cebu City, Philippines

Vicente Sotto Memorial Medical Center, Cebu City, Philippines

J. J. G. Guerrero et al. (eds.), *Mycology in a Changing Planet: Applications and Perspectives from Southeast Asia*, https://doi.org/10.1007/978-981-95-2640-6_8

Introduction

Infectious diseases are a major cause of morbidity in Southeast Asia and are widely recognized as a public health concern. Tuberculosis, dengue, malaria, and seasonal infections are widespread in the region. With the rapidly increasing prevalence of acquired immunodeficiency syndrome (AIDS), the number of opportunistic infections (OIs) is also on the rise (Batac & Denning, 2017). Some of the most common OIs are due to fungal organisms, which may be part of the normal flora of mucosal sites. *Candida albicans*, for instance, is a normal colonizer of the oral cavity and gastrointestinal tract (Handog & Dayrit, 2005). Talaromycosis, previously known as penicilliosis, is also increasingly being recognized as an emerging OI in Southeast Asia (Chakrabarti & Slavin, 2011).

Fungal genera that commonly cause infections in humans include *Aspergillus*; *Candida*; dermatophytes such as *Trichophyton*, *Epidermophyton*, and *Microsporum*; *Pneumocystis*; and *Cryptococcus* (Notarte et al., 2023). All these species are endemic to Southeast Asia. Except for the dermatophytes, all these species are included in the World Health Organization's (WHO) list of priority fungal pathogens. *Cryptococcus neoformans* and *Pneumocystis jirovecii* are known to cause invasive diseases in immunocompromised individuals, such as people living with human immunodeficiency virus infection (PLHIV). In contrast, *Candida* spp. and *Aspergillus fumigatus* are known to cause severe disease in other patient groups (WHO, 2022).

Culture remains the gold standard in the diagnosis of fungal infections, as it allows the direct visualization of fungal morphology; however, this may not always be available in certain settings (Challa et al., 2014). A survey performed by Chindamporn et al. (2018) showed that most laboratories in Southeast Asia rely on microscopic and culture methods, and there is almost no universal access to more advanced diagnostic tests in countries such as Indonesia, the Philippines, and Thailand (Table 8.1). Furthermore, depending on the quality of the samples, some cultures yield negative results, which should not be used as the sole basis to completely rule out a mycotic infection (Borman et al., 2022).

Molecular methods are being developed to shorten the turnaround time for samples obtained from patients suspected of mycotic infections. The WHO recommends the development of rapid, accurate, and affordable diagnostic tests, particularly point-of-care (POC) and biomarker-based tests, that may be rolled out in low- to middle-income countries (LMICs). This includes the development of rapid diagnostics for susceptibility testing and those that are applicable for use among different patient subpopulations (WHO, 2022). In Southeast Asia, such techniques may be introduced to diagnostic laboratories, which usually lack the resources for extensive fungal studies (Chindamporn et al., 2018). The WHO further recognizes the threat of antimicrobial-resistant (AMR) fungal organisms and has therefore included in its research priorities several themes related to the diagnosis of priority fungal pathogens (Bertagnolio et al., 2024).

Table 8.1 Usual laboratory practices employed in Southeast Asia for fungal diagnosis

Direct microscopy
Potassium hydroxide (KOH) method
Wet smear
Histopathologic techniques
Routine paraffin processing with H and E staining
Cytologic examination (cell block and smears)
Microbiologic methods
Culture
Sensitivity testing
Laboratory tests with limited availability
Special stains (e.g., GMS, PAS)
Immunohistochemistry
MALDI-TOF
Nucleic acid amplification tests

Overview of Histopathologic Techniques

The timely detection and diagnosis of fungal infections is of the utmost importance, especially in immunocompromised individuals. Laboratory samples, including biopsy specimens, that may harbor fungal organisms must be carefully evaluated, as antifungal therapy may be initiated depending upon the provisional species identification. Particular attention should be given to body sites that are normally sterile, as the presence of fungal elements is practically diagnostic of a disseminated or opportunistic infection (Borman et al., 2022).

Tissue samples are usually submitted for the evaluation of a possible malignancy. Due to overlapping clinical signs and symptoms, infectious diseases may mimic certain neoplasms. Laboratory personnel working in the histopathology laboratory, including pathologists and medical laboratory scientists, should be knowledgeable in identifying possible infectious processes in a biopsy sample, as this will help guide patient management. One widely recognized benefit of being able to identify the presence of cancer-mimicking infections in tissue is that it prevents a more radical surgical procedure, and the medical team may opt instead for pharmacologic management of the patient. From the author's experience, this is a relatively common scenario in many healthcare training institutions, wherein infectious agents are detected on mass-forming lesions. However, the extent to which infectious agents, let alone fungal infections, mimic malignancies remains understudied.

On occasion, tissues are outright submitted to determine if there are microscopic features suggestive of an infection. Morphologic features that are compatible with infections include granuloma formation and necrosis, which are changes consistent with the host immune response (Guarner & Brandt, 2011). These morphologic features also help physicians and other healthcare professionals understand host-pathogen interactions, as the pattern and progression of infection may vary from acute neutrophilic inflammation to a chronic lymphoplasmacytic response.

Histopathologic detection is a rapid and cost-effective technique in the presumptive diagnosis of fungal infection. In general, laboratory staff might be able to detect fungal spores and hyphae in histologic slides routinely stained with hematoxylin and eosin (H and E; Figs. 8.1 and 8.2). The H and E slide is useful in highlighting the inflammatory processes in a tissue sample, including the presence of necrotic material and hemorrhage. Pigmented fungi are relatively easy to identify on H- and E-stained slides (Guarner & Brandt, 2011).

Other stains include the Papanicolau stain, GMS, PAS, and calcofluor white. The Papanicolau stain is sufficient to detect fungal elements in smears from exfoliative gynecologic samples (Fig. 8.3). Stains such as Gomori methenamine silver (GMS) stain, periodic acid-Schiff (PAS), and Calcofluor white stain may prove useful in

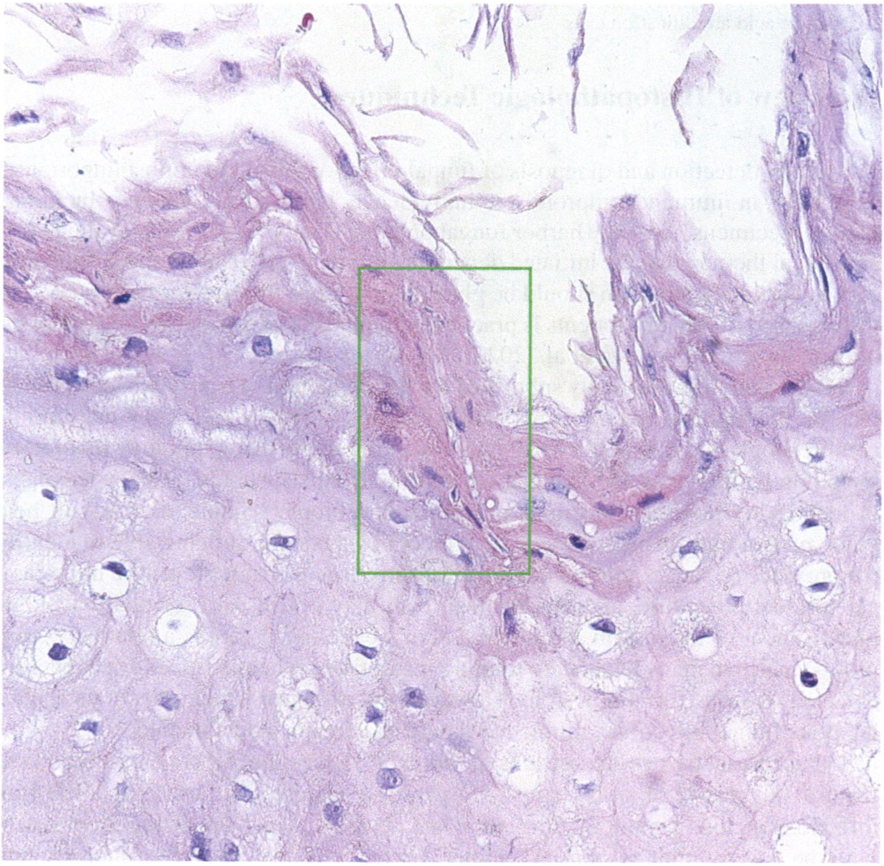

Fig. 8.1 In this biopsy section from the skin, careful evaluation of the slide revealed the subtle presence of fungal hyphae (enclosed in the green box). Further laboratory identification of the fungus was not possible as no other samples were submitted for additional laboratory tests (H and E, 400×; photomicrograph part of the author's archives)

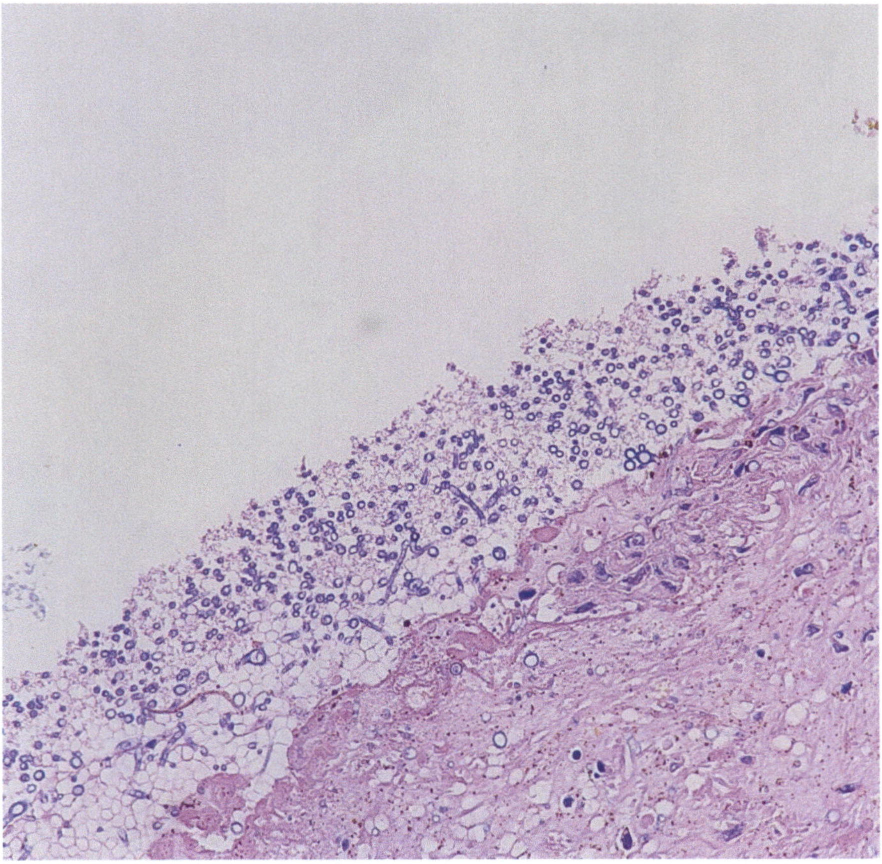

Fig. 8.2 In this biopsy section from a sample labeled "ulcer," the fungal hyphae and yeast forms are readily identifiable. Further laboratory identification of the fungus was not possible as no other samples were submitted for additional microbiology tests (H and E, 100×; photomicrograph part of the author's archives)

highlighting the fungal elements. GMS and PAS can be used to stain carbohydrates in the cell walls of fungi, although PAS can also stain mucoproteins and mucin in tissues, limiting its utility. In GMS, the tissue is stained green, providing a stark contrast to the dark brown to black stain of the fungal elements (Fig. 8.4). However, members of the *Mucorales* might be pale-staining in GMS (Guarner & Brandt, 2011).

Other stains may highlight specific cellular substances. Mucin stains can be used to highlight the presence of mucopolysaccharides of certain encapsulated organisms, e.g., *Cryptococcus*. In select cases, the Fontana-Masson stain, which detects melanin pigment, may also be employed to detect fungal organisms (Howell, 2023).

The final pathologic report must describe in detail the morphology of the fungus seen, including other findings that are related to the host immune reaction and invasion of the tissue submitted. A list indicating the possible fungi associated with the

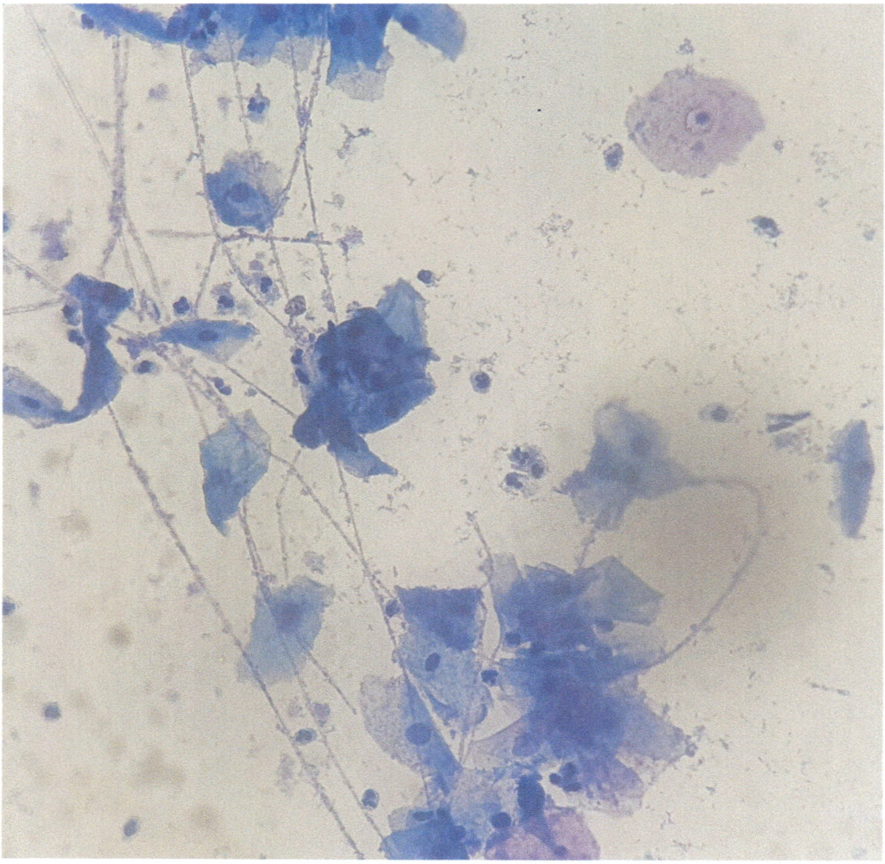

Fig. 8.3 Fungal elements, particularly *Candida* spp., are readily detectable on pap smears. Conventional smears are readily available in most laboratories (H and E, 400×; photomicrograph part of the author's archives)

described histologic features, including parasites that may appear similar to fungal elements, must be included (Guarner & Brandt, 2011).

Examples of Applications of Histopathologic Techniques in Select Fungal Groups

Dermatophyte Infections

The highly infectious dermatophytes cause superficial infections in a wide range of hosts, including humans and other mammals. In the Philippines, initial epidemiologic studies on mycoses focused on dermatophytosis, as the isolation and

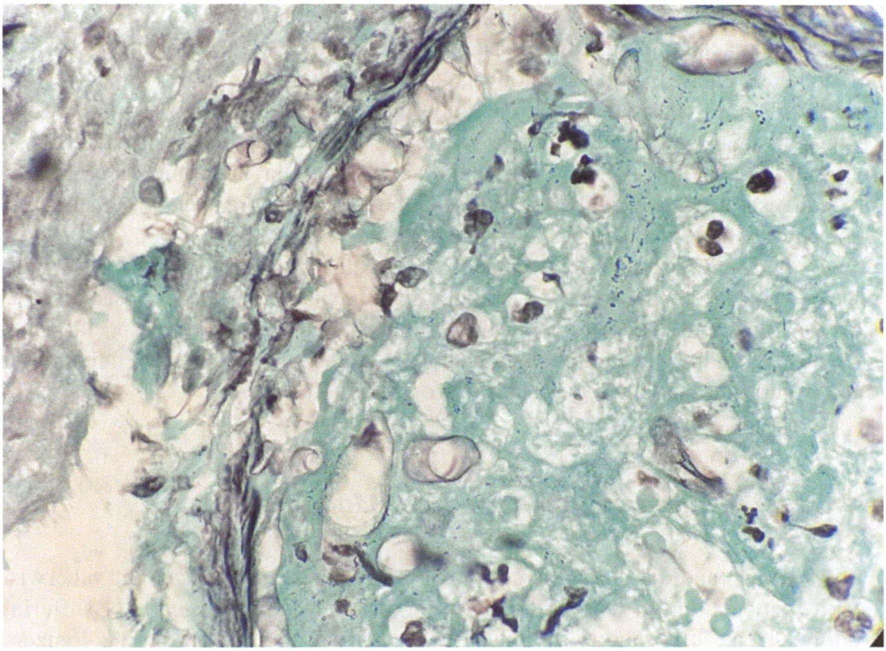

Fig. 8.4 Intestinal biopsy specimen, taken from a 37-year-old diabetic patient with invasive intestinal mucormycosis. The silver stain readily highlights the fungal structures within the mucosal biopsy sample (GMS, 400×; courtesy of Dr. John Anthony Tindoc, Silliman University Medical Center)

identification of fungi were limited to skin, hair, and nail samples (Batac & Denning, 2017). Dermatophytosis in humans is classified based on the anatomic location, i.e., tinea barbae, tinea corporis, tinea pedis, tinea cruris, tinea unguium, tinea manuum, tinea faciei, and tinea capitis. Each dermatophyte may involve more than one anatomic site. Skin lesions may be directly examined using dermoscopy, although species identification may not be possible (Howell, 2023; Moskaluk & VandeWoude, 2022).

Skin and nail scrapings are suitable samples for screening for the presence of dermatophytes. Potassium hydroxide (KOH), which is cheap and readily accessible, can be applied to these samples in order to detect the presence of fungal elements. However, a disadvantage to this technique is that it cannot differentiate between living and dead cells. Species identification is also not possible using this method (Moskaluk & VandeWoude, 2022). In addition, experience with the use of KOH is required, as artifacts may mimic fungal forms (Knoll et al., 2023). Sensitivity may be enhanced by using a fluorescent stain such as calcofluor white (Howell, 2023).

Histopathologic techniques may be employed to evaluate the presence of invasive dermatophytosis. Although fungal spores and hyphae are usually detected in the stratum corneum, certain infections (e.g., *Trichophyton rubrum*) may be seen in the dermis and subcutaneous tissues. Aside from the presence of fungal elements, there are characteristic, albeit non-specific, changes in the skin associated with dermatophytes. These include keratotic changes, a spongiotic pattern of dermatitis, neutrophils at the basal layers of the epidermis, and eosinophils within the dermis (Moskaluk & VandeWoude, 2022). However, the diagnosis of dermatophytosis relies mainly on clinical grounds. Usually, observation via direct microscopy is sufficient to establish the diagnosis, precluding the need for more extensive histopathologic studies such as H and E and special staining.

An important consideration to make when evaluating skin samples for the presence of fungal elements is that dermatophytosis may be seen superimposed on other skin disorders and even on neoplastic entities. As such, aside from special stains such as PAS and GMS, immunohistochemistry studies may prove beneficial to rule out a concurrent malignancy, such as mycoses fungoides.

Detection of Fungal Organisms from Mucosal Sites

Vulvovaginal candidiasis is a mycosis that is usually seen in young and otherwise healthy patients. In 2017, around 1.4 million women were estimated to be suffering from recurrent vulvovaginal candidiasis. *Candida* species are common colonizers, i.e., commensals of mucosal tissue and skin. The presence of isolates consistent with *Candida* may not be significant in tissue biopsies taken from commonly colonized sites. Nevertheless, the presence of this organism may be significant in immunocompromised patients and may be seen in the form of esophageal candidiasis, peritonitis, and even candidemia (Handog & Dayrit, 2005).

Similar to what is done to detect dermatophytes, mucosal samples may be subjected to KOH preparations for the presumptive identification of *Candida* species. Likewise, calcofluor white may be used to increase the sensitivity of direct examination of samples prepared using KOH. In tissue samples, the presence of budding yeasts, hyphae, and pseudohyphae indicates *Candida* infection, and these may be detected on routine H and E examination or with the use of special stains such as PAS and GMS (Howell, 2023).

Exceptionally, immunohistochemical methods (IHCs), using antibodies targeted toward specific *Candida* antigens, may be employed to confirm the presence of fungal elements. Cross-reaction might occur among different fungal species; therefore, IHCs may be limited in their ability to determine the species of a fungus. Commercial IHC antibodies are available for *Candida* and other fungi, such as *Aspergillus* and *Mucor* (Howell, 2023).

With the advent of tests such as matrix-assisted laser desorption/ionization time-of-flight mass spectrometry (MALDI-TOF), which allows the identification of fungi within hours, fungal IHCs are not locally procured. Several referral centers are already offering MALDI-TOF as part of their microbiology services. These include institutions in the capital, Metro Manila, and one in Metro Cebu. Also available are in situ hybridization methods and nucleic acid amplification techniques, but most centers still employ culture methods or MALDI-TOF for fungal identification.

Conventional pap smear is widespread in most laboratories in Southeast Asia, and anatomic pathologists are competent in the evaluation of these preparations. Despite its low sensitivity, the ease with which samples can be collected, even by general practitioners, and the simple manner of preparation, allows most clinics and diagnostic laboratories to offer this service. Liquid-based biopsy has not yet gained wide acceptance, despite the improved sensitivity, mainly because of the high cost associated with it.

Opportunistic Fungal Infections

Immunocompromised patients, patients with hematological malignancies, and individuals with diabetes mellitus are at risk of acquiring mycotic infections (Howell, 2023). These include *Mucor, Aspergillus,* and *Cryptococcus*, which are present in the environment. *Mucor and Aspergillus* exist in the environment as molds, whereas *Cryptococcus* is found in yeast forms. Mucormycosis, in general, is increasingly being detected in patients with diabetes mellitus, and one such case was reported locally (Fig. 8.4; Thompson 3rd et al., 2024). In these infections, noninvasive mycetomas or fungus balls may form in the affected sites. These organisms, particularly *Aspergillus*, may result in disseminated bloodstream infections.

Aspergillosis is a major fungal infection in people with comorbidities. Chronic pulmonary aspergillosis (CPA), commonly affects patients with concurrent tuberculosis. Allergic bronchopulmonary aspergillosis (ABPA) is seen in patients with asthma (Batac & Denning, 2017). Various species of *Aspergillus* have varying susceptibilities to antifungal drugs; therefore, accurate species identification is best performed using molecular methods.

Cryptococcus is an opportunistic infection that may cause meningitis in AIDS patients with a CD4 count of less than 100 cells per μL. The yeast form may be readily detected in the cerebrospinal fluid using the classic India ink method, wherein the mucopolysaccharide capsule of the organism gives it the appearance of a clear halo. This capsule contains starch-like compounds, which are amenable to being stained by mucicarmine or Alcian blue on tissue sections (Howell, 2023). However, white blood cells and fat droplets may resemble *C. neoformans*, making the diagnosis difficult. The use of KOH may dissolve leukocytes and other native cells, assisting in the recognition of bona fide yeast forms (Knoll et al., 2023). Mucicarmine and Fontana-Masson stains may also assist in the detection of *Cryptococcus* (Guarner & Brandt, 2011).

The various stains discussed in the preceding sections are summarized in Table 8.2.

Table 8.2 Summary of the most commonly available stains available in the region

Stain	Applications	Limitations
Hematoxylin and eosin (H and E)	Available in most laboratories in the region, especially those with histopathology services Can be used to detect the presence of most fungal elements	Does not highlight fungal walls, making them difficult to distinguish from the background Cannot be used to identify specific species
Silver stains (e.g., GMS)	For screening of fungal infections; may be combined with H and E to visualize host reaction	*Mucor* may appear pale
Periodic acid-Schiff (PAS)	Binds to the glycogen in the wall of fungi, allowing them to be highlighted for screening	Some tissues may be glycogen-rich, limiting the utility of PAS
Mucin stains (e.g., mucicarmine)	Stains the capsule of *Cryptococcus*	Also stain native mucin, which may preclude evaluation
Fontana-Masson stain	Stains melanin in certain fungi; able to stain *Cryptococcus*	Melanin pigment may be present in native tissue
Immunohistochemistry (IHC)	Can target specific antigens in various cells, allowing increased sensitivity and specificity Several chromogens are available	Expensive and might not be suitable in low-resource settings IHCs for fungi are not readily available

Adapted from Guarner and Brandt (2011)

Challenges in Histopathologic Diagnosis of Fungal Infections

In a survey conducted by the Asia Fungal Working Group (AFWG), as many as 87% of physicians have access to histopathology services in countries such as Indonesia, the Philippines, Singapore, and Thailand. However, several challenges related to diagnostics are encountered, including lack of formal training in mycology, lack of non-culture based tests, and absence of formal procedures to guide patient management (Tan et al., 2020). These challenges are discussed in more detail below.

One key aspect of laboratory testing is the issuance of standardized reports. The use of a standardized histology reporting template for fungal infections may be beneficial in relaying clinically significant results while acknowledging the inherent limitations of tissue diagnosis of fungi. Knoll et al. (2023) suggest that mycology reports include the presence or absence of yeast forms, septate or non-septate hyphae, pigmentation, cellular location of the organisms, and any other specific structures being present.

Such a template, however, demands that the laboratory staff be sufficiently knowledgeable in fungal morphology. It is unfortunate that a lack of knowledge in mycology was found to be a major source of misclassification errors in the study by Sangoi et al. (2009). This lack of knowledge in mycology not only affects the histopathology services but also the microbiology services of a laboratory, as Borman et al. (2022) noted that the diagnostic sensitivity of culture studies might be affected by difficulties in the interpretation of results, which relies on the expertise of the laboratory staff.

Use of these templates also varies from one country to another. In the Philippines, for instance, structured biopsy reports are not routinely employed. Practices vary from one laboratory to another, with certain laboratories settling only for a report of the presence or absence of fungal structures. Tertiary and referral centers, on the other hand, especially those located in metropolitan areas, are able to provide a more comprehensive report, sometimes integrating information from other sections of the laboratory, such as microbiology, clinical microscopy, and serology. Such variations in practices highlight not only the lack of access to mycology services but also the number of laboratory staff with subspecialty training in mycology.

Other challenges that the histopathology laboratory may encounter when it receives biopsy samples for microscopic evaluation are related to the preanalytical phase, i.e., the paucity of tissue fragments and procedure-related distortion of the specimen. For biopsy samples obtained via invasive procedures (e.g., bronchoscopy), the amount of tissue submitted may be very limited, not only for histologic evaluation but also for other sections of the laboratory, such as the microbiology unit.

The quality of the processed slides also plays a significant role in fungal identification on biopsy samples (Challa et al., 2014). Two commonly used methods in the preparation of samples for histopathologic evaluation are tissue homogenization and dicing. A study by Borman et al. (2022) showed that the former can significantly reduce the recovery rates of organisms such as *Mucor* and *Aspergillus*. When the tissue sample submitted to the laboratory is limited, the method of sample preparation should be carefully considered to optimize the yield of fungal elements. Locally, if the biopsy samples are small, they are processed in their entirety, with no homogenization or dicing. These may account for false-positive results, as depending on the number of tissue levels, fungal structures may or may not be apparent. The rate of false positive results, however, remains unknown, as no formal studies have been conducted in this aspect.

If a fungal infection is suspected in solid organs, the goal of the laboratory staff is to determine the presence of fungal elements within the tissue sample, i.e., hyphae and yeasts. It is possible for the pathologist with specialty training in microbiology to narrow down on certain species, but culture and molecular techniques remain the most optimal methods of determining the exact species of a fungus. Otherwise, histopathology is usually not able to provide the fungal genus and species (Guarner & Brandt, 2011). In certain instances when the culture is negative, biopsy findings may be beneficial in the detection of certain mycoses (Challa et al., 2014).

Overlapping morphologic features and the presence of non-fungal mimickers, including normal human tissue structures, may make the diagnosis difficult. For instance, when staining with GMS is performed, neurosecretory granules may be mistaken for yeast cells. Basement membrane may also be confused with fungal hyphae (Guarner & Brandt, 2011). Another pitfall, albeit uncommon, is the presence of dual infections. The study by Challa et al. (2014) showed occasional discordance between culture results and histopathologic evaluation. Some of the discordant results are due to the presence of two distinct fungal organisms, in which a fungus was not seen or recognized on biopsy but grew in culture media.

Table 8.3 Some of the challenges in Southeast Asian laboratories in terms of fungal histopathology

Lack of facilities offering histopathologic and microbiologic services
Reliance on sending out samples to referral laboratories, precluding initiatives to establish local mycology-related services
Lack of laboratory personnel with subspecialty training in diagnostic mycology, including:
Poor processing techniques
Lack of expertise in fungal morphology
No standardized reporting template available
Few diagnostic courses available for laboratory staff
Difficulty in obtaining tissues samples from viscera

Aside from the morphologic similarities among certain fungal species, the presence of normal colonizers should also be considered, especially in mucosal sites. As stated earlier, *Candida* is part of the normal flora, and therefore, the laboratory staff should give due attention to the host tissue response. The importance of a good patient history cannot be overemphasized, as the presence of *Candida* hyphae in the mucosal biopsy of a neutropenic patient is most likely indicative of a true opportunistic infection. On the other hand, some patients may self-medicate with steroids, and this alters the clinical appearance of certain lesions, which prevents the collection of good samples (Howell, 2023).

Table 8.3 summarizes the challenges encountered by Southeast Asian laboratories in terms of fungal diagnosis through histopathologic services.

Toward a More Responsive Fungal Histopathology Service in Southeast Asia

Although the preceding section focused mainly on the lack of standardized reporting and procedures related to fungal biopsy diagnosis, it is evident that these are just part and parcel of the greater challenge in fungal histopathology, i.e., limited access, inadequate resources, and lack of properly trained personnel.

There are opportunities in the region for the enhancement of mycology services, especially in low- and middle-income countries. In particular, the Southeast Asian Ministers of Education Tropical Medicine and Public Health Network (SEAMEO TROPMED) offers short and diploma courses in various fields related to tropical medicine. Some of these courses include microbiology and parasitology. Although mycology-related trainings are usually offered as short courses—with the most recent one being held last November 2024, the network can serve as the lead convenor in order to promote and enhance the capacity of laboratory staff in various techniques related to fungal detection and diagnosis.

Laboratory staff members trained in mycology are in the best position to lobby for more comprehensive laboratory testing. Accordingly, the author has seen remarkable progress in certain laboratories when a subspecialist is present. The subspecialist is able to design testing packages and protocols, do the costing, communicate with clinicians regarding optimal sample collection, and provide reliable results. Although this might take several years in the making, this will definitely lessen the number of send-outs to referral laboratories, which can be costly and time-consuming.

Furthermore, the presence of trained laboratory staff will also push the field of mycology forward, as they can spearhead research projects that will help establish the epidemiology of fungal infections, elucidate the pathogenesis of certain diseases, and, more importantly, determine the best treatment options through the conduct of multidisciplinary trials involving other healthcare professionals.

As mentioned earlier, most laboratories are able to offer basic microbiology and histopathology services already. A careful consideration of the current status of mycology testing in the various countries in Southeast Asia can help push laboratories towards expanding their services, ultimately providing patients with a more comprehensive evaluation of their infections.

Conclusion

A number of laboratory techniques are available or being developed to screen for fungal infections, especially for their detection among immunocompromised individuals. These include procedures from different sections of the laboratory, including microbiology, clinical microscopy, immunology, and histopathology. The histopathology section plays a key role in the detection of invasive fungal infections in tissue samples. For certain organisms, fungal elements may be readily detected with simple, affordable, and readily available methods, such as KOH and H and E stains. With the use of special stains, including GMS, PAS, and stains for other substances, the sensitivity of tissue detection may be increased. Laboratory staff may be able to provide a presumptive differential list of fungal species, but this must be correlated with other pertinent information, including the patient's history and ancillary findings. A template is ideal in order to provide a meaningful report. Challenges related to pre-analytical and analytical factors may be encountered. The laboratory must exercise due prudence in the provision of an initial diagnosis to ensure that the appropriate treatment is instituted for the patient. Training opportunities to empower staff in the diagnosis of fungi in tissue are available in the region, and this should be maximized in order to provide comprehensive and responsive diagnostic mycology services.

References

Batac, M. C. R., & Denning, D. (2017). Serious fungal infections in The Philippines. *European Journal of Clinical Microbiology & Infectious Diseases, 36*(6), 937–941. https://doi.org/10.1007/s10096-017-2918-7

Bertagnolio, S., Dobreva, Z., Centner, C. M., et al. (2024). WHO global research priorities for antimicrobial resistance in human health. *The Lancet. Microbe, 5*(11), 100902. https://doi.org/10.1016/S2666-5247(24)00134-4

Borman, A. M., Mohammed, S., Palmer, M. D., Childs, N., & Johnson, E. M. (2022). The importance of appropriate processing and direct microscopic examination for the timely diagnosis and management of invasive infections caused by filamentous fungi. *Medical Mycology, 60*(12), myac081. https://doi.org/10.1093/mmy/myac081

Chakrabarti, A., & Slavin, M. A. (2011). Endemic fungal infections in the Asia-Pacific region. *Medical Mycology, 49*(4), 337–344. https://doi.org/10.3109/13693786.2010.551426. Epub 2011 Jan 24. PMID: 21254966.

Challa, S., Pamidi, U., Uppin, S. G., Uppin, M. S., & Vemu, L. (2014). Diagnostic accuracy of morphologic identification of filamentous fungi in paraffin embedded tissue sections: Correlation of histological and culture diagnosis. *Indian Journal of Pathology & Microbiology, 57*(4), 583–587. https://doi.org/10.4103/0377-4929.142673

Chindamporn, A., Chakrabarti, A., Li, R., Sun, P. L., Tan, B. H., Chua, M., Wahyuningsih, R., Patel, A., Liu, Z., Chen, Y. C., & Chayakulkeeree, M. (2018). Survey of laboratory practices for diagnosis of fungal infection in seven Asian countries: An Asia Fungal Working Group (AFWG) initiative. *Medical Mycology, 56*(4), 416–425. https://doi.org/10.1093/mmy/myx066

Guarner, J., & Brandt, M. E. (2011). Histopathologic diagnosis of fungal infections in the 21st century. *Clinical Microbiology Reviews, 24*(2), 247–280. https://doi.org/10.1128/CMR.00053-10

Handog, E. B., & Dayrit, J. F. (2005). Mycology in The Philippines, revisited. *Nihon Ishinkin Gakkai zasshi = Japanese Journal of Medical Mycology, 46*(2), 71–76. https://doi.org/10.3314/jjmm.46.71

Howell, S. A. (2023). Dermatopathology and the diagnosis of fungal infections. *British Journal of Biomedical Science, 80*, 11314. https://doi.org/10.3389/bjbs.2023.11314

Knoll, M. A., Steixner, S., & Lass-Flörl, C. (2023). How to use direct microscopy for diagnosing fungal infections. *Clinical Microbiology and Infection, 29*(8), 1031–1038. https://doi.org/10.1016/j.cmi.2023.05.012

Moskaluk, A. E., & VandeWoude, S. (2022). Current topics in dermatophyte classification and clinical diagnosis. *Pathogens (Basel, Switzerland), 11*(9), 957. https://doi.org/10.3390/pathogens11090957

Notarte, K. I. R., Pastrana, A. M., Ver, A. T. M., et al. (2023). Mycosis in The Philippines: Epidemiology, clinical presentation, diagnostics and interventions. In J. J. G. Guerrero, T. U. Dalisay, et al. (Eds.), *Developments in microbiology: Mycology in the tropics* (pp. 213–233). Academic. https://doi.org/10.1016/B978-0-323-99489-7.00005-6

Sangoi, A. R., Rogers, W. M., Longacre, T. A., Montoya, J. G., Baron, E. J., & Banaei, N. (2009). Challenges and pitfalls of morphologic identification of fungal infections in histologic and cytologic specimens: A ten-year retrospective review at a single institution. *American Journal of Clinical Pathology, 131*(3), 364–375. https://doi.org/10.1309/AJCP99OOOZSNISCZ

Tan, B. H., Chakrabarti, A., Patel, A., Chua, M. M. M., Sun, P. L., Liu, Z., Rotjanapan, P., Li, R., Wahyuningsih, R., Chayakulkeeree, M., & Chen, Y. C. (2020). Asia Fungal Working Group (AFWG). Clinicians' challenges in managing patients with invasive fungal diseases in seven Asian countries: An Asia Fungal Working Group (AFWG) Survey. *International Journal of Infectious Diseases, 95*, 471–480. https://doi.org/10.1016/j.ijid.2020.01.007. Epub 2020 Jan 13. PMID: 31945491.

Thompson, G. R., 3rd, Chen, S. C., Alfouzan, W. A., Izumikawa, K., Colombo, A. L., & Maertens, J. (2024). A global perspective of the changing epidemiology of invasive fungal disease and real-world experience with the use of isavuconazole. *Medical Mycology, 62*(9), myae083. https://doi.org/10.1093/mmy/myae083

World Health Organization. (2022). *WHO fungal priority pathogens list to guide research, development and public health action*. WHO. Licence: CC BY-NC-SA 3.0 IGO.

Part III
Mycology and the Challenges of Agriculture

Chapter 9
Mycology in Enhancing Agriculture: From Crop Protection to Crop Yield Improvement

Mari Neila P. Seco, Kris Lord T. Santos, Victor L. Corbita, Diana P. Paguntalan, Mark Christian Felipe R. Redillas, Latiffah Zakaria, Melissa P. Montecalvo, Angelbert D. Cortes, Billy Joel M. Almarinez, Thaddeus M. Carvajal, Nico Bravo, Jonathan Jaime G. Guerrero, Divina M. Amalin, Christian Joseph R. Cumagun, Joy M. Jamago, and Mark Angelo O. Balendres

Abstract Fungi are a diverse group of eukaryotic organisms that play an important role in agricultural ecosystems. While many fungi are known for causing plant diseases, others are highly beneficial, promoting soil fertility, breaking down environmental pollutants, plant growth, nutrient cycling, and natural pest and disease control. In a changing planet where consumers demand food production systems to use safer, cost-effective, and environmentally friendlier strategies, technologies, and applications, fungi have become indispensable. This chapter discusses the multi-pronged agroecological and beneficial functions of fungi, from promoting crop growth to protecting crop health. This chapter also highlights the applications and significance of beneficial fungi in enhancing agriculture in the Philippines, Malaysia, and Indonesia, which strongly rely on agriculture and allied industries. Beneficial fungi are crucial for addressing environmental and agricultural challenges, includ-

M. N. P. Seco
Department of Pest Management, College of Agriculture and Food Science, Visayas State University, Baybay City, Leyte, Philippines

K. L. T. Santos
Department of Biology, College of Science, De La Salle University, Manila, Philippines

Institute of Biological Control, De La Salle University-Laguna Campus, LTI Spine Road Barangays Biñan and Malamig, Biñan City, Laguna, Philippines

V. L. Corbita
Department of Forestry, College of Forestry and Environmental Science, Caraga State University, Ampayon, Butuan City, Philippines

J. J. G. Guerrero et al. (eds.), *Mycology in a Changing Planet: Applications and Perspectives from Southeast Asia*, https://doi.org/10.1007/978-981-95-2640-6_9

ing climate change, biodiversity loss, and ecosystem degradation. As research advances, fungi will be increasingly important in ecosystem restoration, climate change mitigation efforts, and sustainable agriculture in Southeast Asia and beyond.

Keywords Arbuscular mycorrhizal fungi · Biological control · Ectomycorrhizal fungi · Entomopathogenic fungi · *Trichoderma*

D. P. Paguntalan
Division of Biological Sciences, College of Arts and Sciences, University of the Philippines Visayas, Miagao, Iloilo, Philippines

M. C. F. R. Redillas · M. A. O. Balendres (✉)
Department of Biology, College of Science, De La Salle University, Manila, Philippines

Plant and Soil Health Research Unit, College of Science, De La Salle University, Manila, Philippines
e-mail: mark.angelo.balendres@dlsu.edu.ph

L. Zakaria
School of Biological Sciences, Universiti Sains Malaysia, Penang, Malaysia

M. P. Montecalvo
National Crop Protection Center, College of Agriculture and Food Science, University of the Philippines Los Baños, Los Baños, Laguna, Philippines

A. D. Cortes
Department of Biological Sciences, College of Arts and Sciences, Cavite State University, Indang, Cavite, Philippines

B. J. M. Almarinez · T. M. Carvajal · D. M. Amalin
Department of Biology, College of Science, De La Salle University, Manila, Philippines

Institute of Biological Control, De La Salle University-Laguna Campus, LTI Spine Road Barangays Biñan and Malamig, Biñan City, Laguna, Philippines

Plant and Soil Health Research Unit, College of Science, De La Salle University, Manila, Philippines

N. Bravo
Department of Biology, College of Science, De La Salle University, Manila, Philippines

Plant and Soil Health Research Unit, College of Science, De La Salle University, Manila, Philippines

National University Manila, Manila, Philippines

J. J. G. Guerrero
College of Medicine, University of the Philippines Manila, Manila, Philippines

C. J. R. Cumagun
Parma Research and Extension Center, University of Idaho, Parma, Idaho, USA

J. M. Jamago
Department of Agronomy and Plant Breeding, College of Agriculture, Central Mindanao University, Maramag, Bukidnon, Philippines

Introduction

Fungi are a diverse group of eukaryotic organisms that play an important role in agricultural ecosystems. While many fungi are known for causing plant diseases, others are highly beneficial, promoting soil fertility, plant growth, nutrient cycling, and natural pest and disease control. Central to the valuable role of fungi is their capacity to break down organic matter and facilitate nutrient recycling. By recycling nutrients, fungi help maintain soil fertility and ecosystem productivity. Species of basidiomycetes and ascomycetes can break down lignin, a complex polymer in wood (Treseder & Lennon, 2015), allowing them to support the overall nutrient dynamics, e.g., in forests (Arnstadt et al., 2016; Fukasawa, 2021). For instance, *Phanerochaete chrysosporium*, a white-rot fungus, is widely regarded as the model laboratory species for studying lignin degradation as it can colonize wood quickly, has strong ligninolytic abilities, is fast-growing, and is easy to propagate (Li et al., 2022). Another fungus of global importance is the *Trichoderma*, a saprophyte genus. It is a beneficial soil microorganism that improves soil quality by enhancing nutrient composition (Singh et al., 2015) and promoting plant growth (Azarmi et al., 2011; Zhang et al., 2019). By interacting with other microbes in the rhizosphere (Jaroszuk-Ściseł et al., 2019), *Trichoderma* helps shape the soil microbiome (Asghar & Kataoka, 2021; Wu et al., 2022a, b), making it a valuable fungus in sustainable agriculture.

Mycorrhizal fungi form mutualistic associations with approximately 90% of terrestrial plant species, including many important crops such as wheat, maize, and legumes (Smith & Read, 2008a, b). The two most common types of mycorrhizal associations are arbuscular mycorrhizal fungi (AMF) and ectomycorrhizal fungi (EMF). The members of the Glomeromycotina (syn. Glomeromycota) are fungi that penetrate the root cells and form structures called arbuscules. These structures facilitate the mutual exchange of various nutrients between plant cells and fungi. They are regarded as the most important structures of AMF (Pimprikar & Gutjahr, 2018). They form mutualistic associations with terrestrial plants, improving nutrient and water absorption. They are especially effective in enhancing phosphorus uptake, which is often limited in soils, leading to improved biomass and plant growth (Naseer et al., 2022; Wu et al., 2022a, b), higher crop yields (Gao et al., 2020; Kang et al., 2023), and biotic and abiotic stress tolerance (Begum et al., 2019). Aside from phosphorus, AMFs are also known to transfer considerable amounts of nitrogen to their plant host. The extent, however, differs across varying plant-fungus associations (Gao et al., 2021). AMF species that have been widely studied and developed as commercial inoculants for plant growth enhancement include those in *Glomus, Gigaspora, Rhizophagus, Funneliformis*, and *Claroideoglomus* (Basiru & Hijri, 2022). EMF, in contrast, forms a sheath around the roots of host plants, mainly trees in temperate forests. By promoting tree growth and population dynamics (Anthony et al., 2022; Bennett et al., 2017), enhancing water and nutrient uptake (Tedersoo & Bahram, 2019), and contributing to carbon sequestration potential (Wallander et al., 2011), EMF plays an important role in forest ecosystems.

Fungi are also known for their ability to degrade environmental pollutants, including hydrocarbons, pesticides, and industrial chemicals, making them invaluable for cleaning contaminated environments (Harms et al., 2011) due to their intra- and extracellular enzymatic systems that are low in substrate specificity (El-Gendi et al., 2021) and their capacity to secrete acids (Deshmukh et al., 2016). Known fungi that can degrade pesticides include those in the genera *Trametes* (Hu et al., 2020), *Fusarium* (Bhatt et al., 2020), *Pleurotus* (Purnomo et al., 2010), *Ganoderma*, *Rhizopus*, and *Penicillium* (Akpasi et al., 2023). Some species in the genera *Aspergillus, Penicillium,* and *Trichoderma* (Khatua et al., 2024; Malachová et al., 2020) are known for plastic polymer degradation. Some species in the genera *Bjerkandera* and *Phanerochaete* (Rodarte-Morales et al., 2011) are known for the degradation of pharmaceuticals. Finally, filamentous fungi, including *Trichoderma* (Singh et al., 2015), *Penicillium* (Su et al., 2010), *Aspergillus* (Saravanan et al., 2021), and *Gloeophylum* (Achal et al., 2011), are known for heavy metal remediation among many others. The potential of many fungal genera harnesses numerous novel enzyme-based bioremediation strategies. In addition, other fungal species have been used in ecosystem restoration efforts, such as in regenerative agriculture. Mycorrhizal fungi have been applied to restore degraded lands and have been shown to improve mineral acquisition (phosphorus, nitrogen, and other minerals that are critical for plant growth) and maintain ecosystem stability and functioning (Asmelash et al., 2016; Kalamulla et al., 2022). This approach is auspicious in combating soil degradation in many areas around the globe.

Endophytic fungi colonize plant tissues without damaging the host and have diverse ecological functions. They have gained much research interest as novel sources of many important compounds and substances. They have been viable alternative sources of cytotoxic compounds such as those belonging to the genera *Alternaria, Aspergillus, Botryodiplodia, Botrytis, Cladosporium, Ectostroma, Fusarium, Metarhizium, Monochaetia, Mucor, Ozonium, Papulaspora, Periconia, Pestalotia, Pestalotiopsis, Phyllosticta, Pithomyces*, and *Taxomyces* (Uzma et al., 2018), antibacterial substances from important genera including *Diaporthe*, *Xylaria*, *Chaetomium*, *Talaromyces*, and *Paraphaeosphaeria* (Deshmukh et al., 2022). Several species of endophytic *Trichoderma* have been demonstrated to improve plant defense systems and resistance against abiotic stress (Azad & Kaminskyj, 2016; Lombardi et al., 2018). Other fungi of similar nature that are used as effective natural pest control agents are those in the genus *Beauveria* (Zimmermann, 2007) and *Metarhizium* (Butt et al., 2016a, b). Their use in biocontrol is promising in several agricultural settings (discussed in succeeding sections of this chapter).

As the demand for agricultural products increases, many farmers use chemical pesticides and synthetic fertilizers to improve the quality and quantity of their crops. However, excessive use of synthetic pesticides and fertilizers has proven to have caused great damage to the environment, causing soil and water pollution, resistant plant diseases, and residual pesticides causing contamination (Lu et al., 2015). Diverting back to conventional ways of cultivation has led to an increase in green agricultural development that minimizes the use of chemicals for pesticides and fertilizers. Several fungal genera are utilized as a supplement or alternative to increase crop yield due to their capacity for biological control and nutrient recycling. The demand for biological control programs, including integrated pest or disease management (IPM or IDM), has increased in recent years as the future of crop protection has increased (Stenberg et al., 2021). This chapter discusses fungi's multipronged agroecological beneficial functions, from promoting crop growth to protecting crop health. It highlights their versatility and significance in Southeast Asia. How beneficial fungi are utilized in some Southeast Asian countries (the Philippines, Malaysia, and Indonesia) that strongly rely on agriculture and allied industries is also presented.

Fungi in Plant Disease Management

Controlling plant diseases using natural means has become an increasingly adopted and favored strategy in sustainable agriculture. This is due to the unique features of beneficial fungi, which include safety, limited host range or target specificity, pose no harmful residues, are relatively friendly to the environment, and are easy to apply (Rajapakse et al., 2016; Thambugala et al., 2020). Among the beneficial fungi are the endophytic fungi, which live in plant tissues, and most do not harm the plants. These endophytes primarily produce bioactive compounds that may be essential for biotic stress resistance against plant pathogens (Panchalingam et al., 2022; Ratnaweera & de Silva, 2017; Singh & Kumar, 2023; Stone et al., 2004; Tian et al., 2022). Besides, these fungi use several methods like antibiosis, mycoparasitism, and induced host resistance to stop or inhibit the growth of pathogens (Manathunga et al., 2024; Muhammad et al., 2024). Antibiosis is the process by which some strains produce compounds that inhibit the growth of pathogenic fungi. For example, *Chaetomium seminudum* is reported to produce chaetosemins B, which have antifungal action against *Gibberella saubinetti* and *Magnaporthe oryzae* (Wen et al., 2022). The same is true for *Fusarium* sp. from maize, which can produce caffeine that has antifungal activity against *Alternaria alternata* (Fan & Shi, 2024). Moreover, *Curvularia lunata* and *Hypocreales* sp. produce prussic acid against rice diseases (Jena et al., 2024).

Trichoderma species are among the most successful agents in controlling root rots, wilts, and foliar diseases (Dutta et al., 2022; Singh et al., 2014). They are well-known for their ability to produce enzymes that can attack the cell walls of pathogens and take up nutrients and space that fungi would otherwise use (Guzmán-Guzmán et al., 2023). For instance, *T. asperellum* is active against *Phytophthora megakarya* by entering the sporocysts, coiling around the hyphae, forming appressoria, and producing hydrolytic enzymes, like laminarinase (Tondje et al., 2007). Also, a *T. harzianum* strain has been shown to attack soilborne wood-decaying fungi (Alshammari et al., 2024). Whereas *T. asperellum* species have displayed antagonistic effects to a variety of pathogens in maize, e.g., *Ustilago maydis* (Cuervo-Parra et al., 2022), in rice (Nurhayati et al., 2021), in tomato (El-Komy et al., 2015), and *Pythium myriotylum* in cocoyam, which causes root rot disease (Mbarga et al.,

2012). Chan et al. (2023) reported that the fast-growing capabilities of *T. harzianum* have resulted in the growth inhibition of cosmopolitan opportunistic fungal phytopathogens *R. microporus* (causative agent of root rot disease) and *Lasiodiplodia theobromae* (causative agent for black root rot), both >50%. However, *T. harzianum* has been described as inhibiting the growth of *P. noxius* (causative agent of brown tea root disease) by >80%. Antibiosis is another mechanism of *Trichoderma* species against plant pathogenic fungi, whereby the secretion of antagonistic substances inhibits plant pathogen growth (Yao et al., 2023).

Trichoderma viridae, T. koningii, T. asperrelum, T. parareesei, and *T. virens* have been proven to inhibit the growth of a cosmopolitan phytopathogenic fungus *F. oxysporum* that causes crown rot, head blight, and scab on cereal grains on different plants crops (Naher et al., 2019). The inoculation of *Trichoderma* species into crop plants has been shown to induce systemic resistance. *Trichoderma* species can improve the plant's self-defense mechanism by regulating elicitors and effectors (Yao et al., 2023). Cell wall-degrading enzymes produced by the fungi also contribute to the crop's resistance to a pathogen. Volatile organic compounds (VOCs) produced by *T. asperelloides* inhibited the growth of fungal infections, activated the defense responses, and promoted the plant growth and development of *Arabidopsis thaliana* (Chan et al., 2023).

Similar to *Trichoderma* species, *Penicillium* species, in addition to their known biological activities, are also recognized as antagonists against plant pathogens via various methods of action, including inducing resistance in plants (Hossain et al., 2007), synthesis of antimicrobial compounds, and development of mycoparasitic contacts (Nicoletti et al., 2007; Yang et al., 2008). The culture filtrate of *Penicillium janczewskii* was shown to possess an inhibitory effect against *R. solani* (Nicoletti et al., 2007), *Sclerotinia* (Akem, 1985), and *Macrophomina phaseolina* (Khan & Javaid, 2022). Antifungal compounds citrinin and emodin have been attributed to the antagonistic behavior of *P. citrinum* against several fungal plant pathogens (Luo et al., 2019). Some strains of *Aspergillus flavus, A. terreus,* and *Paecilomyces maximus* effectively inhibited *Ilyonectria destructans.* At the same time, *Mucor racemosus* was effective against *Botrytis cinerea* and *Neoscytalidium dimidiatum* (Haleem et al., 2024). Nevertheless, some strains of *Aspergillus* species can produce aflatoxins, and some *Mucor* species are post-harvest pathogens. These must be considered when choosing biological control agents against plant disease-causing fungi.

Mycorrhizal fungi are common soil-borne organisms with which most terrestrial plants (Smith & Read, 2010) form symbiotic relationships. These fungi provide benefits to their host plants, including higher nutrient uptake, better stress tolerance, and fungal pathogen protection (Azcón-Aguilar & Barea, 1997; Jung et al., 2012; Linderman, 1994; Sarkar et al., 2016; Talukder et al., 2024). It is supposed that pathogenic and AMF compete for shared resources in the roots, e.g., infection sites, space, and photosynthate, thus resulting in microbial alterations in the mycorrhizosphere, lignification of the root endodermal cells, respectively, better nutrition of the plant, competition for photosynthates, and activation of the plant defense mechanisms like phytoalexins (Singh et al., 2000; Whipps, 2004). The application of mycorrhizal fungi has been proven to boost the disease resistance of various horticultural crops to help them combat soil-borne pathogens (Azcón-Aguilar & Barea, 1997; Dar & Reshi, 2017) and root-infecting fungi (Sharma et al., 1992). For instance, the mycorrhizal fungus *Glomus mosseae* mitigated the symptoms of *Phytophthora* blight caused by *Phytophthora capsica* in pepper (Ozgonen & Erkilic, 2007) and damping off in eggplant, tomato, and ash gourd (*Benincasa hispida)* seedlings (Talukder et al., 2019, 2024). It is also antagonistic to *Sclerotium rolfsii* in peanut (Krishna & Bagyaraj, 1983), *Verticillium* wilt in eggplant (Matsubara et al., 1995), *Phytophthora nicotianae* var. *parasitica* in tomato (Trotta et al., 1996), *R. solani* in pea (Karagiannidis et al., 2002), and *Fusarium* wilt in tomato (Akköprü & Demir, 2005). Other species of *Glomus,* such as *G. etunicatum* and *G. intraradices,* have successfully controlled *R. solani* in potato (Yao et al., 2002).

Metarhizium anisopliae has also shown promising potential in inhibiting the growth of *Colletotrichum gloeosporioides* and *Pestalotia mangiferae* in mango (Visalakshy et al., 2017). Gupta et al. (2021) claimed that *M. brunneum* and *Beauveria bassiana* could trigger systemic plant immunity and increase resistance to disease by activating the defense mechanism of the host plant, resulting in high ROS production and the expression of genes related to defenses. The action of *B. bassiana* might be a response to the release of certain compounds by plants, thereby contributing to its role in biocontrol by increasing plant growth and disease resistance (Sui et al., 2023).

Fungi in Insect Pest Management

Protecting crops using entomopathogenic fungi (EPF) that can infect and kill insect pests is not new. Historical records show that fungi have been known to grow in insects (Steinhaus, 1975). However, the first definitive record of a fungus infecting an insect was documented in 1835 when Augusto Bassi showed that a disease of silkworms (*Bombyx mori* L.) was due to an EPF that was later named *Beauveria bassiana* (Balsamo) Vuillemin (Bassi, 1835). A few decades later, the first field trials of a mass-produced EPF formulation (conidia of *Metarhizium anisopliae* (Mechnikov) Sorokin harvested from sterilized brewer's mash and mixed with sand granules were carried out to control the sugar-beet weevil *Asproparthenis punctiventris* Germar (Vega et al., 2009). However, despite the potential for insect pest management, research on EPF stalled due to the surge in the use of cheap and fast-acting synthetic chemical pesticides in the 1940s. Recently, there has been renewed interest in using safer methods for insect pest control due to the documented adverse effects of chemical insecticides on beneficial insects, humans, and the environment (Ahmad et al., 2024a, b). EPF, being highly specific to target organisms and posing little risk to humans and the environment, has been sought as an alternative to conventional chemical insecticides and is expected to become a more prominent component of many IPM practices for different crops (Skinner et al., 2014).

There have been approximately 1000 species of fungi and fungal-like organisms reported as entomopathogenic (St. Leger & Wang, 2010). These species include 476

Ascomycota, 474 Entomophtoromycota, 339 Microsporidia, 238 Basidiomycota, 65 Chytridiomycota, and 12 species of fungal-like Oomycetes (Litwin et al., 2020 citing Araújo & Hughes, 2016). While many species have been reported, only a handful have been used extensively for field applications and commercialized. Out of the 171 registered mycoinsecticides and mycoacaricides developed worldwide, about 80% are based only on two genera of Ascomycete fungi, *Beauveria* and *Metarhizium,* with *B. bassiana* (33.9%) and *M. anisopliae* (33.9%) being the most common species (de Faria & Wraight, 2007). Almost half of these products are produced in South America (de Faria & Wraight, 2007). However, the use and the number of commercially available EPF formulations are expected to increase due to a global goal to reduce the use of chemical insecticides and promote biopesticides (Butt et al., 2016a, b). These formulations are usually available as wettable powders or oil formulations. Wettable powders are the most common, and the technical powder (the fungal spore and/or mycelia) is mixed with fillers, dispersants, and surfactants (Burges, 1998). On the other hand, oil formulations are preferred to improve shelf life and increase efficiency by promoting the survival and viability of conidia (Moore et al., 1995).

While the use of EPF in agriculture has become popular, the exact nature or mechanism of how they incapacitate their arthropod hosts remains unclear. For instance, genomic studies on *B. bassiana* and other EPF revealed that pathogenicity developed through convergent evolution across different species and that many aspects of virulence correlate with many species-specific genes (Xiao et al., 2012). The exact mechanisms that mediate pathogenicity and virulence seem to vary across EPF species and their respective hosts because of different coevolutionary relationships (Butt et al., 2016a, b). Even so, EPF seemingly follows a typical infection process. Inoculum comes from the soil, which is the natural habitat of most EPF (Behie & Bidochka, 2014). The infective spores first adhere to the host's cuticle, germinate, and then penetrate host tissues using specialized structures (Sinha et al., 2016). Once established, EPF proliferates inside the host and sporulates after emergence, creating a new source of inoculum. The mycosed insect cadavers usually serve as sources of isolates for later propagation.

After isolation, the EPF needs to be cultured for mass propagation. This is a crucial step that involves consideration of complex factors that affect fungal growth, such as the nature of the substrate, nutritional additives, and others that can affect the virulence and viability of spores that will later be harvested (Feng et al., 1994; Sahayaraj & Namasivayam, 2008). The most typical production method involves the propagation of EPF on grains or vegetable materials like rice, corn grits, and cobs. However, this is not practical for mass production. Submerged fermentation, which forms mycelial mats that can be dried and stored, is often used for mass production (Mascarin et al., 2024). In this procedure, stabilizing the fungal spores before application is very important. The dried mycelial mats or spores are often processed by mixing with protectants to lengthen their shelf life (Tanada & Kaya, 1993). Microencapsulation of spores using biopolymers can also improve formulations under field conditions (Baldiviezo et al., 2023).

In addition to their direct effects on their arthropod hosts, EPF has also been reported to have other beneficial effects on crops that can benefit agricultural production. EPF can have diverse ecological roles based on their complex relationships with their arthropod hosts and plants, with which they can form endophytic relationships (Jaber & Enkerli, 2017). *Beauveria bassiana* and *M. anisopliae* have been reported to colonize the tissues of maize as endophytes and were observed to promote significant growth in the host plant (Liu et al., 2022). These two EPFs have also been reported to be able to colonize cassava roots, with *M. anisopliae* reportedly being more stable and able to persist as an endophyte than *B. bassiana* (Greenfield et al., 2016). Endophytic EPF has been observed to promote plant growth by resisting insects and some pathogenic microbes (Moonjely et al., 2016). In addition, some EPFs have also been shown to help plants sequester nitrogen from insects for additional nutrition (Behie & Bidochka, 2014). These benefits make EPF excellent agricultural inputs for pest control and improve overall crop health.

Fungi in Enhancing Crop Yield

One of the notable groups of fungi with widespread use in organic agriculture is the arbuscular mycorrhizal fungi (AMF) (see previous section), which act as biofertilizers and biopesticides (Smith & Read, 2008a, b), promoting long-term soil health and fertility (Mahdi et al., 2010; Subashini et al., 2007). Plants with arbuscular mycorrhiza associations often show increased nutrient uptake (Lehmann et al., 2014; Lehmann & Rillig, 2015) and improved growth and yield production (Hoeksema et al., 2010; Pellegrino et al., 2015). For instance, millet plants inoculated with 40–50 AMF spores/20 g soil resulted in plants with a 20% increase in shoot dry matter (Sharif et al., 2011). The most prominent feature of AMF is the acquisition of immobile nutrients beyond the range of plant roots via their hyphae (Smith & Read, 2008a, b), including macro- and micronutrients (Govindarajulu et al., 2005; Marschner & Dell, 1994). Several studies have documented the beneficial effect of AMF in increasing nutrient uptake and consequently increasing the growth and yield of many crops (Liu et al., 2012; Tian et al., 2013). Some of the reported AMF species include *Rhizophagus clarus,* which increased soybean and cotton yield under field conditions (Cely et al., 2016), and *R. intraradices,* which increased corn grain yield under different edaphoclimatic conditions in Brazil (Stoffel et al., 2020). *Glomusetunicatum* increased vegetative growth, chlorophyll, and nutrient levels in maize more than in non-mycorrhizal maize plants (Sadhana, 2014). *Glomus intraradices* mycoremediate zinc from the *Medicago truncatula* environment, which led to enhanced growth and development (Hildebrandt et al., 2007). The application of *Glomus mosseae* enhanced corn seedlings' biomass production and height because the inoculum increased the soil's total N, P, K, and organic matter (Wu et al., 2005). Daei et al. (2009) found that *G. mosseae* enhanced wheat yield even under stress conditions. Zhang et al. (2011) also reported that *G. mosseae,* when inoculated with *Mortierella* spp., substantially impacted the dry weights of shoots and roots of *Kosteletzkya virginica* in salt stress. *Acaulosporalacunosa* and *Glomusconstrictum* enhanced the foliar nutritional status of onions (*Allium porrum*) (Hart & Forsythe, 2012).

Phosphate-solubilizing fungi (PSF) are another beneficial fungi that economically enhance the growth, yield, and quality of important crops. The genera commonly reported in the rhizosphere include *Aspergillus, Penicillium,* and *Trichoderma* (Wakelin et al., 2004). In a study conducted by Richa et al. (2007), they reported on the improved growth and phosphorus uptake of corn with the inoculation of two *Aspergillus* species—*A. tubingensis* and *A. niger.* Patil et al. (2012) also studied corn inoculated with *Penicillium bilaiae* and *Penicillium* spp. They noted a significant increase in plant height, number of leaves, dry matter production, cob length, grain weight, and grain yield. In Australia and Canada, bioinoculants made of *P. bilaiae* showed beneficial effects on wheat crops (Wakelin et al., 2004). There are also reports in wheat crops showing higher growth and yield when inoculated with *P. oxalicum* (Singh & Reddy, 2011) and increased uptake of N, P, and K with the inoculation of *A. awamori* and *A. niger.* In Colombia, a biofertilizer based on *P. janthinellum* has been used in rice cultivation, which has resulted in higher rice production (Moreno et al., 2007). Mittal et al. (2008) showed a significant increase in shoot height, seed number, and seed weight of chickpea plants following inoculation of *A. awamori* and *P. citrinum.* Kapri and Tewari (2010) also reported an increase in fresh and dry weight of shoot and root, length of shoot and root in phosphate-deficient soil of chickpea plants after the inoculation of P-solubilizing *Trichoderma* sp. Promwee et al. (2014) evaluated the growth of rubber trees after the inoculation of *T. harzianum* and rock phosphates, which resulted in increased fresh and dry weight of shoot and root.

Moreover, Yadav et al. (2011) demonstrated that co-inoculation with *T. harzianum* and *A. niger* significantly increased the growth of chickpea plants. Manivannan et al. (2012) observed increased cowpea productivity in *Aspergillus*-enriched vermicompost, while Saber et al. (2009) demonstrated that inoculation of green gram seedlings with *A. niger* and *Penicillium* sp. in the presence of rock phosphate (RP) and calcium superphosphate (CP) greatly improved plant growth, seed yield, and phosphorus uptake. Chakraborty et al. (2010) reported a considerable improvement in soybean growth following the application of *A. niger*, *A. mellus*, and *A. clavatus.* Similarly, Islam et al. (2014) found that soybean stem length, fresh weight, dry weight, plant height, leaf area, and chlorophyll content significantly increased with the inoculation of *Aspergillus* spp. The primary mechanism of PSF for stimulating plant development is to increase plant P acquisition efficiency, allowing plants to convert insoluble P forms into an accessible form (orthophosphate) (Kalayu, 2019). They also absorb phosphorus from a larger area by developing an expanded network surrounding the root system (Mehrvarz et al., 2008). Several fungal species can also produce phosphatases such as phytase (Aseri et al., 2009; Dodor & Tabatabai, 2003; Khan et al., 2009; Kumar & Patel, 2018; Santana et al., 2016; Satyaprakash et al., 2017; Selvi et al., 2017; Tarafdar et al., 2003), which hydrolyze organic forms of phosphate compounds, releasing inorganic phosphorus that plants mobilize.

Fungi and Soil Health

Recent studies have highlighted that fungi play a more significant role than bacteria in soil priming and aggregate formation. However, these functions vary among species (Husain et al., 2024). For instance, *Funneliformis mosseae* and *Rhizophagus intraradices* initiate microaggregation by binding to organo-mineral particles, followed by macroaggregate formation (Jiménez-Martínez et al., 2024). In corn, the absence of fertilizers such as diammonium phosphate (DAP) and urea enhanced macroaggregate stability (>0.25 mm), as fertilizer application was shown to reduce AMF activity, impacting soil structure (Gómez-Leyva et al., 2023). Manure and *T. harzianum* application in corn fields also improved soil fertility by increasing soil organic carbon, total nitrogen, bacterial communities, and aggregate stability. However, long-term grazing exclusion adversely affected microbial diversity and fungal communities, potentially hindering aggregation processes (Ju et al., 2024) and affecting plant-AM interaction (Xu et al., 2024). Angulo et al. (2024) compared 29 fungal isolates and found that aggregation is species-dependent and influenced by soil moisture, with a positive correlation between fungal biomass and soil aggregation. Additionally, AMF contributed to drought tolerance through its multifaceted roles in soil aggregation, microbial biomass, respiration, and nutrient cycling (Tang et al., 2024).

AMF has also been widely reported to enhance drought and salinity tolerance in plants (Abrar et al., 2024; Arcidiacono et al., 2024; Basyal, 2024; Chávez-González et al., 2024; Chen et al., 2024; Gou et al., 2024; Singh et al., 2024). However, non-AMF fungi have also demonstrated significant contributions. For instance, inoculation with the root endophyte *Piriformospora indica* in watermelon mitigated water stress by reducing leaf temperature, improving leaf gas exchange parameters, increasing leaf pigment concentration, enhancing relative water content, and decreasing the accumulation of reactive oxygen species (ROS) (Jyothymol et al., 2024). In rice, *P. indica* colonization enhanced growth in drought-susceptible and drought-tolerant varieties by increasing cell membrane stability, photosynthetic rate, transpiration rate, and stomatal conductance (Mohan et al., 2024). Similarly, in date palms, *P. indica* improved salinity tolerance by reinforcing the cortical cells of the roots (Ahmad et al., 2024a, b). Additionally, *P. indica* enhanced soybean resilience to saline-alkaline stress by improving ROS scavenging, osmotic adjustment, and photosynthetic capacity (Zhu et al., 2024). The co-application of *P. indica* and biochar has also been reported to enhance the quality of coastal saline soils (Wu et al., 2024).

The potential of fungi, particularly *P. indica*, to enhance plant tolerance to drought and salinity stress underscores the broader role of fungal symbiosis in addressing environmental and climate change challenges. Similarly, AMF and biochar have been explored as solutions to improve plant resilience and soil quality in saline environments. Saline soils significantly limit the amount of arable land available for cultivation. However, the use of AMF has shown promise in extending these areas for agricultural use by enhancing soil quality (Bencherif et al., 2023; Li et al.,

2023; Masrahi et al., 2023). Other recent studies have focused on the role of AMF in altering the rhizosphere and microbial communities, resulting in improved plant growth under saline conditions (Fan et al., 2024; Wen et al., 2024; Wu et al., 2024; Zhang et al., 2024). Additionally, the co-application of biochar and AMF has been investigated, with positive outcomes (Ashraf & Chen, 2023; Hammer et al., 2015; Wen et al., 2024; Wu et al., 2024). Biochar enhances soil structure by improving water infiltration, mitigating the harmful effects of high sodium levels, and balancing soil ion concentrations through cation exchange (Khan et al., 2024). This leads to improving the Na/K ratio and compounds with the beneficial effects of AMF (Hammer et al., 2015). Soil amendments using biochar have also affected the synergy between bacteria and fungi, enhancing soil quality (Ding et al., 2024). In summary, the intricate interactions between soil fungi, particularly AMF and *Piriformospora indica*, are pivotal in improving soil structure, microbial diversity, and plant resilience under abiotic stresses such as drought and salinity. The co-application of biochar further amplifies these effects by enhancing soil physical properties and ion balance, ultimately contributing to improved plant growth and soil health.

Fungi in Philippine Agriculture in the Philippines

In the Philippines, fungal species, particularly *Trichoderma* and mycorrhizal fungi, have been recognized as sustainable solutions to agricultural challenges, enhancing crop yields and improving soil health. *Trichoderma* species, including *T. viride*, *T. aureoviride*, *T. hamatum*, *T. harzianum*, *T. koningii*, *T. glaucum*, *T. piluliferum*, *T. reesei*, *T. pseudokoningii*, and *T. ghanense*, are effective biological control agents and biofertilizers that improve plant health, controlling pests and boosting productivity (Cumagun, 2012). Research has demonstrated that *Trichoderma* is a valuable tool for crop protection, such as controlling damping-off pathogens (Cuevas et al., 2005; Lopez et al., 2015). Cuevas (2006) also demonstrated that *T. pseudokoningii* improved early flowering and fruit production in tomato and enhanced nutrient uptake (P, K, Ca, Zn, and Fe). Additionally, applying *T. pseudokoningii* in rice fields increased the yield by 18% before transplanting (Cuevas, 2006). *Trichoderma* reduced club root disease in crucifers, leading to a threefold increase in farmers' income in Buguias, Benguet (Cuevas et al., 2011). *T. harzianum* combined with hot water treatment also controlled crown rot in banana, reduced decay, and improved fruit quality (Alvindia & Acda, 2012). In addition, *Trichoderma* protected coconut seedlings from pests like *Brontispa longissima* (Cuevas et al., 2023) and lanzones scale insects (Silva et al., 2019). Integrating salt with *Trichoderma* inoculant aided in the recovery of mature coconut trees damaged by coconut scale insects (Cuevas et al., 2023). In addition to pest control, *Trichoderma* acted as a compost activator, enhancing crop residue decomposition in rice-based systems (Cumagun et al., 2009). Strains like *T. harzianum* strain no. 94–016, with strong cellulolytic and competitive abilities, enhanced crop residue decomposition, organic matter, and soil health in rice-based systems. This has resulted in a 30–50% reduction in fertilizer use and a 20% increase in rice and corn yields (Cuevas, 1997). The findings of Lopez et al. (2015) and Organo et al. (2022) further supported the role of *Trichoderma*-based compost activators in accelerating decomposition and improving composting efficiency. With the benefits of using *Trichoderma*, organic and Good Agricultural Practices (GAP) producers use *T. koningii* as a biofungicide and soil enhancer, with an annual sales growth of 29.7%, the adoption rate of 0.03%, and 4–92% yield improvement in Benguet in 2018 (Launio et al., 2020).

Fungal species are also known to promote plant growth through efficient nutrient uptake. Vesicular arbuscular mycorrhizae (VAM) can reduce fertilizer use by up to 80% while providing plant disease resistance (PhilRice, 2002b). Mykorich® and Mykocap®, arbuscular mycorrhizae-based biofertilizers, can replace 60–85% of the recommended fertilizer for optimal growth and enhanced cacao (*Theobroma cacao*) seedling growth and soil properties in acidic agroforests (Victoria & Aggangan, 2023). Combining AMF with chemical fertilizers improved plant growth and soil physicochemical characteristics. Integrated nutrient management for tomato, using 50% chemical fertilizer combined with vermicompost and mycorrhizae, boosted productivity and profitability (Bernados et al., 2024). Additionally, zinc-solubilizing bacterial isolates from VAM root inoculants alleviated zinc deficiency in rice and improved plant growth (Nepomuceno et al., 2020). Biofertilizers (Mykorich, Mycogroe, MykoPlus, Mykovam, and VAMRI) with mycorrhizae as active ingredients also improved soil and crop health (BIOTECH UPLB, 2023).

Another group of beneficial fungi, entomopathogenic fungi (EPF), are highly effective biocontrol agents for managing key agricultural pests. In the Philippines, Gabriel (1968) documented the natural occurrence of various EPF species, including *Entomophthora, Coelomomyces acariformes, Hypocrella, Myriangium, Cordyceps, Torrubiella flava, Laboulbenia, Amphoromorpha, Aschersonia, Aspergillus, Beauveria, Cephalosporium, Metarhizium, Oospora obducens, Spicaria, Isaria, Fusarium, Nectria, Microcera, Peziotrichum, Septobasidium*, causing epizootics in major insect pests. More recently, Montecalvo and Navasero (2024) identified native *Metarhizium rileyi* against onion and fall armyworms in corn. Potent strains of *M. rileyi* were isolated from these pests (Montecalvo et al., 2022; Montecalvo & Navasero, 2020), and along with *Beauveria bassiana* and *Metarhizium anisopliae*, they were found to be effective against various life stages of the fall armyworm (Montecalvo & Navasero, 2021). Additionally, *M. anisopliae* has been proven effective against oriental migratory locusts (Santiago et al., 2001), mites (Sanjaya et al., 2013), and rice black bugs (PhilRice, 2002a). *M. anisopliae* and *B. bassiana* have also been suggested for managing *Brontispa* infestations in coconut palms (Alfiler et al., 2009). In addition to pest control, EPF strains inhibited the growth of *Pseudocercospora fuligena*, reducing tomato leaf mold severity and boosting yield (Yago et al., 2016). Biopesticides of *B. bassiana* and *Isaria fumosorosea* also reduced twig blight and citrus rind borer damage in citrus trees (Yago et al., 2020).

Efforts to isolate beneficial fungi from various sources and explore their novel use have expanded their potential applications in agriculture. In banana farms, promising weed endophytes (*L. theobromae*, *T. asperellum*, and *Ceratobasidium* species) reduced *Fusarium* wilt in Cavendish bananas (Catambacan & Cumagun, 2021). Further studies by Puig and Cumagun (2019) highlighted endophytic *Schizophyllum commune* CGP119, which controlled *Fusarium* wilt in GCTCV 219 banana by producing enzymes like amylase, cellulase, protease, lipase, and laccase. Other effective antagonists included *Fusarium* CGP150, *Penicillium* CGP116, *Trichoderma* CGP106, and *Pestalotiopsis* CGP117. Additionally, Alvindia and Natsuaki (2008) found that fungal epiphytes (*Clonostachys byssicola*, *Curvularia pallescens*, *Penicillium oxalicum*, *T. harzianum*) from banana fruit surfaces limited crown rot pathogens and significantly lowered rot incidence in postharvest bananas compared to fungicide-treated or untreated fruits. In addition, fungal species degrade toxic compounds (e.g., pollutants, pesticides), offering a sustainable solution for environmental contamination in agricultural systems. Guerrero et al. (2022) highlighted the potential of fungi in serving as biological control agents and plastic degraders, as well as remediating heavy metals, hydrocarbons, and pesticides. Fungal species like *Neodeightonia subglobosa* and *Sclerotium hydrophilum* in Taal Lake degraded butachlor effectively (Carascal et al., 2017). Additionally, *Trichoderma* combined with rice straw compost improved rice yields and reduced copper and lead concentrations in contaminated mined areas in Marinduque (Cuevas & Banaay, 2022).

A breakthrough in mycology in the Philippines was development of a fungus-based nematode control product, *Paecilomyces lilacinus*, pioneered by National Scientist Dr. Romulo Davide (Nature, 1987). This was the first commercial biological control for nematodes manufactured in the Philippines. This eco-friendly alternative to chemical nematicides showed 50–90% nematode mortality in mungbean. It was effective against *Radopholus similis* in Cavendish bananas and the golden potato cyst nematode. There are also products such as BioSpark, a *Trichoderma*-based product for controlling damping-off in vegetables and tropical fruit diseases, which has been commercialized in the Philippines (Cumagun, 2012), and a few other *Trichoderma*-based products in the market. Aside from these, yeast-based products were developed, including Bio-Y (for pre- and post-harvest disease control) and Hi-Yeast (for pathogens affecting various crops) (BIOTECH UPLB, 2023). Product registration is regulated by the Department of Agriculture (DA), Fertilizer and Pesticide Authority, and the Bureau of Agriculture and Fisheries Product Standards. Similarly, Filipino farmers and stakeholders can acquire beneficial fungi like *M. anisopliae*, *B. bassiana*, and *Trichoderma* at no cost from the DA Regional Crop Protection Centers.

Fungi in Enhancing Agriculture in Indonesia

Indonesia is the world's largest archipelagic country, consisting of 17,504 islands by the sea (Djunarsjah & Putra, 2021). It is an agricultural country that offers a wide diversity of tropical products and agricultural commodities, including oil palm, rubber, rice, cassava, corn, coffee, and cacao (Food and Agriculture Organization, 2018). However, invasion of pests, disease outbreaks, and poor soil conditions, which lead to low agricultural yield, remain a problem. Synthetic chemicals are used as interventions, but prolonged use can cause environmental compromises. Meanwhile, indigenous fungi can serve as an alternative farming input to protect and enhance the growth and development of these cash crops through disease suppression caused by phytopathogenic infections and the production of secondary metabolites. The application of these indigenous fungi in agriculture was also evaluated in Indonesia.

As biological control agents, fungi effectively suppressed the growth of pathogenic *Fusarium verticillioides* by 50% in corn seedlings. They consistently showed a positive effect (≥90%) on seed viability and vigor (Mirsam et al., 2021). In South Sumatra, the pathogenesis of endophytic fungi was evaluated against a new invasive fall armyworm, *Spodoptera frugiperda*, which causes corn yield losses of 18 million tons per year (Herlinda et al., 2020). The larvae of *S. frugiperda* are difficult to control by contact as they emerge from the leaf midrib when feeding and immediately hide in the corn stalk (Gustianingtyas et al., 2021). Endophytic *Beauveria bassiana* and *Metarhizium anisophliae* caused adverse effects on fungal-treated corn seedlings on developing *S. frugiperda* or the fall armyworm. The fungal-colonized young corn increased the mortality rate of all larval instars of *S. frugiperda* up to 51.33%. It decreased the last instar development to the pupal stage, the adult emergence, the eggs laid, and the percentage of hatched eggs (Sari et al., 2022). Similar results were previously recorded by Gustianingtyas et al. (2021), where *Beauveria* isolates induced larval mortality by up to 29.33% and reduced the emergence of adult armyworms by up to 44%. Also, the entomopathogenic fungi *Metarhizium* spp. caused up to 78.67% mortality in the larvae and suppressed the emergence of adult worms up to 81.2%, causing morphological abnormalities and shape malformation (Herlinda et al., 2020).

Culvularia lunata is an endophytic fungus that was found to have insecticidal activity against *S. frugiperda*, causing up to 17.33% mortality of the larvae (Herlinda et al., 2021). In Lampung Province, *B. bassiana* and *Aspergillus oryzae* species effectively suppressed the larval growth of *S. litura* in corn seedlings (Fitriana et al., 2021). Moreover, three entomopathogenic fungal species, namely, *M. anisopliae, Cordyceps militaris, and B. bassiana,* were reported as biocontrol agents to manage the populations of insect pests in the Indonesian oil palm industry. First, *Metarhizium* was used to control *Oryctes* larvae, which showed 88.75% mortality and prevented termites in the field. Second, *Cordyceps* was also found effective in reducing the field moth population of *Setothosea asigna* by infecting up to 80% of the pupae. Third, *Beauveria* in an effervescent formulation demonstrated a better efficacy on *Darnatrima* larvae control (Priwiratama & Susanto, 2014). Moreover, *Trichoderma* was used to control basal stem rot disease caused by *Ganoderma* sp. and was found to extend the life of oil palms up to 2–3 years. However, as the roots developed, *Trichoderma* could no longer protect the oil palm from *Ganoderma* infection (Priwiratama & Susanto, 2014). Aside from *Trichoderma* species, other antagonistic fungi, such as *Verticillium lecanii, Aureobasidium pullulans, Muscodor albus,* and *Acremonium breve,* were reported to control citrus and subtropical fruit diseases by producing volatile organic compounds (e.g., alcohols, esters, ketones, acids, and lipids) that exhibit inhibitory effects against the pathogens (Dwiastuti et al., 2021).

On the other hand, fungi are also known for their plant growth-promotion abilities, facilitating plant nutrient uptake and soil-borne disease protection. AMFs, natural symbionts of many terrestrial plants, are widely known as biofertilizers. In Indonesia, AMF application has been shown to support oil palm (*Elaeis guineensis*) cultivation by facilitating nutrient absorption, enhancing water availability, improving physicochemical properties (e.g., pH, organic carbon content, available P, and cationic capacity exchange), and establishing beneficial soil microorganisms (Zebua, 2024). The AMF colonization rate in treated plants increased with the corn-sorghum cropping sequence system's dry biomass (root, shoot, and grain) in Lombok Province, Indonesia (Astiko et al., 2019). Aside from increasing the soil P availability, AMF is also capable of improving the growth and yield of corn in calcareous soil in the East Nusa Tenggara Timur Province, Indonesia (Ishaq et al., 2021) and red chili (*Capsicum annuum*) in andosol soil in the Highlands of West Bandung, Indonesia (Rosliani et al., 2023). Interestingly, the effect of AMF as a bioinoculant can be improved when it is coupled with either a co-inoculant or a soil conditioner. When AMF was coupled with *Rhizobium* species, Wangiyana & Farida (2019) recorded a significant increase in soybean yield components, such as grain yield, weight of 100 grains, grain number, and total biomass per clump of soybean. Likewise, adding biochar as a soil conditioner to AMF inoculation improved the growth and productivity of rice, its nutrient uptake, and the associated physicochemical soil properties (Mulyadi & Jiang, 2023). A greenhouse experiment in Lampung, Indonesia, revealed that cellulolytic fungi (*Aspergillus fumigatus*) and ligninolytic fungi (*Geotrichum* sp.) induced the compost material, resulting in an optimal effect on vegetative growth of red chili plants, such as plant height, dry weight, and relative water content (Irawan et al., 2019).

Fungi in Enhancing Agriculture in Malaysia

The agriculture industry in Malaysia contributes approximately 8.9% of the gross domestic product. It largely depends on oil palm, cash crops, and rubber (Statista, 2024). Both industrial- and small-scale agriculture sectors rely highly on synthetic chemicals to control pathogens and pests affecting crops. A heavy reliance on chemicals can lead to health and environmental problems as well as pathogen and pest resistance. One of the possible solutions is the use of beneficial fungi that have the potential to protect economically important crops from pathogens and pests. As the leading industrial crop in the country, it is evident that oil palms receive great attention for protection, prevention, and management against pests and diseases using beneficial fungi.

Basal stem rot caused by *Ganoderma boninense* is the most serious disease of oil palm, and Integrated *Ganoderma* Management (IGM) was introduced to manage the disease. In IGM, biological control is integrated with chemical and cultural methods to mitigate the effect of basal stem rot (BSR) disease (http://sawitsecure.mpob.gov.my/disease-control-and-management/). The incorporation of biological control in IGM relies on the recognition that a reduced occurrence of disease is attained through the presence of natural competitors that regulate the population of pathogenic fungi (Bivi et al., 2010). To strengthen biological control against BSR, several formulations based on beneficial microbes have been developed and commercialized by the Malaysian Palm Oil Board. Among the beneficial fungi developed into biological control were endophytic *Trichoderma,* which colonizes the roots, protects oil palm against BSR, and stimulates root growth. Treatment of oil palm seedlings using endophytic *Trichoderma virens* triggered induced systemic resistance and supported the seedlings' vegetative growth (Paudzai et al., 2019). Endophytic *Hendersonia* also colonizes oil palm roots and has a strong antagonistic effect against *G. boninense* (Idris et al., 2012).

In Malaysia, entomopathogenic fungi have been proven effective against insect pests such as beetles, moths, and aphid-infested crops. Some of these entomopathogenic fungi are *Beauveria bassiana*, *Metarhizium anisopliae, Isaria fumosorosea* (syn. *Paecilomyces fumosoroseus*), *Paecilomyces carneus*, and *P. farinosus* (Table 9.1). Commercialized products based on *Metarhizium anisopliae* and *Beauveria bassiana* are available. The biocontrol products are used to control various types of insect pests. Infection by entomopathogenic fungi is through airborne conidia that land on the exoskeleton of the insect pest. The hyphae can penetrate the insect's body cavity directly due to the production of extracellular enzymes that degrade the cuticle. Further hyphae development in the insects' body cavity will eventually kill the host (Fernandes et al., 2012a, b).

Root infection by the mycorrhizal fungi increases the active absorption surface. It stimulates nutrient and water uptake even under water stress. Mycorrhizal association also increases the host plant's ability to suppress disease (Cardoso & Kuyper,

Table 9.1 Some of the entomopathogenic fungi tested as biocontrol agents against several pests

Beneficial fungi (biocontrol agent)	Pest	Infested plant	References
Beauveria bassiana, *Paecilomyces fumosoroseus*	Diamondback moth (*Plutella xylostella*)	Cabbage	Ibrahim and Low (1993)
Paecilomyces carneus, *P. Farinosus*	Bagworm moth (*Pteroma pendula*)	Oil palm	Bakeri et al. (2009)
Beauveria bassiana	Tiger moth (*Atteva sciodoxa*)	*Eurycoma longifolia*	Abood et al. (2010)
Metarhizium anisopliae var. *major*	Rhinoceros beetle (*Oryctes rhinoceros*)	Oil palm fronds	Moslim et al. (2009, 2011)
Isaria fumosorosea (syn. *Paecilomyces umosoroseus*) and *Metarhizium anisopliae*	Tiger moth (*Atteva sciodoxa*)	*Eurycoma longifolia*	Sajap et al. (2014)
Metarhizium anisopliae	Aphid (*Aphis gossypii*)	Chilli and Brinjal	Latiff et al. (2022)

2006). Since the 1980s, mycorrhizal studies have become prominent in Malaysia, where most research is conducted at Universiti Putra Malaysia. Since then, other universities and research institutes have also studied mycorrhizal in various crops. Some of the mycorrhizal fungi that have been identified and tested as biofertilizers as well as biocontrol agents are *Glomus mosseae*, *Glomus intraradices*, *Glomus clarum*, and *Scutellospora calospora* (Masri et al., 1998; Sundram et al., 2015). From the findings, many mycorrhiza-based products have been developed and commercialized. These products commonly combine mycorrhiza fungi, other beneficial fungi, and bacteria. Mycorrhiza-based products are used in several agricultural industries, including oil palm, rubber, and fruit crops, due to the ability of the mycorrhiza fungi to facilitate the uptake of nutrients and water that enhance plant growth. Mycorrhiza can also resist soil-borne pathogens and reduce the development of plant disease.

The mechanism of fungal herbicides involves spore germination, penetration, and establishment in the weed host. The mycelial growth in the host prevents water and nutrient supply. In addition, fungal pectinases break through the host cell wall, rupturing the polysaccharide layers. Some fungi produce phytotoxin that inhibits mitosis and affects seedling growth (Jiang et al., 2008). Fungal potential to reduce weed infestation in Malaysia has been assessed on barnyard grass (*Echinocloa crus-galli*), itch grass (*Rottboellia cochinchinensis*), and goosegrass (*Eleusine indica*). *Exserohilum longirostratum* and *E. monoceras* have been found effective against infestation of barnyard grass, which was a noxious weed in rice fields (Hailmi et al., 2011; Juraimi et al., 2005; Ng et al., 2011). *Exserohilum longirostratum* was found to be pathogenic to barnyard grass in field conditions. As for *E. monoceras,* the fungus could suppress the growth of barnyard grass (Hailmi et al., 2011; Juraimi et al., 2005).

Exserohilum prolatum and *E. longirostratum* have been reported as potential mycoherbicides to control itch grass (*Rottboellia cochinchinensis*). The weed occurs in agricultural areas and infests annual and perennial crops (Alloub et al., 2009; Hailmi et al., 2017). The potential of *Bipolaris bicolor* to control goosegrass (*Eleusine indica*), a noxious weed in oil palm plantations, was reported by Fakri et al. (2022). The results also suggest that *B. bicolor* can be used in integrated weed control management in combination with Ametryn, a non-selective herbicide, as the fungus and herbicide acted synergistically and inhibited the growth of goosegrass seedlings (Fakri et al., 2022).

Conclusion and Future Directions

The global importance of fungi is emphasized in this chapter in the examples of symbiotic partnerships, their role as endophytes, their involvement in nutrient cycling, plant disease, and insect pest management. In the tropics, beneficial fungi also enhance plant resilience to environmental stressors by stimulating plant immune response and improving defense efficiency (Cui et al., 2021) against herbivores and pathogens. Beyond their roles in nutrient cycling and plant health, fungi are also known for their ability to degrade environmental pollutants.

Soil microorganisms, particularly fungi, are also critical to determining soil quality. This chapter has presented studies that underscore the potential of fungal symbiosis and soil amendments as sustainable strategies for mitigating environmental stress in agricultural systems, especially in saline and marginal soils. Furthermore, the natural ability of the fungi to recycle nutrients from the soil is beneficial to the increased growth of crops, reducing the reliance on fertilizers by increasing nitrogen-fixing activities in the rhizosphere.

Fungi are pivotal in attaining sustainable agriculture and food security (Basiru et al., 2020; Rillig et al., 2016; Thirkell et al., 2017). Their different characteristics and abilities make them essential in maintaining healthy soils, boosting plant development, and lowering the need for chemical fertilizers. Fungi are also known as colonizers of plant tissues without causing damage or morphological changes in their host—as endophytes. These fungi can infect and kill pests without harming non-target organisms, making them safe and sustainable options for pest management. Hence, using fungi as biological control agents against plant pathogens, insect pests, and weeds can decrease the need for chemical pesticides and herbicides.

Despite a long experimental history of developing fungi as biocontrol agents, there has been limited commercial success in Southeast Asian countries, particularly in mycoherbicide products. Extensive field testing, mass production, and formulation studies are still required for commercial success. Although fungal-based biocontrol products take longer to develop and achieve significant results, they are still a step towards more sustainable agricultural practices.

To expand the use of fungi in enhancing agriculture, it is also vital to understand their mechanisms and physiology, explore their application across crops and under varying field conditions, and assess stakeholder acceptance. Successful biological control requires understanding plant-pathogen-environment interactions and addressing economic and social factors (He et al., 2021). For instance, *Trichoderma* adoption is hindered by high costs, limited information, slow effects, and inadequate marketing (Launio et al., 2020), necessitating continuous R&D and supportive policies to overcome these challenges and fully unlock the potential and opportunities of beneficial fungi for sustainable agriculture in the Southeast Asian region and beyond.

References

Abood, F. K., et al. (2010). Pathogenicity of *Beauveria bassiana* against the Tiger Moth, *Atteva sciodoxa* (Lepidoptera: Yponomeutidae). *Journal of Entomology, 7*(1), 19–32. https://doi.org/10.3923/je.2010.19.32

Abrar, M., Zhu, Y., Rehman, M. M. U., Batool, A., Duan, H. X., Ashraf, U., et al. (2024). Functionality of arbuscular mycorrhizal fungi varies across different growth stages of maize under drought conditions. *Plant Physiology and Biochemistry, 213*, 108839.

Achal, V., Kumari, D., & Pan, X. (2011). Bioremediation of chromium contaminated soil by a brown-rot fungus, *Gloeophyllum sepiarium*. *Research Journal of Microbiology, 6*, 166–171. https://doi.org/10.3923/jm.2011.166.171

Ahmad, M. F., Ahmad, F. A., Alsayegh, A. A., Zeyaullah, M., AlShahrani, A. M., Muzammil, K., Saati, A. A., Wahab, S., Elbendary, E. Y., Kambal, N., Abdelrahman, M. H., & Hussain, S. (2024a). Pesticides impacts on human health and the environment with their mechanisms of action and possible countermeasures. *Heliyon, 10*(7), e29128. https://doi.org/10.1016/j.heliyon.2024.e29128

Ahmad, M., Abdul Aziz, M., Sabeem, M., Kutty, M. S., Sivasankaran, S. K., Brini, F., et al. (2024b). Date palm transcriptome analysis provides new insights on changes in response to high salt stress of colonized roots with the endophytic fungus *Piriformospora indica*. *Frontiers in Plant Science, 15*, 1400215.

Akem, C. N. (1985). *Inhibition of sclerotinia minor by penicillium citrinum; A potential biological control agent*. PhD thesis, Oklahoma State University. https://shareok.org/handle/11244/15878

Akköprü, A., & Demir, S. (2005). Biological control of *Fusarium* wilt in tomato caused by *Fusarium oxysporum* f. sp. *lycopersici* by AMF Glomus intraradices and some rhizobacteria. *Journal of Phytopathology, 153*(9), 544–550.

Akpasi, S. O., Anekwe, I. M. S., Tetteh, E. K., Amune, U. O., Shoyiga, H. O., Mahlangu, T. P., & Kiambi, S. L. (2023). Mycoremediation as a potentially promising technology: Current status and prospects—a review. *Applied Sciences, 13*(8), 4978. https://doi.org/10.3390/app13084978

Alfiler, A. R. R., Orense, J. C., Imperial, M. L. R., & Zipagan, M. B. (2009). Control of *Brontispa longissima* (Gestro) (Coleoptera: Chrysomelidae) infestations in young coconut palms using *Metarhizium anisopliae* (Mets.) Sorokin and *Beauveria bassiana* (Balsamo) Vuillemin. *Physiological Entomology, 23*(2), 198.

Alloub, H., Juraimi, A. S., Kadir, J., Sastroutomo, S., & Begum, M. (2009). Field efficacy of – *Exserohilum prolatum* a potential mycoherbicide for biological control of itchgrass (*Rottboellia cochinchinensis*). *Journal of Biological Sciences, 9*(2), 119–127.

Alshammari, W., Bairum, R., Sulieman, A. M., Alshammari, N., & Elamin, H. (2024). In vitro and in vivo study of antagonistic and biocontrol of *Trichoderma harzianum* strains against wood decay pathogens. *Polish Journal of Environmental Studies, 33*(1), 515–521.

Alvindia, D. G., & Acda, M. A. (2012). An integrated approach with *Trichoderma harzianum* DGA01 and hot water treatment on control of crown rot disease and retention of overall quality in banana. *Biocontrol Science and Technology, 22*(9), 1021–1033.

Alvindia, D. G., & Natsuaki, K. T. (2008). Evaluation of fungal epiphytes isolated from banana fruit surfaces for biocontrol of banana crown rot disease. *Crop Protection, 27*(8), 1200–1207.

Angulo, V., Bleichrodt, R. J., Dijksterhuis, J., Erktan, A., Hefting, M. M., Kraak, B., & Kowalchuk, G. A. (2024). Enhancement of soil aggregation and physical properties through fungal amendments under varying moisture conditions. *Environmental Microbiology, 26*(5), e16627.

Anthony, M. A., Crowther, T. W., van der Linde, S., Suz, L. M., Bidartondo, M. I., Cox, F., Schaub, M., Rautio, P., Ferretti, M., Vesterdal, L., De Vos, B., Dettwiler, M., Eickenscheidt, N., Schmitz, A., Meesenburg, H., Andreae, H., Jacob, F., Dietrich, H. P., Waldner, P., Gessler, A., Frey, B., Schramm, O., van den Bulk, P., Hensen, A., & Averill, C. (2022). Forest tree growth is linked to mycorrhizal fungal composition and function across Europe. *The ISME Journal, 16*(5), 1327–1336. https://doi.org/10.1038/s41396-021-01159-7

Araújo, J. P. M., & Hughes, D. P. (2016). Chapter 1: Diversity of Entomopathogenic fungi: Which groups conquered the insect body? In B. Lovett & R. J. St. Leger (Eds.), *Advances in genetics* (pp. 1–39). Academic.

Arcidiacono, M., Pellegrino, E., Nuti, M., & Ercoli, L. (2024). Field inoculation by arbuscular mycorrhizal fungi with contrasting life-history strategies differently affects tomato nutrient uptake and residue decomposition dynamics. *Plant and Soil, 500*(1), 105–127.

Arnstadt, T., Hoppe, B., Kahl, T., Kellner, H., Krüger, D., Bauhus, J., & Hofrichter, M. (2016). Dynamics of fungal community composition, decomposition, and resulting deadwood properties in logs of *Fagus sylvatica, Picea abies*, and *Pinus sylvestris*. *Forest Ecology and Management, 382*, 129–142. https://doi.org/10.1016/j.foreco.2016.10.004

Aseri, G. K., Jain, N., & Tarafdar, J. C. (2009). Hydrolysis of organic phosphate forms by phosphatases and phytase producing fungi of arid and semi-arid soils of India. *American-Eurasian Journal of Agriculture and Environment Science, 5*, 564–570.

Asghar, W., & Kataoka, R. (2021). Effect of co-application of *Trichoderma* spp. with organic composts on plant growth enhancement, soil enzymes, and fungal community in soil. *Archives of Microbiology, 203*, 4281–4291. https://doi.org/10.1007/s00203-021-02227-6

Ashraf, F. A. I. S. A. L., & Chen, Y. O. U. Y. U. A. N. (2023). Synergistic effects of biochar and arbuscular mycorrhizal fungi on enhancing *Elymus elymoides* growth in saline coastal soil. *Pakistan Journal of Botany, 55*, 119–126.

Asmelash, F., Bekele, T., & Birhane, E. (2016). The potential role of arbuscular mycorrhizal fungi in the restoration of degraded lands. *Frontiers in Microbiology, 7*, 1095. https://doi.org/10.3389/fmicb.2016.01095

Astiko, W., Wangiyana, W., & Susilowati, L. E. (2019). Indigenous Mycorrhizal seed-coating inoculation on plant growth and yield, and NP-uptake and availability on Maizesorghum cropping sequence in Lombok's Drylands. *Pertanika Journal of Tropical Agricultural Science, 42*(3), 1131.

Azad, K., & Kaminskyj, S. (2016). A fungal endophyte strategy for mitigating the effect of salt and drought stress on plant growth. *Symbiosis, 68*, 73–78. https://doi.org/10.1007/s12155-016-9711-8

Azarmi, R., Hajieghrari, B., & Giglou, A. (2011). Effect of *Trichoderma* isolates on tomato seedling growth response and nutrient uptake. *African Journal of Biotechnology, 10*, 5850–5855.

Azcón-Aguilar, C., & Barea, J. M. (1997). Arbuscular mycorrhizas and biological control of soil-borne plant pathogens—An overview of the mechanisms involved. *Mycorrhiza, 6*(6), 457–464. https://doi.org/10.1007/s005720050147

Bakeri, S. A., Ali, S. R. A., Tajuddin, N. S., & Kamaruzzaman, N. E. (2009). Efficacy of entomopathogenic fungi, *Paecilomyces* spp., In controlling the oil palm bag worm, pteroma pendula (Joannis). *Journal of Oil Palm Research, 21*, 693–699.

Baldiviezo, L. V., Nieva, L. B., Pedrini, N., & Cardozo, R. M. (2023). Microencapsulation of a native strain of the entomopathogenic fungus *Beauveria bassiana* and bioinsecticide activity against Pyrethroid-resistant *Triatoma infestans* to vector control of Chagas disease in the Argentine Gran Chaco Region. *Tropical Medicine and Infectious Disease, 8*(5), 245. https://doi.org/10.3390/tropicalmed8050245

Basiru, S., & Hijri, M. (2022). The potential applications of commercial arbuscular mycorrhizal fungal inoculants and their ecological consequences. *Microorganisms, 10*(10), 1897. https://doi.org/10.3390/microorganisms10101897

Basiru, S., Mwanza, H. P., & Hijri, M. (2020). Analysis of Arbuscular Mycorrhizal fungal inoculant benchmarks. *Microorganisms, 9*(1), 81. https://doi.org/10.3390/microorganisms9010081

Bassi, A. (1835). *Del mal del segno, calcinaccio o moscardino: Malattia che affligge i bachi da seta*. Dalla tipografia Orcesi.

Basyal, B. (2024). Plant-Arbuscular Mycorrhizal fungi association under drought stress. In *Arbuscular Mycorrhizal fungi in sustainable agriculture: Nutrient and crop management* (pp. 211–221). Springer Nature Singapore.

Begum, N., Qin, C., Ahanger, M. A., Raza, S., Khan, M. I., Ashraf, M., Ahmed, N., & Zhang, L. (2019). Role of arbuscular mycorrhizal fungi in plant growth regulation: Implications in abiotic stress tolerance. *Frontiers in Plant Science, 10*, 1068. https://doi.org/10.3389/fpls.2019.01068

Behie, S. W., & Bidochka, M. J. (2014). Ubiquity of insect-derived nitrogen transfer to plants by endophytic insect-pathogenic fungi: An additional branch of the soil nitrogen cycle. *Applied and Environmental Microbiology, 80*(5), 1553–1560. https://doi.org/10.1128/AEM.03338-13

Bencherif, K., Laruelle, F., Tisserant, B., Dalpé, Y., & Lounés-Hadj Sahraoui, A. (2023). Engineering approach for production of arbuscular mycorrhizal inoculum adapted to saline soil management. *Stress, 3*(2), 404–423.

Bennett, J. A., Maherali, H., Reinhart, K. O., Lekberg, Y., Hart, M. M., & Klironomos, J. (2017). Plant-soil feedbacks and mycorrhizal type influence temperate forest population dynamics. *Science, 355*(6321), 181–184. https://doi.org/10.1126/science.aai8212

Bernados, L. C., Espineli, J. P., Anarna, J. A., & Aggangan, N. S. (2024). Increasing tomato productivity through integrated nutrient sources and inoculation with arbuscular mycorrhizal fungi and *Azospirillum* spp. *Horticulturae, 10*(10), 1056.

Bhatt, P., Zhang, W., Lin, Z., Pang, S., Huang, Y., & Chen, S. (2020). Biodegradation of allethrin by a novel fungus *Fusarium proliferatum* strain CF2, isolated from contaminated soils. *Microorganisms, 8*(4), 593. https://doi.org/10.3390/microorganisms8040593

BIOTECH UPLB. (2023). *Products and technologies*. Accessed at https://biotech.uplb.edu.ph/products-and-technologies/

Bivi, M. R., Farhana, M. S., & Khairulmazmi, A. (2010). Control of *Ganoderma boninense*: A causal agent of basal stem rot disease in oil palm with endophyte. *International Journal of Agriculture and Biology, 12*(6), 833.

Burges, H. D. (1998). Formulation of mycoinsecticides. In H. D. Burges (Ed.), *Formulation of microbial biopesticides: Beneficial microorganisms, nematodes and seed treatments* (pp. 131–185). Springer.

Butt, T. M., Coates, C. J., Dubovskiy, I. M., & Ratcliffe, N. A. (2016a). Chapter 9: Entomopathogenic fungi: New insights into host–pathogen interactions. In B. Lovett & R. J. St. Leger (Eds.), (pp. 307–364). Academic.

Butt, T. M., Jackson, C., & Magan, N. (Eds.). (2016b). *Fungi as biocontrol agents: Progress, problems and potential*. CABI.

Carascal, M. B., del Rosario, M. J. G., Notarte, K. I. R., Huyop, F., Yaguchi, T., & dela Cruz, T. E. E. (2017). Butachlor biodegradation potential of fungi isolated from submerged wood and surface water collected in Taal Lake, Philippines. *Philippine Science Letters, 10*, 81–88.

Cardoso, I. M., & Kuyper, T. W. (2006). Mycorrhizas and tropical soil fertility. *Agriculture, Ecosystems & Environment, 116*, 72.

Catambacan, D. G., & Cumagun, C. J. R. (2021). Weed-associated fungal endophytes as biocontrol agents of *Fusarium oxysporum* f. sp. *cubense* TR4 in Cavendish banana. *Journal of Fungi, 7*(3), 224.

Cely, M. V. T., De Oliveira, A. G., De Freitas, V. F., De Luca, M. B., Barazetti, A. R., Dos Santos, I. M. O., Gionco, B., Garcia, G. V., Prete, C. E. C., & Andrade, G. (2016). Inoculant of Arbuscular Mycorrhizal fungi (*Rhizophagus clarus*) increase yield of soybean and cotton under Field conditions. *Frontiers in Microbiology, 7*. https://doi.org/10.3389/fmicb.2016.00720

Chakraborty, B. N., Chakraborty, U., Saha, A., Sunar, K., & Dey, P. L. (2010). Evaluation of phosphate solubilizers from soils of North Bengal and their diversity analysis. *World Journal of Agricultural Sciences, 6*(2), 195–200.

Chan, M. E., Tan, J. Y., Lee, Y. Y., Lee, D., Fong, Y. K., Mutwil, M., Wong, J. Y., & Hong, Y. (2023). Locally isolated *Trichoderma harzianum* species have broad spectrum biocontrol activities against the wood rot fungal species through both volatile inhibition and Mycoparasitism. *Journal of Fungi, 9*(6), 675. https://doi.org/10.3390/jof9060675

Chávez-González, J. D., Flores-Núñez, V. M., Merino-Espinoza, I. U., & Partida-Martínez, L. P. (2024). Desert plants, arbuscular mycorrhizal fungi and associated bacteria: Exploring the diversity and role of symbiosis under drought. *Environmental Microbiology Reports, 16*(4), e13300.

Chen, X., Aili, Y., Ma, X., Wang, H., & Dawuti, M. (2024). Mycorrhizal fungal colonization promotes apparent growth and physiology of *Alhagi sparsifolia* seedlings under salt or drought stress at vulnerable developmental stage. *Plant Growth Regulation, 102*(2), 267–278.

Cuervo-Parra, J. A., Pérez España, V. H., Zavala-González, E. A., Peralta-Gil, M., Aparicio Burgos, J. E., & Romero-Cortes, T. (2022). *Trichoderma Asperellum* strains as potential biological control agents against *Fusarium verticillioides* and *Ustilago maydis* in maize. *Biocontrol Science and Technology, 32*(5), 624–647. https://doi.org/10.1080/09583157.2022.2042196

Cuevas, V. C. (1997). Rapid composting technology in the philippines: Its role in producing good-quality fertilizers. *Food Fert. Tech. Center Extn. Bull., 444*, 1–13.

Cuevas, V. C. (2006). Soil inoculation with *Trichoderma pseudokoningii* Rifai enhances yield of rice. *The Philippine Journal of Science, 135*(1), 31.

Cuevas, V. C., & Banaay, C. G. B. (2022). *In situ* bioremediation and crop growth promotion using *Trichoderma* microbial inoculant (TMI) ameliorate the effects of Cu contamination in lowland rice paddies. *The Philippine Journal of Science, 151*(3), 1255–1265.

Cuevas, V. C., Sinohin, A. M., & Orajay, J. I. (2005). Performance of selected Philippine species of *Trichoderma* as biocontrol agents of damping off pathogens and as growth enhancer of vegetables in farmers' field. *Philippine Agricultural Science, 88*(1), 63–71.

Cuevas, V. C., Lagman, C. A., Jr., & Cuevas, A. C. (2011). Potential impacts of the use of *Trichoderma* spp. on farmers' profit in the field control of club root disease of crucifers caused by *Plasmodiophora brassicae* Wor. *Philippine Agricultural Science, 94*(2), 171–178.

Cuevas, V., Caasi-Lit, M., Lit, I., Jr., Alviola, M., Banaay, C. G., & Salazar, A. (2023). Recovery of coconut trees from scale insect infestation with salt and *Trichoderma* microbial inoculant (TMI). *Journal of Environmental Science and Management, 26*(2), 71–82.

Cui, R., Lu, X., Chen, X., Malik, W. A., Wang, D., Wang, J., et al. (2021). A novel raffinose biological pathway is observed by symbionts of cotton *Verticillium dahliae* to improve salt tolerance genetically on cotton. *Journal of Agronomy and Crop Science, 207*, 956–969. https://doi.org/10.1111/jac.12556

Cumagun, C. J. R. (2012). Managing plant diseases and promoting sustainability and productivity with *Trichoderma*: The Philippine experience (review articles). *Journal of Agricultural Science and Technology, 14*, 699–714.

Cumagun, C. J. R., Manalo, J. O., Salcedo-Bacalangco, N. A., & Ilag, L. L. (2009). Cellulose decomposing ability of *Trichoderma* in relation to their saprophytic survival. *Archives of Phytopathology and Plant Protection, 42*(7), 698–704.

Daei, G., Ardekani, M., Rejali, F., Teimuri, S., & Miransari, M. (2009). Alleviation of salinity stress on wheat yield, yield components, and nutrient uptake using arbuscular mycorrhizal fungi under field conditions. *Journal of Plant Physiology, 166*, 617–625.

Dar, M. H., & Reshi, Z. A. (2017). Vesicular Arbuscular Mycorrhizal (VAM) fungi- as a major biocontrol agent in modern sustainable agriculture system. *Russian Agricultural Sciences, 43*(2), 138–143. https://doi.org/10.3103/S1068367417020057

de Faria, M. R., & Wraight, S. P. (2007). Mycoinsecticides and Mycoacaricides: A comprehensive list with worldwide coverage and international classification of formulation types. *Biological Control, 43*(3), 237–256. https://doi.org/10.1016/j.biocontrol.2007.08.001

Deshmukh R., Khardenavis A. A., Purohit H. J. (2016). Diverse metabolic capacities of fungi for bioremediation. *Indian J. Microbiol. 56*, 247–264. https://doi.org/10.1007/s12088-016-0584-6

Deshmukh, S. K., Dufossé, L., Chhipa, H., Saxena, S., Mahajan, G. B., & Gupta, M. K. (2022). Fungal endophytes: A potential source of antibacterial compounds. *Journal of Fungi (Basel), 8*(2), 164. https://doi.org/10.3390/jof8020164

Ding, Y., Gao, X., Shu, D., Siddique, K. H., Song, X., Wu, P., et al. (2024). Enhancing soil health and nutrient cycling through soil amendments: Improving the synergy of bacteria and fungi. *Science of the Total Environment, 923*, 171332.

Djunarsjah, E., & Putra, A. P. (2021). May. The concept of an archipelagic province in Indonesia. In *IOP conference series: Earth and environmental science* (Vol. 777, No. 1, p. 012040). IOP Publishing.

Dodor, D. E., & Tabatabai, M. A. (2003). Effect of cropping systems on phosphatases in soils. *Journal of Plant Nutrition and Soil Science, 166*(1), 7–13.

Dutta, P., Deb, L., & Pandey, A. K. (2022). *Trichoderma*-from lab bench to field application: Looking back over 50 years. *Frontiers in Agronomy, 4*. https://doi.org/10.3389/fagro.2022.932839

Dwiastuti, M. E., Soesanto, L., Aji, T. G., Devy, N. F., & Hardiyanto. (2021). Biological control strategy for postharvest diseases of citrus, apples, grapes and strawberries fruits and application in Indonesia. *Egyptian Journal Of Biological Pest Control, 31*, 1–12.

El-Komy, M. H., Saleh, A. A., Eranthodi, A., & Molan, Y. Y. (2015). Characterization of novel *Trichoderma asperellum* isolates to select effective biocontrol agents against tomato *Fusarium* wilt. *The Plant Pathology Journal, 31*(1), 50.

El-Gendi, H., Saleh, A. K., Badierah, R., Redwan, E. M., El-Maradny, Y. A., & El-Fakharany, E. M. (2021). A comprehensive insight into fungal enzymes: Structure, classification, and their role in mankind's challenges. *Journal of Fungi (Basel), 8*(1), 23. https://doi.org/10.3390/jof8010023

Fakri, M. A., Ghazali, N. F., Sapak, Z., Muhammad Saiful, A. H., & Seng, C. T. (2022). Potential of *Bipolaris bicolor* in combination with ametryn for inhibition of glyphosate-resistant goosegrass (*Eleusine indica*) biotypes. *International Journal of Agriculture and Biology, 27*, 123–129. https://doi.org/10.17957/IJAB/15.1908

Fan, Y., & Shi, B. (2024). Endophytic fungi from the four staple crops and their secondary metabolites. *International Journal of Molecular Sciences, 25*(11), 6057.

Fan, L., Zhang, P., Cao, F., Liu, X., Ji, M., & Xie, M. (2024). Effects of AMF on maize yield and soil microbial community in sandy and saline soils. *Plants, 13*(15), 2056.

Feng, M. G., Poprawski, T. J., & Khachatourians, G. G. (1994). Production, formulation and application of the entomopathogenic fungus *Beauveria bassiana* for insect control: Current status. *Biocontrol Science and Technology, 4*(1), 3–34. https://doi.org/10.1080/09583159409355309

Fernandes, E. G., Valério, H. M., Feltrin, T., & Van Der Sand, S. T. (2012a). Variability in the production of extracellular enzymes by entomopathogenic fungi grown on different substrates. *Brazilian Journal of Microbiology, 43*, 827–833.

Fernandes, E. G., Valério, H. M., Feltrin, T., & Van Der Sand, S. T. (2012b). Variability fertility. *Agriculture, Ecosystems and Environment, 116*, 72–84. https://doi.org/10.1016/j.agee.2006.03.011

Fitriana, Y., Suharjo, R., Swibawa, I. G., Semenguk, B., Pasaribu, L. T., Hartaman, M., Rwandini, R. A., Indriyati, I., Purnomo, P., & Solikhin, S. (2021). *Aspergillus oryzae* and *Beauveria bassiana* as entomopathogenic fungi of *Spodoptera litura* Fabricius (Lepidoptera: Noctuidae) infesting corn in Lampung, Indonesia. *Egyptian Journal Of Biological Pest Control, 31*, 1–12.

Food and Agriculture Organization. (2018). *Country factsheet on small family farms – Indonesia*. Retrieved on February 13, 2025 from https://www.fao.org/agrifood-economics/publications/detail/en/c/1132206/

Fukasawa, Y. (2021). Ecological impacts of fungal wood decay types: A review of current knowledge and future research directions. *Ecological Research, 36*, 910–931. https://doi.org/10.1111/1440-1703.12260

Gabriel, B. P. (1968). Entomogenous microorganisms in The Philippines: New and past records. *Physiological Entomology, 1*(2), 97–130.

Gao, X., Guo, H., Zhang, Q., et al. (2020). Arbuscular mycorrhizal fungi (AMF) enhanced the growth, yield, fiber quality, and phosphorus regulation in upland cotton (*Gossypium hirsutum* L.). *Scientific Reports, 10*, 2084. https://doi.org/10.1038/s41598-020-59180-3

Gao, W., Zaynur, T., Sang, Y., & Ma, X. (2021). Effect of arbuscular mycorrhizal fungi on nitrogen absorption of plants: A review. *Chinese Agricultural Sciences Bulletin, 37*, 53–58.

Gómez-Leyva, J. F., Segura-Castruita, M. A., Hernández-Cuevas, L. V., & Íñiguez-Rivas, M. (2023). Arbuscular Mycorrhizal fungi associated with maize (*Zea mays* L.) in the formation and stability of aggregates in two types of soil. *Microorganisms, 11*(11), 2615.

Gou, X., Hu, Y., Ni, H., Wang, X., Qiu, L., Chang, X., et al. (2024). Arbuscular mycorrhizal fungi alleviate erosional soil nitrogen loss by regulating nitrogen cycling genes and enzymes in experimental agro-ecosystems. *Science of the Total Environment, 906*, 167425.

Govindarajulu, M., Pfeffer, P. E., Jin, H. R., Abubaker, J., Douds, D. D., Allen, J. W., Bucking, H., Lammers, P. J., & Shachar-Hill, Y. (2005). Nitrogen transfer in the arbuscular mycorrhizal symbiosis. *Nature, 435*, 819–823.

Greenfield, M., Gómez-Jiménez, M. I., Ortiz, V., Vega, F. E., Kramer, M., & Parsa, S. (2016). *Beauveria bassiana* and *Metarhizium anisopliae* endophytically colonize cassava roots following soil drench inoculation. *Biological Control, 95*, 40–48. https://doi.org/10.1016/j.biocontrol.2016.01.002

Guerrero, J. J. G., Malonzo, C. A., Regalado, R. R. H., & Dolooiras-Laraños, A. D. (2022). Environmental mycology in The Philippines. In T. U. Dalisay, M. P. De Leon, M. A. O. Balendres, K. I. R. Notarte, & T. E. E. D. Cruz (Eds.), *Mycology in the tropics: Updates on Philippine Fungi*. Elsevier.

Gupta, R., Gupta, R., Keppanan, R., Leibman-Markus, M., David, D. R., Elad, Y., Ment, D., & Bar, M. (2021). The entomopathogenic fungi *Metarhizium brunneum* and *Beauveria bassiana* promote systemic immunity and confer resistance to a broad range of pests and pathogens in tomato. *Phytopathology*. https://doi.org/10.1094/phyto-08-21-0343-r

Gustianingtyas, M., Herlinda, S., & Suwandi, S. (2021). The endophytic fungi from South Sumatra (Indonesia) and their pathogenecity against the new invasive fall armyworm, Spodoptera frugiperda. *Biodiversitas Journal of Biological Diversity, 22*(2). https://doi.org/10.13057/biodiv/d220262

Guzmán-Guzmán, P., Kumar, A., de los Santos-Villalobos, S., Parra-Cota, F. I., del Orozco-Mosqueda, M. C., Fadiji, A. E., Hyder, S., Babalola, O. O., & Santoyo, G. (2023). *Trichoderma* species: Our best fungal allies in the biocontrol of plant diseases—A review. *Plants, 12*(3), Article 3. https://doi.org/10.3390/plants12030432

Hailmi, M. S., Kadir, J., Abdullah, T. A., Suhaili, Z., Mohamad, N. M., Norhayati, N., & Salmah, M. (2011). Potential of *Exserohilum monoceras* as a bioherbicide for controlling *Echinochloa crus-galli* (rumput sambau). *Journal of Agrobiotechnology, 2*, 1–15.

Hailmi, S., Jugah, K., Fadzlina, N. A. N., & Gooi, P. S. (2017). Effect of selected pesticides on survivality of *Exserohilum longirostratum*, a bioherbicide for itchgrass. *Journal of Agrobiotechnology, 8*, 43–48.

Haleem, R. A., Nashat, L. H., & Ali, S. H. (2024). Molecular identification and antimicrobial potential of Endophytic fungi against some grapevine pathogens. *bioRxiv*, 2024–08.

Hammer, E. C., Forstreuter, M., Rillig, M. C., Kohler, J. (2015). Biochar increases arbuscular mycorrhizal plant growth enhancement andameliorates salinity stress. *Applied soil ecology, 96*, 114–121.

Harms, H., Schlosser, D., & Wick, L. Y. (2011). Untapped potential: Exploiting fungi in bioremediation of hazardous chemicals. *Nature Reviews Microbiology, 9*(3), 177–192. https://doi.org/10.1038/nrmicro2528

Hart, M. M., & Forsythe, J. A. (2012). Using arbuscular mycorrhizal fungi to improve the nutrient quality of crops; nutritional benefits in addition to phosphorus. *Scientia Horticulturae, 148*, 206–214.

He, D. C., He, M. H., Amalin, D. M., Liu, W., Alvindia, D. G., & Zhan, J. (2021). Biological control of plant diseases: An evolutionary and eco-economic consideration. *Pathogens, 10*(10), 1311.

Herlinda, S., Noni, O. C. T. A. R. I. A. T. I., Suwandi, S., & Hasbi, H. (2020). Exploring entomopathogenic fungi from South Sumatra (Indonesia) soil and their pathogenicity against a new invasive maize pest, *Spodoptera frugiperda*. *Biodiversitas Journal of Biological Diversity, 21*(7), 2955.

Herlinda, S., Gustianingtyas, M., Suwandi, S., Suharjo, R., Sari, J. M. P., & Lestari, R. P. (2021). Endophytic fungi confirmed as entomopathogens of the new invasive pest, the fall armyworm, *Spodoptera frugiperda* (JE Smith) (Lepidoptera: Noctuidae), infesting maize in South Sumatra, Indonesia. *Egyptian Journal Of Biological Pest Control, 31*, 1–13.

Hildebrandt, U., Regvar, M., & Bothe, H. (2007). Arbuscular mycorrhiza and heavy metal tolerance. *Phytochemistry, 68*, 139–146.

Hoeksema, J. D., Chaudhary, V. B., Gehring, C. A., Johnson, N. C., Karst, J., Koide, R. T., Pringle, A., Zabinski, C., Bever, J. D., & Moore, J. C. (2010). A meta-analysis of context-dependency in plant response to inoculation with mycorrhizal fungi. *Ecology Letters, 13*, 394–407.

Hossain, M. M., Sultana, F., Kubota, M., Koyama, H., & Hyakumachi, M. (2007). The plant growth-promoting fungus *Penicillium simplicissimum* GP17-2 induces resistance in Arabidopsis thaliana by activation of multiple defense signals. *Plant and Cell Physiology, 48*(12), 1724–1736. https://doi.org/10.1093/pcp/pcm144

Hu, K., Peris, A., Torán, J., Eljarrat, E., Sarrà, M., Blánquez, P., & Caminal, G. (2020). Exploring the degradation capability of *Trametes versicolor*on selected hydrophobic pesticides through setting sights simultaneously on culture broth and biological matrix. *Chemosphere, 250*, Article 126293. https://doi.org/10.1016/j.chemosphere.2020.126293

Husain, H., Keitel, C., & Dijkstra, F. A. (2024). Fungi are more important than bacteria for soil carbon loss through priming effects and carbon protection through aggregation. *Applied Soil Ecology, 195*, 105245.

Ibrahim, Y. B., & Low, W. (1993). Potential of mass production and field efficacy of isolates of the entomopathogenic fungi *Beauveria bassiana* and *Paecilomyces fumosoroseus* against *Plutella xylostella*. *International Journal of Pest Management, 39*(10), 288–292. https://doi.org/10.1080/09670879309371807

Idris, A. S., Nurrashyeda, R., Maizatul, S. M., Madihah, A. Z., Tarmizi, A. M., Kushairi, A., Wan Azha, W. M., & Tony Peng, S. H. (2012). *Biofertiliser Hendersonia GanoEF as biological control of Ganoderma in oil palm*. MPOB Information Series TT No. 50.

Irawan, B., Zulkifli, Z., Handayani, T. T., & Hadi, S. (2019). Effect of induced compost by cellulolitic (*Aspergillus fumigatus*) and ligninolitic (*Geotrichum* sp.) fungi inoculum application on vegetative growth of red chili (*Capsicum annuum* L.). *Journal of Pure and Applied Microbiology, 13*(2), 815–821.

Ishaq, L., Tae, A. A., Airthur, M. A., & Bako, P. O. (2021). Effect of single and mixed inoculation of arbuscular mycorrhizal fungi and phosphorus fertilizer application on corn growth in calcareous soil. *Biodiversitas Journal of Biological Diversity, 22*(4), 1920.

Islam, S., Akanda, A. M., Sultana, F., & Hossain, M. M. (2014). Chilli rhizosphere fungus *Aspergillus* spp. PPA1 promotes vegetative growth of cucumber (*Cucumis sativus*) plants upon root colonisation. *Archives of Phytopathology and Plant Protection, 47*, 1231–1238.

Jaber, L. R., & Enkerli, J. (2017). Fungal entomopathogens as endophytes: Can they promote plant growth? *Biocontrol Science and Technology, 27*(1), 28–41. https://doi.org/10.1080/09583157.2016.1243227

Jaroszuk-Ściseł, J., Tyśkiewicz, R., Nowak, A., Ozimek, E., Majewska, M., Hanaka, A., Tyśkiewicz, K., Pawlik, A., & Janusz, G. (2019). Phytohormones (Auxin, Gibberellin) and ACC deaminase in vitro synthesized by the Mycoparasitic *Trichoderma* DEMTkZ3A0 strain and changes in the level of auxin and plant resistance markers in wheat seedlings inoculated with this Strain Conidia. *International Journal of Molecular Sciences, 20*(19), 4923. https://doi.org/10.3390/ijms20194923

Jena, R., Mukherjee, A. K., Swain, H., Samanta, S., & Adak, T. (2024). Isolation of endophytic fungi from wild rice species for disease management and growth promotion in cultivated rice (*Oryza sativa* L.). *Biocontrol Science and Technology, 34*(5), 411–437. https://doi.org/10.1080/09583157.2024.2351812

Jiang, S. J., Qiang, S., Zhu, Y. Z., & Dong, Y. F. (2008). Isolation and phytotoxicity of a metabolite from *Curvularia eragrostidis* and characterisation of its modes of action. *Annals of Applied Biology, 152*, 103–111. https://doi.org/10.1111/j.1744-7348.2007.00202.x

Jiménez-Martínez, A., del Carmen Gutiérrez-Castorena, M., Montaño, N. M., Gutiérrez-Castorena, E. V., Alarcón, A., & Gavito, M. E. (2024). Micromorphology and thematic micromapping reveal differences in the soil structuring traits of three arbuscular mycorrhizal fungi. *Pedobiologia, 104*, 150953.

Ju, W., Fang, L., Shen, G., Delgado-Baquerizo, M., Chen, J., Zhou, G., et al. (2024). New perspectives on microbiome and nutrient sequestration in soil aggregates during long-term grazing exclusion. *Global Change Biology, 30*(1), e17027.

Jung, S. C., Martinez-Medina, A., Lopez-Raez, J. A., & Pozo, M. J. (2012). Mycorrhiza-induced resistance and priming of plant defenses. *Journal of Chemical Ecology, 38*(6), 651–664. https://doi.org/10.1007/s10886-012-0134-6

Juraimi, A. S., Tasrif, A., Kadir, J., Napis, S., & Sastroutomo, S. S. (2005). Phytotoxicity and field efficacy of *Exserohilum longirostra* Jc/min the control of barnyard grass ecotypes (*Echinochloa crus-galli* var. crus-galli (l.) Beauv). *Biotropia, 24*, 20–29.

Jyothymol, C. P., Kutty, M. S., Pradeepkumar, T., Parvathi, M. S., & Rashmi, C. R. (2024). *Piriformospora indica* improves water stress tolerance in watermelon (*Citrullus lanatus* (Thunb.) Matsum & Nakai). *Plant Physiology Reports, 29*, 1–13.

Kalamulla, R., Karunarathna, S. C., Tibpromma, S., Galappaththi, M. C. A., Suwannarach, N., Stephenson, S. L., Asad, S., Salem, Z. S., & Yapa, N. (2022). Arbuscular mycorrhizal fungi in sustainable agriculture. *Sustainability, 14*(19), 12250. https://doi.org/10.3390/su141912250

Kalayu, G. (2019). Phosphate solubilizing microorganisms: Promising approach as biofertilizers. *International Journal of Agronomy, 2019*, 1–7. https://doi.org/10.1155/2019/4917256

Kang, J., Yang, W., Liu, S., Xuan, N., Shao, Y., Geng, Y., Afzal, M., Zhang, Y., Yue, S., Mushtaq, R., & Chen, G. (2023). Arbuscular mycorrhizal fungi increased peanut (*Arachis hypogaea* L.) yield by changing the rhizosphere microbial community structure in saline-alkali soil. *Frontiers in Microbiology, 14*, 1303979. https://doi.org/10.3389/fmicb.2023.1303979

Kapri, A., & Tewari, L. (2010). Phosphate solubilization potential and phosphatase activity of rhizospheric Trichoderma spp. *Brazilian Journal of Microbiology, 41*. https://doi.org/10.1590/S1517-83822010005000001

Karagiannidis, N., Bletsos, F., & Stavropoulos, N. (2002). Effect of Verticillium wilt (*Verticillium dahliae* Kleb.) and mycorrhiza (*Glomus mosseae*) on root colonization, growth and nutrient uptake in tomato and eggplant seedlings. *Scientia Horticulturae, 94*(1–2), 145–156.

Khan, I. H., & Javaid, A. (2022). DNA cleavage of the fungal pathogen and production of antifungal compounds are the possible mechanisms of action of biocontrol agent *Penicillium italicum* against *Macrophomina phaseolina. Mycologia, 114*(1), 24–34. https://doi.org/10.1080/00275514.2021.1990627

Khan, A., Jilani, V., Akhtar, M. S., Naqvi, S. M. S., & Rasheed, M. (2009). Phosphorus solubilizing bacteria: Occurrence, mechanisms and their role in crop production. *Journal of Agricultural and Biological Science, 1*, 48–58.

Khan, S., Irshad, S., Mehmood, K., Hasnain, Z., Nawaz, M., Rais, A., et al. (2024). Biochar production and characteristics, its impacts on soil health, crop production, and yield enhancement: A review. *Plants, 13*(2), 166.

Khatua, S., Simal-Gandara, J., & Acharya, K. (2024). Myco-remediation of plastic pollution: Current knowledge and future prospects. *Biodegradation, 35*, 249–279. https://doi.org/10.1007/s10532-023-10053-2

Krishna, K. R., & Bagyaraj, D. J. (1983). Changes in the free amino nitrogen and protein fractions of groundnut caused by inoculation with VA mycorrhiza. *Annals of Botany, 51*(3), 399–401.

Kumar, A., & Patel, H. (2018). Role of microbes in phosphorus availability and acquisition by plants. *International Journal of Current Microbiology and Applied Sciences, 7*(5), 1344–1347.

Latiff, N., Omar, D., Sajap, A. S., Awang, R. M., Rajimin, R., Muning, M., & Peng, T. L. (2022). Effectiveness of isolates of *Metarhizium anisopliae* against *Aphis gossypii* (Hemiptera: Aphididae) on *Capsicum annum* and *Solanum melongena. International Journal of Agriculture and Biology, 28*, 1–6. https://doi.org/10.17957/IJAB/15.1969

Launio, C. C., Labon, K. O., Bañez, A. A., & Batani, R. S. (2020). Adoption and economic analysis of using biological control in Philippine highland farms: Case of *Trichoderma koningii* strain KA. *Crop Protection, 136*, 105177.

Lehmann, A., & Rillig, M. C. (2015). Arbuscular mycorrhizal contribution to copper, manganese and iron nutrient concentrations in crops – A meta-analysis. *Soil Biology & Biochemistry, 81*, 147–158.

Lehmann, A., Veresoglou, S. D., Leifheit, E. F., & Rillig, M. C. (2014). Arbuscular mycorrhizal influence on zinc nutrition in crop plants – A meta-analysis. *Soil Biology & Biochemistry, 69*, 123–131.

Li, T., Cui, L., Song, X., et al. (2022). Wood decay fungi: An analysis of worldwide research. *Journal of Soils and Sediments, 22*, 1688–1702. https://doi.org/10.1007/s11368-022-03225-9

Li, X., Wang, Y., Guo, P., Zhang, Z., Cui, X., Hao, B., & Guo, W. (2023). Arbuscular mycorrhizal fungi facilitate *Astragalus adsurgens* growth and stress tolerance in cadmium and lead contaminated saline soil by regulating rhizosphere bacterial community. *Applied Soil Ecology, 187*, 104842.

Linderman, R. G. (1994). Role of VAM fungi in biocontrol. In *Mycorrhizae and plant health*. https://cir.nii.ac.jp/crid/1571417125329059968

Litwin, A., Nowak, M., & Różalska, S. (2020). Entomopathogenic fungi: Unconventional applications. *Reviews in Environmental Science and Biotechnology, 19*(1), 23–42. https://doi.org/10.1007/s11157-020-09525-1

Liu, X., Li, Y., Han, B., Zhang, Q., Zhou, K., Zhang, X., & Hashemi, M. (2012). Yield response of continuous soybean to one-season crop disturbance in a previous continuous soybean field in Northeast China. *Field Crops Research, 138*, 52–56.

Liu, Y., Yang, Y., & Wang, B. (2022). Entomopathogenic fungi *Beauveria bassiana* and *Metarhizium anisopliae* play roles of maize (*Zea mays*) growth promoter. *Scientific Reports, 12*, 15706. https://doi.org/10.1038/s41598-022-19899-7

Lombardi, N., et al. (2018). Root exudates of stressed plants stimulate and attract *Trichoderma* soil fungi. *Molecular Plant-Microbe Interactions (MPMI), 31*, 982–994. https://doi.org/10.1094/MPMI-02-18-0031-R

Lopez, L. L. M. A., Reyes, R. G., & Alvindia, D. G. (2015). Evaluation of two species of *Trichoderma* as compost activator and bio-control agents. *Journal of Agricultural Technology, 11*(2), 525–537.

Lu, Y., Song, S., Wang, R., Liu, Z., Meng, J., Sweetman, A. J., et al. (2015). Impacts of soil and water pollution on food safety and health risks in China. *Environment International, 77*, 5–15. https://doi.org/10.1016/j.envint.2014.12.010

Luo, H., Qing, Z., Deng, Y., Deng, Z., Tang, X., Feng, B., & Lin, W. (2019). Two polyketides produced by Endophytic *Penicillium citrinum* DBR-9 from medicinal plant *Stephania kwangsiensis* and their antifungal activity against plant pathogenic fungi. *Natural Product Communications, 14*(5), 1934578X1984679. https://doi.org/10.1177/1934578X19846795

Mahdi, S., Hassan, G., Samoon, S., Rather, H., Dar, S., & Zehra, B. (2010). Bio-fertilizers in organic agriculture. *Journal of Phytology, 2010*(2), 42–54.

Malachová, K., Novotný, Č., Adamus, G., Lotti, N., Rybková, Z., Soccio, M., et al. (2020). Ability of *Trichoderma hamatum* isolated from plastics-polluted environments to attack petroleum-based, synthetic polymer films. *Processes, 8*, 467. https://doi.org/10.3390/pr8040467

Manathunga, K. K., Gunasekara, N. W., Meegahakumbura, M. K., Ratnaweera, P. B., Faraj, T. K., & Wanasinghe, D. N. (2024). Exploring Endophytic fungi as natural antagonists against fungal pathogens of food crops. *Journal of Fungi, 10*(9), 606.

Manivannan, N., Thajuddin, N., Daniel, T., & Gunasekaran, M. (2012). Effect of *Aspergillus* sp. as a bioinoculant in vermicompost with special reference to phosphate enrichment. *American-Eurasian Journal of Sustainable Agriculture, 6*, 205–208.

Marschner, H., & Dell, B. (1994). Nutrient uptake in mycorrhizal symbiosis. *Plant and Soil, 159*, 89–102.

Mascarin, G. M., Golo, P. S., de Souza, R.-S. C., Muniz, E. R., de Oliveira, F. A., Kobori, N. N., & Fernandes, É. K. K. (2024). Advances in submerged liquid fermentation and formulation of entomopathogenic fungi. *Applied Microbiology and Biotechnology, 108*(1), 451. https://doi.org/10.1007/s00253-024-13287-z

Masrahi, A. S., Alasmari, A., Shahin, M. G., Qumsani, A. T., Oraby, H. F., & Awad-Allah, M. M. (2023). Role of arbuscular mycorrhizal fungi and phosphate solubilizing bacteria in improving yield, yield components, and nutrients uptake of barley under salinity soil. *Agriculture, 13*(3), 537.

Masri, M., Azizah, H., Razi, I. M., & Mamat, A. S. (1998). Arbuscular mycorrhiza enchances growth and reduces nursery period of mangosteen (*Garcinia mangostana* L.) seedlings. *Journal of Tropical Agriculture and Food Science, 26*, 7–16.

Matsubara, Y., Tamura, H., & Harada, T. (1995). Growth enhancement and Verticillium wilt control by vesicular-arbuscular mycorrhizal fungus inoculation in eggplant. *Journal of the Japanese Society for Horticultural Science, 64*(3), 555–561.

Mbarga, J. B., Ten Hoopen, G. M., Kuaté, J., Adiobo, A., Ngonkeu, M. E. L., Ambang, Z., Akoa, A., Tondje, P. R., & Begoude, B. A. D. (2012). *Trichoderma asperellum*: A potential biocontrol agent for *Pythium myriotylum*, causal agent of cocoyam (*Xanthosoma sagittifolium*) root rot disease in Cameroon. *Crop Protection, 36*, 18–22.

Mehrvarz, S., Chaichi, M. R., & Alikhani, H. A. (2008). Effects of phosphate solubilizingmicroorganisms and phosphorus chemical fertilizer on yield and yield components of barely (*Hordeum vulgare* L.). *American-Eurasian Journal of Agricultural and Environmental Science, 3*, 822–828.

Mirsam, H., Kalqutny, S. H., Aqil, M., Azrai, M., Pakki, S., Muis, A., Djaenuddin, N., & Rauf, A. W. (2021). Indigenous fungi from corn as a potential plant growth promoter and its role in *Fusarium verticillioides* suppression on corn. *Heliyon, 7*(9), e07926.

Mittal, V., Singh, O., Nayyar, H., Kaur, J., & Tewari, R. (2008). Stimulatory effect of phosphate solubilizing fungal strains (*Aspergillus awamori* and *Penicillium citrinum*) on the yield of chickpea (*Cicer arietinum* cv GPF2). *Soil Biology and Biochemistry, 40*, 718–727.

Mohan, S. L., Beena, R., & Joy, M. (2024). An improvement in water stress tolerance in rice by altering morpho-physiological and biochemical mechanisms using root colonizing endophyte Piriformospora indica. *Vegetos, 38*, 1–12.

Montecalvo, M. P., & Navasero, M. M. (2020). Effect of entomopathogenic fungus *Metarhizium* (*Nomuraea*) *rileyi* (Farl.) Samson on the third instar larvae of the onion armyworm, *Spodoptera exigua* Hübner (Lepidoptera: Noctuidae), under laboratory conditions. *Philippine Agricultural Science, 103*(2), 159–164.

Montecalvo, M. P., & Navasero, M. M. (2021). Comparative virulence of *Beauveria bassiana* (Bals.) Vuill. and *Metarhizium anisopliae* (Metchnikoff) Sorokin to *Spodoptera frugiperda* (JE Smith) (Lepidoptera: Noctuidae). *Journal of the International Society for Southeast Asian Agricultural Sciences, 27*(1), 15–26.

Montecalvo, M. P., & Navasero, M. M. (2024). Occurrence and characterization of *Metarhizium rileyi* (Farl.) Kepler, SA Rehner and Humber from invasive armyworm species in The Philippines with potential as a biopesticide. *Journal of the International Society for Southeast Asian Agricultural Sciences, 30*(1), 80–91.

Montecalvo, M. P., Navasero, M. M., & Navasero, M. V. (2022). Lethal effect of native *Metarhizium rileyi* (Farlow) Samson isolate to invasive fall armyworm, *Spodoptera frugiperda* (JE Smith), infesting corn in The Philippines. *International Journal of Agricultural Technology, 18*(1), 257–270.

Moonjely, S., Barelli, L., & Bidochka, M. J. (2016). Insect pathogenic fungi as endophytes. *Advances in Genetics, 94*, 107–135. https://doi.org/10.1016/bs.adgen.2015.12.004

Moore, D., Bateman, R. P., Carey, M., & Prior, C. (1995). Long-term storage of *Metarhizium flavoviride* conidia in oil formulations for the control of locusts and grasshoppers. *Biocontrol Science and Technology, 5*(2), 193–200. https://doi.org/10.1080/09583159550039918

Moreno, N., Moreno, L., & Uribe, D. (2007). Biofertilizantes para la agricultura en Colombia. In M. L. Izaguirre-Mayoral, C. Labanderay, & J. Sanjuán (Eds.), *Biofertilizantes en Iberoamérica: Una Visión Técnica, Científica y Empresarial* (pp. 38–45). Editorial Universitaria.

Moslim, R., et al. (2009). Pathogenicity of granule formulations of *Metarhizium anisopliae* against the larvae of the oil palm rhinoceros beetle, *Oryctes rhinoceros* (L.). *Journal of Oil Palm Research, 21*, 602–612.

Moslim, R., et al. (2011). Trap for the auto dissemination of *Metarhizium anisopliae* in the management of rhinoceros beetle, *Oryctes rhinoceros. Journal of Oil Palm Research, 23*, 1011–1017.

Muhammad, M., Basit, A., Ali, K., Ahmad, H., Li, W., Khan, A., & Mohamed, H. I. (2024). A review on endophytic fungi: A potent reservoir of bioactive metabolites with special emphasis on blight disease management. *Archives of Microbiology, 206*(3), 129. https://doi.org/10.1007/s00203-023-03828-x

Mulyadi, & Jiang, L. (2023). The combined application of biochar and Arbuscular Mycorrhizal Fungi (AMF) enhanced the physical and chemical properties of soil and rice productivity in Indonesia. *Sustainability, 15*(12), 9782.

Naher, L., Syawani, N., Amieza, N., Kamarudin, A. B., & Karim, S. M. R. (2019). Trichoderma species diversity in rhizosphere soils and potential antagonism with Fusarium oxysporum. *Bioscience Journal*, 13–26. https://doi.org/10.14393/bj-v35n1a2019-41605

Naseer, M., Zhu, Y., Li, F. M., Yang, Y. M., Wang, S., & Xiong, Y. C. (2022). Nano-enabled improvements of growth and colonization rate in wheat inoculated with arbuscular mycorrhizal fungi. *Environmental Pollution, 295*, 118724–118712. https://doi.org/10.1016/j.envpol.2021.118724

Nature. (1987). *'BIOCON' controls nematodes biologically*. Accessed at https://www.nature.com/articles/nbt0887-772.pdf

Nepomuceno, R. A., Brown, C. B., Gargarino, A. M. P., Pedro, M. S., & Brown, M. B. (2020). Growth enhancement of rice (*Oryza sativa* L.) by zinc-solubilizing bacteria isolated from Vesicular-Arbuscular Mycorrhizal Root Inoculant (VAMRI). *Philippine Journal of Crop Science, 45*, 34–40.

Ng, S. C., Kadir, J., Hailmi, M. S., & Rahim, A. A. (2011). Efficacy of *Exserohilum longirostratum* on barnyard grass (*Echinochloa crus-galli* spp. *crusgalli*) under field conditions. *Biocontrol Science and Technology, 21*(4), 449–460. https://doi.org/10.1080/09583157.2011.554972

Nicoletti, R., Lopez-Gresa, M. P., Manzo, E., Carella, A., & Ciavatta, M. L. (2007). Production and fungitoxic activity of Sch 642305, a secondary metabolite of *Penicillium canescens. Mycopathologia, 163*(5), 295–301. https://doi.org/10.1007/s11046-007-9015-x

Nurhayati, Y., Suryanti, S., & Wibowo, A. (2021). In vitro evaluation of Trichoderma asperellum isolate UGM-LHAF against Rhizoctonia solani causing sheath blight disease of rice. *Jurnal Perlindungan Tanaman Indonesia, 25*(1), 64–73.

Organo, N. D., Granada, S. M. J. M., Pineda, H. G. S., Sandro, J. M., Nguyen, V. H., Gummert, M. (2022). Assessing the potential of a Trichoderma-based compost activator to hasten the decomposition of incorporated rice straw. *Scientific Reports, 12*(1), 448.

Ozgonen, H., & Erkilic, A. (2007). Growth enhancement and *Phytophthora* blight (*Phytophthora capsici* Leonian) control by arbuscular mycorrhizal fungal inoculation in pepper. *Crop Protection, 26*(11), 1682–1688.

Panchalingam, H., Powell, D., Adra, C., Foster, K., Tomlin, R., Quigley, B. L., Nyari, S., Hayes, R. A., Shapcott, A., & Kurtböke, D. İ. (2022). Assessing the various antagonistic mechanisms of Trichoderma strains against the brown root rot pathogen Pyrrhoderma noxium infecting heritage fig trees. *Journal of Fungi, 8*(10), 1105.

Patil, P. M., Kuligod, V. B., Hebsur, N. S., Patil, C. R., & Kulkarni, G. N. (2012). Effect of phosphate solubilizing fungi and phosphorus levels on growth, yield and nutrient content in maize (*Zea mays*). *Karnataka Journal of Agricultural Sciences, 25*, 58–62.

Paudzai, F. A. M., Sundram, S., Yusof, M. T., Angel, L. P. L., Hashim, A. M., & Abdullah, S. N. A. (2019). Induced systemic resistance and promotion of plant growth in oil palm seedlings by endophytic *Trichoderma virens. Journal of Oil Palm Research, 31*(4), 572–581.

Pellegrino, E., Opik, M., Bonari, E., & Ercoli, L. (2015). Responses of wheat to arbuscular mycorrhizal fungi: a meta-analysis of field studies from 1975 to 2013. *Soil Biology & Biochemistry, 84*, 210–217.

PhilRice. (2002a). *Metarhizium*: Microbial control agent for rice black bug. Rice Technology Bulletin 44.

PhilRice. (2002b). Pagpaparami ng VAM (Vesicular arbuscular mycorrhizae). Rice Technology Bulletin 61.

Pimprikar, P., & Gutjahr, C. (2018). Transcriptional regulation of arbuscular mycorrhiza development. *Plant Cell Physiology, 59*, 673–690. https://doi.org/10.1093/pcp/pcy024

Priwiratama, H., & Susanto, A. (2014). Utilization of fungi for the biological control of insect pests and Ganoderma disease in the Indonesian oil palm industry. *Journal of Agricultural Science and Technology. A, 4*(2A), 103.

Promwee, A., Issarakraisila, M., Intana, W., Chamswarng, C., & Yenjit, P. (2014). Phosphate solubilization and growth promotion of rubber tree (*Hevea brasiliensis* Muell. Arg.) by *Trichoderma* strains. *The Journal of Agricultural Science, 6*, 8. https://doi.org/10.5539/jas.v6n9p8

Puig, C. G., & Cumagun, C. J. R. (2019). Rainforest fungal endophytes for the bio-enhancement of banana toward *Fusarium oxysporum* f. sp. *cubense* Tropical Race 4. *Archives of Phytopathology and Plant Protection, 52*(7–8), 776–794.

Purnomo, A. S., Mori, T., Kamei, I., Nishii, T., & Kondo, R. (2010). Application of mushroom waste medium from *Pleurotus ostreatus* for bioremediation of DDT-contaminated soil. *International Biodeterioration & Biodegradation, 64*, 397–402. https://doi.org/10.1016/j.ibiod.2010.04.007

Rajapakse, R. H. S., Ratnasekera, D., & Abeysinghe, S. (2016). Biopesticides research: Current status and future trends in Sri Lanka. In *Agriculturally important microorganisms: Commercialization and regulatory requirements in Asia* (pp. 219–234).

Ratnaweera, P. B., & de Silva, E. D. (2017). Endophytic fungi: A remarkable source of biologically active secondary metabolites. *Endophytes: Crop Productivity and Protection, 2*, 191–212.

Richa, G., Khosla, B., & Reddy, M. S. (2007). Improvement of maize plant growth by phosphate solubilizing fungi in rock phosphate amended soils. *World Journal of Agricultural Sciences, 3*, 481–484.

Rillig, M. C., Sosa-Hernàndez, M. A., Roy, J., Aguilar-Trigueros, C. A., Vàlyi, K., & Lehmann, A. (2016). Towards an integrated mycorrhizal technology: Harnessing mycorrhiza for sustainable intensification in agriculture. *Frontiers in Plant Science, 7*, 1–5.

Rodarte-Morales, A. I., Feijoo, G., Moreira, M. T., & Lema, J. M. (2011). Degradation of selected pharmaceutical and personal care products (PPCPs) by white-rot fungi. *World Journal of Microbiology and Biotechnology, 27*, 1839–1846. https://doi.org/10.1007/s11274-010-0642-x

Rosliani, R., Prathama, M., Sulastrini, I., Sugiartini, E., & Santosa, A. D. (2023). The effect of arbuscular mycorrhizae inoculation on growth and yield of red chili on the highland andisol soil. In *IOP conference series: Earth and environmental science* (Vol. 1287, No. 1, p. 012005). IOP Publishing.

Saber, W. I. A., Ghanem, K. M., & El hersh, M.S. (2009). Rock phosphate solubilisation by 2 isolates of *Aspergillus niger* and *Penicillium* sp and their promotion to mungbean plants. *Research Journal of Microbiology, 4*, 235–250.

Sadhana, B. (2014). Arbuscular mycorrhizal fungi (AMF) as a biofertilizer. A review. *International Journal of Current Microbiology and Applied Sciences, 3*, 384–400.

Sahayaraj, K., & Namasivayam, S. K. R. (2008). Mass production of entomopathogenic fungi using agricultural products and by products. *African Journal of Biotechnology, 7*(12), 1907–1910. https://doi.org/10.5897/AJB07.778

Sajap, A. S., Rozihawati, Z., Omar, D., & Lau, W. H. (2014). *Isaria fumosorosea* and *Metarhizium anisopliae* for controlling *Atteva sciodoxa* (Lepidoptera: Yponomeutidae), a pest of *Eurycoma longifolia*. *Journal of Tropical Forest Science, 26*(1), 84–91.

Sanjaya, Y., Ocampo, V. R., & Caoili, B. L. (2013). Infection process of entomopathogenic fungi *Metarhizium anisopliae* in the *Tetranychus kanzawai* (Kishida) (Tetranychidae: Acarina). *AGRIVITA Journal of Agricultural Science, 35*(1), 64–72.

Santana, E. B., Marques, E. L. S., & Dias, J. C. T. (2016). Effects of phosphate-solubilizing bacteria, native microorganisms and rock dust on *Jatropha curcas* L. growth. *Genetics and Molecular Research, 15*(4). https://doi.org/10.4238/gmr.15048729

Santiago, D. R., Castillo, A. G., Arapan, R. S., Navasero, M. V., & Eusebio, J. E. (2001). Efficacy of *Metarhizium anisopliae* (metsch.) Sor. against the oriental migratory locust, *Locusta migratoria manilensis* meyen. *Philippine Agricultural Science, 84*(1), 26–34.

Saravanan, A., Kumar, P. S., Govarthanan, M., George, C. S., Vaishnavi, S., Moulishwaran, B., et al. (2021). Adsorption characteristics of magnetic nanoparticles coated mixed fungal biomass for toxic Cr(VI) ions in aquatic environment. *Chemosphere, 267*, 129226. https://doi.org/10.1016/j.chemosphere.2020.129226

Sari, J. M. P., Herlinda, S., & Suwandi, S. (2022). Endophytic fungi from South Sumatra (Indonesia) in seed-treated corn seedlings affecting development of the fall armyworm, *Spodoptera frugiperda* JE Smith (Lepidoptera: Noctuidae). *Egyptian Journal Of Biological Pest Control, 32*(1), 103.

Sarkar, A., Asaeda, T., Wang, Q., & Rashid, M. H. (2016). Arbuscular Mycorrhizal Association for Growth and Nutrients Assimilation of *Pharagmites japonica* and *Polygonum cuspidatum* Plants Growing on River Bank Soil. *Communications in Soil Science and Plant Analysis, 47*(1), 87–100. https://doi.org/10.1080/00103624.2015.1108432

Satyaprakash, M., Nikitha, T., Reddi, E. U. B., Sadhana, B., & Vani, S. S. (2017). A review on phosphorous and phosphate solubilising bacteria and their role in plant nutrition. *International Journal of Current Microbiology and Applied Sciences, 6*, 2133–2144.

Selvi, K. B., Paul, J. J. A., Vijaya, V., & Saraswathi, K. (2017). Analyzing the efficacy of phosphate solubilizing microorganisms by enrichment culture techniques. *Biochemistry and Molecular Biology Journal, 3*, 1.

Sharif, M., Ahmad, E., Sarir, M. S., Muhammad, D., Shafi, M., & Bakht, J. (2011). Response of different crops to arbuscular mycorrhiza fungal inoculation in phosphorus-deficient soil. *Communications in Soil Science and Plant Analysis, 42*(19), 2299–2309.

Sharma, A. K., Johri, B. N., & Gianinazzi, S. (1992). Vesicular-arbuscular mycorrhizae in relation to plant disease. *World Journal of Microbiology & Biotechnology, 8*(6), 559–563. https://doi.org/10.1007/BF01238788

Silva, B. B., Banaay, C. G., & Salamanez, K. (2019). *Trichoderma*-induced systemic resistance against the scale insect (*Unaspis mabilis* Lit & Barbecho) in lanzones (*Lansium domesticum* Corr.). *Agriculture and Forestry/Poljoprivreda i šumarstv, 65*(2), 59.

Singh, V. K., & Kumar, A. (2023). Secondary metabolites from endophytic fungi: Production, methods of analysis, and diverse pharmaceutical potential. *Symbiosis, 90*(2), 111–125.

Singh, H., & Reddy, M. S. (2011). Effect of inoculation with phosphate solubilising fungus on growth and nutrient uptake of wheat and maize plants fertilized with rock phosphate in alkaline soils. *European Journal of Soil Biology, 47*, 30–34.

Singh, R., Adholeya, A., & Mukerji, K. G. (2000). Mycorrhiza in control of soil borne pathogens. In K. G. Mukerji, B. P. Chamola, & J. Singh (Eds.), *Mycorrhizal biology* (pp. 173–196). Springer US. https://doi.org/10.1007/978-1-4615-4265-0_12

Singh, A., Sarma, B. K., Singh, H. B., & Upadhyay, R. S. (2014). Trichoderma: A silent worker of plant rhizosphere. In *Biotechnology and biology of Trichoderma* (pp. 533–542). Elsevier. https://www.sciencedirect.com/science/article/pii/B9780444595768000400

Singh, B. N., Singh, A., Singh, G. S., & Dwivedi, P. (2015). Potential role of *Trichoderma asperellum* T42 strain in growth of pea plant for sustainable agriculture. *Journal of Pure and Applied Microbiology, 9*, 1069–1074.

Singh, M., Sharma, J. G., & Giri, B. (2024). Augmentative role of Arbuscular Mycorrhizal fungi, *Piriformospora indica*, and plant growth-promoting bacteria in mitigating salinity stress in maize (*Zea mays* L.). *Journal of Plant Growth Regulation, 43*(4), 1195–1215.

Sinha, K. K., Choudhary, A. K., & Kumari, P. (2016). Chapter 15: Entomopathogenic fungi. In Omkar (Ed.), *Ecofriendly pest management for food security* (pp. 475–505). Academic.

Skinner, M., Parker, B. L., & Kim, J. S. (2014). Chapter 10: Role of Entomopathogenic fungi in integrated Pest management. In D. P. Abrol (Ed.), *Integrated Pest management* (pp. 169–191). Academic.

Smith, S. E., & Read, D. J. (2008a). *Mycorrhizal symbiosis* (3rd ed.). Academic.

Smith, S. E., & Read, D. J. (2008b). *Mycorrhizal symbiosis*. Academic.

Smith, S. E., & Read, D. J. (2010). *Mycorrhizal symbiosis*. Academic. https://books.google.com/books?hl=en&lr=&id=qLciOJaG0C4C&oi=fnd&pg=PP1&dq=Smith+SE,+Read+DJ.

+Mycorrhizal+symbiosis.+Academic+press%3B+2010.&ots=zstWkVQxoP&sig=ozjvH4Wqa1R4snu742g1GpnY1uw

St. Leger, R. J., & Wang, C. (2010). Genetic engineering of fungal biocontrol agents to achieve greater efficacy against insect pests. *Applied Microbiology and Biotechnology, 85*(4), 901–907. https://doi.org/10.1007/s00253-009-2306-z

Statista. (2024). *Agriculture in Malaysia – Statistics & facts*. https://www.statista.com/topics/10680/agriculture-in-malaysia/#topicOverview. Accessed 13 Aug 2024.

Steinhaus, E. A. (1975). *Disease in a minor chord: Being a Semihistorical and semibiographical account of a period in science when one could be happily yet seriously concerned with the diseases of lowly animals without backbones, especially the insects*. Ohio State University Press.

Stenberg, J. A., Sundh, I., Becher, P. G., et al. (2021). When is it biological control? A framework of definitions, mechanisms, and classifications. *Journal of Pest Science, 94*, 665–676. https://doi.org/10.1007/s10340-021-01354-7

Stoffel, S. C. G., Soares, C. R. F. S., Meyer, E., Lovato, P. E., & Giachini, A. J. (2020). Yield increase of corn inoculated with a commercial arbuscular mycorrhizal inoculant in Brazil. *Ciência Rural, 50*(7), e20200109. https://doi.org/10.1590/0103-8478cr20200109

Stone, J. K., Polishook, J. D., & White, J. F. (2004). Endophytic fungi. *Biodiversity of Fungi: Inventory and Monitoring Methods, 241*, 270.

Su, S., Zeng, X., Bai, L., Jiang, X., & Li, L. (2010). Bioaccumulation and biovolatilization of pentavalent arsenic by *Penicillium janthinellum*, *Fusarium oxysporum*, and *Trichoderma asperellum* under laboratory conditions. *Current Microbiology, 61*, 261–266. https://doi.org/10.1007/s00284-010-9605-6

Subashini, H. D., Malarvannan, S., & Kumaran, P. (2007). Effect of biofertilizers (N-fixers) on the yield of Rice varieties at Puducherry, India. *Asian Research Journal of Agriculture, 2007*(1), 146–150.

Sui, L., Lu, Y., Zhou, L., Li, N., Li, Q., & Zhang, Z. (2023). Endophytic *Beauveria bassiana* promotes plant biomass growth and suppresses pathogen damage by directional recruitment. *Frontiers in Microbiology, 14*, 1227269.

Sundram, S., Meon, S., Seman, I. A., & Othman, R. (2015). Application of arbuscular mycorrhizal fungi with *Pseudomonas aeruginosa* UPMP3 reduces the development of *Ganoderma* basal stem rot disease in oil palm seedlings. *Mycorrhiza, 25*, 387–397. https://doi.org/10.1007/s00572-014-0620-5

Talukder, M. R., Sarkar, A., & Rashid, M. H. (2019). The role of arbuscular mycorrhizal fungi in the bioprotection of ash gourd (*Benincasa hispida*) against damping off disease. *Fundamental and Applied Agriculture, 4*(1), Article 1. https://doi.org/10.5455/faa.6098

Talukder, M. R., Sarkar, A., Rashid, M. H., & Hossain, M. M. (2024). Biocontrol of damping off disease in Brinjal (*Solanum melongena*) and tomato (*Solanum lycopersicum*) by Arbuscular Mycorrhiza. *OnLine Journal of Biological Sciences, 24*(4), 633–642. https://doi.org/10.3844/ojbsci.2024.633.642

Tanada, Y., & Kaya, H. K. (1993). Chapter 10: Fungal infections. In Y. Tanada & H. K. Kaya (Eds.), *Insect pathology* (pp. 318–387). Academic.

Tang, B., Man, J., Lehmann, A., & Rillig, M. C. (2024). Arbuscular mycorrhizal fungi attenuate negative impact of drought on soil functions. *Global Change Biology, 30*(7), e17409.

Tarafdar, J. C., Bareja, M., & Panwar, J. (2003). Efficiency of some phosphatase producing soil-fungi. *Indian Journal of Microbiology, 43*, 27–32.

Tedersoo, L., & Bahram, M. (2019). Mycorrhizal types differ in ecophysiology and alter plant nutrition and soil processes. *Biological Reviews of the Cambridge Philosophical Society, 94*(5), 1857–1880. https://doi.org/10.1111/brv.12538

Thambugala, K. M., Daranagama, D. A., Phillips, A. J., Kannangara, S. D., & Promputtha, I. (2020). Fungi vs. fungi in biocontrol: An overview of fungal antagonists applied against fungal plant pathogens. *Frontiers in Cellular and Infection Microbiology, 10*, 604923.

Thirkell, T. J., Charters, M. D., Elliott, A. J., Sait, S. M., & Field, K. J. (2017). Are mycorrhizal fungi our sustainable saviours? Considerations for achieving food security. *Journal of Ecology, 105*, 921–929.

Tian, H., Drijber, R., Zhang, J., & Li, X. (2013). Impact of long-term nitrogen fertilization and rotation with soybean on the diversity and phosphorus metabolism of indigenous arbuscular mycorrhizal fungi within the roots of maize (*Zea mays* L.). *Agriculture, Ecosystems and Environment, 164*, 53–61.

Tian, Y., Zhang, D., Cai, P., Lin, H., Ying, H., Hu, Q.-N., & Wu, A. (2022). Elimination of *Fusarium* mycotoxin deoxynivalenol (DON) via microbial and enzymatic strategies: Current status and future perspectives. *Trends in Food Science & Technology, 124*, 96–107.

Tondje, P. R., Roberts, D. P., Bon, M. C., Widmer, T., Samuels, G. J., Ismaiel, A., Begoude, A. D., Tchana, T., Nyemb-Tshomb, E., & Ndoumbe-Nkeng, M. (2007). Isolation and identification of mycoparasitic isolates of *Trichoderma asperellum* with potential for suppression of black pod disease of cacao in Cameroon. *Biological Control, 43*(2), 202–212.

Treseder, K. K., & Lennon, J. T. (2015). Fungal traits that drive ecosystem dynamics on land. *Microbiology and Molecular Biology Reviews, 79*(2), 243–262. https://doi.org/10.1128/MMBR.00001-15

Trotta, A., Varese, G. C., Gnavi, E., Fusconi, A., Sampo, S., & Berta, G. (1996). Interactions between the soilborne root pathogen *Phytophthora nicotianae* var. *parasitica* and the arbuscular mycorrhizal fungus *Glomus mosseae* in tomato plants. *Plant and Soil, 185*, 199–209.

Uzma, F., Mohan, C. D., Hashem, A., Konappa, N. M., Rangappa, S., Kamath, P. V., et al. (2018). Endophytic fungi—Alternative sources of cytotoxic compounds: A review. *Frontiers in Pharmacology, 9*, 309. https://doi.org/10.3389/fphar.2018.00309

Vega, F. E., Goettel, M. S., Blackwell, M., Chandler, D., Jackson, M. A., Keller, S., Koike, M., Maniania, N. K., Monzón, A., Ownley, B. H., Pell, J. K., Rangel, D. E. N., & Roy, H. E. (2009). Fungal entomopathogens: New insights on their ecology. *Fungal Ecology, 2*(4), 149–159. https://doi.org/10.1016/j.funeco.2009.05.001

Victoria, K. S., & Aggangan, N. S. (2023). Effects of bio-and chemical fertilizers on growth and soil microbial population of cacao (*Theobroma cacao* L.) seedlings. *The Philippine Journal of Science, 152*(4), 1495–1505.

Visalakshy, P. N., Frenita, L., Darshana, C. N., Swathi, C., & Krisnamoorthy, A. (2017). Antagonistic effect of entomopathogenic fungus, *Metarhizium anisopliae* (IIHR strain) on plant pathogens, *Colletotrichum gloeosporioides, Pestalotia mangiferae* and *Botrydipliodia theobromae*. *Pest Management in Horticultural Ecosystems, 23*(2), 179–181.

Wakelin, S., Warren, R., Harvey, P., & Ryder, M. (2004). Phosphate solubilization by *Penicillium* spp. closely associated with wheat roots. *Biology and Fertility of Soils, 40*, 36–43.

Wallander, H., Ekblad, A., & Bergh, J. (2011). Growth and carbon sequestration by ectomycorrhizal fungi in intensively fertilized Norway spruce forests. *Forest Ecology and Management, 262*(6), 999–1007. https://doi.org/10.1016/j.foreco.2011.05.035

Wen, J., Okyere, S. K., Wang, S., Wang, J., Xie, L., Ran, Y., & Hu, Y. (2022). Endophytic fungi: An effective alternative source of plant-derived bioactive compounds for pharmacological studies. *Journal of Fungi, 8*(2), 205.

Wen, Y., Wu, R., Qi, D., Xu, T., Chang, W., Li, K., et al. (2024). The effect of AMF combined with biochar on plant growth and soil quality under saline-alkali stress: Insights from microbial community analysis. *Ecotoxicology and Environmental Safety, 281*, 116592.

Whipps, J. M. (2004). Prospects and limitations for mycorrhizas in biocontrol of root pathogens. *Canadian Journal of Botany, 82*(8), 1198–1227. https://doi.org/10.1139/b04-082

Wu, S. C., Cao, Z. H., Li, Z. G., Cheung, K. C., & Wong, M. H. (2005). Effects of biofertilizer containing N- fixer, P and K solubilizers and AM fungi on maize growth: a greenhouse trial. *Geoderma, 125*, 155–166. https://doi.org/10.1016/j.geoderma.2004.07.003

Wu, J., et al. (2022a). Beneficial effect on the soil microenvironment of *Trichoderma* applied after fumigation for cucumber production. *PLoS ONE, 17*, Article e0266347. https://doi.org/10.1371/journal.pone.0266347

Wu, S., Shi, Z., Chen, X., Gao, J., & Wang, X. (2022b). Arbuscular mycorrhizal fungi increase crop yields by improving biomass under rainfed conditions: A meta-analysis. *PeerJ, 10*, e12861. https://doi.org/10.7717/peerj.12861

Wu, Q., Ning, K., Liu, B., Zheng, X., Li, C., Li, X., et al. (2024). Co-application of biochars and *Piriformospora indica* improved the quality of coastal saline soil and promoted the growth of forage. *Frontiers in Plant Science, 15*, 1434097.

Wangiyana W, Farida N (2019) Application bio-fertilizers to increase yields of zero-tillage soybean of two varieties under different planting distances in dry season on vertisol land of Central Lombok, Indonesia. AIP Conference Proceedings 2199, 040009.

Xiao, G., Ying, S.-H., Zheng, P., Wang, Z.-L., Zhang, S., Xie, X.-Q., Shang, Y., St. Leger, R. J., Zhao, G.-P., Wang, C., & Feng, M.-G. (2012). Genomic perspectives on the evolution of fungal entomopathogenicity in *Beauveria bassiana*. *Scientific Reports, 2*(1), 483. https://doi.org/10.1038/srep00483

Xu, T., Johnson, D., & Bardgett, R. D. (2024). Defoliation modifies the response of arbuscular mycorrhizal fungi to drought in temperate grassland. *Soil Biology and Biochemistry, 192*, 109386.

Yadav, J., Verma, J. P., & Tiwari, K. N. (2011). Plant growth promoting activities of fungi and their effect on chickpea plant growth. *Asian Journal of Biological Sciences, 4*, 291–299.

Yago, J. I., Agrimor, A. D., Baligad, J. L. P., Irabagon, E. T., Villanueva, L. M., & Maghirang, R. G. (2016). Extracellular cell wall-degrading enzyma as mycopesticide derived from *Beauveria bassiana* against tomato black lead mold (*Pseudocercospora fuligena*). *Philippine Journal of Crop Science, 42*(1), 26–27.

Yago, J. I., Chung, K. R., & Bating, R. A. (2020). Efficacy evaluation of biopesticides derived from entomopathogenic fungi against rind borer and twig blight disease of citrus. *International Journal of Agriculture and Biological Sciences*, 27–47.

Yang, L., Xie, J., Jiang, D., Fu, Y., Li, G., & Lin, F. (2008). Antifungal substances produced by *Penicillium oxalicum* strain PY-1—Potential antibiotics against plant pathogenic fungi. *World Journal of Microbiology and Biotechnology, 24*(7), 909–915. https://doi.org/10.1007/s11274-007-9626-x

Yao, M., Tweddell, R., & Désilets, H. (2002). Effect of two vesicular-arbuscular mycorrhizal fungi on the growth of micropropagated potato plantlets and on the extent of disease caused by *Rhizoctonia solani*. *Mycorrhiza, 12*(5), 235–242. https://doi.org/10.1007/s00572-002-0176-7

Yao, X., Guo, H., Zhang, K., Zhao, M., Ruan, J., & Chen, J. (2023). *Trichoderma* and its role in biological control of plant fungal and nematode disease. *Frontiers in Microbiology, 14*. https://doi.org/10.3389/fmicb.2023.1160551

Zebua, A. C. (2024). Application of FMA on soil chemical properties in oil palm (*Elaeis guineensis* Jacq.) plants. *Jurnal Agronomi Tanaman Tropika (Juatika), 6*(3), 894–903.

Zhang, H., Wu, X., Li, G., & Qin, P. (2011). Interactions between arbuscular mycorrhizal fungi and phosphate-solubilizing fungus (*Mortierella* sp.) and their effects on *Kostelelzkya virginica* growth and enzyme activities of rhizosphere and bulk soils at different salinities. *Biology and Fertility of Soils, 47*, 543–554.

Zhang, F., Xu, X., Huo, Y., & Xiao, Y. (2019). *Trichoderma*-inoculation and mowing synergistically altered soil available nutrients, rhizosphere chemical compounds, and soil microbial community, potentially driving alfalfa growth. *Frontiers in Microbiology, 9*, 3241. https://doi.org/10.3389/fmicb.2018.03241

Zhang, Z., Ding, S., Diao, F., Jia, B., Shi, Z., & Guo, W. (2024). AMF symbiosis drives the rhizosphere microbiome to synergistically improve herbage growth in saline–alkaline soils. *Land Degradation & Development, 35*(11), 3663–3674.

Zhu, S., Shi, F., Li, H., Ding, Y., Chang, W., Ping, Y., & Song, F. (2024). *Piriformospora indica* alleviates soda saline-alkaline stress in *Glycine max* by modulating plant metabolism. *Frontiers in Plant Science, 15*, 1406542.

Zimmermann, G. (2007). Review on safety of the entomopathogenic fungi *Beauveria bassiana* and *Beauveria brongniartii*. *Biocontrol Science and Technology, 17*(5–6), 553–596. https://doi.org/10.1080/09583150701625608

Chapter 10
Fungi in the Postharvest of Agricultural Commodities

Reynaldo Peja Jr., Deborah Anne Dimayacyac, Jonathan Barcelo, Rachele De Torres, Julie Aiza Mandap, Junrey Amas, Searle S. Duay, Latiffah Zakaria, Christian Joseph Cumagun, Gianfranco Romanazzi, and Mark Angelo O. Balendres

Abstract Postharvest diseases caused by fungal pathogens are among the major contributors to postharvest losses. An estimated $150 billion is lost due to postharvest constraints globally. Fresh fruits, vegetables, and stored cereal grains are the most affected commodities by postharvest losses. Postharvest diseases are often associated with "latent" or "quiescent" infections, wherein the pathogen infects the host at an earlier stage (typically before harvest) but enters a state of dormancy until favorable conditions occur. Latent infections are more commonly associated with postharvest diseases in fruits. However, they are also reported in vegetables such as those incited by *Colletotrichum* species and *Botrytis cinerea*. On cereal grains, toxigenic fungi (e.g., *Aspergillus* species) are often contaminated during storage and marketing. This chapter discusses common postharvest diseases caused by fungi in fruits, vegetables, and cereal grains and the various disease control strategies used. Finally, this chapter examines emerging and innovative approaches for future postharvest disease management, including, but not limited to, genomics, edible coat-

R. Peja Jr.
Department of Biological Sciences, Faculty of Natural and Mathematical Sciences, Visayas State University, Baybay City, Philippines

D. A. Dimayacyac
Department of Plant Pathology, Georg-August University, Göttingen, Germany

J. Barcelo
School of Nursing Allied Health and Biological Sciences, Saint Louis University, Baguio City, Philippines

R. De Torres · J. A. Mandap
Institute of Weed Science, Entomology and Plant Pathology, College of Agriculture and Food Science, University of the Philippines Los Baños, Los Baños, Laguna, Philippines

J. Amas
School of Biological Sciences, University of Western Australia, Perth, Western Australia, Australia

J. J. G. Guerrero et al. (eds.), *Mycology in a Changing Planet: Applications and Perspectives from Southeast Asia*, https://doi.org/10.1007/978-981-95-2640-6_10

ings, nanotechnology, plant chemistry, and RNA interference. Continued exploration of these technologies and their adoption (either singly or in combination) may hold promise for the future of postharvest disease management.

Keywords Latent infection · *Colletotrichum* · Mycotoxin · Seed contamination · Food security

Introduction

Fresh produce such as fruits and vegetables serves as an indispensable part of the human diet and, in parallel, holds a pivotal role in global commerce and economic development (Schreinemachers et al., 2018). However, being highly perishable due to its high moisture and nutrient content, fleshy tissue, and higher respiration rate, these commodities face persistent challenges from abiotic and biotic stresses, which are limited not only during cultivation but also throughout the postharvest period. Food loss can occur postharvest, especially during handling, storage processing, marketing, and consumer delivery (Prusky, 2011; Sneha et al., 2024). This could lead to decreased marketability of agricultural products, leading to significant economic loss. One major contributor to economic loss during postharvest is fungal contamination (Altomare et al., 2021; Davies et al., 2021; Gatto et al., 2011; Matrose et al., 2021). During postharvest stages, fungal contaminants may remain on fruits or grains and become activated when the environmental conditions become more favorable (Birhanu et al., 2014). Several fungal species, such as *Aspergillus* spp., *Rhizopus* spp., and *Penicillium* spp., are capable of producing conidia/spores that allow pathogens to persist and thrive in postharvest conditions (Amenu et al., 2024). However, there is a diversity of pathogenic fungi that can infect several crops. Once activated, fungal pathogens produce enzymes that destroy plant tissues, leading to rotting and decay (Matrose et al., 2021; Sneha et al., 2024). In some cases, some

S. S. Duay
Department of Chemistry, College of Science, De La Salle University, Manila, Philippines

L. Zakaria
School of Biological Sciences, Universiti Sains Malaysia, Penang, Malaysia

C. J. Cumagun
Parma Research and Extension Center, University of Idaho, Parma, Idaho, USA

G. Romanazzi
Department of Agricultural, Food and Environmental Sciences, Marche Polytechnic University, Ancona, Italy

M. A. O. Balendres (✉)
Department of Biology, College of Science, De La Salle University, Manila, Philippines

Plant and Soil Health Research Unit, Center for Natural Science and Environmental Research, De La Salle University, Manila, Philippines
e-mail: mark.angelo.balendres@dlsu.edu.ph

fungal contaminants such as *Fusarium* sp., *Aspergillus* sp., *Penicillium* sp., and *Alternaria* sp. cause mycotoxin contamination in fruits, vegetables, and other agricultural products (Dwiastuti et al., 2021). Some of these mycotoxins include aflatoxins, *Alternaria* toxins, ochratoxin A, patulin, and citrinin, which are difficult to remove during postharvest stages (Liu et al., 2020a, b; Oztekin et al., 2023).

Due to physiological deterioration, postharvest damage, and diseases, more economic losses are realized during the postharvest phase. Moreover, these losses are incurred as significant investments are allotted to postharvest processes (harvesting, processing, storage, and transport) to ensure that the produce reaches the consumers at optimal conditions (Huang et al., 2021; Moradinezhad et al., 2023). Postharvest losses are realized globally, ranging between 28% and 55%, with developing countries being highly affected due to the absence of proper storage facilities and transportation (Karoney et al., 2024a, b; Kitinoja et al., 2019; Tripathi et al., 2022). In vegetable production, postharvest losses range from 30% to 40% and have been caused by various aspects concerning mechanical and biological damages, with the latter estimated to cause the highest rate of losses (Tek & Budak, 2022). Meanwhile, yearly postharvest fruit losses reach 210,000 metric tons, appraised at about US$ 90 million (Rajapaksha et al., 2021).

Postharvest diseases are often classified based on how infections are initiated. The first classification is referred to as "latent" or "quiescent" infections, wherein the pathogen infects the host at an earlier stage (typically before harvest) but enters a state of dormancy until favorable conditions such as physiological changes (ethylene production, cuticular modifications, and cell-wall softening, and reduction of antifungal compounds) allow the reactivation of infection often after harvest or during storage (Prusky et al., 2013). Latent infections are more commonly associated with postharvest diseases in fruits but are also reported in vegetables such as those incited by *Colletotrichum* sp. and *Botrytis cinerea*. The second classification of postharvest diseases is those that begin during or after harvest wherein the fungal pathogens enter through surface wounds inflicted by physical, insect, or mechanical damage; examples are diseases caused by *Pythium* and *Rhizopus* species (Coates & Johnson, 1997; Tripathi et al., 2022). However, there are still postharvest pathogens reported to direct penetration of the cuticle or entry through stomata, lenticels, or abscission of scar tissue of the produce, such as *Sclerotinia* sp. and some species of *Colletotrichum* (Singh & Sharma, 2018).

This chapter discusses common postharvest diseases in fruits, vegetables, and cereal grains caused by fungi, alongside an overview of current disease control strategies. This chapter also examines emerging and innovative approaches for future postharvest disease management, including, but not limited to, genomics, edible coatings, nanotechnology, plant chemistry, and RNA interference. These advancements, individually or combined, have significant potential for enhancing postharvest disease control. This chapter emphasizes the importance of continued research and the potential for these cutting-edge technologies to address the evolving challenges of fungal disease management on a changing planet.

Postharvest Fungi in Fruits

Fungal phytopathogens are estimated to account for more than 50% of postharvest loss in fruit production (Zhang et al., 2017). From accelerating fruit senescence to causing fruit decay, these pathogens contribute to economic losses due to quality deterioration and fruit wastage and pose health threats as some fungi have the ability to produce mycotoxins. Many of these pathogens exhibit diverse adaptive lifestyle patterns that facilitate successful host colonization, including biotrophic, hemibiotrophic, and necrotrophic strategies, and some even alternate between them. While some infect multiple hosts, others are host specific. Several species cause disease and manifest symptoms closely linked to fruit ripeness, often remaining dormant when the fruit is unripe. Table 10.1 lists the common fungal species associated with postharvest diseases in fruits.

Alternaria spp. are fungal pathogens responsible for postharvest diseases in various fruits, including apples, pears, and citrus (Table 10.1). The infection process typically begins in the field, where spores colonize the fruit surface, often entering through natural openings such as stomata, lenticels, or wounds caused by mechanical damage, insects, or environmental stress (Güney et al., 2023; Wang et al., 2021). Once inside the host, the fungus can remain latent until favorable conditions, such as high humidity or temperature fluctuations during storage, trigger disease development (Vilaplana et al., 2017). *Alternaria* spp. are primarily necrotrophic. It colonizes and kills host tissues for nutrients, which results in symptoms like black rot, black spot, and *Alternaria* rot (Kobiler et al., 2011; Qi et al., 2024; Rizwana et al., 2021). However, some studies suggest it may exhibit hemibiotrophic behavior, initially establishing a biotrophic relationship before transitioning to necrotrophy (Mincuzzi et al., 2022). In *A. alternata*, infected fruits have been shown to have outwardly asymptomatic fruits, only revealing internal black rot when the fruit is opened (Güney et al., 2023). Members of this genus also pose significant health risks as they produce mycotoxins. *Alternaria* species can produce mycotoxins such as alternariol monomethyl ether and tenuazonic acid at room temperature. Even under cold storage conditions, alternariol can be produced after prolonged storage periods exceeding 30 days (Mao et al., 2023). Field infections are common, as spores can survive on plant debris or soil, spreading via wind, rain, or contaminated equipment (Li et al., 2017). Postharvest infections often occur when fruits are stored under suboptimal conditions, such as high humidity or poor ventilation, which favor fungal growth (Cancino et al., 2023).

Aspergillus spp., particularly *A. niger* and *A. flavus*, are prominent postharvest pathogens causing significant decay in fruits such as mango, citrus, and dates. These fungi primarily infect fruits through wounds, insect damage, or natural openings, thriving in warm, humid storage conditions (Bashir et al., 2020; Javadpour et al., 2018). Unlike *Alternaria*, which often initiates latent field infections, *Aspergillus* infections frequently occur postharvest. However, field contamination via spores on damaged fruit surfaces can predispose produce to rot during storage (Barkai-Golan, 2001). These pathogens are predominantly necrotrophic, secreting enzymes like

Table 10.1 Common fungal species associated with postharvest diseases in fruits

Genus	Causal species	Disease	Fruit crop	References
Alternaria	*A. alternata*	Black rot	Apple, pear, pomegranate, yellow dragon fruit, cherry fruits, cherry tomatoes	Vilaplana et al. (2017), Mincuzzi et al. (2022), Cancino et al. (2023), and Qi et al. (2024)
		Internal black rot	Lemon	Güney et al. (2023)
		Black spot	Persimmon	Kobiler et al. (2011)
		Alternaria rot	Citrus fruits, blueberry, tomatoes, peaches, nectarines, cherry	Cara et al. (2022), Wang et al. (2021)
		Alternaria spot	Apple, tomato	Rizwana et al. (2021), Cabrefiga et al. (2022)
		Postharvest rot	Kiwi fruit	Li et al. (2017)
	A. arborescens	Alternaria rot	Blueberry	Wang et al. (2021)
	A. tenuissima	Black spot	Black rot	Cancino et al. (2023)
		Alternaria fruit rot	Strawberry	Sun et al. (2023)
	A. citri	Alternaria brown spot	Citrus fruits	Umer et al. (2021)
	A. solani	Postharvest rot	Tomato	Bhalerao et al. (2019)
Aspergillus	*A. niger*	Postharvest decay	Mango	Javadpour et al. (2018)
		Black rot	Dates, grape, tomato	Barkai-Golan et al. (1993), Storari et al. (2012)
	A. sect. *nigri*	Postharvest rot	Mango, pomegranate	Abdullah et al. (2016), Mincuzzi et al. (2022)
	A. flavus	Postharvest decay	Lemon, watermelon	Kotan et al. (2009), Odelade and Oladeji (2020)
	A. fumigatus	Postharvest decay	Citrus fruits (lime, lemon, sweet orange)	Bashir et al. (2020)

(continued)

Table 10.1 (continued)

Genus	Causal species	Disease	Fruit crop	References
Botryosphaeria	*B. dothidea*	Postharvest rot	Plum, quince	Vučković et al. (2023), Yuan et al. (2024)
		Ring rot	Apple	Li, Liu, et al. (2024a), Huang et al. (2024)
	B. wangensis	Postharvest rot	Plum	Wang et al. (2025)
Botryodiplodia	*B. theobromae*	Crown rot	Banana	Renganathan and Muthukumar (2012)
Botrytis	*B. cinerea*	Gray mold	Strawberry, raspberry, cherry, grapes, apple, pear, cherry, peach, plum, persimmon, citrus, kiwifruit, tomato, pepper, brinjal, cucumber, squash, melon, pumpkin	Elmer and Michailides (2007a, b), Kozhar and Peever (2018), Singh and Sharma (2018), Petrasch et al. (2019)
Colletotrichum	*C. gloeosporioides*	Anthracnose	Avocado, banana; citrus fruits, guava, mango, papaya, soursop	Kefialew and Ayalew (2008), Zhou et al. (2016), Siddiqui and Ali (2014), González-Ruíz et al. (2021), and Peralta-Ruiz et al. (2023)
		Bitter rot	Pome and stone fruits	Ahmadi-Afzadi et al. (2013), Wenneker and Thomma (2020)
		Ripe rot	Grapes	Hsieh et al. (2023)
	C. fruticola	Bitter rot	Pear	Li et al. (2013)
	C. musae	Crown rot	Banana	Su et al. (2011)
Diaporthe	*D. ambigua, D. eres, D. nobilis, D. novem, D. passiflorae, D. rudis*	Fruit rot	Kiwifruit	Auger et al. (2013), Díaz et al. (2014), Li et al. (2019), and Liu et al. (2021)
	D. arctii	Postharvest decay	Pear	Bennett et al. (2024)
	D. passiflorae	Brown rot	Passion fruit	Wang et al. (2023a, b, c)

(continued)

Table 10.1 (continued)

Genus	Causal species	Disease	Fruit crop	References
Fusarium	*F. solani, F. semitectum, F. oxysporum, F. subglutinans and F. equiseti*	Soft rot	Tomato	Abu-Bakar et al. (2013)
	F. musae, F. pallidoroseum, and F. incarnatum	Crown rot	Banana	Triest and Hendrickx (2016), Conti Taguali et al. (2024)
	F. proliferatum and *F. sacchari*	Postharvest rot	Banana	Conti Taguali et al. (2024)
	F. incarnatum	Fruit rot	Musk melon	Wonglom and Sunpapao (2020)
	F. moniliforme	Black heart	Pineapple	Barkai-Golan (2005)
Geotrichum	*G. candidum*	Sour rot	Tomato	Thomidis et al. (2021)
Lasiodiplodia	*L. theobromae*	Fruit rot	Avocado, guava, longan, watermelon, passion fruit, persimmon	Garibaldi et al. (2012), Palou et al. (2013), Chen et al. (2021), Zee et al. (2021), Khuna et al. (2024), and Weber et al. (2025)
		Stem end rot	Citrus, mango, papaya	Zhang (2014), Zhao et al. (2015), Suasa-ard et al. (2019), and Samithri et al. (2020)
	L. pseudotheobromae	Fruit rot	Jujube, citrus	Chen et al. (2021), Guo et al. (2023a, b)
		Postharvest decay	Grapefruit strawberry	Olmedo et al. (2023), Zhang et al. (2024a, b)
		Soft rot	Plum	Wang et al. (2024)
	L. irainensis	Postharvest decay	*Capsicum* spp.	Akhtar and Thiruppathi (2025)
	L. hormozganensis and *L. laeliocattleyae*	Crown rot	Banana	da Silva França et al. (2024)

(continued)

Table 10.1 (continued)

Genus	Causal species	Disease	Fruit crop	References
Monilinia	*M. laxa, M. fructigena,* and *M. fructicola*	Brown rot	Stone fruit	Martini and Mari (2014), Garcia-Benitez et al. (2017), and Garcia-Benitez et al. (2020)
Mucor	*M. piriformis*	Soft rot	Apple, pear, tomato, strawberry, raspberry, mandarin fruit	Li et al. (2014), López et al. (2016), and Saito et al. (2016)
	M. irregularis	Soft rot	Papaya	Cruz-Lachica et al. (2017)
Neoscytalidium	*N. dimidiatum*	Fruit rot	Pear	Derviş et al. (2024)
Neofabraea	*Neofabraea alba*	Bull's eye rot	Apple	Köhl et al. (2018)
Penicillium	*P. digitatum*	Green mold	Citrus fruits	Palou (2014), Ghanei Ghooshkhaneh et al. (2018)
	P. italicum	Blue mold	Citrus fruits	Papoutsis et al. (2019), Bhatta (2022)
	P. expansum	Postharvest decay	Apple, pome	Luciano-Rosario et al. (2020), Yu et al. (2020)
Phomopsis	*P. citri*	Stem end rot	Citrus fruits	Cerioni et al. (2017), Chen et al. (2022a, b)
	P. longicolla	Fruit rot	Kiwifruit	Liu et al. (2020a, b)
	P. longanae Chi	Pericarp browning	Longan	Chen et al. (2014)
Rhizopus	*R. stolonifer*	Watery white rot	Apple, avocado, citrus fruit, cherry, grape, jackfruit, loquat, papaya, peach, pear, pepper, squash, strawberry, tomato,	Bautista-Baños et al. (2014), Singh and Sharma (2018), and Liu et al. (2024)
Sclerotinia	*S. sclerotorium*	Fruit rot	Cherry, kiwifruit, blueberry, papaya	Lee et al. (2015), Lopez et al. (2015), Rafiq and Fatima (2023), and Ruan et al. (2023)
		White mold	Cucumber	Sadek et al. (2022)

(continued)

Table 10.1 (continued)

Genus	Causal species	Disease	Fruit crop	References
Trichothecium	*T. roseum*	Fruit rot	Cantaloupe, passion fruit, tomato	Guo et al. (2007), Dal Bello (2008)
		Pink rot	Avocado, pepper	Lin et al. (2016), Sharma et al. (2016)
		Brown rot	Peach	May-De Mio et al. (2014)
		Black spot	Apple	Dai et al. (2019)

pectinases and cellulases to macerate host tissues, leading to symptoms such as black rot of *A. niger* or aflatoxin-contaminated lesions of *A. flavus* (Storari et al., 2012; Odelade & Oladeji, 2020). Among species, *A. flavus* is paramount due to its production of aflatoxins. These carcinogenic mycotoxins persist even in processed foods, posing severe health risks (Kotan et al., 2009). *A. niger*, while less toxigenic, causes widespread postharvest decay in mango and pomegranate, which is often compounded by its resilience to conventional fungicides (Abdullah et al., 2016; Mincuzzi et al., 2022). Meanwhile, some species exhibit thermotolerance. *A. fumigatus*, though only reported in citrus fruits, can grow at elevated temperatures of up to 50 °C, which causes decay and spoilage in poorly managed storage environments (Bashir et al., 2020).

Members of the genus *Botryosphaeria*, particularly *B. dothidea* and *B. wangensis*, are important postharvest pathogens that cause rots in apples, plums, and quince. These fungi typically infect fruits through wounds or natural openings, often during preharvest stages. However, symptoms manifest postharvest under favorable storage conditions (Vučković et al., 2023; Yuan et al., 2024). *B. dothidea* is notorious for causing ring rot in apples, characterized by concentric, sunken lesions that severely reduce marketability (Huang et al., 2024; Li et al., 2024a). Similarly, *B. wangensis* induces postharvest rot in plums, of which the fruit displays brown, circular lesions with the peel surrounding the lesions transitioning from cyan to red (Wang et al., 2025; Yuan et al., 2024). The infection process involves the secretion of cell wall-degrading enzymes, which macerate fruit tissues to facilitate fungal colonization. Damages are exacerbated by high humidity and suboptimal storage temperatures, which accelerate disease progression. Notably, *Botryosphaeria* spp. can remain latent in asymptomatic fruits, which makes early detection challenging (Navarro et al., 2022). Postharvest losses are significant, as infected fruits are unmarketable and prone to secondary infections by other pathogens (Navarro et al., 2022).

Lasiodiplodia theobromae (syn. *Botryodiplodia theobromae*) is a notable postharvest pathogen causing crown rot in bananas and stem end rot in mango. This fungus infects fruits through wounds or natural openings, often during harvesting or handling, and thrives in warm, humid storage conditions. Crown rot manifests as dark, sunken lesions at the fruit's crown, leading to rapid decay and significant postharvest losses (Renganathan & Muthukumar, 2012). The pathogen secretes enzymes that degrade cell walls and facilitate tissue maceration and secondary infections by other microbes.

Botrytis cinerea, which causes gray mold, is one of the most destructive postharvest fungal pathogens, second to *Magnaporthe* sp. in terms of economic impact (Zhang et al., 2021a, b). It infects a wide range of fruits, including strawberries, grapes, apples, and kiwifruit, which results in significant losses during storage and transport (Kozhar & Peever, 2018; Singh & Sharma, 2018). Gray mold causes up to 20–25% postharvest losses globally, especially in perishable fruits like strawberries and grapes (Petrasch et al., 2019). The fungus employs a necrotrophic lifestyle. The infection begins with spore germination on the fruit surface, often facilitated by wounds, natural openings, or senescing tissues. The fungus secretes cell wall-degrading enzymes (e.g., polygalacturonases and pectinases) and toxins, which macerate host tissues and suppress plant defenses (Petrasch et al., 2019; van Kan, 2003). The pathogen exploits the host's ripening processes in ripe fruits, such as cell wall softening and sugar accumulation, to establish infection. Unripe fruits, however, often resist infection due to their robust cell walls and active defense mechanisms (Petrasch et al., 2019). Infections by *B. cinerea* can initiate in the field, where spores colonize flowers, stems, or immature fruits, remaining latent until favorable conditions arise. Postharvest, the pathogen thrives in high-humidity environments, such as storage facilities, where it rapidly spreads to adjacent fruits. This dual-phase infection strategy makes *B. cinerea* particularly challenging to control (Droby & Lichter, 2007; Petrasch et al., 2019).

Another important genus is *Colletotrichum*, and members are associated with anthracnose, bitter rot, ripe rot, and crown rot in many fruits (Table 10.1). Among the most common species, *C. gloeosporioides, C. fruticola*, and *C. musae* are major contributors to postharvest spoilage in various fruit crops. The infection process of *Colletotrichum* spp. follows a typical hemibiotrophic lifestyle. The pathogen infects the host by producing conidia germinating and forming appressoria that facilitate penetration into the host tissue (Kefialew & Ayalew, 2008; Zhou et al., 2016). Upon landing on the fruit surface, *Colletotrichum* spores germinate and develop appressoria, which generate turgor pressure to breach the cuticle and epidermal layers of the fruit (Bordoh et al., 2020; Peralta-Ruiz et al., 2023). The fungus remains asymptomatic within the host tissue, establishing a quiescent infection. During this stage, it remains dormant until favorable conditions, such as fruit ripening, trigger disease progression (da Silva & Michereff, 2013; Zakaria, 2021). As the fruit ripens, the fungus transitions to a necrotrophic lifestyle, secreting enzymes and toxins that degrade cell walls and lead to visible disease symptoms such as lesions, decay, and fruit rot (Zakaria, 2021). *C. gloeosporioides* causes anthracnose in avocado, mango, papaya, and citrus fruits, *C. musae* crown rot in banana, and *C. fructicola* bitter rot in apple and pear (González-Ruíz et al., 2021; Li et al., 2013; Siddiqui & Ali, 2014; Su et al., 2011). Infections by *Colletotrichum* species can occur both in the field and postharvest, with the primary mode of infection varying based on environmental conditions and host susceptibility (Worku et al., 2025). Cross-contamination can occur during handling, packing, or transportation, especially if fruits are wounded (Zakaria, 2021).

Diaporthe spp. are opportunistic pathogens that infect fruits through wounds, natural openings, or latent infections that remain asymptomatic until favorable conditions trigger disease development (Auger et al., 2013; Li et al., 2019). The infection process of *Diaporthe* typically begins during the preharvest stage, where the fungus establishes itself as an endophyte in plant tissues. It can remain dormant until stress factors, such as mechanical injuries or environmental changes, activate its pathogenic phase. The fungi penetrate fruit tissues through wounds or lenticels, leading to rot, discoloration, and fruit collapse (Díaz et al., 2017; Liu et al., 2021). Most *Diaporthe* species exhibit a hemibiotrophic lifestyle, which allows them to establish persistent infections that intensify postharvest losses (Bennett et al., 2024).

Several *Fusarium* species are well-known soil pathogens that cause postharvest diseases in fruits. These fungi typically infect fruits through wounds, natural openings, or contact with contaminated surfaces. Once inside the host, they colonize tissues, disrupting cellular integrity and leading to visible symptoms such as soft rot, discoloration, and tissue breakdown. Infection often starts in the field but progresses during storage and transport, where environmental conditions favor fungal growth (Abu-Bakar et al., 2013; Triest & Hendrickx, 2016). For instance, *F. musae* is responsible for crown rot in bananas. The fungus enters through wounds sustained during harvesting, with symptoms developing during postharvest storage. The initial signs include mycelial growth on the crown and internal spread into the peduncle, leading to blackening and fruit deterioration (Triest & Hendrickx, 2016; Conti Taguali et al., 2024). In tomatoes, *F. oxysporum* is a major cause of soft rot. The pathogen gains entry through wounds or root tips. It rapidly colonizes the vascular system, resulting in tissue maceration and a characteristic water-soaked appearance. High humidity and poor storage conditions accelerate disease progression, making it a major concern in postharvest management (Abu-Bakar et al., 2013). Other species also cause significant postharvest damage. *F. incarnatum* is associated with fruit rot in musk melon, often characterized by softening and discoloration (Wonglom & Sunpapao, 2020). Meanwhile, *F. moniliforme* leads to black heart disease in pineapples, which remains undetected until the fruit is cut open, revealing internal browning and decay (Barkai-Golan, 2005). During infection, these pathogens may produce various mycotoxins that can contaminate fruits and pose health risks to consumers. Necrotic lesions in banana caused by *F. proliferatum* and *F. sacchari* have been detected of fumonisins and enniatins (Conti Taguali et al., 2024), and trichothecenes, specifically T-2 toxin, have been detected in tomato fruits infected with *F. acuminatum* (Petrasch et al., 2019). Hence, *Fusarium* spp. not only deteriorate fruit quality but also pose significant health risks.

Postharvest Fungi in Vegetables

Various causes of postharvest vegetable loss exist, with fungal infection and contamination being the most notable. Fungal diseases are of major concern as pathogens can sporulate and have the possibility to withstand postharvest treatments targeted to eliminate the present pathogen; thus, resisting these treatments would allow the pathogen to start infecting the produce while on its way to commercial endpoint (Tripathi & Dubey, 2004; Troncoso-Rojas & Tiznado-Hernández, 2014). It is important to assess both the quantity and quality of losses in vegetables, as diseases, particularly those caused by fungi, not only decrease the number of marketable crops but can also diminish their market value. During storage and transportation, fruits and vegetables provide an ideal environment for fungal pathogens to thrive, leading to spoilage. This deterioration renders the produce unsuitable for sale, resulting in substantial postharvest financial losses.

The damage caused by fungal pathogens extends beyond physical and direct economic impact, as it can render the produce unsafe for consumption due to potential health hazards. Several fungal species are known to produce mycotoxins under certain conditions, and the greatest risk of mycotoxin contamination occurs when infected produce is used in the production of processed foods or animal feed (Coates & Johnson, 1997; Sanzani et al., 2016; Balendres et al., 2019). The majority of fungi during postharvest belong to the phylum Ascomycota, and the anamorph (asexual) stage of the fungus is often more encountered compared to the teleomorph (sexual) stage (Coates & Johnson, 1997). Generally, important fungal genera of anamorphic postharvest pathogens in fruits and vegetables include *Alternaria, Aspergillus, Botrytis, Colletotrichum, Fusarium, Lasiodiplodia, Phomopsis, Rhizoctonia, Sclerotium,* and *Sclerotinia*. Other fungal species from other phylum are also of postharvest importance, *Phytophthora* and *Pythium* from Phylum *Oomycota,* causing cottony leaks in cucurbits, *Rhizopus* sp. and *Mucor* sp. from Phylum *Zygomycota* causing rapid soft rot in various vegetables, and *Sclerotium rolfsii* and *Rhizoctonia solani* from Phylum Basidiomycota, which may lead to substantial losses in tomato and potato, but development of symptoms intensifies after harvest (Coates & Johnson, 1997). The aforementioned fungal pathogens generally degrade the infected produce through the secretion of cell-wall-degrading enzymes and the toxic secondary metabolites produced, which are known to produce mycotoxins (Moradinezhad & Ranjbar, 2023).

With the ever-changing world, most fungal pathogens have expanded their host range beyond their originally associated plant host. Ecological changes, globalization, and elevated agricultural trade all contributed to the spread of these pathogens, enabling them to acclimatize and adapt to new geographical areas and hosts. In the case of postharvest diseases, a wide range of fungal pathogens contribute to postharvest diseases in vegetables. However, some genera of major importance and will be further discussed in this section include *Botrytis cinerea* (Gray mold), *Alternaria* sp. (Alternaria rot), *Aspergillus niger* (Black mold rot), and *Colletotrichum* sp. (Anthracnose and fruit rot) (Tripathi et al., 2022).

Botrytis cinerea *(Gray Mold)*

Botrytis cinerea is a ubiquitous, necrotrophic pathogen with a wide global distribution and host range, with the reported number of hosts increased from over 200 (Jarvis, 1977) to more than 1400 (CABI, 2022a, b; Elad et al., 2016). This pathogen particularly causes a disastrous impact on the majority of crops, which led to its consideration as a "high risk" pathogen due to its resistance development connected with its high genetic diversity, abundant sporulation, polycyclic nature, wide host range, and requires intensive fungicide management for control (Elad et al., 2004). Moreover, this pathogen is considered one of the most significant pathogens in vegetable diseases, especially on salad vegetables, some solanaceous crops, beans, and cucurbits, leading to significant losses due to soft rot during storage, transport, and at the market (CABI, 2022a, b; Cheung et al., 2020).

Destruction by *B. cinerea* is evident in mature tissues of the infected host. However, this pathogen gains entry at an earlier stage. Upon successful establishment of *B. cinerea* on an infected host, the pathogen undergoes a brief biotrophic phase and then transitions into its necrotrophic phase, being triggered by biochemical changes mainly leading to host ripening (Neri et al., 2014; Williamson et al., 2007). At this stage, the fungus secretes virulence factors, oxalic acid, cell-wall-degrading enzymes, and plant hormone analogs, disrupting the host metabolism and immune defenses (Sasanuma & Suzuki, 2016). These virulence factors and disruption of plant immune response, in turn, lead to the symptoms manifesting as brown lesions, water-soaked, then progressing to be grayish and necrotic. At advanced stages of infection, obvious signs and sporulation of *B. cinerea* are present, covering the surface of old infected tissue, which is typical for this disease (Williamson et al., 2007). The infection spreads to adjacent plant tissues, allowing gray mold proliferation to nearby plants or fruits. Significant damage occurs after the harvest of crops that initially appear healthy. As these crops are subjected to storage or transport, the extent of losses becomes apparent. However, infections in some cases are possible to occur at later stages, with evidence indicating that some *Botrytis* species can be present inside the plant tissues without abruptly triggering plant response, leading to the appearance of symptoms (Elad et al., 2016). In terms of pathogen spread, dispersal is mainly through airborne conidia brought about by the wind. Other reports of dispersal include water splashes, insect adhesion, and contaminated agricultural machinery and tools (Blanco et al., 2006; Fermaud & Le Menn, 1992).

This pathogen causes both qualitative and quantitative losses. However, since *B. cinerea* has a broad host range and is found in vast areas of the world, exact estimates of losses on vegetables are hard to quantify as records are scarcely documented, and most of which are focused on grapes and strawberries (Elmer & Michailides, 2007a, b; Mertely et al., 2000). In onions, 7–30 % yield losses were recorded in untreated fields (De Visser, 1996; Shoemaker & Lorbeer, 1971). Rotting of mature tomato fruits occurs at postharvest with a symptom manifesting as a "ghost spot," which renders the produce unvendable (Williamson et al., 2007). Generally, losses can exceed 50% and require frequent use of chemical treatments. The global cost for crop protection for this pathogen amounts to $310 million, mainly for the grapevine industry in France and Italy and vegetables for China, Japan, Spain, Germany, Australia, Chile, and the USA (Elad et al., 2016).

Alternaria alternata *(Alternaria Rot)*

Several species of *Alternaria* are important pathogens that cause plant diseases and postharvest rot, eventually leading to substantial agricultural losses (Deshpande, 2002). This fungal genus, being a hardy fungus that withstands extreme conditions, can thrive at even lower temperatures, making spoilage of produce realized even at refrigerated transport and storage (Ostry, 2008; Troncoso-Rojas & Tiznado-Hernández, 2014). Besides these, human and animal health are also affected due to its capacity to produce secondary metabolites—mycotoxins, reported to play a role in mutations, DNA damage, and chromosomal aberrations (Escrivá et al., 2017). There are well over 70 secondary metabolites reportedly produced by *Alternaria* that render toxicity to plants, but only a small portion of these have been chemically analyzed to be toxic to humans and animals (Dall'Asta, 2014).

Among the *Alternaria* species, *A. alternata* is the most significant mycotoxin-producing species and most commonly associated with the decay of vegetables and fruits in many areas globally, implicating the fact that *A. alternata* is well adjusted to varying environmental conditions (CABI, 2022a, b; Dall'Asta, 2014; Weidenbörner, 2001). Postharvest losses brought about by *A. alternata* are well documented in numerous fruit crops like pome fruits, mangoes, kiwi, and melons, but also have notable effects on vegetable crops, including fruit rot in chili peppers (Anand et al., 2009), brinjal, and potato (Balodi et al., 2024), and are widely reported to be a major decay biotic factor in tomatoes (Wang et al., 2010; Ventura-Aguilar et al., 2021). Infection hastens when tomato fruits have already sustained injuries or debilitation during extended storage. Typical symptoms in tomatoes manifest as dark brown to black, slightly sunken lesions with firm texture, and expansion of several centimeters in diameter throughout the infection progression (Yang et al., 2020). These symptoms are mostly found near the peduncle of the blossom end of the fruit, leading to fruit rotting, thus limiting its marketability (Prusky et al., 2013). Reports in India indicated that in the solanaceous crops affected, tomato was most susceptible compared to brinjal and potato, with disease indices of 98.99%, 82.36%, and 61.19%, respectively.

On top of factors such as the growing season, harvesting conditions, handling, transport, mode of storage, and marketing conditions, there are also several factors during the postharvest phase that influence *A. alternata* infection. Another factor is the fruit ripening. Prusky et al. (1996) reported that the natural resistance of fruit and vegetables to fungal pathogen attack declines with storage duration and ripeness as the natural antifungal compounds decline to sub-fungitoxic levels, coinciding with the development of disease symptoms in the fruit.

Aspergillus niger *(Black Mold Rot)*

Aspergillus sp. has a wide global distribution and a wide range of substrates. This species is well known as plant pathogens that cause agricultural and economic losses in plants and also causes spoilage in postharvest in fruits, cereals, and vegetables (Bennett, 2010; Mahfooz et al., 2017; Samson et al., 2010). In previous research, the three main *Aspergillus* species commonly associated with causing disease in agricultural crops are *A. flavus,* and *A. niger*. Over time, other species have been reported as phytopathogens, especially with the mandate of the one fungus, one name concept, and succeeding taxonomic revisions (Taylor, 2011). However, the *A. niger* (*Aspergillus* section *Nigri*) species complex is reported to be the most prevalent species causing spoilage and responsible for postharvest decay of fresh fruits and vegetables, emphasizing its high potential to cause rots in mangoes, grapes, tomatoes, and most devastating in onions throughout the world (Gautam et al., 2011). This species is often associated with plants as an opportunistic pathogen as it primarily requires an existing wound for successful entry, infection, and colonization of the host (Sexton & Howlett, 2006). Infection is usually initiated in the field, where the mycelium of the pathogen is limited to specific tissues of the plant, seed, or fruit, and later on affects infected produce (Shu et al., 2015). The compelling effect of mycotoxin production is realized at the postharvest stage, predominantly during storage. Some species of *Aspergillus* are capable of infecting entire organs (seeds and fruits) through the production and action of mycotoxins such as aflatoxins (Afs), ochratoxin A (OTA), fumonisins (FBs), and Patulin (PAT) (Amaike & Keller, 2011; Scott, 2012; Snini et al., 2014; Sanzani et al., 2015).

Aspergillus niger has a widespread distribution and is present in most kinds of soil. It has the capacity to withstand a wide range of pH, salinity, alkalinity, and even in the presence of heavy metal soil pollutants (Ahmad et al., 2006; Kis-Papo et al., 2003).

Symptoms of black mold rot in onions manifest as black discoloration at the neck, with superficial lesions on outer scales and black mycelial streaking. For the signs, conidia are evident beneath the dry outer scale (Toit & Schwartz, 2011). It was reported that contaminated soil and seeds are the main inoculum sources for this disease (Köycü & Özer, 1997; Tyson & Fullerton, 2004). Disease spread is through the bulb-to-bulb contact, by mechanical means, or through wind dissemination of air-borne conidia. Germination of conidia is favored by high relative humidity (RH) at a range of 70–80%, and optimum temperature for growth ranges from 28 °C to 34 °C but is inhibited at below 17 °C and above 47 °C (Musa et al., 1973; Sumner, 1995).

Incidences of black mold in onions have been widely documented globally. Black mold disease becomes apparent during storage, transport, and at the market, causing massive losses as the bulbs discolor with tissues already disintegrating (Zakaria, 2024). Spoilage was reported to reach as high as 80%, and losses can be as high as 60% during summer storage (Qadri et al., 1982; Tanaka, 1991). Moreover, it was reported that black mold was the major disease found during onion storage in the tropics under ambient conditions (Hayden et al., 1994).

Colletotrichum *spp. (Anthracnose and Fruit Rot)*

Colletotrichum is a large fungal genus comprising different species informally gathered in 15 species complexes (Talinhas & Baroncelli, 2021). Species under this genus are the most prevalent fungal pathogen causing disease in subtropical and tropical crops both in the field, during postharvest storage, and in the market. Attack of this pathogen, like other postharvest pathogens, lowers the shelf life of produce and renders these commodities unvendable (Siddiqui, 2014). Since *Colletotrichum* infection is predominant in subtropical and tropical regions of the world, this pathogen is less present in temperate latitudes due to its onset requirements, including warm conditions (25–28 °C), high relative humidity, leaf wetness, and light intensity (Banya et al., 2020; Tripathy et al., 2021). This pathogen is widely distributed, which approximately infects about 470 different host genera and is responsible for more than 50% of losses incurred in fruits and vegetables (Awang et al., 2011; Paull et al., 1997).

Initiation of infection of *Colletotrichum* species follows the same pattern, wherein conidia spread from infected vegetal matter or through an infected planting tool to neighboring susceptible tissues and adhere to the vegetal material with a hemicellulosic mucilage, germinates, and infect with a specialized structure called appressoria, or otherwise, infection can proceed by entry/penetrance through stomata, lenticels, wounds, or abscission or scar tissue (de Oliveira Filho et al., 2021; Nazzaro et al., 2017). However, specific host-pathogen interactions and infection strategies have been reported in *Colletotrichum,* which follows two main strategies that act with dependence on the species infecting, host, and tissue type: (i) intracellular hemibiotrophy, or (ii) subcuticular, intramural necrotrophy (Banya et al., 2020; Wharton & Diéguez-Uribeondo, 2004). The first strategy involves a brief biotrophic stage (~24–72 h), which encompasses intracellular primary hyphae formation, and then followed by a necrotrophic phase, where secondary ramified narrow hyphae spread throughout the host tissues (Latunde-Dada et al., 1996; Wharton & Julian, 1996). Physiological changes, or "signals" related to fruit ripening, changes in cell wall integrity, ethylene production, fluctuations in environmental pH, changes in organic compound content, and decline in antifungal compounds (phytoalexins, polyphenols), all lead to the transition to necrotrophic phase (Alkan et al., 2015; O'Connell et al., 2012; Prusky et al., 2016). The second strategy begins with a shorter (24 h) asymptomatic phase where the pathogen establishes itself with the periclinal and anticlinal walls of the host's epidermal cells and is then followed by rapid colonization in host tissues, resulting in distinct disease symptoms. Species that employ the second strategy are reportedly observed in *C. truncatum* (syn. *C. capsici*) (Pring et al., 1995; Roberts & Snow, 1984).

Symptoms of anthracnose highly vary depending on plant tissue; these can be seen on vegetative plant parts, such as leaves, blossoms, and developing fruit. However, the highest impact of this disease is evident during postharvest storage due to fruit damage (Ciofini et al., 2022). Symptoms of fruits generally appear to be dark lesions covered with conidial masses, wherein the conidial color depends on the species infecting them (de Oliveira Filho et al., 2021). Economic losses are realized due to reduced quality and aesthetic standards of the produce (Lakshmi et al., 2011). Anthracnose disease has caused huge losses in vegetable crops, with reports about solanaceous crops. In chili peppers, as it also affects immature fruits in the field and mature fruits at harvest, storage, and transit (given that conditions are favorable), losses are reported to reach up to 50% in Thailand (Pakdeevaraporn et al., 2005), 12–50% in India (Bagri et al., 2004; Sharma & Shenoy, 2014), 15% in Korea (Kim & Park, 1988), and 50% in Malaysia (Sariah, 1994).

Postharvest Fungi in Grains/Cereals

Cereal grains, such as maize, rice, wheat, sorghum, millet, barley, oats, rye, and millets, are monocotyledonous crops in the Poaceae family. Cereal production and utilization were largely attributed to maize and wheat (FAO, 2024). These crops have been produced for consumption since the advent of agriculture. Cereals are renowned for their long-term storability, contributing to their status as primary staple food in numerous countries, other than being utilized as animal feed and raw material for various industrial products (Batey, 2017). The Food and Agriculture Organization (FAO) initially projected a record high of global cereal production for 2023–2024 at 13.2 million tons, surpassing the 2022–23 record by 1.2% (World Grain, 2024). However, due to adverse weather conditions, such as hot and dry climates in major cereal-producing countries, including the EU, Mexico, and Ukraine, the production forecast led to a 9.4 million tons downturn by mid-2024 (FAO, 2024).

Cereal grain loss occurs not only in the field but also in storage due to fungal contamination, mycotoxins, stored grain pests, or unfavorable conditions, particularly in traditional storage facilities. Fungal invasion represents the largest fraction of causing grain loss, mainly through direct grain yield reduction, spoilage that renders grain and derived cereal products unsuitable for consumption (Magan et al., 2004; Pitt & Hocking, 2009), and the production of mycotoxins (Miller, 1995; Munkvold et al., 2019). Early studies accounted for more than 150 species of filamentous fungi and yeasts isolated from cereal grains (Christensen & Kaufmann, 1969). Fungi that attack cereals are broadly categorized as field fungi or storage fungi, with their distinction often defined based on their capacity to thrive at 18% grain moisture, although this criterion may not apply to all fungi (Christensen, 1957; Munkvold et al., 2019). The development of storage fungi is influenced primarily by temperature and moisture conditions available to the grain. For instance, fungal growth is promoted in stored rice grains when the storage environment allows for continuous production of CO_2 and water, generating heat and creating a conducive condition for fungal proliferation (Atungulu et al., 2019).

Storage fungi, such as *Aspergillus* species, colonize grains targeting primarily the embryo or germ (Harmon & Pfleger, 1974), initiating invasion even at the lower limit of available moisture (Sauer et al., 1992). These fungi penetrate the seed coat actively through their specialized structures, eventually colonizing the endosperm and all parts of the seed, including the embryo (Agarwal & Sinclair, 1997; Harman 1983). Embryo death can occur before grain discoloration becomes apparent.

Likewise, grains can become contaminated with mycotoxins before the appearance of visible mold growth (USDA, 2016). Once sporulation is dense on the grain surface, mustiness or off-odors may develop, along with the grain's webbing and "caking." Fungal species that are known to invade cereal embryos include *A. glaucus*, *A restrictus*, *A. candidatus*, *A. ochraceus*, *A. versicolor*, and *Penicillium viridicatum* (Christensen & Kaufmann, 1974; Tsuruta et al., 1981).

The two major genera of fungi in stored cereals are *Aspergillus* and *Penicillium,* as consistently reported in classical seed pathology books and current literature (Martin et al., 2022; Agarwal & Sinclair, 1997; Neergaard, 1977). In general, *Aspergillus* spp. infect cereal grains with 13–16% moisture content (MC) and stored at 65–85% relative humidity (RH), whereas *Penicillium* spp. are mostly favored by 16–20% MC and higher RH ranges of 80–90% (Meronuck, 1987). Other less relevant fungal genera/species reported in cereal grains include *Mucor*, *Chaetomium* (Martin et al., 2022), *Curvularia lunata, Rhizoctonia solani, A. alternata, Rhizopus stolonifer,* and *Fusarium solani* (Bhalerao & Chavan, 2017). Fungal species capable of infecting stored cereal grains are generally similar, though the predominant species vary by region and season, influenced by the prevailing field and storage conditions. For example, *A. ochraceous, A. flavus,* and *A. parasiticus* are often associated with Turkish rice grains (Aydin et al., 2011), *A. flavus* and *A. niger* in India (Reddy et al., 2008), and *P. citrinum* and *A. candidus* in South Korea (Park et al., 2005).

Aspergillus spp. are opportunistic invaders, with *A. flavus* being the most important storage fungus in cereal crops due to aflatoxin production, especially in maize (Munkvold et al., 2019; Nishimwe et al., 2020). Several outbreaks of acute and fatal toxicoses in humans have been documented due to exposure to aflatoxins (CAST, 2003; Lewis et al., 2005). Many *Aspergillus* species generally cause olive green discoloration of germ tissues and whole grains, exacerbated by insect activity such as maize weevils and Indianmeal moths (Mandap et al., 2024). Over 100 *Aspergillus* species have been associated with crops (USDA, 2016), with the most significant species found in cereal grains being *A. flavus, A. parasiticus, A. niger, A. candidus, A. glaucus,* and *A. ochraceous* (Magnoli et al., 2006; Sauer et al., 1992). Other species reported include *A. restrictus*, *A. versicolor*, and *A. clavatus* (Hocking, 2006; Magan & Lacey, 1984).

Penicillium spp. are also opportunistic saprophytes with cosmopolitan distribution. It causes known blue-eye damage in maize, resulting from the heavy sporulation of *Penicillium* spp. between the germ and pericarp (Hocking, 2006; Roberts & Wu, 1981). In rice grains, *Penicillium* spp. infection leads to "yellow rice," which has been linked to causing *Shoshin-kakke* (heart-attacking paralysis) in Japan (Kushiro, 2015). Species of *Penicillium* in maize, rice, and wheat include *P. citrinum, P. expansum, P. oxalicum, P. griseofulvum, P. aurantiogriseum, P. viridicatum, P. tricolor,* and *P. freii* (Mills et al., 1995; Munkvold et al., 2019; Pitt & Hocking, 2009). Among these, *P. verrucosum* is the most reported species in wheat grains, predominantly occurring in temperate regions (Cairns-Fuller et al., 2005) and deemed important for producing ochratoxin A and citrin (Jacobsen, 2014).

In wheat, aside from *Aspergillus* and *Penicillium* species, *Fusarium* is commonly reported to proliferate in storage (Birck et al., 2006). Although *Fusarium* generally requires higher moisture content (>18% MC) and relative humidity (>80% RH) for optimal growth, some *Fusarium* species can persist even under drier conditions. *Fusarium* is regarded as the most damaging fungi in wheat and is responsible for the disease Fusarium Head Blight (FHB). This disease is caused by a species complex of over 16 *Fusarium* species, with *F. graminearum* (Dweba et al., 2017) and *F. culmorum* (Schmidt et al., 2016) being the most prominent pathogen species.

Fungal colonization in the whole bin or bulk storage tends to be stratified due to uneven moisture distribution throughout the bin (Qi et al., 2023; Semeniuk, 1954). Since different fungal species are favored at different moisture levels, a hot spot (localized high-temperature zones) may develop in an area of the bin where grain moisture and temperature fluctuate in response to changes in ambient conditions, respiratory heat of insects, or initial mold growth, leading to the succession of several fungal species (Bothast, 1978; Jian & Jayas, 2012). For example, in a metal bin of maize, *P. palitans* grows alone in the upper 300 mm of maize, creating a fungal mat and encouraging the development of *A. flavus* below the colonized layer (Semeniuk, 1954). In some cases, when a hot spot reaches a moisture level of over 80% RH, the growth of Mucorales fungi such as *Absidia, Mucor,* or *Rhizopus* is favored (Nishimwe et al., 2020). Hot spots in bulk maize from several warehouses have been determined to be mainly caused by *Aspergillus* species (Qi et al., 2023). Protecting the entire grain container from fungal growth requires regular monitoring of storage conditions, particularly moisture levels, and temperature, and checking the bin for potential occurrences of hot spots.

Management of Postharvest Fungi

Several postharvest management approaches have been adopted to combat harvest fungal contamination during handling, storage, transport, and marketing. During postharvest, preventing fungal growth is the first step to controlling the infection in agricultural products (Liu et al., 2020a, b). In many cases, growers manage postharvest fungal contamination through physical methods, chemical methods, biological control, or a combination of these approaches (Eshel et al., 2009; Sneha et al., 2024). The different methods are also combined to improve their efficiency in controlling the growth of postharvest fungal contaminants (Liu et al., 2020a, b), depending on the type of agricultural product, type of fungal pathogen, and availability of resources.

Cultural or Physical Approaches

Several growers have preferred cultural and physical methods of managing postharvest fungal contamination due to the absence of chemical residues and minimal environmental impact (Usall et al., 2016). Generally, cultural and physical approaches involve controlling environmental conditions during postharvest, such as temperature and humidity (Sneha et al., 2024). Cultural approaches may involve proper harvesting techniques, proper drying techniques, and various processing operations such as manual selection or sorting, cleaning or washing, and partial removal of contaminated agricultural products (Mohapatra et al., 2017). Various factors promote the growth and spread of fungal pathogens during postharvest stages, including temperature, relative humidity, carbon dioxide levels, nutrient availability, and physical damage (Sneha et al., 2024). Hence, the main purpose of cultural and physical approaches is to change these conditions to make the environment less conducive to fungal contaminants. Water content and temperature can be easily controlled during food storage (Liu et al., 2020a, b). By controlling temperature and humidity, the growth of fungal pathogens in agricultural products during postharvest is prevented. However, physical methods can directly induce damage to fungal contaminants, thereby decreasing the incidence of fungal contamination of agricultural products during postharvest stages.

Cold storage, controlled, and modified atmospheres are primarily used to maintain commodity quality after harvest. However, they may also be employed to delay the development of pathogens causing postharvest diseases (Usall et al., 2016). Cold storage is achieved by storing fruits at low temperatures after harvest. When cold storage or refrigeration may prove inadequate in lengthening the shelf life of perishables, a controlled atmosphere can be used. This is done by lowering oxygen levels and increasing carbon dioxide concentrations in the storage facilities (Arshad et al., 2023). On the other hand, modified atmosphere packaging refers to placing fruits or vegetables in packaging where the internal conditions, particularly the atmospheric gases, are modified (Kargwal et al., 2020).

Some physical approaches involve heat treatment using hot air or hot water (Abubakar et al., 2020; Dai et al., 2021; de Moura et al., 2024; Eshel et al., 2009; Jin et al., 2016; Scherrer Montero et al., 2010; Schirra et al., 2011), irradiation using UV-C light (Ou et al., 2016; Sari et al., 2016; Sripong et al., 2015), bagging (Lamine Senghor et al., 2007), cold atmospheric plasma treatment (Guo et al., 2023a, b; Yinxin et al., 2022), use of light (e.g., pulsed light, LED light) (Huo et al., 2024; Wang et al., 2023a, b, c), and manipulation of carbon dioxide levels (Heilborn et al., 2023). Heat treatment involves the melting of epicuticular waxes that seal the natural openings and prevent the entry of pathogens into the fruits (de Moura et al., 2024) and induces the defense response mechanism of fruits (Jin et al., 2016; Sripong et al., 2015). However, it is recommended that heat treatment should be combined with other treatments to improve its efficiency (Schirra et al., 2011). Heat treatment is widely used to disinfect crops against various pests and diseases (Lurie & Pedreschi, 2014). It can be done through hot water dipping, where the commodities are dipped or immersed in water heated at a set temperature for a fixed amount of time (Kadam et al., 2020). In a study by Alvindia and Acda (2015), it was demonstrated that immersing 'Carabao' mangoes in hot water at 53 °C for 20 min resulted in a significant reduction in the severity of anthracnose (*C. gloeosporioides*) and stem end rot (*L. theobromae*). Specifically, the severity of anthracnose decreased by 48.71–52.63%, while stem end rot severity was reduced by 48–60.86%.

Meanwhile, hot air is also reported to increase the antifungal compounds of fruits, thereby aiding in the control of postharvest fungal diseases (Wan et al., 2020). This was shown in the study by Liu et al. (2022), where forced air heat treatment of papaya elicited disease resistance against *C. gloeosporioides* by delaying ripening and inducing defense from pathogen attacks.

Exposure to different sources of ionizing and non-ionizing radiation is another type of physical treatment for postharvest diseases. The ability of the non-ionizing radiation UV-C to severely impair microorganisms has made it an attractive tool for extending the shelf-life of perishables and reducing their decay. Janisiewicz et al. (2021) showed that 222 nm UV-C light can kill the fungal pathogens of strawberry *B. cinerea*, *P. expansum*, and various *Colletotrichum* species. They also showed that irradiation of fruits with a 30-s dose of 222 nm UV-C light effectively controlled anthracnose (*Colletotrichum* spp.) (Janisiewicz et al., 2021). In bananas, a UVC irradiation dose of 0.01 kJ m^{-2} has demonstrated significant effectiveness in controlling crown rot (Mohamed et al., 2017). Similarly, ionizing radiation also has the potential to inactivate pathogens (Piri et al., 2011). Jeong et al. (2015) observed that postharvest fungal pathogens *B. cinerea*, *P. expansum*, and *R. stolonifer* may be inactivated by ionizing gamma, X-ray, or e-beam irradiation. Irradiation using UV light (UV-C) not only inactivates fungal pathogens (Dai et al., 2012) but also promotes and increases the resistance of *Ananas comosus* against *Candida tropicalis* by promoting the activity of phenylalanine ammonia-lyase, which is responsible for increasing the total phenolic content of the fruit (Ou et al., 2016). A similar observation on increased defense-related enzymes induced by UV-C was reported in mangoes (Sripong et al., 2015). Furthermore, exposing fruits to UV-C induces modifications in the cell walls of plants (Duarte-Molina et al., 2016) and increases the antioxidant activity in fruit peels (Sari et al., 2016).

Bagging, on the other hand, reduces fruit decay and promotes the growth of healthy fungal communities in fruits (Gao et al., 2022; Khwantongyim et al., 2021), thereby preventing the growth of pathogenic fungi during postharvest stages. Cold plasma treatment destroys the structure of fungal spores and hyphae and neutralizes the gene expression-related synthesis of the cell wall (Yinxin et al., 2022). Aside from reducing the activity of fungal pathogens, cold plasma treatment also degrades mycotoxins such as ochratoxin A and deoxynivalenol in rice (Guo et al., 2023a, b). Pulsed light was reported to delay the softening of fruits, leading to increased shelf-life (Duarte-Molina et al., 2016). In addition, pulsed light can also increase the antioxidant capacity of fruits and improve the fungal community during cold storage (Wang et al., 2023a, b, c). Blue LED light exhibited antifungal activities against *A. carbonarius* by destroying cell wall integrity and permeability (Huo et al., 2024). Similarly, elevated levels of CO_2 can cause cell membrane malfunction (Heilborn et al., 2023). While each physical method has specific mechanisms that lead to the inhibition of fungal pathogens, physical methods combined with other methods (Jin et al., 2016; Sripong et al., 2015).

Biological (Microbial) Approaches

Biological approaches to managing postharvest fungal contaminants involve using microorganisms such as yeast, fungi, and bacteria to suppress fungal growth and disease progression in fruits, vegetables, and other agricultural products (Sharma et al., 2009). Compared to chemical approaches, biological approaches are less harmful (Fenta et al., 2023). Hence, biological approaches are a potential substitute for chemical methods (Liu et al., 2020a, b). The presence of wounds and the high water content of fruits and vegetables during postharvest stages facilitate the infection and growth of pathogenic fungi (Bano et al., 2023). Similarly, postharvest fungal contamination of seeds and grains happens as temperature and moisture content become more favorable (Mohapatra et al., 2017). According to Sneha et al. (2024), biological approaches involve direct inoculation of microbial culture, postharvest spray, dipping agricultural products in liquid formulations, or fumigation of cell-free supernatant or fermentation broths. Bacteria can form colonies on surfaces of agricultural products such as fruits, vegetables, and grains that result in competition for nutrients and space (Zhou et al., 1999). Furthermore, bacteria produce several antifungal compounds in cell-free supernatants.

Biological control methods may be used with edible coatings (Miranda et al., 2024). Because of their film-forming properties that allow them to create a physical barrier around the commodity, edible coatings are used for preservation (Miranda et al., 2024). Edible coatings (EC) may be obtained from several agricultural products, including corn, potatoes, rice, wheat, soybeans, and milk (Miranda et al., 2024). Popescu et al. (2022) showed that ECs containing essential oils can reduce the coated strawberries' microbial load (yeasts and molds) during storage.

Direct microbial inoculation prevents the progression of the infection caused by fungal contaminants through colonization of the surface of fruit peels and pulp, which suppresses fungal growth (Li et al., 2022). When directly inoculated into fresh fruits, several *Bacillus* species have been reported to exhibit antifungal activities against *P. digitatum* in apples (Khadiri et al., 2023) and citrus (Khadiri et al., 2023; Li et al., 2022; Tian et al., 2020), *A. flavus* in maize (Einloft et al., 2021), *Aspergillus parasiticus* in pistachio (Siahmoshteh et al., 2017), *C. gloeosporioides* in mango (Lamine Senghor et al., 2007), *Monilinia laxa* in apples (Khadiri et al., 2023), *Geotrichum candidum* (Taha et al., 2023), and *A. alternata* in tomatoes (Gorai et al., 2021; Taha et al., 2023). *Pseudomonas* spp. have also exhibited antifungal activities against *Geotrichum candidum* in tomato (Taha et al., 2023), *M. fructicola,* and *R. stolonifera* in peaches (Zhou et al., 1999). Lactobacilli exhibited inhibitory activities against *Botrytis cinerea* in strawberries (De Simone et al., 2024).

Bacterial cell-free supernatants have also exhibited significant antifungal properties when applied to fruits. Cell-free supernatants can be applied through direct dipping (De Simone et al., 2024; Dhundale et al., 2018), spraying, or fumigation (Sharma et al., 2021). In several studies, applying the cell-free supernatant of *Bacillus* spp. is effective against *P. digitatum* in mandarin oranges (Li et al., 2024b;

Tian et al., 2020), *C. acutatum* in tamarillo fruits (Arroyave-Toro et al., 2017), *A. ochracerus* in cereal flour (Zhao et al., 2023), and *C. gloeosporiodes* and *P. expansum* in loquat fruits (Ye et al., 2021). The application of the cell-free supernatant of *Pseudomonas* spp. was also reported to inhibit the infection caused by *C. gloeosporioides* in mango (Archana et al., 2021), *M. fructicola* (Shan et al., 2024), and *M. fructigena* in peaches (Aiello et al., 2019). Lactobacilli cell-free supernatant inhibited fruit decay caused by *P. digitatum* and *P. italicum* in lemons (Volentini et al., 2023), while *Streptomyces* spp. was effective against *C. musae* in bananas (Zeng et al., 2024).

Some fungal species can also act as biocontrol agents during postharvest stages through space and nutrient competition against fungal pathogens (He et al., 2024; Nally et al., 2012; Pesce et al., 2018; Spadaro & Droby, 2016). Direct inoculation of *Wickerhamomyces anomalus* and *Meyerozyma guilliermondii* can reduce the disease incidence caused by *C. gloeosporioides* in papaya fruit (Lima et al., 2013). In another study, *W. anomalus* and *Metschnikowia pulcherrima* exhibited antagonistic effects against *B. cinerea* in tomatoes, while *Candida lusitaniae* and *C. oleophila* exhibited antagonistic effects against *P. expansum* (Fernandez-San Millan et al., 2021). *Saccharomyces cerevisiae* demonstrated the ability to inhibit *Botrytis cinerea* in grapes (Nally et al., 2012) and decreased the severity of rotting in grapes caused by *L. theobromae* and *Neofusicoccum parvum* (Carvalho Castro et al., 2020). Direct inoculation of *Trichoderma guizhouense*, *T. harzianum*, and *T. koningiopsis* in Delta Valencia orange fruits exhibited antagonistic activities against *P. digitatum*, *C. gloeosporioides,* and *A. alternata*. Epiphytic yeast strains also inhibited the growth of *P. digitatum* in kinnow fruits during storage (Habiba et al., 2019).

Several mechanisms are involved in the antifungal activities of biocontrol agents such as bacteria and fungi. The direct inoculation of bacteria results in nutrient and space competition (Zhou et al., 1999). *Trichoderma* spp. rapidly spreads in the microenvironment, deprives nutrients of the pathogen's conidia, and blocks the germination of the fungal pathogen (Ferreira et al., 2020). Microbial agents also produce several compounds that exhibit wide-spectrum antifungal activities. The cell-free supernatant extracts of *Bacillus* spp. were attributed to volatile organic compounds such as aldehydes, ketones, alcohols, esters, hydrocarbons, acids, pyrazines, and terpenes (Calvo et al., 2020; Santos et al., 2023; Poulaki & Tjamos, 2023), mycolytic enzymes such as chitinase, β-1,3-glucanase, and protease (Abdelmoteleb et al., 2017), and lipopeptides such as surfactins, fengycins, bacillaene, bacilysin, and iturin A inhibits fungal growth (Arroyave-Toro et al., 2017; Chen et al., 2016; Khadiri et al., 2023; Li et al., 2024a; Trupo et al., 2023). *Lactobacillus* species inhibit fungal growth due to the production of organic acids that decrease the pH of substrates of fungal pathogens (Saragoca et al., 2024). In addition, *Lactobacillus* spp. also increases the shelf-life of fruits and vegetables due to the production of bacteriocin (Agriopoulou et al., 2020). On the other hand, fungi produce polyketides, aromatic compounds, non-ribosomal peptides, antibiotics, and heterocyclic metabolites that exhibit antifungal activities (Islam et al., 2024). Enzymes produced by *Trichoderma* species cause morphological changes and lysis of the pathogen's conidia (Ferreira et al., 2020).

The microorganism should be non-pathogenic, be able to survive in various postharvest conditions, not produce harmful metabolites to humans, exhibit wide-spectrum antifungal activity, be genetically stable, effective in low concentrations, and compatible with other physical or chemical treatment methods (Ferreira et al., 2020; Sharma et al., 2009). Hence, some biological control agents exhibit higher antifungal activities when combined with other methods (Eshel et al., 2009; Sharma et al., 2009). Lyousfi et al. (2022) reported that combining *B. amyloliquefaciens*, *Alcaligenes faecalis*, and sodium bicarbonate led to the inhibition of the growth and spore germination of *M. fructigena* in nectarines without modifying the quality parameters of the fruit. Other methods have combined yeast and bacteria (Lyousfi et al., 2022) or biocontrol agents with chemical methods (Usall et al., 2008). Yeast was also combined with UV-C radiation against the parasitic fungus *Chalara paradoxa* (De Seyn.) Sacc in pineapples (*Ananas comosus*) (Ou et al., 2016), which led to high chitinase and β-1,3-glucanase activity.

Botanical Approaches

Botanical approaches involve the application of plant extracts that exhibit antifungal properties. These extracts can be obtained using various solvents such as water, methanol, ethanol, ethyl acetate, or ether. The botanical extracts contain several bioactive compounds that protect plants from various pathogens. Since the extract is plant-based, it is generally considered a safer alternative to chemical fungicides (Pazolini et al., 2016). Plant extracts or plant-derived metabolites are applied through direct spraying (Kushaha et al., 2024), direct dipping (Pangallo et al., 2021), or using them as components of composite coatings of harvested fruits (Alvarez et al., 2023; Bordoh et al., 2022). Generally, botanical approaches are considered to be safer compared to chemical approaches (Sernaite et al., 2020). In addition, plant extracts are considered generally recognized as safe (GRAS) and are widely accepted by consumers (Seshadri et al., 2020). Botanical approaches involve the use of plant extracts that contain antifungal metabolites such as phenolic compounds (Fadda et al., 2021; Hernández et al., 2021; Rongai et al., 2015; Wang et al., 2018), terpenes (López-Velázquez et al., 2021; Souza et al., 2024; Zulu et al., 2023), alkaloids (Du et al., 2017; López-Velázquez et al., 2021), and coumarins (López-Velázquez et al., 2021). The antifungal activity of the ethanol extracts from the leaves and branches of *Flourensia cernua*, *F. microphylla*, and *F. retinophylla* has been attributed to sesquiterpenes, diterpenes, fatty acid, esters, polyols, amides, amines, benzofuranes, and dioxolane (de Rodriguez et al., 2017).

Crude plant extracts and their fractions have exhibited variable in vitro and in vivo antifungal activities against fungal pathogens. In several studies, the in vitro antifungal properties of crude extracts are first investigated, followed by in vivo experiments (Bello et al., 2022; Elshikh & Al Farraj, 2024; España et al., 2017; Maj et al., 2024). Very few studies, however, have applied botanical extracts in actual postharvest stages. The efficacy of botanical extracts against fungal pathogens is highly influenced by their chemical composition (Atnafu et al., 2024; Maj et al., 2024) and actual postharvest conditions such as temperature and humidity (Gupta & Jain, 2012). The conditions during experiments, however, may not be similar to the environmental conditions in postharvest stages. The antifungal activity of plant extracts in actual postharvest settings, however, cannot be predicted by in vitro experiments (Fadda et al., 2021).

Nevertheless, some studies have reported the antifungal activities of plant extracts when applied directly to fruits. In the study of Worku et al. (2025), the *Ruta chalepensis* extract inhibited the growth of *C. gloeosporioides* in mango when combined with heat treatment. The 20% crude extract of Neem (*Azadirachta indica* J., Meliaceae) also inhibited the spoilage of mango caused by fungal pathogens (Gupta & Jain, 2012). The oil extract of cinnamon (*Cinnamomum cassia*) exhibited an antagonistic effect against *Botrytis cinerea* in apples (Sernaite et al., 2020). Pomegranate peel extract exhibited antifungal activities against *P. digitatum* and *P. italicum* in oranges (Pangallo et al., 2021). Canola and Indian mustard extracts also inhibited the pathogenic effects of *M. fructicola* in peaches (Pazolini et al., 2016). In general, the higher antioxidant activity of plant extracts was attributed to their antifungal activities (de Rodriguez et al., 2017).

Specific compounds from botanical extracts have been reported to exhibit antifungal activities when applied to fruits. Among the bioactive compounds of plant extracts, the most commonly reported antifungal metabolites were essential oils and phenolic compounds. Essential oils can be obtained from citrus plants (Simas et al., 2017; Souza et al., 2024), herbs (Siddique et al., 2025; Yan et al., 2020), spices (Souza et al., 2024; Yan et al., 2020; Zulu et al., 2023), and medicinal plants (Rguez et al., 2018). The essential oils from *Mentha piperita* were reported to inhibit the growth of *F. sambucinum* in peppers (*Capsicum pubescens*) (Pérez-Vázquez et al., 2021). *Eucalyptus camaldulensis* essential oils inhibited the growth of *Aspergillus niger* and *Penicillium griseofulvum* in tomatoes (Siddique et al., 2025), while the essential oil of *Cupressus sempervirens* has been reported to inhibit *B. cinerea* in tomato during postharvest stages (Rguez et al., 2018). Five days post-inoculation, the *Cupressus sempervirens* essential oils exhibited 54% inhibition of fungal rotting in tomato (Rguez et al., 2018). Lastly, essential oils obtained from *Cinnamomum loureirii* and *Evolvulus alsinoide* exhibited antifungal activities against rot fungi that are inoculated in guava (Seshadri et al., 2020). The antifungal activities of essential oils are related to their ability to disrupt plasma membrane, alter hyphal morphology, and stimulate endogenous ROS (radical oxygen species) production (Yan et al., 2020), thereby inhibiting mycelial growth and spore germination (Sneha et al., 2024).

On the other hand, phenolic extracts from myrtle leaves contain hydroxybenzoic acids, hydrolyzable tannins, and flavonols that inhibit the decay and lesion formation caused by *P. digitatum* and *P. italicum* in Pterostilbene and piceatannol inhibited the effects of *Botrytis cinerea* in grapes (Xu et al., 2018a, b). When applied to peach, ferulic acid was active against *M. fructicola*, *A. alternata*, and *B. cinerea* (Hernández et al., 2021). Phenolic compounds from plant extracts interfere with binding of protein molecules, act as chelating agents, and weaken the permeability barrier of the cell membrane (Rongai et al., 2015). Overall, plant extracts offer safer alternatives to chemical fungicides. However, additional studies are required to determine their efficiency in actual postharvest environments.

Chemical Approaches

Chemical approaches are the primary choice for controlling fungal contamination and subsequent decay of agricultural products such as fruits and vegetables during postharvest stages (Abdelmoteleb et al., 2017; Khadiri et al., 2023). However, the frequent use of synthetic chemical fungicides has resulted in the accumulation of chemical residues in food (Dwiastuti et al., 2021; Wang et al., 2022), which led to public health issues regarding food safety and environmental pollution (Sharma et al., 2009). Furthermore, chemical fungicides have resulted in antifungal resistance among fungal pathogens (Islam et al., 2024; Wang et al., 2022), which leads to reduced control efficacy or control failure (Li et al., 2024a). Fungal resistance, however, depends on the natural adaptability of the fungal species (Bi & Ma, 2016). These issues have led to the restriction or withdrawal of several chemical fungicide treatments during postharvest (Schirra et al., 2011).

The most commonly used chemical fungicides in various agricultural products are thiabendazole and imazalil (Altieri et al., 2013; Cerioni et al., 2017; Sharma et al., 2009). Several fungicides, such as benomyl, captan, maneb, thiophanate methyl, and chlorothalonil, have been discontinued and replaced by less toxic fungicides (Wisniewski et al., 2016). The new chemical fungicides pyrimethanil, fludioxonil, and azoxystrobin exhibited lower risks and have been introduced as postharvest management to fungal contamination in citrus fruits (Cerioni et al., 2017). Pymimethanil is an anilinopyrimidine that inhibits the biosynthesis of methionine, affecting cystathionine β-lyase, while fludioxonil is a phenylpyrrole that inhibits the transport-associated phosphorylation of glucose, as well as prevents glycerol synthesis (Vischetti et al., 2023). Pyrimethanil and fludioxonil have exhibited antifungal activities against *Botrytis cinerea* in strawberries. Fludioxonil, pyrimethanil, imazalil, benomyl, and thiabendazole are fungicides used to control the growth of *P. expansum* (Yu et al., 2020).

The synthetic fungicide prochloraz is also used to control postharvest fungal contamination during storage in citrus fruits (Li et al., 2024a, b). Aside from chemical fungicides, several studies have also investigated the application of salts such as calcium chloride (Muñoz et al., 2025), sodium carbonate (Lyousfi et al., 2022; Miller et al., 2011; Usall et al., 2008), and zinc oxide (Ingle et al., 2024; Sardella et al., 2017) in combination with biological or physical approaches. Disinfectants such as chlorine dioxide, hydrogen peroxide, peroxyacetic acid, and mixtures of hydrogen peroxide and peroxyacetic acid have also been applied against postharvest fungal pathogens (Miller et al., 2011). Recently, the application of nanoparticles to control fungal growth has also been explored by several studies (Ingle et al., 2024; Shende et al., 2024).

Chemical fungicides can be applied via dipping, spraying, or fumigation (Dubey et al., 2011). In the study of Altieri et al. (2013), dipping oranges to 1000 mg/kg aqueous solution of Imazalil significantly prevented fruit decay caused by *P. digitatum*, *P. italicum*, *Geotrichum candidum*, *P.* spp., *A. citri*, *L. natalensis*, or *Phomopsis citri*. Chemical fungicides such as azoxystrobin, fludioxonil, and difenoconazole have been applied to potatoes via spraying (Kirk et al., 2013). Direct spraying of benzalkonium chloride, disodium-ethylenediaminetetraacetic acid, and sodium hypochlorite was the most effective in combating *A. parasiticus* growth during storage of hazelnuts, although the activities of the chemicals may decrease with longer incubation hours (Basaran, 2010). Foliar spraying of copper sulfate and chlorothalonil has also controlled the growth of *P. palmivora* in papaya (Vawdrey et al., 2015). Overall, chemical fungicides exhibit high effectiveness against fungal pathogens. However, current research is focused on non-chemical alternatives due to the toxicity of chemical fungicides.

Research Prospects and Way Forward

Genomics Approach

Genomics has been considered a powerful tool for understanding the intricacies involved in host-pathogen interactions (Neik et al., 2020), providing a multitude of possibilities for developing viable solutions to manage plant diseases, including postharvest diseases (Belay & James Caleb, 2022). One of the most important applications of genomics in plant disease management is developing host genetic resistance. Genome sequencing enables the discovery of genetic variants linked to disease resistance through genome-wide association studies (GWAS) or QTL mapping (Song et al., 2023). Breeders can then use this information to select the best-performing progenies with robust disease resistance characteristics and other important traits. Furthermore, the availability of genome-wide molecular markers from whole genome sequencing has allowed the implementation of genomic selection (GS), accelerating the development of resistant cultivars (Amas et al., 2021). This approach showed promising outcomes in breeding for resistance against postharvest diseases in strawberry, peaches, and pepper (Fu et al., 2022; Hong et al., 2020; Petrasch et al., 2021). More recently, advances in pangenomics (Bayer et al., 2021) have significantly improved the discovery of genomic variants linked to disease resistance traits by integrating multiple genomes (Amas et al., 2023), providing breeders with a comprehensive genomic toolkit to develop better cultivars.

Genomics has also played a vital role in characterizing many plant pathogens, including postharvest fungal pathogens (Chen et al., 2022a, b), which are challenging to detect owing to their ability to undergo a quiescent phase (Balendres et al., 2020). Genome sequencing can reveal key genetic features that can be used to develop early detection systems, such as highly sensitive PCR-based techniques (Harpaz et al., 2022). Moreover, the advances in sequencing platforms, including the availability of portable long-read sequencing devices (i.e., MinION nanopore sequencing), are streamlining direct genome sequencing to rapidly identify pathogens (Phannareth et al., 2021). This is particularly crucial for managing transboundary diseases, which can potentially become global biosecurity concerns. The availability of complete genome sequences for fungal pathogens also enables the identification of genes involved in pathogenicity, providing potential targets for novel chemistries (Xu et al., 2024).

Metagenomics is an emerging approach to managing postharvest diseases by analyzing the collective genomic material in healthy and infected samples. It can detect shifts in microbial communities that indicate progressing deterioration and sensitivity to certain fungal pathogens (Kusstatscher et al., 2019). This approach is a novel way to monitor disease development during storage, preventing an unprecedented disease outbreak. Moreover, new species of microorganisms with biological control (biological control agents or BCA) characteristics can be identified from metagenomic analyses (Zhang et al., 2021a, b). Genes responsible for their biocontrol properties, including those encoding for antimicrobial compounds, may be isolated and functionally studied, potentially producing effective BCA products against postharvest pathogens.

Advances in genomics have been the cornerstone of the emergence of powerful biotechnological tools, including genome editing and RNA interference (RNAi) (Mao et al., 2019; Zhao et al., 2024). Genome editing using *CRISPR-Cas* (clustered regularly interspaced short palindromic repeats-CRISPR-associated protein) systems has become one of the most popular biotechnological tools for gene manipulation for various applications (Wang & Doudna, 2023). For managing plant diseases, this tool can be used to manipulate susceptible alleles or knock down susceptibility genes in the host, resulting in enhanced resistance (Sun et al., 2018). Similarly, it can be deployed to edit pathogenicity or virulence genes in the pathogens to make them less or non-infectious (Ropero-Pérez et al., 2024). In some cases, modifying genes indirectly involved in pathogen recognition but participating in other resistance mechanisms, including metabolite production, could lead to increased resistance to diseases (Wang et al., 2023a, b, c). Moreover, multiplexed CRISPR technologies can manipulate several genes in a single reaction, resulting in higher editing efficiency and greater chances of obtaining desirable phenotypes in both the host and pathogen (Sun et al., 2018; Zhang et al., 2023).

Another emerging biotechnological advance showing great potential for fungal postharvest disease control is the spray-induced gene silencing (SIGS) technique based on RNAi technology (de Oliveira Filho et al., 2021). In this strategy, double-stranded RNAs (dsRNAs), capable of hijacking the fungal cellular machinery, are directly applied to the harvested products (i.e., fruits). This effectively attenuates the pathogen's growth, achieving high levels of disease control. SIGS has been successfully used for managing several postharvest pathogens, including *B. cinerea* (Wang et al., 2016) and *Fusarium* spp. (Gu et al., 2019). Because RNAi is based on a natural process conserved in most living organisms, it is considered a safer alternative to synthetic fungicides. Furthermore, it is highly specific to the target pathogen and can be applied straightforwardly, making large-scale treatments convenient (de Oliveira Filho et al., 2021).

The prospects of leveraging genomics for plant disease control have significantly expanded with the surge of sequencing information from numerous hosts and pathogens. The challenge now lies in our ability to efficiently manage this wealth of information to develop sustainable solutions to solve issues surrounding crop protection. Artificial intelligence (AI) has been recognized as a powerful tool for addressing this challenge with its ability to analyze huge datasets (Jafar et al., 2024). AI has been increasingly used for detecting disorders in horticultural products such as diseased fruits and vegetables, preventing further yield and economic losses (Dolatabadian et al., 2025). Together with genomics, AI can infer gene networks involved in host-pathogen interactions, providing insights into the molecular mechanisms involved in resistance or susceptibility (Sperschneider, 2020). This can inform strategies for designing effective control measures, including developing broad-spectrum fungicides, enhancing host resistance, or deploying gene drives to curb pathogen virulence (Tek & Budak, 2022). Moreover, machine and deep learning methods, which are forms of AI, can integrate complex phenotyping data, including hyperspectral images of diseases, with genomic information to increase the power of genomic prediction (Upadhyaya et al., 2024). This has the potential to enhance the efficiency of GS for developing resistant cultivars, ultimately providing growers with sustainable genetic solutions to manage plant diseases, including postharvest diseases.

Plant Chemistry

The utilization of plant metabolites as antifungal agents against postharvest diseases in crops has been extensively discussed in various review papers (Deresa & Diriba, 2023; Khetabi et al., 2022; Matrose et al., 2021; Pusztahelyi et al., 2015). These metabolites are usually identified from plants' essential oils (Eosinophil). The metabolite components are then studied using various in vitro and in vivo techniques to determine their fungistatic and fungicidal activities. An example is an in vitro study of EO components from Mediterranean aromatic plants. Citral, carvacrol, and thymol exhibited antifungal activities against different postharvest fruit rot fungi (Camele et al., 2012). The use of such metabolites to prevent or treat postharvest fungal diseases offers a more environmentally friendly approach compared to the use of synthetic fungicides. Furthermore, since these metabolites come from natural plant sources and are commonly tested for their toxicity in humans, their use reduces concerns about food safety and health risks.

The most promising application of plant metabolites as a postharvest treatment is to mix EOs containing them with natural edible films or coatings. Natural edible films or coatings can be wrapped around food products to extend their shelf life and may be consumed with the food (Pavlath & Orts, 2009; Summo & De Angelis, 2022). Using such films and coatings helps in addressing the volatility of EOs. Commonly, a base material that is classified as generally recognized as safe (GRAS) by the US Food and Drug Administration is combined with either the metabolite itself or the EO that contains the metabolites. In a review by Cortés et al. (2022), it was identified that cinnamaldehyde, 2-nonanone, eugenol, thymol, carvacrol, and citral are the metabolites that are commonly studied for active packaging. EOs that contain multiple plant metabolites can also be combined with natural coatings. An example would be *Aloe vera* gel mixed with garlic essential oil, which increased the resistance of postharvest bananas against anthracnose (Khaliq et al., 2019). In their in vivo study, the gel combined with 0.1% of the EO reduced anthracnose disease incidence and severity. An attempt was made to formulate a natural coating base mixed with antifungal EOs to control the growth of *B. cinerea* in *Rosa* sp. (Herrera-Romero et al., 2017). However, none of the explored combinations presented significant growth inhibition.

Polysaccharides such as chitosan and pullulan are used as natural coating bases that, when applied to fruits, can decrease the fruit's respiration rate, inhibit fungal growth, and delay ripening (Du et al., 1997). Related studies presented the ability of chitosan to prolong the shelf life of various citrus fruits (Arnon et al., 2014; Chien et al., 2007; Contreras-Oliva et al., 2012; Xu et al., 2018a, b). A coating of 1.5% chitosan and crude hairy fig (*Ficus hirta* Vahl.) extract improved the postharvest quality of navel oranges by decreasing the respiration rate (Chen et al., 2018). In Brazil, biodegradable films extracted from tilapia skin were used to coat strawberries that were previously immersed in different plant extracts (catuaba, guarana, sibipiruna, and barbatimão) (Gardenal et al., 2024). However, the physicochemical analyses revealed no significant difference between the film-coated strawberries with and without the plant extracts. Nonetheless, the crude extracts showed antifungal activity against *B. cinerea*, suggesting its potential alternative to prevent mold development in fruits. Waxes from natural sources can also be used to develop edible coatings. These sources include candelilla, rice bran, and carnauba (Campos et al., 2011). Carnauba wax and thyme essential oil were mixed to develop a bell pepper packaging system (López-Velásquez et al., 2024). Both the EO alone and the combination of the wax and EO can inhibit the growth of gray mold in bell peppers. However, the new bell pepper packaging system was superior in maintaining the fruit's firmness, reducing weight loss, and enhancing the fruit's appearance.

Doing sensory analysis upon application of edible coatings is important in the palatability or attractiveness of the fruit for consumption. For example, the incorporation of carvacrol into polyvinyl alcohol coating was recommended only as a preharvest treatment due to the negative effect of the taste and odor of carvacrol (Sapper et al., 2020). This is despite its ability to reduce the incidence and severity of black and blue molds on *Golden Delicious* apples.

The essential oils can also be encapsulated into a nanocapsule before being incorporated into the coating. A multisystem coating composed of encapsulated EO from *Thymus vulgaris* and pullulan polysaccharide was formulated by Piña-Barrera et al. (2019). This coating prevented the growth of microorganisms, resulting in maintained color, firmness, titrable acidity, and total soluble solid content of table grapes (*Vitis vinifera* L.). Metabolites that are mainly hydrophobic, such as eugenol, can be nanoencapsulated to protect them against degradation and improve their physical and chemical stability (Piña-Barrera et al., 2022). Polymeric nanoparticles based on preformed polymers were incorporated with alginate as the edible coating.

A technique called nanoprecipitation was used to add eugenol to the mix. The produced coating resulted in delayed ripening of tomatoes and protection of tomatoes against *Colletotrichum gloeosporioides*. Nanoencapsulation of eugenol allowed the even distribution of eugenol on the fruit surface and its gradual release (Piña-Barrera et al., 2022). Chitosan nanoemulsion was used to encapsulate thymol, *Hippophae rhamnoides* oil (HEO), or a mixture of both to create a chitosan composite film (Zhang et al., 2024a, b). The film that contains both thymol and HEO provided both antimicrobial and antioxidant activities, including less nutrient loss, to strawberries. Synergism between the polymer coating and EO is also possible. For example, the addition of thyme essential oil to chitosan nanoparticles exhibited synergistic effects in reducing the incidence of *C. gloesporoides* in avocados (Correa-Pacheco et al., 2017).

A list of commercially available botanical fungicides was compiled in the review paper of Deresa and Diriba (2023). However, these fungicides, such as seed treatment, are applied pre-harvest or as part of the planting process. As of writing, no commercial product for postharvest application is available. There is also no available study on why none of this developed technology for postharvest disease prevention in crops is translated commercially. We can only surmise that the challenges in adopting botanical fungicides for pre-harvest application apply here. For example, the effectivity of the botanical fungicides is generally less than the available synthetic fungicides. Scaling the in vitro and in vivo studies to a larger population of postharvest crops must also be studied further. Compared with the cost of producing synthetic fungicides, extracting the metabolites or EOs from plants, and incorporating them into the edible films may be larger. Thus, more studies must be conducted on improving the benefit-cost ratio of producing such materials. Farmers, suppliers, and local markets may also lack awareness of the potential of this technology to prolong the shelf life of their produce postharvest. More importantly, local and international agricultural organizations must lobby for the advantages of these technologies to the national governments. In response, there must be support from the government, both in terms of funding and national policies, in realizing the technology transfer to the market.

Nanotechnology

A more recent advancement in postharvest pathology is the use of nanotechnology. Nanotechnology focuses on designing and utilizing materials, devices, and systems on an atomic or molecular scale—typically between 1 and 100 nanometers (Saini et al., 2010). The technology may be used to develop nanocoating, nano-bubbles, and nano-packing (Mahmoud & Gaber, 2024), which may all be used to prolong the shelf life of agricultural products. All three may also inhibit pathogen infection or enhance antimicrobial capacity (Shan et al., 2023). In 2024, a study by Yin et al. showed that HLDP nanoprotectants prepared from a hydrophilic and lipophilic diblock polymer (HLDP) have the ability to control mango anthracnose by direct pathogen inhibition and amplified plant immune responses. Sadek et al. (2022) observed that sulfur nanoparticles (S-NPs) and copper nanoparticles (Cu-NPs) effectively suppressed fungal (*B. cinerea* and *Sclerotinia sclerotiorum*) infections in cucumbers.

References

Abdelmoteleb, A., Troncoso-Rojas, R., Gonzalez-Soto, T., & González-Mendoza, D. (2017). Antifungical activity of autochthonous *Bacillus subtilis* isolated from *Prosopis juliflora* against phytopathogenic fungi. *Mycobiology, 45*(4), 385–391. https://doi.org/10.5941/myco.2017.45.4.385

Abdullah, Q., Mahmoud, A., & Al-harethi, A. (2016). Isolation and identification of fungal post-harvest rot of some fruits in Yemen. *PSM Microbiology, 1*(1), 36–44.

Abubakar, M. M., Norida, M., Rafii, M. Y., & Nakasha, J. J. (2020). Effects of post-harvest hot water treatments on the fungi contamination, physiology and quality of Rock Melon Fruit. *Australian Journal of Crop Science, 14*(7), 1081–1087. https://doi.org/10.21475/ajcs.20.14.07.p2218

Abu-Bakar, A. I., Naur Ain Izzati, M. Z., & Umi Kalsom, Y. (2013). Diversity of *Fusarium* species associated with post-harvest fruit rot disease of tomato. *Sains Malaysiana, 42*(7), 911–920.

Agarwal, V. K. S., & Sinclair, J. B. (1997). *Principles of seed pathology* (2nd ed., 442p). CRC Press.

Agriopoulou, S., Stamatelopoulou, E., Sachadyn-Król, M., & Varzakas, T. (2020). Lactic acid bacteria as antibacterial agents to extend the shelf life of fresh and minimally processed fruits and vegetables: Quality and safety aspects. *Microorganisms, 8*(6), 952. https://doi.org/10.3390/microorganisms8060952

Ahmad, I., Ansari, M. I., & Aqil, F. (2006). Biosorption of Ni, Cr and Cd by metal-tolerant *Aspergillus niger* and *Penicillium* sp., using single and multi-metal solution. *Indian Journal of Experimental Biology, 44*(1), 73–76.

Aiello, D., Restuccia, C., Stefani, E., Vitale, A., & Cirvilleri, G. (2019). Postharvest biocontrol ability of *Pseudomonas synxantha* against *Monilinia fructicola* and *Monilinia fructigena* on stone fruit. *Postharvest Biology and Technology, 149*, 83-89.

Akhtar, R., & Thiruppathi, S. K. (2025). First report of *Lasiodiplodia iraniensis* causing post-harvest decay in chillies (*Capsicum* spp.). *Physiological and Molecular Plant Pathology, 136*, 102580.

Alkan, N., Friedlander, G., Ment, D., Prusky, D., & Fluhr, R. (2015). Simultaneous transcriptome analysis of *Colletotrichum gloeosporioides* and tomato fruit pathosystem reveals novel fungal pathogenicity and fruit defense strategies. *New Phytologist, 205*(2), 801–815. https://doi.org/10.1111/nph.13175

Altieri, G., Di Renzo, G. C., Genovese, F., Calandra, M., & Strano, M. C. (2013). A new method for the postharvest application of imazalil fungicide to citrus fruit. *Biosystems Engineering, 115*(4), 434–443. https://doi.org/10.1016/j.biosystemseng.2013.04.008

Altomare, C., Logrieco, A. F., & Gallo, A. (2021). Mycotoxins and mycotoxigenic fungi: Risk and management. A challenge for future Global Food Safety and Security. *Encyclopedia of Mycology*, 64–93. https://doi.org/10.1016/b978-0-12-819990-9.00032-9

Alvarez, M., Pérez-Gago, M., Taberner, V., Settier-Ramírez, L., Martínez-Blay, V., & Palou, L. (2023). Postharvest application of novel bio-based antifungal composite edible coatings to reduce sour rot and quality losses of 'valencia' oranges. *Coatings, 13*(8), 1412. https://doi.org/10.3390/coatings13081412

Alvindia, D. and Acda, M. (2015). Revisiting the efficacy of hot water treatment in managing anthracnose and stem-end rot diseases of mango cv. 'Carabao'. Crop Protection. 67. https://doi.org/10.1016/j.cropro.2014.09.016

Amaike, S., & Keller, N. P. (2011). Aspergillus flavus. *Annual Review of Phytopathology, 49*, 107–133.

Amas, J., Anderson, R., Edwards, D., Cowling, W., & Batley, J. (2021). Status and advances in mining for blackleg (*Leptosphaeria maculans*) quantitative resistance (QR) in oilseed rape (*Brassica napus*). *Theoretical and Applied Genetics, 134*(10), 3123–3145. https://doi.org/10.1007/S00122-021-03877-0

Amas, J. C., Bayer, P. E., Hong Tan, W., Tirnaz, S., Thomas, W. J. W., Edwards, D., & Batley, J. (2023). Comparative pangenome analyses provide insights into the evolution of *Brassica rapa* resistance gene analogues (RGAs). *Plant Biotechnology Journal, 21*(10), 2100–2112. https://doi.org/10.1111/PBI.14116

Amenu, D., Desalegn, R., Nugusa, A., Tolera, C., & Tafesse, T. (2024). Evaluation of antifungal activity of plant extracts against plant pathogen associated with sweet orange (*Citrus sinensis* L.), in Jimma, western Ethiopia. *Discover Applied Sciences, 6*(7). https://doi.org/10.1007/s42452-024-06054-2

Anand, T., Bhaskaran, R., Raguchander, T., Samiyappan, R., Prakasam, V., & Gopalakrishnan, C. (2009). Defence responses of chilli fruits to *Colletotrichum capsici* and *Alternaria alternata*. *Biologia plantarum, 53*(3), 553–559.

Archana, T. J., Gogoi, R., Kaur, C., Varghese, E., Sharma, R. R., Srivastav, M., Tomar, M., Kumar, M., & Kumar, A. (2021). Bacterial volatile mediated suppression of postharvest anthracnose and quality enhancement in Mango. *Postharvest Biology and Technology, 177*, 111525. https://doi.org/10.1016/j.postharvbio.2021.111525

Arnon, H., Zaitsev, Y., Porat, R., & Poverenov, E. (2014). Effects of carboxymethyl cellulose and chitosan bilayer edible coating on postharvest quality of citrus fruit. *Postharvest Biology and Technology, 87*, 21–26. https://doi.org/10.1016/j.postharvbio.2013.08.007

Arroyave-Toro, J. J., Mosquera, S., & Villegas-Escobar, V. (2017). Biocontrol activity of *Bacillus subtilis* EA-CB0015 cells and lipopeptides against postharvest fungal pathogens. *Biological Control, 114*, 195–200. https://doi.org/10.1016/j.biocontrol.2017.08.014

Arshad, M. O., Chauhan, Y., Singh, P., Srivastav, P., Gupta, M., & Patwa, N. (2023). Advancements in controlled atmosphere storage technology—A review. In S. Yadav, A. Haleem, P. K. Arora, & H. Kumar (Eds.), *Proceedings of second international conference in mechanical and energy technology* (Smart innovation, systems and technologies) (Vol. 290). Springer. https://doi.org/10.1007/978-981-19-0108-9_42

Atnafu, B., Abedeta, C., Lemessa, F., Mohammed, A., Oufensou, S., & Chala, A. (2024). Chemical composition of selected aromatic plant essential oils and their antifungal efficacy against toxigenic fungi associated with maize (*Zea mays* L). *Cogent Food and Agriculture, 10*(1). https://doi.org/10.1080/23311932.2024.2329116

Atungulu, G. G., Kolb, R. E., Karcher, J., & Mohammadi Shad, Z. (2019). Postharvest technology: Rice storage and cooling conservation. In J. Bao (Ed.), *Rice chemistry and technology* (pp. 517–555). Elsevier Inc. in cooperation with AACC International. https://doi.org/10.1016/B978-0-12-811508-4.00016-2

Auger, J., Pérez, I., & Esterio, M. (2013). *Diaporthe ambigua* associated with post-harvest fruit rot of kiwifruit in Chile. *Plant Disease, 97*(6), 843–843.

Awang, Y., Ghani, M. A. A., Sijam, K., & Mohamad, R. B. (2011). Effect of calcium chloride on anthracnose disease and postharvest quality of red-flesh dragon fruit (*Hylocereus polyrhizus*). *African Journal of Microbiology Research, 5*(29), 5250–5259.

Aydin, A., Aksu, H., & Gunsen, U. (2011). Mycotoxin levels and incidence of mould in Turkish rice. *Environmental Monitoring and Assessment, 178*(1–4), 271–280. https://doi.org/10.1007/s10661-010-1688-9

Bagri, R. K., Choudhary, S. L., & Rai, P. K. (2004). Management of fruit rot of chilli with different plant products. *Indian Phytopathology, 57*(1), 107–109.

Balendres, M. A., Karlovsky, P., & Cumagun, C. J. (2019). Mycotoxigenic fungi and mycotoxins in agricultural crop commodities in The Philippines a review. *Food, 8*, 249.

Balendres, M. A., Mendoza, J. V., & Dela Cueva, F. (2020). Characteristics of *Colletotrichum musae* PHBN0002 and the susceptibility of popular banana cultivars to postharvest anthracnose. *Indian Phytopathology, 73*(1), 57–64. https://doi.org/10.1007/S42360-019-00178-X/METRICS

Balodi, R., Raghavendra, K. V., Singh, P. K., et al. (2024). *Alternaria alternata* species complex impairing solanaceous vegetables in northern parts of India: An emerging problem in *Solanum lycopersicum* L. *3 Biotech, 14*(218). https://doi.org/10.1007/s13205-024-04059-7

Bano, A., Gupta, A., Prusty, M. R., & Kumar, M. (2023). Elicitation of fruit fungi infection and its protective response to improve the postharvest quality of fruits. *Stress, 3*(1), 231–255. https://doi.org/10.3390/stresses3010018

Banya, M., Garg, S., & Lal Meena, N. (2020). A review: Chilli anthracnose, its spread and management. *Journal of Pharmacognosy and Phytochemistry, 9*, 1432–1438.

Barkai-Golan, R. (2001). *Postharvest diseases of fruits and vegetables: Development and control*. Elsevier.

Barkai-Golan, R. (2005). *Postharvest diseases of fruit and vegetables*. A division of Reed Elsevier India, Pvt. Limited..

Barkai-Golan, R., Padova, R., Ross, I., Lapidot, M., Davidson, H., & Copel, A. (1993). Combined hot water and radiation treatments to control decay of tomato fruits. *Scientia horticulturae, 56*(2), 101–105.

Basaran, P. (2010). Antifungal effect of acids and surface active compounds for post-harvest control of aspergillus parasiticus growth on hazelnut. *Journal of Food Processing and Preservation, 35*(2), 236–246. https://doi.org/10.1111/j.1745-4549.2009.00442.x

Bashir, M. A., Hamza, F., & Pukuma, M. S. (2020). Postharvest fungal spoilage of some citrus fruits. *Biotechnology and Bioengineering, 7*(1), 10–14.

Batey, I. (2017). The diversity of uses for cereal grains. In C. W. Wrigley & I. L. Batey (Eds.), *Cereal grains: Assessing and managing quality* (2nd ed., pp. 45–56). Woodhead Publishing Limited. https://doi.org/10.1016/B978-0-08-100719-8.00003-6

Bautista-Baños, S., Bosquez-Molina, E., & Barrera-Necha, L. L. (2014). *Rhizopus stolonifer* (soft rot). In *Postharvest decay* (pp. 1–44). Academic.

Bayer, P. E., Petereit, J., Danilevicz, M. F., Anderson, R., Batley, J., & Edwards, D. (2021). The application of pangenomics and machine learning in genomic selection in plants. *The Plant Genome, 14*(3), e20112. https://doi.org/10.1002/TPG2.20112

Belay, Z. A., & James Caleb, O. (2022). Role of integrated omics in unravelling fruit stress and defence responses during postharvest: A review. *Food Chemistry: Molecular Sciences, 5*, 100118. https://doi.org/10.1016/J.FOCHMS.2022.100118

Bello, T. S., Adebola, M. O., & Chikwendu, C. S. (2022). Comparison of the antifungal activities of selected botanical extracts against the *carica papaya* fruit rot pathogen (*Cladosporium herbarum*). *Archives of Phytopathology and Plant Protection, 55*(12), 1464–1482. https://doi.org/10.1080/03235408.2022.2104589

Bennett, J. W. (2010). An overview of the genus *aspergillus*. In M. Machida & K. Gomi (Eds.), *Aspergilli: Genomics, medical aspects, biotechnology, and research methods* (pp. 1–17). Caister Academic Press.

Bennett, J. S., Evans, B., Barnes, C., & Collum, T. D. (2024). First report of *Diaporthe* sp. from the *D. arctii* species complex causing postharvest decay of European pear in West Virginia, United States. *Plant Disease*. https://doi.org/10.1094/PDIS-09-23-1794-PDN

Bhalerao, V., & Chavan, A. (2017). Isolation of filamentous fungi in post-harvest cereal grain during the storage and their effect on seed health. *International Journal of Advanced Research, 5*(8), 2171–2177. https://doi.org/10.21474/IJAR01/5283

Bhalerao, J. B., Chavan, R. A., Dharbale, B. B., Kendre, A. H., & Mete, V. S. (2019). Study on in-vitro efficacy of botanicals and chemicals against Alternaria solani associated with post-harvest rot of tomato (*Lycopersicon esculentum* Mill.). *Journal of Pharmacognosy and Phytochemistry, 8*(5), 2045–2049.

Bhatta, U. K. (2022). Alternative management approaches of citrus diseases caused by *Penicillium digitatum* (green mold) and *Penicillium italicum* (blue mold). *Frontiers in Plant Science, 12*, 833328.

Bi, Q., & Ma, Z. (2016). Sensitivity, resistance stability, and cross-resistance of *Plasmopara viticola* to four different fungicides. *Crop Protection, 89*, 265–272. https://doi.org/10.1016/j.cropro.2016.07.030

Birck, N. M., Lorini, I., & Scussel, V. M. (2006). Fungus and mycotoxins in wheat grain at post harvest. In *9th international working conference on stored product protection* (pp. 198–205).

Birhanu, S., Akhtar, M. S., & Muleta, D. (2014). Management of post-harvest fruit spoilage fungi by some potential spice extracts. *Archives of Phytopathology and Plant Protection, 47*(17), 2124–2140. https://doi.org/10.1080/03235408.2013.869891

Blanco, C., de los Santos, B., & Romero, F. (2006). Relationship between concentrations of *Botrytis cinerea* conidia in air, environmental conditions, and the incidence of grey mold in strawberry flowers and fruits. *European Journal of Plant Pathology, 114*(4), 415–425.

Bordoh, P. K., Ali, A., Dickinson, M., Siddiqui, Y., & Romanazzi, G. (2020). A review on the management of postharvest anthracnose in dragon fruits caused by *Colletotrichum* spp. *Crop Protection, 130*, 105067.

Bordoh, P. K., Ali, A., Dickinson, M., Siddiqui, Y., & Ansah, F. A. (2022). Bioefficacy of composite medicinal plant extracts and gum arabic on improving postharvest quality in dragon fruit. *International Journal of Food Science, 2022*, 1–17. https://doi.org/10.1155/2022/3889563

Bothast, R. J. (1978). Fungal deterioration and related phenomena in cereals, legumes and oilseeds. In H. O. Hultin & M. Milner (Eds.), *Post-harvest biology and biotechnology* (462pp). Food Nutrition Press Inc.

CABI International. (2022a). *Alternaria alternata (Alternaria leaf spot)*. PlantwisePlus Knowledge Bank. https://doi.org/10.1079/pwkb.species.4480

CABI International. (2022b). *Botrytis cinerea (Grey mold rot)*. CABI Compendium. https://www.cabidigitallibrary.org/doi/abs/10.1079/cabicompendium.9611

Cabrefiga, J., Salomon, M. V., & Vilardell, P. (2022). Improvement of *Alternaria* leaf blotch and fruit spot of apple control through the management of primary inoculum. *Microorganisms, 11*(1), 101.

Cairns-Fuller, V., Aldred, D., & Magan, N. (2005). Water, temperature and gas composition interactions affect growth and ochratoxin A production by isolates of *Penicillium verrucosum* on wheat grain. *Journal of Applied Microbiology, 99*(5), 1215–1221. https://doi.org/10.1111/j.13652672.2005.02695.x

Calvo, H., Mendiara, I., Arias, E., Gracia, A. P., Blanco, D., & Venturini, M. E. (2020). Antifungal activity of the volatile organic compounds produced by *Bacillus velezensis* strains against postharvest fungal pathogens. *Postharvest Biology and Technology, 166*, 111208. https://doi.org/10.1016/j.postharvbio.2020.111208

Camele, I., Altieri, L., De Martino, L., De Feo, V., Mancini, E., & Rana, G. L. (2012). *In vitro* control of post-harvest fruit rot fungi by some plant essential oil components. *International Journal of Molecular Sciences, 13*, 2290–2300. https://doi.org/10.3390/ijms13022290

Campos, C., Gerschenson, L. N., & Flores, S. (2011). Development of edible films and coatings with antimicrobial activity. *Food Bioprocess Technology, 4*, 849–875. https://doi.org/10.1007/s11947-010-0434-1

Cancino, S., Lolas, M., Galdós, L., Hernández, Y., Ferrada, E., Riveros, P., et al. (2023). Occurrence of *Alternaria alternata* and *A. tenuissima* causing black rot in cherry fruits (*Prunus avium*) in Central Chile. *Plant Disease, 107*(12), 4032.

Cara, M., Toska, M., Frasheri, D., Baroncelli, R., & Sanzani, S. M. (2022). *Alternaria* species causing pomegranate and citrus fruit rots in Albania. *Journal of Plant Diseases and Protection, 129*(5), 1095–1104.

Carvalho Castro, A. P., Tavares, P. F., Araújo, C. P., da Paz, C. D., & Gava, C. A. T. (2020). Semi-commercial field evaluation of yeast formulations for control of mango postharvest decay caused by Botryosphaeriacean fungi in organic production. *International Journal of Fruit Science, 20*(2), 207–220. https://doi.org/10.1080/15538362.2019.1613469

CAST. (2003). *Mycotoxins: Risks in plant, animal, human systems* (Task force report no. 38). CAST.

Cerioni, L., Bennasar, P. B., Lazarte, D., Sepulveda, M., Smilanick, J. L., Ramallo, J., & Rapisarda, V. A. (2017). Conventional and reduced-risk fungicides to control postharvest *Diplodia* and *Phomopsis* stem-end rot on Lemons. *Scientia Horticulturae, 225*, 783–787. https://doi.org/10.1016/j.scienta.2017.08.008

Chen, Y., Lin, H., Jiang, Y., Zhang, S., Lin, Y., & Wang, Z. (2014). *Phomopsis longanae* Chi-induced pericarp browning and disease development of harvested longan fruit in association with energy status. *Postharvest Biology and Technology, 93*, 24–28.

Chen, X., Zhang, Y., Fu, X., Li, Y., & Wang, Q. (2016). Isolation and characterization of *Bacillus amyloliquefaciens* PG12 for the biological control of Apple Ring Rot. *Postharvest Biology and Technology, 115*, 113–121. https://doi.org/10.1016/j.postharvbio.2015.12.021

Chen, C., Cai, N., Chen, J., Peng, X., & Wan, C. (2018). Chitosan-based coating enriched with hairy Fig (*Ficus hirta* Vahl.) fruit extract for "newhall" navel orange preservation. *Coatings*. https://doi.org/10.3390/coatings8120445

Chen, J., Zhu, Z., Fu, Y., Cheng, J., Xie, J., & Lin, Y. (2021). Identification of *Lasiodiplodia pseudotheobromae* causing fruit rot of citrus in China. *Plants, 10*(2), 202.

Chen, C., Cai, N., Wan, C., Kai, W., & Chen, J. (2022a). Carvacrol delays *Phomopsis* stem-end rot development in pummelo fruit in relation to maintaining energy status and antioxidant system. *Food Chemistry, 372*, 131239.

Chen, Y., Zhang, Z., Tian, S., & Li, B. (2022b). Application of -omic technologies in postharvest pathology: Recent advances and perspectives. *Current Opinion in Food Science, 45*, 100820. https://doi.org/10.1016/J.COFS.2022.100820

Cheung, N., Tian, L., Liu, X., & Li, X. (2020). The destructive fungal pathogen *Botrytis cinerea*—Insights from gene studies with mutant analysis. *Pathogens, 9*(11), 923. https://doi.org/10.3390/pathogens9110923

Chien, P.-J., Sheu, F., & Lin, H.-R. (2007). Coating citrus (*Murcott tangor*) fruit with low molecular weight chitosan increases postharvest quality and shelf life. *Food Chemistry, 100*, 1160–1164. https://doi.org/10.1016/j.foodchem.2005.10.068

Christensen, C. M. (1957). Deterioration of stored grain by fungi. *The Botanical Review, 23*(2), 108–134. https://doi.org/10.1007/BF02930521

Christensen, C. M., & Kaufmann, H. H. (1969). *Grain storage: The role of fungi in quality loss* (292pp). University of Minnesota Press.

Christensen, C. M., & Kaufmann, H. H. (1974). Microflora. In C. M. Christensen (Ed.), *Storage of cereal Grain and their products* (pp. 219–240). American Association of Cereal Chemists, Inc, St. Paul.

Ciofini, A., Negrini, F., Baroncelli, R., & Baraldi, E. (2022). Management of post-harvest anthracnose: Current approaches and future perspectives. *Plants, 11*(1856). https://doi.org/10.3390/plants11141856

Coates, L. M., & Johnson, G. (1997). Postharvest pathology of fruit and vegetables. In J. Brown & H. Ogle (Eds.), *Plant pathogens and plant diseases*.

Conti Taguali, S., Riolo, M., Dopazo, V., Meca, G., & Cacciola, S. O. (2024). Characterization of mycotoxins produced by two Fusarium species responsible for postharvest rot of banana fruit. *Journal of Plant Pathology, 106*(4), 1785–1800.

Contreras-Oliva, A., Perez-Gago, M. B., & Rojas-Argudo, C. (2012). Effects of chitosan coatings on physicochemical and nutritional quality of clementine mandarins cv. 'Oronules'. *Food Science and Technology International, 18*, 303–315. https://doi.org/10.1177/1082013211427953

Correa-Pacheco, Z. N., Bautista-Baños, S., Valle-Marquina, M. Á., & Hernández-López, M. (2017). The effect of nanostructured chitosan and chitosan-thyme essential oil coatings on *Colletotrichum gloeosporioides* growth *in vitro* and on cv Hass avocado and fruit quality. *Journal of Phytopathology, 165*, 297–305. https://doi.org/10.1111/jph.12562

Cortés, L. A., Herrera, A. O., & Castellanos, D. A. (2022). Natural plant-based compounds applied in antimicrobial active packaging and storage of berries. *Journal of Food Processing & Preservation, 46*, e16995. https://doi.org/10.1111/jfpp.16995

Cruz-Lachica, I., Márquez-Zequera, I., García-Estrada, R. S., Carrillo-Fasio, J. A., León-Félix, J., & Allende-Molar, R. (2017). Identification of mucoralean fungi causing soft rot in papaya (*Carica papaya* L.) fruit in Mexico. *Revista mexicana de fitopatología, 35*(3), 397–417.

da Silva França, K. R., de Medeiros Ferro, M. M., Feijó, F. M., da Silva, S. J. C., de Melo, M. P., de Andrade Lima, G. S., & Assunção, I. P. (2024). First report of *Lasiodiplodia hormozganensis* and *Lasiodiplodia laeliocattleyae* causing crown rot disease in banana fruits in Brazil. *Crop Protection, 184*, 106784.

da Silva, C. D. F. B., & Michereff, S. J. (2013). Biology of *Colletotrichum* spp. and epidemiology of the anthracnose in tropical fruit trees. *Revista Caatinga, 26*(4), 130–138.

Dai, P., Liang, X., Wang, Y., Gleason, M. L., Zhang, R., & Sun, G. (2019). High humidity and age-dependent fruit susceptibility promote development of *Trichothecium* black spot on apple. *Plant Disease, 103*(2), 259–267.

Dai, K., Han, P., Zou, X., Jiang, S., Xu, F., Wang, H., Wei, Y., & Shao, X. (2021). Hot air treatment reduces postharvest decay in Chinese bayberries during storage by affecting fungal community composition. *Food Research International, 140*, 110021. https://doi.org/10.1016/j.foodres.2020.110021

Dai, T., Vrahas, M. S., Murray, C. K., & Hamblin, M. R. (2012). Ultraviolet C irradiation: an alternative antimicrobial approach to localized infections?. Expert review of anti-infective therapy, 10(2), 185–195.

Dal Bello, G. (2008). First report of *Trichothecium roseum* causing postharvest fruit rot of tomato in Argentina. *Australasian Plant Disease Notes, 3*(1), 103–104.

Dall'Asta, C. (2014). Mycotoxins from *Alternaria*. In *Advances in molecular toxicology* (Vol. 8, pp. 107–121). Elsevier. https://doi.org/10.1016/B978-0-444-63406-1.00003-9

Davies, C. R., Wohlgemuth, F., Young, T., Violet, J., Dickinson, M., Sanders, J.-W., Vallieres, C., & Avery, S. V. (2021). Evolving challenges and strategies for fungal control in the Food Supply Chain. *Fungal Biology Reviews, 36*, 15–26. https://doi.org/10.1016/j.fbr.2021.01.003

de Moura, A. P., Alves, T. R., dos Silva, S. J. L., da Silva, I. V., de Souza, J. J., Negreiros, A. M., Evangelista, L. F., Freires, A. L., Figueiredo, F. R., & de Queiroz Ambrósio, M. M. (2024). Thermotherapy in the management of Galia melon rot caused by *Fusarium falciforme* and *Fusarium sulawesiense*. *European Journal of Plant Pathology, 169*(4), 715–725. https://doi.org/10.1007/s10658-024-02866-5

de Oliveira Filho, J. G., Silva, G. D. C., Cipriano, L., Gomes, M., & Egea, M. B. (2021). Control of postharvest fungal diseases in fruits using external application of RNAi. *Journal of Food Science, 86*(8), 3341–3348. https://doi.org/10.1111/1750-3841.15816. Epub 2021 Jul 16. PMID: 34272735.

De Rodríguez, D. J., Salas-Méndez, E. D. J., Rodríguez-García, R., Hernández-Castillo, F. D., Díaz-Jiménez, M. L. V., Sáenz-Galindo, A., González-Morales, S., Flores-López, M.L., Villarreal-Quintanilla, J.A., Peña-Ramos, F.M., & Carrillo-Lomelí, D. A. (2017). Antifungal activity in vitro of ethanol and aqueous extracts of leaves and branches of *Flourensia* spp. against postharvest fungi. *Industrial Crops and Products, 107*, 499–508.

De Simone, N., Scauro, A., Fatchurrahman, D., Amodio, M. L., Capozzi, V., Colelli, G., Spano, G., Fragasso, M., & Russo, P. (2024). Probiotic *Lactiplantibacillus plantarum* strains showing anti-botrytis activity: A food-grade approach to improve the overall quality of strawberry in post-harvest. *Postharvest Biology and Technology, 218*, 113125. https://doi.org/10.1016/j.postharvbio.2024.113125

De Visser, C. L. M. (1996). Field evaluation of a supervised control system for *Botrytis* leaf blight in spring-sown onions in The Netherlands. *European Journal of Plant Pathology, 102*(8), 795–805.

Deresa, E. M., & Diriba, T. F. (2023). Phytochemicals as alternative fungicides for controlling plant diseases: A comprehensive review of their efficacy, commercial representatives, advantages, challenges for adoption, and possible solutions. *Heliyon, 9*, e13810. https://doi.org/10.1016/j.heliyon.2023.e13810

Derviş, S., Zholdoshbekova, S., Güney, İ. G., & Özer, G. (2024). *Neoscytalidium dimidiatum*: A newly identified postharvest pathogen of pears and its implications for pome fruits. *Journal of Phytopathology, 172*(3), e13322.

Deshpande, S. E. (2002). Fungal toxins. In *Handbook of food toxicology* (pp. 387–456). CRC Press.

Dhundale, V., Hemke, V., Desai, D., & Dhundale, P. (2018). Evaluation and exploration of lactic acid bacteria for preservation and extending the shelf life of fruit. *International Journal of Fruit Science, 18*(4), 355–368. https://doi.org/10.1080/15538362.2018.1435331

Díaz, G. A., Latorre, B. A., Jara, S., Ferrada, E., Naranjo, P., Rodríguez, J., & Zoffoli, J. P. (2014). First report of *Diaporthe novem* causing postharvest rot of kiwifruit during controlled atmosphere storage in Chile. *Plant Disease, 98*(9), 1274–1274.

Díaz, G. A., Latorre, B. A., Lolas, M., Ferrada, E., Naranjo, P., & Zoffoli, J. P. (2017). Identification and characterization of *Diaporthe ambigua, D. australafricana, D. novem,* and *D. rudis* causing a postharvest fruit rot in kiwifruit. *Plant Disease, 101*(8), 1402–1410.

Dolatabadian, A., Neik, T. X., Danilevicz, M. F., Upadhyaya, S. R., Batley, J., & Edwards, D. (2025). Image-based crop disease detection using machine learning. *Plant Pathology, 74*(1), 18–38. https://doi.org/10.1111/PPA.14006

Droby, S., & Lichter, A. (2007). Post-harvest Botrytis infection: etiology, development and management. In Botrytis: Biology, pathology and control (pp. 349–367). Dordrecht: Springer Netherlands.

Du, J. M., Gemma, H., & Ihawori, S. (1997). Effects of chitosan coating on the storage of peach, Japanese pear, and kiwifruit. *Journal of the Japanese Society for Horticultural Science, 66*, 15–22. https://doi.org/10.2503/jjshs.66.15

Du, R., Liu, J., Sun, P., Li, H., & Wang, J. (2017). Inhibitory effect and mechanism of *Tagetes erecta* L. fungicide on *Fusarium oxysporum* f. sp. niveum. *Scientific Reports, 7*(1). https://doi.org/10.1038/s41598-017-14937-1

Duarte-Molina, F., Gómez, P. L., Castro, M. A., & Alzamora, S. M. (2016). Storage quality of Strawberry Fruit treated by pulsed light: Fungal decay, water loss and mechanical properties. *Innovative Food Science and Emerging Technologies, 34*, 267–274. https://doi.org/10.1016/j.ifset.2016.01.019

Dubey, N. K., Singh, P., Prakash, B., & Mishra, P. K. (2011). Plant chemicals in post harvest technology for management of fungal, mycotoxin and insect contamination of food commodities. In *Plant defence: Biological control* (pp. 195–208). https://doi.org/10.1007/978-94-007-1933-0_8

Dweba, C. C., Figlan, S., Shimelis, H. A., Motaung, T. E., Sydenham, S., Mwadzingeni, L., & Tsilo, T. J. (2017). *Fusarium* head blight of wheat: Pathogenesis and control strategies. *Crop Protection, 91*, 114–122. https://doi.org/10.1016/j.cropro.2016.10.002

Dwiastuti, M. E., Soesanto, L., Aji, T. G., Devy, N. F., & Hardiyanto. (2021). Biological control strategy for postharvest diseases of citrus, apples, grapes and strawberries fruits and application in Indonesia. *Egyptian Journal of Biological Pest Control, 31*(1), 141. https://doi.org/10.1186/s41938-021-00488-1

Einloft, T. C., Bolzan de Oliveira, P., Radünz, L. L., & Dionello, R. G. (2021). Biocontrol capabilities of three *Bacillus* isolates towards aflatoxin B1 producer *A. flavus* in vitro and on maize grains. *Food Control, 125*, 107978. https://doi.org/10.1016/j.foodcont.2021.107978

Elad, Y., Vivier, M., & Fillinger, S. (2016). *Botrytis*, the good, the bad, and the ugly. In S. Fillinger & Y. Elad (Eds.), *Botrytis – The fungus, the pathogen and its management in agricultural systems* (pp. 1–16). Springer. https://doi.org/10.1007/978-3-319-23371-0_1

Elad Y, Williamson B, Tudzynski P, Delen N (eds) (2004) *Botrytis*: biology, pathology and control. Kluwer Academic Publishers (Springer), Dordrecht, 416 pp

Elmer, P. A. G., & Michailides, T. J. (2007a). Epidemiology of *Botrytis cinerea* in orchard and vine crops. In Y. Elad, B. Williamson, P. Tudzynski, & N. Delen (Eds.), *Botrytis: Biology, pathology, and control* (pp. 243–272). Kluwer Academic Publishers.

Elmer, P. A., & Michailides, T. J. (2007b). Epidemiology of *Botrytis cinerea* in orchard and vine crops. In *Botrytis: Biology, pathology and control* (pp. 243–272). Springer Netherlands.

Elshikh, M. S., & Al Farraj, D. A. (2024). Postharvest Disease Control in banana using organic extract from the brown alga *Turbinaria ornata. Physiological and Molecular Plant Pathology, 131*, 102283. https://doi.org/10.1016/j.pmpp.2024.102283

Escrivá, L., Oueslati, S., Font, G., & Manyes, L. (2017). Alternaria mycotoxins in food and feed: An overview. *Journal of Food Quality, 2017*, Article ID 1569748. https://doi.org/10.1155/2017/1569748

Eshel, D., Regev, R., Orenstein, J., Droby, S., & Gan-Mor, S. (2009). Combining physical, chemical and biological methods for synergistic control of postharvest diseases: A case study of black root rot of Carrot. *Postharvest Biology and Technology, 54*(1), 48–52. https://doi.org/10.1016/j.postharvbio.2009.04.011

España, M. D., Arboleda, J. W., Ribeiro, J. A., Abdelnur, P. V., & Guzman, J. D. (2017). Eucalyptus leaf byproduct inhibits the anthracnose-causing fungus *Colletotrichum gloeosporioides*. *Industrial Crops and Products, 108*, 793–797. https://doi.org/10.1016/j.indcrop.2017.08.002

Fadda, A., Sarais, G., Lai, C., Sale, L., & Mulas, M. (2021). Control of postharvest diseases caused by *penicillium* spp. with myrtle leaf phenolic extracts: *in vitro* and *in vivo* study on Mandarin fruit during storage. *Journal of the Science of Food and Agriculture, 101*(10), 4229–4240. https://doi.org/10.1002/jsfa.11062

FAO [Food and Agriculture Organization]. (2024). *World food situation*. Available at https://www.fao.org/worldfoodsituation/csdb/en. Accessed 26 Aug 2024.

Fenta, L., Mekonnen, H., & Kabtimer, N. (2023). The exploitation of microbial antagonists against postharvest plant pathogens. *Microorganisms, 11*(4), 1044. https://doi.org/10.3390/microorganisms11041044

Fermaud, M., & le Menn, R. (1992). Transmission of *Botrytis cinerea* to grapes by grape berry moth larvae. *Phytopathology, 82*(12), 1393–1398.

Fernandez-San Millan, A., Larraya, L., Farran, I., Ancin, M., & Veramendi, J. (2021). Successful biocontrol of major postharvest and soil-borne plant pathogenic fungi by antagonistic yeasts. *Biological Control, 160*, 104683. https://doi.org/10.1016/j.biocontrol.2021.104683

Ferreira, F. V., Herrmann-Andrade, A. M., Calabrese, C. D., Bello, F., Vázquez, D., & Musumeci, M. A. (2020). Effectiveness of *trichoderma* strains isolated from the rhizosphere of citrus tree to control *Alternaria alternata*, *Colletotrichum gloeosporioides* and *Penicillium digitatum* A21 resistant to pyrimethanil in post-harvest oranges (*Citrus sinensis* L. (Osbeck)). *Journal of Applied Microbiology, 129*(3), 712–727. https://doi.org/10.1111/jam.14657

Fu, W., da Linge, C. S., Lawton, J. M., Gasic, K., Fu, W., da Linge, C. S., Lawton, J. M., & Gasic, K. (2022). Feasibility of genomic prediction for brown rot (*Monilinia* spp.) resistance in peach. *Fruit Research, 2*(1), 1–12. https://doi.org/10.48130/FRURES-2022-0002

Gao, C., Zhang, Y., Li, H., Gao, Q., Cheng, Y., Ogunyemi, S. O., & Guan, J. (2022). Fruit bagging reduces the postharvest decay and alters the diversity of fruit surface fungal community in 'yali' pear. *BMC Microbiology, 22*(1). https://doi.org/10.1186/s12866-022-02653-4

Garcia-Benitez, C., Melgarejo, P., & De Cal, A. (2017). Fruit maturity and post-harvest environmental conditions influence the pre-penetration stages of *Monilinia* infections in peaches. *International Journal of Food Microbiology, 241*, 117–122.

Garcia-Benitez, C., Casals, C., Usall, J., Sánchez-Ramos, I., Melgarejo, P., & De Cal, A. (2020). Impact of postharvest handling on preharvest latent infections caused by *Monilinia* spp. in nectarines. *Journal of Fungi, 6*(4), 266.

Gardenal, A. C., Leite, A. R. Z., Cardoso, F. A. R., Mello, J. C. P., Marques, L. L. M., & Perdoncini, M. R. F. G. (2024). Strawberries treated with biodegradable film containing plant extracts. *Brazilian Journal of Biology, 84*, e276874. https://doi.org/10.1590/1519-6984.276874

Garibaldi, A., Bertetti, D., Amatulli, M. T., Cardinale, J., & Gullino, M. L. (2012). First report of postharvest fruit rot in avocado (*Persea americana*) caused by *Lasiodiplodia theobromae* in Italy. *Plant Disease, 96*(3), 460–460.

Gatto, M. A., Ippolito, A., Linsalata, V., Cascarano, N. A., Nigro, F., Vanadia, S., & Di Venere, D. (2011). Activity of extracts from wild edible herbs against postharvest fungal diseases of fruit and vegetables. *Postharvest Biology and Technology, 61*(1), 72–82. https://doi.org/10.1016/j.postharvbio.2011.02.005

Gautam, A. K., Sharma, S., Avasthi, S., & Bhadauria, R. (2011). Diversity, pathogenicity and toxicology of *A. niger*: An important spoilage fungi. *Research Journal of Microbiology, 6*, 270–280. https://doi.org/10.3923/jm.2011.270.280

Ghanei Ghooshkhaneh, N., Golzarian, M. R., & Mamarabadi, M. (2018). Detection and classification of citrus green mold caused by *Penicillium digitatum* using multispectral imaging. *Journal of the Science of Food and Agriculture, 98*(9), 3542–3550.

González-Ruíz, A. V., Palomino-Hermosillo, Y. A., Balois-Morales, R., Ochoa-Jiménez, V. A., Casas-Junco, P. P., López-Guzmán, G. G., et al. (2021). Pathogenic fungi associated with soursop fruits (*Annona muricata* L.) during postharvest in Nayarit, Mexico. *Horticulturae, 7*(11), 471.

Gorai, P. S., Ghosh, R., Konra, S., & Mandal, N. C. (2021). Biological control of early blight disease of potato caused by Alternaria alternata EBP3 by an endophytic bacterial strain *Bacillus velezensis* SEB1. *Biological Control, 156*, 104551. https://doi.org/10.1016/j.biocontrol.2021.104551

Gu, K. X., Song, X. S., Xiao, X. M., Duan, X. X., Wang, J. X., Duan, Y. B., Hou, Y. P., & Zhou, M. G. (2019). A β2-tubulin dsRNA derived from *Fusarium asiaticum* confers plant resistance to multiple phytopathogens and reduces fungicide resistance. *Pesticide Biochemistry and Physiology, 153*, 36–46. https://doi.org/10.1016/J.PESTBP.2018.10.005

Güney, İ. G., Tekin, F., Günen, T. U., Özer, G., & Derviş, S. (2023). *Alternaria alternata* causing inner black rot of lemon (*Citrus limon*) fruits in Turkey: Genetic diversity and characterisation. *Physiological and Molecular Plant Pathology, 125*, 101998.

Guo, Y., Liu, L., Zhao, J., & Bi, Y. (2007). Use of silicon oxide and sodium silicate for controlling *Trichothecium roseum* postharvest rot in Chinese cantaloupe (*Cucumis melo* L.). *International Journal of Food Science & Technology, 42*(8), 1012–1018.

Guo, J., He, Z., Ma, C., Li, W., Wang, J., Lin, F., Liu, X., & Li, L. (2023a). Evaluation of cold plasma for decontamination of molds and mycotoxins in Rice Grain. *Food Chemistry, 402*, 134159. https://doi.org/10.1016/j.foodchem.2022.134159

Guo, M., Zhang, N., Wei, S., Yang, C., & Cui, C. (2023b). First report of *Lasiodiplodia pseudotheobromae* causing postharvest fruit rot on Indian jujube in China. *Plant Disease, 107*(10), 3307.

Gupta, N., & Jain, S. K. (2012). Storage behavior of mango as affected by post harvest application of plant extracts and storage conditions. *Journal of Food Science and Technology, 51*(10), 2499–2507. https://doi.org/10.1007/s13197-012-0774-0

Habiba, Noreen, R., Ali, S. A., Hasan, K. A., Sultana, V., Ara, J., & Ehteshamul-Haque, S. (2019). Evaluation of biocontrol potential of epiphytic yeast against postharvest penicillium digitatum rot of stored kinnow fruit (*Citrus reticulata*) and their effect on its physiochemical properties. *Postharvest Biology and Technology, 148*, 38–48. https://doi.org/10.1016/j.postharvbio.2018.10.007

Harman, G. (1983). Mechanisms of seed infection and pathogenesis. *Phytopathology, 73*, 326–328. https://doi.org/10.1094/Phyto-73-32

Harmon, G., & Pfleger, F. (1974). Pathogenicity and infection sites of *Aspergillus* species in stored seeds. *Phytopathology, 64*, 1339–1344. https://doi.org/10.1094/Phyto-64-1339

Harpaz, D., Duanis-Assaf, D., Alkan, N., & Eltzov, E. (2022). Detection of postharvest pathogenic fungi by RNA markers in high-throughput platform on streptavidin plate. *Postharvest Biology and Technology, 183*, 111728. https://doi.org/10.1016/J.POSTHARVBIO.2021.111728

Hayden, N. J., Maude, R. B., & Protor, F. J. (1994). Strategies for the control of black mold (*Aspegillus niger*) on stored tropical onion. *Acta Horticulture, 358*, 271–274.

He, Y., Degraeve, P., & Oulahal, N. (2024). Bioprotective yeasts: Potential to limit postharvest spoilage and to extend shelf life or improve microbial safety of Processed Foods. *Heliyon, 10*(3). https://doi.org/10.1016/j.heliyon.2024.e24929

Heidler von Heilborn, D., Reinmüller, J., Yurkov, A., Stehle, P., Moeller, R., & Lipski, A. (2023). Fungi under Modified Atmosphere—The Effects of CO2 Stress on Cell Membranes and Description of New Yeast *Stenotrophomyces fumitoleransgen.* nov., sp. nov. *Journal of Fungi, 9*(10), 1031.

Herrera-Romero, I., Ruales, C., Caviedes, M., & Leon-Reyes, A. (2017). Postharvest evaluation of natural coatings and antifungal agents to control *Botrytis cinerea* in Rosa sp. Phytoparasitica, 45(1), 9–20.

Hernández, A., Ruiz-Moyano, S., Galván, A. I., Merchán, A. V., Pérez Nevado, F., Aranda, E., Serradilla, M. J., de Córdoba, M., & Martín, A. (2021). Anti-fungal activity of phenolic sweet orange peel extract for controlling fungi responsible for post-harvest fruit decay. *Fungal Biology, 125*(2), 143–152. https://doi.org/10.1016/j.funbio.2020.05.005

Hocking, A. D. (2006). *Aspergillus* and related teleomorphs. In C. d. W. Blackburn (Ed.), *Technology and nutrition, food spoilage microorganisms* (Woodhead publishing series in food science) (pp. 451–487). Woodhead Publishing. https://doi.org/10.1533/9781845691417.4.451. ISBN 9781855739666.

Hong, J. P., Ro, N., Lee, H. Y., Kim, G. W., Kwon, J. K., Yamamoto, E., & Kang, B. C. (2020). Genomic selection for prediction of fruit-related traits in pepper (*Capsicum* spp.). *Frontiers in Plant Science, 11*, 570871. https://doi.org/10.3389/FPLS.2020.570871/BIBTEX

Hsieh, T. F., Shen, Y. M., Huang, J. H., Tsai, J. N., Lu, M. T., & Lin, C. P. (2023). Insights into grape ripe rot: A focus on the *Colletotrichum gloeosporioides* species complex and its management strategies. *Plants, 12*(15), 2873.

Huang, X., Ren, J., Li, P., Feng, S., Dong, P., & Ren, M. (2021). Potential of microbial endophytes to enhance the resistance to postharvest diseases of fruit and vegetables. *Journal of the Science of Food and Agriculture, 101*(5), 1744–1757. https://doi.org/10.1002/jsfa.10840

Huang, Y., Li, J., Shan, X., Wang, H., & Duan, Y. (2024). Bioactivities evaluation of an endophytic bacterial strain *Bacillus tequilensis* QNF2 inhibiting apple ring rot caused by *Botryosphaeria dothidea* on postharvest apple fruits. *Food Microbiology, 123*, 104590.

Huo, X., Tian, X., Liu, Z., Wang, L., Kong, Q., Wang, D., & Ren, X. (2024). Combination of LED blue light with peppermint essential oil emulsion for the postharvest storage of Shine Muscat grape to control *Aspergillus carbonarius*. *Postharvest Biology and Technology, 218*, 113175. https://doi.org/10.1016/j.postharvbio.2024.113175

Ingle, P. U., Rai, M., Golińska, P., & Gade, A. K. (2024). Phytomediated zinc oxide and sulfur nanoparticles for management of soft-rot causing pathogenic fungi in ginger. *Biocatalysis and Agricultural Biotechnology, 58*, 103229. https://doi.org/10.1016/j.bcab.2024.103229

Islam, T., Danishuddin, Tamane, N., Barai, H. R., & Haque, M. A. (2024). Resistance mechanisms of plant pathogenic fungi to fungicide, environmental impacts of fungicides, and Sustainable Solutions. *Plants, 13*(19), 2737. https://doi.org/10.3390/plants13192737

Jacobsen, B. J. (2014). Good agricultural and harvest practices to reduce mycotoxin contamination in wheat in temperate countries. In J. F. Leslie & A. F. Logrieco (Eds.), *Mycotoxin reduction in grain chains* (pp. 289–308).

Jafar, A., Bibi, N., Naqvi, R. A., Sadeghi-Niaraki, A., & Jeong, D. (2024). Revolutionizing agriculture with artificial intelligence: Plant disease detection methods, applications, and their limitations. *Frontiers in Plant Science, 15*, 1356260. https://doi.org/10.3389/FPLS.2024.1356260/BIBTEX

Janisiewicz, W., Takeda, F., Evans, B., & Camp, M. (2021). Potential of far ultraviolet (UV) 222 nm light for management of strawberry fungal pathogens. *Crop Protection., 150*, 105791. https://doi.org/10.1016/j.cropro.2021.105791

Jarvis, W. R. (1977). *Botrytinia and Botrytis species: Taxonomy, physiology, and pathogenicity, a guide to the literature* (Monograph No. 15). Canada Department of Agriculture.

Javadpour, S., Golestani, A., Rastegar, S., & Dastjer, M. M. (2018). Postharvest control of *Aspergillus niger* in mangos by means of essential oil. *Advances in Horticultural Science, 32*(3), 389–398.

Jeong, R. D., Shin, E. J., Chu, E. H., & Park, H. J. (2015). Effects of ionizing radiation on postharvest fungal pathogens. *Plant Pathology Journal, 31*(2), 176–180. https://doi.org/10.5423/PPJ.NT.03.2015.0040. Epub 2015 Jun 30. PMID: 26060436; PMCID: PMC4453998.

Jian, F., & Jayas, D. S. (2012). The ecosystem approach to grain storage. *Agricultural Research, 1*, 148–156. https://doi.org/10.1007/s40003-012-0017-7

Jin, P., Zheng, C., Huang, Y., Wang, X., Luo, Z., & Zheng, Y. (2016). Hot air treatment activates defense responses and induces resistance against *Botrytis cinerea* in strawberry fruit. *Journal of Integrative Agriculture, 15*(11), 2658–2665. https://doi.org/10.1016/s2095-3119(16)61387-4

Kadam, S. K., Ahirrao, P. P., Relekar, P. P., Khalangre, A. B., & Nigade, S. D. (2020). Hot Water Treatment for Fruits and Vegetables: A Review. *Trends in Biosciences, 13*(12), 773–781.

Kargwal, R., Garg, M. K., Singh, V., Garg, R., & Kumar, N. (2020). Principles of modified atmosphere packaging for shelf life extension of fruits and vegetables: An overview of storage conditions. *International Journal of Chemical Studies, 8*, 2245–2252. https://doi.org/10.22271/chemi.2020.v8.i3af.9545

Karoney, E. M., Molelekoa, T., Bill, M., Siyoum, N., & Korsten, L. (2024a). Global research network analysis of fresh produce postharvest technology: Innovative trends for loss reduction. *Postharvest Biology and Technology, 212*, 112642. https://doi.org/10.1016/j.postharvbio.2023.112642

Karoney, E., Molelekoa, T., Bill, M., Siyoum, N., & Korsten, L. (2024b). Global research network analysis of fresh produce postharvest technology: Innovative trends for loss reduction. *Postharvest Biology and Technology, 208*, 112642. https://doi.org/10.1016/j.postharvbio.2023.112642

Kefialew, Y., & Ayalew, A. (2008). Postharvest biological control of anthracnose (Colletotrichum gloeosporioides) on mango (Mangifera indica). *Postharvest Biology and Technology, 50*(1), 8–11.

Khadiri, M., Boubaker, H., Askarne, L., Ezrari, S., Radouane, N., Farhaoui, A., El Hamss, H., Tahiri, A., Barka, E. A., & Lahlali, R. (2023). Bacillus cereus B8W8 an effective bacterial antagonist against major postharvest fungal pathogens of Fruit. *Postharvest Biology and Technology, 200*, 112315. https://doi.org/10.1016/j.postharvbio.2023.112315

Khaliq, G., Abbas, H. T., Ali, I., & Waseem, M. (2019). *Aloe vera* gel enriched with garlic essential oil effectively controls anthracnose disease and maintains postharvest quality of banana fruit during storage. *Horticulture, Environment, and Biotechnology, 60*, 659–669. https://doi.org/10.1007/s13580-019-00159-z

Khetabi, A. E., Lahlali, R., Ezrari, S., Radouane, N., Lyousfi, N., Banani, H., Askarne, L., Tahiri, A., Ghadraoui, L. E., Belmalha, S., & Barka, E. A. (2022). Role of plant extracts and essential oils in fighting against postharvest fruit pathogens and extending fruit shelf life: A review. *Trends in Food Science and Technology, 120*, 402–417. https://doi.org/10.1016/j.tifs.2022.01.009

Khuna, S., Kumla, J., & Suwannarach, N. (2024). First report of post-harvest fruit rot on watermelon caused by *Lasiodiplodia theobromae* in Northern Thailand. *Canadian Journal of Plant Pathology, 46*(3), 265–274.

Khwantongyim, P., Wansee, S., Lu, X., Zhang, W., & Sun, G. (2021). Variations in the community structure of fungal microbiota associated with Apple fruit shaped by fruit bagging-based practice. *Journal of Fungi, 7*(9), 764. https://doi.org/10.3390/jof7090764

Kim, C. H., & Park, K. S. (1988). A predictive model of disease progression of red-pepper anthracnose. *Korean Journal of Pathology, 4*, 325–331.

Kirk, W. W., Gachango, E., Schafer, R., & Wharton, P. S. (2013). Effects of in-season crop-protection combined with postharvest applied fungicide on suppression of potato storage diseases caused by *Fusarium* pathogens. *Crop Protection, 51*, 77–84. https://doi.org/10.1016/j.cropro.2013.04.009

Kis-Papo, T., Oren, A., Wasser, S. P., & Nevo, E. (2003). Survival of filamentous fungi in hypersaline Dead Sea water. *FEMS Microbiology Ecology, 45*(2), 183–190.

Kitinoja, L., Odeyemi, O., Dubey, N., Musanase, S., & Gill, G. S. (2019). Commodity system assessment studies on the postharvest handling and marketing of tomatoes in Nigeria, Rwanda, and Maharashtra, India. *Journal of Horticultural Postharvest Research, 2*, 1–14. https://doi.org/10.22077/JHPR.2019.2060.1040

Kobiler, I., Akerman, M., Huberman, L., & Prusky, D. (2011). Integration of pre-and postharvest treatments for the control of black spot caused by *Alternaria alternata* in stored persimmon fruit. *Postharvest Biology and Technology, 59*(2), 166–171.

Köhl, J., Wenneker, M., Groenenboom-de Haas, B. H., Anbergen, R., Goossen-van de Geijn, H. M., Lombaers-van der Plas, C. H., et al. (2018). Dynamics of post-harvest pathogens *Neofabraea* spp. and *Cadophora* spp. in plant residues in Dutch apple and pear orchards. *Plant Pathology, 67*(6), 1264–1277.

Kotan, R., Dikbas, N., & Bostan, H. (2009). Biological control of post harvest disease caused by *Aspergillus flavus* on stored lemon fruits. *African Journal of Biotechnology, 8*(2), 209–214.

Köycü, N. D., & Özer, N. (1997). Determination of seedborne fungi in onion and their transmission to onion sets. *Phytoparasitica, 25*(1), 25.

Kozhar, O., & Peever, T. L. (2018). How does *Botrytis cinerea* infect red raspberry? *Phytopathology, 108*(11), 1287–1298.

Kushaha, T. M., Mkindi, A. G., Mbega, E. R., Stevenson, P. C., & Belmain, S. R. (2024). Botanical extracts control the fungal pathogen *Colletotrichum boninense* in smallholder production of Common Bean. *Phytopathology Research, 6*(1). https://doi.org/10.1186/s42483-024-00235-y

Kushiro, M. (2015). Historical review of researches on yellow rice and mycotoxigenic fungi adherent to rice in Japan. *Mycotoxins, 65*(1), 19–23. https://doi.org/10.2520/myco.65.19

Kusstatscher, P., Zachow, C., Harms, K., Maier, J., Eigner, H., Berg, G., & Cernava, T. (2019). Microbiome-driven identification of microbial indicators for postharvest diseases of sugar beets. *Microbiome, 7*(1), 1–12. https://doi.org/10.1186/S40168-019-0728-0/TABLES/1

Lakshmi, B. K. M., Reddy, P. N., & Prasad, R. D. (2011). Cross-infection potential of *Colletotrichum gloeosporioides* Penz. isolates causing anthracnose in subtropical fruit crops. *Tropical Agricultural Research, 22*, 183–193.

Lamine Senghor, A., Liang, W. J., & Ho, W. C. (2007). Integrated control of*colletotrichum gloeosporioides*on mango fruit in Taiwan by the combination of*bacillussubtilis*and fruit bagging. *Biocontrol Science and Technology, 17*(8), 865–870. https://doi.org/10.1080/09583150701527409

Latunde-Dada, A. O., O'Connell, R. J., Nash, C., Pring, R. J., Lucas, J. A., & Bailey, J. A. (1996). Infection process and identity of the hemibiotrophic anthracnose fungus (*Colletotrichum destructivum*) from cowpea (*Vigna unguiculata*). *Mycological Research, 100*(10), 1133–1141. https://doi.org/10.1016/S0953-7562(96)80050-8

Lee, J. H., Kwon, Y. H., Kwack, Y. B., & Kwak, Y. S. (2015). Report of postharvest rot of kiwifruit in Korea caused by *Sclerotinia sclerotiorum. International Journal of Food Microbiology, 206*, 81–83.

Lewis, L., Onsongo, M., Njapau, H., Schurz-Rogers, H., Luber, G., Kieszak, S., Nyamongo, J., Backer, L., Dahiye, A. M., & Misore, A. (2005). Aflatoxin contamination of commercial maize products during an outbreak of acute aflatoxicosis in eastern and Central Kenya. *Environmental Health Perspective, 113*(12), 1763.

Li, H. N., Jiang, J. J., Hong, N., Wang, G. P., & Xu, W. X. (2013). First report of *Colletotrichum fructicola* causing bitter rot of pear (*Pyrus bretschneideri*) in China. *Plant Disease, 97*(7), 1000–1000.

Li, J., Gaskins, V. L., Yan, H. J., Luo, Y. G., & Jurick, W. M., II. (2014). First report of *Mucor* rot on stored 'Gala' apple fruit caused by *Mucor piriformis* in Pennsylvania. *Plant Disease, 98*(8), 1157–1157.

Li, L., Pan, H., Liu, W., Chen, M. Y., & Zhong, C. H. (2017). First report of *Alternaria alternata* causing postharvest rot of kiwifruit in China. *Plant Disease, 101*(6), 1046.

Li, H., Yu, S., Tang, W., Miao, M., & Liu, Y. (2019). First report of *Diaporthe passiflorae* and *Diaporthe nobilis* causing a postharvest kiwifruit rot in Sichuan province, China. *Plant Disease, 103*(4), 771–771.

Li, Y., Xia, M., He, P., Yang, Q., Wu, Y., He, P., Ahmed, A., Li, X., Wang, Y., Munir, S., & He, Y. (2022). Developing *Penicillium digitatum* management strategies on post-harvest citrus fruits with metabolic components and colonization of bacillus subtilis L1-21. *Journal of Fungi, 8*(1), 80. https://doi.org/10.3390/jof8010080

Li, P., Liu, J., Li, Y., Yang, T., Ge, X., & Tian, P. (2024a). Natural compound berberine inhibits apple ring rot fungus *Botryosphaeria dothidea. International Journal of Pest Management, 70*(4), 961–970.

Li, S., Hu, J., Ning, S., Li, W., Jiang, R., Huang, J., & Li, Y. (2024b). *Bacillus velezensis* HY19 as a sustainable preservative in post-harvest citrus (*Citrus reticulata* Blanco L.) fruit management. *Food Control, 155*, 110068. https://doi.org/10.1016/j.foodcont.2023.110068

Lima, J. R., Gondim, D. M. F., Oliveira, J. T. A., Oliveira, F. S. A., Gonçalves, L. R. B., & Viana, F. M. P. (2013). Use of killer yeast in the management of postharvest papaya anthracnose. *Postharvest Biology and Technology, 83*, 58–64. https://doi.org/10.1016/j.postharvbio.2013.03.014

Lin, S. H., Wang, J. H., & Wu, A. B. (2016). A post-harvest pink rot of pepper caused by *Trichothecium roseum* in China. *Plant Disease, 100*(10), 2164–2164.

Liu, Q., Chen, Q., Liu, H., Du, Y., Jiao, W., Sun, F., & Fu, M. (2024). *Rhizopus stolonifer* and related control strategies in postharvest fruit: A review. *Heliyon, 10*(8).

Liu, H., Pang, L., Lu, X., Wang, R. C., & Zhou, Q. (2020a). First report of *Phomopsis longicolla* associated with postharvest fruit rot of kiwifruit in China. *Plant Disease, 104*(2), 579.

Liu, Y., Galani Yamdeu, J. H., Gong, Y. Y., & Orfila, C. (2020b). A review of postharvest approaches to reduce fungal and mycotoxin contamination of foods. *Comprehensive Reviews in Food Science and Food Safety, 19*(4), 1521–1560. https://doi.org/10.1111/1541-4337.12562

Liu, J., Guo, X., Zhang, H., Cao, Y., & Sun, Q. (2021). First report of postharvest fruit rot disease of hardy kiwifruit caused by *Diaporthe eres* in China. *Plant Disease, 105*(11), 3745.

Liu, B., Xue, M., Zhou, J., Zhang, H., Ren, L., & Fan, J. (2022). Hot air treatment elicits disease resistance against *Colletotrichum gloeosporioides* and improves the quality of papaya by Metabolomic profiling. *BioMed Research International, 2022*, 5162845. https://doi.org/10.1155/2022/5162845. Retraction in: Biomed Res Int. 2023;2023:9862038. 10.1155/2023/9862038. PMID: 35968234; PMCID: PMC9371874.

Lopez, M. B., Spadaro, D., & Gullino, M. L. (2015). First report of *Sclerotinia sclerotiorum* causing postharvest *Sclerotinia* rot on highbush blueberry in Europe. *Plant Disease, 99*(11), 1648–1648.

López, S. N., Sangorrín, M. P., & Pildain, M. B. (2016). Fruit rot of sweet cherries and raspberries caused by *Penicillium crustosum* and *Mucor piriformis* in South Patagonia, Argentina. *Canadian Journal of Plant Pathology, 38*(4), 511–516.

López-Velásquez, J. G., Barraza-López, F. J., Vega-García, M. O., López-López, M. E., Gutiérrez-Dorado, R., Chaidez-Gastelum, D. C., & Ayón-Reyna, L. E. (2024). Microstructural and physicochemical quality maintenance in green bell pepper infected with *Botrytis cinerea* and treated with thyme essential oil combined with carnauba wax. *Journal of Food Science, 89*, 2943–2955. https://doi.org/10.1111/1750-3841.17041

López-Velázquez, J. G., Delgado-Vargas, F., Ayón-Reyna, L. E., López-Angulo, G., Bautista-Baños, S., Uriarte-Gastelum, Y. G., López-López, M. E., & Vega-García, M. O. (2021). Postharvest application of partitioned plant extracts from Sinaloa, Mexico for controlling papaya pathogenic fungus *Colletotrichum gloeosporioides*. *Journal of Plant Pathology, 103*(3), 831–842. https://doi.org/10.1007/s42161-021-00838-w

Luciano-Rosario, D., Keller, N. P., & Jurick, W. M. (2020). *Penicillium expansum*: Biology, omics, and management tools for a global postharvest pathogen causing blue mould of pome fruit. *Molecular Plant Pathology, 21*(11), 1391–1404.

Lurie, S., & Pedreschi, R. (2014). Fundamental aspects of postharvest heat treatments. *Horticulture Research, 1*, 14030. https://doi.org/10.1038/hortres.2014.30

Lyousfi, N., Letrib, C., Legrifi, I., Blenzar, A., El Khetabi, A., El Hamss, H., Belabess, Z., Barka, E. A., & Lahlali, R. (2022). Combination of sodium bicarbonate (SBC) with bacterial antagonists for the control of brown rot disease of Fruit. *Journal of Fungi, 8*(6), 636. https://doi.org/10.3390/jof8060636

Magan, N., & Lacey, J. (1984). Effects of gas composition and water activity on growth of field and storage fungi and their interactions. *Transactions of the British Mycological Society, 82*(2), 305–314. https://doi.org/10.1016/S0007-1536(84)80074-1

Magan, N., Sanchis, V., & Aldred, D. (2004). Role of spoilage fungi in seed deterioration. In *Fungal biotechnology in agricultural, food and environmental applications* (pp. 311–323). Marcel Dekker.

Magnoli, C., Hallak, C., Astoreca, A., Ponsone, L., Chiacchiera, S., & Dalcero, A. M. (2006). Occurrence of ochratoxin A producing fungi in commercial corn kernels in Argentina. *Mycopathologia, 161*(1), 53–58. https://doi.org/10.1007/s11046-005-0237-5

Mahfooz, S., Singh, S. P., Mishra, N., & Mishra, A. (2017). A comparison of microsatellites in phytopathogenic *Aspergillus* species in order to develop markers for the assessment of genetic diversity among its isolates. *Frontiers in Microbiology, 8*, 1774. https://doi.org/10.3389/fmicb.2017.01774

Mahmoud, G. A. E., & Gaber, S. H. (2024). Nanotechnology applications in postharvest disease management. In N. Benkeblia (Ed.), *Recent advances in postharvest technologies* (Vol. 2). Springer. https://doi.org/10.1007/978-3-031-65816-7_8

Maj, W., Pertile, G., Różalska, S., Skic, K., & Frąc, M. (2024). Comprehensive antifungal investigation of natural plant extracts against *Neosartorya* spp. (*aspergillus* spp.) of agriculturally significant microbiological contaminants and shaping their metabolic profile. *Scientific Reports, 14*(1). https://doi.org/10.1038/s41598-024-58791-4

Mandap, J. A. L., Hellmich, R. L., Busman, M., Maier, D. E., & Munkvold, G. P. (2024). Protection from stored grain insects using transgenic maize hybrids and implications for *Aspergillus flavus* and aflatoxin contamination. *Journal of Stored Products Research, 106*, 102258.

Mao, Y., Botella, J. R., Liu, Y., & Zhu, J. K. (2019). Gene editing in plants: Progress and challenges. *National Science Review, 6*(3), 421. https://doi.org/10.1093/NSR/NWZ005

Mao, X., Chen, A., Qu, J., Luo, P., You, Y., Gao, Y., et al. (2023). New insights into in mycotoxins production in *Alternaria* infected apple during postharvest storage. *Postharvest Biology and Technology, 198*, 112238.

Martín, I., Gálvez, L., Guasch, L., & Palmero, D. (2022). Fungal pathogens and seed storage in the dry state. *Plants, 11*(22), 3167

Martini, C., & Mari, M. (2014). *Monilinia fructicola, Monilinia laxa* (*Monilinia* rot, brown rot). In *Postharvest decay* (pp. 233–265). Academic.

Matrose, N. A., Obikeze, K., Belay, Z. A., & Caleb, O. J. (2021). Plant extracts and other natural compounds as alternatives for post-harvest management of Fruit Fungal Pathogens: A review. *Food Bioscience, 41*, 100840. https://doi.org/10.1016/j.fbio.2020.100840

May-De Mio, L. L., Negri, G., & Michailides, T. J. (2014). Effect of *Trichothecium roseum*, lime sulphur and phosphites to control blossom blight and brown rot on peach. *Canadian Journal of Plant Pathology, 36*(4), 428–437.

Meronuck, R. (1987). The significance of fungi in cereal grains. *Plant Disease, 71*, 287–292.

Mertely, J. C., Chandler, C. K., Xiao, C. L., & Legard, D. E. (2000). Comparison of sanitation and fungicides for management of *Botrytis* fruit rot of strawberry. *Plant Disease, 84*(11), 1197–1202.

Miller, J. D. (1995). Fungi and mycotoxins in grains: Implication for stored products research. *The Journal of Stored Products Research, 31*(1), 1–16.

Miller, J. S., Hamm, P. B., Olsen, N., Geary, B. D., & Johnson, D. A. (2011). Effect of post-harvest fungicides and disinfestants on the suppression of silver scurf on potatoes in storage. *American Journal of Potato Research, 88*(5), 413–423. https://doi.org/10.1007/s12230-011-9207-2

Mills, J. T., Seifert, K. A., Frisvad, J. C., & Abramson, D. (1995). Nephrotoxigenic *Penicillium* species occurring on farm-stored cereal grains in western Canada. *Mycopathologia, 130*(1), 23–28. https://doi.org/10.1007/BF01104345. PMID: 7666923.

Mincuzzi, A., Sanzani, S. M., Palou, L., Ragni, M., & Ippolito, A. (2022). Postharvest rot of pomegranate fruit in southern Italy: Characterization of the main pathogens. *Journal of Fungi, 8*(5), 475.

Miranda, M., Bai, J., Pilon, L., Torres, R., Casals, C., Solsona, C., & Teixidó, N. (2024). Fundamentals of edible coatings and combination with biocontrol agents: A strategy to improve postharvest fruit preservation. *Food, 13*(18), 2980. https://doi.org/10.3390/foods13182980

Mohamed, N. T., Ding, P., Kadir, J., & Ghazali, H. (2017). Potential of UVC germicidal irradiation in suppressing crown rot disease, retaining postharvest quality and antioxidant capacity of *Musa* AAA Berangan during fruit ripening. *Food Sci Nutr., 5*(5), 967–980.

Mohapatra, D., Kumar, S., Kotwaliwale, N., & Singh, K. K. (2017). Critical factors responsible for fungi growth in stored food grains and non-chemical approaches for their control. *Industrial Crops and Products, 108*, 162–182. https://doi.org/10.1016/j.indcrop.2017.06.039

Moradinezhad, F., & Ranjbar, A. (2023). Advances in postharvest diseases management of fruits and vegetables: A review. *Horticulturae, 9*(1099). https://doi.org/10.3390/horticulturae9101099

Munkvold, G. P., Arias, S., Taschl, I., & Gruber-Dorninger, C. (2019). Chapter 9: Mycotoxins in corn: Occurrence, impacts, and management. In S. O. Serna-Saldivar (Ed.), *Corn* (3rd ed., pp. 235–287). AACC International Press. https://doi.org/10.1016/B978-0-12-811971-6.00009-7. ISBN 9780128119716.

Muñoz, M., Behnke, L. E., Bridges, W. C., Schnabel, G., & Faust, J. E. (2025). Postharvest calcium chloride dips. An effective strategy to reduce botrytis blight severity and increase petal strength in cut roses. *Postharvest Biology and Technology, 219*, 113292. https://doi.org/10.1016/j.postharvbio.2024.113292

Musa, S. K., Habish, H. A., Abdalla, A. A., & Adlan, A. B. (1973). Problems of onion storage in Sudan. *Tropical Science, 15*, 319–327.

Nally, M. C., Pesce, V. M., Maturano, Y. P., Muñoz, C. J., Combina, M., Toro, M. E., de Figueroa, L. I. C., & Vazquez, F. (2012). Biocontrol of *Botrytis cinerea* in table grapes by non-pathogenic indigenous saccharomyces cerevisiae yeasts isolated from viticultural environments in Argentina. *Postharvest Biology and Technology, 64*(1), 40–48. https://doi.org/10.1016/j.postharvbio.2011.09.009

Navarro, B. L., Edwards Molina, J. P., & Nogueira Júnior, A. F. (2022). Penetration by Botryosphaeriaceae species in avocado, guava and persimmon fruit during postharvest. *Journal of Phytopathology, 170*(1), 57–68.

Nazzaro, F., Fratianni, F., Coppola, R., & De Feo, V. (2017). Essential oils and antifungal activity. *Pharmaceuticals, 10*(4), 86. https://doi.org/10.3390/ph10040086

Neergaard, P. (1977). *Seed pathology* (Vol. 1 and 2, p. 839). The MacMillan Press.

Neik, T. X., Amas, J., Barbetti, M., Edwards, D., & Batley, J. (2020). Understanding host–pathogen interactions in *Brassica napus* in the omics era. *Plants, 9*(10), 1336. https://doi.org/10.3390/PLANTS9101336

Neri, F., Cappellin, L., Spadoni, A., Cameldi, I., Alarcon, A. A., Aprea, E., Romano, A., Gasperi, F., & Biasioli, F. (2014). Role of strawberry volatile organic compounds in the development of *Botrytis cinerea* infection. *Plant Pathology, 64*(4), 709–717. https://doi.org/10.1111/ppa.12287

Nishimwe, K., Mandap, J. A. L., & Munkvold, G. M. (2020). Advances in understanding fungal contamination in cereals. In D. E. Maier (Ed.), *Advances in postharvest management of cereals and grains* (478 pp). Burleigh Dodds Science Publishing.

O'Connell, R. J., Thon, M. R., Hacquard, S., Amyotte, S. G., Kleemann, J., Torres, M. F., Damm, U., Buiate, E., Epstein, L., Alkan, N., & Others. (2012). Lifestyle transitions in plant pathogenic *Colletotrichum* fungi deciphered by genome and transcriptome analyses. *Nature Genetics, 44*(10), 1060–1065. https://doi.org/10.1038/ng.2372

Odelade, K. A., & Oladeji, O. S. (2020). Isolation of phytopathogenic fungi associated with the post-harvest deterioration of watermelon fruits. *Scientific African, 8*, e00366.

Olmedo, G. M., Zhang, J., Zhao, W., Mattia, M., Rosskopf, E. N., Ritenour, M., et al. (2023). Application of Thymol vapors to control postharvest decay caused by *Penicillium digitatum* and *Lasiodiplodia theobromae* in grapefruit. *Food, 12*(19), 3637.

Ostry, V. (2008). *Alternaria* mycotoxins: An overview of chemical characterization, producers, toxicity, analysis, and occurrence in foodstuffs. *World Mycotoxin Journal, 1*(2), 175–188.

Ou, C., Liu, Y., Wang, W., & Dong, D. (2016). Integration of UV-C with antagonistic yeast treatment for controlling post-harvest disease and maintaining fruit quality of ananas comosus. *BioControl, 61*(5), 591–603. https://doi.org/10.1007/s10526-016-9740-5

Oztekin, S., Dikmetas, D. N., Devecioglu, D., Acar, E. G., & Karbancioglu-Guler, F. (2023). Recent insights into the use of antagonistic yeasts for sustainable biomanagement of postharvest pathogenic and mycotoxigenic fungi in fruits with their prevention strategies against Mycotoxins. *Journal of Agricultural and Food Chemistry, 71*(26), 9923–9950. https://doi.org/10.1021/acs.jafc.3c00315

Pakdeevaraporn, P., Wasee, S., Taylor, P. W. J., & Mongkolporn, O. (2005). Inheritance of resistance to anthracnose caused by *Colletotrichum capsici* in *Capsicum*. *Plant Breeding, 124*(2), 206–208. https://doi.org/10.1111/j.1439-0523.2004.01050.x

Palou, L. (2014). *Penicillium digitatum, Penicillium italicum* (green mold, blue mold). In Postharvest decay (pp. 45–102). Academic Press.

Palou, L., Montesinos-Herrero, C., Besada, C., & Taberner, V. (2013). Postharvest Fruit rot of Persimmon (*Diospyros kaki*) in S pain Caused by *Lasiodiplodia theobromae* and *Neofusicoccum* spp. *Journal of Phytopathology, 161*(9), 625–631.

Pangallo, S., Li Destri Nicosia, M. G., Scibetta, S., Strano, M. C., Cacciola, S. O., Belgacem, I., Agosteo, G. E., & Schena, L. (2021). Preharvest and postharvest applications of a pomegranate peel extract to control citrus fruit decay during storage and shelf life. *Plant Disease, 105*(4), 1013–1018. https://doi.org/10.1094/pdis-01-20-0178-re

Papoutsis, K., Mathioudakis, M. M., Hasperué, J. H., & Ziogas, V. (2019). Non-chemical treatments for preventing the postharvest fungal rotting of citrus caused by *Penicillium digitatum* (green mold) and *Penicillium italicum* (blue mold). *Trends in Food Science & Technology, 86*, 479–491.

Park, J. W., Choi, S. Y., Hwang, H. J., & Kim, Y. B. (2005). Fungal mycoflora and mycotoxins in Korean polished rice destined for humans. *The International Journal of Food Microbiology, 103*, 305–314. https://doi.org/10.1016/j.ijfoodmicro.2005.02.001

Paull, R. E., Nishijima, W., Reyes, M., & Cavaletto, C. (1997). Postharvest handling and losses during marketing of papaya (*Carica papaya*). *Postharvest Biology and Technology, 11*(3), 165–179. https://doi.org/10.1016/S0925-5214(97)00052-0

Pavlath, A. E., & Orts, W. (2009). Edible films and coatings: Why, what, and how. In K. C. Huber & M. E. Embuscado (Eds.), *Edible films and coatings for food applications* (pp. 1–23). Springer.

Pazolini, K., dos Santos, I., Giaretta, R. D., Marcondes, M. M., Reiner, D. A., & Citadin, I. (2016). The use of brassica extracts and thermotherapy for the postharvest control of brown rot in peach. *Scientia Horticulturae, 209*, 41–46. https://doi.org/10.1016/j.scienta.2016.06.008

Peralta-Ruiz, Y., Rossi, C., Grande-Tovar, C. D., & Chaves-López, C. (2023). Green management of postharvest anthracnose caused by *Colletotrichum gloeosporioides*. *Journal of Fungi, 9*(6), 623.

Pérez-Vázquez, M. A., Pacheco-Hernández, Y., Lozoya-Gloria, E., Mosso-González, C., Ramírez-García, S. A., Romero-Arenas, O., & Villa-Ruano, N. (2021). Peppermint essential oil and its major volatiles as protective agents against soft rot caused by *fusarium sambucinum* in Cera Pepper (*capsicum pubescens*). *Chemistry and Biodiversity, 19*(1). https://doi.org/10.1002/cbdv.202100835

Pesce, V. M., Nally, M. C., Carrizo, G. P., Rojo, C., Pérez, B. A., Toro, M. E., Castellanos de Figueroa, L. I., & Vazquez, F. (2018). Antifungal activity of native yeasts from different microenvironments against *Colletotrichum gloeosporioides* on ripe olive fruits. *Biological Control, 120*, 43–51. https://doi.org/10.1016/j.biocontrol.2017.03.005

Petrasch, S., Knapp, S. J., Van Kan, J. A., & Blanco-Ulate, B. (2019). Grey mould of strawberry, a devastating disease caused by the ubiquitous necrotrophic fungal pathogen *Botrytis cinerea*. *Molecular Plant Pathology, 20*(6), 877–892.

Petrasch, S., Mesquida-Pesci, S. D., Pincot, D. D. A., Feldmann, M. J., López, C. M., Famula, R., Hardigan, M. A., Cole, G. S., Knapp, S. J., & Blanco-Ulate, B. (2021). Genomic prediction of strawberry resistance to postharvest fruit decay caused by the fungal pathogen *Botrytis cinerea*. *G3: Genes|Genomes|Genetics, 12*(1), jkab378. https://doi.org/10.1093/G3JOURNAL/JKAB378

Phannareth, T., Nunziata, S. O., Stulberg, M. J., Galvez, M. E., & Rivera, Y. (2021). Comparison of nanopore sequencing protocols and real-time analysis for Phytopathogen diagnostics. *Plant Health Progress, 22*(1), 31–36. https://doi.org/10.1094/PHP-02-20-0013-RS/ASSET/IMAGES/LARGE/PHP-02-20-0013-RSF4-1616092604040.JPEG

Piña-Barrera, A. M., Álvarez-Román, R., Báez-González, J. G., Amaya-Guerra, C. A., Rivas-Morales, C., Gallardo-Rivera, C. T., & Galindo-Rodríguez, S. A. (2019). Application of a multisystem coating based on polymeric Nanocapsules containing essential oil of *Thymus vulgaris* L. to increase the shelf life of table grapes (*Vitis vinifera* L.). *IEEE Transactions on Nanobioscience, 18*(4), 549–557. https://doi.org/10.1109/TNB.2019.2941931

Piña-Barrera, A. M., Pérez, M. S. R., Román, R. A., Gonzáles, J. G. B., Guerra, C. A. A., & Rodríguez, S. A. G. (2022). Edible alginate-based coating in combination with nanoencapsulated eugenol and its preservative effect on the shelf life of tomato (*Solanum lycopersicum*). *Characterization and Application of Nanomaterials, 5*(2), 47–57. https://doi.org/10.24294/can.v5i2.1686

Piri, I., Babayan, M., Tavassoli, A., & Javaheri, M. (2011). The use of gamma irradiation in agriculture. *African Journal of Microbiology Research, 5*, 5806–5811.

Pitt, J. I., & Hocking, A. D. (2009). *Aspergillus* and related teleomorphs. In J. I. Pitt & A. D. Hocking (Eds.), *Fungi and food spoilage* (3rd ed., 520pp). Springer.

Popescu, P.-A., Palade, L. M., Nicolae, I.-C., Popa, E. E., Miteluţ, A. C., Drăghici, M. C., Matei, F., & Popa, M. E. (2022). Chitosan-based edible coatings containing essential oils to preserve the shelf life and postharvest quality parameters of organic strawberries and apples during cold storage. *Food, 11*(21), 3317. https://doi.org/10.3390/foods11213317

Poulaki, E. G., & Tjamos, S. E. (2023). *Bacillus* species: Factories of plant protective volatile organic compounds. *Journal of Applied Microbiology, 134*(3). https://doi.org/10.1093/jambio/lxad037

Pring, R. J., Nash, C., Zakaria, M., & Bailey, J. A. (1995). Infection process and host range of *Colletotrichum capsici. Physiological and Molecular Plant Pathology, 46*(3), 137–152.

Prusky, D. (1996). Pathogen quiescence in postharvest diseases. *Annual Review of Phytopathology, 34*, 413–434.

Prusky, D. (2011). Reduction of the incidence of postharvest quality losses, and future prospects. *Food Security, 3*(4), 463–474. https://doi.org/10.1007/s12571-011-0147-y

Prusky, D., Alkan, N., Mengiste, T., & Fluhr, R. (2013). Quiescent and necrotrophic lifestyle choice during postharvest disease development. *Annual Review of Phytopathology, 51*, 155–176.

Prusky, D., Barad, S., Ment, D., & Bi, F. (2016). The pH modulation by fungal secreted molecules: A mechanism affecting pathogenicity by postharvest pathogens. *Israel Journal of Plant Sciences, 63*(1), 22–30.

Pusztahelyi, T., Holb, I. J., & Pócsi, I. (2015). Secondary metabolites in fungus-plant interactions. *Frontiers in Plant Science, 6*, 573. https://doi.org/10.3389/fpls.2015.00573

Qadri, S. M. H., Srivastava, K. J., Bhonde, S. R., Pandey, U. B., & Bhagchandani, P. M. (1982). Fungicidal bioassay against certain important pathogens of onion (*Allium cepa* L.). *Pesticides, 16*, 11–16.

Qi, Z., Tian, L., Zhang, H., Lei, Y., & Tang, F. (2023). Fungal community analysis of hot spots in bulk maize under different storage conditions. *LWT, 182*, 114819. https://doi.org/10.1016/j.lwt.2023.114819. ISSN 0023-6438.

Qi, C., Zhang, H., Chen, W., & Liu, W. (2024). Curcumin: An innovative approach for postharvest control of *Alternaria alternata* induced black rot in cherry tomatoes. *Fungal Biology, 128*(2), 1691–1697.

Rafiq, R., & Fatima, H. (2023). Morphological characterization of Sclerotinia sclerotiorum causing fruit rot of papaya and its management using biopesticides. *Journal of Agriculture and Biology, 1*(2), 47–52.

Rajapaksha, L., Gunathilake, D. M. C. C., Pathirana, S. M., et al. (2021). Reducing post-harvest losses in fruits and vegetables for ensuring food security – Case of Sri Lanka. *MOJ Food Process Technols, 9*(1), 7–16. https://doi.org/10.15406/mojfpt.2021.09.00255

Reddy, K. R. N., Reddy, C. S., Abbas, H. K., Abel, C. A., & Muralidharan, K. (2008). Mycotoxigenic fungi, mycotoxins, and management of rice grains. *Toxin Reviews, 27*(3–4), 287–317. https://doi.org/10.1080/15569540802432308

Renganathan, P., & Muthukumar, A. (2012). Evaluation of fungicides against *Colletotrichum musae* and *Botryodiplodia theobromae*, the cause of banana crown rot. *Indian Journal of Plant Protection, 40*(3), 240–242.

Rguez, S., Djébali, N., Ben Slimene, I., Abid, G., Hammemi, M., Chenenaoui, S., Bachkouel, S., Daami-Remadi, M., Ksouri, R., & Hamrouni-Sellami, I. (2018). *Cupressus sempervirens* essential oils and their major compounds successfully control postharvest grey mould disease of tomato. *Industrial Crops and Products, 123*, 135–141. https://doi.org/10.1016/j.indcrop.2018.06.060

Rizwana, H., Bokahri, N. A., Alsahli, S. A., Al Showiman, A. S., Alzahrani, R. M., & Aldehaish, H. A. (2021). Postharvest disease management of *Alternaria* spots on tomato fruit by *Annona muricata* fruit extracts. *Saudi Journal of Biological Sciences, 28*(4), 2236–2244.

Roberts, R. G., & Snow, J. P. (1984). Histopathology of cotton boll rot caused by *Colletotrichum capsici. Phytopathology, 74*(4), 390–397.

Roberts, R. G., & Wu, H. C. (1981). *Penicillium* mold on stored corn. *Cereal Chemistry, 58*(4), 357–361.

Rongai, D., Pulcini, P., Pesce, B., & Milano, F. (2015). Antifungal activity of some botanical extracts on *Fusarium oxysporum. Open Life Sciences, 10*(1). https://doi.org/10.1515/biol-2015-0040

Ropero-Pérez, C., Marcos, J. F., Manzanares, P., & Garrigues, S. (2024). Increasing the efficiency of CRISPR/Cas9-mediated genome editing in the citrus postharvest pathogen *Penicillium digitatum. Fungal Biology and Biotechnology, 11*(1), 1–11. https://doi.org/10.1186/S40694-024-00179-0/FIGURES/3

Ruan, R., Huang, K., Luo, H., Zhang, C., Xi, D., Pei, J., & Liu, H. (2023). Occurrence and characterization of *Sclerotinia sclerotiorum* causing fruit rot on sweet cherry in Southern China. *Plants, 12*(24), 4165.

Sadek, M. E., Shabana, Y. M., Sayed-Ahmed, K., & Abou Tabl, A. H. (2022). Antifungal activities of sulfur and copper nanoparticles against cucumber postharvest diseases caused by *Botrytis cinerea* and *Sclerotinia sclerotiorum*. *Journal of Fungi, 8*(4), 412. https://doi.org/10.3390/jof8040412

Saini, R., Saini, S., & Sharma, S. (2010). Nanotechnology: The future medicine. *Journal of Cutaneous and Aesthetic Surgery, 3*(1), 32–33. https://doi.org/10.4103/0974-2077.63301. PMID: 20606992; PMCID: PMC2890134.

Saito, S., Michailides, T. J., & Xiao, C. L. (2016). *Mucor* rot—An emerging postharvest disease of mandarin fruit caused by *Mucor piriformis* and other Mucor spp. In california. *Plant Disease, 100*(6), 1054–1063.

Samithri, Y. A. S., Karunanayake, K. O. L. C., & Kulasinghe, A. A. (2020). In vitro study of selected essential oils against *Colletotrichum* sp. and *Lasiodiplodia* sp. causing postharvest diseases in papaya. *Ceylon Journal of Science, 49*(5), 1–11.

Samson, R. A., Houbraken, J., Thrane, U., Frisvald, J. C., & Andersen, B. (2010). *Food and indoor fungi* (CBS Laboratory manual series). CBS-KNAW Fungal Diversity Centre.

Santos, J. B., de Cruz, J., Geraldo, L. C., Dias, E. G., Queiroz, P. R., Monnerat, R. G., Borges, M., Blassioli-Moraes, M. C., & Blum, L. E. (2023). Detection and evaluation of volatile and non-volatile antifungal compounds produced by *Bacillus* spp. strains. *Microbiological Research, 275*, 127465. https://doi.org/10.1016/j.micres.2023.127465

Sanzani, S. M., Reverberi, M., Fanelli, C., & Ippolito, A. (2015). Detection of ochratoxin A using molecular beacons and real-time PCR thermal cycler. *Toxins, 7*(3), 812–820.

Sanzani, S. M., Miazzi, M. M., Di Rienzo, V., Fanelli, V., Gambacorta, G., Taurino, M. R., & Montemurro, C. (2016). A rapid assay to detect toxigenic *Penicillium* spp. contamination in wine and musts. *Toxins, 8*(8), 235.

Sapper, M., Martin-Esparza, M. E., Chiralt, A., & Martinez, C. G. (2020). Antifungal polyvinyl alcohol coatings incorporating Carvacrol for the postharvest preservation of *Golden delicious* apple. *Coatings, 10*, 1027. https://doi.org/10.3390/coatings10111027

Saragoça, A., Canha, H., Varanda, C. M., Materatski, P., Cordeiro, A. I., & Gama, J. (2024). Lactic acid bacteria: A sustainable solution against phytopathogenic agents. *Environmental Microbiology Reports, 16*(6), e70021.

Sardella, D., Gatt, R., & Valdramidis, V. P. (2017). Physiological effects and mode of action of zno nanoparticles against postharvest fungal contaminants. *Food Research International, 101*, 274–279. https://doi.org/10.1016/j.foodres.2017.08.019

Sari, L. K., Setha, S., & Naradisorn, M. (2016). Effect of UV-C irradiation on postharvest quality of 'phulae' pineapple. *Scientia Horticulturae, 213*, 314–320. https://doi.org/10.1016/j.scienta.2016.09.049

Sariah, M. (1994). Incidence of *Colletotrichum* spp. on chilli in Malaysia and pathogenicity of *C. gloeosporioides*. *Biotropia, 54*, 103–120.

Sasanuma, I., & Suzuki, T. (2016). Effect of calcium on cell-wall degrading enzymes of *Botrytis cinerea. Bioscience, Biotechnology, and Biochemistry, 80*(9), 1730–1736. https://doi.org/10.1080/09168451.2016.1146064

Sauer, D. B., Meronuck, R. A., & Christensen, C. M. (1992). Microflora. In D. B. Sauer (Ed.), *Storage of cereal grains and their products* (4th ed., pp. 313–340). American Association of Cereal Chemists, Inc.

Scherrer Montero, C. R., Antes, R. B., Schwarz, L. L., dos Santos, L. C., dos Santos, R. P., & Bender, R. J. (2010). Complementary physical and chemical treatments as an alternative to fungicide use to control postharvest decay incidence and fruit quality of Montenegrina tangerines. *Crop Protection, 29*(10), 1076–1083. https://doi.org/10.1016/j.cropro.2010.06.014

Schirra, M., D'Aquino, S., Cabras, P., & Angioni, A. (2011). Control of postharvest diseases of fruit by heat and fungicides: Efficacy, residue levels, and residue persistence. A review. *Journal of Agricultural and Food Chemistry, 59*(16), 8531–8542. https://doi.org/10.1021/jf201899t

Schmidt, M., Horstmann, S., De Colli, L., Danaher, M., Speer, K., Zannini, E., & Arendt, E. K. (2016). Impact of fungal contamination of wheat on grain quality criteria. *The Journal of Cereal Science, 69*, 95–103. https://doi.org/10.1016/j.jcs.2016.02.010

Schreinemachers, P., Simmons, E. B., & Wopereis, M. C. S. (2018). Tapping the economic and nutritional power of vegetables. *Global Food Security, 16*, 36–45.

Scott, P. M. (2012). Recent research on fumonisins: a review. *Food Additives & Contaminants: Part A, 29*(2), 242–248.

Šernaitė, L., Rasiukevičiūtė, N., Dambrauskienė, E., Viškelis, P., & Valiuškaitė, A. (2020). Biocontrol of strawberry pathogen *Botrytis cinerea* using plant extracts and essential oils.

Semeniuk, G. (1954). Microflora. In J. A. Anderson & A. W. Alcock (Eds.), *Storage of cereal grains and their products* (572pp). American Association of Cereal Chemists.

Seshadri, V. D., Balasubramanian, B., Al-Dhabi, N. A., Esmail, G. A., & Arasu, M. V. (2020). Essential oils of *Cinnamomum loureirii* and *Evolvulus alsinoides* protect guava fruits from spoilage bacteria, fungi and insect (pseudococcus longispinus). *Industrial Crops and Products, 154*, 112629. https://doi.org/10.1016/j.indcrop.2020.112629

Sexton, A. C., & Howlett, B. J. (2006). Parallels in fungal pathogenesis on plant and animal hosts. *Eukaryotic Cell, 5*, 1941–1949. https://doi.org/10.1128/EC.00277-06

Shan, Y., Li, T., Qu, H., Duan, X., Farag, M., Xiao, J., Gao, H., & Jiang, Y. (2023). Nanopreservation: An emerging postharvest technology for quality maintenance and shelf life extension of fresh fruit and vegetable. *Food Frontiers, 4*. https://doi.org/10.1002/fft2.201

Shan, X., Zhang, C., Li, J., Duan, Y., & Huang, Y. (2024). Pseudomonas protegens controlled brown rot (*Monilinia fructicola*) on peach fruit in postharvest by suppressing fungal development and inducing plant defense-related genes. *Postharvest Biology and Technology, 213*, 112971. https://doi.org/10.1016/j.postharvbio.2024.112971

Sharma, G., & Shenoy, B. D. (2014). *Colletotrichum fruticola* and *C. siamense* are involved in chilli anthracnose in India. *Archives of Phytopathology and Plant Protection, 47*(11), 1179–1194.

Sharma, R. R., Singh, D., & Singh, R. (2009). Biological control of postharvest diseases of fruits and vegetables by microbial antagonists: A review. *Biological Control, 50*(3), 205–221. https://doi.org/10.1016/j.biocontrol.2009.05.001

Sharma, G., Maymon, M., & Freeman, S. (2016). First detailed report of *Trichothecium roseum* causing post-harvest pink rot of avocado in Israel. *Plant Disease, 100*(4), 856–856.

Sharma, A., Sharma, I. M., Sharma, M., Sharma, K., & Sharma, A. (2021). Effectiveness of fungal, bacterial and yeast antagonists for management of Mango Anthracnose (*Colletotrichum gloeosporioides*). *Egyptian Journal of Biological Pest Control, 31*(1). https://doi.org/10.1186/s41938-021-00480-9

Shende, S. S., Gade, A. K., Minkina, T. M., Ingle, P. U., Rajput, V. D., Sushkova, S. N., Mandzhieva, S. S., Rai, M., & Wong, M. H. (2024). Exploring sustainable management by using green nanosilver to combat three post-harvest pathogenic fungi in crops. *Discover Nano, 19*(1). https://doi.org/10.1186/s11671-024-03986-x

Shoemaker, P. B., & Lorbeer, J. W. (1971). Spray volume, interval, and fungicide rate for control of *Botrytis* leaf blight of onion. *Plant Disease Reporter, 55*(6), 565–569.

Shu, X., Livingston, D. P., III, Franks, R. G., Boston, R. S., Woloshuk, C. P., & Payne, G. A. (2015). Tissue-specific gene expression in maize seeds during colonization by *Aspergillus flavus* and *Fusarium verticillioides*. *Molecular Plant Pathology, 16*(6), 662–674.

Siahmoshteh, F., Siciliano, I., Banani, H., Hamidi-Esfahani, Z., Razzaghi-Abyaneh, M., Gullino, M. L., & Spadaro, D. (2017). Efficacy of *Bacillus subtilis* and *Bacillus amyloliquefaciens* in the control of *Aspergillus parasiticus* growth and aflatoxins production on pistachio. *International Journal of Food Microbiology, 254*, 47–53. https://doi.org/10.1016/j.ijfoodmicro.2017.05.011

Siddique, Y., Masih, R., Ahmad, I., & Sadiq, M. B. (2025). Control of postharvest fungal contamination in rice seeds and cherries using *eucalyptus camaldulensis* essential oil and modified sweet potato starch. *Archives of Phytopathology and Plant Protection*, 1–14. https://doi.org/10.1080/03235408.2024.2375807

Siddiqui, Y. (2014). *Postharvest decay: Colletotrichum gloeosporioides* (Anthracnose). In *Postharvest decay* (pp. 337–371). Elsevier. https://doi.org/10.1016/B978-0-12-411552-1.00011-9

Siddiqui, Y., & Ali, A. (2014). *Colletotrichum gloeosporioides* (Anthracnose). In *Postharvest decay* (pp. 337–371). Academic.

Simas, D. L. R., de Amorim, S. H. B. M., Goulart, F. R. V., Alviano, C. S., Alviano, D. S., & da Silva, A. J. (2017). *Citrus* species essential oils and their components can inhibit or stimulate fungal growth in fruit. *Industrial Crops and Products, 98*, 108–115. https://doi.org/10.1016/j.indcrop.2017.01.026

Singh, D., & Sharma, R. R. (2018). Postharvest diseases of fruits and vegetables and their management. In *Postharvest disinfection of fruits and vegetables* (pp. 1–52). Academic. https://doi.org/10.1016/b978-0-12-812698-1.00001-7

Sneha, K. B., Indra, N., Vanitha, S., Saranya, S., & Ramalakshmi, A. (2024). Exploring the non-chemical alternatives for the management of post-harvest fungal diseases of major tropical fruits- mango, banana and Papaya. *Physiological and Molecular Plant Pathology, 134*, 102460. https://doi.org/10.1016/j.pmpp.2024.102460

Snini, E. P., Tadrist, S., Laffite, J., Jamin, E. L., Oswald, I. P., & Puel, O. (2014). The gene PatG involved in the biosynthesis pathway of patulin, a food-borne mycotoxin, encodes a 6-methylsalicylic acid decarboxylase. *International Journal of Food Microbiology, 171*, 77–83.

Song, B., Ning, W., Wei, D., Jiang, M., Zhu, K., Wang, X., Edwards, D., Odeny, D. A., & Cheng, S. (2023). Plant genome resequencing and population genomics: Current status and future prospects. *Molecular Plant, 16*(8), 1252–1268. https://doi.org/10.1016/J.MOLP.2023.07.009/ASSET/49A7C524-4BC7-4EC9-821B-4A6677857016/MAIN.ASSETS/GR3.JPG

Souza, E. J., Kringel, D. H., Lima Costa, I. H., Hackbart, H. C., Cantillano, R. F., Ueno, B., Dias, A. R., & da Zavareze, E. (2024). Antifungal potential of essential oils from different botanical sources against *penicillium digitatum*: Chemical composition and antifungal mechanisms of action by direct contact and volatile. *Natural Product Research*, 1–9. https://doi.org/10.1080/14786419.2024.2405865

Spadaro, D., & Droby, S. (2016). Development of biocontrol products for postharvest diseases of fruit: The importance of elucidating the mechanisms of action of yeast antagonists. *Trends in Food Science and Technology, 47*, 39–49. https://doi.org/10.1016/j.tifs.2015.11.003

Sperschneider, J. (2020). Machine learning in plant–pathogen interactions: Empowering biological predictions from field scale to genome scale. *New Phytologist, 228*(1), 35–41. https://doi.org/10.1111/NPH.15771

Sripong, K., Jitareerat, P., Tsuyumu, S., Uthairatanakij, A., Srilaong, V., Wongs-Aree, C., Ma, G., Zhang, L., & Kato, M. (2015). Combined treatment with hot water and UV-C elicits disease resistance against anthracnose and improves the quality of harvested mangoes. *Crop Protection, 77*, 1–8. https://doi.org/10.1016/j.cropro.2015.07.004

Storari, M., Broggini, G. A., Bigler, L., Cordano, E., Eccel, E., De Filippi, R., ... & Pertot, I. (2012). Risk assessment of the occurrence of black aspergilli on grapes grown in an alpine region under a climate change scenario. *European journal of plant pathology, 134*(3), 631–645.

Su, Y. Y., Noireung, P., Liu, F., Hyde, K. D., Moslem, M. A., Bahkali, A. H., et al. (2011). Epitypification of Colletotrichum musae, the causative agent of banana anthracnose. *Mycoscience, 52*(6), 376–382.

Suasa-ard, S., Eakjamnong, W., & Dethoup, T. (2019). A novel biological control agent against postharvest mango disease caused by *Lasiodioplodia theobromae*. *European Journal of Plant Pathology, 155*, 583–592.

Summo, C., & De Angelis, D. (2022). The importance of edible films and coatings for sustainable food development. *Food, 11*, 3221. https://doi.org/10.3390/foods11203221

Sumner, D. R. (1995). Diseases of bulbs caused by fungi—Black mold. In H. F. Schwartz & S. K. Mohan (Eds.), *Compendium of onion and garlic diseases* (pp. 26–27). APS Press.

Sun, Q., Lin, L., Liu, D., Wu, D., Fang, Y., Wu, J., & Wang, Y. (2018). CRISPR/Cas9-mediated multiplex genome editing of the BnWRKY11 and BnWRKY70 genes in *Brassica napus* L. *International Journal of Molecular Sciences, 19*(9), 2716. https://doi.org/10.3390/IJMS19092716

Sun, X., Wang, C., Gao, X., Wu, X., & Fu, Y. (2023). Characterization of *Alternaria* species associated with black spot of strawberry in Dandong, China. *Agronomy, 13*(4), 1014.

Taha, N. A., Elsharkawy, M. M., Shoughy, A. A., El-Kazzaz, M. K., & Khedr, A. A. (2023). Biological control of postharvest tomato fruit rots using *Bacillus* spp. and *Pseudomonas* spp. *Egyptian Journal of Biological Pest Control, 33*(1), 106. https://doi.org/10.1186/s41938-023-00752-6

Talhinhas, P., & Baroncelli, R. (2021). *Colletotrichum species* and complexes: geographic distribution, host range and conservation status. *Fungal Diversity, 110*(1), 109–198.

Tanaka, K. (1991). Studies on the black mould disease of onion bulbs caused by *Aspergillus niger* van tieghem. *Bulletin of the Faculty of Agriculture, Saga University, 70*, 1–54.

Taylor, J. W. (2011). One fungus = one name: DNA and fungal nomenclature twenty years after PCR. *IMA Fungus, 2*, 113–120.

Tek, M. I., & Budak, K. (2022). A new approach to develop resistant cultivars against the plant pathogens: CRISPR drives. *Frontiers in Plant Science, 13*, 889497. https://doi.org/10.3389/FPLS.2022.889497/BIBTEX

Thomidis, T., Prodromou, I., Farmakis, A., & Zambounis, A. (2021). Effect of temperature on the growth of *Geotrichum candidum* and chemical control of sour rot on tomatoes. *Tropical Plant Pathology, 46*, 545–552.

Tian, Z., Chen, C., Chen, K., Liu, P., Fan, Q., Zhao, J., & Long, C. (2020). Biocontrol and the mechanisms of bacillus sp. w176 against postharvest green mold in citrus. *Postharvest Biology and Technology, 159*, 111022. https://doi.org/10.1016/j.postharvbio.2019.111022

Toit, L. J., & Schwartz, H. F. (2011). *Storage fungal diseases* [PDF document]. Onion ipmPIPE Diagnostic Pocket Series. Retrieved from http://mtvernon.wsu.edu/path_team/Onion-Card-Storage-Fungal-Diseases.pdf

Triest, D., & Hendrickx, M. (2016). Postharvest disease of banana caused by *Fusarium musae*: A public health concern? *PLoS Pathogens, 12*(11), e1005940.

Tripathi, P., & Dubey, N. K. (2004). Exploitation of natural products as an alternative strategy to control postharvest fungal rotting of fruit and vegetables. *Postharvest Biology and Technology, 32*(3), 235–245. https://doi.org/10.1016/j.postharvbio.2003.11.005

Tripathi, A. N., Tiwari, S. K., & Behera, T. K. (2022). Postharvest diseases of fruits and vegetables: Diagnosis and management. In M. Ahiduzzaman (Ed.), *Postharvest technology: Recent advances, new perspectives, and applications* (pp. 109–127). Springer. https://doi.org/10.1007/978-3-030-92093-9

Tripathy, A. P., Satapathy, R. R., & Rout, M. (2021). Management of *Colletotrichum gloeosporioides* causing papaya anthracnose with best resulting plant extract and bio-agents. *The Pharma Innovation Journal, 10*, 629–630.

Troncoso-Rojas, R., & Tiznado-Hernández, M. E. (2014). Chapter 5: *Alternaria alternata* (Black Rot, Black Spot). In *Postharvest decay* (pp. 147–187). Academic.

Trupo, M., Magarelli, R. A., Martino, M., Larocca, V., Giorgianni, A., & Ambrico, A. (2023). Crude lipopeptides from culture of *Bacillus subtilis* strain ET-1 against *Podosphaera xanthii* on *Cucumis melo*. *Journal of Natural Pesticide Research, 4*, 100032. https://doi.org/10.1016/j.napere.2023.100032

Tsuruta, O., Gohara, S., & Saito, M. (1981). Scanning electron microscopic observations of a fungal invasion of corn kernels. *Transactions of the Mycological Society of Japan, 22*, 121–126.

Tyson, J., & Fullerton, R. (2004). Effect of soil-borne inoculum on incidence of onion black mould (*Aspergillus niger*). *New Zealand Plant Protection, 57*, 138–141.

Umer, M., Mubeen, M., Ateeq, M., Shad, M. A., Atiq, M. N., Kaleem, M. M., Iqbal, S., Shaikh, A.A., Ullah, I., Khan, M., & Abbas, A. (2021). Etiology, epidemiology and management of citrus black rot caused by *Alternaria citri*: an outlook. *Plant Prot, 5*(2), 105–115.

Upadhyaya, S. R., Danilevicz, M. F., Dolatabadian, A., Neik, T. X., Zhang, F., Al-Mamun, H. A., Bennamoun, M., Batley, J., & Edwards, D. (2024). Genomics-based plant disease resistance prediction using machine learning. *Plant Pathology, 73*(9), 2298–2309. https://doi.org/10.1111/PPA.13988

Usall, J., Smilanick, J., Palou, L., Denis-Arrue, N., Teixidó, N., Torres, R., & Viñas, I. (2008). Preventive and curative activity of combined treatments of sodium carbonates and *Pantoea agglomerans* CPA-2 to control postharvest green mold of citrus fruit. *Postharvest Biology and Technology, 50*(1), 1–7. https://doi.org/10.1016/j.postharvbio.2008.03.001

Usall, J., Ippolito, A., Sisquella, M., & Neri, F. (2016). Physical treatments to control postharvest diseases of fresh fruits and vegetables. *Postharvest Biology and Technology, 122*, 30–40. https://doi.org/10.1016/j.postharvbio.2016.05.002

USDA [United States Department of Agriculture]. (2016). *Grain fungal diseases and mycotoxin reference* (p. 45). USDA.

van Kan, J. A. (2003). Infection strategies of *Botrytis cinerea*. In *VIII international symposium on postharvest physiology of ornamental plants* (Vol. 669, pp. 77–90).

Vawdrey, L. L., Male, M., & Grice, K. R. E. (2015). Field and laboratory evaluation of fungicides for the control of *Phytophthora* fruit rot of papaya in Far North Queensland, Australia. *Crop Protection, 67*, 116–120. https://doi.org/10.1016/j.cropro.2014.10.003

Ventura-Aguilar, R. I., Bautista-Baños, S., Hernández-López, M., & Llamas-Lara, A. (2021). Detection of *Alternaria alternata* in tomato juice and fresh fruit by the production of its biomass, respiration, and volatile compounds. *International Journal of Food Microbiology, 342*, 109092. https://doi.org/10.1016/j.ijfoodmicro.2021.109092

Vilaplana, R., Páez, D., & Valencia-Chamorro, S. (2017). Control of black rot caused by *Alternaria alternata* in yellow pitahaya (*Selenicereus megalanthus*) through hot water dips. *LWT-Food Science and Technology, 82*, 162–169.

Vischetti, C., Feliziani, E., Landi, L., De Bernardi, A., Marini, E., & Romanazzi, G. (2023). Effectiveness of four synthetic fungicides in the control of post-harvest gray mold of strawberry and analyses of residues on fruit. *Agronomy, 14*(1), 65. https://doi.org/10.3390/agronomy14010065

Volentini, S. I., Olmedo, G. M., Grillo-Puertas, M., Rapisarda, V. A., Hebert, E. M., Cerioni, L., & Villegas, J. M. (2023). Biological control of Green and blue molds on postharvest lemon by lactic acid bacteria. *Biological Control, 185*, 105303. https://doi.org/10.1016/j.biocontrol.2023.105303

Vučković, N., Vico, I., & Duduk, N. (2023). First report of *Botryosphaeria dothidea* causing postharvest rot of quince fruits in Serbia. *Journal of Plant Pathology, 105*(2), 605–605.

Wan, C., Kahramanoğlu, İ., Chen, J., Gan, Z., & Chen, C. (2020). Effects of hot air treatments on postharvest storage of Newhall navel Orange. *Plants (Basel), 9*(2), 170. https://doi.org/10.3390/plants9020170. PMID: 32024051; PMCID: PMC7076504.

Wang, J. Y., & Doudna, J. A. (2023). CRISPR technology: A decade of genome editing is only the beginning. *Science, 379*(6629). https://doi.org/10.1126/SCIENCE.ADD8643/ASSET/F32937F7-FE8D-4FDA-B0EC-7DB7B79A1B94/ASSETS/IMAGES/LARGE/SCIENCE.ADD8643-F4.JPG

Wang, M., Weiberg, A., Lin, F. M., Thomma, B. P. H. J., Da Huang, H., & Jin, H. (2016). Bidirectional cross-kingdom RNAi and fungal uptake of external RNAs confer plant protection. *Nature Plants, 2*(10), 1–10. https://doi.org/10.1038/nplants.2016.151

Wang, K., Jiang, S., Pu, T., Fan, L., Su, F., & Ye, M. (2018). Antifungal activity of phenolic monoterpenes and structure-related compounds against plant pathogenic fungi. *Natural Product Research, 33*(10), 1423–1430. https://doi.org/10.1080/14786419.2017.1419232

Wang, F., Saito, S., Michailides, T. J., & Xiao, C. L. (2021). Postharvest use of natamycin to control *Alternaria* rot on blueberry fruit caused by *Alternaria alternata* and *A. arborescens*. *Postharvest Biology and Technology, 172*, 111383.

Wang, X., Zhang, X., Sun, M., Wang, L., Zou, Y., Fu, L., Han, C., Li, A., Li, L., & Zhu, C. (2022). Impact of vanillin on postharvest disease control of Apple. *Frontiers in Microbiology, 13*. https://doi.org/10.3389/fmicb.2022.979737

Wang, H., Chen, H., Lin, Y., Li, M., Liu, Q., Lin, Y., et al. (2023a). Insights into the isolation, identification, and biological characterization analysis of and novel control strategies for *Diaporthe passiflorae* in postharvest passion fruit. *Journal of Fungi, 9*(10), 1034.

Wang, R., Liu, K., Tang, B., Su, D., He, X., Deng, H., Wu, M., Bouzayen, M., Grierson, D., & Liu, M. (2023b). The MADS-box protein SlTAGL1 regulates a ripening-associated SlDQD/SDH2 involved in flavonoid biosynthesis and resistance against *Botrytis cinerea* in post-harvest tomato fruit. *The Plant Journal, 115*(6), 1746–1757. https://doi.org/10.1111/TPJ.16354

Wang, Y., Zheng, Y., Shang, J., Wu, D., Zhou, A., Cai, M., Gao, H., & Yang, K. (2023c). Pulsed light reduces postharvest losses of Chinese bayberries by affecting fungal microbiota during Cold storage. *Food Control, 146*, 109524. https://doi.org/10.1016/j.foodcont.2022.109524

Wang, Y. R., Huang, B. Y., Usman, H. M., Javed, K., Al-Otibi, F., Hyde, K. D., & Wang, Y. (2025). First report of *Botryosphaeria wangensis, Colletotrichum nymphaeae, Diaporthe eres*, and *Geotrichum candidum* causing postharvest fruit rot of plums (Prunus salicina) in China. *Crop Protection, 190*, 107089.

Wang, Y. R., Song, Q. J., Shang, L., Wang, X. C., Muhae-Ud-Din, G., Wang, Y., & Chen, H. (2024). First report of Lasiodiplodia pseudotheobromae causing soft rot of plum in China. *Plant Disease, 108*(4), 1097.

Wang, Y., Yu, T., Xia, J., Yu, D., Wang, J., & Zheng, X. (2010). Biocontrol of postharvest gray mold of cherry tomatoes with the marine yeast *Rhodosporidium paludigenum. Biological control, 53*(2), 178–182.

Weber, S., Lugo-Duque, M. A., & Gazis, R. (2025). First report of postharvest fruit rot on passion fruit (*Passiflora edulis*) caused by *Lasiodiplodia theobromae* in the USA. *Plant Disease, 109*(3), 722.

Weidenbörner, M. (2001). *Encyclopedia of food mycotoxins*. Springer.

Wenneker, M., & Thomma, B. P. (2020). Latent postharvest pathogens of pome fruit and their management: From single measures to a systems intervention approach. *European Journal of Plant Pathology, 156*(3), 663–681.

Wharton, P. S., & Diéguez-Uribeondo, J. (2004). The biology of *Colletotrichum acutatum. Anales del Jardín Botánico de Madrid, 61*, 3–22. https://doi.org/10.3989/ajbm.2004.v61.i1.3

Wharton, P. S., & Julian, A. M. (1996). A cytological study of compatible and incompatible interactions between *Sorghum bicolor* and *Colletotrichum sublineolum. New Phytologist, 134*(1), 25–34. https://doi.org/10.1111/j.1469-8137.1996.tb01144.x

Williamson, B., Tudzynski, B., Tudzynski, P., & Van Kan, J. A. L. (2007). *Botrytis cinerea*: The cause of grey mold disease. *Molecular Plant Pathology, 8*(5), 561–580. https://doi.org/10.1111/j.1364-3703.2007.00417.x

Wisniewski, M., Droby, S., Norelli, J., Liu, J., & Schena, L. (2016). Alternative management technologies for postharvest disease control: The journey from simplicity to complexity. *Postharvest Biology and Technology, 122*, 3–10. https://doi.org/10.1016/j.postharvbio.2016.05.012

Wonglom, P., & Sunpapao, A. (2020). *Fusarium incarnatum* is associated with postharvest fruit rot of muskmelon (*Cucumis melo*). *Journal of Phytopathology, 168*(4), 204–210.

Worku, S., Alemu, K., & Tsedaley, B. (2025). Combining plant extracts and hot water treatments for the management of Postharvest Mango Anthracnose (*Colletotrichum gloeosporioides*). *Scientific Reports, 15*(1), 5074. https://doi.org/10.1038/s41598-025-89587-9

World Grain. (2024). *FAO projects record 2023–24 cereal output*. Available at https://www.world--grain.com/articles/19577-fao-projects-record-2023-24-cereal-output. Accessed 26 Aug 2024.

Xu, D., Deng, Y., Han, T., Jiang, L., Xi, P., Wang, Q., Jiang, Z., & Gao, L. (2018a). In vitro and in vivo effectiveness of phenolic compounds for the control of postharvest gray mold of table grapes. *Postharvest Biology and Technology, 139*, 106–114. https://doi.org/10.1016/j.postharvbio.2017.08.019

Xu, D., Qin, H., & Ren, D. (2018b). Prolonged preservation of tangerine fruits using chitosan/montmorillonite composite coating. *Postharvest Biology and Technology, 143*, 50–57. https://doi.org/10.1016/j.postharvbio.2018.04.013

Xu, R., Wu, J., Han, X., Wang, Z., Lou, Y., Gao, Y., Shang, S., Song, Z., Song, J., & Li, J. (2024). Discovery of natural rosin-based preservative candidates to control *Phytophthora capsici* for postharvest disease management of solanaceae vegetables. *Postharvest Biology and Technology, 218*, 113133. https://doi.org/10.1016/J.POSTHARVBIO.2024.113133

Yan, J., Wu, H., Shi, F., Wang, H., Chen, K., Feng, J., & Jia, W. (2020). Antifungal activity screening for Mint and thyme essential oils against *rhizopus stolonifer* and their application in postharvest preservation of strawberry and Peach Fruits. *Journal of Applied Microbiology, 130*(6), 1993–2007. https://doi.org/10.1111/jam.14932

Yang, J., Jing, X., Wang, T., Di, J., Chen, L., & Wang, Y. (2020). Overactivation of glutamate-consuming pathways in L-glutamate-treated tomato fruits leads to resistance against *Alternaria alternata*. *Postharvest Biology and Technology, 169*, 111311. https://doi.org/10.1016/j.postharvbio.2020.111311

Ye, W. Q., Sun, Y. F., Tang, Y. J., & Zhou, W. W. (2021). Biocontrol potential of a broad-spectrum antifungal strain *Bacillus amyloliquefaciens* B4 for Postharvest Loquat Fruit Storage. *Postharvest Biology and Technology, 174*, 111439. https://doi.org/10.1016/j.postharvbio.2020.111439

Yinxin, L., Can, Z., Menglu, H., Cui, S., Jinping, C., Jingyu, W., & Huang, L. (2022). Effect of cold atmospheric plasma on the gray mold rot of postharvest mulberry fruit. *Food Control, 137*, 108906. https://doi.org/10.1016/j.foodcont.2022.108906

Yu, L., Qiao, N., Zhao, J., Zhang, H., Tian, F., Zhai, Q., & Chen, W. (2020). Postharvest control of *Penicillium expansum* in fruits: A review. *Food Bioscience, 36*, 100633. https://doi.org/10.1016/j.fbio.2020.100633

Yuan, X., Qi, Y., Wu, Z., Yang, C., & Cui, C. (2024). First report of *Botryosphaeria dothidea* causing postharvest fruit rot of plum in China. *Plant Disease, 108*(4), 1111.

Zakaria, L. (2021). Diversity of *Colletotrichum* species associated with anthracnose disease in tropical fruit crops—A review. *Agriculture, 11*(4), 297.

Zakaria, L. (2024). An overview of *Aspergillus* species associated with plant diseases. *Pathogens, 13*(9), 813. https://doi.org/10.3390/pathogens13090813

Zee, K. Y., Asib, N., & Ismail, S. I. (2021). First report of *Lasiodiplodia theobromae* causing postharvest fruit rot on guava (Psidium guajava) in Malaysia. *Plant Disease, 105*(9), 2716.

Zeng, W., Feng, J., Wei, Y., Chen, Y., Zhang, M., Zhou, D., Qi, D., Zhang, L., Xie, J., & Wang, W. (2024). Biocontrol efficiency and potential mechanism of streptomyces distatochromogenes XT34 against postharvest anthracnose caused by *Colletotrichum musae* on Banana fruit. *Postharvest Biology and Technology, 212*, 112899. https://doi.org/10.1016/j.postharvbio.2024.112899

Zhang, J. (2014). *Lasiodiplodia theobromae* in citrus fruit (*Diplodia* stem-end rot). In *Postharvest decay* (pp. 309–335). Academic.

Zhang, H., Mahunu, G. K., Castoria, R., Apaliya, M. T., & Yang, Q. (2017). Augmentation of biocontrol agents with physical methods against postharvest diseases of fruits and vegetables. *Trends in Food Science & Technology, 69*, 36–45.

Zhang, H., Serwah Boateng, N. A., Ngolong Ngea, G. L., Shi, Y., Lin, H., Yang, Q., Wang, K., Zhang, X., Zhao, L., & Droby, S. (2021a). Unravelling the fruit microbiome: The key for developing effective biological control strategies for postharvest diseases. *Comprehensive Reviews in Food Science and Food Safety, 20*(5), 4906–4930. https://doi.org/10.1111/1541-4337.12783

Zhang, Z. Q., Chen, T., Li, B. Q., Qin, G. Z., & Tian, S. P. (2021b). Molecular basis of pathogenesis of postharvest pathogenic fungi and control strategy in fruits: Progress and prospect. *Molecular Horticulture, 1*, 1–10.

Zhang, Y., Zheng, L., & Xie, K. (2023). CRISPR/dCas9-mediated gene silencing in two plant fungal pathogens. *MSphere, 8*(1). https://doi.org/10.1128/MSPHERE.00594-22/SUPPL_FILE/MSPHERE.00594-22-S0007.DOCX

Zhang, J., Yan, J., Bai, J., Hu, C., Pan, T., Carrillo, Y. J., et al. (2024a). First report of *Lasiodiplodia pseudotheobromae* causing postharvest decay of strawberries in Florida. *Plant Disease, 108*(2), 519.

Zhang, Y., Zhang, K., Bao, Z., Hao, J., Ma, X., Jia, C., Liu, M., Wei, D., Yang, S., & Qin, J. (2024b). A novel preservative film with a pleated surface structure and dual bioactivity properties for application in strawberry preservation due to its efficient apoptosis of pathogenic fungal cells. *Journal of Agricultural and Food Chemistry, 72*, 18027–18044. https://doi.org/10.1021/acs.jafc.4c04579

Zhao, W., Bai, J., McCollum, G., & Baldwin, E. (2015). High incidence of preharvest colonization of huanglongbing-symptomatic *Citrus sinensis* fruit by *Lasiodiplodia theobromae* (*Diplodia natalensis*) and exacerbation of postharvest fruit decay by that fungus. *Applied and environmental microbiology, 81*(1), 364–372.

Zhao, Z., Liu, D., Ruan, L., Wang, T., & Liang, Z. (2023). Antifungal mechanism of *Bacillus amyloliquefaciens* SC-B15 and its application in cereal mildewproof and grape preservation. *Food Bioscience, 56*, 103287. https://doi.org/10.1016/j.fbio.2023.103287

Zhao, J.-H., Liu, Q.-Y., Xie, Z.-M., & Guo, H.-S. (2024). Exploring the challenges of RNAi-based strategies for crop protection. *Advanced Biotechnology, 2*(3), 1–10. https://doi.org/10.1007/S44307-024-00031-X

Zhou, T., Northover, J., & Schneider, K. E. (1999). Biological control of postharvest diseases of peach with phyllosphere isolates of *Pseudomonas syringae*. *Canadian Journal of Plant Pathology, 21*(4), 375–381. https://doi.org/10.1080/07060669909501174

Zhou, Y., Li, S., & Zeng, K. (2016). Exogenous nitric oxide-induced postharvest disease resistance in citrus fruit to *Colletotrichum gloeosporioides*. *Journal of the Science of Food and Agriculture, 96*(2), 505–512.

Zulu, L., Gao, H., Zhu, Y., Wu, H., Xie, Y., Liu, X., Yao, H., & Rao, Q. (2023). Antifungal effects of seven plant essential oils against *Penicillium digitatum*. *Chemical and Biological Technologies in Agriculture, 10*(1). https://doi.org/10.1186/s40538-023-00434-3

Chapter 11
Veterinary Mycology: Challenges and Applications

Cherry P. Fernandez-Colorado, Rohani B. Cena-Navarro, and Andrea Gail M. Villavicencio

Abstract Veterinary mycology is an important field of veterinary science that focuses on the identification, diagnosis, and treatment of infectious and zoonotic fungal diseases that pose risks to both animal and public health. The emergence of new and resistant fungal pathogens is influenced by human-animal-environmental interactions. In Southeast Asia, the environmental conditions with warm, humid climate, dense biodiversity, urbanization, and habitat disruption can influence the growth and transmission of these novel fungal pathogens. This chapter highlights the importance of fungal pathogens, zoonotic transmission, and the impact of environmental factors and climate change in the transmission and evolution of fungal diseases. It describes the growing concerns and challenges of antifungal resistance and the increasing cases of fungal diseases, especially those with zoonotic potential. In response to these challenges, this chapter reviews recent and current applications of advanced diagnostic techniques used for the timely identification of fungal pathogens, improving the accuracy and timeliness of disease reporting, treatment, and control. Development of antifungal drugs is also addressed in this chapter with more emphasis on the importance of a collaborative response, that is, a One Health approach involving veterinary, medical, and environmental health to foster more effective surveillance systems and intervention strategies. Moreover, this chapter explores the future outlook of interdisciplinary collaboration when it comes to fungal disease monitoring, antifungal drug development, and sustainable research

C. P. Fernandez-Colorado (✉)
Department of Veterinary Paraclinical Sciences, College of Veterinary Medicine, University of the Philippines Los Baños, Los Baños, Philippines
e-mail: cpfernandezcolorado@up.edu.ph

R. B. Cena-Navarro
National Institute of Molecular Biology and Biotechnology, National Institutes of Health, University of the Philippines Manila, Manila, Philippines

A. G. M. Villavicencio
Department of Medical Microbiology, College of Public Health, University of the Philippines Manila, Manila, Philippines

J. J. G. Guerrero et al. (eds.), *Mycology in a Changing Planet: Applications and Perspectives from Southeast Asia*, https://doi.org/10.1007/978-981-95-2640-6_11

funding to mitigate the growing threat of fungal diseases and strengthen animal, human, and environmental health.

Keywords Veterinary mycology · Zoonosis · Antifungal resistance · Southeast Asia · One Health

Introduction

Veterinary mycology is the field of veterinary science that is involved in the study of fungi that cause infections in animals, classification, pathogenicity, diagnostics, as well as treatment. There are a variety of fungal pathogens that include yeasts, molds, as well as dimorphic fungi, some of which are zoonotic in character and can be transmitted between animals and human hosts (Ainsworth, 2019). Some of these fungal pathogens have also developed resistance toward antifungal agents; hence, early detection and effective antifungal treatments are crucial in combating these infections (Verweij et al., 2016). Enhancing diagnostics, treatments, and prevention strategies is crucial in a bid to secure both animal and human health.

Within the One Health framework, the interconnectedness extends beyond animal and human health to environmental health. Many fungi that infect animals are zoonotic, and environmental factors also influence the spread of fungal diseases in animals and humans. Veterinary mycology is also involved in preventing outbreaks, planning antifungal therapies, and minimizing risks to human health through examination of animal fungal pathogens. These fungal pathogens also affect animals that are economically important. The effects that these pathogens have on cattle, poultry, and other domestic animals, as well as measures on how to counteract them, are dealt with in this chapter.

The diverse applications and challenges of veterinary mycology, novel diagnostic methods, and novel therapeutic approaches to combat fungal infections in animals are also discussed in this chapter. Also in consideration is the role of environmental drivers and climate change as well as future directions in veterinary mycology with emphasis on multidisciplinary and multidimensional teamwork, enhanced surveillance measures, and new therapy development in a bid to boost veterinary management of fungal infections.

Important Animal Fungal Diseases in Southeast Asia

Southeast Asia is a region in which animal fungal infections are a serious threat not only to animals but also to humans. The environmental conditions (humid, warm climatic conditions) in the region are a perfect breeding ground for fungal pathogens. Although increasingly important, animal fungal infections are not as much in the forefront as are those from bacteria or viruses. In this section, we discuss the key animal fungal infections and pathogens in Southeast Asia, focusing on their cause, clinical signs, and symptoms, as well as diagnostic procedures and treatment regimens. Knowledge of these infections is important in formulating effective prevention as well as control measures to secure animal and human health in Southeast Asia.

Emerging and Re-emerging Fungal Pathogens

The emergence and re-emergence of pathogenic fungi are attributed to several factors, such as climatic changes, deforestation, urbanization, and increasing numbers of immunocompromised hosts. As a result, new and increasing cases of fungal infections are continuously being reported. Fortunately, developments in diagnosis and epidemiological studies allowed awareness of the existence of animal fungal diseases in Southeast Asia. Understanding the disease profile, proper diagnosis, and management can aid in the control and prevention of fungal diseases. Table 11.1 summarizes the disease profile of fungal diseases in animals already reported in Southeast Asia.

Zoonotic Fungal Pathogens

The incidence of fungal infections, including histoplasmosis and talaromycosis (penicilliosis), in Southeast Asia has been recorded as early as 1972 in humans, especially in patients with sexually transmitted diseases such as HIV (Chakrabarti & Slavin, 2011). Aside from humans, it has been established previously that there are also several reports of fungal diseases in animals in Southeast Asia. With these, it can be asserted that there is an overlap between the fungal species that infects both humans and animals, making zoonotic infection a possible mode of transmission. One of the most concerning zoonotic fungal pathogens in Southeast Asia, particularly in tropical and subtropical regions, is *Cryptoccocus gattii,* which is associated with respiratory and neurological signs in animals and is characterized by meningoencephalitis in humans (Seyedmousavi et al., 2015). The direct transmission of *Sporothrix schenckii*, the causative agent of sporotrichosis, in humans through scratches or bites from felines was reported in Malaysia and Thailand (Kundu et al., 2024). Moreover, *Talaromyces marneffei* (formerly known as *Penicillium marneffei*), a dimorphic fungus endemic to Southeast Asia, was also reported to infect a wide range of animals, including rodents, and can cause severe systemic mycosis in immunocompromised humans (Nguyen et al., 2024). *Microsporum canis* and *Trichophyton mentagrophytes*, isolated from pets and stray animals across Southeast Asia, have also been reported to have zoonotic potential infecting humans via direct exposure to infected animals or contaminated environments (Patouillat et al., 2024).

In northern Thailand, *H. capsulatum* has been detected via PCR in environmental samples with bat guano and avian droppings, where there have been reports of

Table 11.1 Emerging and re-emerging animal fungal diseases in Southeast Asia

Disease/fungal pathogen	Animals affected	Clinical signs and symptoms	Diagnosis	Treatment/ management
Aspergillosis (*Aspergillus* spp.)	Poultry Horses Cattle Canine	Horses: Guttural pouch mycosis (characterized by epistaxis, unilateral purulent nasal discharge, and abnormal head position); sporadic abortion Cattle: Mastitis, pneumonia, sporadic abortion Dogs: Rhinitis Poultry: Brooder pneumonia (periorbital swelling and ocular discharge and inflammation; necrotic foci in organs upon necropsy)	Routine microscopy (10% KOH) Histopathological stains (periodic acid-Schiff, Grocott's silver, or methenamine silver) Isolation and identification: Require multiple isolation as it is a common contaminant Molecular methods such as antigen detection and PCR	*Topical:* Enilconazole, miconazole, and terbinafine *Systemic:* Fluconazole, itraconazole, ketoconazole, and 5-fluorocytosine Amphotericin B: Both topical and systemic applications
Candidiasis (*Candida* spp.)	Poultry Cattle Dogs Cats Pigs	*Poultry* (*C. albicans*): Thrush (crop mycosis) wherein the upper gastrointestinal tract develops white plaques and "Turkish towel" appearance in the crop. *Dogs*: Exfoliative dermatitis in the muzzle, scrotum, and feet; peritonitis. Pigs: Dermatologic lesions *Cattle*: Mycotic mastitis *Foals and calves*: Gastroesophageal lesions	Routine microscopy and isolation, demonstration of germ tube formation, and chlamydospore production Histopathological samples: Blastospores and pseudo hyphae Immunofluorescence and immunohistochemical staining (*C. albicans*) ELISA; PCR	In poultry (orally via feed or drinking water): Amphotericin B, fluconazole, copper sulfate, and nystatin Debriding is done in dermatologic cases along with topical administration of povidone-iodine to disinfect, and copper sulfate, chlorhexidine, or selenium sulfide

(continued)

Table 11.1 (continued)

Disease/fungal pathogen	Animals affected	Clinical signs and symptoms	Diagnosis	Treatment/ management
Dermatophytosis (*Trichophyton, Microsporum, Epidermophyton* spp.)	Cattle Horse Sheep Poultry Pigs Dogs Cats	*Cattle*: Ringworm, with lesions appearing to be circular which may or may not produce large plaques or "asbestos" *Horses*: "Moth-eaten appearance" *Sheep* (*T. verrucosum*): "Club-lamb fungus" *Pigs* (*Microsporum* spp.): Small, circular inflammations forming crusts *Poultry*: Scaly, necrotizing inflammation; white comb or "flavus" (*M. gallinae*) Dogs, cats: Skin inflammation which may progress into a "wet eczema"	Routine microscopy: Two types of hair invasion—Ectothrix and endothrix. Chlorazol black staining for hyphae identification Histopathological slides (PAS, GMS, and calcofluor white (CFW) stains): Dead and degenerating mycelium *Microsporum* spp.: Hair and scale samples from the margin of the lesion transported in brown paper Molecular methods are also available for dermatophytes	*Trichophyton*: Hair clipping of the affected area followed by topical application of antifungals *Microsporum*: Systemic antifungals are preferred; itraconazole and ketoconazole are usually administered via pulse therapy
Histoplasmosis (*Histoplasma capsulatum*)	Canine Feline Bats Cattle Equine	*Pulmonary*: Coughing, dyspnea, and ulceration of oral and nasal mucosa, *systemic*: Fever, anorexia, and lethargy	Routine microscopy (wet or gram stain): Intracellular gram-positive yeast cells Serological tests: Passive hemagglutination, immunodiffusion, CFT, and ELISA, and skin hypersensitivity test (histofarcin test) Antigen detection, PCR	Dogs and cats: Itraconazole Large animals, horses: Amphotericin B
Mycetoma (various fungal species such as *Madurella*, *Pseudallescheria*, and *Scedosporium*)	Horses Cats Dogs Goats Cattle	Subcutaneous nodules with purulent exudate containing granules (mainly the trunk and hind legs)	Fluorescence microscopy using calcofluor white Direct microscopy or crushed granules with modified acid-fast staining Histopathology (H&E stain, Gram-Weigert stain, PAS, and Gridley stains) Serological tests to detect antibodies and antigens, PCR	Surgical removal Thiabendazole powder or antimicrobial to prevent secondary infections

(continued)

Table 11.1 (continued)

Disease/fungal pathogen	Animals affected	Clinical signs and symptoms	Diagnosis	Treatment/ management
Oomycosis (*Pythium insidiosum)*	Horse Dogs Cattle Sheep	*Cutaneous form*: Granulomatous lesions on the face and limbs *Intestinal form*: Vomiting, anorexia, intermittent diarrhea, and segmental thickening of various gastrointestinal organs	Wet mount microscopy with 10% KOH Histopathological staining (H&E, GMS, and PAS): Granulomas showing immune cells and fungi Serological tests to detect antibodies, agglutination techniques, PCR	Amputation of affected legs (cutaneous infections) or resectioning (gastrointestinal cases) and itraconazole therapy (2–3 months) post-surgery
Sporotrichosis (*Sporothrix schenckii)*	Dogs Cats Equine Goats Cattle Fowl Swine	*Dogs, horses*: Subcutaneous nodules which eventually ulcerates and becomes purulent and spread toward the neighboring lymph nodes are commonly seen in dogs *Cats*: Initially show tiny, puncturing-type of wounds in the head, base of tail, and legs, and may become nodular. Systemic infection is common in cats, affecting various internal organs as well as the central nervous system	Direct examination of wet mounts using 10% KOH Histopathological slides (stained with H&E, GMS, or PAS): Show spherical cells and PAS-positive capsules. Granuloma divided into three zones Sporotrichin test: a skin test for delayed type of hypersensitivity Serological tests, PCR	Potassium iodide (continued for 2–3 months) or sodium iodide

Reference: Samanta (2015). Veterinary Mycology (No. 11606). Springer India

histoplasmosis in humans. This suggests that there is a possibility that animals can indirectly infect humans via the environment (Norkaew et al., 2013). Other records of histoplasmosis in bats have also been reported in other Southeast Asian countries, namely Malaysia and Thailand, which also have human cases of histoplasmosis (Gugnani & Denning, 2023). In 2017, the first case of feline cutaneous histoplasmosis was confirmed in Thailand (Larsuprom et al., 2017). Evidence of dermatophytes, specifically *Trichophyton* spp., in both humans and animals has also been reported in 2019, namely in Malaysia, Thailand, and Vietnam (Taghipour et al., 2019). Additionally, cases of dermatophytosis in companion animals, which humans are in constant contact with, have been rising in tropical Southeast Asian countries such as Indonesia (Tan et al., 2021). An isolate of *Pythium insidiosum* from humans has also been reported in Thailand and was described to be very similar to human and animal isolates, showing an antigenic relationship between human and animal *P. insidiosum* (Mendoza et al., 1987; Schurko et al., 2003a, b). Human sporotrichosis cases in Malaysia, Thailand, and Laos also report exposure to animals prior to fungal infection (Tovikkai et al., 2020). Continuing reports of incidence of cryptococcal infections in Malaysia, Laos, Vietnam, and Singapore in humans have also shown light to its zoonotic potential (Abraham, 2020; Singh et al., 2018). While there is no established evidence of penicilliosis in animals in Southeast Asian countries to date, the likelihood of zoonotic occurrence can be considered in the future. Globally, eumycetoma infections caused by filamentous fungi, causing subcutaneous infections, are neglected despite it being a multifactorial mycosis affecting both humans and animals. Despite this, the underlying fact remains that eumycetoma infections are also zoonotic and require a multi-sectoral approach for control and prevention measures (Siddig & Ahmed, 2024). Similarly, aforementioned zoonotic fungal diseases should also consider a One Health approach involving various sectors in order to control and prevent the incidence of these infections. Surveillance of both human and animal cases as well as correlation between exposures and incidences should also be strengthened.

Challenges in Veterinary Mycology

Global health concerns on fungal infections are increasing and have been challenging both human and animal health. However, there is very limited information on the prevalence and epidemiology of fungal diseases, including their transmission, spread, and antifungal resistance patterns (WHO, 2022). Knowledge gaps attributed to lack of funding, underreported emerging fungal diseases and a low priority for fungal infections often complicate diagnosis and treatment of fungal infections in animals (Baker & Denning, 2022; WHO, 2022).

Diagnosis and Treatment Challenges

The environmental conditions in subtropical and tropical areas of Southeast Asian countries are warm and humid, which fosters fungal growth, leading to an extremely difficult and challenging diagnosis (Cheng et al., 2021). Moreover, fungal diseases such as aspergillosis, cryptococcosis, and dermatophytosis present with nonspecific symptoms. This is demonstrated in cattle and poultry with aspergillosis and histoplasmosis due to vague clinical signs that resemble bacterial or viral infections (Arunmozhi Balajee et al., 2021). In addition, the symptoms of fungal diseases, which often overlap or mimic those of other serious conditions, nonspecific/

non-localizing clinical presentations, and the host's physical conditions and immunity (Colombo et al., 2017; Badiee & Hashemizadeh, 2014), are some of the factors making diagnosis more challenging.

The significant challenges in the diagnosis of fungal infections in Southeast Asia are attributed to limited knowledge or awareness, lack of laboratory facilities catering diagnostic methods, and even the limited access to antifungal drugs. Fungal infections in animals are frequently underreported and underdiagnosed, which is associated with the lack of knowledge and skills among veterinarians. Advanced diagnostic techniques, confusing clinical symptoms, limited availability and high costs of antifungal treatments, and limited research on fungal diseases also contribute to diagnostic and treatment challenges. Moreover, co-infections with other bacteria, viruses, or parasites have been reported to also cause difficulties in diagnosis and treatment. Mycotic mastitis and aspergillosis in bovines have been found to co-occur with bacterial infections (Fatima, 2021), sporothricosis has been linked to other immunosuppressive diseases in dogs and cats (Eissa, 2024), and most fungal skin infections, including *Malassezia dermatitis*, frequently manifest with bacterial coinfections (Hoffman et al., 2023). Most recently, a fatal case of disseminated protothecosis caused by *P. wickherhamii* was complicated by coinfection with hemoparasites, *Hepatozoon* and *Ehrlichia canis,* causing chronic progression of the disease (Aicardi et al., 2025).

On the other hand, access to diagnostic methods such as fungal culture and advanced molecular diagnostic methods in rural areas is often limited, resulting in misdiagnosis and delayed treatment (Kaewrak et al., 2022). A recent survey has shown significant differences in the diagnostic capabilities of laboratory mycology services across Asia-Pacific in terms of molecular and serological assays for fungal identification (Salmanton-García et al., 2023). Furthermore, the absence of structured mycology laboratories in many Southeast Asian countries contributes to misdiagnosis and treatment delays (Chakrabarti, 2020). On the other hand, epidemiological data on surveillance of invasive fungal infections (IFIs) are limited and necessitate the importance of rapid diagnostic tools (Chen et al., 2022). Because of these diagnostic challenges, veterinarians may experience difficulty in getting efficient or accurate treatment.

Another challenge is the unavailability or limited access to antifungal drugs which impairs treatment against fungal infections. For instance, itraconazole and amphotericin B, commonly used antifungals, are expensive and most of the time unavailable or not readily accessible in veterinary clinics in low-resource regions (Kaewrak et al., 2022). Moreover, more antimycotics were approved for use in crops as compared to antifungals available for animal and human treatments against fungal infections (Fisher et al., 2018). Nevertheless, polyenes and azoles are the most widely used antifungals in both animals and humans and the latter for the control of fungal pathogens in the environment (Price et al., 2015; Bermas & Geddes-McAlister, 2020).

Access to diagnostic techniques like microscopic examination, fungal culture, identification of antifungal resistance, and molecular methods may be lacking, if not limited, especially in low-resource areas should also be considered. Advanced molecular methods like PCR remain costly and accessible in these areas with limited resources as compared to traditional diagnostic methods such as microscopy and culture techniques (Lamoth et al., 2016). It is therefore important to focus on enhancing or improving the laboratory capacity and diagnostic capabilities by increasing veterinary training and easy access to antifungal drugs to improve diagnosis and treatment of animal fungal infections in Southeast Asia (Jongwattanapisan et al., 2021). Meanwhile, the lack of diagnostic facilities, particularly in low-resourced regions, results in empirical antifungal treatments where broad-spectrum antifungals are used without confirming the specific fungal pathogen, further contributing to resistance development (Cheng et al., 2021).

Challenges on Antifungal Resistance

Antifungal resistance is the term used to describe the decreased susceptibility of fungal pathogens to antifungal drugs, resulting in treatment failures and recurring fungal infections. Due to the overuse and misuse of some antifungal drugs in animals, the emergence of antifungal resistance in veterinary medicine has become a growing concern in Southeast Asian countries. With the increasing global concerns on antifungal resistance, several studies demonstrated that the emergence of antifungal resistance was attributed to several mechanisms. The overexpression of efflux pumps such as ATP-binding cassette (ABC) transporters is one of the reported primary mechanisms of resistance, where the efficacy of antifungal agents was reduced (Cheng et al., 2021). Moreover, in a recent study conducted by Hanafiah et al. (2023), mutations in target genes, such as ERG11 in *Candida* species and CYP51A in *Aspergillus*, resulted in reduced azole antifungal susceptibility. Meanwhile, the reduced drug binding affinity of *Candida* and *Aspergillus* species was attributed to the mutations in the ERG11 gene, which encodes lanosterol 14α-demethylase (Morio et al., 2010; Perlin et al., 2017), and FKS1 gene (Krysan et al., 2019), resulting in resistance to azoles and echinocandin, respectively. These mutations reduce drug effectiveness, necessitating alternative treatment strategies. Moreover, ATP-binding cassette (ABC) transporter genes (CDR1 and CDR2) and major facilitator superfamily (MFS) transporters are examples of efflux pumps that can be overexpressed to remove antifungal medications from fungal cells, lowering intracellular drug concentrations (Kanafani & Perfect, 2008; Cowen et al., 2014). In addition to these factors are biofilms that protect fungi from antifungals, and these biofilms are considered potential factors contributing to antifungal resistance (Taff et al., 2013; Harding et al., 2017). Moreover, antifungal agents can be enzymatically broken down by fungi, as demonstrated by the sterol 14α-demethylase mutations in the Cyp51A, which affect the effectiveness of medications (Nascimento et al., 2003; Snelders et al., 2012). Antifungal resistance is also influenced by oxidative stress (Kwon-Chung & Sugui, 2013), cellular stress, and drug exposure (Perlin et al., 2017). These genetic adaptations allow fungi to survive prolonged antifungal exposure, making infections harder to treat and increasing the risk of transmission between animals and humans (Kaewrak et al., 2022).

In Southeast Asia, the unregulated and widespread use of many antifungal drugs is rampant, allowing the increasing number of antifungal resistance cases (Jongwattanapisan et al., 2021). Moreover, treatment of fungal infections is challenging in animals due to the emergence of resistant fungal strains in the environment (Snelders et al., 2008; Verweij et al., 2016; Kaewrak et al., 2022). Among the significant factors that contribute to antifungal resistance are also self-medication by pet owners or farmers, improper prescription practices by veterinarians, as well as the failure to complete the medication (White et al., 2017; Tang et al., 2021). According to Bowman et al. (2020), a lot of animal owners use human antifungal drugs or give their pets the wrong dosages of antifungals without first seeing a veterinarian. This results in subtherapeutic exposure and resistance selection, and partially eradicated fungal infections may re-emerge with resistant characteristics (White et al., 2017).

There have been reports of transmission through contaminated environments, farms, and veterinary clinics. Furthermore, antifungal contamination of the environment, especially in aquaculture and agricultural settings, encourages the development of resistance in fungi (Chowdhary et al., 2013). Antifungal residues persisting in animal feed, bedding materials, and the environment sustain selective pressure on fungal populations, promoting resistance development (Buil et al., 2019; Fisher et al., 2022). Moreover, a study has shown the increasing resistance to azoles in agricultural settings, which are commonly used to treat fungal infections in companion and livestock animals (Berger et al., 2017). Moreover, antifungal resistance is further aggravated by inadequate biosecurity, sanitation procedures, improper disinfection methods, and a lack of personal protective equipment (Bowman et al., 2020). The presence of resistant fungal strains in soil and water sources further facilitates their spread to animals and humans, creating a One Health challenge that requires coordinated intervention (Miller et al., 2020).

Indeed, the growing number of resistant fungal strains makes treatment more difficult and increases the risk of zoonotic infections and socio-economic impacts. Considerable efforts have been made to discover novel therapeutic approaches; however, antifungal resistance is still an emerging concern in veterinary medicine and significantly impacting the treatment of fungal infections in animals. Alternative antifungal treatments like natural products derived from indigenous plants or herbs can be promising alternative antifungal treatments (Jongwattanapisan et al., 2021). The efficacy and safety of traditional herbal remedies, often used as alternative treatments, are poorly studied (Cheng et al., 2021). Additionally, there are inconsistencies in managing fungal infections across different regions in Southeast Asia because of the lack of standardized treatment guidelines (Hanafiah et al., 2023). In order to combat the emergence of antifungal resistance, the conduct of antimicrobial stewardship programs will greatly influence responsible antifungal use and improve diagnostic capabilities in Southeast Asia (Hanafiah et al., 2023). Additionally, stricter surveillance of resistant fungal strains and stronger regulations on antifungal use in veterinary medicine will be useful tools in controlling the spread of antifungal resistance in the region (Miller et al., 2020). With this growing concern, it is imperative to address this in a multifaceted way through the development of point-of-care diagnostic techniques, new or novel antifungal agents, and enhanced public awareness and campaigns on antifungal resistance.

Zoonotic Potential

There is an increasing concern about zoonotic fungal diseases such as blastomycosis, coccidioidomycosis, cryptococcosis, histoplasmosis, dermatophytosis, talaromycosis, sporotrichosis, and *Malassezia* infection, which cause severe infections and threaten both veterinary and human medicine (Garcia-Rubio et al., 2017; MSD Veterinary Manual, 2025). As zoonotic fungal infections continue to emerge in Southeast Asia (Table 11.1), increasing public health concerns on the persistence of antifungal-resistant fungal strains have also been reported due to the risk of transmission to human populations (Bowman et al., 2020). Various fungal species have been reported, and some of these pathogens have the zoonotic potential of spillovers affecting humans. Moreover, detection of antifungal residues in animal waste products can also contribute to fungal infections and eventually spread within humans (Berger et al., 2017). Additionally, subsequent zoonotic transmission of fungal diseases can also be linked to contaminated agricultural and livestock environments. For example, the use of fungicides can lead to the emergence of azole-resistant *Aspergillus fumigatus*, a zoonotic fungal pathogen (Verweij et al., 2016). Due to the increased cases of zoonotic fungal infections in Southeast Asia, the risk of transmission is also high between humans and animals (Hanafiah et al., 2023). Therefore, the zoonotic potential of these fungal pathogens should be recognized, and urgent actions for a One Health approach are needed to mitigate the risks of transmission.

Mycology and One Health

A holistic strategy to address these emerging and zoonotic fungal pathogens and the increasing emergence of antifungal resistance is the One Health approach which is often understudied. The interconnectedness or interdisciplinary collaboration between clinicians, veterinarians, and environmentalists makes a One Health approach to combat fungal diseases and overcome antifungal resistance (Woods et al., 2023).

In Southeast Asia, the transmission of zoonotic fungal pathogens is not only influenced by human-animal interactions but also by environmental factors such as urbanization, climate change, and habitat destruction. The disruption of the environment may affect the natural balance with the rising global temperatures and shifting precipitation patterns allowing fungal pathogens to thrive and spread more easily in animals and humans and favoring fungal proliferation into new geographic regions.

This poses high transmission risk and exposes humans and animals to invasive fungal pathogens by inhaling fungal cells and spores, leading to systemic and fatal infections (Woods et al., 2023).

Reports on the increasing number of confined fungal species in one region have now been appearing in new regions. In addition, climate change can also support the proliferation of fungal species, allowing fungi to evade host immune responses as well as antifungal treatments (Garcia-Solache & Casadevall, 2010). Due to shifting climate conditions, deforestation and urbanization can also influence the increasing cases of *Talaromyces marneffei* infections in rodents and domestic animals, resulting in zoonotic infections (Siengsanan-Lamont & Blacksell, 2018; Hyde et al., 2020). Furthermore, climate-induced stress in animals also weakens their immune responses, making them more susceptible to opportunistic fungi like *Candida* and *Aspergillus*, which cause respiratory and systemic infections (Seidel et al., 2024). Some agricultural fungi in Indonesia were also reported to cause important mycotic diseases in animal livestock (Ahmad & Gholib, 2016). On the other hand, the production of aflatoxins by *Aspergillus* spp. and fumonisins by *Fusarium* spp., which are prevalent among Southeast Asian countries, has also been reported to thrive in warm and humid environments (Benkerroum, 2020). The production of these toxins affects cattle, poultry, and swine, resulting in immunosuppression and increased fungal disease outbreaks (Paterson & Lima, 2011).

There is also the need to monitor our wildlife for fungal infections. The interconnected nature of zoonotic fungal transmission has been demonstrated in surveillance studies conducted in Vietnam, where this study identified important fungal pathogens not only in domesticated animals and wet market environments but also in wildlife (Pham-Thanh et al., 2022). This surveillance showed that the identified fungal pathogens were shared between humans and animals (Pham-Thanh et al., 2022), suggesting a combined response to surveillance and disease management is important. Additionally, human-animal interactions have been documented in stray animals with cases of *Sporothrix schenckii* (Salazar-Hamm & Torres-Cruz, 2024). On the other hand, azole-resistant *Aspergillus fumigatus* (Chowdhary et al., 2013) and *Candida auris* (Casadevall et al., 2019) can be associated with warmer temperatures and climate change, respectively, aside from the increased use of antifungal agents in veterinary medicine. This may raise zoonotic risks, complicate treatment, and highlight the urgent need to conduct surveillance and management strategies. Moreover, One Health approach can also be implemented in new antifungals for testing by integrating agricultural applications with clinical use avoiding dual drivers of resistance (Woods et al., 2023). With the integration of data on surveillance from human medicine, veterinary medicine and environment, this One Health approach can improve early detection, enhance diagnostic capacity, and elevate targeted interventions or treatments against fungal infections.

Limited Research Funding and Expertise in Veterinary Mycology

The limited funding or resources available on fungal diseases of animals is one of the major challenges in veterinary mycology. Less prioritization of fungal diseases has hindered the development of improved diagnostic methods, surveillance studies, and novel antifungal treatments (White et al., 2017; Hanafiah et al., 2023). The lack or limited number of trained mycologists and diagnostic laboratories or infrastructure gaps are another challenge in veterinary medicine, which is limiting the ability to accurately diagnose fungal infections in animals (Tang et al., 2021). Additionally, few research institutions in Southeast Asia offer specialized training in veterinary mycology (Kaewrak et al., 2022).

Meanwhile, bacterial and viral diseases are often the priority of research funding institutions due to their immediate economic and public health impact, allowing minimal resources allocated to mycological research (Cheng et al., 2021). This limits the ability of researchers to conduct various research studies on fungal pathogens, allowing gaps in epidemiological surveillance, transmission, and even antifungal susceptibility testing (Jongwattanapisan et al., 2021). Thus, in order to address these research gaps, there is a need to increase investment to enhance stronger government support for mycological research, extensive training programs, and scholarships to advance fungal disease diagnosis and treatment.

Diagnostic Approaches

In recent years, the diagnostic methods for fungal infections affecting animals have changed dramatically. Though fungal infections in animals across Southeast Asia experienced significant diagnostic challenges due to the region's humid or tropical climate, which supports fungal growth, the development of several advanced diagnostic techniques has been proven to be more specific and sensitive in detecting fungal pathogens in animals. Visualization of fungal specimens in samples can be performed with the use of potassium hydroxide (KOH) or through staining with the use of lactophenol cotton blue (Johansen et al., 2010). These techniques have been performed in combination with fungal culture and molecular techniques such as polymerase chain reaction (PCR). Recently, identification of *Microsporum canis* and *Sporothrix schenckii* from hair plucks, skin scrapings, and lesion swabs from cats in Malaysia has been successfully performed using lactophenol cotton blue staining, fungal culture using Saboroud's dextrose agar (SDA), and PCR targeting the internal transcribed spacer region (ITS) and β-tubulin gene (Mohd-Khlubi et al., 2024). *Aspergillus* spp. was also found to be the causative pathogen in broiler chickens with respiratory distress in Indonesia using KOH, methylene blue staining, and fungal culture using SDA methods (Kaler et al., 2024). On the other hand, *Aspergillus* spp. in poultry, dermatophytes in ruminants, and *Histoplasma* spp. in horses were identified using $NaOH_2$/KOH 10–20% solution and SDA fungal culture methods (Ahmad & Gholib, 2016). Histopathology is another method that can be used to identify fungal infections in laboratory animals. Using the staining methods such as periodic acid-Schiff (PAS) and Grocott-Gomori Methenamine Silver (GMS), fungal infections can be distinguished from bacterial or viral diseases (Shourie et al., 2021). In addition, fungal hyphae, spores, or yeast cells can be identified through histopathological analysis, which aids in distinguishing fungal infections from bacterial or viral illnesses (Shourie et al., 2021).

Animal enclosures, zoos, and their environments may also serve as reservoirs for various pathogenic, zoonotic, or opportunistic fungi, impacting both animal and human health. Considerable efforts have been conducted to identify fungal contaminants in the environment. Some studies were conducted in captive animal environments, and these studies from Thailand, Malaysia, Laos, Indonesia, and the Philippines have identified various fungi, including *Aspergillus* spp., *Penicillium* spp., *Candida* spp., and *Fusarium* spp. A study in Malaysia assessed the fungal contamination in captive wildlife enclosures and identified *Penicillium* and *Candida* species, which are known opportunistic pathogens, from the environmental samples using a passive air sampling technique with SDA plates for fungal culture followed by wet mount and lactophenol cotton blue staining to visualize the fungal structures (Omar et al., 2020). Another study in Thailand also reported fungal diversity in dung samples of wild and domestic animals using conventional methods (microscopy and culture methods) on the identification of overfungi such as *Alternaria, Curvularia, and Fusarium* (Jeamjitt et al., 2006). Similarly, Piasai and Manoch (2009) investigated 16 genera and 20 species of fungi in wildlife dung from barking deer, buffalo, cows, and elephants using microscopy and culture techniques.

The gold standard for fungal identification is still the culture-based methods, such as SDA and chromogenic media, and direct microscopy for morphological and biochemical characterization. In low-resource settings, these methods still play a crucial role in fungal identification and disease management but are time-consuming, making growing fastidious fungi a challenge. Furthermore, conventional culture methods give low sensitivity and are difficult to differentiate closely related fungal species. Hence, molecular diagnostics such as PCR and MALDI-TOF mass spectrometry can be integrated into the diagnostic plan. Advanced diagnosis of both human and animal fungal infections in developed countries is well established. However, fungal-specific molecular diagnostic methods across Southeast Asia remain limited. As compared to conventional diagnostic methods like microscopy and culture techniques, molecular diagnostics like PCR offer higher sensitivity and specificity. However, a study reported that 37 out of 219 (16.9%) laboratories across seven Asian countries only, including Indonesia, Thailand, and the Philippines, are performing fungal PCR testing (Chindamporn et al., 2018). Only a few or limited studies have developed advanced fungal diagnostic techniques in Southeast Asia. Studies across the region have primarily focused on zoonotic pathogens; however, these studies utilized fungal-specific PCR. A multiplex PCR assay was developed to identify *Pythium insidiosum* in animals. In addition, the study also used this method to genotype the fungal pathogen (Rujirawat et al., 2017). PCR, histology, and immunohistochemistry were used to identify *Batrachochytrium dendrobatidis* in captive frogs in Thai zoos (Techangamsuwan et al., 2017). On the other hand, PCR-based diagnosis of *Blastocystis* sp. in companion animals and livestock was also evaluated across Malaysia, Thailand, and Indonesia, and the dynamics of the *Sporothrix schenckii* epidemic in cats in Thailand was assessed using genotypic approaches (Yingchanakiat et al., 2023).

Next-generation sequencing (NGS) is an advanced and sophisticated diagnostic tool for high-throughput identification of fungal pathogens in animals. NGS and metabarcoding of the ITS1 region was used to identify fungal diversity in soil samples in Thailand (Suwannarach et al., 2021). Moreover, NGS can be used to characterize resistance genes in pathogenic fungi, making it an invaluable tool in antifungal resistance surveillance (Jiang et al., 2022; Naik et al., 2024). However, NGS-based fungal identification in animals is an area yet to be fully explored, as fungal genome reference databases are still limited. Although this method has been commonly used in developed countries, its utilization is limited, particularly in Southeast Asian countries where veterinary mycology is already constrained by limited resources.

Treatment and Prevention

Current Therapeutic Regimen

With the growing concern about antifungal resistance, there is a critical need to develop new antifungal drugs and more stringent animal husbandry protocols such as proper hygiene, housing, nutrition, and biosecurity measures. These factors significantly contributed to the emergence and control of fungal disease outbreaks. Accumulation of organic wastes, poor ventilation, and high humidity favor the growth of fungal pathogens such as *Aspergillus* spp., and *Candida* spp. (Ainsworth & Austwick, 1973). The improper use of antibiotics and other treatment drugs in husbandry can disrupt normal microbiota and predispose animals to opportunistic fungal infections, suppressing immune responses (Bond et al., 2020). The selection of treatment for fungal diseases depends on the species, the type of mycosis, and the availability of other treatments. Factors such as the health status of the animal should also be considered, as they may exacerbate the condition, especially with immunocompromised animals. Despite the need for antifungal drugs for animals, only a few antifungals are approved for veterinary use. Table 11.2 shows the common classes of antifungal drugs used for animals.

Several classes of fungal treatment for animals are available depending on the type of mycosis. Griseofulvin is one of the oldest antifungal agents and has been widely used to treat dermatophytosis in dogs and cats (Moriello, 2004) and other livestock and wild animals. Polyenes are systemic, broad-spectrum antifungal drugs that target and disrupt fungal cell membranes by binding to ergosterol and forming pores (Carolus et al., 2020; Ngece et al., 2024). Commonly used polyenes for

Table 11.2 Common classes of antifungal drugs in veterinary medicine

Chemical classes	Drugs	Formulation
Polyenes	Amphotericin B Natamycin	Injectable lotion 3% ((Rochette et al., 2003), intravenous solution (Kahn, 2005), Suspension (Rochette et al., 2003)
Azoles	Clotrimazole	Topical solution (Stanton et al., 2018), intranasal (Smith et al., 1998)
	Miconazole	Topical, intranasal (Kahn, 2005)
	Fluconazole	Oral preparation (Kukanich et al., 2023)
	Itraconazole	Oral preparation (Lee et al., 2024),
	Ketoconazole	Oral preparation (Kahn, 2005)
Allylamines	Terbinafine	Tablet or granules (Williams et al., 2011)
Benzimidazoles	Thiabendazole	4% Solution (Rochette et al., 2003), ointment, otic preparation (Kahn, 2005)
Pyrimidine	Flucytosine	Capsule, intravenous solution (Kahn, 2005)
Others	Griseofulvin	7.5% Powder feed additive; bolus, liquid powder (Rochette et al., 2003)

animals are amphotericin B and natamycin. Amphotericin B has been used for deep mycosis and is known to be effective against *Candida* and *Cryptococcus*. However, cumulative doses of amphotericin must be used with caution due to its possible nephrotoxic effects (Chan et al., 2023). Like other polyenes drugs, natamycin has been used for many years to treat ringworm in animals, particularly horses and cattle (Oldenkamp, 1979a; Oldenkamp, 1979b).

Azoles are broad-class synthetic compounds against fungi that block ergosterol biosynthesis, a component of the cytoplasmic membrane of the fungi (Hector, 2005). Clotrimazole and miconazole are antifungal azole treatments of choice for superficial mycosis. It has been used to treat *Aspergillus* infection in companion (pets) and livestock animals. However, studies have shown mechanisms that increase resistance against azoles in veterinary use (Castelo-Branco et al., 2020; van Dijk et al., 2024). Fluconazole has been used for localized and disseminated mycoses caused by many fungal pathogens, including coccidioidomycosis in dogs (Carter et al., 2022). It is commonly used for urinary tract infections other than those caused by *Aspergillus* sp. (Hector, 2005). Itraconazole has a broad spectrum of activity against many fungal pathogens in veterinary medicine, including dermatophytes in dogs and cats (Moriello, 2004). Although there are reports of treatment failure in itraconazole, combination therapy with itraconazole with potassium iodide was associated with higher effectivity in cats (Reis et al., 2024) as well as itraconazole with acylhydrazone (AH) (Dib Ferreira Gremião et al., 2024). Ketoconazole is one of the oldest antifungal drugs for veterinary use but is less potent than other azole drugs. Therefore, it is commonly used in combination with other antifungal agents.

Allylamines is another class of antifungal that act on fungal cell membranes that specifically inhibit squalene epoxidase, which is important for fungal sterol biosynthesis (Hammoudi Halat et al., 2022). Terbinafine is a common treatment for microsporosis (Kotnik et al., 2001) and Malassezia dermatitis (Berger et al., 2012). Benzimidazoles are known for their broad spectrum of activity with antiparasitic and antifungal activities. Thiabendazole has been used to treat Malassezia in dogs and is applied topically in otic preparation (Rochette et al., 2003). Another class of antifungal agents are pyrimidines, which act by inhibiting RNA and DNA synthesis (Basha & Goudgaon, 2021). A member of this class is flucytosine, which is commonly combined with amphotericin B due to its synergistic effects (Rochette et al., 2003). Animal studies using a murine model have also shown the effectiveness of the flucytosine + amphotericin B combination against cryptococcal meningitis (Larsen et al., 1996).

Antifungal treatments can either target topical or systemic infection. For topical treatments, 1% clotrimazole has been shown to have 94% success rate for the treatment of mycotic rhinitis-rhinosinusitis in dogs (Hazuchova et al., 2017) and enilconazole with miconazole/chlorhexidine shampoo as a treatment for dermatophytosis in dogs and cats (Moriello, 2004). For wild and domestic animals, mycosis is commonly treated with topical agents containing C-alkyl esters, specifically targeting hooves, claws, and nails (Meyer, 2014). One study has demonstrated the effectiveness of amphotericin B with dimethyl sulfoxide (DMSO) as a treatment for equine cutaneous pythiosis (Dias et al., 2012). For systemic infections, common treatment includes itraconazole and nystatin (Elad, 2018), as well as amphotericin B, ketoconazole, and fluconazole (Foy & Trepanier, 2010). However, due to limited licensed products for animals, veterinarians frequently use off-label human-approved antifungals (Seyedmousavi et al., 2018).

Other than treatments, another known effective method is improving animal husbandry protocols that contribute to the overall prevention and control of fungal diseases in animals, especially livestock animals. Good husbandry practices in feed preparation and storage, water and sanitation, disinfection, waste management, and improved health monitoring are ways that can prevent the development of fungal diseases.

Emerging Therapies

Effective prevention and control is a way to mitigate the spread of fungal diseases in animals. Although mycology diseases are relatively common in domestic and livestock animals, treatment can be challenging with the emergence of anti-fungal resistant pathogens. Advancements in research have led to the development of antifungals, vaccines, and alternative therapies. The use of vaccines has been used for some animals and is also available in other countries. However, limited information is still known about the efficacy of the vaccines. Vaccines have been developed for dermatophytosis in cattle, dogs, and cats (Lund & DeBoer, 2008). Experimental vaccines were evaluated for coccidiomycosis in dogs and are promising (Cabañes, 2023).

Probiotics were also reported to be effective against various pathogens, including fungi. Probiotic strains have inhibitory effects in vitro against *Candida albicans, Candida tropicalis, and Candida parasilosis* (Poon & Hui, 2023). While these have not been evaluated in animals, their use for veterinary medicine is promising.

Future Directions of Veterinary Mycology

With the emergence of new pathogens and the increase in stability and virulence of pathogens, treatment becomes challenging. Therefore, several key areas need attention to address the challenges in veterinary mycology.

Enhancement of diagnostics is needed for early detection and treatment of fungal infection. The need to develop new diagnostic tools is essential for rapid and accurate diagnosis. Anti-fungal resistance monitoring is also needed to characterize mechanisms and develop new treatment protocols. Determining the prevalence of fungal infections in animals is also warranted to determine new interventions. One Health highlights the importance of the interconnectedness of humans, animals, and the environment. The varying weather conditions also drive the expansion of geographic ranges of fungi and the emergence of new pathogens, which can impact humans, animals, and the environment. Reports have shown that with the new emergence of fungi, common antifungals that have been used for humans, animals, and crops have developed resistance to drugs (Kakeya, 2022). Therefore, collaborative efforts between multiple stakeholders to combat these challenges, including prioritizing research efforts, are warranted.

In 2022, the World Health Organization released a list of priority diseases, called the fungal priority pathogens list (FPPL), that pose a global threat to public health. The list is divided into three categories: critical, high, and medium priority, depending on their impact and anti-fungal resistance risk. The list also serves as a guide to which research efforts should be focused in a given country. However, while these pathogens in the list are considered priorities, careful consideration must be made for pathogens that are endemic and are more of concern in specific local regions.

In veterinary medicine, research on fungal pathogens is limited. Several tools are already available to support studies, such as genomics, nanotechnology, and even artificial intelligence (AI). With the growing need to expand the knowledge and address research gaps in mycology research, the use of advanced technology is essential. The use of genomics has been widely used to characterize fungal pathogens and discover functional genomes (Gervais & Shapiro, 2024). The use of nanotechnology has also been revolutionary in the fields of agriculture (Nizamani et al., 2024) and veterinary medicine (Alghuthaymi et al., 2021). The emerging use of AI in various applications is becoming popular, including its use in the field of mycology, especially in diagnostics. Several studies have evaluated AI in diagnostics for human health (Xu et al., 2023; Bulińska et al., 2024) and very few in veterinary medicine (Hwang et al., 2022). All these new technologies can help in the overall approach in diagnostics and monitoring of fungal diseases.

Finally, increasing awareness and education about fungi affecting animals is also needed, especially among veterinarians. Training and skill development is essential for early detection, accurate diagnosis, and treatment in animals.

Conclusion

Veterinary mycology plays a critical role in understanding the fungal diseases not only affecting animals but also those with zoonotic potential. Southeast Asia is considered a "hotspot" due to its climate, which may contribute to the emergence of new pathogens. The development of advanced technologies has improved the early and accurate detection of fungi and its timely disease reporting. The increasing number of antifungal resistances calls for active surveillance and collaboration between animal and human health in developing innovative treatments.

References

Abraham, O. C. (2020). Cryptococcosis in Asia. In A. Chakrabarti (Ed.), *Clinical practice of medical mycology in Asia*. Springer. https://doi.org/10.1007/978-981-13-9459-1_18

Ahmad, R. Z., & Gholib, D. (2016). Important mycotic diseases in animal livestock caused by some agricultural fungi in Indonesia. *International Journal of Tropical Veterinary and Biomedical Research, 1*(1), 38–44. https://doi.org/10.21157/IJTVBR.V1I1.5079

Aicardi, V., Giannini, L., & Rossi, A. (2025). Fatal disseminated protothecosis in a dog co-infected with *Hepatozoon*and *Ehrlichia canis*. *Journal of Veterinary Internal Medicine, 39*(1), 120–130.

Ainsworth, G. C. (2019). *Introduction to the history of mycology*. Cambridge University Press.

Ainsworth, G. C., & Austwick, P. K. C. (1973). *Fungal diseases of animals*. Commonwealth Agricultural Bureaux.

Alghuthaymi, M. A., Hassan, A. A., Kalia, A., Sayed El Ahl, R. M. H., El Hamaky, A. A. M., Oleksak, P., Kuca, K., & Abd-Elsalam, K. A. (2021). Antifungal nano-therapy in veterinary medicine: Current status and future prospects. *Journal of Fungi (Basel), 7*(7), 494. https://doi.org/10.3390/jof7070494

Arunmozhi Balajee, S., Borman, A. M., Brandt, M. E., et al. (2021). Diagnosing and treating invasive fungal infections in veterinary medicine: Challenges and advances. *Veterinary Mycology Journal, 35*(4), 215–230.

Badiee, P., & Hashemizadeh, Z. (2014). Opportunistic invasive fungal infections: Diagnosis & clinical management. *The Indian Journal of Medical Research, 139*(2), 195–204.

Baker, S., & Denning, D. W. (2022). Global burden of fungal infections and the need for urgent action. *The Lancet Infectious Diseases, 22*(10), e277–e282. https://doi.org/10.1016/S1473-3099(22)00293-4

Basha, J., & Goudgaon, N. M. (2021). A comprehensive review on pyrimidine analogs—Versatile scaffold with medicinal and biological potential. *Journal of Molecular Structure, 1246*, 131168. https://doi.org/10.1016/j.molstruc.2021.131168

Benkerroum, N. (2020). Chronic and acute toxicities of aflatoxins: Mechanisms of action. *International Journal of Environmental Research and Public Health, 17*(2), 423. https://doi.org/10.3390/ijerph17020423

Berger, D. J., Lewis, T. P., Schick, A. E., & Stone, R. T. (2012). Comparison of once-daily versus twice-weekly terbinafine administration for the treatment of canine Malassezia

dermatitis: A pilot study. *Veterinary Dermatology, 23*(5), 418–e79. https://doi.org/10.1111/j.1365-3164.2012.01074.x
Berger, S., El Chazli, Y., Babu, A. F., & Coste, A. T. (2017). Azole resistance in *aspergillus fumigatus*: A consequence of antifungal use in agriculture? *Frontiers in Microbiology, 8*, 1024. https://doi.org/10.3389/fmicb.2017.01024
Bermas, A., & Geddes-McAlister, J. (2020). Combating the evolution of antifungal resistance in *Cryptococcus neoformans. Molecular Microbiology, 114*, 721–734. https://doi.org/10.1111/mmi.14565
Bond, R., Morris, D. O., Guillot, J., et al. (2020). Biology, diagnosis and treatment of Malassezia dermatitis in dogs and cats. *Clinical, Cosmetic and Investigational Dermatology, 13*, 463–472. https://doi.org/10.2147/CCID.S245345
Bowman, J. C., et al. (2020). Global surveillance of antifungal resistance. *Clinical Microbiology Reviews, 33*(2), e00111–e00119.
Buil, J. B., et al. (2019). Azole-resistant Aspergillus fumigatus in the environment: The Netherlands as a model country? *Frontiers in Microbiology, 10*, 2520. https://doi.org/10.3389/fmicb.2019.02520
Bulińska, B., Mazur-Milecka, M., Sławińska, M., Rumiński, J., & Nowicki, R. J. (2024). Artificial intelligence in the diagnosis of onychomycosis—Literature review. *Journal of Fungi, 10*(8), 534. https://doi.org/10.3390/jof10080534
Cabañes, F. J. (2023). A promising candidate vaccine for coccidioidomycosis. *Revista Iberoamericana de Micología, 40*(1), 1–2. https://doi.org/10.1016/j.riam.2022.03.002
Carolus, H., Pierson, S., Lagrou, K., & Van Dijck, P. (2020). Amphotericin B and other polyenes: Discovery, clinical use, mode of action, and drug resistance. *Journal of Fungi (Basel), 6*(4), 321. https://doi.org/10.3390/jof6040321
Carter, T. D., Fuller, B. D., Shaver, S. L., & Foy, D. S. (2022). Retrospective evaluation of the efficacy of fluconazole for the treatment of coccidioidomycosis in dogs: 49 cases (2015–2020). *Journal of the American Veterinary Medical Association, 260*(10), 1200–1205. https://doi.org/10.2460/javma.22.01.0006
Casadevall, A., Kontoyiannis, D. P., & Robert, V. (2019). On the emergence of Candida auris: Climate change, Azoles, Swamps, and Birds. *mBio, 10*(4), e01397–e01319. https://doi.org/10.1128/mBio.01397-19
Castelo-Branco, D. S. C. M., Paiva, M. A. N., Teixeira, C. E. C., Caetano, É. P., Guedes, G. M. M., Cordeiro, R. A., Brilhante, R. S. N., Rocha, M. F. G., & Sidrim, J. J. C. (2020). Azole resistance in *Candida* from animals calls for the one health approach to tackle the emergence of antimicrobial resistance. *Medical Mycology, 58*(7), 896–905. https://doi.org/10.1093/mmy/myz135
Chakrabarti, A. (2020). *Difficulties faced in Asian countries for the diagnosis of fungal infections and possible solutions* (pp. 199–205). *Springer*. https://doi.org/10.1007/978-981-13-9459-1_13
Chakrabarti, A., & Slavin, M. A. (2011). Endemic fungal infections in the Asia-Pacific region. *Medical Mycology, 49*(4), 337–344. https://doi.org/10.3109/13693786.2010.551426
Chan, J. C., Dear, J., Palm, C., & Reagan, K. (2023). Incidence of acute kidney injury in dogs with systemic mycotic infections treated with amphotericin B (1996–2020). *Journal of Veterinary Internal Medicine,* 37(3), 1030–1037. https://doi.org/10.1111/jvim.16728 *Erratum in:* Chan, J. C., Dear, J., Palm, C., & Reagan, K. (2023). Journal of Veterinary Internal Medicine, *37*(5), 1945. https://doi.org/10.1111/jvim.16817.
Chen, Y.-C., Chayakulkeeree, M., Chakrabarti, A., Gan, G. G., Kwong, Y.-L., Liu, W.-L., Tan, B. H., & Todi, S. (2022). Unmet needs and practical solutions in the management of invasive mould infections in Asia. *Journal of Antimicrobial Chemotherapy, 77*(10), 2579–2585. https://doi.org/10.1093/jac/dkac251
Cheng, X., et al. (2021). Mechanisms of antifungal resistance in veterinary medicine: A southeast Asian perspective. *Journal of Veterinary Mycology, 35*(2), 112–125.
Chindamporn, A., Chakrabarti, A., Li, R., Sun, P. L., Tan, B. H., Chua, M., Wahyuningsih, R., Patel, A., Liu, Z., Chen, Y. C., & Chayakulkeeree, M. (2018). Survey of laboratory practices

for diagnosis of fungal infection in seven Asian countries: An Asia Fungal Working Group (AFWG) initiative. *Medical Mycology, 56*(4), 416–425. https://doi.org/10.1093/mmy/myx066

Chowdhary, A., Sharma, C., & Hagen, F. (2013). Environmental reservoirs of azole-resistant *Aspergillus fumigatus*. *Clinical Microbiology Reviews, 26*(4), 667–681.

Colombo, A. L., de Almeida, J. N., Rosenthal, B. D., & Nucci, M. (2017). Epidemiology and treatment of hematogenous candidiasis: A Brazilian perspective. *Brazilian Journal of Microbiology, 48*(3), 401–413. https://doi.org/10.1016/j.bjm.2017.01.002

Cowen, L. E., Sanglard, D., Howard, S. J., Rogers, P. D., & Perlin, D. S. (2014). Mechanisms of antifungal drug resistance. *Cold Spring Harbor Perspectives in Medicine, 5*(7), a019752. https://doi.org/10.1101/cshperspect.a019752

Dias, D. P. M., Dória, R. G. S., Pereira, R. N., Canola, P. A., & Di Filippo, P. A. (2012). Topical treatment using amphotericin B and DMSO for an atypically located equine cutaneous pythiosis. *Acta Scientiae Veterinariae, 40*(4), 1–8. https://www.cabdirect.org/cabdirect/abstract/20123339861

Dib Ferreira Gremião, I., Pereira-Oliveira, G. R., Pereira, S. A., Corrêa, M. L., Borba-Santos, L. P., Viçosa, A. L., Garg, A., Haranahalli, K., Dasilva, D., Pereira de Sa, N., Matos, G. S., Silva, V., Lazzarini, C., Fernandes, C. M., Miranda, K., Artunduaga Bonilla, J. J., Nunes, A. L., Nimrichter, L., Ojima, I., Mallamo, J., McCarthy, J. B., & Del Poeta, M. (2024). Combination therapy of itraconazole and an acylhydrazone derivative (D13) for the treatment of sporotrichosis in cats. *Microbiology Spectrum, 12*(6), e0396723. https://doi.org/10.1128/spectrum.03967-23

Eissa, M. (2024). Sporotrichosis in dogs and cats: Clinical and epidemiological perspectives. *Veterinary Mycology, 12*(1), 88–102.

Elad, D. (2018). Therapy of non-dermatophytic mycoses in animals. *Journal of Fungi, 4*(4), 120. https://doi.org/10.3390/JOF4040120

Fatima, N. (2021). Role of fungal pathogens in bovine mastitis: An emerging concern. *International Journal of Veterinary Sciences and Animal Husbandry, 6*(3), 45–52.

Fisher, M. C., Hawkins, N. J., Sanglard, D., & Gurr, S. J. (2018). Worldwide emergence of resistance to antifungal drugs challenges human health and food security. *Science, 360*(6390), 739–742. https://doi.org/10.1126/science.aap7999

Fisher, M. C., Alastruey-Izquierdo, A., Berman, J., Bicanic, T., Bignell, E. M., Bowyer, P., Bromley, M., Brüggemann, R., Garber, G., Cornely, O. A., Gurr, S. J., Harrison, T. S., Kuijper, E., Rhodes, J., Sheppard, D. C., Warris, A., White, P. L., Xu, J., Zwaan, B., & Verweij, P. E. (2022). Tackling the emerging threat of antifungal resistance to human health. *Nature Reviews. Microbiology, 20*(9), 557–571. https://doi.org/10.1038/s41579-022-00720-1

Foy, D. S., & Trepanier, L. A. (2010). Antifungal treatment of small animal veterinary patients. *Veterinary Clinics of North America: Small Animal Practice, 40*(6), 1171–1188. https://doi.org/10.1016/j.cvsm.2010.07.006

Garcia-Rubio, R., et al. (2017). The fungal cell wall: Candida, Cryptococcus, and aspergillus species. *Frontiers in Microbiology, 8*, 299. https://doi.org/10.3389/fmicb.2019.02993

Garcia-Solache, M. A., & Casadevall, A. (2010). Global warming will bring new fungal diseases for mammals. *mBio, 1*(1), e00061–e00010. https://doi.org/10.1128/mBio.00061-10

Gervais, N. C., & Shapiro, R. S. (2024). Discovering the hidden function in fungal genomes. *Nature Communications, 15*(1), 8219. https://doi.org/10.1038/s41467-024-52568-z

Gugnani, H. C., & Denning, D. W. (2023). Infection of bats with *Histoplasma* species. *Medical Mycology, 61*(8), myad080.

Hammoudi Halat, D., Younes, S., Mourad, N., & Rahal, M. (2022). Allylamines, benzylamines, and fungal cell permeability: A review of mechanistic effects and usefulness against fungal pathogens. *Membranes, 12*(12), 1171. https://doi.org/10.3390/membranes12121171

Hanafiah, R., et al. (2023). Emerging antifungal resistance in zoonotic fungi: Implications for Southeast Asia. *Tropical Medicine and Infectious Disease, 8*(1), 45–58.

Harding, M. W., Marques, L. L. R., Howard, R. J., & Olson, M. E. (2017). Biofilm morphologies and their role in antifungal resistance. *Journal of Fungi, 3*(2), 12. https://doi.org/10.3390/jof3020012
Hazuchova, K., Neiger, R., & Stengel, C. (2017). Topical treatment of mycotic rhinitis-rhinosinusitis in dogs with meticulous debridement and 1% clotrimazole cream: 64 cases (2007–2014). *Journal of the American Veterinary Medical Association, 250*(3), 309–315. https://doi.org/10.2460/javma.250.3.309
Hector, R. F. (2005). An overview of antifungal drugs and their use for treatment of deep and superficial mycoses in animals. *Clinical Techniques in Small Animal Practice, 20*(4), 240–249. https://doi.org/10.1053/j.ctsap.2005.07.005
Hoffman, R. L., Chen, S., & Patel, M. R. (2023). Malassezia dermatitis and bacterial co-infections in dermatology. *Veterinary Dermatology, 34*(2), 56–67.
Hwang, S., Shin, H. K., Park, J. M., et al. (2022). Classification of dog skin diseases using deep learning with images captured from multispectral imaging device. *Molecular & Cellular Toxicology, 18*, 299–309. https://doi.org/10.1007/s13273-022-00249-7
Hyde, K. D., Dong, Y., Phookamsak, R., et al. (2020). Fungal diversity notes 1151–1276: Taxonomic and phylogenetic contributions on genera and species of fungal taxa. *Fungal Diversity, 100*, 5–277. https://doi.org/10.1007/s13225-020-00439-5
Jeamjitt, O., Manoch, L., Visarathanonth, N., & Chamswarng, C. (2006). Diversity and distribution of hyphomycetes from dung in Thailand. *Kasetsart Journal: Natural Sciences, 40*(4), 890–901. http://kasetsartjournal.ku.ac.th/kuj_files/2008/A080402154114.pdf
Jiang, S., Chen, Y., Han, S., Lv, L., & Li, L. (2022). Next-generation sequencing applications for the study of fungal pathogens. *Microorganisms, 10*(10), 1882. https://doi.org/10.3390/microorganisms10101882
Johansen, M. V., Sithithaworn, P., Bergquist, R., & Utzinger, J. (2010). Towards improved diagnosis of zoonotic trematode infections in Southeast Asia. *Advances in Parasitology, 73*, 267–315. https://doi.org/10.1016/S0065-308X(10)73007-4
Jongwattanapisan, T., et al. (2021). Antifungal stewardship and resistance management in southeast Asian veterinary practice. *Asian Journal of Animal Sciences, 19*(3), 221–234.
Kaewrak, P., et al. (2022). Environmental reservoirs of antifungal resistance in southeast Asian agriculture. *Veterinary Research and Policy, 14*(4), 287–301.
Kahn, C. M. (2005). *The Merck veterinary manual* (9th ed.). Merck.
Kakeya, H. (2022). Medical mycology seen through a one health approach. *Medical Mycology Journal, 63*(1), 11–15. https://doi.org/10.3314/mmj.22.001
Kaler, I. G. B. U., Suarjana, I. G. K., Berata, I. K., Suardana, I. B. K., & Suratma, N. A. (2024). Aspergillosis in broiler chickens in Tunjuk Village, Tabanan District, Tabanan regency, Bali. *Veterinary Science and Medicine Journal, 6*(4), 10. https://doi.org/10.24843/vsmj.2024.v06.i04.p10
Kanafani, Z. A., & Perfect, J. R. (2008). Resistance to antifungal agents: Mechanisms and clinical impact. *Clinical Infectious Diseases, 46*(1), 120–128. https://doi.org/10.1086/524071
Kotnik, T., Kozuh Erzen, N., Kuzner, J., & Drobnic-Kosorok, M. (2001). Terbinafine hydrochloride treatment of *Microsporum canis*-experimentally induced ringworm in cats. *Veterinary Microbiology, 83*(2), 161–168. https://doi.org/10.1016/s0378-1135(01)00412-6
Krysan, D. J., Didone, L., & Aksenov, S. (2019). Mechanisms of echinocandin resistance in *Candida* and *Aspergillus*species. *Clinical Microbiology Reviews, 32*(4), e00135–e00118. https://doi.org/10.1128/CMR.00135-18
Kukanich, K., Kukanich, B., & Magnin, G. (2023). Oral fluconazole has variable pharmacokinetics in dogs. *Journal of Veterinary Pharmacology and Therapeutics, 46*(2), 71–76. https://doi.org/10.1111/jvp.13101
Kundu, R., Bansal, Y., & Singla, N. (2024). The zoonotic potential of fungal pathogens: Another dimension of the one health approach. *Diagnostics, 14*(18), 2050. https://doi.org/10.3390/diagnostics14182050

Kwon-Chung, K. J., & Sugui, J. A. (2013). Aspergillus fumigatus--what makes the species a ubiquitous human fungal pathogen? *PLoS Pathogens, 9*(12), e1003743. https://doi.org/10.1371/journal.ppat.1003743

Lamoth, F., Lockhart, S. R., Berkow, E. L., & Calandra, T. (2016). Challenges in the molecular diagnosis of fungal infections: A veterinary perspective. *Journal of Fungal Pathology, 12*(3), 145–159.

Larsen, R. A., Bauer, M., Weiner, J. M., Diamond, D. M., Leal, M. E., Ding, J. C., Rinaldi, M. G., & Graybill, J. R. (1996). Effect of fluconazole on fungicidal activity of flucytosine in murine cryptococcal meningitis. *Antimicrobial Agents and Chemotherapy, 40*(9), 2178–2182. https://doi.org/10.1128/AAC.40.9.2178

Larsuprom, L., Duangkaew, L., Kasorndorkbua, C., Chen, C., Chindamporn, A., & Worasilchai, N. (2017). Feline cutaneous histoplasmosis: The first case report from Thailand. *Medical Mycology Case Reports, 18*, 28–30.

Lee, H., Koo, Y., Yun, T., Lee, D., Chae, Y., Min, K. D., Kim, H., Yang, M. P., & Kang, B. T. (2024). A single-blind randomised study comparing the efficacy of fluconazole and itraconazole for the treatment of *Malassezia* dermatitis in client-owned dogs. *Veterinary Dermatology, 35*(3), 284–295. https://doi.org/10.1111/vde.13233

Lund, A., & DeBoer, D. J. (2008). Immunoprophylaxis of Dermatophytosis in animals. *Mycopathologia, 166*, 407–424. https://doi.org/10.1007/s11046-008-9111-6

Mendoza, L., Kaufman, L., & Standard, P. (1987). Antigenic relationship between the animal and human pathogen *Pythium insidiosum* and nonpathogenic *Pythium* species. *Journal of Clinical Microbiology, 25*(11), 2159–2162. https://doi.org/10.1128/jcm.25.11.2159-2162.1987

Meyer, H. (2014). U.S. patent application no. 14/301,368.

Miller, J. D., et al. (2020). Fungal zoonoses and antifungal resistance: A one health perspective. *Mycopathologia, 185*(5), 761–774.

Mohd-Khlubi, A. N., Omar, S., Khairani-Bejo, S., & Megat Abdul Rani, P. A. (2024). *Microsporum canis* and *Sporothrix schenckii*: Fungi causing skin infections in cats. *Pertanika Journal of Tropical Agricultural Science, 47*(3), 645–658. https://doi.org/10.47836/pjtas.47.3.04

Moriello, K. A. (2004). Treatment of dermatophytosis in dogs and cats: Review of published studies. *Veterinary Dermatology, 15*(2), 99–107. https://doi.org/10.1111/j.1365-3164.2004.00361.x

Morio, F., Pagniez, F., Besse, B., Gay-Andrieu, F., Miegeville, M., & Le Pape, P. (2010). Deciphering azole resistance mechanisms in *Candida albicans* clinical isolates. *Antimicrobial Agents and Chemotherapy, 54*(4), 1442–1450. https://doi.org/10.1128/AAC.01370-09

MSD Veterinary Manual. (2025). *Global zoonotic diseases: Fungal diseases*. Retrieved April 2, 2025, from https://www.msdvetmanual.com/multimedia/table/global-zoonotic-diseases-fungal-diseases.

Naik, S., Kashyap, D., Deep, J., Darwish, S., Cross, J., Mansoor, E., Garg, V. K., & Honnavar, P. (2024). Utilizing next-generation sequencing: Advancements in the diagnosis of fungal infections. *Diagnostics, 14*(15), 1664. https://doi.org/10.3390/diagnostics14151664

Nascimento, A. M., Goldman, G. H., Park, S., Marras, S. A., Delmas, G., Oza, U., et al. (2003). Multiple resistance mechanisms among aspergillus fumigatus mutants with high-level resistance to itraconazole. *Antimicrobial Agents and Chemotherapy, 47*(5), 1719–1726. https://doi.org/10.1128/AAC.47.5.1719-1726.2003

Ngece, K., Ntondini, T. L., Khwaza, V., Paca, A. M., & Aderibigbe, B. A. (2024). Polyene-based derivatives with antifungal activities. *Pharmaceutics, 16*(8), 1065. https://doi.org/10.3390/pharmaceutics16081065

Nguyen, T. T., Mai, T. N., Dang-Xuan, S., Nguyen-Viet, H., Unger, F., & Lee, H. S. (2024). Emerging zoonotic diseases in Southeast Asia in the period 2011–2022: A systematic literature review. *The Veterinary Quarterly, 44*(1), 1–15. https://doi.org/10.1080/01652176.2023.2300965

Nizamani, M. M., Hughes, A. C., Zhang, H. L., & Wang, Y. (2024). Revolutionizing agriculture with nanotechnology: Innovative approaches in fungal disease management and plant health monitoring. *Science of the Total Environment, 928*, 172473. https://doi.org/10.1016/j.scitotenv.2024.172473

Norkaew, T., Ohno, H., Sriburee, P., Tanabe, K., Tharavichitkul, P., Takarn, P., et al. (2013). Detection of environmental sources of *Histoplasma capsulatum* in Chiang Mai, Thailand, by nested PCR. *Mycopathologia, 176*, 395–402.

Oldenkamp, E. P. (1979a). Treatment of ringworm in horses with natamycin. *Equine Veterinary Journal, 11*(1), 36–38. https://doi.org/10.1111/j.2042-3306.1979.tb01294.x

Oldenkamp, E. P. (1979b). Natamycin treatment of ringworm in cattle in the United Kingdom. *Veterinary Record, 105*(24), 5554–5556.

Omar, S., Jalaludin, F. A., Yee, J. M., Kamarudin, Z., Jayaseelan, K., Khlubi, A. N. M., Madaki, Y. L., Hassan, H., Ramli, M. N., Topani, R., & Che-Amat, A. (2020). Mycological isolation from animal enclosures and environments in National Wildlife Rescue Centre and National zoo, Malaysia. *Journal of Veterinary Medical Science, 82*(8), 1236–1242. https://doi.org/10.1292/jvms.20-0229

Paterson, R. R. M., & Lima, N. (2011). Further mycotoxin effects from climate change. *Food Research International, 44*(9), 2555–2566. https://doi.org/10.1016/j.foodres.2011.05.038

Patouillat, L., Hambuckers, A., Subrata, S. A., et al. (2024). Zoonotic pathogens in wild Asian primates: A systematic review highlighting research gaps. *Frontiers in Veterinary Science, 11*, 1386180. https://doi.org/10.3389/fvets.2024.1386180

Perlin, D. S., Rautemaa-Richardson, R., & Alastruey-Izquierdo, A. (2017). The global problem of antifungal resistance: Prevalence, mechanisms, and management. *The Lancet Infectious Diseases, 17*(12), e383–e392. https://doi.org/10.1016/S1473-3099(17)30316-X

Pham-Thanh, L., Nhu, T. V., Nguyen, T. V., Tran, K. V., Nguyen, K. C., Nguyen, H. T., Hoa, N. T., & Padungtod, P. (2022). Zoonotic pathogens and diseases detected in Vietnam, 2020–2021. *One health (Amsterdam, Netherlands), 14*, 100398. https://doi.org/10.1016/j.onehlt.2022.100398

Piasai, O., & Manoch, L. (2009). Coprophilous ascomycetes from Phu Luang wildlife sanctuary and Khao Yai National Park in Thailand. *Kasetsart Journal: Natural Sciences, 43*, 34–40. https://www.cabdirect.org/cabdirect/abstract/20103085077

Poon, Y., & Hui, M. (2023). Inhibitory effect of lactobacilli supernatants on biofilm and filamentation of Candida albicans, Candida tropicalis, and Candida parapsilosis. *Frontiers in Microbiology, 14*, 1105949. https://doi.org/10.3389/fmicb.2023.1105949

Price, C. L., Parker, J. E., Warrilow, A. G., Kelly, D. E., & Kelly, S. L. (2015). Azole fungicides – Understanding resistance mechanisms in agricultural fungal pathogens. *Pest Management Science, 71*(8), 1054–1058. https://doi.org/10.1002/ps.4029

Reis, E. G. D., Pereira, S. A., Miranda, L. H. M., Oliveira, R. V. C., Quintana, M. S. B., Viana, P. G., Figueiredo, A. B. F., Honorato, C. C. D. S., Pereira-Oliveira, G. R., Silva, J. N., Schubach, T. M. P., & Gremião, I. D. F. (2024). A randomized clinical trial comparing itraconazole and a combination therapy with itraconazole and potassium iodide for the treatment of feline sporotrichosis. *Journal of Fungi, 10*(2), 101. https://doi.org/10.3390/jof10020101

Rochette, F., Engelen, M., & Vanden Bossche, H. (2003). Antifungal agents of use in animal health—Practical applications. *Journal of Veterinary Pharmacology and Therapeutics, 26*(1), 31–53. https://doi.org/10.1046/j.1365-2885.2003.00457.x

Rujirawat, T., Sridapan, T., Lohnoo, T., Yingyong, W., Kumsang, Y., Sae-Chew, P., Tonpitak, W., & Krajaejun, T. (2017). Single nucleotide polymorphism-based multiplex PCR for identification and genotyping of the oomycete *Pythium insidiosum* from humans, animals, and the environment. *Infection, Genetics and Evolution, 54*, 429–436. https://doi.org/10.1016/j.meegid.2017.08.004

Salazar-Hamm, P., & Torres-Cruz, T. (2024). The impact of climate change on human fungal pathogen distribution and disease incidence. *Current Clinical Microbiology Reports, 11*, 1–13. https://doi.org/10.1007/s40588-024-00224-x

Salmanton-García, J., Au, W. Y., Hoenigl, M., Chai, L. Y. A., Badali, H., Basher, A., Brockhoff, R. A., Chen, S. C.-A., Chindamporn, A., Chowdhary, A., Heath, C. H., Jabeen, K., Lee, J., Matar, M. J., Taj-Aldeen, S. J., Tan, B. H., Ueno, K., Wahyuningsih, R., Chakrabarti, A., &

Cornely, O. A. (2023). The current state of laboratory mycology in Asia/Pacific: A survey from the European Confederation of Medical Mycology (ECMM) and International Society for Human and Animal Mycology (ISHAM). *International Journal of Antimicrobial Agents, 61*(3), 106718. https://doi.org/10.1016/j.ijantimicag.2023.106718

Samanta, I. (2015). *Veterinary mycology* (Vol. No. 11606). Springer.

Schurko, A., Mendoza, L., de Cock, A. W. A. M., & Klassen, G. R. (2003a). Evidence for geographic clusters: Molecular genetic differences among strains of *Pythium insidiosum* from Asia, Australia, and the Americas are explored. *Mycologia, 95*(2), 200–208. https://doi.org/10.1080/15572536.2004.11833105

Schurko, A. M., Mendoza, L., Lévesque, C. A., & Klassen, G. R. (2003b). A molecular phylogeny of *Pythium insidiosum. Mycological Research, 107*(5), 537–544. https://doi.org/10.1017/S0953756203007663

Seidel, D., Wurster, S., Jenks, J. D., Sati, H., Gangneux, J.-P., Egger, M., Alastruey-Izquierdo, A., Ford, N. P., Chowdhary, A., Sprute, R., Cornely, O., Thompson, G. R., Hoenigl, M., & Kontoyiannis, D. P. (2024). Impact of climate change and natural disasters on fungal infections. *The Lancet Microbe, 5*(6), e594–e605. https://doi.org/10.1016/S2666-5247(24)00039-9

Seyedmousavi, S., Guillot, J., Tolooe, A., Verweij, P. E., & de Hoog, G. S. (2015). Neglected fungal zoonoses: Hidden threats to man and animals. *Clinical Microbiology and Infection: The Official Publication of the European Society of Clinical Microbiology and Infectious Diseases, 21*(5), 416–425. https://doi.org/10.1016/j.cmi.2015.02.031

Seyedmousavi, S., Bosco, S. M. G., de Hoog, S., Ebel, F., Elad, D., Gomes, R. R., Jacobsen, I. D., Jensen, H. E., Martel, A., Mignon, B., Pasmans, F., Piecková, E., Rodrigues, A. M., Singh, K., Vicente, V. A., Wibbelt, G., Wiederhold, N. P., & Guillot, J. (2018). Fungal infections in animals: A patchwork of different situations. *Medical Mycology, 56*(suppl_1), 165–187. https://doi.org/10.1093/mmy/myx104

Shourie, A., Singh, R., & Singh, A. (2021). *Histopathologic diagnosis of fungal infections of lab animals* (pp. 73–91). Springer. https://doi.org/10.1007/978-3-030-69507-1_6

Siddig, E. E., & Ahmed, A. (2024). The urgent need for developing and implementing a multisectoral one health strategy for the surveillance, prevention, and control of eumycetoma. *IJID One Health, 100048, 100048.*

Siengsanan-Lamont, J., & Blacksell, S. D. (2018). A review of laboratory-acquired infections in the Asia-Pacific: Understanding risk and the need for improved biosafety for veterinary and zoonotic diseases. *Tropical Medicine and Infectious Disease, 3*(2), 36. https://doi.org/10.3390/tropicalmed3020036

Singh, K., Ilkit, M., Shokohi, T., Tolooe, A., Malik, R., & Seyedmousavi, S. (2018). Cryptococcosis: Emergence of *Cryptococcus gattii* in animals and zoonotic potential. In S. Seyedmousavi, G. de Hoog, J. Guillot, & P. Verweij (Eds.), *Emerging and epizootic fungal infections in animals* (pp. xx–xx). Springer. https://doi.org/10.1007/978-3-319-72093-7_12

Smith, S. A., Andrews, G., & Biller, D. S. (1998). Management of nasal aspergillosis in a dog with a single, noninvasive intranasal infusion of clotrimazole. *Journal of the American Animal Hospital Association, 34*(6), 487–492. https://doi.org/10.5326/15473317-34-6-487

Snelders, E., van der Lee, H. A., Kuijpers, J., Rijs, A. J., Varga, J., Samson, R. A., Mellado, E., Donders, A. R., Melchers, W. J., & Verweij, P. E. (2008). Emergence of azole resistance in aspergillus fumigatus and spread of a single resistance mechanism. *PLoS Medicine, 5*(11), e219. https://doi.org/10.1371/journal.pmed.0050219

Snelders, E., van der Lee, H. A., Kuijpers, J., Rijs, A. J., Varga, J., Samson, R. A., et al. (2012). Emergence of azole resistance in *aspergillus fumigatus* and spread of a single resistance mechanism. *PLoS Medicine, 9*(11), e1001183. https://doi.org/10.1371/journal.pmed.1001183

Stanton, J. A., Miller, M. L., Johnson, P., Davignon, D. L., & Barr, S. C. (2018). Treatment of canine sinonasal aspergillosis with clotrimazole infusion in patients with cribriform plate lysis. *Journal of Small Animal Practice, 59*(7), 411–414. https://doi.org/10.1111/jsap.12835

Suwannarach, N., Kumla, J., In-On, A., Lumyong, S., & Lumyong, S. (2021). Soil Metabarcoding offers a new tool for the investigation and hunting of truffles in northern Thailand. *Journal of Fungi, 7*(4), 293. https://doi.org/10.3390/JOF7040293

Taff, H. T., Nett, J. E., Zarnowski, R., Ross, K. M., Sanchez, H., Cain, M. T., et al. (2013). A *Candida albicans* biofilm-induced pathway for matrix glucan delivery: Implications for drug resistance. *PLoS Pathogens, 9*(4), e1003562. https://doi.org/10.1371/journal.ppat.1003562

Taghipour, S., Pchelin, I. M., Zarei Mahmoudabadi, A., Ansari, S., Katiraee, F., Rafiei, A., et al. (2019). *Trichophyton mentagrophytes* and *T. Interdigitale* genotypes are associated with particular geographic areas and clinical manifestations. *Mycoses, 62*(11), 1084–1091.

Tan, Y. F., Widyarini, S., Yanuartono, Y., Nururrozi, A., & Indarjulianto, S. (2021). Canine dermatophytosis in Indonesia, a tropical country, compared to the non-tropical countries. *Indonesian Journal of Veterinary Sciences, 2*(2), 58–67.

Tang, J. L., et al. (2021). Antifungal stewardship in veterinary medicine: An emerging necessity. *Veterinary Microbiology, 258*, 109078.

Techangamsuwan, S., Sommanustweechai, A., Kamolnorranart, S., Siriaroonrat, B., Khonsue, W., & Pirarat, N. (2017). Emerging chytrid fungal pathogen, *Batrachochytrium dendrobatidis*, in zoo amphibians in Thailand. *Acta Veterinaria-Beograd, 67*(4), 525–539. https://doi.org/10.1515/acve-2017-0042

Tovikkai, D., Maitrisathit, W., Srisuttiyakorn, C., Vanichanan, J., Thammahong, A., & Suankratay, C. (2020). Sporotrichosis: The case series in Thailand and literature review in Southeast Asia. *Medical Mycology Case Reports, 27*, 59–63.

van Dijk, M. A. M., Buil, J. B., Tehupeiory-Kooreman, M., Broekhuizen, M. J., Broens, E. M., Wagenaar, J. A., & Verweij, P. E. (2024). Azole resistance in veterinary clinical *aspergillus fumigatus* isolates in The Netherlands. *Mycopathologia, 189*(4), 50. https://doi.org/10.1007/s11046-024-00850-5

Verweij, P. E., Chowdhary, A., Melchers, W. J., & Meis, J. F. (2016). Azole resistance in aspergillus fumigatus: Can we retain the clinical use of Mold-active antifungal azoles? *Clinical Infectious Diseases: An Official Publication of the Infectious Diseases Society of America, 62*(3), 362–368. https://doi.org/10.1093/cid/civ885

White, P. L., et al. (2017). A national strategy to diagnose fungal infections in the United Kingdom. *Journal of Medical Microbiology, 66*(1), 11–17.

Williams, M. M., Davis, E. G., & KuKanich, B. (2011). Pharmacokinetics of oral terbinafine in horses and greyhound dogs. *Journal of Veterinary Pharmacology and Therapeutics, 34*(3), 232–237. https://doi.org/10.1111/j.1365-2885.2010.01213.x

Woods, M., McAlister, J. A., & Geddes-McAlister, J. (2023). A one health approach to overcoming fungal disease and antifungal resistance. *WIREs Mechanisms of Disease, 15*(4), e1610. https://doi.org/10.1002/wsbm.1610

World Health Organization. (2022). *Fungal priority pathogens list to guide research, development and public health action*. World Health Organization. https://www.who.int/publications/i/item/9789240060241

Xu, J., Luo, Y., Wang, J., Tu, W., Yi, X., Xu, X., Song, Y., Tang, Y., Hua, X., Yu, Y., Yin, H., Yang, Q., & Huang, W. E. (2023). Artificial intelligence-aided rapid and accurate identification of clinical fungal infections by single-cell Raman spectroscopy. *Frontiers in Microbiology, 14*, 1125676. https://doi.org/10.3389/fmicb.2023.1125676

Yingchanakiat, K., Limsivilai, O., Sunpongsri, S., Niyomtham, W., Lugsomya, K., & Yurayart, C. (2023). Phenotypic and genotypic characterization and antifungal susceptibility of *Sporothrix schenckii sensu stricto* isolated from a feline Sporotrichosis outbreak in Bangkok, Thailand. *Journal of Fungi (Basel, Switzerland), 9*(5), 590. https://doi.org/10.3390/jof9050590

Chapter 12
From Mushroom Farms to Food Tables: Fungi for Food Security

Jeane A. dela Cruz, James Kennard S. Jacob, Marjone S. Beltran, and Thomas Edison E. dela Cruz

Abstract Fungi, particularly edible mushrooms, have long been a component of human diet, providing a diverse array of flavors, textures, and nutritional advantages. Edible mushrooms are rich sources of high-quality protein, dietary fiber, essential amino acids, a variety of vitamins and minerals, antioxidants, and other bioactive compounds. This book chapter focuses on edible mushrooms, both on their commercial cultivation and the nutritional and therapeutic advantages they offer. We also examined traditional and contemporary culinary traditions in relation to mushrooms. Furthermore, we provide insights on how edible mushrooms can aid in addressing food security through their efficient, low-cost, substrate-based cultivation, which makes them accessible and sustainable.

Keywords Edible fungi · Food items · Fruiting body production · Mushroom farming · Tropical fungi

---

J. A. dela Cruz
The Graduate School, University of Santo Tomas, Manila, Philippines

Department of Biological Sciences, College of Science, University of Santo Tomas, Manila, Philippines

J. K. S. Jacob
Department of Biological Sciences, College of Arts and Sciences, Isabela State University, Isabela, Philippines

M. S. Beltran
Fungal Biodiversity, Ecogenomics and Systematics-Metabolomics (FBeS) Group, Research Center for the Natural and Applied Sciences, University of Santo Tomas, Manila, Philippines

T. E. E. dela Cruz (✉)
The Graduate School, University of Santo Tomas, Manila, Philippines

Department of Biological Sciences, College of Science, University of Santo Tomas, Manila, Philippines

Fungal Biodiversity, Ecogenomics and Systematics-Metabolomics (FBeS) Group, Research Center for the Natural and Applied Sciences, University of Santo Tomas, Manila, Philippines
e-mail: tedelacruz@ust.edu.ph

J. J. G. Guerrero et al. (eds.), *Mycology in a Changing Planet: Applications and Perspectives from Southeast Asia*, https://doi.org/10.1007/978-981-95-2640-6_12

Introduction

Fungi are recognized as a diverse kingdom that includes single-celled yeasts, filamentous molds, and mushrooms and play vital ecological roles as decomposers, mutualistic symbionts, and pathogens in various ecosystems. Many species possess distinctive properties that make them indispensable in numerous culinary, industrial, and biotechnological applications. Among these are the edible mushrooms, which have garnered significant attention due to their multifaceted value, not only as a gourmet food but also for their remarkable nutritional, medicinal, and therapeutic properties. Edible mushrooms, such as *Lentinula edodes* (shiitake), *Pleurotus ostreatus* (oyster mushroom), and *Agaricus bisporus* (button mushroom), are rich sources of essential macro- and micronutrients, which contribute to overall human health and well-being (Chang & Miles, 2004). Many species of these edible fungi contain biologically active compounds with antioxidant, anti-inflammatory, anticancer, and immune-modulatory properties, and thus, further enhance their potential as natural functional foods and therapeutic agents. These benefits are corroborated by numerous studies as reported by O'Callaghan et al. (2015) and Wasser (2010). In addition to edible fruiting bodies of fungi, filamentous molds, particularly those under the genera *Aspergillus*, *Penicillium*, and *Rhizopus*, are used in the production of a wide range of fermented foods and beverages through enzymatic activity, enhancing nutrient bioavailability and contributing to the distinctive flavors, textures, and nutritional profiles defining many culinary staples worldwide (Pouris et al., 2024). These fungi are also critical in the fermentation processes of traditional and modern food production. For example, *Aspergillus oryzae* is essential in the production of soy-based fermented products like miso, soy sauce, and tempeh, while *Penicillium* species are indispensable in the maturation of cheeses such as Roquefort, Camembert, and Brie. The biochemical activities of these molds during fermentation—ranging from the breakdown of complex carbohydrates and proteins to the synthesis of unique metabolites—are key to the sensory and nutritional qualities of these foods, influencing both taste and health benefits (Hawksworth, 2012). Many of these fungal species are classified as Generally Recognized As Safe (GRAS), supporting their widespread use in food production. The well-defined metabolic pathways involved in these processes have been well-documented in studies conducted by Ito and Matsuyama et al. (2021) and Arora and Ramesh (2017).

Aside from filamentous fungi, yeasts are also widely used in the food industry. *Saccharomyces cerevisiae* is well known for its fermentation abilities and metabolic versatility, which make it ideal for baking, brewing, and winemaking. Interestingly, its use dates back to ancient civilizations like the Egyptians and Sumerians, who used yeast to ferment bread and alcoholic beverages (Harutyunyan & Malfeito-Ferreira, 2022). Over time, *S. cerevisiae* has been very useful not just in making food and drinks but also in molecular biology and biotechnology. Today, it plays a key role in the large-scale production of foods, beverages, and even bioethanol.

As research on fungi continues to unfold, the economic significance of edible mushrooms is increasingly recognized to address global challenges in food security, environmental sustainability, and renewable resource utilization (Hogan et al., 2015). This chapter will look into these edible mushrooms and how their nutritional benefits, cultivation, and innovative food applications can support and address food security challenges by providing affordable, nutrient-rich sources that support dietary diversity and environmental sustainability.

Challenges and Threats to Food Security

Global Population Growth

Food security, a fundamental human right, is becoming more complex due to urbanization reducing available arable land, climate change negatively affecting agricultural productivity, dietary transitions changing food demand, lack of resources, and international migration altering population distribution, placing immense pressure on food systems and disrupting food distribution systems. The Rome Declaration on World Food Security (1996) defines food security as a state in which "all people, at all times, have physical and economic access to sufficient, safe, and nutritious food that meets their dietary needs and food preferences for an active and healthy life" (FAO and ECO-RCCFS, 2025). This requires that food production and distribution, whether local or imported, should allow all individuals to have economic and physical means to access nutritious food, clean water, sanitation, and healthcare. Furthermore, food access must remain stable despite economic crises, political instability, and climate-related disturbances. This definition encompasses four key dimensions: food availability, access, utilization, and stability.

The global population reached 8 billion and is projected to increase to 9.8 billion by 2050 (United Nations, 2024). The rapidly growing population places immense pressure on agricultural systems to meet increasing food demand, particularly in areas such as Sub-Saharan Africa and South Asia, where rapid population growth is coupled with severe food insecurity. Although recent trends show a slowdown due to declining fertility rates, this demographic remains a huge challenge in achieving one of the 2030 Sustainable Development Goals, particularly in eradicating poverty and hunger. In response, modern agriculture has turned to intensified agricultural practices to maximize yields. Intensified agricultural practices often entail an aggressive pursuit of higher productivity through methods that can have significant environmental and economic costs. While such practices may involve technology, they frequently rely on the extensive use of chemical fertilizers and pesticides. The indiscriminate use of synthetic fertilizers and pesticides has been shown to cause soil degradation, water pollution, and harm to the environment and agroecological systems (Khatri et al., 2023).

Agricultural practices such as monoculture farming maximize land use by cultivating a single crop, commonly seen in rice terraces, sugarcane fields, and banana plantations (Belete & Yadete, 2023). However, this reduces biodiversity and increases crop vulnerability to pests and diseases. It also quickly depletes soil nutrients, as the land isn't given time to naturally recover. To manage pests and diseases, pesticides are again used, offering quick solutions but leading to pest resistance and more severe outbreaks. This results in heavier financial burdens for agricultural operations. Other agricultural practices, such as excessive irrigation and repeated tilling, can contribute to soil erosion, salinization, and the depletion of groundwater resources (Mallareddy et al., 2023. In many cases, the drive to intensify agriculture prioritizes short-term gains over long-term sustainability, putting the environment and rural livelihoods at risk. When these issues occur simultaneously, they can severely impact and continue to threaten food security.

Antimicrobial Resistance Among Crop Pathogens

Despite advancements in agriculture, technology, and policy, the world continues to struggle with hunger, malnutrition, and food inequality (FAO, 2025). Food insecurity continues to exist due to unequal access to technology, poor infrastructure and logistics, poverty and inequality, and the effect of climate change. Food security also faces a major challenge as crop diseases persist and continuously evolve, making conventional fungicides and pesticides increasingly ineffective due to resistance. Globally, 20–30% of total crop yields are lost due to pathogens and pests, with fungal pathogens playing a significantly large role (Savary et al., 2019). While fungi have a narrower host range than some pests, such as insect and vertebrate pests, their widespread dispersal enables global crop infestations (Bebber et al., 2014). Global average yield losses due to pests and diseases for major crops are estimated for wheat to be 21.5% (10.1–28.1%), rice: 30.3% (24.6–40.9%), maize: 22.6% (19.5–41.4%), potato: 17.2% (8.1–21%), and soybean: 21.4% (11–32.4%) (Ristainoa et al., 2021; Savary et al., 2019). Hence, farmers resort to chemical interventions to minimize the losses.

The historical impact of crop pathogens on food security is well documented, and such pathogens have had devastating consequences that shaped human history. One of the most infamous examples is the Irish Potato Famine (1845–1852), caused by the oomycete *Phytophthora infestans*. The famine resulted in mass starvation, millions of deaths, and large-scale migration from Ireland (Collins & Zimran, 2019). The crisis was exacerbated by Ireland's dependence on a single potato variety, the Lumper, demonstrating the vulnerability of these varieties to plant disease outbreaks. The use of single crop varieties or monocropping reduces crop genetic diversity. This makes the plant vulnerable to plant pathogens and reduces soil microbial diversity, further exacerbating disease susceptibility (Belete & Yadete, 2023). Other notable pathogens include *Puccinia polysora*, which causes corn rust (Sun et al., 2021); *Magnaporthe oryzae*, the rice blast fungus responsible for up to

30% of global rice yield loss (Nalley et al., 2016); and *Puccinia graminis*, the wheat stem rust pathogen, which continues to threaten global wheat production (Lidwell-Durnin & Lapthorn, 2020). Nearly two centuries after the Irish Potato Famine, *Phytophthora infestans* remains the most destructive disease of potato globally (Fry et al., 2015; Haverkort et al., 2008).

As chemical control strategies have overtaken modern agriculture, fungal pathogens have also evolved resistance mechanisms to evade fungicides, diminishing the effectiveness of these chemicals. Antimicrobial resistance (AMR) is an emerging threat arising from the widespread misuse and overuse of antimicrobials in agriculture and food production (Aslam et al., 2024). For example, *Zymoseptoria tritici*, the causal agent of septoria leaf blotch in wheat (Fones et al., 2017), has developed resistance to multiple fungicide classes, including azoles and SDHI fungicides (McDonald et al., 2019; Omrane et al., 2017). Similarly, *Botrytis cinerea*, responsible for gray mold in fruits and vegetables, has shown resistance to several fungicide groups, leading to frequent treatment failures in vineyards and greenhouse crops like tomatoes and strawberries (Shao et al., 2021; Sofianos et al., 2023). Another major concern is the resistance of the environmental saprobe and a causative agent of fruit rot, *Aspergillus fumigatus*, to azole fungicides, as it is also an opportunistic human pathogen (Behnsen et al., 2008). Azoles used for crop protection and those used as the first line of treatment in human aspergillosis have identical mechanisms of action (Fisher et al., 2018). Thereby, cross-resistance poses a risk for both agriculture and public health (Berger et al., 2017; Fisher et al., 2018). Since agricultural and therapeutic azoles work the same way, resistance from crop applications can erode their clinical effectiveness. Likewise, as soil unintentionally becomes the reservoir for agricultural chemical residues, fungicide applications also alter soil microbial diversity and distribution (Ma et al., 2021; Roman et al., 2021) and reduce community stability, leading to poor soil health and resilience.

The management of AMR among plant diseases can be achieved through plant breeding of disease-resistant and climate-resilient crops, which can reduce dependency on antimicrobials while keeping yield losses at bay. For example, the breeding of IR64, a widely cultivated high-yielding rice variety that is resistant to *M. oryzae*, exemplifies the use of resistant crop varieties (Fujita et al., 2022). Despite the development of resistant varieties, *M. oryzae* evolves rapidly, overcoming single-gene or vertical resistance due to its high genetic variability and adaptability. This is a typical challenge for resistant varieties, not just in rice (Haque et al., 2021; Fujita et al., 2022). To withstand this, researchers make use of horizontal resistance, a more durable form of resistance (Haque et al., 2021). Horizontal resistance offers broad-spectrum protection as it uses multiple genes, providing long-term resistance. Unlike vertical resistance, which can be easily overcome by the pathogen, horizontal resistance offers a more stable genetic makeup that strengthens defense across various environmental and pathogenic conditions. However, breeding plant varieties is time-consuming, especially through conventional breeding. It can take a decade or more to develop a stable, high-performing variety. In addition, the complexity of identifying and combining the right set of genes makes the process more difficult (Haque et al., 2021).

Globalization and Climate Change

Globalization has significantly improved the availability and accessibility of food through global trade, positively contributing to food security. Importation allows for a range of food products that may not be locally available, improving dietary diversity and nutrition. Additionally, globalization facilitates the transfer of agricultural technologies, knowledge, and resources, driving innovative practices in food production. An example is the development of drought-tolerant maize varieties, a collaboration between the International Maize and Wheat Improvement Center (CIMMYT) and National Agricultural Research Organizations (NAROs) in Africa (Simtowe et al., 2019). This initiative resulted in increased yields and reduced crop failure, helping build resilience amid climate change. However, globalization also accelerates the spread of crop pathogens and AMR due to the global movement of people, animals, and agricultural products (Arcilla et al., 2017). International trade and travel enable pathogens to rapidly reach new geographic regions through contaminated plant products, soil, and even clothing or footwear (Bebber et al., 2014). Fungal spores and soilborne pathogens can be inadvertently introduced to new environments, threatening local agricultural systems. For instance, *M. oryzae* has originated in Southeast Asia but has spread to other parts of the world through human activities (Duan et al., 2024).

Lastly, food security is profoundly impacted by climate change. Rising global temperatures, altered precipitation patterns, and extreme weather events significantly affect agricultural productivity, leading to food shortages and hunger. In addition, climate change influences pathogen distribution and behavior. Research indicates that warmer temperatures are driving pests and pathogens toward higher latitudes, allowing them to spread into regions that were previously unaffected (Bebber et al., 2013; Lahlali et al., 2024). Higher humidity favors spore germination and fungal diseases such as anthracnose, while lowering plant host resistance to pathogens (Lim et al., 2023).

As the global population continues to increase, ensuring a food-secure world remains a challenge. The pressing and combined threats of crop pathogens, antimicrobial resistance, globalization, and climate change require urgent response and science-based actions. A multi-pronged approach, involving genetic crop diversity, sustainable agricultural practices, improved disease surveillance, and international cooperation, is crucial for mitigating the risks and securing a stable, nutritious, and accessible food supply for future generations. For instance, integrated pest management (IPM) offers a holistic strategy that integrates biological, cultural, physical, and chemical tools in a way that minimizes risks to human health and the environment (Angon et al., 2023). This requires the use of chemical pesticides only as a last resort and applying them in minimal quantities while promoting crop rotation, the use of resistant crop varieties, and the conservation of agroecological systems. When implemented effectively, IPM reduces the likelihood of pest outbreaks, slows the development of pesticide resistance, and supports the long-term sustainability of agricultural systems.

Mushroom Production Process

With the challenges of food security aggravated by globalization, climate change, booming population, and rise of AMR, mushroom cultivation can help address food security by providing alternative food sources. Mushrooms are rich in high-quality protein, dietary fiber, essential amino acids, vitamins, minerals, antioxidants, and other bioactive compounds, while requiring low inputs and exhibiting rapid growth. Consumers are increasingly turning to toward mushroom consumption due to their distinct flavors and numerous health benefits. Mushrooms have long been a staple in human diets across various cultures in many countries. Before mushroom cultivation was practiced, people foraged in woods and fields, a practice limited by seasonal availability, unpredictable yields, risk of misidentification, and the unsustainable depletion of wild mushroom populations. However, with the growing demand for consistent and high-quality produce, mushroom cultivation was developed into a highly controlled, science-based production.

Mushroom production and processing have become a financially significant industry generating billions, with the global market valued at $56 billion in 2022 and projected to reach $136 billion by 2032, showing an annual growth rate of 9.5% from 2023 to 2032. Despite being fungi, they are considered as vegetable in the market (IndustryArc, 2024). The global mushroom market is widely distributed across various end-use industries, with food and culinary applications accounting for 42%, pharmaceuticals for 34%, and cosmetics for 24%. This highlights the versatility and multifaceted value of mushrooms across different sectors. In addition to their long history of medicinal use and health benefits, mushrooms are also known for being low in calories while providing minimal dietary fiber and essential nutrients, thus making them an underrated yet valuable food (IndustryArc, 2024; Khan et al., 2024).

At this point, we present the process of edible mushroom production on farms, focusing on key stages that ensure successful cultivation. It begins with substrate selection, where suitable materials such as agricultural waste or compost are meticulously chosen and properly prepared to support mushroom growth. This is followed by the inoculation stage, during which mushroom spawn is introduced into the substrate to initiate colonization. In the growth phase, the mycelium spreads throughout the substrate and eventually produces fruiting bodies. Once the mushrooms reach maturity, they are harvested at the optimal time to ensure quality and yield. Post-harvest processing involves cleaning, packaging, and storing the mushrooms properly to maintain freshness and extend shelf life. These carefully managed stages ultimately ensure that mushrooms reach the consumer's table ready for preparation and consumption. Additionally, we also emphasize factors contributing to the growth of mushroom farming, the role of innovation in the industry, and the future potential of mushroom cultivation.

Substrate Selection and Preparation

In crop production, land preparation and the purchase of viable seeds are typically done. Similarly, mushroom cultivation begins with the selection and preparation of substrates. Substrates are organic materials on which mushrooms grow, thereby providing essential nutrients for their mycelium. The substrate not only provides food for the mycelium but also influences factors such as moisture retention, aeration, and pH, which are important for proper mushroom development. The mycelium, the thread-like vegetative structure of a fungus, spreads throughout the substrate, breaking down organic matter and facilitating the development of the mushroom fruiting bodies. Choosing an appropriate substrate is a prerequisite for good mushroom growth, and this varies based on the type of mushroom being cultivated. Wild mushrooms can thrive in a wide range of natural environments, such as grasslands, rotting wood, leaf litter, and soils; however, commercial mushroom production relies on controlled conditions and carefully selected substrates to ensure consistent growth, yield, and quality. The typical substrates are lignocellulosic wastes such as sawdust, straw of grain crops, logs, corn cobs, bagasse, water lilies, banana leaves, and other agricultural by-products (Subedi et al., 2023). Several factors, such as the nutritional requirements of the specific mushroom species and the desired growth rate, are considered in selecting specific substrates.

Oyster mushrooms (*Pleurotus ostreatus*) are commonly cultivated on substrates like rice straw, sawdust, and banana leaves due to their adaptability and rapid growth. Muswati et al. (2021) grew these mushrooms on modified substrates of cotton waste and mixed substrates of cotton husk, wheat straw, and baobab fruit shells and found out that it can enhance mycelial colonization, yield, and biological efficiency, thereby optimizing mushroom production. Other researchers have explored the use of alternative substrates, such as corncob substrates, which improved substrate efficiency by promoting inoculation and flushing, increasing mushroom weight, and enhancing biological efficiency. The corncob substrate performed comparably to sawdust in fruiting body yield and marketable harvest, suggesting its viability as an alternative substrate for oyster mushroom cultivation (Mangilog et al., 2024). Similarly, straw mushrooms (*Volvariella volvacea*) thrive on agricultural waste materials such as rice straw, banana leaves, or water lily stems. In some cases, mushroom farms utilize other organic by-products, including corn husks and sugarcane bagasse, to promote sustainable and cost-effective production (Sarma & Paul, 2024).

Other research used cucurbit-based media such as squash and winter melon for the cultivation of tiger sawgill mushroom (*Lentinus tigrinus*), discovering that squash decoction gelatin and all cucurbit species supported the greatest mycelial growth, while winter melon decoction enhanced antioxidant activity and phenolic content. The study reveals that cucurbit decoctions can be used as promising culture media for Tiger Sawgill mushrooms production (Liwanag et al., 2020). Other agricultural wastes can also be used to cultivate *L. tigrinus* and other mushrooms (De Leon et al., 2013a, b; 2017).

Once the substrate is selected, it must be prepared free of competing microorganisms or contaminants that could hinder the growth of the mushroom spores. Sterilization or pasteurization is commonly done to prevent contamination. Sterilization is done by heating the substrate to temperatures of 121 °C (250 °F) or higher, typically using a pressure cooker or autoclave, to eradicate any harmful bacteria, molds, or other fungi present in the material. Pasteurization, on the other hand, is a slightly less intense heating process, usually done at 60–80 °C (140–176 °F), that removes harmful pathogens while retaining beneficial microorganisms (Grimm et al., 2024). In either case, the substrate must be allowed to cool to a suitable temperature before it is inoculated with mushroom spores or mycelium. Furthermore, innovative practices such as the use of locally sourced agricultural waste as a substrate not only reduce costs but also support sustainability by recycling organic materials that would otherwise go to waste.

Advancements in technology have significantly improved substrate preparation. Automated pasteurization and sterilization systems allow for higher efficiency and precision, enabling mushroom farms to produce large quantities of substrate with consistent quality (Balan et al., 2022). Grimm et al. (2024) tested various pasteurization and sterilization techniques on oyster mushroom growth, yield, and resource efficiency. Results showed that autoclave sterilization increased dry yields by more than half compared to pasteurization, while hot water pasteurization matched autoclaving in fresh yield and significantly outperformed other pasteurization methods. Hot air pasteurization, used in countries like Mexico, Japan, and several European nations, is the most energy- and water-efficient method—using 75% less energy than autoclaving and 85% less water than hot water pasteurization—making it a sustainable option for mushroom cultivation, particularly in small-scale production systems (Grimm et al., 2024).

Inoculum Preparation, Colonization, and Cultivation

Following substrate preparation, the next step in mushroom production is inoculation or the process by which spores or mycelium are introduced to the substrate. There are two main categories of inoculation methods in mushroom cultivation: solid inoculum and liquid inoculum. Solid inoculum includes grain spawn, sawdust spawn, mushroom plugs, and other lignocellulosic materials such as corn cobs with mycelial growth. Grain spawn is widely used due to its high nutrient content and fast colonization rate, while sawdust spawn is ideal for inoculating logs and outdoor substrates. Plugs or seed spawn, typically wooden dowels impregnated with mycelium, are commonly used in small-scale or log-based cultivation. On the other hand, liquid culture serves as a sterile and scalable liquid inoculum option, where mycelium is suspended in a nutrient-rich solution, allowing for precise and rapid inoculation. In recent years, liquid culture has gained popularity due to its ability to promote faster colonization and higher yields (Kumla et al., 2020). Local farmers with small-scale production often use sterilized grain seeds such as corn, rice, or sorghum because of their inability to keep liquid inoculum sterile. Although grain-based inoculum can be less prone to contamination, it has a shorter shelf life and may result in slower colonization compared to other spawn types but offers a cheaper and more convenient way of inoculum preparation (Ghimire et al., 2021).

Once the spawn is introduced, the substrate is placed in growing rooms where it is maintained under controlled conditions. Factors such as temperature, humidity, and airflow are carefully monitored to establish an optimal environment for mycelium colonization. During the colonization phase, the mycelium spreads throughout the substrate, breaking down the organic material to make nutrients available and preparing it for fruiting. Signs of successful colonization include the appearance of dense, white mycelial growth covering the substrate. This stage can take several weeks, depending on the type of mushroom and conditions in the growing area. Mycelium growth is a delicate process that requires careful attention to environmental conditions. Temperature dictates the growth of most mushroom species that require specific ranges to let the mycelia spread throughout the substrate (Bellettini et al., 2016). The optimal temperature ranges for various *Pleurotus* species vary across different growth stages. Mycelial growth generally occurs between 5–35 °C, with most species favoring 10–25 °C. Primordia formation typically requires 10–27 °C, while fruiting body development occurs best at 20–30 °C. The ideal fruiting temperature varies among species, ranging from 5 to 30 °C, depending on environmental conditions and specific cultivation requirements (Bellettini et al., 2016).

Humidity levels contribute to mushroom growth as high humidity promotes mycelium growth and prevents the substrate from drying out (Azevedo et al., 2017). Most fungi can tolerate a humidity range of 20–70%, but *Pleurotus* species require higher humidity for optimal growth. During spawn running, 60–75% humidity is ideal, while mycelial stimulation thrives at 85–97%. High humidity (80–95%) is essential for pinning and fruiting, with successful cultivation achieved by maintaining these levels through regular water spraying. If humidity drops too low, the substrate and mycelium can dry out, causing stunted growth, poor colonization, and lower mushroom yields. On the other hand, excessively high humidity can create overly wet conditions that promote contamination by molds and bacteria, which can harm the crop and reduce its quality. Therefore, maintaining optimal humidity is vital to balance moisture requirements while minimizing the risk of contamination.

Unlike traditional cultivation methods that rely on natural environmental conditions, controlled environments like climate-controlled growing rooms allow for precise regulation of temperature, humidity, and light, leading to more consistent and higher-quality mushroom production (Ogwu et al., 2025). Mushroom farms can ensure consistent and predictable mycelium growth. In addition, using advanced air filtration and sterilization systems helps keep contamination at bay, allowing mushrooms to grow without any unwanted interference (Sangeeta et al., 2024). In the study of Setiawati et al. (2021), a temperature and humidity control system based on the Internet of Things (IoT) was developed and tested for oyster mushroom cultivation. IoT is a powerful tool that uses a network of interconnected devices that collect and share data using the internet. This technology was used to address the high failure rates in mushroom growth caused by excessive heat and low humidity in conventional growing structures. This system enabled real-time monitoring and automatic regulation of environmental conditions, helping to improve the success and efficiency of mushroom cultivation.

Fruiting and Harvesting

After a few weeks to months of incubation, following the right temperature, humidity, and nutrients, the substrate will be fully colonized by mycelium, and the fruiting stage follows (Kobukata et al., 2024). Fruiting is the process by which the mycelium produces basidiocarp, the fruiting structures of mushrooms, which are naked to the eye. In the wild, this natural transformation is typically triggered by changes in climatic factors such as temperature, humidity, and light exposure. However, in a controlled farming environment, mushroom growers have mastered these conditions, holding the skill of making mushrooms grow and emerge precisely whenever wanted. To initiate fruiting, mushroom farmers adjust environmental factors such as temperature, humidity, and light (Chen et al., 2021). For instance, the temperature is often lowered to simulate the change in seasons that would naturally prompt fruiting. Similarly, mushrooms require high humidity and clean air to produce healthy fruiting bodies. Light is also a crucial factor in mushroom development, although mushrooms do not utilize light in the same way that plants do. Instead, light helps signal the transition from mycelial growth to fruiting. As the mycelium transitions to fruiting, tiny structures called primordia (or pins) begin to form on the substrate's surface. These small, white structures will eventually grow into fully mature mushrooms. The process of fruiting can take several days to weeks, depending on the type of mushroom and environmental conditions (Bellettini et al., 2016).

Mushroom farming has gained traction even in micro agribusinesses in the Philippines, particularly in backyard settings, due to its low investment requirements and the availability of agricultural byproducts as substrates. Materials such as bamboo poles and nylon ropes are used to hang the mushroom bags, providing sturdy support for vertical cultivation. Polyvinyl chloride (PVC) tubes serve as channels to efficiently deliver water to the growing area. Rubber bands secure the bags and substrates in place, while PEP plastics cover and protect the fruiting bodies, helping maintain proper humidity and cleanliness during growth. Watering is typically done using ordinary hoses and pails to keep the mushrooms hydrated. Some farms also use natural substances such as molasses to promote the growth of their mushrooms. These materials are good for urban setups (Philippine Rice Research Institute, 2018).

The mushrooms are harvested once they reached full maturity or size. Harvesting is a delicate process that requires care to avoid damaging the surrounding substrate or the mushrooms themselves. Most mushrooms are harvested by hand, with workers carefully twisting or cutting the mushrooms at the base to avoid disturbing the mycelium. In some larger-scale operations, mechanized harvesters are used to collect mushrooms more efficiently.

The right timing and proper harvesting ensure high-quality mushrooms. They should be collected at peak maturity, as those harvested too early may not be fully developed, while those left too long may begin to release spores, reducing their quality and shelf life. Harvesting time is when the mushrooms become firm, their caps begin to flatten, and their color changes, as these often indicate freshness and maturity. Some mushroom varieties like white button mushrooms are harvested in multiple flushes, meaning several cycles of fruiting that occur before the substrate is depleted.

Post-harvest Processing and Packaging

Many might imagine how mushrooms end up in cans or packaging. This is because of the careful processing and packing after harvest. Once harvested, mushrooms require careful post-harvest handling to ensure they reach consumers in the best possible condition. Mushrooms are highly perishable, with a shelf life that only typically lasts from 3 to 7 days. Effective post-harvest practices such as packaging and processing help improve food security by prolonging mushroom freshness, minimizing spoilage, and providing a consistent source of nutrients.

Post-harvest processing begins with cleaning. Mushrooms are typically wiped clean to remove any substrate or dirt. Unlike vegetables, they should not be washed with water, as mushrooms act like sponges and absorb excess moisture, which can lead to spoilage. This is due to their porous structure and high surface area. After cleaning, mushrooms undergo sorting according to size, shape, and quality, followed by grading to ensure that mushrooms meet market standards (Dawadi et al., 2022).

Proper packaging extends the shelf life and protects the products during transport. Most commercially grown mushrooms are packaged in perforated plastic or cardboard containers, which allow for airflow while maintaining humidity. Some markets also use innovative packaging, such as biodegradable materials or vacuum-sealed containers, to reduce environmental impact. Mushrooms that are not sold fresh may undergo further processing, such as drying, freezing, and canning, to extend their shelf life and provide a value-added product. Drying mushrooms allows their use in a variety of meals, including stews, sauces, and soups. Mushroom chips, which are dried in dryers, are also very popular among consumers of different ages. In commercial production, automated drying machines are used (Sangeeta et al., 2024). On the other hand, Ansari et al. (2024) used different methods of drying mushrooms and suggested the use of solar drying, tray drying, vacuum drying, freeze drying, and heat pump as it has the potential to maintain the quality of dried mushrooms while extending their shelf life, considering also cost-effectiveness, capacity to minimize drying time, and limit damage to the product.

Additionally, sliced or pre-packaged mushrooms are increasingly common in supermarkets, providing consumers with a convenient and ready-to-use option. These mushrooms are often pre-washed and stored in controlled packaging to maintain freshness. Canned mushrooms, on the other hand, undergo a brief blanching process, typically lasting 2–5 min in hot water, which helps eliminate microorganisms and slow down spoilage, thereby preserving their flavor and firmness before sealing. This process helps extend their shelf life, making them a practical choice for long-term storage and various culinary applications. Some innovations are developed showcasing products that are ecofriendly.

After packaging, mushrooms are transported to distribution centers, where they are shipped to retailers or directly to consumers. Given their perishable nature, mushrooms are typically transported in refrigerated vehicles to maintain freshness during transit. The global distribution of mushrooms has expanded significantly in recent years, with major exporting countries like China and the Netherlands. With meticulous post-harvest handling, mushrooms maintain their freshness and flavor, making them suitable for various culinary applications. Proper storage and transportation from farm to consumer play a crucial role in preserving their quality. These measures ensure that mushrooms retain their desirable characteristics, enhancing the overall dining experience. Effective post-harvest management contributes to the consistent availability of high-quality mushrooms in the market. Equally important is traceability in the supply chain, which ensures food safety by allowing quick identification and removal of contaminated batches if issues arise. This enhances consumer trust and supports regulatory compliance.

Contaminations in Mushroom Farms

While mushrooms can be easily cultivated, farmers who use makeshift materials such as recycled plastic bags, untreated wooden racks, and old sacks often face contamination issues caused by houseflies, rodents, and other external factors. These materials, though affordable, may harbor microbes or residual chemicals if not properly cleaned or sterilized. Storing substrates or tools in open, unsanitary areas also exposes them to airborne contaminants, increasing the risk of contamination in the mushroom beds. To cope, they have used sticky traps and organic fertilizers made with molasses and other natural compounds (Philippine Rice Research Institute, 2018; Hyde et al., 2019). Local farms also have their trade secrets, refined through experience. Some farmers use biological control in managing mushroom contaminants. Plant-derived essential oils like oregano (*Origanum vulgare*) exhibit strong antifungal properties, making it effective against common mushroom pathogens such as *Trichoderma harzianum* and *Verticillium fungicola* (Gea et al., 2021). The antimicrobial and antifungal compounds in oregano, particularly carvacrol and thymol, help inhibit the growth of contaminating organisms by breaking the outer layer of their cells, providing a natural alternative to synthetic chemicals in mushroom cultivation (Basak & Guha, 2018). This natural remedy does not harm mushrooms because of the unique composition and toughness of their cell structure. Incorporating oregano essential oil into pest management strategies can aid in controlling fungal diseases and insect pests like *Megaselia halterata*, enhancing overall farm productivity (Leyva-López et al., 2017). This method is also a practical choice for many small-scale mushroom farmers due to its wide availability and low cost. They offer an alternative for controlling contaminants and pests, making them both accessible and environmentally friendly. While the effectiveness of these methods can vary depending on factors such as concentration, application method, and environmental conditions, proper use and continued innovation can enhance their reliability. With growing interest in sustainable farming, these natural approaches hold great promise for safer, more resilient mushroom production.

Nutritional Content of Edible Mushrooms

Fruiting bodies of edible mushrooms are rich in nutritional value, culinary versatility, and health-promoting properties. These edible mushrooms serve as an excellent source of essential nutrients while being low in calories, making them a valuable component of a balanced diet (Kalač, 2016). Due to their unique biochemical composition, edible mushrooms contribute to human health by providing proteins, dietary fiber, vitamins, minerals, and bioactive compounds with antioxidant and immunomodulatory effects (Nowacka et al., 2015; Valverde et al., 2015; Wasser, 2010). Kalaras et al. (2017) support the role of mushrooms in improving cardiovascular health, boosting immunity, and offering potential neuroprotective effects. Mushrooms have distinct biochemical pathways that allow them to produce ergothioneine, beta-glucans, and other specialized metabolites with antioxidant, anti-inflammatory, and immune-enhancing functions. Their ability to decompose complex organic matter in their environment also allows them to accumulate minerals like selenium, potassium, and copper from substrates. This metabolic versatility is what sets mushrooms apart as both a nutritional and medicinal food source. Listed below are the nutritional profiles of edible mushrooms.

Proteins Mushrooms contain high-quality proteins that provide essential amino acids necessary for numerous physiological functions, including enzyme production, tissue repair, and immune responses. High-quality proteins are those that contain all the essential amino acids in appropriate proportions required by the human body and are easily digestible and bioavailable. This makes them especially valuable in supporting growth, maintenance, and overall health. Certain species, such as *Lentinula edodes* (shiitake), *Pleurotus ostreatus* (oyster mushroom), and *Agaricus bisporus* (button mushroom), have been recognized for their significant protein content. Shiitake mushrooms contain all nine essential amino acids, making them a great natural source of complete protein (Mattila et al., 2001). The protein composition of mushrooms is comparable to that of legumes and some cereals, making them an important dietary protein source, especially in vegetarian and vegan diets (Ghosh et al., 2009). Moreover, mushroom proteins often have better digestibility and are less likely to contain antinutritional factors such as phytic acid and lectins that are commonly found in legumes and cereals, giving them a unique advantage in terms of bioavailability and digestive tolerance.

Dietary Fiber Mushrooms are also a valuable source of dietary fiber, primarily composed of non-starch polysaccharides, such as β-glucans, chitin, hemicellulose, mannans, and xylans, which contribute to digestive health and metabolic regulation. Among these, β-glucans are glucose polymers, while chitin is a linear polymer of N-acetylglucosamine, structurally similar to cellulose. These polysaccharides function as prebiotics, selectively stimulating the growth of beneficial gut microbiota, thereby enhancing gut barrier function and reducing inflammation (Cheung, 2013). The consumption of mushrooms has been associated with improved bowel function, cholesterol reduction, and glycemic control, making them particularly beneficial for individuals with metabolic disorders such as diabetes and obesity (Jayachandran et al., 2017). Jayachandran et al. (2017) suggest that β-glucans from mushrooms can enhance immune responses by modulating gut microbiota and promoting the production of short-chain fatty acids (SCFAs), which contribute to overall gut homeostasis. SCFAs, especially acetate, propionate, and butyrate, help maintain gut health by providing energy to intestinal cells, strengthening the gut lining, supporting beneficial bacteria, reducing inflammation, and preventing the growth of harmful microorganisms.

Vitamins and Minerals Mushrooms are abundant in vitamins and essential minerals that play crucial roles in maintaining human health. They are particularly rich in B vitamins, including riboflavin (B2), niacin (B3), and pantothenic acid (B5), which are essential for energy metabolism, nervous system function, and red blood cell production. Mushroom-derived B vitamins support key enzymatic reactions involved in converting carbohydrates, fats, and proteins into usable energy. For example, riboflavin and niacin are vital components of coenzymes (FAD and NAD) used in cellular respiration, while pantothenic acid is part of coenzyme A, essential in the Krebs cycle and fatty acid metabolism (Cheung, 2010). Additionally, mushrooms are one of the few natural alternative sources of vitamin D, as they can synthesize ergocalciferol (vitamin D_2) when exposed to ultraviolet (UV) light. This makes them an important dietary source of vitamin D, especially for individuals with limited sun exposure. Significant vitamin D_2 production typically occurs when mushrooms are exposed to UV-B light (280–315 nm) for 15–60 min, depending on the intensity and duration of exposure, with the gill side facing up to maximize conversion of ergosterol to vitamin D_2 (Cardwell et al., 2018). Minerals such as selenium, potassium, and copper are also abundant in mushrooms. Selenium acts as an essential antioxidant that supports immune function and thyroid health, while potassium contributes to cardiovascular health by regulating blood pressure (Valverde et al., 2015). Copper, a key cofactor in enzymatic reactions, plays a vital role in iron metabolism, neurotransmitter synthesis, and antioxidant defense mechanisms (Dubost et al., 2007). These micronutrients collectively enhance the nutritional value of mushrooms and contribute to their widespread recognition as functional foods.

Antioxidants and Other Bioactive Compounds Mushrooms contain a variety of bioactive compounds with strong antioxidant properties, including ergothioneine, glutathione, and polyphenols. Ergothioneine, a sulfur-containing amino acid found in high concentrations in mushrooms, has been shown to protect cells from oxidative stress and inflammation, potentially reducing the risk of neurodegenerative diseases such as Alzheimer's and Parkinson's disease (Kalaras et al., 2017). Glutathione, another potent antioxidant, plays a crucial role in cellular detoxification and immune function (Dubost et al., 2007). The antioxidant potential of mushrooms has also been widely studied. Recent studies demonstrate that mushrooms possess significantly higher levels of ergothioneine compared to other dietary sources, making them an important dietary antioxidant (Kalaras et al., 2017). Additionally, studies suggest that regular consumption of mushrooms may contribute to the prevention of chronic diseases, including cardiovascular diseases and certain types of cancer, by reducing oxidative damage to cells (Dubost et al., 2007). Beta-glucans in shiitake mushrooms can aid cancer treatments and improve survival rates in advanced gastric cancer patients (Vetvicka & Vetvickova, 2007). This anticancer activity occurs through the activation of immune pathways, which trigger cytokine release and enhance the body's ability to detect and destroy tumor cells. Other polysaccharides in mushrooms have been found to stimulate the activity of immune cells, macrophages, and free radicals that can stop the growth and spread of tumor cells and cause existing tumor cells to die.

Culinary Applications of Edible Mushrooms

Mushrooms are widely celebrated for their culinary versatility, offering a diverse range of flavors, textures, and aromas that enhance various cuisines across the globe. Their ability to absorb and complement different seasonings, along with their rich umami profile, makes them indispensable in gourmet cooking and everyday meals alike. Many edible mushrooms contain umami-enhancing compounds such as glutamates and nucleotides, which naturally boost flavor without the need for artificial additives or excessive salt (Feeney et al., 2014). Different species of edible mushrooms are prized for their distinct sensory characteristics, such as the meaty texture of oyster mushrooms, the rich umami flavor of shiitake, and the delicate, earthy aroma of button mushrooms, all of which influence their application in various culinary traditions. Presented here are some mushrooms commonly used in cooking in various culinary practices.

Agaricus bisporus **(White Button, Cremini, and Portobello Mushrooms)** The most cultivated mushroom worldwide, *Agaricus bisporus*, is available in different stages of maturity—white button (youngest), cremini (mid-aged), and portobello (fully mature). White button mushrooms are mild in flavor and widely used in salads, soups, pasta, and sauces. Cremini mushrooms have a deeper, earthier taste and are preferred for sautéing, roasting, and adding richness to stews. The mature portobello mushroom, with its meaty texture, is often grilled or used as a natural alternative in burgers (Botella-Martinez et al., 2023). The ability of *Agaricus bisporus* to enhance umami flavors makes it a staple ingredient in many Western and Asian cuisines, such as in beef stroganoff or Korean bulgogi with mushrooms (Feeney et al., 2014).

Pleurotus ostreatus **(Oyster Mushroom)** Oyster mushrooms (*Pleurotus ostreatus*) are known for their delicate, slightly sweet flavor and tender texture. Their versatility allows them to be used in stir-fries, stews, and soups, where they readily absorb seasonings and enhance the dish's overall taste. Studies have highlighted their application in sustainable, alternative diets due to their high protein content and ability to mimic the texture of seafood and meat when cooked properly (González et al., 2021). They are considered sustainable because they can be easily cultivated on agricultural waste and require minimal water and energy inputs. Their rapid cooking time and compatibility with various seasonings make them a favorite in both home cooking and professional kitchens.

Lentinula edodes **(Shiitake Mushroom)** Shiitake mushrooms (*Lentinula edodes*) are renowned for their robust umami flavor, characterized by a smoky, savory depth that enhances soups, broths, and stir-fried dishes. A staple in East Asian cuisines, shiitake mushrooms are widely used in miso soup, ramen, and braised dishes, imparting a rich aroma and depth of taste. Additionally, dried shiitake mushrooms have an even more concentrated umami profile, as the drying process reduces water content and concentrates natural umami substances, making them ideal for intensifying broths and sauces (Barros et al., 2008).

Cantharellus cibarius **(Chanterelle Mushroom)** The chanterelle mushroom (*Cantharellus cibarius*) is a prized wild mushroom known for its bright golden hue, fruity aroma, and peppery taste. This gourmet mushroom is commonly featured in French and Scandinavian cuisines, where it is sautéed in butter and paired with delicate proteins such as poultry, fish, and pasta dishes. Due to its unique flavor, chanterelles are often used in fine dining to enhance risottos, omelets, and creamy sauces (Boa, 2004), as seen in dishes like risotto *aux girolles* (chanterelle risotto), omelette *aux chanterelles* (chanterelle omelet), and the Swedish *kantarellsås* (chanterelle cream sauce), typically served with meat or fish. It has identified the presence of carotenoids and phenolic compounds in chanterelles, contributing not only to their distinctive color and flavor but also to their antioxidant properties (Barros et al., 2008).

Tuber melanosporum **(Black Truffle)** Black truffles (*Tuber melanosporum*), one of the most expensive and sought-after fungi, are revered for their pungent aroma and complex earthy flavor. Found primarily in European truffle-producing regions such as France and Italy, black truffles are used sparingly to elevate dishes, often shaved over pasta, risotto, eggs, or infused into oils and butters. Their high content of volatile aromatic compounds, including sulfur-based molecules, gives them their characteristic intense fragrance (Frauendorfer et al., 2019). Black truffles are typically cultivated in symbiosis with the roots of specific trees such as oaks and hazelnuts, which support their subterranean growth. The culinary applications of black truffles extend beyond traditional European cuisine, as chefs worldwide incorporate them into modern gastronomy to enhance sensory experiences, such as in truffle sushi rolls, truffle aioli on gourmet burgers, and truffle foam garnishes.

Fungi and Fermented Foods

Beyond their culinary and medicinal significance, fungi are also used in the fermentation of various foods and beverages. Fermentation, driven by microbial activity, enhances the flavor, texture, and nutritional properties of foods while also extending their shelf life. Certain filamentous fungi play a key role in both traditional and modern fermentation processes by producing enzymes and metabolites that catalyze biochemical transformations in raw ingredients. These beneficial changes include the development of flavor compounds such as esters, alcohols, and organic acids; improvement in digestibility through the breakdown of proteins and starches; and the synthesis of bioactive compounds with antioxidant or antimicrobial properties. Notable fungal enzymes involved in these processes include amylases, proteases, and lipases, which break down starches, proteins, and fats, respectively. Additionally, metabolites such as kojic acid, citric acid, and β-glucans contribute to the nutritional enhancement and preservation of fermented products. Several studies have investigated the biochemical mechanisms and health benefits associated with fungal-fermented foods (Machida et al., 2008; Giraud et al., 2010).

Fungal fermentation is particularly prominent in East Asian, European, and Southeast Asian cuisines, where specific fungal species are responsible for the production of alcoholic beverages, fermented soy products, cheeses, and functional foods. The ability of fungi to break down complex carbohydrates, proteins, and lipids into bioavailable compounds has made them indispensable in the global food industry. Additionally, many fungal-fermented foods exhibit probiotic-like properties and contain bioactive compounds that contribute to human health (Wang et al., 2019). Examples of functional foods produced through fungal fermentation include tempeh, a fermented soybean product rich in nutrients and digestible proteins. Specific bioactive compounds found in fungal-fermented foods include ergothioneine, lovastatin, and γ-aminobutyric acid, which have been linked to antioxidant, cholesterol-lowering, and neuroprotective effects, respectively (Chen et al., 2012). The following section outlines notable fungal species commonly used in the fermentation of foods across different cultures and industries.

Aspergillus oryzae **(Koji Mold)** *Aspergillus oryzae*, commonly known as koji mold, is a filamentous fungus widely utilized in East Asian fermentation processes. It produces a range of hydrolytic enzymes, including amylases, proteases, and lipases, which break down starches and proteins into fermentable sugars and amino acids. This enzymatic activity is essential to produce staple fermented products such as sake (Japanese rice wine), soy sauce, miso, and amazake (Kitamoto, 2015). In sake brewing, *A. oryzae* converts rice starches into simple sugars, which are then fermented by yeast to produce alcohol. Studies have also highlighted the genetic and metabolic pathways that enable *A. oryzae* to optimize fermentation efficiency and flavor development (Jin et al., 2021). Specifically, the organism exhibits an expansion of genes encoding hydrolytic enzymes like amylases and proteases, along with the duplication of α-amylase genes, which enhances its ability to degrade complex substrates during fermentation.

Penicillium roqueforti **(Blue Cheese Mold)** *Penicillium roqueforti* is a critical fungal species in the production of blue-veined cheeses, such as Roquefort, Stilton, and Gorgonzola. This mold is responsible for the characteristic blue-green marbling in these cheeses, which results from the production of secondary metabolites such as roquefortine C that contribute to their distinctive sharp and tangy flavors (Giraud et al., 2010). Roquefortine C is formed when the amino acids tryptophan and histidine are joined together by non-ribosomal peptide synthetase and then shaped into a ring structure through an oxidation process. The mold also produces lipolytic and proteolytic enzymes that break down fats and proteins, leading to the creamy texture and pungent aroma of blue cheeses. Giraud et al. (2010) have further elucidated the role of *P. roqueforti* in enhancing the sensory and textural qualities of fermented dairy products.

Monascus purpureus **(Red Yeast Rice Fungus)** *Monascus purpureus* is a traditional fungus used in East Asian fermentation, particularly in the production of red yeast rice. This fungus ferments rice, imparting a deep red pigment due to the production of monacolins, including monacolin K, which has been found to lower cholesterol levels by inhibiting HMG-CoA reductase, an enzyme involved in cholesterol synthesis (Wang et al., 2019). Red yeast rice is commonly used as a natural food coloring agent and a functional food ingredient in Chinese and Japanese cuisines. Pataková (2013) and Zhu et al. (2019) has demonstrated its potential health benefits, particularly in cardiovascular disease prevention. However, the safety and regulatory status of *M. purpureus* products depend on the presence of citrinin, a mycotoxin that can be produced during fermentation (Lachenmeier et al., 2012). The production of citrinin is influenced by several factors, including strain selection, fermentation conditions such as temperature and pH, substrate composition, and oxygen availability. Using non-citrinin-producing strains and optimizing environmental parameters during fermentation can significantly reduce or eliminate citrinin synthesis.

Rhizopus oligosporus **(Tempeh Fermentation Fungus)** *Rhizopus oligosporus* is a key fungal species in the fermentation of tempeh, a traditional Indonesian soybean-based food. This fungus forms a dense mycelial network that binds soybean cotyledons together, creating a firm, cake-like texture. During fermentation, *R. oligosporus* produces proteolytic enzymes that enhance the digestibility of soy proteins and increase the bioavailability of essential nutrients (Nout & Kiers, 2005). Several reports also revealed that tempeh fermentation by *R. oligosporus* also reduces antinutritional factors such as phytic acid, making soy-based foods more nutritionally accessible (Nout & Kiers, 2005). In addition to phytic acid, *R. oligosporus* helps break down tannins and trypsin inhibitors, which can otherwise impair protein digestibility and mineral absorption. This enzymatic activity enhances the bioavailability of key nutrients like iron, zinc, and amino acids, improving the overall nutritional profile of tempeh. Additionally, this fungus exhibits antimicrobial properties that inhibit the growth of pathogenic bacteria, contributing to food safety and preservation.

Neurospora intermedia **(Oncom Production Fungus)** *Neurospora intermedia* is a lesser-known but culturally significant fungus used in the fermentation of oncom, a traditional fermented food from Indonesia. Oncom is produced by fermenting byproducts such as peanut press cake or coconut meal, resulting in a protein-rich, textured food like tempeh. *N. intermedia* fermentation has been shown to enhance the bioavailability of B vitamins and improve the functional properties of alternative proteins (Dianatdar et al., 2025). This fungus has been utilized for its potential application in sustainable food systems and due to its ability to ferment agricultural waste products efficiently (Maini Rekdal et al., 2024). Examples of agricultural waste degraded by *N. intermedia* include cassava pulp, corn stover, sugarcane bagasse, and other lignocellulosic residues. These byproducts, often considered low-value or waste, are effectively converted into nutrient-rich biomass and value-added food ingredients through fungal fermentation.

Safety Considerations in Fungal Consumption

While fungi contribute significantly to nutrition, medicine, and food fermentation, it is important to recognize that not all fungi are safe for consumption. Certain wild mushrooms contain toxic compounds that can cause severe poisoning or even be fatal if misidentified. For example, *Amanita phalloides* (death cap) contains lethal amatoxins that damage the liver and kidneys, often leading to death if not treated promptly. This species is especially dangerous because it closely resembles edible mushrooms, increasing the risk of accidental poisoning. Similarly, *Gyromitra esculenta* (false morel) produces gyromitrin, a toxin that can cause seizures and liver failure. Though consumed in some areas after processing, its unpredictable toxicity makes it a significant health risk (He et al., 2022). Similarly, some fungi used in fermentation may produce secondary metabolites that require careful monitoring to ensure food safety. Furthermore, it is also crucial to emphasize the importance of mycotoxin regulation and fungal species identification to mitigate risks associated with fungal consumption (Benjamin, 1995; Alshannaq & Yu, 2017).

Mushroom foraging is a common practice in many cultures, but it carries inherent risks due to the presence of toxic and sometimes lethal species. Many poisonous mushrooms closely resemble edible varieties, making misidentification a serious hazard. Among the poisonous mushrooms are *Amanita phalloides* (Death Cap), which contains amatoxins that inhibit RNA polymerase II, leading to severe liver and kidney damage (Enjalbert et al., 2002); *Cortinarius* sp., which produce orellanine, a nephrotoxin that can cause irreversible kidney failure (Karlson-Stiber & Persson, 2003); and *Gyromitra esculenta* (False Morel), which contains gyromitrin that hydrolyzes into monomethylhydrazine, a compound that affects the central nervous system and can be carcinogenic (Michelot & Toth, 1991). Given these risks, mycologists and public health agencies strongly recommend proper species identification using field guides, DNA barcoding, or expert consultation before consuming any wild mushrooms. Accurate mushroom identification is challenging due to similar physical traits, specimen damage, and limited expert availability. While DNA barcoding improves precision, it is constrained by resource requirements, DNA quality, and the need for specialized equipment. Even expert consultation can be subjective because it relies on morphology-based identification and limited knowledge, particularly when dealing with cryptic or processed specimens. Therefore, multi-method verification is essential for safety. The safest approach is to rely on commercially cultivated mushrooms such as *Agaricus bisporus*, *Lentinula edodes*, and *Pleurotus ostreatus*, which are grown under controlled conditions and free from toxic contaminants.

Although fungi play a crucial role in food fermentation, some species can produce harmful mycotoxins under certain conditions. Mycotoxin contamination is a major concern in food safety and is regulated by organizations such as the European Food Safety Authority (EFSA) and the US Food and Drug Administration (FDA). Fungal growth and subsequent mycotoxin production during fermentation are influenced by environmental factors such as temperature, humidity, and pH levels in the growth medium. Furthermore, the specific fungal strain involved and the properties of the substrate significantly impact the extent of mycotoxin contamination. Common mycotoxins associated with fungal fermentation include citrinin, which is produced by *Monascus purpureus*. Citrinin is a nephrotoxin that can accumulate in red yeast rice and pose health risks if present at high levels (Farawahida et al., 2022). Quality control measures, such as controlled fermentation conditions and strain selection, are essential to minimize citrinin production (Lachenmeier et al., 2012). Aflatoxins are produced by *Aspergillus flavus* and *Aspergillus parasiticus*, which can contaminate grains, nuts, and fermented soy products if storage conditions are inadequate (Wild & Gong, 2010). Aflatoxins primarily affect the liver and can cause acute poisoning symptoms, such as nausea, vomiting, abdominal pain, liver necrosis, edema, and, in severe cases, acute hepatic failure and death. Chronic exposure is linked to immunosuppression, liver cirrhosis, growth retardation, and a significantly increased risk of hepatocellular carcinoma and gallbladder cancer. On the other hand, patulin, a mycotoxin produced by *Penicillium* species, is commonly found in spoiled fruits and fruit juices and can lead to gastrointestinal distress and immune suppression (Puel et al., 2010). To ensure safety, industrial food producers implement stringent mycotoxin screening, good manufacturing practices (GMP), and post-harvest interventions such as UV treatment, ozone exposure, and microbial detoxification to reduce mycotoxin contamination in fungal-fermented foods (Loi et al., 2020).

Food safety authorities worldwide also enforce strict regulations to control mycotoxin levels in edible fungi and fermented products. For example, the Codex Alimentarius sets permissible limits for mycotoxins in food, and routine testing is conducted in many countries to prevent contaminated products from reaching consumers (https://www.fao.org/fao-who-codexalimentarius/about-codex/en/). Public awareness campaigns also emphasize the importance of sourcing fungi-based foods from reputable suppliers and avoiding wild mushroom consumption unless identified by experts. In Nepal, coordinated efforts involving government ministries and consumer associations use mass media, posters, jingles, and in-person outreach to raise awareness about mycotoxin risks and safe practices. Similarly, the UP-RISE campaign around World Food Safety Day 2024 highlighted mycotoxin impacts and mitigation strategies through television, radio, and social media to inform the public, especially in Africa, where contamination threatens staple crops (Nepal project and UP-RISE campaign) (Leslie et al., 2023; UP-RISE, 2024). These campaigns emphasize trusted sources and clear messaging to improve knowledge and encourage safe food handling.

As research advances, biotechnology is being employed to enhance the safety of fungal products. Genome sequencing of beneficial fungi, genetic modification of fermentation strains, and real-time mycotoxin detection methods are being developed to improve food safety standards and reduce risks associated with fungal consumption (Zhang et al., 2019). Genome sequencing of beneficial fungi provides a comprehensive understanding of their genetic makeup, which helps in identifying genes responsible for beneficial traits, such as enhanced enzyme production or improved stress tolerance.

Concluding Points

Fungi offer numerous nutritional and functional benefits and can be used to address issues related to food security. The future of mushroom farming holds great promise due to its low environmental footprint, efficient conversion of agricultural waste into nutrient-rich food, minimal resource requirements, and its ability to provide high nutritional yield with bioactive compounds that support food security and sustainability. With further research, technological advancement, and increased consumer demand, mushroom production will continue to thrive. The industry's focus on sustainability, efficient farming practices, and innovation ensures that mushrooms will remain a part of diverse diets and part of the global food system, providing nutritious and sustainable food options. With the growing demand for fungi and mushrooms as functional foods, their importance in food production continues to rise, particularly through fermentation processes. Fermentation not only enhances the flavor and nutritional value of foods but also allows fungi to produce beneficial bioactive compounds. However, it is important to address toxicity risks and mycotoxin contamination to ensure these products remain safe for consumption. Proper identification of wild mushrooms, stringent quality control measures during fungal fermentation, and adherence to food safety regulations are crucial to maximize the benefits of fungi while minimizing potential hazards. Future research in mycology, food science, and biotechnology will further enhance the safety and efficacy of fungal-derived foods, ensuring their sustainable incorporation into human nutrition.

References

Abdul Halim Farawahida, Jon Palmer, Steve Flint, (2022) Monascus spp. and citrinin: Identification, selection of Monascusspp. isolates, occurrence, detection and reduction of citrinin during the fermentation of red fermented rice,*International Journal of Food Microbiology, 379*, 109829, https://doi.org/10.1016/j.ijfoodmicro.2022.109829

Alshannaq A, Yu JH. (2017) Occurrence, Toxicity, and Analysis of Major Mycotoxins in Food. *Int J Environ Res PublicHealth. 14*(6):632. Jun 13. https://doi.org/10.3390/ijerph14060632

Angon, P. B., Mondal, S., Jahan, I., Datto, M., Antu, U. B., Ayshi, F. J., & Islam, M. S. (2023). Integrated Pest Management (IPM) in agriculture and its role in maintaining ecological balance and biodiversity. *Advances in Agriculture, 2023*, 1–19. https://doi.org/10.1155/2023/5546373

Ansari, A., Guria, A., Lakra, N. H., & Rai, P. (2024). Dehydration and quality of dried mushrooms: A review. *International Journal of Bio-Resource and Stress Management, 15*(Sep, 9), 01–12. https://doi.org/10.23910/1.2024.5554a

Arcilla, M. S., van Hattem, J. M., Haverkate, M. R., Bootsma, M. C. J., van Genderen, P. J. J., Goorhuis, A., Grobusch, M. P., Lashof, A. M. O., Molhoek, N., Schultsz, C., Stobberingh, E. E., Verbrugh, H. A., de Jong, M. D., Melles, D. C., & Penders, J. (2017). Import and spread of extended-spectrum β-lactamase-producing Enterobacteriaceae by international travellers (COMBAT study): A prospective, multicentre cohort study. *The Lancet Infectious Diseases, 17*(1), 78–85. https://doi.org/10.1016/S1473-3099(16)30319-X

Arora, D. K., & Ramesh, S. (2017). Molds in food and fermentation. *International Journal of Food Science and Technology, 52*(3), 899–915.

Aslam, B., Asghar, R., Muzammil, S., Shafique, M., Siddique, A. B., Khurshid, M., Ijaz, M., Rasool, M. H., Chaudhry, T. H., Aamir, A., & Baloch, Z. (2024). AMR and sustainable development goals: At a crossroads. *Globalization and Health, 20*(1), 73. https://doi.org/10.1186/s12992-024-01046-8

Azevedo, S., Cunha, L. M., Oliveira, J. C., Mahajan, P. V., & Fonseca, S. C. (2017). Modelling the influence of time, temperature and relative humidity conditions on the mass loss rate of fresh oyster mushrooms. *Journal of Food Engineering, 212*, 108–112. https://doi.org/10.1016/j.jfoodeng.2017.05.026

Balan, V., Zhu, W., Krishnamoorthy, H., Benhaddou, D., Mowrer, J., Husain, H., & Eskandari, A. (2022). Challenges and opportunities in producing high-quality edible mushrooms from lignocellulosic biomass in a small scale. *Applied Microbiology and Biotechnology, 106*(4), 1355–1374. https://doi.org/10.1007/s00253-021-11749-2

Barros, L., Baptista, P., Estevinho, L. M., & Ferreira, I. C. (2008). Effect of fruiting body maturity stage on chemical composition and antioxidant activity of wild edible mushrooms. *Food Chemistry, 103*(4), 1458–1464.

Basak S., Guha P. (2018). A review on antifungal activity and mode of action of essential oils and their delivery asnano-sized oil droplets in food system. *Journal of Food Science and Technology, 55*(12), 4701–4710

Bebber, D. P., Ramotowski, M. A. T., & Gurr, S. J. (2013). Crop pests and pathogens move polewards in a warming world. *Nature Climate Change, 3*(11), 985–988. https://doi.org/10.1038/nclimate1990

Bebber, D. P., Holmes, T., & Gurr, S. J. (2014). The global spread of crop pests and pathogens. *Global Ecology and Biogeography, 23*(12), 1398–1407. https://doi.org/10.1111/geb.12214

Behnsen, J., Hartmann, A., Schmaler, J., Gehrke, A., Brakhage, A. A., & Zipfel, P. F. (2008). The opportunistic human pathogenic fungus *aspergillus fumigatus* evades the host complement system. *Infection and Immunity, 76*(2), 820–827. https://doi.org/10.1128/IAI.01037-07

Belete, T., & Yadete, E. (2023). Effect of mono cropping on soil health and fertility management for sustainable agriculture practices: A review. *Journal of Plant Sciences, 11*(6), 192–197. https://doi.org/10.11648/j.jps.20231106.1

Bellettini, M. B., Fiorda, F. A., Maieves, H. A., Teixeira, G. L., Ávila, S., Hornung, P. S., Júnior, A. M., & Ribani, R. H. (2016). Factors affecting mushroom Pleurotus spp. *Saudi Journal of Biological Sciences, 26*(4), 633–646. https://doi.org/10.1016/j.sjbs.2016.12.005

Benjamin, D. R. (1995). *Mushrooms: Poisons and panaceas—A handbook for naturalists, mycologists, and physicians*. W. H. Freeman.

Berger, S., El Chazli, Y., Babu, A. F., & Coste, A. T. (2017). Azole resistance in *Aspergillus fumigatus*: A consequence of antifungal use in agriculture? *Frontiers in Microbiology, 8*(JUN), 1–6. https://doi.org/10.3389/fmicb.2017.01024

Boa, E. (2004). *Wild edible fungi: A global overview of their use and importance to people*. Food and Agriculture Organization of the United Nations.

Botella-Martínez C, Muñoz-Tebar N, Lucas-González R, Pérez-Álvarez JA, Fernández-López J, Viuda-Martos M. (2023)Assessment of Chemical, Physico-Chemical and Sensory Properties of Low-Sodium Beef Burgers Formulated withFlours from Different Mushroom Types. *Foods. Sep 27;12*(19):3591. https://doi.org/10.3390/foods12193591

Cardwell, G., Bornman, J. F., James, A. P., & Black, L. J. (2018). A review of mushrooms as a potential source of dietary vitamin D. *Nutrients, 10*(10), 1498. https://doi.org/10.3390/nu10101498

Chang, S. T., & Miles, P. G. (2004). *Mushrooms: Cultivation, nutritional value, medicinal effect, and environmental impact*. CRC Press.

Chen, S.-Y., Ho, K.-J., Hsieh, Y.-J., Wang, L.-T., & Mau, J.-L. (2012). Contents of lovastatin, γ-aminobutyric acid and ergothioneine in mushroom fruiting bodies and mycelia. *LWT, 47*(2), 274–278. https://doi.org/10.1016/j.lwt.2012.01.019

Chen, L., Qian, L., Zhang, X., Li, J., Zhang, Z., & Chen, X. (2021). Research progress on indoor environment of mushroom factory. *International Journal of Agricultural and Biological Engineering, 14*(6), 25–32. https://doi.org/10.25165/j.ijabe.20221501.6872

Cheung, P. C. K. (2010). The nutritional and health benefits of mushrooms. *British Journal of Nutrition, 104*(2), 1–4.

Cheung, P. C. K. (2013). Mini-review on edible mushroom β-glucans and their bioactivities. *Food Hydrocolloids, 25*(4), 202–207.

Collins, W. J., & Zimran, A. (2019). The economic assimilation of Irish Famine migrants to the United States. *Explorations in Economic History¸, 74*, 101302. https://doi.org/10.1016/j.eeh.2019.101302

Dawadi, E., Magar, P. B., Bhandari, S., Subedi, S., Shrestha, S., & Shrestha, J. (2022). Nutritional and post-harvest quality preservation of mushrooms: A review. *Heliyon, 8*(12), e12093. https://doi.org/10.1016/j.heliyon.2022.e12093

De Leon, A.M., Reyes, R.G., dela Cruz, T.E.E. (2013a). Enriched cultivation of three wild strains of Lentinustigrinus (Bull.) Fr. using agricultural wastes. *International Journal of Agricultural Technology 9* (5): 1199–1214.

De Leon, A.M., Reyes, R.G., dela Cruz, T.E.E. (2013b). Lentinus squarrosulus (mont.) Singer and PolyporusgrammocephalusBerk.: newly domesticated, wild edible macrofungi from the Philippines. *Philippine Agricultural Scientists 96*(4): 411–418.

De Leon, A.M., Orpilla, J.O.V., Cruz, K.V., Dulay, R.M.R., Kalaw, S.P., dela Cruz, T.E.E. (2017). Optimization of mycelial growth and mycochemical screening of Lentinus sajor-caju (Fr.) from Banaue, Ifugao Province, Philippines. *International Journal of Agricultural Technology 13*(7.3): 2549–2567

Duan, G., Liu, Y., Zheng, C., Yu, K., Xie, J., Wang, B., Zheng, H., Tang, W., Bao, J., Wang, Z., & Chen, M. (2024). Chinese populations of *Magnaporthe oryzae* serving as a source of human-mediated gene flow to Asian countries: A population genomic analysis. *Journal of Fungi, 10*(11), 739. https://doi.org/10.3390/jof10110739

Dubost, N. J., Beelman, R. B., Peterson, D., & Royse, D. J. (2007). Identification and quantification of ergothioneine in cultivated mushrooms by liquid chromatography-mass spectroscopy. *International Journal of Medicinal Mushrooms, 9*(3), 215–222. https://doi.org/10.1615/IntJMedMushr.v8.i3.30

Enjalbert, F., Rapior, S., Nouguier-Soulé, J., Guillon, S., Amouroux, N., & Cabot, C. (2002). Treatment of amatoxin poisoning: 20-year retrospective analysis. *Journal of Toxicology: Clinical Toxicology, 40*(6), 715–757. https://doi.org/10.1081/CLT-120014645

FAO and ECO-RCCFS. (2025). *Overview of food security in economic cooperation organization countries, 2022*. FAO. https://doi.org/10.4060/cd3902en

Farzaneh Dianatdar, Zahra Etemadifar, Mohammad J. Taherzadeh. (2025) Production of B vitamins and carotenoids asactive biomolecules by edible filamentous fungus neurospora intermedia, *Food Bioscience, 69*,106850, https://doi.org/10.1016/j.fbio.2025.106850

Feeney, M. J., Miller, A. M., Roupas, P., & Beelman, R. B. (2014). Mushrooms—Biologically distinct and nutritionally unique: Exploring a "third food kingdom". *Nutrition Today, 49*(6), 301–307. https://doi.org/10.1097/NT.0000000000000063

Fisher, M. C., Hawkins, N. J., Sanglard, D., & Gurr, S. J. (2018). Worldwide emergence of resistance to antifungal drugs challenges human health and food security. *Science (New York, N.Y.), 360*(6390), 739–742. https://doi.org/10.1126/SCIENCE.AAP7999

Food and Agriculture Organization of the United Nations. (1996). Rome Declaration on World Food Security andWorld Food Summit Plan of Action: *World Food Summit, 13–17 November 1996, Rome, Italy* (p. 43). FAO. http://digitallibrary.un.org/record/195568

Fones, H. N., Fisher, M. C., & Gurr, S. J. (2017). Emerging fungal threats to plants and animals challenge agriculture and ecosystem resilience. *Microbiology Spectrum, 5*(2). https://doi.org/10.1128/MICROBIOLSPEC.FUNK-0027-2016

Frauendorfer, F., Schieberle, P., & Grosch, W. (2019). Volatile compounds in black truffles (*tuber melanosporum*) and their contribution to aroma. *European Food Research and Technology, 245*(4), 785–794.

Fry, W. E., Birch, P. R. J., Judelson, H. S., Grünwald, N. J., Danies, G., Everts, K. L., Gevens, A. J., Gugino, B. K., Johnson, D. A., Johnson, S. B., McGrath, M. T., Myers, K. L., Ristaino, J. B., Roberts, P. D., Secor, G., & Smart, C. D. (2015). Five reasons to consider *Phytophthora infestans* a reemerging pathogen. *Phytopathology, 105*(7), 966–981. https://doi.org/10.1094/PHYTO-01-15-0005-FI

Fujita, D., Tagle, A. G., Koide, Y., Simon, E. V., Fukuta, Y., Ishimaru, T., & Kobayashi, N. (2022). Characterization of QTLS for grain weight from new plant type rice cultivars through the development of near-isogenic lines with an IR 64 background. *Euphytica, 218*(5). https://doi.org/10.1007/s10681-022-03008-w

Gea, F. J., Navarro, M. J., Santos, M., Diánez, F., & Carrasco, J. (2021). Control of fungal diseases in mushroom crops while dealing with fungicide resistance: A review. *Microorganisms, 9*(3), 585–585. https://doi.org/10.3390/microorganisms9030585

Ghimire, A., Pandey, K. R., Joshi, Y. R., & Subedi, S. (2021). Major fungal contaminants of mushrooms and their management. *International Journal of Applied Sciences and Biotechnology, 9*(2), 80–93. https://doi.org/10.3126/ijasbt.v9i2.37513
Ghosh, S., Sarkar, T., & Pati, B. R. (2009). Nutritional and medicinal importance of edible mushrooms: A review. *International Journal of Food Science & Technology, 44*(3), 245–252.
Giraud, F., Giraud, T., Aguileta, G., Fournier, E., Samson, R., Cruveiller, S., & Brygoo, Y. (2010). Microsatellite loci for *Penicillium roqueforti*, a species used in blue-veined cheese making. *Food Microbiology, 27*(7), 1048–1052.
González, A., Combet, E., & Ros-Baró, M. (2021). Culinary applications of edible mushrooms in sustainable diets: A review. *International Journal of Gastronomy and Food Science, 26*, 100382.
Grimm, D., Sonntag, E., & Rahmann, G. (2024). Evaluation of different pasteurization and sterilization methods for oyster mushroom substrates. *Journal of Microbiology Biotechnology and Food Sciences, 13*(5), e10428–e10428. https://doi.org/10.55251/jmbfs.10428
Haque, M. A., Rafii, M. Y., Yusoff, M. M., Ali, N. S., Yusuff, O., Datta, D. R., Anisuzzaman, M., & Ikbal, M. F. (2021). Recent advances in Rice varietal development for durable resistance to biotic and abiotic stresses through marker-assisted gene pyramiding. *Sustainability, 13*(19), 10806. https://doi.org/10.3390/su131910806
Harutyunyan, M., & Malfeito-Ferreira, M. (2022). The rise of wine among ancient civilizations across the Mediterranean Basin. *Heritage, 5*(2), 788–812. https://doi.org/10.3390/heritage5020043
Haverkort, A. J., Boonekamp, P. M., Hutten, R., Jacobsen, E., Lotz, L. A. P., Kessel, G. J. T., Visser, R. G. F., & Van Der Vossen, E. A. G. (2008). Societal costs of late blight in potato and prospects of durable resistance through cisgenic modification. *Potato Research, 51*(1), 47–57. https://doi.org/10.1007/s11540-008-9089-y
Hawksworth, D. L. (2012). Global species numbers of fungi: Are tropical studies and molecular approaches contributing to a more robust estimate? *Biodiversity and Conservation, 21*(9), 2425–2433.
He, M.-Q., Wang, M.-Q., Chen, Z.-H., Deng, W.-Q., Li, T.-H., Vizzini, A., Jeewon, R., Hyde, K. D., & Zhao, R.-L. (2022). Potential benefits and harms: A review of poisonous mushrooms in the world. *Fungal Biology Reviews, 42*, 56–68. https://doi.org/10.1016/j.fbr.2022.06.002
Hogan, J. S., Feingold, B. S., & Bostock, R. M. (2015). The role of fungi in biomanufacturing: A review. *Fungal Biology, 119*(3), 193–207. https://doi.org/10.1016/j.fgb.2014.10.003
Hyde, K. D., Xu, J., Rapior, S., Jeewon, R., Lumyong, S., Niego, A. G. T., Abeywickrama, P. D., Aluthmuhandiram, J. V. S., Brahamanage, R. S., Brooks, S., Chaiyasen, A., Chethana, K. W. T., Chomnunti, P., Chepkirui, C., Chuankid, B., de Silva, N. I., Doilom, M., Faulds, C., Gentekaki, E., et al. (2019). The amazing potential of fungi: 50 ways we can exploit fungi industrially. *Fungal Diversity, 97*(1), 1–136. https://doi.org/10.1007/s13225-019-00430-9
IndustryArc. (2024). Mushroom market-by type by distribution channel (Offline & Online), By Geography – Global Opportunity Analysis & Industry Forecast, 2024–2030. Industryarc.com; *IndustryARC*. https://www.industryarc.com/Report/16679/mushroom-market.html
Ito, K. & Matsuyama, A. (2021). Koji molds for japanese soy sauce brewing: Characteristics and key enzymes. *Journal of Fungi, 7*, 658. https://doi.org/10.3390/jof7080658
Jayachandran, M., Xiao, J., & Xu, B. (2017). A critical review on health-promoting benefits of edible mushrooms through gut microbiota. *Food & Function, 8*(8), 3563–3583. https://doi.org/10.1039/C7FO00602B
Jin, F.-J., Hu, S., Wang, B.-T., & Jin, L. (2021). Advances in genetic engineering technology and its application in the industrial fungus *Aspergillus oryzae*. *Frontiers in Microbiology, 12*, 644404. https://doi.org/10.3389/fmicb.2021.644404
Kalač, P. (2016). *Edible mushrooms: Chemical composition and nutritional value*. Academic Press.
Kalaras, M. D., Beelman, R. B., Elias, R. J., & Wisniewski, S. R. (2017). Mushrooms: A rich source of the antioxidants ergothioneine and glutathione. *Food Chemistry, 233*, 429–433.
Karlson-Stiber, C., & Persson, H. (2003). Cytotoxic fungi—An overview. *Toxicon, 42*(4), 339–349.

Khan, A., Murad, W., Salahuddin, N., Ali, S., Shah, S. S., Halim, S. A., Khalid, A., Kashtoh, H., Khan, A., & Al-Harrasi, A. (2024). Contribution of mushroom farming to mitigating food scarcity: Current status, challenges and potential future prospects in Pakistan. *Heliyon, 10*(23), e40362–e40362. https://doi.org/10.1016/j.heliyon.2024.e40362

Khatri, P., Kumar, P., Shakya, K. S., Kirlas, M. C., & Tiwari, K. K. (2023). Understanding the intertwined nature of rising multiple risks in modern agriculture and food system. *Environment, Development and Sustainability, 26*(9), 24107–24150. https://doi.org/10.1007/s10668-023-03638-7

Kitamoto, K. (2015). Cell biology of the koji mold Aspergillus oryzae. *Bioscience, Biotechnology, and Biochemistry,79*(6), 863–869. https://doi.org/10.1080/09168451.2014.998611

Kobukata, T., Nakazawa, T., Yamasaki, F., Sugano, J., Oh, M., Kawauchi, M., Sakamoto, M., & Honda, Y. (2024). Identification of two genes essential for basidiospore formation during the postmeiotic stages in *Pleurotus ostreatus*. *Fungal Genetics and Biology, 172*, 103890–103890. https://doi.org/10.1016/j.fgb.2024.103890

Kumla, J., Suwannarach, N., Sujarit, K., Penkhrue, W., Kakumyan, P., Jatuwong, K., Vadthanarat, S., & Lumyong, S. (2020). Cultivation of mushrooms and their Lignocellulolytic enzyme production through the utilization of agro-industrial waste. *Molecules, 25*(12), 2811. https://doi.org/10.3390/molecules25122811

Lachenmeier, D. W., Monakhova, Y. B., Kuballa, T., & Reusch, H. (2012). Identification and risk assessment of mycotoxins in red yeast rice food supplements. *Mycotoxin Research, 28*(1), 9–15.

Lahlali, R., Taoussi, M., Laasli, S. E., Gachara, G., Ezzouggari, R., Belabess, Z., Aberkani, K., Assouguem, A., Meddich, A., El Jarroudi, M., & Barka, E. A. (2024). Effects of climate change on plant pathogens and host-pathogen interactions. *Crop and Environment, 3*(3), 159–170. https://doi.org/10.1016/j.crope.2024.05.003

Leslie, J. F., Morris, J. B., Gurung, J. K., Harvey, J., Ayalew, A., Baker, R., & Zhang, G. (2023). Mycotoxin communications: Managing messages for different audiences. *Frontiers in Sustainable Food Systems, 6*, 1095256. https://doi.org/10.3389/fsufs.2022.1095256

Leyva-López, N., Gutiérrez-Grijalva, E., Vazquez-Olivo, G., & Heredia, J. (2017). Essential oils of oregano: Biological activity beyond their antimicrobial properties. *Molecules, 22*(6), 989–989. https://doi.org/10.3390/molecules22060989

Lidwell-Durnin, J., & Lapthorn, A. (2020). The threat to global food security from wheat rust: Ethical and historical issues in fighting crop diseases and preserving genetic diversity. *Global Food Security¸, 26*, 100446. https://doi.org/10.1016/j.gfs.2020.100446

Lim, J. A., Yaacob, J. S., Mohd Rasli, S. R. A., Eyahmalay, J. E., El Enshasy, H. A., & Zakaria, M. R. S. (2023). Mitigating the repercussions of climate change on diseases affecting important crop commodities in Southeast Asia, for food security and environmental sustainability—A review. *Frontiers in Sustainable Food Systems, 6*, 1030540. https://doi.org/10.3389/fsufs.2022.1030540

Liwanag, E. J., Dulay, R. M., & Kalaw, S. (2020). Mycelial growth of Philippine mushroom Lentinus tigrinus in selected cucurbit-based media and its antioxidant activity. *Asian Journal of Agriculture and Biology, 8*(3), 323–329. https://doi.org/10.35495/ajab.2019.12.577

Loi, M., Fanelli, F., Liuzzi, V. C., Logrieco, A. F., & Mulè, G. (2020). Mycotoxin biotransformation by native and commercial enzymes: Present and future perspectives. *Toxins, 12*(11), 629.

Ma, G., Gao, X., Nan, J., Zhang, T., Xie, X., & Cai, Q. (2021). Fungicides alter the distribution and diversity of bacterial and fungal communities in ginseng fields. *Bioengineered, 12*(1), 8043–8056. https://doi.org/10.1080/21655979.2021.1982277

Machida, M., Yamada, O., & Gomi, K. (2008). Genomics of Aspergillus oryzae: Learning from the history of kojimold and exploration of its future. *DNA Research, 15*(4), 173–183. https://doi.org/10.1093/dnares/dsn020

Maini Rekdal, V., Villalobos-Escobedo, J.M., Rodriguez-Valeron, N. et al. (2024) Neurospora intermedia from a traditionalfermented food enables waste-to-food conversion. *Nat Microbiol 9*, 2666–2683. https://doi.org/10.1038/s41564-024-01799-3

Mallareddy, M., Thirumalaikumar, R., Balasubramanian, P., Naseeruddin, R., Nithya, N., Mariadoss, A., Eazhilkrishna, N., Choudhary, A. K., Deiveegan, M., Subramanian, E., Padmaja, B., & Vijayakumar, S. (2023). Maximizing water use efficiency in rice farming: A comprehensive review of innovative irrigation management technologies. *Water, 15*(10), 1802. https://doi.org/10.3390/w15101802

Mangilog, K., Mendez, J. F., & Soriano, R. (2024). Maize residue as a viable substrate for farm scale cultivation of oyster mushroom (*Pleurotus ostreatus*). *Advances in Agriculture, 3*(1), 9–18. https://doi.org/10.1155/2015/213251

Mattila, P., Könkö, K., Eurola, M., Pihlava, J. M., Astola, J., Vahteristo, L., Hietaniemi, V., Kumpulainen, J.,Valtonen, M., & Piironen, V. (2001). Contents of vitamins, mineral elements, and some phenolic compounds incultivated mushrooms. *Journal of Agricultural and Food Chemistry, 49*(5), 2343–2348. https://doi.org/10.1021/jf001525d

McDonald, M. C., Renkin, M., Spackman, M., Orchard, B., Croll, D., Solomon, P. S., & Milgate, A. (2019). Rapid parallel evolution of azole fungicide resistance in Australian populations of the wheat pathogen *Zymoseptoria tritici*. *Applied and Environmental Microbiology, 85*(4), e01908-18. https://doi.org/10.1128/AEM.01908-18

Michelot, D., & Toth, B. (1991). Poisoning by *Gyromitra esculenta* (false morel) mushrooms: A review. *Journal of Applied Toxicology, 11*(4), 235–243.

Muswati, C., Simango, K., Tapfumaneyi, L., Mutetwa, M., & Ngezimana, W. (2021). The effects of different substrate combinations on growth and yield of oyster mushroom (*Pleurotus ostreatus*). *International Journal of Agronomy, 2021*, 1–10. https://doi.org/10.1155/2021/9962285

Nout, M. J. R., & Kiers, J. L. (2005). Tempe fermentation, innovation and functionality: Update into the thirdmillennium. *Journal of Applied Microbiology, 98*(4), 789–805. https://doi.org/10.1111/j.1365-2672.2004.02471.x

Nowacka, N., Nowak, R., Drozd, M., Olech, M., Los, R., & Malm, A. (2015). Antibacterial, antiradical potential andphenolic compounds of thirty-one Polish mushrooms. *PLOS ONE, 10*(10), e0140355. https://doi.org/10.1371/journal.pone.0140355

Nalley, L., Tsiboe, F., Durand-Morat, A., Shew, A., & Thoma, G. (2016). Economic and environmental impact of riceblast pathogen (Magnaporthe oryzae) alleviation in the United States. *PLOS ONE, 11*(12), e0167295. https://doi.org/10.1371/journal.pone.0167295

O'Callaghan, Y. C., O'Brien, N. M., Kenny, O., Harrington, T., Brunton, N., & Smyth, T. J. (2015). Anti-inflammatoryeffects of wild Irish mushroom extracts in RAW264.7 mouse macrophage cells. *Journal of Medicinal Food, 18*(2),202–207. https://doi.org/10.1089/jmf.2014.0012

Ogwu, M. C., Chime, A. O., Sawyer, W. E., Izah, S. C., & Ovuru, K. F. (2025). Sustainable agriculture and environmental monitoring through edible mushrooms. In S. C. Izah, M. C. Ogwu, & M. Akram (Eds.), *Bioactive compounds in edible mushrooms* (Reference series in phytochemistry) (Vol. 1–28). Springer. https://doi.org/10.1007/978-3-031-52642-8_20-1

Omrane, S., Audéon, C., Ignace, A., Duplaix, C., Aouini, L., Kema, G., Walker, A.-S., & Fillinger, S. (2017). Plasticity of the MFS1 promoter leads to multidrug resistance in the wheat pathogen *Zymoseptoria tritici*. *MSphere, 2*(5), 1–19. https://doi.org/10.1128/msphere.00393-17

Patakovà, P. (2013). Monascus secondary metabolites: Production and biological activity. *Journal of Industrial Microbiology & Biotechnology, 40*(2), 169–181. https://doi.org/10.1007/s10295-012-1216-8

Philippine Rice Research Institute. (2018). *Rice-based mushroom production manual – Pinoy rice*. Pinoy Rice Knowledge Bank. https://www.pinoyrice.com/wp-content/uploads/rice-based-mushroom-production-manual.pdf

Pouris, J., Kolyva, F., Bratakou, S., Vogiatzi, C. A., Chaniotis, D., & Beloukas, A. (2024). The role of fungi in food production and processing. *Applied Sciences, 14*(12), 5046–5046. https://doi.org/10.3390/app14125046

Puel, O., Galtier, P., & Oswald, I. P. (2010). Biosynthesis and toxicological effects of patulin. *Toxins, 2*(4), 613–631. https://doi.org/10.3390/toxins2040613

Ristainoa, J. B., Andersonb, P. K., Bebber, D. P., Braumane, K. A., Cunniffef, N. J., & Fedoroff, N. V. (2021). The persistent threat of emerging plant disease pandemics to global food security.

Proceedings. National Academy of Sciences. United States of America, 118(23), e2022239118. https://doi.org/10.1073/pnas.2022239118

Roman, D. L., Voiculescu, D. I., Filip, M., Ostafe, V., & Isvoran, A. (2021). Effects of triazole fungicides on soil microbiota and on the activities of enzymes found in soil: A review. *Agriculture (Switzerland), 11*(9), 1–18. https://doi.org/10.3390/agriculture11090893

Sangeeta, S., Sharma, D., Ramniwas, S., Mugabi, R., Uddin, J., & Nayik, G. A. (2024). Revolutionizing mushroom processing: Innovative techniques and technologies. *Food Chemistry X, 23*, 101774–101774. https://doi.org/10.1016/j.fochx.2024.101774

Sarma, H. H., & Paul, A. (2024). Turning waste into wealth: Exploring strategies for effective agricultural waste management. *Vigyan Varta, 5*(5), 322–330. https://www.vigyanvarta.in/adminpanel/upload_doc/VV_0524_70.pdf

Savary, S., Willocquet, L., Pethybridge, S. J., Esker, P., McRoberts, N. & Nelson, A. (2019). The global burden of pathogens and pests on major food crops. *Nature Ecology & Evolution, 3*, 430–439. https://doi.org/10.1038/s41559-018-0793-y

Setiawati, D. A., Utomo, S. G., Murad, M., & Putra, G. M. D. (2021). Design of temperature and humidity control system on oyster mushroom plant house based on internet of things (IoT). *IOP Conference Series: Earth and Environmental Science, 712*(1), 012002. https://doi.org/10.1088/1755-1315/712/1/012002

Shao, W., Zhao, Y., & Ma, Z. (2021). Advances in understanding fungicide resistance in botrytis cinerea in China. *Phytopathology, 111*(3), 455–463. https://doi.org/10.1094/PHYTO-07-20-0313-IA

Simtowe, F., Amondo, E., Marenya, P., Rahut, D., Sonder, K., & Erenstein, O. (2019). Impacts of drought-tolerant maize varieties on productivity, risk, and resource use: Evidence from Uganda. *Land Use Policy, 88*(July), 104091. https://doi.org/10.1016/j.landusepol.2019.104091

Sofianos, G., Samaras, A., & Karaoglanidis, G. (2023). Multiple and multidrug resistance in Botrytis cinerea: Molecular mechanisms of MLR/MDR strains in Greece and effects of co-existence of different resistance mechanisms on fungicide sensitivity. *Frontiers in Plant Science, 14*(October), 1–13. https://doi.org/10.3389/fpls.2023.1273193

Subedi, S., Kunwar, N., Pandey, K. R., & Joshi, Y. R. (2023). Performance of oyster mushroom (*Pleurotus ostreatus*) on paddy straw, water hyacinth and their combinations. *Heliyon, 9*(8), e19051–e19051. https://doi.org/10.1016/j.heliyon.2023.e19051

Sun, Q., Li, L., Guo, F., Zhang, K., Dong, J., Luo, Y., & Ma, Z. (2021). Southern corn rust caused by *Puccinia polysora* Underw: A review. *Phytopathology Research, 3*, 25. https://doi.org/10.1186/s42483-021-00102-0

United Nations, DESA Population Division. (2024). World population prospects in 2024. https://population.un.org/wpp/graphs?loc=900&type=Demographic%20Profiles&category=Line%20Charts

UP-RISE. (2024). *World food safety day 2024: Did you know campaign | UP-RISE food safety*. UP-RISE Food Safety. https://uprisefoodsafety.org/world-food-safety-day-2024-did-you-know/

Valverde, M. E., Hernández-Pérez, T., & Paredes-López, O. (2015). Edible mushrooms: Improving human health and promoting quality life. *International Journal of Microbiology, 2015*, 376387. https://doi.org/10.1155/2015/376387

Větvička, V., & Větvičková, J. (2007). Physiological effects of different types of beta-glucan. Biomedical Papers ofthe Medical Faculty of the University Palacký, Olomouc, *Czech Republic, 151*(2), 225–231. https://doi.org/10.5507/bp.2007.038

Wang, C., Chen, W., & Ning, Z. (2019). Monascus fermentation and its beneficial effects on human health. *Journal of Functional Foods, 57*, 202–213.

Wasser, S. P. (2010). Medicinal mushroom science: Current perspectives, advances, evidences, and challenges. *International Journal of Medicinal Mushrooms, 12*(1), 1–16. https://doi.org/10.1615/IntJMedMushr.v12.i1.10

Wild, C. P., & Gong, Y. Y. (2010). Mycotoxins and human disease: A largely ignored global health issue. *Carcinogenesis, 31*(1), 71–82. https://doi.org/10.1093/carcin/bgp264

Zhang, J., Chen, H., Chen, W., Leng, X., Ren, Z., Chen, Y., & Deng, L. (2019). Advances in biosensors and molecular-based methods for mycotoxin detection. *Toxins, 11*(11), 635.

Zhu B, Qi F, Wu J, Yin G, Hua J, Zhang Q, Qin L. (2019) Red Yeast Rice: A Systematic Review of the Traditional Uses,Chemistry, Pharmacology, and Quality Control of an Important Chinese Folk Medicine. *Front Pharmacol. Dec 2;10*:1449. https://doi.org/10.3389/fphar.2019.01449

Chapter 13
Fungal-Based Products in the Southeast Asian Market: Applications, Emerging Trends, Consumer Demands, and Key Challenges in the Agro-industrial Sector

Diana P. Paguntalan, Jonathan Jaime G. Guerrero, and Mark Angelo O. Balendres

Abstract The Southeast Asian market has seen a growing demand for commercial fungal products, driven by their diverse applications in agriculture, medicine, and food production. These products derived from beneficial fungi are increasingly used as food, biofertilizers, biopesticides, and nutritional supplements. With a rising focus on sustainable farming and natural solutions, the region has embraced fungal products as ecofriendly alternatives to chemical inputs, particularly for crop production and crop protection. The growth is supported by the rich biodiversity of fungi in the area, alongside advancements in biotechnology, making fungal products a promising sector in Southeast Asia's agro-industrial economic landscape. This chapter presents common commercial fungal-based agricultural products in the Southeast Asian market, as well as trends, demands, challenges, and opportunities for growth and expansion.

Keywords Biofertilizers · Biopesticides · Ecofriendly alternatives · Mycopesticides

D. P. Paguntalan
Division of Biological Sciences, College of Arts and Sciences, University of the Philippines Visayas, Iloilo, Philippines

J. J. G. Guerrero
College of Medicine, University of the Philippines Manila, Manila, Philippines

M. A. O. Balendres (✉)
Department of Biology, College of Science, De La Salle University, Manila, Philippines
e-mail: mark.angelo.balendres@dlsu.edu.ph

J. J. G. Guerrero et al. (eds.), *Mycology in a Changing Planet: Applications and Perspectives from Southeast Asia*, https://doi.org/10.1007/978-981-95-2640-6_13

Introduction

Fungi have long been valued for their medicinal properties (Niego et al., 2023a, b). The fungal products are increasingly utilized as functional foods, dietary supplements, and sources of bioactive compounds (Gupta et al., 2018; Li et al., 2023). Functional foods, defined as foods that provide health benefits beyond basic nutrition, represent a growing industry (Kavya et al., 2024). Fungi are also used in agriculture to improve crop yield and protect plants from pests and diseases (Koul et al., 2022). Niego et al. (2023a, b) estimated that the global value of fungi is around USD 54.57 trillion. This highlights the growing recognition of fungi's multifaceted roles in various sectors, increased commercial interest, and market growth. The Southeast Asian market has seen a growing demand for commercial fungal products, driven by their diverse applications in agriculture, medicine, and food production (Dai et al., 2021; Liu et al., 2023; Sudheer et al., 2023).

The markets for bio-based agricultural products, including biopesticides and biofertilizers made up of fungi, are growing (Mordor Intelligence, 2025) due to increasing consumer demand for organic and more sustainably produced food products and the pressing need to reduce dependence on agrochemicals to address environmental concerns (Rajapakse et al., 2016; Leite et al., 2022; Joshi & Gauraha, 2022). For instance, the global market value of biofertilizers alone is expected to reach USD 5.2 billion by 2028, almost double the value in 2022 (2.8 billion) (Kumar et al., 2024). In the Asia Pacific region, which includes Australia, China, India, Indonesia, Japan, Philippines, Thailand, and Vietnam, Mordor Intelligence (2025) reports growth of the size of the market to reach USD 1.21 billion by 2030.

This chapter provides an overview of Southeast Asia's rapidly growing fungi-based product market, highlighting market trends, consumer demands, and some challenges faced within the agro-industrial sector.

Food Additives and Functional Foods

Various microorganisms are used to produce single-cell protein (SCP), fungi being one. The term refers to any protein from microbial sources that substitutes protein-rich foods (Li et al., 2024). Today, many companies produce microbial proteins for use in food applications, as reflected by the increasing number of patents in this field. For instance, Marlow Foods (UK) made available seven patents of meat alternatives utilizing filamentous fungi (Amara & Abd El-Baky, 2023). They can serve as a protein supplement in animal feed, both in aquaculture and poultry farming (Kelechi & Ukaegbu-Obi, 2016; Li et al., 2024). Based on the Global Market Insights Inc. (2022), the SCP market exceeded USD 8 billion in 2021 and is expected to grow at a Compound Annual Growth Rate (CAGR) of over 9% from 2022 to

2030, surpassing USD 18.5 billion by 2030. Malaysia led the ASEAN market with a value of USD 9.7 million in 2020.

Aside from SCP, mushroom production has significantly contributed to global microbial protein production, which reached approximately 11.9 million tons annually in 2019 (FAOSTAT, 2020). Hence, mycoprotein refers to protein-rich food derived from fungal biomass, which serves as a meat alternative. Edible mushrooms serve as low-calorie functional foods and nutraceuticals (Lu et al., 2020), providing health benefits and a unique flavor profile that appeals to consumers. The global mushroom industry is categorized into cultivated (54%), medicinal (38%), and wild fungi (8%). In 2020, the total world mushroom production reached 42.59 million tons, with China contributing 93.93% of this output (FAOSTAT, 2022). The yield in Asia was about seven million tons, and a 3.2 million kg/ha yield for mushroom and truffle production in 2018. The global market value of mushrooms for food 2020 was USD 28.16 billion, representing 62% of total mushroom production. Known for its low-fat content and high protein and fiber (Souza Filho et al., 2018), it has been shown to benefit blood cholesterol and insulin levels (Turnbull & Ward, 1995; Cherta-Murillo et al., 2020). Economically significant fungi include *Agaricus bisporus*, valued globally at USD 16.69 billion in 2022 (Market Data Forecast, 2022), and *Hericium* species, valued at USD 963.4 million in exports in 2021 (Tridge, 2022). *Lentinula edodes* (Shiitake) had a market value of USD 4.31 billion in 2022. It is expected to grow to USD 4.7 billion by 2030 (The Brainy Insights, 2022). *Monascus purpureus* is used to produce red yeast rice, a staple Asian food valued at USD 380.6 million in 2022 (Nguyen et al., 2017; Persistence Market Research, 2022). The highly prized *Ophiocordyceps sinensis*, known for its numerous nutraceutical products, had a market value of USD 5–11 billion in 2018 (He, 2018). Meanwhile, there are around 2000 species of edible wild fungi, of which 470 also have medicinal properties (Li et al., 2021), making up 8% of global mushroom production, with a market value of USD 3.63 billion in 2020 (Royse et al., 2017). The Southeast Asia mushroom market is expected to grow at a CAGR of 7.02% between 2024 and 2032 (IMARC Group, 2023). This growth is mainly driven by the rising demand for organic food products and innovations in mushroom packaging technologies (IMARC Group, 2023).

Another important fungal product essential in food processing and industrial applications is the organic acids produced during fermentation. Key acids like citric, gluconic, and lactic acids are commonly derived from *Aspergillus niger*. In contrast, others, such as fumaric and itaconic acids, come from various fungal species (Copetti, 2019). Citric acid dominates the market, with a value of USD 2.89 billion in 2022, followed by fumaric acid (USD 654.4 million in 2020), gluconic acid (over USD 50 million in 2017), itaconic acid (USD 98.4 million in 2021), and lactic acid (USD 1.1 billion in 2020) (Global Market Insights, 2022b).

Fermented Foods and Beverages

Southeast Asia is known for its cultural and ethnic diversity and indigenous fermented foods. Alcoholic beverages play an important role in the cultural practices of Southeast Asians, where social gatherings frequently involve the generous consumption of alcohol. Yeasts are crucial in alcoholic beverage production, including beer, wine, and distilled spirits (Walker & Stewart, 2016). They convert sugars into alcohol and carbon dioxide through fermentation while producing secondary metabolites that contribute to the taste and aroma of beverages. For example, *Saccharomyces cerevisiae* is an important yeast species used in beer and wine production (Lodolo et al., 2008), while some have reported using *S. eubayanus* (Sampaio, 2018). Dark spirits like rum, whiskey, and brandy are aged in barrels, gaining distinct flavors. Overall, the alcohol production that involves yeast fermentation significantly impacts the global economy, with alcoholic beverages valued at USD 1108.45 billion (Niego et al., 2023a, b). In Southeast Asia, especially in non-Muslim countries, some of the popularly consumed beverages include *brem* in Bali for Hindu rituals (Kruger, 2014), *sra peang* in Cambodia for ceremonial alliances (Yamamoto & Matsumoto, 2011), and *ciu* in Java for traditional medicine (Beers, 2012). Also, various rice-based alcoholic beverages like *tapuy* in the Philippines and *lao-lao in* Laos are popular (Nagai & Tamang, 2010; Culloty, 2010). Additionally, vinegar is produced from further fermentation of alcoholic beverages, and traditional kinds of vinegar and spirits like *arak* and *tuak* are widely used across the region (Lim-Castillo, 2006).

Aside from alcoholic beverages, fermented foods in Southeast Asia, as categorized by Surya (2024), include fermented starch products such as the famous *Tapai* (*ragi* in Indonesia and Malaysia, *look paeng* in Thailand, or *bubod* in the Philippines), commonly fermented by molds (e.g., *Aspergillus oryzae*, *Rhizopus oryzae*, *Amylomyces rouxii*, and *Mucor* sp.) and yeasts (e.g., *S. cerevisiae*, *S. fibuliger*, and *Endomycopsis burtonii*) (Lamang, 2019). Other categories include fermented fruit and vegetable products such as *nata de coco* from the Philippines, and other non-legume and soybean-based products like *tempeh* and *kecap manis* (sweet soy sauce) from Indonesia, *thua nao* from Thailand, *sieng* from Laos and Cambodia, and *pepok* from Myanmar (Nagai & Tamang, 2010). Dairy products are also fermented, like the *dadih* from Indonesia and *tairu* from Malaysia, which are yogurt-like fermented foods (Arnold et al., 2021). Traditionally prepared cheese is also produced, such as the *dangke* chees of Indonesia (Setiarto, 2025) and *kesong puti* in the Philippines (Aquino et al., 2011). Lastly, meat-based fermented products are also popular in Southeast Asia, such as *naem* or *nham* in Thailand and Laos cuisine (Laxananil et al., 2009), *Sai krok Isaan* in the northern part of Thailand (Yongsmith & Malaphan, 2016), *urutan* in Bali, Indonesia (Darmayanti et al., 2014), and *bekamal* in Java, Indonesia (Prastujati et al., 2022).

Other fermented food products in Southeast Asia include fermented tea, eggs, chili paste, coffee, and by-products from the vegetable oil industries (Surya, 2024). In Northern Thailand, fermented tea products like *miang* are popular, with tea leaves fermented by filamentous fungi and bacteria (Khanongnuch et al., 2017). Myanmar also has a traditional fermented tea called *lahpet*, which is often used in dishes (Huang et al., 2016). Salted duck eggs, originating from China, are popular in the Philippines and Indonesia (Sumekar & Al-Baarri, 2020). Sriracha sauce, made from fermented chili peppers, is a well-known Thai condiment (Watts et al., 2018). In Indonesia, *oncom* is a nutritious fermented food made from oil press cake by-products, with black and red varieties using different molds. Finally, civet coffee, or *kopi luwak*, is a unique and expensive fermented coffee produced from beans eaten and excreted by luwak (*Paradoxorus hermaphroditus*), a cat-like animal (Muzaifa et al., 2019).

Food Enhancers and Colorants

Soy sauce, a key food enhancer, is produced using *Aspergillus* molds through the fermentation of soybeans, wheat, and salt, which produced a global market value of USD 39.5 billion in 2019 (Verified Market Research, 2022. Miso, another soybean-based condiment, involves similar fungal species, with a market value of USD 67.8 million in 2022 (Verified Market Research, 2022. *Tempeh*, a staple food in Indonesia made from fermented soybeans using *Rhizopus* species, reached a market value of USD 4.53 billion in 2021 (The Business Research Company, 2022). In the food colorants sector, natural pigments from fungi like *Penicillium, Monascus*, and some marine fungi are becoming popular (Lin & Xu, 2022). The natural food colorant market was valued at USD 1.6 billion in 2020 (Mordor Intelligence, 2022). For instance, astaxanthin and carotenoids have high market demand (Nair et al., 2023). Other fungal pigments, such as those from *Monascus* and *Neurospora*, are used in the food and textile industries (Gmoser et al., 2018). However, the share of fungal-based products in these markets is unclear. Food-grade pigments from filamentous fungi, such as *Monascus* pigments, Arpink red™ from *Penicillium oxalicum*, riboflavin from *Ashbya gossypii*, and lycopene and β-carotene from *Blakeslea trispora*, exist in the market (Park et al., 2000).

Cosmetics

Natural biomolecules synthesized by fungi are widely used in cosmetic products for their beneficial properties, including anti-aging, antioxidants, skin revitalization, whitening, and hair care effects (Badalyan et al., 2022; Usman et al., 2023). Key compounds such as polysaccharides, glycoproteins, phenolic compounds, terpenoids, and lipids are highly valued for enhancing skin health (Ślusarczyk et al., 2021). Various mushroom species have been included as ingredients in topical creams, lotions, ointments, serums, and facial products for cosmetic purposes. These mushrooms, which have traditional uses in East Asian cultures, include

Shiitake (*Lentinula edodes*), Maitake (*Grifola frondosa*), Reishi or Lingzhi (*Ganoderma lucidum*), Fu Ling (*Wolfiporia extensa*), Yartsa Gunbu (*Cordyceps sinensis*), cauliflower mushroom (*Sparassis latifolia*), and jelly fungi (*Tremella* spp.) (Wu et al., 2016; Liu, 2002; Zhang et al., 2002). For example, Shiitake mushrooms have diverse cosmetic applications, including an exfoliant, an anti-inflammatory agent, and a skin brightener, promoting faster skin renewal and improved elasticity (Wu et al., 2016). Reishi mushrooms have been incorporated into anti-aging skincare products since the 1980s, with the Japanese brand Menard being one of the first to do so (Wu et al., 2016). Glucans, primarily from yeast, had a global market value of USD 403.8 million in 2020 (Markets and Markets, 2022). Lentinan from *L. edodes* was valued at USD ten million in 2022 (Market Watch, 2023). Ergothioneine, an antioxidant from edible mushrooms like *A. bisporus* and *Pleurotus* (Kalač, 2016), had a market value of USD 15.02 million in 2021 (Market Data Forecast, 2022). Kojic acid, produced by *Aspergillus* species and used for its skin-lightening effects and popularly incorporated in products like lotions, soaps, and skin creams (Ola et al., 2019), had a global market value of USD 36 million in 2022 (Market Watch, 2022).

Aside from topical applications, nutricosmetic products are also popular as oral applications in the form of nutritional supplements where several ingredients are combined with mushroom extracts and other mushroom-derived components (Barel et al., 2014) and are known to have the same beneficial influences on beauty and health (Wu et al., 2016).

Antibacterials and Antifungals

Fungi have been used for medicinal purposes since ancient times, valued for their nutritional properties and ability to produce bioactive secondary metabolites. These compounds have led to the development of many important pharmaceuticals (Bills & Gloer, 2016). These fungal metabolites, particularly from species of the Ascomycota and Basidiomycota (Sandargo et al., 2019), have been vital in creating drugs, most notably anti-infective agents, cardiovascular drugs, and immunomodulatory medications.

Penicillin discovery from *Penicillium rubens* is one of the medical breakthroughs that has revolutionized the treatment of several bacterial infections and is still widely used. (Sawant et al., 2023). The penicillin market was estimated to be worth USD 1.96 billion in 2022 (Market Watch, 2022). Cephalosporins, another group of β-lactam antibiotics initially discovered from *Acremonium* species, have been developed through five generations against drug-resistant bacteria (Lin & Kuck, 2022). Carbapenems, first derived from *Streptomyces* species, were introduced as potent antibiotics against resistant bacteria (McGowan et al., 1999). Fusidic acid, a fungal metabolite from *Ramularia coccinea*, is a topical agent for treating skin infections (Fernandes, 2016).

On the other hand, antifungal agents such as echinocandins (Huttel, 2021) and griseofulvin (Gupta et al., 2017) have shown therapeutic properties and economic value. Echinocandins target invasive fungal infections, while griseofulvin targets superficial fungal infections, reaching a global market size of USD 515.20 million in 2021 (Data Bridge Market Research, 2022) and USD 6.2 billion by 2027 (Global Market Insights, 2022a), respectively. Overall, fungal-derived drugs, despite their slow and meticulous development process, have become indispensable in modern medicine, offering vast economic and therapeutic benefits.

Traditional Medicine and Other Drugs

In Southeast Asia, several traditional fungal-based medicines (Paterson, 2008) have evolved into popular commercial products in both local and international markets. *Ganoderma lucidum* (Reishi or Lingzhi), highly regarded for its immune-boosting and anti-aging properties, is widely sold in the form of capsules, powders, teas, and extracts (Ekiz et al., 2023). These products, often marketed as supplements, claim to enhance overall well-being, reduce stress, and promote longevity (Zhu et al., 1998). Brands like Thailand, Vietnam, and Indonesia offer Reishi products, with many exported globally (Chang & Miles, 2004; Bhosle et al., 2010). *Cordyceps sinensis* (Yartsa Gunbu) is another highly sought-after valuable medicinal and edible fungus that is extensively utilized for promoting overall health and preventing diseases (Wang et al., 2024). *Cordyceps* is commonly available in capsules, liquid extracts, and even infused into herbal drinks, often targeting athletes or individuals looking to improve stamina and vitality (Shweta et al., 2023). The demand for *Ophiocordyceps sinensis* in Chinese markets is substantial, with annual production estimated at 100 to 150 tons (Zha et al., 2018).

Other important drugs where fungi play a significant role are cardiovascular health, immunosuppression, and nematicide treatments. Statins, such as pravastatin, lovastatin, and their synthetic derivatives like atorvastatin, are widely used for lowering cholesterol to prevent cardiovascular diseases (Rea, 2008), with the global statin market expected to reach USD 17.5 billion by 2027 (IMARC Group, 2022). Immunosuppressive agents, including cyclosporin A derived from *Tolypocladium inflatum* (Yang et al., 2018), are found vital for organ transplantation, while fingolimod (Gilenya) is used to treat multiple sclerosis (Costa Novo & Felgueiras, 2021). Emodepside, derived from a fungal endophyte (Helaly et al., 2018), is a nematicide used in veterinary medicine and shows potential for treating human parasitic infections (Hubner et al., 2021). These drugs demonstrate the medicinal potential of fungal-derived compounds and represent significant market value in the global pharmaceutical industry.

Mycopesticides and Biofertilizers

Fungi-based biological control agents, known as mycopesticides, have been used for years in agriculture and urban environments, making up roughly 50% of microbial biopesticides (Mascarin et al., 2019). *Mycar* is the first mycopesticide product registered in 1981 that utilized *Hirsutella thompsonii* to combat spider mites (Rosas-Acevedo et al., 2003). Mycoinsecticides provide a safer and more environmentally friendly alternative to chemical pesticides, posing less risk to humans and pets (Bamisile et al., 2021). Prominent fungal genera such as *Akanthomyces*, *Beauveria*, and *Metarhizium* are widely used as biological control agents targeting insect pests. In 2022, the global biopesticides market was valued at USD 5.5 billion (Research and Markets, 2022), with mycopesticides accounting for approximately 10% of this value, or USD 550 million (Zaki et al., 2020). Additionally, strobilurins, bioactive compounds discovered initially in *Strobilurus tenacellus*, have been extracted from various fungi and have led to the development of agricultural fungicides, such as azoxystrobin and pyraclostrobin (Nofiani et al., 2018).

Mycorrhiza-based biofertilizers are natural fertilizers that enhance plant nutrient uptake through fungal symbiosis (Barea et al., 2011; Masrahi et al., 2023). Arbuscular mycorrhizal fungi (AMF) and ectomycorrhizal fungi, such as *Pisolithus tinctorius*, *Rhizophagus*, and *Trichoderma*, are key species and are highly valued. In Japan, for instance, AMF production was 25 tons in 2011, priced at USD100 for 1 kg (Yokoyama, 2023). Examples of well-studied AMF species include *Glomus intraradices*, *Glomus mosseae*, *Glomus etunicatum*, *Glomus claroideum*, *Glomus aggregatum*, *Glomus versiforme*, *Glomus geosporum*, *Glomus fasciculatum*, and *Glomus intraradices* (Ferreyra-Suarez et al., 2024). In the Philippines, soil-based biofertilizers with effective species of fungi that assist crops in absorbing water and nutrients have been made available in the market (Aguilar et al., 2018). The global biofertilizer market is distributed across various categories, with *Rhizobium* representing 18%, *Azotobacter* 17%, *Azospirillum* 16%, blue-green algae 13%, phosphate-solubilizing bacteria 15%, mycorrhiza 17%, and other biofertilizers accounting for 4% (Sansinenea, 2021).

Fungal Enzymes

Fungal enzymes enhance industrial processes by improving efficiency. At the same time, biofertilizers promote sustainable farming practices by fostering plant growth and soil health (Singh et al., 2025). The capacity of fungi to release beneficial enzymes is valuable for food production. Many endophytic fungi have been explored for important enzymes. Examples of such enzymes are lipases, which improve the taste of many dairy products (Singh & Mukhopadhyay, 2012), pectinases, which are important enzymes for fruit juice extraction (Tapre & Jain, 2014); proteases for their importance in brewing and baking (Saxena et al., 2001), and many others, such as lactases, hemicellulases, pullulanases, glucanases, α-amylases, and β-amylases, which are used in several food processing (Kavanagh, 2017; Pouris et al., 2024). With its rich culinary heritage, Southeast Asia has adopted fungal enzymes in traditional foods like soy sauce, miso, and tempeh. In Thailand and Vietnam, where fermented foods are dietary staples, fungal enzymes are part of efficient production processes, offering cleaner alternatives to chemical additives and aligning with the preferences of health-conscious consumers (Waché et al., 2018).

Market Trends and Consumer Demands for Fungal-Based Agricultural Products

Market trends for fungal-based products in the Southeast Asian Market are primarily driven by biofertilizers and biopesticides, which include mycopesticides. Market trends for mycopesticides, described as "living preparations that use fungal cells, such as spores and hyphae, as active ingredients" (Jiang & Wang, 2023), have been influenced by several factors, including agricultural intensification, sustainable agriculture, and the regulatory environment. With the need to produce more food, the intensification of crop production has increased the use of pesticides to reduce losses brought by destructive pests and diseases. However, the use of pesticides in pest and disease management has deleterious effects on non-target organisms (Skovmand, 2007; Crowder & Harwood, 2014; Tscharntke et al., 2007; Sann et al., 2018). Biopesticides, including mycopesticides, are a more sustainable, ecofriendly alternative to reducing pesticide inputs (Sann et al., 2018; Rajapakse et al., 2016; Bateman et al., 2018). On the other hand, consumer demands align well with market trends. Consumers have become more aware of the impact and footprints of pesticide and agrochemical use in crop production, resulting in adopting alternative pest and disease control strategies (Skovmand, 2007). For instance, using biological control agents like *Beauveria bassiana* and *Metarhizium anisopliae* has been considered a safer approach to pest management (Rajapakse et al., 2016; Leite et al., 2022).

Nevertheless, the adoption of mycopesticides in Southeast Asian countries is limited and still challenged by several factors, including the economics of the mycopesticides, education, and training of the users, and external factors that include regulatory frameworks and market competition (Skovmand, 2007; Istriningsih et al., 2024; Sackman, 2013), to name some. For instance, the study by Istriningsih et al. (2024) found that despite the benefits of biopesticides, farmers' adoption of the technology in Indonesia is low. One factor identified to have affected this is economic consideration or pricing strategy.

In Southeast Asia, biological control agents are categorized into attractant, botanical, growth stimulator, microbial, natural product, product mix, semiochemical, and others (Bateman et al., 2014). Biological control agents categorized as "microbial" dominate (174 of the 471) on the commercial product list, which includes bacteria and fungi (Bateman et al., 2014). Fungi-based products were second to bacteria (generally represented by *Bacillus* species). However, they have been commercialized in some Southeast Asian countries as microbial or product mixes (Table 13.1) (Bateman et al., 2014). Most registered products have *B. bassiana* and *M. anisopliae* as key components.

Seventy-seven (77) *Trichoderma*-based commercialized biocontrol agents are available (Tyśkiewicz et al., 2022). The main species used for these commercialized products are *T. viride* and *T. harzianum*. Products registered in India include Anushka, Bhoomika, Biofort, Jaivaka, and Monitor. In contrast, Predatox, Tanus, Pardella, and Ecotrich WP are registered for use in Brazil (Tyśkiewicz et al., 2022). Some 77 *Trichoderma*-based commercialized products are registered in India, Brazil, Chile, Canada, EU member states, the USA, New Zealand, Australia, Uganda, Uruguay, and Poland (Tyśkiewicz et al., 2022). Interestingly, none has

Table 13.1 Pest diseases against which BCA have been registered in Indonesia, Malaysia, Philippines, and Vietnam (Status as of October 2013)

Pests and diseases	Biocontrol agent (AI)	Category	Country registered
Cocoa pod borer	*Beauveria bassiana*	Microbial	Indonesia
Coffee berry borer	*Beauveria bassiana*	Microbial	Indonesia
Cutworm	*Metarhizium anisopliae*	Product mix	Indonesia
Mirid bug	*Beauveria bassiana*	Microbial	Indonesia
White root disease	*Trichoderma koningii*	Microbial	Indonesia
Rice bug	*Beauveria bassiana*	Microbial	Indonesia
Rhinoceros beetle	*Metarhizium anisopliae*	Product mix	Indonesia
Rhinoceros beetle	*Metarhizium anisopliae* Var. *Majus* (St-01)	Microbial	Malaysia
Aphid	*Beauveria bassiana* strain GHA	Microbial	Philippines
Mealybugs	*Beauveria bassiana* strain GHA	Microbial	Philippines
Mosquito	*Beauveria bassiana* strain GHA	Microbial	Philippines
Phytophthora (blight, rot)	*Trichoderma* spp.	Microbial	Philippines
Psyllids	*Beauveria bassiana* strain GHA	Microbial	Philippines
Thrips	*Beauveria bassiana* strain GHA	Microbial	Philippines
Whitefly	*Beauveria bassiana* strain GHA	Microbial	Philippines
Black Rot	*Trichoderma* spp.	Microbial	Vietnam
Buklworm	*Beauveria bassiana*	Microbial	Vietnam
Brown Plant Hopper	*Beauveria bassiana*	Product mix	Vietnam
Brown Plant Hopper	*Metarhizium anisopliae* var. *anisopliae*	Microbial	Vietnam
Diamondback Moth	*Beauveria bassiana*	Microbial	Vietnam
Mold	*Trichoderma viride*	Microbial	Vietnam

Source: ASEAN Sustainable Agrifood Systems (Biocontrol) Project; Bateman et al. (2014)

been reported (Tyśkiewicz et al., 2022) as a registered commercialized product in the Southeast Asian region. Registered *Trichoderma*-based commercialized products are in demand for managing plant diseases caused by fungi in the genus *Fusarium*, *Rhizoctonia*, *Alternaria*, *Colletotrichum*, and fungi-like organisms in the genus *Phytophthora* and *Pythium.* Regarding protected crops, each product is only allowed for use in certain crops with some exemptions (e.g., India, Brazil, Chile), where most products have been permitted for use on all crops. For instance, Predatox is registered in Brazil for lettuce, soybeans, and other beans. At the same time, Pardella can be used in all crops.

While the global market share for mycopesticides is relatively low compared with other biological control agent products (Bergin, 2014; Chen, 2017; Zaki et al., 2020), both mycopesticides and biofertilizers are experiencing market growth, resulting from consumer demand for more sustainable agricultural food products. This trend is expected to continue as consumers' environmental concerns and health consciousness rise, supportive government policies encourage farmers, and pricing strategies favor farmers and growers.

Challenges and Opportunities in the Agricultural Sector

Multiple fungal products and commodities bolster the Southeast Asian economy but are also hampered by a few notorious fungal species that damage crops, post-harvest commodities, and marketed food products. The region's diverse but generally warm, humid climate, as well as the changes brought by global warming, provides an ideal environment for fungal growth, leading to devastating plant diseases and post-harvest spoilage (Chakraborty et al., 2021; Lim et al., 2023; Pangga et al., 2023; Phan et al., 2021). In the Philippines, Indonesia, Thailand, and Vietnam, agriculture is vital to the economy and is particularly vulnerable to fungal infestations. At the same time, despite fungi's potential in biotechnology and food production, their utilization remains underdeveloped due to limited investment and market awareness (Jenol et al., 2024.

In the field, a few species, such as *Magnaporthe oryzae*, threaten staple crops in the region. This pathogen is responsible for significant yield losses in rice-producing countries such as Vietnam and the Philippines (Phi, 2023; Simkhada & Thapa, 2022). Lesions on leaves, stems, and panicles reduce photosynthetic capabilities and thus lead to reduced grain quality and production. Efforts to combat rice blast include the development of resistant rice varieties, crop rotation, and applying fungicides (Bui & Nguyen, 2024). However, the rapidly evolving fungus, owing to the rise of resistant strains and the overuse and improper use of fungicides, presents an ongoing challenge, curbing economic growth and development.

Fusarium species cause vascular wilt in banana, including Panama disease (Foc Tropical Race 4), which threatens the Cavendish banana variety grown in the Philippines and Indonesia. *Fusarium verticillioides*, and *F. graminearum* contaminate maize and wheat with mycotoxins, such as fumonisin and deoxynivalenol

(DON), posing health risks to humans and livestock when consumed. The persistence of these fungi in soil and plant residues makes their management particularly difficult. Cash crops are important economic drivers of many countries; thus, damage to them results in great economic loss.

Post-harvest fungal contamination is another primary concern, especially with *Aspergillus* and *Penicillium* species, which produce mycotoxins that compromise production and marketing but also extend to losses due to health and safety reasons. *Aspergillus flavus* and *A. parasiticus* contaminate peanuts, corn, and cassava with aflatoxins. These potent carcinogens have been linked to liver cancer in countries like Thailand and the Philippines (Gómez-Salazar et al., 2023). Inappropriate storage conditions, such as high humidity and inadequate ventilation, further exacerbate fungal contamination, highlighting the need for improved post-harvest handling and storage infrastructure (Alvito et al., 2021; Owolabi et al., 2022; Sun et al., 2023).

Despite fungi's negative impact on agriculture, they also hold great potential for food and pharmaceutical industries. Edible fungi such as shiitake (*Lentinula edodes*) and oyster mushrooms (*Pleurotus* spp.) are widely consumed in some parts of Southeast Asia (Fung & Tan, 2021) and a good source of income for many farmers (Raman et al., 2021), yet their production and marketing remain low compared to other agricultural products. Additionally, fungal-based biofertilizers and biopesticides offer sustainable alternatives to chemical inputs (Ferreyra-Suarez et al., 2024). However, their adoption in mainstream agriculture is still limited due to a lack of awareness, regulatory hurdles, and insufficient funding for research and development.

Fungal threats extend beyond field infections to food processing and storage facilities. Mold growth on stored grains, nuts, and dried fruits can lead to spoilage, economic losses, and foodborne illnesses. Countries that export large quantities of palm oil, spices, and dried commodities face challenges in maintaining fungal-free products for international trade. Strict quality control measures, including antifungal treatments and improved storage conditions, are necessary to meet food safety standards and prevent economic losses.

Addressing the challenges posed by fungi in Southeast Asian agriculture requires an integrated approach, combining resistant crop varieties, biocontrol methods, improved post-harvest management, and stricter agricultural and food safety regulations. At the same time, greater investment in fungal biotechnology, including the commercial production of edible fungi, biofertilizers, and medicinal compounds, could help turn fungal research into an economic opportunity. Climate change, which alters temperature and humidity patterns, may further influence fungal prevalence and severity in the future. Therefore, continuous research, farmer education, and international collaboration are essential to developing sustainable solutions that ensure the region's food security and economic resilience.

References

Aguilar, E., Elleva, L. I., Realin, D., Garcia, G., Divina, F., & Aggangan, N. (2018). Arbuscular mycorrhizal fungi increased root-mycorrhizal association and enhanced seedling growth of abaca, papaya, and sugarcane. *Journal of the International Society for Southeast Asian Agricultural Sciences, 24*(1), 22–29.

Alvito, P., Barcelo, J., De Meester, J. O. H. A. N., Rito, E. L. I. A. S., & Suman, M. (2021). Mitigation of mycotoxins during food processing: Sharing experience among Europe and South East Asia. *Science & Technology of Cereals, Oils & Foods*, 6.

Amara, A. A., & Abd El-Baky, N. (2023). Fungi as a source of edible proteins and animal feed. *Journal of Fungi, 9*(1), 73. https://doi.org/10.3390/jof9010073

Aquino, E. M., Tapay, N. E., & Barraquio, V. L. (2011). A case study of the indigenous technology for making white soft cheese kesong puti in Lumban, Laguna, Philippines. *Philippine Journal of Veterinary and Animal Sciences, 37*(1), 89–100.

Arnold, M., Rajagukguk, Y. V., & Gramza-Michalowska, A. (2021). Characterization of Dadih: Traditional fermented buffalo milk of Minangkabau. *Beverages, 7*(3), 60. https://doi.org/10.3390/beverages7030060

Badalyan, S. M., Barkhudaryan, A., & Rapior, S. (2022). Medicinal macrofungi as cosmeceuticals: A review. *International Journal of Medicinal Mushrooms, 24*, 1–13. https://doi.org/10.1615/IntJMedMushrooms.2022043124

Bamisile, B., Akutse, K., Siddiqui, J., & Xu, Y. (2021). Model application of entomopathogenic fungi as alternatives to chemical pesticides: Prospects, challenges, and insights for next-generation sustainable agriculture. *Frontiers in Plant Science, 12*, 741804. https://doi.org/10.3389/fpls.2021.741804

Barea, J. M., Palenzuela, J., Cornejo, P., et al. (2011). Ecological and functional roles of mycorrhizas in semi-arid ecosystems of Southeast Spain. *Journal of Arid Environments, 75*(12), 1292–1301. https://doi.org/10.1016/j.jaridenv.2011.07.004

Barel, A., Paye, M., & Maibach, H. (2014). Use of food supplements as nutricosmetics in health and fitness. *Handbook of Cosmetic Science and Technology*, 583–596. https://doi.org/10.1201/b16716-56

Bateman, R., Ginting, S., Moltmann, J., & Jakel, T. (2014). *ASEAN guidelines on the regulation, use, and trade of biological control agents (BCA)*. German Cooperation.

Bateman, M. L., Day, R. K., Luke, B., Edgington, S., Kuhlmann, U., & Cock, M. J. (2018). Assessment of potential biopesticide options for managing fall armyworm (Spodoptera frugiperda) in Africa. *Journal of Applied Entomology, 142*(9), 805–819.

Beers, S. J. (2012). *Jamu: The ancient Indonesian art of herbal healing*. Tuttle Publishing.

Bergin, J. (2014). *Agricultural biotechnology: Emerging technologies and global markets* (p. 7215). BCC Res.

Bhosle, S., Ranadive, K., Bapat, G., Garad, S., Deshpande, G., & Vaidya, J. (2010). Taxonomy and diversity of *Ganoderma* from the western parts of Maharashtra (India). *Mycosphere, 1*(3), 249–262.

Bills, G. F., & Gloer, J. B. (2016). Biologically active secondary metabolites from fungi. *Microbiology Spectrum, 4*(5). https://doi.org/10.1128/microbiolspec.funk-0009-2016

Bui, L. T., & Nguyen, P. H. (2024). Assessing impacts on rice production and quantifying economic cost losses with PM2. 5–based damage–A case of the Long Xuyen Quadrangle, Vietnam Mekong Delta. *Computers and Electronics in Agriculture, 217*, 108570.

Chakraborty, A., Singh, J., & Upadhyay, H. (2021). Fungal diversity and ecology: Major challenges for crop production in present climate scenario. In *Recent trends in mycological research: Volume 1: Agricultural and medical perspective* (pp. 89–107).

Chang, S. T., & Miles, P. G. (2004). *Mushrooms: Cultivation, nutritional value, medicinal effect, and environmental impact* (2nd ed.). CRC Press.

Chen, J. (2017). *Global markets for biopesticides* (p. 7215). BCC Res.

Cherta-Murillo, A., Lett, A. M., Frampton, J., Chambers, E. S., Finnigan, T. J. A., & Frost, G. S. (2020). Effects of mycoprotein on glycaemic control and energy intake in humans: A systematic review. *The British Journal of Nutrition, 123*(12), 1321–1332. https://doi.org/10.1017/S0007114520000756. Epub 2020 Feb 26.

Copetti, M. V. (2019). Fungi as industrial producers of food ingredients. *Current Opinion in Food Science, 25*, 52–56. https://doi.org/10.1016/j.cofs.2019.02.006

Costa Novo, J., & Felgueiras, H. (2021). Neuro-ophthalmologic manifestations of multiple sclerosis other than acute optic neuritis. *Multiple Sclerosis and Related Disorders, 48*, 102730. https://doi.org/10.1016/j.msard.2020.102730

Crowder, D. W., & Harwood, J. D. (2014). Promoting biological control in a rapidly changing world. *Biological Control, 75*, 1–7.

Culloty, D. (2010). *Food from northern Laos: The boat landing cookbook.* Galangal Press.

Dai, X., Sharma, M., & Chen, J. (Eds.). (2021). *Fungi in sustainable food production.* Springer International Publishing.

Darmayanti, L. P. T., Antara, N. S., & Duniaji, A. S. (2014). Physicochemical characteristic and protein profile of fermented Urutan (Balinese Sausage). *International Journal on Advanced Science, Engineering and Information Technology, 4*, 112–116.

Data Bridge Market Research. (2022). Global echinocandins market–Industry trends and forecast to 2029. https://www.databridgemarketresearch.com/reports/global-echinocandins-market.

Ekiz, E., Oz, E., Abd El-Aty, A. M., Proestos, C., Brennan, C., Zeng, M., et al. (2023). Exploring the potential medicinal benefits of *Ganoderma lucidum*: From metabolic disorders to coronavirus infections. *Food, 12*(7), 1512. https://doi.org/10.3390/foods12071512

FAOSTAT Food and Agriculture Data. 2020. [(accessed on October 2024)]. Available online: http://www.fao.org/faostat/en/#home

FAOSTAT. (2022). Mushroom and truffles. https://www.fao.org/faostat/en/#search/Mushrooms%20and%20truffles. Accessed October 2024.

Fernandes, P. (2016). Fusidic acid: A bacterial elongation factor inhibitor for the oral treatment of acute and chronic staphylococcal infections. *Cold Spring Harbor Perspectives in Medicine, 6*(5), a025437. https://doi.org/10.1101/cshperspect.a025437

Ferreyra-Suarez, D., García-Depraect, O., & Castro-Muñoz, R. (2024). A review on fungal-based biopesticides and biofertilizers production. *Ecotoxicology and Environmental Safety, 283*, 116945.

Fung, S. Y., & Tan, C. S. (2021). Mushroom market in Malaysia. In *Advances in macrofungi* (pp. 41–50). CRC Press.

Global Market Insights. (2022a). *Onychomycosis treatment market size*. https://www.gminsights.com/industry-analysis/onychomycosis-treatment-market.

Global Market Insights. (2022b). *Gluconic acid market*. https://www.gminsights.com/industry-analysis/gluconic-acid-market.

Global Market Insights Inc. (2022). Single cell protein market share | industry statistics—2030. Available online: https://www.gminsights.com/industry-analysis/single-cell-protein-market.

Gmoser, R., Ferreira, J. A., Lundin, M., Taherzadeh, M. J., & Lennartsson, P. R. (2018). Pigment production by the edible filamentous fungus *Neurospora intermedia. Fermentation, 4*(1), 11. https://doi.org/10.3390/fermentation4010011

Gómez-Salazar, J. A., Ruiz-Hernández, K., Martínez-Miranda, M. M., & Castro-Ríos, K. (2023). Postharvest strategies for decontamination of aflatoxins in cereals. *Food Reviews International, 39*(7), 3635–3662.

Gupta, A. K., Foley, K. A., & Versteeg, S. G. (2017). New antifungal agents and new formulations against dermatophytes. *Mycopathologia, 182*(1–2), 127–141. https://doi.org/10.1007/s11046-016-0045-0

Gupta, V. K., Treichel, H., Shapaval, V., de Oliveira, L. A., Tuohy, M. G., & Wiley, J. (Eds.). (2018). *Microbial functional foods and nutraceuticals*. Wiley.

He, J. (2018). Harvest and trade of caterpillar mushroom (*Ophiocordyceps sinensis*) and the implications for sustainable use in the Tibet region of Southwest China. *Journal of Ethnopharmacology, 221*, 86–90. https://doi.org/10.1016/j.jep.2018.04.022

Helaly, S. E., Thongbai, B., & Stadler, M. (2018). Diversity of biologically active secondary metabolites from endophytic and saprotrophic fungi of the ascomycete order Xylariales. *Natural Product Reports, 35*(10), 992–1014. https://doi.org/10.1039/C8NP00010G

Huang, C., Liu, C., & Xiao, X. (2016). Quality characteristics of a pickled tea processed by submerged fermentation. *International Journal of Food Properties, 19*(6), 1194–1206.

Hubner, M. P., Townson, S., Gokool, S., et al. (2021). Evaluation of the in vitro susceptibility of various filarial nematodes to emodepside. *International Journal for Parasitology: Drugs and Drug Resistance, 17*, 27–35. https://doi.org/10.1016/j.ijpddr.2021.07.005

Huttel, W. (2021). Echinocandins: Structural diversity, biosynthesis, and development of antimycotics. *Applied Microbiology and Biotechnology, 105*(1), 55–66. https://doi.org/10.1007/s00253-020-11022-y

IMARC Group. (2022). Statin market: Global industry trends, share, size, growth, opportunity and forecast 2022b–2027. https://www.imarcgroup.com/statin-market

IMARC Group. (2023). Bakery products market: Global industry trends, share, size, growth, opportunity and forecast 2023–2028. https://www.imarcgroup.com/bakery-products-market

Istriningsih, I., Sumarwan, U., Hartoyo, H., & Suharjo, B. (2024). Behavioral intention to use biopesticides among rice farmers in Indonesia. In *E3S web of conferences* (Vol. 595, p. 04005). EDP Sciences.

Jenol, M. A., Chu, P. H., Ramle, I. K., Joyce, L. J. W., Lai-Yee, P., Ibrahim, M. F., et al. (2024). Feasibility of agricultural biomass in Southeast Asia for enzymes production. *Renewable and Sustainable Energy Reviews, 200*, 114601.

Jiang, Y., & Wang, J. (2023). The registration situation and use of mycopesticides in the world. *Journal of Fungi, 9*(9), 940.

Joshi, S. K., & Gauraha, A. K. (2022). Global biofertilizer market: Emerging trends and opportunities. In *Trends of applied microbiology for sustainable economy* (pp. 689–697).

Kalac, P. (2016). Edible mushrooms: chemical composition and nutritional value. *Academic Press*

Kavanagh, K. (2017). *Fungi: Biology and applications* (2nd ed.). Wiley.

Kavya, M., Heeba, S., Anupama, A. R., & Nisha, P. (2024). Functional foods: History, scope and future trends. In *Traditional foods: The reinvented superfoods* (pp. 31–43). Springer.

Kelechi, M., & Ukaegbu-Obi, K. M. (2016). Single cell protein: A resort to global protein challenge and waste management. *Journal of Microbiology and Microbial Technology, 1*, 5–16.

Khanongnuch, C., Unban, K., Kanpiengjai, A., & Saenjum, C. (2017). Recent research advances and ethno-botanical history of miang, a traditional fermented tea (*Camellia sinensis var. assamica*) of northern Thailand. *Journal of Ethnic Foods, 4*(3), 135–144. https://doi.org/10.1016/j.jef.2017.08.006

Koul, B., Chopra, M., & Lamba, S. (2022). Microorganisms as biocontrol agents for sustainable agriculture. In *Relationship between microbes and the environment for sustainable ecosystem services, volume 1* (pp. 45–68). Elsevier.

Kruger, V. (2014). *Balinese food: The traditional cuisine and food culture of Bali*. Tuttle Publishing.

Kumar, D., Kumar, R., Sakshi, D., & Kumar, A. (2024). Commercialization and market perspectives of biofertilizers through advanced approaches. In *Metabolomics, proteomics and gene editing approaches in biofertilizer industry: Volume II* (pp. 339–357). Springer.

Lamang, Y. T. (2019). Tapai: The ancient Malay food in Minangkabau tradition. *Journal of Ethnic Foods, 6*(1), 1–9. https://doi.org/10.1186/s42779-019-0029-z

Laxananil, P., Promchai, R., Wanasen, S., Kamdee, S., Thepkasikul, P., Plengvidhya, V., Visessanguan, W., Valyasevi, R. (2009). Monitoring Lactobacillus plantarum BCC9546 starter culture during fermentation of nham, a traditional Thaiporksausage. *International Journal of Food Microbiology, 129*, 312–315.

Leite, M. O., Alves, D. A., Lecocq, A., Malaquias, J. B., Delalibera, I., Jr., & Jensen, A. B. (2022). Laboratory risk assessment of three entomopathogenic fungi used for pest control toward social bee pollinators. *Microorganisms, 10*(9), 1800.

Li, H., Tian, Y., Menolli, N., et al. (2021). Reviewing the world's edible mushroom species: A new evidence-based classification system. *Comprehensive Reviews in Food Science and Food Safety, 20*, 1982–2014. https://doi.org/10.1111/1541-4337.12708

Li, K., Qiao, K., Xiong, J., Guo, H., & Zhang, Y. (2023). Nutritional values and bio-functional properties of fungal proteins: Applications in foods as a sustainable source. *Food, 12*(24), 4388.

Li, Y. P., Ahmadi, F., Kariman, K., et al. (2024). Recent advances and challenges in single cell protein (SCP) technologies for food and feed production. *npj Science of Food, 8*, 66. https://doi.org/10.1038/s41538-024-00299-2

Lim, J. A., Yaacob, J. S., Mohd Rasli, S. R. A., Eyahmalay, J. E., El Enshasy, H. A., & Zakaria, M. R. S. (2023). Mitigating the repercussions of climate change on diseases affecting important crop commodities in Southeast Asia, for food security and environmental sustainability—A review. *Frontiers in Sustainable Food Systems, 6*, 1030540.

Lim-Castillo, P. (2006). Traditional Philippine vinegars and their role in shaping the culinary culture. In R. Hosking (Ed.), *Authenticity in the kitchen* (pp. xx–xx). Prospect Books.

Lin, X., & Kuck, U. (2022). Cephalosporins as key lead generation beta-lactam antibiotics. *Applied Microbiology and Biotechnology, 106*(20), 8007–8020. https://doi.org/10.1007/s00253-022-12272-8

Lin, L., & Xu, J. (2022). Production of fungal pigments: Molecular processes and their applications. *Journal of Fungi (Basel), 9*(1), 44. https://doi.org/10.3390/jof9010044

Liu, J. (2002). Biologically active substances from mushrooms in Yunnan, China. *Heterocycles, 57*, 157–167.

Liu, M., Li, X., Huai, M., Yang, Y., & Dong, C. (2023). Cultivation, bioactive metabolites, and application of caterpillar mushroom Cordyceps militaris: Current state, issues, and perspectives. In *Fungi and fungal products in human welfare and biotechnology* (pp. 187–210). Springer.

Lodolo, E. J., Kock, J. L. F., Axcell, B. C., & Brooks, M. (2008). The yeast *Saccharomyces cerevisiae*–The main character in beer brewing. *FEMS Yeast Research, 8*(7), 1018–1036. https://doi.org/10.1111/j.1567-1364.2008.00433.x

Lu, H., Lou, H., Hu, J., Liu, Z., & Chen, Q. (2020). Macrofungi: A review of cultivation strategies, bioactivity, andapplication of mushrooms. Comprehensive reviews in food science and food safety, 19(5), 2333-2356

Market Data Forecast. (2022). *Button mushrooms market*. https://www.marketdataforecast.com/market-reports/button-mushrooms-market.

Market Watch. (2022). *Penicillin market*. https://www.marketwatch.com/press-release/penicillin-market-size-share-trends-massive-growth-is-expected-to-grow-at-a-cagr-of-38-cagr-forecast-by-2028-108-pages-report-2022a-07-21.

Market Watch. (2023). *Lentinan market overview 2023 to 2030, future trends and forecast*. Retrieved from https://www.marketwatch.com/press-release/lentinan-market-overview-2023-to-2030-future-trends-and-forecast

Markets and Markets. (2022). *Beta glucan market*. Retrieved from https://www.marketsandmarkets.com/Market-Reports/beta-glucan-market-5191796.html

Mascarin, G. M., Lopes, R. B., Delalibera, I., Jr., Fernandes, E. K. K., Luz, C., & Faria, M. (2019). Current status and perspectives of fungal entomopathogens used for microbial control of arthropod pests in Brazil. *Journal of Invertebrate Pathology, 165*, 46–53. https://doi.org/10.1016/j.jip.2018.01.001

Masrahi, A. S., Alasmari, A., Shahin, M. G., et al. (2023). Role of arbuscular mycorrhizal fungi and phosphate solubilizing bacteria in improving yield, yield components, and nutrient uptake of barley under salinity soil. *Agriculture, 13*(3), 537. https://doi.org/10.3390/agriculture13030537

McGowan, S. J., Holden, M. T. G., Bycroft, B. W., et al. (1999). Molecular genetics of carbapenem antibiotic biosynthesis. *Antonie Van Leeuwenhoek, 75*(1), 135–141. https://doi.org/10.1023/A:1001768428383

Mordor Intelligence. (2022). *Natural food colorants market – Growth, trend, Covid-19 impact, and forecast (2022–2027)*. Retrieved from https://www.mordorintelligence.com/industry-reports/global-natural-food-colorants-market

Mordor Intelligence (2025) *Asia-pacific biofertilizer market size and share analysis-growth trends & forecasts up to 2023*. Accessed via https://www.mordorintelligence.com/industry-reports/asia-pacific-biofertilizer-market, Accessed on 19 Feb 2025.

Muzaifa, M., Hasni, D., & Rahmi, F. (2019). What is kopi luwak? A literature review on production, quality, and problems. *IOP Conference Series: Earth and Environmental Science, 365*(1), 012041.

Nagai, T., & Tamang, J. P. (2010). Fermented legumes: Soybean and non-soybean products. In J. P. Tamang & K. Kailasapathy (Eds.), *Fermented foods and beverages of the world* (pp. xx–xx). CRC Press.

Nair, A., Ahirwar, A., Singh, S., Lodhi, R., Lodhi, A., Rai, A., et al. (2023). Astaxanthin as a king of ketocarotenoids: Structure, synthesis, accumulation, bioavailability, and antioxidant properties. *Marine Drugs, 21*(3), 176. https://doi.org/10.3390/md21030176

Nguyen, T., Karl, M., & Santini, A. (2017). Red yeast rice. *Foods, 6*, 19. https://doi.org/10.3390/foods030019

Niego, A. G. T., Lambert, C., Mortimer, P., Thongklang, N., Rapior, S., Grosse, M., et al. (2023a). The contribution of fungi to the global economy. *Fungal Diversity, 121*(1), 95–137.

Niego, A. G. T., Lambert, C., Mortimer, P., et al. (2023b). The contribution of fungi to the global economy. *Fungal Diversity, 121*, 95–137. https://doi.org/10.1007/s13225-023-00520-9

Nofiani, R., de Mattos-Shipley, K., Lebe, K. E., et al. (2018). Strobilurin biosynthesis in basidiomycete fungi. *Nature Communications, 9*(1), 3940. https://doi.org/10.1038/s41467-018-06202-4

Ola, A. R., Metboki, G., Lay, C. S., Sugi, Y., Rozari, P. D., Darmakusuma, D., & Hakim, E. H. (2019). Single production of kojic acid by Aspergillus flavus and the revision of flufuran. *Molecules, 24*(22), 4200

Owolabi, I. O., Kolawole, O., Jantarabut, P., Elliott, C. T., & Petchkongkaew, A. (2022). The importance and mitigation of mycotoxins and plant toxins in southeast Asian fermented foods. *npj Science of Food, 6*(1), 39.

Pangga, I. B., Macasero, J. B. M., & Villa, J. E. (2023). Epidemiology of fungal plant diseases in The Philippines. In *Mycology in the Tropics* (pp. 189–212).

Park, Y.H., Lee, I.Y., Kim, S.W., Lee, J.H., & Jeong, J.C. (2000). Production method of β-carotene from *Blakeslea trispora*. Korean patent KR20000031738.

Paterson, R. R. M. (2008). Cordyceps–a traditional Chinese medicine and another fungal therapeutic biofactory?. *Phytochemistry, 69*(7), 1469–1495.

Persistence Market Research. (2022). *Red yeast rice market*. https://www.persistencemarketresearch.com/market-research/red-yeast-rice-market.asp.

Phan, L. T. K., Tran, T. M., Audenaert, K., Jacxsens, L., & Eeckhout, M. (2021). Contamination of Fusarium proliferatum and Aspergillus flavus in the rice chain linked to crop seasons, cultivation regions, and traditional agricultural practices in Mekong Delta, Vietnam. *Foods, 10*(9), 2064.

Phi, A. (2023). Current status of Rice blast in Vietnam and future perspectives. *OALib, 10*(11), 1–17.

Pouris, J., Kolyva, F., Bratakou, S., Vogiatzi, C. A., Chaniotis, D., & Beloukas, A. (2024). The role of fungi in food production and processing. *Applied Sciences, 14*(12), 5046. https://doi.org/10.3390/app14125046

Prastujati, A. U., Hilmi, M., Khusna, A., Arief, I. I., Makmur, S., & Maulida, Q. (2022). Isolation and identification of lactic acid bacteria of bekamal (Banyuwangi traditional fermented meat). *IOP Conference Series: Earth and Environmental Science, 1020*(1), 012026.

Rajapakse, R. H. S., Ratnasekera, D., & Abeysinghe, S. (2016). Biopesticides research: Current status and future trends in Sri Lanka. In *Agriculturally important microorganisms: Commercialization and regulatory requirements in Asia* (pp. 219–234).
Raman, J., Jang, K. Y., Oh, Y. L., Oh, M., Im, J. H., Lakshmanan, H., & Sabaratnam, V. (2021). Cultivation and nutritional value of prominent Pleurotus spp.: An overview. *Mycobiology, 49*(1), 1–14.
Rea, P. A. (2008). Statins: From fungus to pharma. *American Scientist*. https://www.americanscientist.org/article/statins-from-fungus-to-pharma
Research and Markets. (2022). *Biopesticides market by type*. Retrieved from https://www.researchandmarkets.com/reports/5116567/biopesticidesmarket-by-type-bioinsecticides
Rosas-Acevedo, J. L., Boucias, D. G., Lezama, R., Sims, K., & Pescador, A. (2003). Exudate from sporulating cultures of *Hirsutella thompsonii* inhibit oviposition by the two-spotted spider mite *Tetranychus urticae*. *Experimental & Applied Acarology, 29*(3–4), 213–225. https://doi.org/10.1023/a:1025801817004
Royse, D. J., Baars, J., & Tan, Q. (2017). Current overview of mushroom production in the world. In C. Z. Diego & A. Pardo-Gimenez (Eds.), *Edible and medicinal mushrooms* (pp. 5–13). Wiley.
Sackman, J. (2013). Report from Southeast Asia. *Pharmaceutical Technology, 37*(6), 32–36.
Sampaio, J. P. (2018). Microbe profile: *Saccharomyces eubayanus*, the missing link to lager beer yeasts. *Microbiology (Reading), 164*(9), 1069–1071. https://doi.org/10.1099/mic.0.000677
Sandargo, B., Chepkirui, C., Cheng, T., et al. (2019). Biological and chemical diversity go hand in hand: Basidiomycota as source of new pharmaceuticals and agrochemicals. *Biotechnology Advances, 37*, 107344. https://doi.org/10.1016/j.biotechadv.2019.01.011
Sann, C., Theodorou, P., Heong, K. L., Villareal, S., Settele, J., Vidal, S., & Westphal, C. (2018). Hopper parasitoids do not significantly benefit from non-crop habitats in rice production landscapes. *Agriculture, Ecosystems & Environment, 254*, 224–232.
Sansinenea, E. (2021). Application of biofertilizers: Current worldwide status. In *Advances in biological inoculants (Vol. 1)*. https://doi.org/10.1016/B978-0-12-821667-5.00004-X
Sawant, A. M., Navale, V. D., & Vamkudoth, K. R. (2023). Isolation and molecular characterization of indigenous *Penicillium chrysogenum/rubens* strain portfolio for penicillin V production. *Microorganisms, 11*(5), 1132. https://doi.org/10.3390/microorganisms11051132
Saxena, R. K., Gupta, R., Saxena, S., & Gulati, R. (2001). Role of fungal enzymes in food processing. In *Applied mycology and biotechnology* (Vol. 1, pp. 353–386). Elsevier.
Setiarto, R. H. B., Nur, N., Romulo, A., & Herlina, V. T. (2025). Dangke: unveiling indonesian traditional fermented cheese from Enrekang, South Sulawesi. *Journal of Ethnic Foods, 12*, 16. https://doi.org/10.1186/s42779-025-00279-1
Shweta, Abdullah, S., Komal, & Kumar, A. (2023). A brief review on the medicinal uses of *Cordyceps militaris*. *Pharmacological Research–Modern Chinese Medicine, 7*, 100228. https://doi.org/10.1016/j.prmcm.2023.100228
Simkhada, K., & Thapa, R. (2022). Rice blast, a major threat to the rice production and its various management techniques. *Turkish Journal of Agriculture-Food Science and Technology, 10*(2), 147–157.
Singh, A. K., & Mukhopadhyay, M. (2012). Overview of fungal lipase: A review. *Applied Biochemistry and Biotechnology, 166*(2), 486–520. https://doi.org/10.1007/s12010-011-9455-3
Singh, S. K., Pachauri, R. K., Khatoon, H., Katiyar, D., & Agnihotri, G. (2025). The role of biofertilizers inenhancing soil and productivity-A review. *International Journal of Plant & Soil Science, 37*(3), 141–161.
Skovmand, O. (2007). Microbial control in Southeast Asia. *Journal of Invertebrate Pathology, 95*(3), 168–174.
Ślusarczyk, J., Adamska, E., & Czerwik-Marcinkowska, J. (2021). Fungi and algae as sources of medicinal and other biologically active compounds: A review. *Nutrients, 13*(9), 3178. https://doi.org/10.3390/nu13093178

Souza Filho, P. F., Nair, R. B., Andersson, D., Lennartsson, P. R., & Taherzadeh, M. J. (2018). Vegan-mycoprotein concentrate from pea-processing industry byproduct using edible filamentous fungi. *Fungal Biology and Biotechnology, 5*, 5.

Sudheer, S., Johny, L., Srivastava, S., & Adholeya, A. (2023). The trade-in-trade: Multifunctionalities, current market and challenges for arbuscular mycorrhizal fungal inoculants. *Symbiosis, 89*(3), 259–272.

Sumekar, W., & Al-Baarri, A. N. M. (2020). Study in agroindustry of salted egg: Length of salting process and marketing reach aspects. *Journal of Applied Food Technology, 7*(1), 25–28.

Sun, L., Li, R., Tai, B., Hussain, S., Wang, G., Liu, X., & Xing, F. (2023). Current status of major mycotoxins contamination in food and feed in Asia— a review. *ACS Food Science & Technology, 3*(2), 231–244.

Surya, R. (2024). Fermented foods of Southeast Asia other than soybean- or seafood-based ones. *Journal of Ethnic Foods, 11*, 27. https://doi.org/10.1186/s42779-024-00241-7

Tapre, A. R., & Jain, R. K. (2014). Pectinases: Enzymes for fruit processing industry. *International Food Research Journal, 21*(1), 271–296.

The Brainy Insights. (2022). *Shiitake mushroom market size by product type*. https://www.thebrainyinsights.com/report/shiitake-mushroom-market-12716.

The Business Research Company. (2022). *Tempeh global market report 2022*. Retrieved from https://www.thebusinessresearchcompany.com/report/tempeh-global-market-report

Tridge. (2022). Lion's mane mushroom. https://www.tridge.com/intelligences/lions-mane-mushroom.\

Tscharntke, T., Bommarco, R., Clough, Y., Crist, T. O., Kleijn, D., Rand, T. A., et al. (2007). Conservation biological control and enemy diversity on a landscape scale. *Biological Control, 43*(3), 294–309.

Turnbull, W. H., & Ward, T. (1995). Mycoprotein reduces glycemia and insulinemia when taken with an oral-glucose-tolerance test. *The American Journal of Clinical Nutrition, 61*, 135–140.

Tyśkiewicz, R., Nowak, A., Ozimek, E., & Jaroszuk-Ściseł, J. (2022). Trichoderma: The current status of its application in agriculture for the biocontrol of fungal phytopathogens and stimulation of plant growth. *International Journal of Molecular Sciences, 23*(4), 2329.

Usman, M. M., Idi, A., Aisami, A., & Maigari, F. U. (2023). A review on endophytic fungi: A natural source of industrial enzymes. *Asian Journal of Plant Biology, 5*(1), 7–11. https://doi.org/10.54987/ajpb.v5i1.820

Verified Market Research. (2022). *Soy sauce market size and forecast*. Retrieved from https://www.verifiedmarketresearch.com/product/soy-sauce-market/

Waché, Y., Do, T. L., Do, T. B. H., Do, T. Y., Haure, M., Ho, P. H., Anal, A., Le, V., Li, W. J., Licandro, H., Lorn, D., Ly-Chatain, M. H., Ly, S., Mahakarnchanakul, W., Dinh Vuong, M., Mith, H., Nguyen, D. H., Nguyen, T. K. C., Nguyen, T. M. T., & Chu-Ky, S. (2018). Prospects for food fermentation in South-East Asia: Topics from the tropical fermentation and biotechnology network at the end of the AsiFood Erasmus+ project. *Frontiers in Microbiology, 9*, 2278. https://doi.org/10.3389/fmicb.2018.02278

Walker, G., & Stewart, G. (2016). *Saccharomyces cerevisiae* in the production of fermented beverages. *Beverages, 2*, 30. https://doi.org/10.3390/beverages2040030

Wang, T., Tang, C., He, H., Cao, Z., Xiao, M., He, M., et al. (2024). Evaluation of *Cordyceps sinensis* quality in 15 production areas using metabolomics and the membership function method. *Journal of Fungi (Basel), 10*(5), 356. https://doi.org/10.3390/jof10050356

Watts, E. G., Janes, M. E., Prinyawiwatkul, W., Shen, Y., Xu, Z., & Johnson, D. (2018). Microbiological changes and their impact on quality characteristics of red hot chili pepper mash during natural fermentation. *International Journal of Food Science and Technology, 53*(8), 1816–1823.

Wu, Y., Choi, M.-H., Li, J., Yang, H., & Shin, H.-J. (2016). Mushroom cosmetics: The present and future. *Cosmetics, 3*(3), 22. https://doi.org/10.3390/cosmetics3030022

Yamamoto, S., & Matsumoto, T. (2011). Rice fermentation starters in Cambodia: Cultural importance and traditional methods of production. *Japanese Journal of Southeast Asian Studies, 49*(2), 192–213.

Yang, X., Feng, P., Yin, Y., Bushley, K., Spatafora, J. W., & Wang, C. (2018). Cyclosporine biosynthesis in *Tolypocladium inflatum* benefits fungal adaptation to the environment. *mBio, 9*(5), e01211–e01218. https://doi.org/10.1128/mBio.01211-18

Yokoyama, T. (2023). Development and use of biofertilizers in Japan: Current situation and future directions. *Asia-Pacific Biofertilizers and Biopesticides Information Platform.*. https://apbb.fftc.org.tw/article/373

Yongsmith, B., & Malaphan, W. (2016). Traditional fermented foods in Thailand. In K. Kristbergsson & J. Oliveire (Eds.), *Traditional foods: General and consumer aspects* (pp. xx–xx). Springer.

Zaki, O., Weekers, F., Thonart, P., et al. (2020). Limiting factors of mycopesticide development. *Biological Control, 144*, 104220. https://doi.org/10.1016/j.biocontrol.2020.104220

Zha, L. S., Huang, S. K., Xiao, Y. P., et al. (2018). An evaluation of common *Cordyceps* (ascomycetes) species found in Chinese markets. *International Journal of Medicinal Mushrooms, 20*(12), 1149–1162. https://doi.org/10.1615/IntJMedMushrooms.2018027330

Zhang, Y., Mills, G. L., & Nair, M. G. (2002). Cyclooxygenase inhibitory and antioxidant compounds from the mycelia of the edible mushroom *Grifola frondosa*. *Journal of Agricultural and Food Chemistry, 50*, 7581–7585.

Zhu, J.-S., Halpern, G. M., & Jones, K. (1998). The scientific rediscovery of an ancient Chinese herbal medicine: *Cordyceps sinensis* part I. *Journal of Alternative and Complementary Medicine, 4*(3), 289–303. https://doi.org/10.1089/acm.1998.4.3-289

Part IV
Mycology in the Spectrum of Biodiversity

Chapter 14
Fun Guys of the Dark: Fungi in Cave Ecosystems and Other Extreme Environments

Marian P. De Leon, Lou Gene B. Sibal, and Noel G. Sabino

Abstract Caves and other extreme environments harbor unique microbial communities, particularly fungi, which exhibit remarkable adaptations to conditions such as nutrient limitation, constant darkness, and high humidity. Cave fungi predominantly from the phyla Ascomycota, Basidiomycota, and Mucoromycota utilize substrates such as bat guano, decomposing organic matter, and mineral surfaces. They play important roles in cave ecosystems by recycling nutrients, maintaining ecological balance, and supporting food webs. These fungi perform various functions, from decomposition to pathogenicity, with some species causing diseases like white-nose syndrome in bats and histoplasmosis in humans. The extreme conditions within caves promote the production of secondary metabolites with antimicrobial, anticancer, and enzymatic properties, making cave fungi as possible sources of novel bioactive compounds. Fungal diversity is affected by both abiotic factors, such as temperature, humidity, and substrate type, and biotic interactions involving bats, insects, and other microorganisms. To investigate these communities, researchers utilize a combination of culture-dependent and omics-based approaches. Beyond cave ecosystems, fungi inhabiting other extreme environments, such as salterns, acidic mangrove sediments, glaciers, and deep-sea ecosystems, exhibit similar stress-tolerance mechanisms, including melanin production, siderophore synthesis, and thermotolerance, which enable survival under extreme environmental conditions. Despite their ecological importance and biotechnological potential, studying cave fungi is faced with notable challenges, including limited accessibility, low cul-

M. P. De Leon (✉) · L. G. B. Sibal
Microbial Culture Collection, Museum of Natural History, University of the Philippines Los Baños, College, Los Baños, Laguna, Philippines
e-mail: mpdeleon1@up.edu.ph

N. G. Sabino
Microbial Culture Collection, Museum of Natural History, University of the Philippines Los Baños, College, Los Baños, Laguna, Philippines

Microbiology Division, Institute of Biological Sciences, College of Arts and Sciences, University of the Philippines Los Baños, College, Los Baños, Laguna, Philippines

J. J. G. Guerrero et al. (eds.), *Mycology in a Changing Planet: Applications and Perspectives from Southeast Asia*, https://doi.org/10.1007/978-981-95-2640-6_14

turability, contamination risks, and ethical concerns regarding the conservation of fragile cave ecosystems. Nonetheless, cave fungi present valuable insights into microbial evolution, environmental stress adaptation, and ecosystem functioning, with potential for applications in drug discovery and biotechnology. In Southeast Asia, particularly the Philippines, fungal diversity in extreme environments like caves remains largely unexplored, emphasizing the need for interdisciplinary, conservation-focused research initiatives. This chapter explores the diversity, ecological functions, and biotechnological potential of fungi in these habitats, emphasizing their significance in light of a changing climate.

Keywords Cave fungi · Extreme environments · Fungal diversity · Secondary metabolites · Ecological adaptation

Introduction

Arguably, a changing planet is most evidently manifested through shifts in climate, as every living organism perceives climate-related changes in some capacity. From a meteorological perspective, climate change refers to the long-term global shift in typical weather and climate patterns (BBC News, 2013). The drivers of climate change are diverse, arising from both anthropogenic and natural causes, with mechanisms ranging from simple to highly complex. These drivers are intricately interconnected, perpetuating the self-reinforcing cycle of climate change. Manifestations of this phenomenon include rising sea levels, global warming, ocean acidification, increased costs of livestock and agricultural products, and a growing incidence of emerging and re-emerging diseases. Among the many organisms affected by and contributing to these changes are the fungi which from a mycological perspective can serve both as contributors to and potential mitigators of climate change's consequences.

On the negative side, global warming—exacerbated by the heavy reliance on fossil fuels—amplifies the threats posed by fungal pathogens to public health through four primary mechanisms as described by Williams et al. (2024). First, global warming creates more hospitable environments for pathogenic fungi, enabling them to expand beyond their historical geographical ranges. For instance, *Coccidioides* spp. G.W. Stiles, which were once confined to regions in South and Central America and the Southwestern United States, have now been detected in Northern parts of the United States. Similarly, the distribution of *Histoplasma* Darling and *Cryptococcus gattii* (Vanbreus and Takashio) Kwon-Chung and Boekhout has expanded, while *Candida auris* Satoh and Makimura has exhibited significant global spread. Initially reported in the Indian subcontinent, South Africa, East Asia, and Venezuela, *C. auris* has now been detected worldwide, except in Antarctica (Lockhart et al., 2023). Its ability to inhabit human skin further facilitates rapid global transmission due to increased globalization and human migration.

Second, global warming intensifies extreme weather events, such as hurricanes, storms, and wildfires, which contribute to the widespread dispersal of pathogenic fungal spores. While fungal spores naturally disperse through atmospheric and environmental processes, abnormal weather patterns accelerate and expand their distribution, increasing human exposure risks. For example, cases of *Coccidioides* infections have been linked to dust storms. Additionally, the destruction caused by severe weather events exacerbates fungal transmission due to compromised hygiene, limited access to healthcare, and overcrowding in emergency shelters.

Third, global warming promotes fungal mutations, as observed in *Cryptococcus deneoformans* (San Felice) Vuill. and *C. auris*. Heat stress has been shown to accelerate antifungal resistance in *Cryptococcus deneoformans* (formerly *C. neoformans* var. *neoformans*) through induced transposon mobilization. These genetic disruptions can alter the expression or function of genes (e.g., ERG11 and AFR1) involved in azole antifungal drug susceptibility (Hu, 2025). Meanwhile, *C. auris* has demonstrated an increased tolerance to higher temperatures beyond its previously recorded range. Such adaptations make fungal infections more persistent, transmissible, and difficult to treat (Leitch, 2024).

Finally, the growing prevalence of fungal infections has led to an increased dependence on antifungal medications, which, when misused or overused, further drives antifungal resistance. This concerning trend is exemplified by three clinically significant pathogens: azole-resistant *Aspergillus fumigatus* Frassen. (Lockhart et al., 2023); pan-resistant *C. auris*, demonstrating resistance to all major antifungal classes including azoles, echinocandins, polyenes, and flucytosine; and *Trichophyton indotineae* Kano, Kimura, Kakurai, Hiruma, Kamata, Suga, and Harada, which shows resistance to both azole and allylamine drugs (Hui et al., 2024). The indiscriminate use of antifungal drugs places selective pressure on fungal populations, favoring the proliferation of resistant strains.

Despite their pathogenic potential, fungi also serve as valuable resources in combating diseases, particularly in the field of antibiotic discovery (Correia et al., 2023; Yurchenko et al., 2025). In Southeast Asia, studies have demonstrated that endophytic fungi—specifically those belonging to the genera *Fusarium*, *Dematophora*, *Acremonium*, and *Trichoderma*—exhibit strong antagonistic activity, primarily against *Escherichia coli* Migula and *Staphylococcus aureus* Rosenbach (Mohd et al., 2021; Chua et al., 2022). Notably, Chua et al. (2022) reported that eight fungal isolates also displayed inhibitory effects against *Bacillus cereus* Frankland and Frankland and *Pseudomonas aeruginosa* Schroeter (Migula), in addition to *E. coli* and *S. aureus*. With advances in epidemiology and medicine, global surveillance has identified a concerning rise in the incidence of emerging and re-emerging diseases. These patterns underscore the profound impact of human activities on planetary health, influencing not only climate, biodiversity, and ecosystems but also the dynamics of infectious diseases.

Emerging diseases are those that exhibit rapid increases in incidence or geographical spread, regardless of prior detection. Examples include Hantavirus pulmonary syndrome, Nipah virus infection, COVID-19, Lyme disease, and pathogenic *E. coli* infections. In contrast, re-emerging diseases refer to infections that were previously controlled or declining but have resurged, such as West Nile fever, Zika virus disease, Ebola hemorrhagic fever, cholera, vancomycin-resistant *S. aureus* (VRSA) infections, and multidrug-resistant tuberculosis.

Both emerging and re-emerging diseases are often zoonotic, transmitted from animals or insects to humans, emphasizing the significance of the human-animal interface as a critical hotspot for disease transmission. Additionally, antimicrobial resistance has rendered many re-emerging bacterial infections more difficult to treat. Key drivers of the increasing prevalence of these diseases include overpopulation, globalization, microbial adaptation, and inadequate animal handling practices (Wang et al., 2021).

In response, global research efforts are intensifying to discover novel bioactive compounds, including antifungal and antibacterial agents. Notably, fungi have long been a source of antibiotics, with species such as *Penicillium chrysogenum* Thom yielding penicillin. The exploration of extreme and underexplored environments is a promising avenue for identifying new fungal-derived compounds with potential therapeutic applications.

The complex interplay between climate change, emerging pathogens, and fungal dynamics underscores the urgency of proactive and sustainable interventions. The rising incidence of emerging and re-emerging diseases, driven by climate change and global interconnectedness, exemplifies a positive feedback loop that, if left unaddressed, will lead to worsening global health crises. Effective mitigation strategies require a multifaceted approach, integrating climate action, disease surveillance, antimicrobial stewardship, and the continued exploration of novel bioactive compounds from fungal sources. Without decisive action, the detrimental consequences of these interconnected challenges will only intensify, further jeopardizing planetary and human health.

Chapter 14 focuses on the fungi from cave ecosystems and other extreme environments relatively unexplored and are now an emerging research field in Southeast Asia. Given their unique conditions, cave fungi can be sources of potentially novel bioactive compounds with promising applications in several fields. This chapter explores their diversity and the initiatives conducted to further harness cave fungi, as well as those from extreme environments, for the changing planet.

Limestone Caves and Fungal Diversity

Limestone Caves

Limestone caves are naturally occurring underground voids primarily composed of calcium carbonate and accessible to humans (Turrini et al., 2024). These environments are typically marked by high salinity, persistent darkness, scarce food sources, low temperatures, and reduced oxygen levels. Water flow is another defining characteristic. Caves with active water flow, known as *wet caves*, continue to form speleothems (Fig. 14.1a), whereas those without active flow, referred to as *dry caves*, no longer exhibit speleothem formation (Fig. 14.1b). Caves serve as vital and unique habitats for many bat species, providing roosting and breeding sites, as well as protection from predators. These underground spaces offer a stable microclimate, with relatively constant temperatures and high humidity, which is crucial for bat hibernation and other social events. Caves also support a diverse community of other animals, including insects, birds, and even some predators, who benefit from the presence of bat colonies (Fig. 14.1c).

Cave organisms are classified into three groups: trogloxenes, troglophiles, and troglobites, each with distinct relationships to the cave environment. Trogloxenes are temporary visitors, troglophiles are transient residents, and troglobites are obligate cave dwellers. Trogloxenes are surface-dwelling organisms that occasionally visit caves for specific purposes like foraging or nesting. They are not adapted to living in caves and may not survive long-term. Examples include ants, some beetles, and crickets. Troglophiles are transient residents that can move in and out of caves but prefer to reside within them. They may visit caves for specific stages of their life, such as roosting, breeding, or hibernation. Examples include bats, woodrats,

Fig. 14.1 Speleothems of Limestone Cave. (**a**) Cavinti Underground River and Cave Complex, Cavinti, Laguna, Philippines; (**b**) Kamantigue Cave, Lobo, Batangas, Philippines; (**c**) Obong Cave, Hinoba-an, Negros Occidental, Philippines

and some salamanders. Troglobites, on one hand, are obligate cave dwellers that have evolved adaptations to survive in caves and cannot survive outside. They often lack eyes and pigmentation due to the absence of sunlight. Examples include cave fish, cave crayfish, and salamanders (Choi et al., 2011).

Like other organisms, microorganisms can also be cave residents, cave-dwelling, or cave transients. Microorganisms may be considered residents if they are present as microflora of resident troglobitic animals and/or found in dark zones of the caves, rarely or seldom reached by tourists or troglophiles or trogloxenes. Troglophilic or trogloxenic microorganisms are usually brought by air, water, humans, and other animals.

Despite extreme conditions, limestone caves support diverse life forms, including bacteria, archaea, viruses, protists, fungi, animals, and plants. The delicate balance between abiotic and biotic factors in caves contributes to their high species diversity but also makes them highly vulnerable ecosystems. Notably, the bulk of this diversity is not contributed by organisms visible to the naked eye but by microorganisms stealthily inhabiting cave floors, walls, water, speleothems, and biological excrements.

These microorganisms play diverse ecological roles: some function as primary producers, others prey on fellow microbes, and some specialize in decomposing organic matter for growth and survival. Unlike larger organisms, cave-dwelling microbes cannot migrate over long distances to seek stable or favorable conditions. To adapt to the challenges of nutrient scarcity and competition, these microbes have evolved unique survival strategies, making them promising subjects for research into novel bioactive compounds. For example, certain microbial species produce compounds that inhibit the growth of competitors, aiding survival in nutrient-scarce environments.

Cave Fungi

Cave ecosystems are connected to surface environments, facilitating the exchange of microorganisms. Water movement, air currents, and human visitors can transport microorganisms, including fungi, between subterranean and surface areas (Gabriel & Northup, 2013). Fungi are key players and the most dominant organisms in cave ecosystems. They can colonize any organic substrata within the caves and can establish different biotic interactions with other cave organisms (Wasti et al., 2020). Zhang et al. (2018) suggest that cave fungi may originate from outside environments, as most genera and species are found in other ecosystems. This suggests that while some fungi may have adapted to cave environments, many may be simply introduced from external sources.

The earliest description of cave fungi was reported by Humboldt in 1794, while the first ecological literature of caves was by Megušar in 1914 (Dobat, 1967; Vanderwolf et al., 2013; Zhang et al., 2018). More than 1150 fungal species in 550 genera have been discovered in caves and mines worldwide by 2017 (Zhang et al.,

2018), and some suspected obligate troglobitic fungi have been reported, such as *Acaulium caviariforme* (Malloch et Hubart) Sandoval-Denis et al., *Aspergillus baeticus* Novakova and Hubka, and *Aspergillus thesauricus* Hubka and Novakova (Vanderwolf et al., 2013). In a recent paper, Savković et al. (2025) reported a total of 1447 sequences corresponding to fungi isolated from various substrata within cave environments around the world, which corresponds to 445 fungal species, members of the 394 genera. The number of fungal species was generated using statistical analysis of sequences in the NCBI database and was much lower than those reported by Zhang et al. (2018), which included the fungal species from mines worldwide. Using the sequences, Savković et al. (2025) reported Ascomycota was the most dominant phylum and Eurotiomycetes the dominant class of fungal dwellers in these environments. The highest species richness is detected for the genus *Penicillium*, followed by *Aspergillus*. On the other hand, the most frequently documented single species was *Pseudogymnoascus destructans* (Blehert and Gargas) Minnis and D. L. Lindner, isolated mostly from hibernating bats and guano, followed by *Penicillium chrysogenum* Thom. Because caves have stable, nutrient-limited, low-competition microhabitats that support unusual or cryptic species, many new fungal taxa have been reported as well (such as *Apiotrichum* and *Cephalotrichum* species).

Kosznik-Kwaśnicka et al. (2022) emphasized the significance of cave fungi. Cave fungi can be decomposers, biodeteriogens, and pathogens but may also serve as sources of food, novel species and bioactive compounds. As saprophytes, cave fungi in bat guano, moonmilk, sediments, and carcasses of animals act as the main decomposers of organic matter in caves, leading to biogeochemical cycling of elements in caves. The mycofloral diversity on bat and arthropod cadavers in Semud Hitam chamber of Gomantong Cave, Borneo, an ecotourism hotspot, was assessed using molecular characterization of the internal transcribed spacer (ITS) gene region to confirm identification. All fungi were Ascomycetes except for one Basidiomycete isolate with *Aspergillus* spp. as the highest occurrence (45.8%), followed by *Penicillium* spp. (25.0%), and *Fusarium* sp. (12.5%). Other fungi such as *Ceratobasidium* sp., *Diaporthe* sp., *Pestalotiopsis* sp., and *Xylaria feejeensis* (Berkeley) Fries were also identified. Wasti et al. (2020) further emphasized that bats and arthropods *play a role in fungal dispersion and introduction in the cave because their exteriors are likely to harbor fungi they are exposed to in the environment* and that cadavers serve as *important substrates for fungal growth and proliferation, perpetuating the role of fungi as important decomposers in caves.*

Fungi are incredibly important in cave environments, acting as the primary decomposers. They are a vital part of the cave food web, breaking down organic matter and making its energy available to other cave-dwelling organisms. Essentially, cave fungi serve as a key food source for various cave creatures. Many cavernicolous insects, for instance, feed directly on fungal spores and mycelia. Beyond insects, fungi are also recognized as an essential food source for a wide range of cave arthropods. Fungi play a significant role in cave environments, particularly those associated with animal droppings. Sustr et al. (2005) observed that fungi on bat guano exhibited a higher biomass compared to other areas of the cave and that bat guano fuels cave food webs (Beck et al., 1976).

However, in many tropical caves, especially dry ones, bat guano stands out as the most common source of organic matter (Decu, 1986; Ferreira & Martins, 1999; Fenolio et al., 2006; Vanderwolf et al., 2013). It often forms the base of the food chain for numerous invertebrate communities. These invertebrates might either eat the guano directly or consume the fungi that grow on it. Beyond bats, other animal droppings, such as those from birds (like the oilbird *Steatornis caripensis* Humboldt in Venezuela) and rodents (like the porcupine *Erethizon dorsatum* Linneaus in eastern North America), can also significantly influence the types of animals and fungi found in caves. Many different kinds of insects are associated with dung in caves, which in turn can impact the fungal populations (Calder & Bleakney, 1965; Herrera, 1995; Moulds, 2006; Moseley, 2007; Ribeiro et al., 2012).

Interestingly, Estrada-Barcenas et al. (2010) found that mites were the most diverse invertebrates in bat guano, with *Sancassania* species being the most abundant microarthropod. These mites feed on microfungi, including the pathogen *Histoplasma capsulatum* Darling. This suggests that mites might act as a natural biological control against this pathogen and generally help regulate the numbers of microfungi in caves.

Insects living on dung in caves are exposed to a wide variety and large quantity of fungi. Some insect species are even known to produce antimicrobial substances to protect themselves from infection. Furthermore, insects can actually introduce fungi into caves by carrying spores both on their bodies and internally. For example, Dickson (1975) noted a positive correlation between invertebrate and fungal populations in Virginia cave sediments. Cave crickets (*Ceuthophilus gracilipes* Haldeman) are also thought to be vectors for spreading dictyostelid cellular slime molds both into and within caves.

Cave fungi are pathogens of cave organisms such as *P. destructans* (formerly *Geomyces destructans*), a causative agent of the white-nose syndrome in bats (Hoyt et al., 2021), while members of genus *Arthrorhynchus* (Ascomycota, Laboulbeniomycetes) as obligate parasites of bat flies (Haelewaters et al., 2018). In recent years, white-nose syndrome has caused a severe epidemic among bat populations in the northeastern US, and it is still spreading quickly. It seems that bats' immune systems are weaker during their hibernation, making them more vulnerable to infection by this fungus. White-nose syndrome (WNS), caused by an introduced fungus, was first discovered in a commercial cave in upstate New York in 2006. Since then, it has rapidly spread across North America, leading to massive bat deaths. The disease has even put several bat species at risk of extinction, despite their previously thriving populations. The fungal infection is most severe in the northeastern USA between January and April, when the amount of fungus on bats is at its highest. While bats typically clear the fungus from their skin after emerging from hibernation in the summer, *P. destructans* can linger in their hibernation sites for extended periods, even without bats present. This persistence is likely due to the fungus origins as a soil-dwelling organism (Langwig et al., 2017). As early as 1965, dermatophytes like *Microsporum gypseum* (E. Bodin) Guiart and Grigoraki and

Trichophyton mentagrophytes (C.P. Robin) R. Blanch. were found in soils and caves where bats lived. These fungi are known to cause skin infections in humans (Kajihiro, 1965; Stomeo et al., 2009a, b). In 2005, a diverse array of *Trichosporon* yeasts were found in Spain's Ardales Cave (Málaga) (Sugita et al., 2005). Simultaneously, new species of *Trichosporon* were identified in bat guano from Japanese caves and categorized into four different serotypes. Two of these, *T. laibachii* and *T. multisporum*, fall into the group I-III serotype. Notably, each of these serotypes has been linked to the development of pneumonitis, a type of lung inflammation. Bats inhabit both Ardales Cave and Murciélagos Cave in Córdoba, Spain, as well as the Japanese caves where these discoveries were made. It is well-known that *Trichosporon* species are commonly isolated from guano (bat droppings), which serves as a thriving environment for numerous yeast types, including some that are still undiscovered (Sugita et al., 2005). Research by Vaughan-Martini et al. (2000) revealed a direct correlation between high counts of blastomycetes (yeasts and yeast-like fungi) in caves and the frequency of visits by humans or animals. This suggests that excessive human presence may have contributed to six out of ten isolated yeast strains being directly associated with human illnesses.

Histoplasmosis, often called "cave disease," is the most common illness linked to caves. One can get it through work or leisure activities that expose one to environments containing the fungus *H. capsulatum*. This fungus is found in many parts of the world, including North, Central, and South America, the Caribbean, Europe, Central and Eastern Asia, Africa, and Australia. Cases of cave-related histoplasmosis have been documented in the Americas, Africa, and Oceania. *H. capsulatum* thrives in soil, and its growth is believed to be boosted by bird and bat droppings. These droppings provide nitrates, which help the fungus produce spores more quickly (Nieves-Rivera et al., 2009; Igreja, 2011). In 2004, Skoulidis and colleagues documented a case where a patient, who had frequent contact with soil in bat caves, contracted an infection from the fungus *Penicillium marneffei* Segretain. It is believed the patient most likely acquired the infection by inhaling airborne spores.

At higher concentrations, cave fungal spores and mycelial fragments impair cave indoor air quality, a health risk for visitors and local guides (Ogórek, 2018). Supporting this point, Abris et al. (2020) assessed the air quality inside Cacupangan Cave, Pangasinan, Philippines, using the Index of Microbial Air Contamination (IMA) for cave systems and found that the cave environment is already threatened by fungal contamination and that management and controls are necessary. Despite these risks, however, caves serve as unique ecological niches. As extreme environments with poor nutrients, caves are sources of novel fungal species with potential bioactive compounds for biotechnological or pharmaceutical applications. Using a combination of morphological observation and multigene phylogenetic analyses, four new species were recovered from bat carcasses, namely, *Mortierella rhinolophicola* Tibpromma, Karun., Karasaki and P.E. Mortimer, *M. multispora* Tibpromma, Karun., Karasaki and P.E. Mortimer, *M. yunnanensis* Tibpromma, Karun., Karasaki and P.E. Mortimer, and *Neocosmospora pallidimors* Tibpromma, Karun., Karasaki and P.E. Mortimer (Karunarathna et al., 2020). A number of infections related to *Neocosmospora* associated with human and animal mycotoxicoses are reported to be increasing and alarming.

Cave fungi act as *biodeteriogens*, meaning they contribute to the degradation of materials in the cave environment, particularly stone and other cave artifacts. They achieve this through both physical and chemical mechanisms, leading to staining, discoloration, and structural damage. In cave walls known for prehistoric wall paintings, sacred objects, murals, and other artifacts in the caves (Bastian et al., 2010), many fungal species have been reported as *biodeteriogens* or the main "culprits" responsible for their biodeterioration. *Biodeterioration induced by fungi is reported for several caves from the UNESCO list of world cultural heritage sites; for example, the visible black fungal stains reported on walls with prehistoric paintings in the Lascaux cave* (Bastian et al., 2010).

Caves are widely recognized as extreme environments characterized by perpetual darkness, which significantly influences their nutrient-deficient conditions. Since most plants, as primary producers, require sunlight for growth, their presence is generally restricted to cave entrances where light is available. For life to persist in the nutrient-poor conditions of caves, energy sources must be available, and energy transfer across trophic levels must be efficient. In cave ecosystems, the excreta of troglophiles and troglobites serve as essential energy sources, with bat guano being the most prevalent due to the substantial bat populations. Numerous organisms, ranging from microorganisms to invertebrates, rely on bat guano for survival. Research suggests that bat guano provides an optimal substrate for fungal growth, offering essential nutrients, moisture, and a stable temperature (Wasti et al., 2020). Consequently, a significant portion of fungal diversity in caves is found within bat guano.

Cave Fungal Diversity

Cave fungal diversity is affected by several factors that include anthropogenic activity, humidity, rainfall, seasonal variations, sunlight exposure, temperature, and proximity to the cave entrance. Notably, increased human disturbance within caves has been associated with higher fungal diversity (Wasti et al., 2020). In general, cave fungi may exist as yeasts, molds, or mushrooms.

Bats play a significant role in shaping the types of fungi found in caves, primarily through two actions: transporting fungal spores and depositing guano and carcasses. As bats move in and out of caves, they can act as vectors, carrying fungal spores with them. This movement helps spread different fungal species throughout the cave environment (Vanderwolf et al., 2013). Moreover, the presence of bat guano and bat carcasses can greatly impact fungal diversity (Min, 1988). For instance, research in a Slovakian cave revealed that the highest diversity of fungal species was found on bat guano compared to other surfaces. In fact, some studies suggest that guano can be the dominant factor determining the entire fungal community within a cave (Novakova, 2009).

In 2013, Vanderwolf and colleagues conducted a global review of published literature to describe fungal diversity in caves and mines. Their findings revealed that cave fungi are mainly distributed among several taxa: Ascomycota, Basidiomycota, Zygomycota, Microsporidiomycota, and Chytridiomycota. Among these taxa, the most frequently reported fungal genera include *Geomyces*, *Histoplasma*, *Mucor*, *Penicillium*, *Trichoderma*, *Fusarium*, *Alternaria*, *Cladosporium*, *Acremonium*, *Paecilomyces*, *Rhizopus*, *Laboulbenia*, *Chrysosporium*, *Chaetomium*, *Mortierella*, *Trichophyton*, *Rhachomyces*, *Humicola*, *Absidia*, *Isaria*, *Phoma*, *Beauveria*, *Aureobasidium*, *Verticillium*, *Gliocladium*, *Cunninghamella*, *Coprinus*, *Geotrichum*, *Microsporum*, *Epicoccum*, *Botrytis*, *Mycena*, *Candida*, *Stachybotrys*, and *Scopulariopsis*.

Understanding cave fungal diversity requires specialized research techniques, which can be broadly categorized into culture-dependent and culture-independent approaches. Culture-dependent methods involve the isolation of culturable cave fungi, which are then identified based on colony morphology, micromorphological characteristics, and genomic data. The internal transcribed spacer (ITS) sequence is commonly used for genus-level identification. However, species-level identification often requires multi-gene analysis due to the presence of cryptic fungal species. Additional markers that can be used alongside the ITS gene include the translation elongation factor 1-alpha (TEF1-α), the second-largest subunit of RNA polymerase II (RPB2), β-tubulin, and the calmodulin gene (CaM). Depending on the target fungal groups, small subunit rRNA (18S rRNA) and large subunit rRNA (28S rRNA) may also be used.

Savković et al. (2025) extracted nucleotide sequences from the NCBI database (https://www.ncbi.nlm.nih.gov/), all fungi related to the cave environment. Using multivariate analyses of 22,407 sequences, the obtained list included a total of 1447 sequences corresponding to fungi isolated from various substrata within cave environments around the world, mostly from Spain, China, the USA, and France, followed by Brazil, Poland, and Slovakia. These sequences correspond to 445 fungal species, members of the 394 genera, with Ascomycota as the most dominant phylum and Eurotiomycetes the dominant class of fungal dwellers in these environments. The highest species richness is detected for the genus *Penicillium*, followed by *Aspergillus*. The most frequently documented single species was *P. destructans*, isolated mostly from hibernating bats and guano, followed by *P. chrysogenum*.

In one of the limestone caves in the Philippines, a partial assessment of culturable cave fungal diversity showed that most fungal isolates belong to the taxa Ascomycota, Mucoromycota, and Basidiomycota, primarily from the genera *Aspergillus*, *Penicillium*, *Moniliella*, and *Trichoderma* (Lopez et al., unpublished manuscript). On a larger scale, commonly isolated cave fungi from karst caves in Southeast Asia include *Aspergillus*, *Penicillium*, *Cladosporium*, *Trichoderma*, and *Talaromyces* (Suetrong et al., 2023). Abris et al. (2020) reported a total of 26 distinct fungal colony forms identified to belong to at least 11 genera based on microscopic observations and cultural morphology, namely, *Aspergillus*, *Penicillium*, *Cladosporium*, *Fusarium*, *Mucor*, *Geotrichum*, *Alternaria*, *Trichoderma*, *Microsporum*, *Rhizopus*, and *Acremonium*. Isolates of *Aspergillus* and *Penicillium* constitute more than half of the total number of identified distinct colonies with *Cladosporium* contributing another 14%. These genera are ubiquitous fungi commonly isolated from natural environments such as soil, plants, and decaying materials. They are saprotrophic and contribute greatly to nutrient cycling in the environment. Samplings conducted during the dry season and those conducted during the wet season showed similar trends, wherein *Aspergillus* and *Penicillium* were the dominant species and sampling points near the entrance had the highest IMA. Previous research from other regions likewise shows the prevalence of *Aspergillus* and *Penicillium* in cave air (Docampo et al., 2011; Proca et al., 2011; Martin-Sanchez et al., 2012). An unpublished research in the same cave has documented the presence of *Aspergillus*, *Penicillium*, and *Cladosporium* from swabs from the dermal layer of bats roosting inside the cave. These may form part of the source propagules in certain areas of the cave. A study by Borda et al. (2004) has identified that bats and bat guano are sources of aerosolized microorganisms inside caves.

In contrast, culture-independent techniques extract genomic DNA directly from environmental samples, bypassing the need for fungal cultivation. Common sample sources include cave water, cave sediment, speleothem swabs, bat guano, and bat rinse (derived from skin/fur). These extracted DNA samples undergo genomic and metagenomic analyses, generating data primarily focused on fungal diversity studies. In general, cave fungal metagenomic studies in Southeast Asia are scarce and mainly concentrated in China and Thailand (Man et al., 2018; Wiseschart et al., 2019). These studies revealed that *Aspergillus* and *Penicillium* are commonly detected fungal genera in karst caves.

Limestone Caves as Potential Sources of Novel Bioactive Compounds

As previously mentioned, the extreme conditions of limestone caves drive microbial production of compounds that aid their survival. These compounds, specifically secondary metabolites, exhibit a wide range of biological functions. For instance, they may serve as antibacterial, antiviral, and antifungal agents, as well as immunosuppressants, anticancer and antitumor agents, industrial enzymes, probiotics, and biosurfactants (Rangseekaew & Pathom-aree, 2019; Jurado et al., 2021; Arndt et al., 2022; Barbosa et al., 2025). In the cave ecosystem, where nutrients are scarce and competition is intense, fungi rely on secondary metabolites as a key survival strategy. Those that produce antimicrobial compounds (such as bactericidal or fungicidal agents) gain a significant advantage by suppressing nearby competing microbes. By inhibiting or killing rival bacteria and fungi, they reduce competition for limited food sources, thereby securing greater access to essential nutrients.

Beyond antimicrobial effects, some secondary metabolites play a crucial role in nutrient acquisition through enzymatic activity. For example, metabolites that break down insoluble or complex substrates (e.g., decaying organic matter or mineral deposits) facilitate the release of trapped nutrients. This enzymatic dissolution allows fungi to exploit otherwise inaccessible resources, sustaining their growth and proliferation even in an environment where nutrients are inherently deficient (Barbosa et al., 2025). Researchers are highly interested in these bioactive compounds, particularly those synthesized by novel, undiscovered microbes in uncharted limestone cave environments. Additionally, understudied endemic cave-dwelling fungal taxa are a key focus, as they are likely to produce previously uncharacterized compounds with significant biotechnological potential.

In the context of cave fungi, bioactive compounds commonly screened exhibited functions such as antibacterial, antitumor, and antioxidant properties, as well as their roles in producing organic acids and industrial enzymes (Barbosa et al., 2025). Among these fungi, recent studies have identified the genera *Aspergillus*, *Penicillium*, *Cladosporium*, *Trichoderma*, *Talaromyces*, and *Xylaria* as notable producers of bioactive compounds. Among these compounds, antimicrobial activity is the most frequently screened potential application in biotechnology (Nuehring, 2023; Suetrong et al., 2023; Barbosa et al., 2025).

The functional determination of these bioactive compounds can be conducted using either culture-based or omics-based methods (Salazar-Hamm et al., 2025). As the names suggest, traditional culture-based methods involve culturing target microbes and screening their bioactive compounds through biochemical assays. In contrast, omics-based approaches, such as proteomics, metabolomics, genomics, and transcriptomics, are employed to discover and characterize these compounds at a molecular level (Salazar-Hamm et al., 2025).

Currently, in Southeast Asia, the screening of bioactive compounds from cave fungi currently relies predominantly on traditional methods, given their cost-effectiveness and proven reliability. However, the limitations of these conventional approaches—particularly the challenges associated with culturing cave fungi, many of which are unculturable or difficult to propagate under laboratory conditions—have prompted a gradual shift toward omics-based techniques. Despite their higher costs and reliance on sophisticated instrumentation, omics methods are increasingly being adopted in the region to overcome these barriers and expand the discovery of novel fungal metabolites.

Fungal Diversity in Other Extreme Environments

Aside from the cave environment, where fungi have been observed growing on cave walls and bat guano despite extreme conditions, other extreme environments—such as those with extremely high or low temperatures and pH levels—raise an important question: Do the detected fungal species in these environments actively grow, or do they merely exist as spores? (Gostinčar et al., 2023; Bueno de Mesquita et al., 2024).

The likely answer is both. Under high-stress conditions, extremophilic fungi typically persist as spores, but when conditions improve or encounter a favorable substrate or microhabitat, they transition into active growth, forming mycelium or other physical structures. A prime example of this is *Cryomyces antarcticus* Selbmann, de Hoog, Mazzaglia, Friedmann and Onofri, which has been observed growing on Antarctica's sandstone (Pacelli et al., 2021).

Due to their unique structural adaptations, spores—both autochthonous and allochthonous—are frequently isolated from extreme environments. This can create challenges for researchers in determining whether a particular fungal species is genuinely thriving in an extreme environment or merely present as a dormant spore passively introduced. However, molecular-level analyses provide valuable insights, whereas fungi that truly grow in extreme environments exhibit specialized genetic adaptations. These adaptations, encoded in their genome, enable them to survive harsh conditions by regulating structural modifications and producing specific enzymes or proteins. Fungi have evolved specialized biosynthetic pathways to overcome environmental stressors. A key example is melanin production through polyketide-derived pathways, which provides protection against UV radiation and osmotic stress (Avalos & Limón, 2022). Similarly, non-ribosomal peptide synthetase (NRPS) systems enable fungi to produce siderophores—small iron-chelating molecules that enhance nutrient acquisition during periods of iron limitation (Bushley & Turgeon, 2010). Additionally, fungi synthesize stress-responsive terpenoids that help maintain membrane stability under thermal stress conditions (Liu et al., 2017). Together, these adaptive mechanisms demonstrate how fungal secondary metabolites serve crucial protective functions in challenging environments.

Table 14.1, adapted from Bueno de Mesquita et al. (2024), outlines key genetic and physiological adaptations of fungi in extreme environments.

Table 14.1 Fungal group adaptations enabling survival in extreme environments characterized by various stressors

Stressors	Adaptations	Fungal groups
High metal content, low pH	Presence of efflux or metal sequestration systems	Sordariomycetes, Eurotiomycetes, Dothideomycetes, Saccharomycetes
Low nutrient content, low temperature level	Production of melanin, cold-shock proteins, and modified membrane lipids, and possess mechanism for dormancy	Sordariomycetes, Eurotiomycetes, Agaricomycetes, Dothideomycetes
High salt content, low water activity	Production of compatible solutes, heat-shock proteins, and melanin	Dothideomycetes, Sordariomycetes, Agaricomycetes, Eurotiomycetes
High temperature	Possess a modified cell membrane, high number of DNA supercoils, and production of heat-shock protein	Eurotiomycetes, Dothideomycetes, Sordariomycetes, Saccharomycetes, Agaricomycetes
Low water activity, high salt content/pH	Production of compatible solutes, melanin, and other alkaline-compatible enzymes	Saccharomycetes, Eurotiomycetes, Sordariomycetes, Dothideomycetes, Agaricomycetes, Chytridiomycetes

Note. Adapted from "Fungal diversity and function in metagenomes sequenced from extreme environments" by Bueno de Mesquita et al. (2024), Fungal Ecology, 72, 101383. https://doi.org/10.1016/j.funeco.2024.101383

These adaptations are observed in extremophilic fungi but not necessarily in extremotolerant fungi. Yarzábal Rodríguez et al. (2024) defined extremophilic fungi as those possessing essential adaptations that allow them to thrive in extreme environments where conditions are detrimental to most fungi. In contrast, extremotolerant fungi are capable of surviving in extreme environments but can also grow in non-extreme conditions. In essence, extremophilic fungi are obligate inhabitants of specific extreme environments, whereas extremotolerant fungi are more versatile, able to persist in a broader range of habitats.

In general, fungi commonly found in extreme environments belong to the imperfect stage of Ascomycota (Wingfield et al., 2023). Recent studies on extremophilic and extremotolerant fungi in Southeast Asia have focused on those inhabiting deep-sea sediments, acidic mangrove sediments, decaying plant matter, compost, salterns, and glaciers.

Regarding environments with elevated to high pH, Zhang et al. (2013) reported the presence of 12 genera of culturable fungi—potentially halophilic/halotolerant and piezophilic/piezotolerant—across 10 fungal families isolated from deep-sea sediments in the West Philippine Sea. This study, however, did not discuss their adaptations to these cold environments. Among the most dominant genera were *Aspergillus, Cladosporium*, and *Penicillium*. These genera are not strictly piezophilic or halophilic but are frequently found in such environments. Furthermore, Wingfield et al. (2023) identified 27 culturable filamentous fungi in man-made salterns in Pattani Province, Thailand. These isolates were classified under the genera *Cladosporium*, *Curvularia*, *Aspergillus*, *Diaporthe*, *Ectophoma*, *Penicillium*, and *Fusarium*. Among these, *Penicillium* sp., *Aspergillus* sp., and *Fusarium* sp. were characterized as halophilic, while *Cladosporium* sp., *Curvularia* sp., *Diaporthe* sp., and *Ectophoma* sp. were identified as halotolerant. The results of both studies indicate that *Aspergillus* and *Penicillium* are common fungal inhabitants of saline to hypersaline environments, consistent with findings from other studies (Wingfield et al., 2023).

In contrast, in low pH environments, Thanh et al. (2019) isolated 105 filamentous fungi from decaying plant organic matter—a low pH microhabitat—collected from northern provinces of Vietnam. With the exception of *A. fumigatus*, all isolates were thermophilic and acidotolerant, distributed across 11 genera, with *Thielavia* being the most dominant. *Thielavia* is a well-documented thermophilic genus (Berka et al., 2011; García-Huante et al., 2017; López-López et al., 2023); however, reports of its presence in low pH environments are scarce. Thanh et al. (2019) described the isolated *Thielavia* spp. as acidotolerant.

In nature, aciduric fungi (acidophilic and acidotolerant) are more prevalent than thermophilic and thermotolerant fungi (Thanh et al., 2019). This could be due to the greater abundance of low pH habitats compared to high-temperature environments, as well as the fact that fungi are inherently more resistant to acidic stress than to heat due to their structural composition. According to Bueno de Mesquita et al. (2024), fungi belonging to the classes Eurotiomycetes, Leotiomycetes, Dothideomycetes, Pezizomycetes, Sordariomycetes, and Saccharomycetes are relatively abundant in extreme environments.

Regarding psychrotolerant and psychrophilic fungi, researchers from China isolated 143 psychrophilic fungi and 1285 psychrotolerant fungal strains from soil samples collected across seven glaciers: Toumingmengke and Qiyi in Gansu Province, Zhadang and Midui in Tibet, Hailuogou in Sichuan Province, Mingyong in Yunnan Province, and Yuzhufeng in Qinghai Province. The psychrophilic fungal isolates primarily belong to the genera *Psychrophila*, *Phoma*, *Tetracladium*, and *Pseudogymnoascus*, while the psychrotolerant isolates are predominantly classified under the order *Helotiales*. These cold-adapted fungi have developed several adaptations, including the production of cold shock proteins, pigments, compatible solutes, antifreeze proteins, and extracellular polysaccharides (Hassan et al., 2016).

Wang et al. (2015) highlighted the significance of psychrophilic and psychrotolerant fungi as potential sources of novel bioactive compounds, noting that relatively few studies had been conducted on them. Since then, various studies have explored their applications, particularly in cold-active enzyme production and bioremediation (Hassan et al., 2016; Kumari et al., 2021). Notable cold-adapted fungal species used for cold-active enzyme production include *Rhodotorula psychrophenolica (= Phenoliferia psychrophenolica* (Margesin and Samp.) Q.M. Wang, F.Y. Bai, M. Groenew. and Boekhout), *R. glacialis (= Phenoliferia glacialis* (Margesin and Samp) Q.M. Wang, F.Y. Bai, M. Groenew. and Boekhout), *Thelebolus microsporus* (Berk. and Broome) Kimbr., *Geomyces* spp., and *Cladosporium* (Hassan et al., 2016). Although filamentous fungi constitute the majority of cold-adapted species, certain yeasts, such as those in the genera *Rhodotorula*, *Mrakiella*, and *Mrakia*, also exhibit cold tolerance (Wang et al., 2015).

Most fungi inhabiting extreme environments possess innate characteristics that are valuable to various industries, particularly their ability to tolerate harsh conditions encountered during the industrial production of bioactive fungal-derived compounds. Notable extremophilic fungi utilized in industrial processes include enzyme-producing strains of *Aspergillus gracilis Bainier*, *Fusarium solani (= Neocosmospora solani* (Mart.) L. Lombard and Crous), *Stachybotrys microspora (= Stachybotrys microsporus* (B.L. Mathur & Sankhla) S.C. Jong & E.E. Davis), *Aspergillus sydowii* (Bainier & Sartory) Thom and Church, *Aspergillus reticulatus* Sklenar, Jurjević, S.W. Peterson and Hubka, and *Trichoderma asperellum* Samuels, Lieckf. and Nirenberg (Yovchevska et al., 2025). Furthermore, since the ability to withstand extreme conditions is conferred by specific adaptations translated from genes, certain fungal species—such as strains of *Saccharomyces cerevisiae* (Desm.) Meyen, *Debaryomyces hansenii* (Zopf) Lodder and Kreger-van Rij, and *Rhodotorula mucilaginosa* (A. Jörg.) F.C. Harrison—can serve as gene donors for enhancing crop tolerance to extreme environmental stresses, including drought, high pH, and heavy metal exposure, through modern gene editing technologies (Gostinčar & Turk, 2012). These attributes underscore their potential in various biotechnological applications.

Challenges and Opportunities on Studying on Fungi from Caves and Extreme Environments

Caves and other extreme environments are simultaneously challenging and incredibly rewarding places to study fungi. The very conditions that make these environments "extreme" also contribute to the potential for discovering novel fungal species, unique ecological interactions, and a deeper understanding of fungal adaptation.

The pursuit of fungal knowledge in caves and extreme environments is faced with difficulties, often demanding specialized approaches and significant perseverance:

1. *Accessibility and Logistical Hurdles*. Many caves, especially deeper or less explored ones, are inherently difficult to access. This can involve challenging terrain, confined spaces, vertical descents, need for specialized caving equipment, and skilled or highly trained individuals ("speleologists"). Similarly, other extreme environments like polar regions, deep-sea hydrothermal vents, or high-altitude ecosystems demand extensive logistical planning, specialized transport, and safety protocols, making sample collection a complex and costly endeavor.

   The Philippines, for one, has over 3000 recorded caves, and a significant number are yet to be discovered, assessed, surveyed, and classified (The Philippine Clearing House Mechanism, 2020). Exploring these caves for their protection, conservation, and preservation requires a collaborative effort from a diverse team of experts, including zoologists, botanists, geologists, hydrologists, microbiologists, entomologists, social scientists/anthropologists, archaeologists, and speleologists.
2. *Harsh Environmental Conditions*. Studying fungi in extreme environments presents significant hurdles for researchers and can compromise sample quality. These extreme conditions present unique challenges, from the need for specialized equipment to the complexities of maintaining sample integrity and cultivating these unique fungi in a laboratory setting. These challenging conditions include:

   - *Low Temperatures and Freezing*: In places like ice caves or polar regions, keeping samples viable and contamination-free during collection and transport is crucial.
   - *High Temperatures and Pressure*: Environments such as deep-sea vents or geothermal caves expose equipment and samples to intense heat and pressure, necessitating specialized, pressure-resistant sampling tools.
   - *Oligotrophy (Nutrient Scarcity)*: Many cave environments are poor in nutrients, leading to slow fungal growth and making isolation in a laboratory difficult.
   - *Low Oxygen Level*: Some deep cave systems or anoxic sediments may harbor anaerobic fungi, requiring more complex, strict anaerobic culturing techniques than standard methods.

- *UV Radiation*: While less common in deep caves, surface-exposed extreme environments like high-altitude deserts can have high UV levels, impacting fungal communities.
- *Low Fungal Biomass and Culturability*: Fungi in some extreme environments, especially oligotrophic ones, might exist at very low densities, making detection and isolation difficult. Furthermore, many fungi from these unique niches are *unculturable* using standard laboratory techniques. They might have highly specific growth requirements or symbiotic relationships that are difficult to replicate ex situ, leading to an underestimation of true diversity.

3. *Contamination Risks*: The pristine nature of many cave environments makes them highly susceptible to contamination from researchers, equipment, or external sources. Strict sterile techniques are paramount but can be challenging to implement in difficult field conditions.
4. *Identification and Characterization*: The discovery of novel or rarely encountered fungal species from extreme environments often means that traditional morphological identification methods are insufficient. This necessitates reliance on molecular techniques (e.g., DNA sequencing of ribosomal RNA genes like ITS, LSU, SSU) for accurate identification and phylogenetic placement. Even then, new species might not have close relatives in existing databases, making their classification challenging.
5. *Ethical and Conservation Concerns*: Many caves are fragile ecosystems, and research activities must be conducted with minimal disturbance. Strict protocols are often in place to protect cave formations, bat populations (especially given concerns like White-Nose syndrome), and other sensitive cave fauna and flora.

Despite the formidable challenges, the study of fungi in caves and extreme environments offers immense opportunities for groundbreaking discoveries and significant advancements in various fields:

1. *Discovery of Novel Biodiversity*: These environments are veritable treasure troves of undiscovered fungal species. The unique selective pressures of extreme conditions drive the evolution of novel adaptations, leading to fungi with previously unknown morphological, physiological, and genetic characteristics. This expands our understanding of fungal evolution and biodiversity on Earth.
2. *Unveiling Novel Metabolic Pathways and Secondary Metabolites*: Fungi surviving in extreme conditions often produce unique enzymes and secondary metabolites to cope with stress, compete for limited resources, or interact with other organisms. These extremophilic fungi could be sources of:

   - *Novel antibiotics*: In the face of increasing antimicrobial resistance, new compounds are desperately needed.
   - *Enzymes for industrial applications*: Extremophilic enzymes (e.g., cold-active, heat-stable, salt-tolerant) have vast potential in biotechnology, bioremediation, and industrial processes.
   - *Bioactive compounds for pharmaceuticals*: New anti-cancer, anti-inflammatory, or immunosuppressive agents could be discovered.

3. *Understanding Adaptation and Survival Mechanisms*: Studying how fungi thrive in environments with limited nutrients, extreme temperatures, high or low pH, or anoxia provides critical insights into the fundamental mechanisms of life under stress. This knowledge can inform astrobiology research, helping us understand the potential for life in extraterrestrial extreme environments.
4. *Ecological Insights*: Fungi serve as fundamental drivers of cave ecosystems, where they facilitate decomposition and nutrient cycling, processes essential for sustaining these isolated environments. Their ecological roles extend beyond mere organic matter breakdown, as they engage in complex interactions with other cave-dwelling organisms. For instance, fungi interact closely with bats, both as pathogens—such as *P. destructans*, the causative agent of White-Nose Syndrome—and as decomposers of bat guano, which forms a critical nutrient base in many caves. These fungal-bat dynamics highlight the delicate balance between parasitism and nutrient recycling in subterranean food webs. Additionally, fungi form mutualistic relationships with bacteria, particularly in oligotrophic cave zones where resources are scarce. Through metabolic exchanges, fungi enhance bacterial nitrogen fixation, while bacteria may aid in breaking down complex organic compounds, collectively sustaining microbial life in otherwise inhospitable conditions. In guano-rich environments, dominant decomposers like *Penicillium* and *Aspergillus* thrive alongside nitrogen-fixing microbes, accelerating nutrient turnover and supporting diverse cave communities. Competition further shapes fungal assemblages, as some species, including antibiotic-producing *Penicillium*, suppress rivals to secure ecological niches. These antagonistic interactions influence species distribution and diversity, particularly in microbe-rich substrates like decaying organic matter or sediment. Such adaptive strategies—ranging from cooperative symbioses to competitive exclusion—enable fungi to endure extreme cave conditions, including perpetual darkness, elevated humidity, and fluctuating nutrient availability. Remarkably, these survival mechanisms mirror those observed in other extreme habitats, reinforcing the versatility of fungi as ecological pioneers in challenging environments.
5. *Biogeographical Studies*: Analyzing fungal communities across different cave systems and extreme environments can shed light on fungal dispersal patterns, endemism, and the factors shaping their global distribution.
6. *Bioremediation Potential*: Fungi capable of thriving in harsh conditions may possess unique abilities to degrade pollutants in contaminated environments. For instance, fungi from heavy metal-rich caves might be able to detoxify contaminated sites.
7. *Conservation and Monitoring*: Fungi can serve as indicators of environmental health. Monitoring fungal communities in caves can provide insights into human impact, climate change effects, and the overall ecological integrity of these sensitive environments.

In conclusion, while the challenges of studying fungi in caves and extreme environments are substantial, the potential rewards are immense. These unique habitats are fertile ground for discovering new life forms, unveiling novel biochemical processes, and expanding our fundamental understanding of life's resilience and adaptability, offering exciting avenues for future research and biotechnological applications. Despite the numerous recorded caves in the Philippines, however, there remains a notable lack of documentation or studies dwelling on filamentous fungi strains from Philippine karstic caves.

References

Abris, M. I., Palanca, M. S., De Leon, M. P., & Banaay, C. G. B. (2020). Microbial air and water quality assessment of a freshwater limestone cave in The Philippines and its implications for ecotourism management. *Journal of Nature Studies, 19*(1), 33–48.

Arndt, O., Stott, M., Vakaloudis, S., Taylor, M., & Sutherland, K. (2022). Isolation, antibacterial screening, and identification of bioactive cave dwelling bacteria in Fiji. *Frontiers in Microbiology, 13*, 1012867. https://doi.org/10.3389/fmicb.2022.1012867

Avalos, J., & Limón, M. C. (2022). Fungal secondary metabolism. *Encyclopedia, 2*(1), 1–13. https://doi.org/10.3390/encyclopedia2010001

Barbosa, R. N., Felipe, M. T. C., Silva, L. F., Silva, E. A., Silva, S. A., Herculano, P. N., Prazeres, J. F. S. A., Lima, J. M. S., Bezerra, J. D. P., Moreira, K. A., Magalhães, O. M. C., & Souza-Motta, C. M. (2025). A review of the biotechnological potential of cave fungi: A toolbox for the future. *Journal of Fungi, 11*(2), 145. https://doi.org/10.3390/jof11020145

Bastian, F., Jurado, V., Nováková, A., Alabouvette, C., & Sáiz-Jiménez, C. (2010). The microbiology of Lascaux cave. *Microbiology, 156*, 644–652.

BBC News. (2013, September 11). *What is climate change? A really simple guide*. BBC News. https://www.bbc.com/news/science-environment-24021772

Beck, B. F., Fram, M., & Carvajal Zamora, J. R. (1976). The Aguas Buenas caves, Puerto Rico: Geology, hydrology and ecology with special reference to the histoplasmosis fungus. *National Speleological Society Bulletin, 38*, 1–16.

Berka, R., Grigoriev, I., Otillar, R., et al. (2011). Comparative genomic analysis of the thermophilic biomass-degrading fungi *Myceliophthora thermophila* and *Thielavia terrestris*. *Nature Biotechnology, 29*(10), 922–927. https://doi.org/10.1038/nbt.1976

Borda, D., Borda, C., & Tamas, T. (2004). Bats, climate, and air microorganisms in a Romanian cave. *Mammalia, 68*(4), 337–343. https://doi.org/10.1515/mamm.2004.033

Bueno de Mesquita, C. P., Vimercati, L., Wu, D., Childress, M., Danz, A., Grupe, A., Haelewaters, D., Hyde, N., Kossmann, T., Oliver, C., Perrotta, C., Young, B., Schmidt, S., Tringe, S., & Quandt, C. A. (2024). Fungal diversity and function in metagenomes sequenced from extreme environments. *Fungal Ecology, 72*, 101383. https://doi.org/10.1016/j.funeco.2024.101383

Bushley, K. E., & Turgeon, B. G. (2010). Phylogenomics reveals subfamilies of fungal nonribosomal peptide synthetases and their evolutionary relationships. *BMC Evolutionary Biology, 10*, 26. https://doi.org/10.1186/1471-2148-10-26

Calder, D. R., & Bleakney, J. S. (1965). Microarthropod ecology of a porcupine inhabited cave in Nova Scotia. *Ecology, 46*(6), 895–899. https://doi.org/10.2307/1934031

Choi, Y. G., Kim, W. R., Woo, K. S., & Cho, Y. B. (2011). A proposal of a new category, Scotophile visitors, within trogloxenes in Korean Limestone Caves. *Journal of Korean Nature., 4*, 191–195. https://doi.org/10.7229/jkn.2011.4.3.191

Chua, R. W., Song, K. P., & Ting, A. (2022). Antimicrobial activities and phytochemical screening of endophytic fungi isolated from *Cymbidium* and *Dendrobium* orchids. *South African Journal of Botany, 151*, 909–918. https://doi.org/10.1016/j.sajb.2022.11.015

Correia, J., Borges, A., Simões, M., & Simões, L. C. (2023). Beyond penicillin: The potential of filamentous fungi for drug discovery in the age of antibiotic resistance. *Antibiotics (Basel, Switzerland), 12*(8), 1250. https://doi.org/10.3390/antibiotics12081250

Decu, V. (1986). Some considerations on the bat guano synusia. *Travaux de l'Institut de Speologie "Emile Racovitza", 25*, 41–51.

Dickson, G. W. (1975). A preliminary study of heterotrophic microorganisms as factors in substrate of troglobitic invertebrates. *NSS Bulletin, 37*, 89–93.

Dobat, K. (1967). Ein bisher unveroffentlichtes botanisches manuskript Alexander von Humboldts: *Plantae subterranae* Europ. 1794 cum Iconibus. *Akademie Der Wissenschaften und der Literatur., 6*, 16–19.

Docampo, S., Mar, T. M., Recio, M., Melgar, M., Garcia-Sanchez, J., & Cabezudo, B. (2011). Fungal spore content of the atmosphere of the cave of Nerja (southern Spain): Diversity and origin. *Science of the Total Environment, 409*, 835–843. https://doi.org/10.1016/j.scitotenv.2010.10.048

Estrada-Barcenas, D. A., Palacios-Vargas, J. G., Estrada-Venegas, E., Klimov, P. B., Martinez-Mena, A., & Taylor, M. L. (2010). Biological activity of the mite *Sancassania* sp. (Acari: Acaridae) from bat guano associated with the pathogenic fungus *Histoplasma capsulatum*. *Memorias do Instituto Oswaldo Cruz, 105*(2), 127–131. https://doi.org/10.1590/S0074-02762010000200003

Fenolio, D. B., Graening, G. O., Collier, B. A., & Stout, J. F. (2006). Coprophagy in a cave-adapted salamander; the importance of bat guano examined through nutritional and stable isotope analyses. *Proceedings: Biological Sciences, 273*(1585), 439–443. https://doi.org/10.1098/rspb.2005.3341

Ferreira, R., & Martins, R. (1999). Trophic structure and natural history of bat guano invertebrate communities, with special reference to Brazilian caves. *Tropical Zoology, 12*(2), 231–252. https://doi.org/10.1080/03946975.1999.10539391

Gabriel, C. R., & Northup, D. E. (2013). Microbial ecology: Caves as an extreme habitat. In N. Cheeptham (Ed.), *Cave microbiomes: A novel resource for drug discovery* (pp. 85–108). Springer Press. https://doi.org/10.1007/978-1-4614-5206-5_5

García-Huante, Y., Cayetano-Cruz, M., Santiago-Hernández, A., et al. (2017). The thermophilic biomass-degrading fungus *Thielavia terrestris* Co3Bag1 produces a hyperthermophilic and thermostable β-1,4-xylanase with exo- and endo-activity. *Extremophiles, 21*(2), 175–186. https://doi.org/10.1007/s00792-016-0893-z

Gostinčar, C., & Turk, M. (2012). Extremotolerant fungi as genetic resources for biotechnology. *Bioengineered, 3*(5), 293–297. https://doi.org/10.4161/bioe.20713

Gostinčar, C., Stajich, J. E., & Gunde-Cimerman, N. (2023). Extremophilic and extremotolerant fungi. *Current Biology, 33*(14), R752–R756. https://doi.org/10.1016/j.cub.2023.06.011

Haelewaters, D., Page, R. A., & Pfister, D. H. (2018). (2018) Laboulbeniales Hyperparasites (fungi, Ascomycota) of bat flies: Independent origins and host associations. *Ecology and Evolution, 8*, 8396–8418.

Hassan, N., Rafiq, M., Hayat, M., Shah, A. A., & Hasan, F. (2016). Psychrophilic and psychrotrophic fungi: A comprehensive review. *Reviews in Environmental Science and Bio/Technology, 15*(2), 147–172. https://doi.org/10.1007/s11157-016-9395-9

Herrera, F. F. (1995). Las comunidades de artropodos del guano del guacharos en la cueva del guacharo, Venezuela. *Boletin de la Sociedad Venezolana de Espeleologia, 29*, 39–46.

Hoyt, J. R., Kilpatrick, A. M., & Langwig, K. E. (2021). Ecology and impacts of white-nose syndrome on bats. *Nature Reviews Microbiology, 2021*(19), 196–210. https://doi.org/10.1038/s41579-020-00493-5

Hu, P. (2025). mSphere of influence: Rapid evolution of pathogenesis and drug resistance in human pathogenic fungi. *mSphere, 10*(3), e0057024. https://doi.org/10.1128/msphere.00570-24

Hui, S. T., Gifford, H., & Rhodes, J. (2024). Emerging antifungal resistance in fungal pathogens. *Current Clinical Microbiology Reports, 11*(2), 43–50. https://doi.org/10.1007/s40588-024-00219-8

Igreja, R. P. (2011). Infectious diseases associated with caves. *Wilderness & Environmental Medicine., 22*(2), 115–121.

Jurado, V., Miller, A. Z., Alias-Villegas, C., & Saiz-Jimenez, C. (2021). Cave microbes as a potential source of drug development in the era of antibiotic resistance. *Microorganisms, 9*(10), 2143. https://doi.org/10.3390/microorganisms9102143

Kajihiro, E. S. (1965). Occurrence of dermatophytes in fresh bat guano. *Applied Microbiology, 13*(5), 720–724.

Karunarathna, S. C., Dong, Y., Karasaki, S., Tibpromma, S., Hyde, K. D., Lumyong, S., Xu, J., Sheng, J., & Mortimer, P. E. (2020). Discovery of novel fungal species and pathogens on bat carcasses in a cave in Yunnan Province. *China. Emerging Microbes & Infections, 9*(1), 1554–1566.

Kosznik-Kwaśnicka, K., Golec, P., Jaroszewicz, W., Lubomska, D., & Piechowicz, L. (2022). Into the unknown: Microbial communities in caves, their role, and potential use. *Microorganisms, 10*(2), 222. https://doi.org/10.3390/microorganisms10020222. PMID: 35208677; PMCID: PMC8877592.

Kumari, M., Padhi, S., Sharma, S., Phukon, L. C., Singh, S. P., & Rai, A. K. (2021). Biotechnological potential of psychrophilic microorganisms as the source of cold-active enzymes in food processing applications. *Biotechnology, 11*(11), 479. https://doi.org/10.1007/s13205-021-03008-y

Langwig, K. E., Hoyt, J. R., Parise, K. L., Frick, W. F., Foster, J. T., & Marm, K. A. (2017). Resistance in persisting bat populations after white-nose syndrome invasion. *Philosophical Transactions of the Royal Society B, 372*(1712), 20160044.

Leitch, C. (2024). *More evidence that climate change can make fungi more dangerous*. Labroots. https://www.labroots.com/trending/earth-and-the-environment/27321/evidence-climate-change-fungi-dangerous

Liu, Y. N., Zhang, T. J., Lu, X. X., Ma, B. L., Ren, A., Shi, L., Jiang, A. L., Yu, H. S., & Zhao, M. W. (2017). Membrane fluidity is involved in the regulation of heat stress induced secondary metabolism in *Ganoderma lucidum. Environmental Microbiology, 19*(4), 1653–1668. https://doi.org/10.1111/1462-2920.13693

Lockhart, S. R., Chowdhary, A., & Gold, J. A. W. (2023). The rapid emergence of antifungal-resistant human-pathogenic fungi. *Nature Reviews Microbiology, 21*(12), 818–832. https://doi.org/10.1038/s41579-023-00960-9

López-López, A., Santiago-Hernández, A., Cayetano-Cruz, M., García-Huante, Y., Campos, J. E., Bustos-Jaimes, I., Marsch-Moreno, R., Cano-Ramírez, C., Benitez-Cardoza, C. G., & Hidalgo-Lara, M. E. (2023). TtCel7A: A native thermophilic Bifunctional cellulose/Xylanase Exogluclanase from the thermophilic biomass-degrading fungus *Thielavia terrestris* Co3Bag1, and its application in enzymatic hydrolysis of Agroindustrial derivatives. *Journal of Fungi (Basel, Switzerland), 9*(2), 152. https://doi.org/10.3390/jof9020152

Man, B., Wang, H., Yun, Y., Xiang, X., Wang, R., Duan, Y., & Cheng, X. (2018). Diversity of fungal communities in Heshang cave of Central China revealed by Mycobiome-sequencing. *Frontiers in Microbiology, 9*, 1400. https://doi.org/10.3389/fmicb.2018.01400

Martin-Sanchez, P. M., Novakova, A., Bastian, F., Alabouvette, C., & Saiz-Jimenez, C. (2012). Use of biocides for the control of fungal outbreaks in subterranean environments: The case of the Lascaux cave in France. *Environmental Science and Technology, 46*, 3762–3770. https://doi.org/10.1021/es2040625

Min, K. H. (1988). Fungus flora of Seongrya cave in Korea. *Transactions of the Mycological Society of Japan, 29*, 479–487.

Mohd, A. Y., Rosli, N., & Ibrahim, M. (2021). Antibacterial activity of endophytic fungi from *Paederia foetida* L. leaves. *Science & Technology Indonesia, 6*(3), 189–195. https://doi.org/10.26554/sti.2021.6.3.189-195

Moseley, M. (2007). Acadian biospeleology: Composition and ecology of cave fauna of Nova Scotia and southern New Brunswick, Canada. *International Journal of Speleology, 36*(1), 1–21. https://doi.org/10.5038/1827-806X.36.1.1

Moulds, T. A. (2006). *The seasonality, diversity, and ecology of cavernicolous guano dependent arthropod ecosystem in southern Australia* (p. 257). PhD Thesis in Environmental Biology,School of Earth and Environmental Sciences, The University of Adelaide.

Nieves-Rivera, A. M., Santos-Flores, C. J., Dugan, F. M., & Miller, T. E. (2009). Guanophilic fungi in three caves of southwestern Puerto Rico. *International Journal of Speleology, 38*(1), 61–70. https://doi.org/10.5038/1827-806X.38.1.7

Novakova, A. (2009). Microscopic fungi isolated from the Domica cave system (Slovak karst National Park, Slovakia). *A review. International Journal of Speleology, 38*(1), 71–82. https://doi.org/10.5038/1827-806X.38.1.8

Nuehring, J. (2023). *Bioactive secondary metabolites in cave dwelling fungi* (honors program thesis, University of Northern Iowa). UNI ScholarWorks. https://scholarworks.uni.edu/hpt/708

Ogórek, R. (2018). Speleomycology of air in Demänovská cave of liberty (Slovakia) and new airborne species for fungal sites. *Journal of Cave & Karst Studies, 80*(3), 153–116.

Pacelli, C., Alessia, C., Siong, L. M., Lorenzo, A., Moeller, R., Fujimori, A., Igor, S., & Silvano, O. (2021). Insights into the survival capabilities of *Cryomyces antarcticus* hydrated colonies after exposure to Fe particle radiation. *Journal of Fungi (Basel, Switzerland), 7*(7), 495. https://doi.org/10.3390/jof7070495

Proca, E., Jurado, V., Martin-Sanchez, P. M., Hermosin, B., Bastian, F., Alabouvette, C., Saiz-Jimenez, C. (2011) Aerobiology: An ecological indicator for early detection and control of fungal outbreaks in caves. *Ecological Indicators, 11*(6), 1594–1598.

Rangseekaew, P., & Pathom-aree, W. (2019). Cave actinobacteria as producers of bioactive metabolites. *Frontiers in Microbiology, 10*, 387. https://doi.org/10.3389/fmicb.2019.00387

Ribeiro, J. M. C., Labruna, M. B., Mans, B. J., Maruyama, S. R., Francischetti, I. M. B., Barizon, G. C., & de Miranda Santos, I. K. F. (2012). The sialotranscriptome of *Antricola delacruzi* female ticks is compatible with non-hematophagous behavior and an alternative source of food. *Insect Biochemistry and Molecular Biology, 42*, 332–342. https://doi.org/10.1016/j.ibmb.2012.01.003

Salazar-Hamm, P. S., Homan, F. E., Good, S. A., Hathaway, J. J. M., Clements, A. E., Haugh, E. G., & Caesar, L. K. (2025). Subterranean marvels: Microbial communities in caves and underground mines and their promise for natural product discovery. *Natural Product Reports, Advance Article.* https://doi.org/10.1039/D4NP00055B

Savković, Ž., Popović, S., & Stupar, M. (2025). Unveiling the subterranean symphony: A comprehensive study of cave fungi revealed through National Center for biotechnology sequences. *Journal of Fungi (Basel)., 11*(4), 286. https://doi.org/10.3390/jof11040286. PMID: 40278107; PMCID: PMC12028181.

Stomeo, F., Portillo, M. C., & Gonzalez, J. M. (2009a). Assessment of bacterial and fungal growth on natural substrates: Consequences for preserving caves with prehistoric paintings. *Current Microbiology, 59*, 321–325. https://doi.org/10.1007/s00284-009-9437-4

Stomeo, F., Ellersdorfer, G., Sterflinger, K., Gonzalez, J., & Saiz-Jimenez, C. (2009b). A preliminary survey on the fungi of Dona Trinidad cave, Ardales, Malaga, Spain. In *Current research topics in applied microbiology and microbial biotechnology* (pp. 153–156). World Scientific.

Suetrong, S., Preedanon, S., Kobmoo, N., Srihom, C., Somrithipol, S., Saengkaewsuk, S., Srikitikulchai, P., Klaysuban, A., Nuankaew, S., Chuaseeharonnachai, C., Chainuwong, B., Muangsong, C., Malimart, K., Rungjindamai, N., Siripornpibul, C., Chareonkunnatum, U., Ploydam, B., Thungprue, N., Tongsima, S., Zhang, Z.-F., Cai, L., & Boonyuen, N. (2023). Unravelling the hidden diversity of cave mycobiota in Thailand's Satun Geopark. *Scientific Reports, 13*(1), Article 19162. https://doi.org/10.1038/s41598-023-43316-2

Sugita, T., Kikuchi, K., Makimura, K., Urata, K., Someya, T., Kamei, K., Niimi, M., & Uehara, Y. (2005). *Trichosporon* species isolated from guano samples obtained from bat-inhabited caves in Japan. *Applied and Environmental Microbiology, 71*, 7626–7629. https://doi.org/10.1128/AEM.71.11.7626-7629.2005

Sustr, V., Elhottova, D., Kristufek, V., Lukesova, A., Novakova, A., Tajovsky, K., & Triska, J. (2005). Ecophysiology of the cave isopod *Mesoniscus graniger* (Frivaldszky 1865)

(Crustacea: Isopoda). *European Journal of Soil Biology, 41*, 69–75. https://doi.org/10.1016/j.ejsobi.2005.09.008

Thanh, V. N., Thuy, N. T., Huong, H. T. T., Hien, D. D., Hang, D. T. M., Anh, D. T. K., Hüttner, S., Larsbrink, J., & Olsson, L. (2019). Surveying of acid-tolerant thermophilic lignocellulolytic fungi in Vietnam reveals surprisingly high genetic diversity. *Scientific Reports, 9*(1), 3674. https://doi.org/10.1038/s41598-019-40213-5

The Philippine Clearing House Mechanism. (2020). http://www.philchm.ph/list-of-classified-caves-of-the-philippines/#:~:text=More%20than%203%2C000%20caves%20have,that%20empties%20into%20Honda%20Bay

Turrini, P., Chebbi, A., Riggio, F., & Visca, P. (2024). The geomicrobiology of limestone, sulfuric acid speleogenetic, and volcanic caves: Basic concepts and future perspectives. *Frontiers in Microbiology, 15*, 1370520. https://doi.org/10.3389/fmicb.2024.1370520

Vanderwolf, K., Malloch, D., McAlpine, D. F., & Forbes, G. J. (2013). A world review on fungi, yeasts, and slime molds in caves. *International Journal of Speleology, 42*(1), 77–96. https://doi.org/10.5038/1827-806X.42.1.9

Vaughan-Martini, A., Angelini, P., & Zacchi, L. (2000). The influence of human and animal visitation on the yeast ecology of three Italian caverns. *Annals of Microbiology, 50*, 133–140.

Wang, M., Jiang, X., Wu, W., Hao, Y., Su, Y., Cai, L., Xiang, M., & Liu, X. (2015). Psychrophilic fungi from the world's roof. *Persoonia, 34*, 100–112. https://doi.org/10.3767/003158515X685878

Wang, W. H., Thitithanyanont, A., Urbina, A. N., & Wang, S. F. (2021). Emerging and re-emerging diseases. *Pathogens (Basel, Switzerland), 10*(7), 827. https://doi.org/10.3390/pathogens10070827

Wasti, I. G., Fui, F. S., Zhi, T. Q., Mun, C. W., Kassim, M. H. S., et al. (2020). Fungi from dead arthropods and bats of Gomantong cave, northern Borneo, Sabah (Malaysia). *Journal of Cave and Karst Studies, 82*(4), 261–275. https://doi.org/10.4311/2019MB0146

Williams, S. L., Toda, M., Chiller, T., Brunkard, J. M., & Litvintseva, A. P. (2024). Effects of climate change on fungal infections. *PLoS Pathogens, 20*(5), e1012219. https://doi.org/10.1371/journal.ppat.1012219

Wingfield, L. K., Jitprasitporn, N., & Che-alee, N. (2023). Isolation and characterization of halophilic and halotolerant fungi from man-made solar salterns in Pattani Province. *Thailand. PLOS ONE, 18*(2), e0281623. https://doi.org/10.1371/journal.pone.0281623

Wiseschart, A., Mhuantong, W., Tangphatsornruang, S., Chantasingh, D., & Pootanakit, K. (2019). Shotgun metagenomic sequencing from Manao-Pee cave, Thailand, reveals insight into the microbial community structure and its metabolic potential. *BMC Microbiology, 19*(1), 144. https://doi.org/10.1186/s12866-019-1521-8

Yarzábal Rodríguez, L. A., Álvarez Gutiérrez, P. E., Gunde-Cimerman, N., Ciancas Jiménez, J. C., Gutiérrez-Cepeda, A., Ocaña, A. M. F., & Batista-García, R. A. (2024). Exploring extremophilic fungi in soil mycobiome for sustainable agriculture amid global change. *Nature Communications, 15*(1), 6951. https://doi.org/10.1038/s41467-024-51223-x

Yovchevska, L., Gocheva, Y., Stoyancheva, G., Miteva-Staleva, J., Dishliyska, V., Abrashev, R., Stamenova, T., Angelova, M., & Krumova, E. (2025). Halophilic fungi—Features and potential applications. *Microorganisms, 13*(1), 175. https://doi.org/10.3390/microorganisms13010175

Yurchenko, E. A., Chingizova, E. A., Aminin, D. L., Pislyagin, E. A., Kolesnikova, S. A., & Dyshlovoy, S. A. (2025). Marine fungi: In search of new antibacterial drugs. *Molecular Biology, 59*(1), 35–50. https://doi.org/10.1134/S0026893324700705

Zhang, X.-Y., Zhang, Y., Xu, X.-Y., & Qi, S.-H. (2013). Diverse deep-sea fungi from the South China Sea and their antimicrobial activity. *Current Microbiology, 67*(5), 525–530. https://doi.org/10.1007/s00284-013-0394-6

Zhang, Z.-F., Zhao, P., & Cai, L. (2018). Origins of cave fungi. *Frontiers in Microbiology, 9*, 1407. https://doi.org/10.3389/fmicb.2018.01407

Chapter 15
Fungi and the Changing Climate

Jon Ray M. Maglonzo, Sittie Aisha B. Macabago, and Thomas Edison E. dela Cruz

Abstract Fungi are primary decomposers, ubiquitous symbionts, or widespread pathogens that influence nutrient cycling and ecosystem maintenance. Thus, fungi, a highly diverse group of organisms, play a critical role in ecosystems and, just like other organisms, are affected by climate change. In this chapter, we will focus on how fungi and fungal-like organisms respond to climatic shifts, particularly to rising temperatures, altered precipitation patterns, and increases in extreme weather events. Key points are the impact of a changing climate on fungal biodiversity, communities, and ecosystem functions. The intricate relationship between fungi and ecosystem services is explored with emphasis on feedback mechanisms that either mitigate or exacerbate the impacts of climate change. Furthermore, the potential consequences of climate-induced shifts in fungal distribution in agriculture and human health highlight the urgent need for integrative research approaches to predict and manage these impacts. Synthesis of current knowledge and emerging research directions in fungal research emphasizes the importance of integrating fun-

J. R. M. Maglonzo
The Graduate School, University of Santo Tomas, Manila, Philippines

Fungal Biodiversity, Ecogenomics and Systematics-Metabolomics (FBeS) Group, Research Center for the Natural and Applied Sciences, University of Santo Tomas, Manila, Philippines

S. A. B. Macabago
Department of Biological Sciences, College of Science, University of Santo Tomas, Manila, Philippines

Pure and Applied Microbiology, Research Center for the Natural and Applied Sciences,, University of Santo Tomas, Manila, Philippines

T. E. E. dela Cruz (✉)
The Graduate School, University of Santo Tomas, Manila, Philippines

Fungal Biodiversity, Ecogenomics and Systematics-Metabolomics (FBeS) Group, Research Center for the Natural and Applied Sciences, University of Santo Tomas, Manila, Philippines

Department of Biological Sciences, College of Science, University of Santo Tomas, Manila, Philippines
e-mail: tedelacruz@ust.edu.ph

J. J. G. Guerrero et al. (eds.), *Mycology in a Changing Planet: Applications and Perspectives from Southeast Asia*, https://doi.org/10.1007/978-981-95-2640-6_15

gal ecology into climate models and global sustainability efforts. Understanding these intricate dynamics is critical for leveraging the role of fungi in ecological processes to formulate better climate mitigation strategies and ensure the maintenance of Earth's biodiversity in an era of rapid environmental change brought by anthropogenically driven climate change.

Keywords Anthropogenic activities · Fungal classification · Fungal diversity · Climate change factors · Impact studies

Introduction

Fungi and other fungal-like organisms (FFLO) are among the most species-rich groups after insects. Current estimates peg its number between 2.2 and 3.8 million species based on species studies, extrapolation using the plant-fungus ratio, and data estimation from environmental sequencing, with 120,000 currently accepted species (Hawksworth & Lücking, 2017). FFLO plays fundamental ecological roles as saprotrophs, mutualists, parasites, and pathogens across diverse ecosystems (Haelewaters et al., 2022). These heterotrophic eukaryotes, characterized by chitinous cell walls and absorptive nutrition, exhibit a wide range of forms, from single-celled yeasts to filamentous mycelial colonies to multicellular networks capable of producing conspicuous macroscopic fruiting structures. Once considered as the mysterious "third kingdom" beyond the plant-animal divide, advances in genetic analysis have revolutionized their classification. As critical components of Earth's biomes, FFLO contribute to ecosystem stability and human industries such as agriculture, forestry, medicine, and biotechnology (Bahram & Netherway, 2022). However, rapid environmental changes driven by anthropogenic activities, particularly global climate change and pollution, threaten the diversity and distribution of FFLO communities and the broader ecological networks they support. Given their integral role in maintaining ecological balance, understanding the impacts of climate change on FFLO is essential for predicting and mitigating potential cascading effects on global biodiversity. In this chapter, we provide a brief presentation on the traditional and modern classification of fungi and outline the drivers of climate change. With this, the current chapter explores the complexities of the impact of climate change on the diversity and ecology of fungi and other fungal-like organisms. Specific examples of how climate change impacts FFLO will be discussed, particularly on the interaction of mycorrhizal fungi and plants, the changing distribution of climate-sensitive lichen species, and the shifting pathogenicity of certain FFLO under changing climate conditions.

Evolutionary Origin and Classification of the Fungi

Evolutionary Origin of Fungi Evidence suggests that the major fungal phyla have ancient origins but leaves us uncertain about the exact timing of the emergence of their ancestral species. Given the structural simplicity of chytrids and the absence of flagella in other fungi, it is proposed that early fungi were likely simple, unicellular organisms that moved through water using a flagellum located at the posterior of the cell (Naranjo-Ortiz & Gabaldón, 2019). This posterior flagellum is a feature shared by both fungi and animals, which places them together in the eukaryotic supergroup known as Opisthokonta (Steenkamp et al., 2006). In contrast, other motile eukaryotes have flagella positioned at the anterior of the cell. Choanoflagellates, a basal group of eukaryotes related to animals, show structural similarities to chytrids, providing additional evidence for a shared evolutionary origin.

Traditional Approach to Fungal Classification Early mycologists aimed to categorize fungi based on shared phenotypic characteristics, and this morphological approach successfully identified the ascomycetes and basidiomycetes, as well as subgroups of closely related genera. In some instances, these early classifications reflected evolutionary relationships because certain broad structural similarities among fungi indicate kinship. However, in many cases, these prior classifications grouped unrelated organisms based on shared traits, such as spore color (Saccardo, 1894), which lacked evolutionary significance. Advances in microscopy and biochemistry helped improve the natural classification of fungi, but it was the introduction of molecular genetic techniques in the 1980s that revolutionized fungal taxonomy and has provided a deeper understanding of their evolution (Guarro et al., 1999; Yang, 2011). Since then, many traditional groups have been reorganized and renamed.

The goal of creating a natural system reflecting evolutionary relationships between species is considered important for several reasons. By understanding the evolutionary relatedness among fungal species, we can gain insights into the biology of a particular fungus by looking at what we know about its relatives. For example, the analyses of gene sequences complement morphological traits in identifying clinically important fungi (Guarro et al., 1999). Molecular methods also facilitate early detection of pathogenic species, which is crucial in implementing appropriate therapeutic plans (Franco-Duarte et al., 2019). Often too, related taxa share common physiological, biological, or ecological features, so knowledge of a broader fungal taxonomy can guide explorations on fungal species.

An important aspect of fungal classification is that the species we encounter today represent only a tiny fraction of all the fungi that have existed over hundreds of millions of years. As a result, the closest relatives of many modern species diverged from their common ancestors a long time ago. Over this vast period, numerous extinctions have wiped out numerous species, leading to gaps in the evolutionary tree of life (Erwin, 2008). This makes it challenging to determine connections between fungal groups solely based on molecular analysis of living organisms.

However, as species and ecosystems re-evolve following extinction of certain lineages, these led to changes in global biogeochemical cycles (Hull, 2015). While biological extinction poses a major challenge for phylogenetic studies in many organisms, the problem is not as relevant in the study of vertebrates because of the rich fossil record. As opposed to the study of fungi, much of the evolutionary history remains unclear, thus limiting our confidence in the finer details of their natural or evolutionary classifications. Despite these limitations, the current understanding of fungal diversity is far more detailed than the textbook view of mycology a generation ago.

A recent review by Naranjo-Ortiz and Gabaldón (2019) presents a taxonomy categorizing the known diversity of true fungi into nine major lineages: Opisthosporidia, Chytridiomycota, Neocallimastigomycota, Blastocladiomycota, Zoopagomycota, Mucoromycota, Glomeromycota, Ascomycota, and Basidiomycota. These lineages together form a monophyletic clade known as the true fungi, which is sister to a group of amoeboid protozoans that includes the Nucleariida (Nuclearia, Micronuclearia, Parvularia) and Funticulida (Fonticula) (Karpov et al., 2014; Spatafora et al., 2017; Tedersoo et al., 2018). This taxonomy can be reconciled with the traditional grouping that includes the zoosporic fungi, zygomycetous fungi, and the Dikarya. Here we provide a brief description of these traditional groupings of fungi. Figure 15.1 shows images of these fungi.

Zoosporic Fungi

All fungi must have descended from a single-celled organism that, at least at some point in its life cycle, was capable of swimming using a posterior flagellum. Several fungal lineages still exhibit this lifestyle, although some have lost this ability through secondary evolution. The evolutionary relationships among these lineages remain unresolved, largely due to their deep divergence, incomplete sampling, and the parasitic nature of several species.

Zygomycetous Fungi

The loss of flagella and the development of hyphal growth enabled a specific group of fungi to colonize terrestrial environments. These land-dwelling fungi encompass most of the described diversity and are included in the subkingdom Dikarya (Basidiomycota + Ascomycota + Entorrhizomycota), along with several lineages collectively known as zygomycetous fungi. The term "zygomycetous" refers to a paraphyletic group (Zygomycota), which is named after the zygospore, a sexual structure common to most of the lineages in this group. Although Glomeromycota lacks zygospores and were historically separated from the "Zygomycota," they are included here for simplicity. Within this context, zygomycetous fungi can be divided

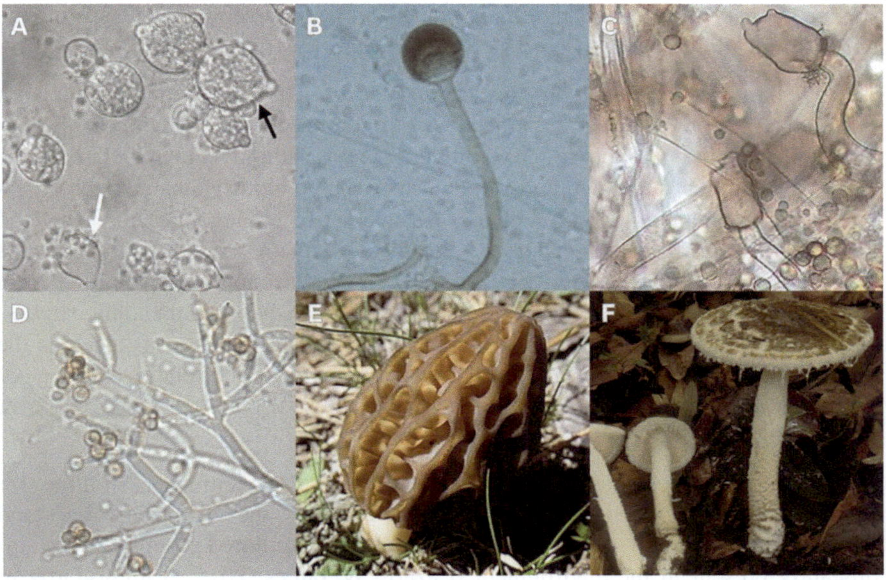

Fig. 15.1 Variation in morphology among traditional groups of fungi. Chytrid Fungi: (**a**) Morphology of *Batrachochytrium dendrobatidis* in culture on tryptone/gelatin-hydrolysate/lactose (TGhL)-broth, showing mature zoosporangia (black arrow) and empty discharged sporangia (white arrow) (van Rooij, 2015). Zygomycetous fungi: (**b**) *Rhizopus microsporus* sporangium containing sporangiospore (Kim et al., 2015) and (**c**) *Mucor plumbeus* sporangiophores and sporangiospores showing the finger-like projections on the columellae (Rodríguez et al., 2021). Ascomycetes (**d**) Conidiophores and Phialides of *Trichoderma caeruloviride*, an ascomycete (Medmyco, 2015), and (**e**) fruiting body of *Morchella deliciosa* (algarsal, 2022). Basidiomycetes: (**f**) fruiting body of *Amanita zangii* (Thongbai et al., 2017)

into two main lineages: one primarily composed of parasites of opisthokonts (Zoopagomycota), and another consisting mostly of plant symbionts and saprotrophs (Glomeromycota + Mucoromycota).

Dikarya

Dikarya is by far the most species-rich and well-studied group of fungi, consisting of two primary phyla: Basidiomycota and Ascomycota. Additionally, a small group of root endophytes has been proposed as a third phylum, Entorrhizomycota, but more recent data seem to support that these root endophytes form a separate group from the Ascomycota and Basidiomycota. Dikarya is defined by a sexual cycle that involves a hyphal fusion without meiosis, resulting in hyphae that contain two separate nuclear populations, known as dikaryotic hyphae. Most members also feature septate hyphae, ergosterol as their membrane sterol, and several lineages having the ability to form multicellular reproductive or vegetative structures.

Phylogenomic Approach to Fungal Classification The phylogenetic framework of fungi has become increasingly robust with genomic data. This wealth of information confirmed the monophyly of Ascomycota and Basidiomycota and clarified the boundaries of their major subphyla, transformed the previously disorganized orders within the "Zygomycota" into a clearly defined set of phylogenetic lineages, and identified the phylogenetic status of Blastocladiomycota, Zoopagomycota, and Mucoromycota, among others (Li et al., 2021). Modern approaches shifted from single-gene-based phylogenetic methods to comprehensive phylogenomic studies, helping resolve some of the most challenging evolutionary questions, such as the placement of *Microsporidia* and the relationships among various clades of the zygomycetous fungi (Naranjo-Ortiz & Gabaldón, 2019; Li et al., 2021; Park & Poulin, 2021).

Environmental sequencing has also opened new and exciting possibilities for fungal taxonomy, leading to the discovery of the Cryptomycota and Archaeorrhizomycetes—two highly diverse, largely uncultured groups that are significantly present in the environment (Lazarus & James, 2015; Naranjo-Ortiz & Gabaldón, 2019; Mariz et al., 2025). These studies suggest that most major fungal lineages include species that have been described using traditional methods (Tedersoo et al., 2018). However, this does not mean we have a comprehensive understanding of fungal diversity, which remains unevenly explored.

Molecular systematics has become the foundation for modern fungal classification at various taxonomic levels. Hibbett and Donoghue (1998) emphasized the significance of DNA-based phylogenetic analyses in fungal classification. Hibbett et al. (2007) were among the first major studies to provide a broad, in-depth examination of higher-level fungal taxonomy. Since then, numerous studies have explored higher-level fungal rankings, though most have been focused on specific taxa.

To date, the most comprehensive work aimed at offering a comprehensive overview of the Kingdom Fungi, covering key concepts of its taxonomy and classification was done by Wijayawardene et al. (2022). In this study, they presented a phylogenomic tree of the fungal phyla. Each phylum and its respective classes were carefully described and presented with illustrations, life modes, distributions, and other notable characteristics. In this study, phylogenetic reconstructions were conducted using both maximum likelihood (ML) and Bayesian methods, utilizing a 300-gene dataset and a taxonomic selection inclusive of early-diverging fungal lineages. In this analysis, 19 fungal phyla are presented (see Table 15.1): Aphelidiomycota, Ascomycota, Basidiobolomycota, Basidiomycota, Blastocladiomycota, Calcarisporiellomycota, Chytridiomycota, Entomophthoromycota, Entorrhizomycota, Glomeromycota, Kickxellomycota, Monoblepharomycota, Mortierellomycota, Mucoromycota, Neocallimastigomycota, Olpidiomycota, Rozellomycota, Sanchytriomycota, and Zoopagomycota. In the phylogenetic trees, *Caulochytriomycota* is placed within *Chytridiomycota*, making it a synonym of the latter, with *Caulochytriomycetes*

Table 15.1 List of phyla based on the 300-gene phylogenomic study by Wijayawardene et al. (2022)

Fungal phyla	References
Aphelidiomycota Tedersoo et al.	Tedersoo et al. (2018)
Ascomycota Caval.-Sm.	Cavalier-Smith (1998)
Basidiobolomycota Doweld	Doweld (2001)
Basidiomycota R.T. Moore	Moore (1980)
Blastocladiomycota T.Y. James	James et al. (2006)
Calcarisporiellomycota Tedersoo et al.	Tedersoo et al. (2018)
Chytridiomycota Doweld	Doweld (2001)
Entomophthoromycota Humber	Humber (2012)
Entorrhizomycota R. Bauer et al.	Bauer et al. (2015)
Glomeromycota C. Walker & A. Schüßler	Schüβler et al. (2001)
Kickxellomycota Tedersoo et al.	Tedersoo et al. (2018)
Monoblepharomycota Doweld	Doweld (2001)
Mortierellomycota Tedersoo et al.	Tedersoo et al. (2018)
Mucoromycota Doweld	Doweld (2001)
Neocallimastigomycota M.J. Powell	James et al. (2006)
Olpidiomycota Doweld	Wijayawardene et al. 2022
Rozellomycota Doweld	Tedersoo et al. (2018) and Vogt-Schilb et al. (2022)
Sanchytriomycota Galindo et al.	Galindo et al. (2021)
Zoopagomycota Gryganskyi et al.	Spatafora et al. (2017)

being considered a class within *Chytridiomycota*. Additionally, a new subphylum, *Sanchytriomycotina* Karpov, is introduced as the sole subphylum within *Sanchytriomycota*. This has increased the previous number of early-diverging fungal phyla from 17 (Voigt et al., 2021) to 19 (Wijayawardene et al., 2022).

Drivers of Climate Change

Climate change is a major driver of ecological change (Williams et al., 2020) through alterations in temperature, precipitation patterns, and extreme weather events, causing significant shifts in species distribution (Sirami et al., 2017), phenology (Inouye, 2022), and community composition (Gornish & Tylianakis, 2013; Bergamin et al., 2024). The current accelerated rate of global warming is mainly attributable to human activities, primarily through greenhouse gas emission pathways (Masson-Delmotte et al., 2018). The steady rise in the concentration of greenhouse gases in the atmosphere, such as carbon dioxide and methane, gradually traps heat via radiative forcing (Trenberth & Fasullo, 2009; Harde, 2017). Radiative forcing refers to the balance between the incoming and outgoing energy to and from Earth, which can be affected by various factors such as solar irradiance, atmospheric greenhouse gas concentrations (e.g. CO_2, CH_4), cloud cover, aerosols, surface albedo, and other "ghost" forcings through elevational, latitudinal, longitudinal, seasonal, and daily variations in atmospheric conditions (Hansen et al., 1997; Intergovernmental Panel On Climate Change, 2014). Evidence from paleoclimate records indicates a direct relationship between the concentration of carbon dioxide in the atmosphere and the temperature (Lüthi et al., 2008; Jouzel, 2012). As fundamental variables in all ecosystems, the interplay between carbon dioxide concentration and temperature has a cascading effect on other components of the Earth's systems that can result in an amplification or forcing effect. An example of this is the release of more carbon from soils and permafrost during warmer periods (Koven et al., 2011; Schneider Von Deimling et al., 2012), lesser carbon dioxide absorption of warmer oceans (Abas & Khan, 2014), and possible reduction in the carbon dioxide sequestration by marine phytoplankton (Beardall et al., 2009). This amplification effect in the rise of global temperature will provide a higher energy input to accelerate the phases of the hydrological cycle including evaporation and surface drying (Trenberth, 2011).

Impact of Climate Change on Fungi and Other Fungal-Like Organisms

All organisms are affected by temperature and water concentration changes (e.g., humidity, evapotranspiration) from organismal to ecosystem levels. With the complexity of climate change and the incredible variations in atmospheric conditions on a regional scale (Salinger, 2005; Xu et al., 2017), climate change's impact on specific groups, such as FFLO, will also differ. Thermotolerant FFLO might have an opportunity to spread and have a wider distribution as suitable conditions for their growth and reproduction increase as the global average temperature becomes warmer. Meanwhile, endophytes and pathogenic FFLO might be affected by the response of their hosts to the changes in climatic conditions (Suryanarayanan & Shaanker, 2021). However, regional variation might also occur, which can lead to colder climates in specific regions, such as that of the forecasted cooler climate in parts of Europe, predominantly in the North Atlantic region (Asbjørnsen & Årthun, 2023). This case of potential regional cooling can result from the weakening of Atlantic Meridional Overturning Circulation (AMOC), a major system of ocean currents in the Atlantic (Jackson et al., 2022). The weakening of parts of the AMOC, particularly the Gulf Stream (Asbjørnsen & Årthun, 2023), will result in less heat being transported northward to North Atlantic Europe, thus resulting in a colder climate in the region. In this scenario, FFLO favoring colder climates will have better chances of growth and reproduction. Therefore, the impact of climate change on FFLO varies widely depending on the species' thermal niche, growth and development, and dispersal and reproduction. We outlined below some of the expected changes brought about by climate change to fungi and other fungal-like organisms.

Impacts on Fungal Phenology and Reproduction Climate change can alter the phenology, reproductive adaptation, and survival of FFLO, commonly by affecting sporulation and dispersal (Vitasse et al., 2021; Norros et al., 2023). Environmental conditions such as temperature, humidity, and microclimatic factors influence fungal development, resulting in the alteration of life cycle stages such as spore production, germination, mycelial growth, reproduction, and survival (Delmas et al., 2024). For macrofungi, many exhibit pronounced delay in fruiting relative to a warming climate, with high interspecific variation (Kauserud et al., 2012). The study of Vogt-Schilb et al. (2022) compared nineteenth-century (1821–1852) and twenty-first-century (2001–2011) herbaria through analyses of phenological traits. Contrasting changes in the emergence of fungal fruit bodies was observed according to ecological guilds, sporocarp life forms, and forest types accompanied long-term climate warming in the Mediterranean (Vogt-Schilb et al., 2022). Early-autumn fruiters, epigeous (above-ground) fungi, and species with cold-climate affinities exhibited fruiting on average 16.4–17.3 days later than those collected during the nineteenth century. In contrast, late fruiters, hypogeous (below-ground) fungi, and species restricted to Mediterranean climates did not change their fruiting dates significantly. Furthermore, concerning guild-specific responses to climate change, saproxylic fungi showed delayed fruiting for 32.5 days. Furthermore, Krah et al. (2023) explored the effect of temperature on the timing and duration of fungal fruiting in major terrestrial biomes. Most biomes exhibited a single prominent fruiting peak each year, though boreal and temperate biomes often have a secondary peak in spring and autumn. Mean temperature and variability were key factors influencing fungal fruiting patterns, while precipitation played a secondary role. The same study also revealed that warming temperatures generally led to lengthened fruiting durations, especially in boreal and temperate biomes. In general, in warmer future climate scenarios, many fungi will have varying growth rates with faster mycelium growth but lower spore production (Damialis et al., 2015).

Impacts on Mycorrhizal Fungi and Plant Interaction The complex symbiosis between mycorrhizal fungi and plants is critical to maintaining terrestrial ecosystem function, affecting nutrient cycling, plant health, and carbon sequestration (Bonfante & Genre, 2010). These important biological interactions may be affected as global warming accelerates and atmospheric carbon dioxide levels increase. However, despite the usual forecasted adverse effects of climate change on the diversity of most organisms, the case of arbuscular mycorrhizal fungi (AMF) may paint a different scenario. AMF are sensitive to colder temperatures than other fungal groups. Thus, rising temperatures could potentially improve AMF-mediated nutrient transfer to plants, which leads to strengthened mutualism (Bunn et al., 2009). Furthermore, higher concentrations of carbon dioxide in the atmosphere may aid in more efficient photosynthesis in plants, which may also help reduce the carbon cost for plants to maintain symbiosis with AMF (Dusenge et al., 2019). Empirical evidence on this is confirmed with wild plants, where an increase in temperature from sub-ambient to ambient boosted mycorrhizal growth, but interestingly, this is not the case with domesticated species (Duarte & Maherali, 2022). Domesticated plants may have been selectively bred for traits that reduce their need for mycorrhizal symbiosis, which resulted in lesser resilience in changing climate conditions, which AMF can mitigate (Martín-Robles et al., 2018). On the other hand, wild plants evolved over extended periods in nutrient-limited or competitive environments, which heavily favors harboring keen symbiosis with AMF for nutrient acquisition (Grassi et al., 2024). The mechanism for this is likely through increased metabolic activity in AMF at higher temperatures, which reinforces the mobilization and transfer of nutrients to the wild plant hosts (Duarte & Maherali, 2022). Relatedly, as elevated carbon dioxide levels stimulate plant growth through the optimization of photosynthesis, the extent to which plants benefit from AMF symbiosis depends on their capacity to allocate carbon to the AMF symbionts.

Impacts on Decomposer Fungi and Carbon Cycling Species richness of soil fungi is predominantly influenced by climate, with mean annual precipitation and distance from the equator being the strongest predictors. Saprotrophic fungi (SF) are the primary decomposers of organic matter. They play a crucial role in carbon flow within ecosystems by breaking down organic matter and facilitating soil respiration (Boddy, 2016). However, climate change may significantly affect the distribution of SF as they are susceptible to changes in temperature, which may induce shifts in SF communities and impact their enzymatic activity. This SF community and diversity alteration may affect global feedback mechanisms (Feng et al., 2022). SF may exhibit differentiated responses to temperature variations, with abrupt decreases in abundance occurring at specific thresholds rather than following a gradual decline (Morera et al., 2022). The most critical thresholds include ~8.86 °C for mean annual temperature, ~22.41 °C for soil temperature, and ~27.97 °C for the maximum temperature of the warmest month (Feng et al., 2022). The findings of Feng et al. (2022) challenge standard linear models of microbial response to climate change, pointing out that the crossing thresholds may lead to suppressing important soil decomposition processes. Additionally, SF is also found to be more susceptible to increases in temperature under arid conditions, which indicates a likely reduction in their bioactivity and resulting carbon flux contributions in dry environments (Berdugo et al., 2020).

Impacts on Fungal Pathogenicity in Humans Dramatic changes in the climatic variables of temperature, precipitation, and extreme weather events generally alter all levels of biodiversity; this includes the potential geographic expansion of pathogenic fungi (Williams et al., 2024). As anthropogenically driven climate change alters these parameters, it is expected to speed up pathogen spread and increase in virulence. Historically, *Coccidiodes* spp. was confined to arid regions of the southwestern regions of the United States. However, due to dramatic changes in the rainfall and drought cycles in the continental US, *Coccidiodes* spp. can now be detected in far northern regions, including Washington (Head et al., 2022). Analogously, *Histoplasma* spp., a fungus that causes histoplasmosis, is detected to expand beyond its traditional range in the Ohio and Mississippi River Valleys driven by the changes in the soil properties as brought by a changing climate (Mazi et al., 2023). Beyond the continental US, other fungal pathogens such as *Cryptococcus gattii*, which was historically restricted to tropical and subtropical ecoregions, have recently been detected in the Pacific Northwest. *C. gattii* is then expected to spread further into Europe as the acceleration in the rise of the global temperature continues (Cogliati, 2021). Besides the gradual alteration of ecological systems, empirical evidence in the increasing frequency of extreme weather events such as dust storms, hurricanes/typhoons, and wildfires may also support the spread of fungal pathogens primarily through expanding their spore dispersal. A concrete example is the increased frequency of dust storms in the southwestern US between 1988 and 2011, which aligned with a proportional increase in the number of *Coccidiodes* spp. (Tong et al., 2017). Furthermore, flooding from hurricanes/typhoons provides a conducive environment for the rapid growth of molds, which increases the risk of respiratory infections, most dangerously for immunocompromised populations (Toda et al., 2023). In the USA, post-disaster scenarios, such as after the devastation of hurricanes, have been found to align with the proliferation of fungal infections, such as the case of necrotizing mucormycosis following a tornado in Joplin, Missouri (Uhrlaß et al., 2022). Most of these reports are from the continental US, where disease spread is well under surveillance; the lack of reports and literature in other ecoregions, such as in Asia and Africa, highlights that the impact of climate change in supporting the geographic spread of fungal pathogens may well be much worse.

Another concern for the spread of fungal pathogens is their strong adaptive capacity in the increasing global temperatures brought by climate change. This high adaptability can increase pathogenicity in their hosts, such as humans. A concrete example of this is the emergence of the multidrug resistant yeast, *Candidozyma auris*, which is hypothesized to be driven by global warming, with distinct lineages appearing simultaneously in multiple continents (Casadevall et al., 2019). Another case is that of *Cryptococcus deneoformans* which exhibited acceleration in genetic mutations and promotion of antifungal resistance due to the selection pressure brought by global warming through heat stress (Gusa et al., 2020). *C. deneoformas* have profound transposon mobilization, which allows it to develop resistance to conventional antifungals rapidly (Gusa et al., 2023). Alarmingly, these findings raise concerns regarding the evolution of other pathogenic fungi in overcoming mammalian thermal barriers which may lead to new and more virulent fungal diseases. The changing climatic conditions brought by anthropogenically driven climate change may have favored the spread of fungal pathogens due to the expansion of suitable habitats, increased frequency of extreme weather events, and their high adaptability to increasing environmental temperatures. Another potential, albeit indirect route, in the further geographic spread of fungal pathogens under the current climate change scenario is the impact of agriculture and human behavior.

Impacts on Fungal Pathogenicity in Plants Besides the empirical evidence on humans' geographical spread of fungal pathogens, mounting evidence is also being established on the role of an anthropogenically driven climate in the spread of fungal plant pathogens (FPP). Among the climatic variables, temperature is considered to be the primary driver of plant disease outbreaks, with global warming supporting the latitudinal spread of FPP such as *Fusarium*, *Alternaria*, and *Venturia* (Delgado-Baquerizo et al., 2020). In China, areas of cropland affected by plant diseases have been recorded to increase significantly over the past 50 years, with warm nights and reduced frosts attributed to driving approximately 20% of the increase (Wang et al., 2021). Furthermore, metabarcoding analyses reveal that soilborne FPP are influenced by rising temperatures, which directly poses a growing risk to agriculture in higher latitudes (Větrovský et al., 2019). With the hosts, plants, mainly affected by changes in temperature and moisture, their FPP is also expected to shift accordingly. Mechanistic models demonstrated that temperature plays a significant role in determining the rates of spore germination, infection, and virulence of FPP (Chaloner et al., 2020). Moisture availability also plays a significant limiting factor in FPP spread, as many foliar pathogens require high humidity or surface water availability for spore germination and host penetration (Prank et al., 2019). Additionally, fungal-like plant pathogens such as *Phytophthora* species indicate evolutionary variability in response to increasing temperatures, suggesting that specific plant pathogenic species may rapidly adapt to warming climates (Chaloner et al., 2020).

Empirical evidence from controlled experiments supplied proof that anthropogenically driven climate change exacerbates plant fungal disease through increasing virulence driven by their adaptability to environmental variations brought by the increasing temperatures. To illustrate, *Fusarium* spp. in salad crops and *Blumeria graminis* in barley are found to increase in severity in alignment with increasing carbon dioxide levels and temperature conditions (Chitarra et al., 2015; Gullino et al., 2017). Higher carbon dioxide concentrations may weaken plant immunity and alter stomatal and trichome densities, which can lead to crops being more vulnerable to FPP (Váry et al., 2015). Additionally, changes in climate conditions can affect disease management strategies in agriculture as experimental evidence points out that the efficacy of fungicides such as mancozeb and azoxystrobin while biocontrol agents such as *Aspergillus quisqualis* also showed enhanced performance in controlling FPP.

Impacts on Lichenized Fungi Lichens are composite organisms composed of lichenized fungi (mycobiont), alga and/or cyanobacteria (photobiont), and other microbes. The mini-ecosystem in these lichen-microbial assemblages is highly sensitive to microclimatic conditions (Sahu et al., 2019; Ellis & Eaton, 2021). Even slight changes in climate may cause shifts in lichen distribution, diversity, and functional traits. Thus, lichens serve as important indicators of climate change due to their high sensitivity to environmental conditions, particularly temperature, humidity, and air quality (Neupane et al., 2017). Lichens as poikilohydric organisms rely on ambient humidity and precipitation for hydration, making them highly responsive to microclimatic changes. Warming related to climate coupled with reduced relative humidity and precipitation can significantly decrease lichen cover, particularly in dryland biocrust communities (Li et al., 2023). This trend underscores the pivotal role of humidity over temperature in determining lichen survival under changing climates (Aptroot et al., 2021). Lichens inhabit climate-sensitive habitats, such as high mountain tropics and Mediterranean biomes, are particularly sensitive to the impacts of changes in climatic conditions of precipitation and temperature (Nelsen & Lumbsch, 2020). In these areas, the effects of climate change may drive local extinctions, particularly for epiphytic species that rely on specific microhabitats under tree canopies (Di Nuzzo et al., 2022). Studies conducted in Mediterranean regions point that even though certain tree morphologies can mitigate adverse effects, they are insufficient to preserve overall lichen diversity and ecosystem functionality (Nelsen & Lumbsch, 2020). The spatial distribution of lichens and their functional traits also correlate significantly with macroclimatic factors of precipitation and temperature. Studies on trebouxioid (*Trebouxia* as photobiont), trentepohlioid (*Trentepohlia* as photobiont), and cyanolichens (cyanobacteria as photobiont) show that the former thrive in colder, drier upland regions, while the latter two prefer wetter and warmer climates, reflecting distinct ecological adaptations (Borgato et al., 2022; Zuo et al., 2023). These differences in lichen distribution patterns and functional traits provide concrete evidence of climatic impacts on ecosystems (Baldauf et al., 2021). Furthermore, different mechanisms work in terms of lichen responses to climate drivers. For instance, warming may promote photosynthesis during favorable periods but reduce overall cover due to lower relative humidity (Ladrón De Guevara et al., 2018). Increased concentrations of atmospheric carbon dioxide are often neutralized by declines in water availability, illustrating the multivariate and complete nature of climate change impacts on lichen diversity (Lange et al., 2007; Matvienko et al., 2024).

Impacts on Fungal-Like Protists (Myxomycetes) The potential impact of climate change on myxomycetes, also referred to as slime molds, is expected to be significant as temperature and moisture are the primary factors regulating their occurrence. A changing climate is then expected to alter their distribution and ecology significantly (Kryvomaz & Stephenson, 2017). However, due to their cryptic life cycle, predicting their resilience to environmental changes remains challenging, as such literature in this area is scarce (Walker & Stephenson, 2016). Myxomycetes may also be less affected by climate change due to the three dormant stages in their life cycle (Kryvomaz & Stephenson, 2017), though strong winds from typhoons positively spread myxomycetes (Cabutaje et al., 2021). However, species confined to restricted microhabitats like alpine snowbanks (Ronikier & Lado, 2015) or limited to small geographic regions like isolated oceanic islands are particularly vulnerable (Kryvomaz et al., 2020).

Concluding Points

FFLOs play a critical role in maintaining the stability of ecosystems and their responses to climate change are yet to be thoroughly explored. As temperature and water availability are important drivers of climate change, the continuous rise in temperatures, shifting precipitation patterns, and increased extreme weather events can impact fungal biodiversity, ecological functions, and interactions with other organisms. Shifts in the biodiversity of FFLOs will incur changes to the ecosystems they support since they are collectively involved in essential ecosystem processes such as carbon and nutrient cycling. While many FFLOs exhibit diverse adaptive strategies in a changing climate, others face significant challenges in survival and reproduction. Many fruticose lichens such as *Usnea* are theorized to be negatively impacted by a warming climate, especially in tropical countries where they are usually found at higher elevations and in mountain tops. The reduction of distribution or potential extinction of these fungi also has an unknown impact on their local ecosystems. On the other hand, some fungal pathogens are expected to shift in their distribution as climate changes. Specific fungal pathogens such as *Fusarium*, *Alternaria*, and *Phytophthora* may become more widespread in distribution as warming climates favor their adaptive and reproductive strategies. These changes in fungal biodiversity may influence carbon cycling, soil health, agriculture, and human health. Addressing these challenges requires interdisciplinary collaboration, long-term research, and integrative conservation efforts to better understand and mitigate the impacts of climate change on fungal communities. Thus, recognizing the role of FFLOs in maintaining ecosystems is imperative for developing holistic climate resilience strategies.

References

Abas, N., & Khan, N. (2014). Carbon conundrum, climate change, CO2 capture and consumptions. *Journal of CO2 Utilization, 8*, 39–48. https://doi.org/10.1016/j.jcou.2014.06.005

algarsal. (2022). Morchella deliciosa [Image]. Wikimedia Commons. https://commons.wikimedia.org/wiki/File:Morchella_deliciosa_young.jpg. Licensed under CC BY 4.0.

Aptroot, A., Stapper, N. J., Košuthová, A., & Van Herk, K. (C. M.). (2021). Lichens as an indicator of climate and global change. In *Climate change* (pp. 483–497). Elsevier. https://doi.org/10.1016/B978-0-12-821575-3.00023-2

Asbjørnsen, H., & Årthun, M. (2023). Deconstructing future AMOC decline at 26.5°N. *Geophysical Research Letters, 50*(14), e2023GL103515. https://doi.org/10.1029/2023GL103515

Bahram, M., & Netherway, T. (2022). Fungi as mediators linking organisms and ecosystems. *FEMS Microbiology Reviews, 46*(2), fuan058. https://doi.org/10.1093/femsre/fuab058

Baldauf, S., Porada, P., Raggio, J., Maestre, F. T., & Tietjen, B. (2021). Relative humidity predominantly determines long-term biocrust-forming lichen cover in drylands under climate change. *Journal of Ecology, 109*(3), 1370–1385. https://doi.org/10.1111/1365-2745.13563

Bauer, R., Garnica, S., Oberwinkler, F., Riess, K., Weiß, M., & Begerow, D. (2015). Entorrhizomycota: A new fungal phylum reveals new perspectives on the evolution of fungi. *PLOS ONE, 10*(7). https://doi.org/10.1371/journal.pone.0128183

Beardall, J., Stojkovic, S., & Larsen, S. (2009). Living in a high CO2 world: Impacts of global climate change on marine phytoplankton. *Plant Ecology & Diversity, 2*(2), 191–205. https://doi.org/10.1080/17550870903271363

Berdugo, M., Delgado-Baquerizo, M., Soliveres, S., Hernández-Clemente, R., Zhao, Y., Gaitán, J. J., Gross, N., Saiz, H., Maire, V., Lehmann, A., Rillig, M. C., Solé, R. V., & Maestre, F. T. (2020). Global ecosystem thresholds driven by aridity. *Science, 367*(6479), 787–790. https://doi.org/10.1126/science.aay5958

Bergamin, R. S., Bastazini, V. A. G., Esquivel-Muelbert, A., Bordin, K. M., Klipel, J., Debastiani, V. J., Vibrans, A. C., Loyola, R., & Müller, S. C. (2024). Elevational shifts in tree community composition in the Brazilian Atlantic Forest related to climate change. *Journal of Vegetation Science, 35*(4), e13289. https://doi.org/10.1111/jvs.13289

Boddy, L. (2016). Fungi, ecosystems, and global change. In *The fungi* (pp. 361–400). Elsevier. https://doi.org/10.1016/B978-0-12-382034-1.00011-6

Bonfante, P., & Genre, A. (2010). Mechanisms underlying beneficial plant–fungus interactions in mycorrhizal symbiosis. *Nature Communications, 1*(1), 48. https://doi.org/10.1038/ncomms1046

Borgato, L., Ertz, D., Van Rossum, F., & Verbeken, A. (2022). The diversity of lichenized Trentepohlioid algal (Ulvophyceae) communities is driven by fungal taxonomy and ecological factors. *Journal of Phycology, 58*(4), 582–602. https://doi.org/10.1111/jpy.13252

Bunn, R., Lekberg, Y., & Zabinski, C. (2009). Arbuscular mycorrhizal fungi ameliorate temperature stress in thermophilic plants. *Ecology, 90*(5), 1378–1388. https://doi.org/10.1890/07-2080.1

Cabutaje, E. M., Pecundo, M. H., & dela Cruz, T. E. E. (2021). Diversity of myxomycetes in typhoon-prone areas: A case study in beach and inland forests of Aurora and Quezon Province, Philippines. *Sydowia, 73*, 113–132. https://doi.org/10.12905/0380.sydowia73-2020-0113

Casadevall, A., Kontoyiannis, D. P., & Robert, V. (2019). On the emergence of *Candida auris*: Climate change, azoles, swamps, and birds. *mBio, 10*(4), e01397–e01319. https://doi.org/10.1128/mBio.01397-19

Cavalier-Smith, T. (1998). A revised six-kingdom system of life. *Biol Rev, 73*(3), 203–266. https://doi.org/10.1017/S0006323198005167

Chaloner, T. M., Gurr, S. J., & Bebber, D. P. (2020). Geometry and evolution of the ecological niche in plant-associated microbes. *Nature Communications, 11*(1), 2955. https://doi.org/10.1038/s41467-020-16778-5

Chitarra, W., Siciliano, I., Ferrocino, I., Gullino, M. L., & Garibaldi, A. (2015). Effect of elevated atmospheric CO2 and temperature on the disease severity of rocket plants caused by Fusarium wilt under Phytotron conditions. *PLoS One, 10*(10), e0140769. https://doi.org/10.1371/journal.pone.0140769

Cogliati, M. (2021). Global warming impact on the expansion of fundamental niche of Cryptococcus gattii VGI in Europe. *Environmental Microbiology Reports, 13*(3), 375–383. https://doi.org/10.1111/1758-2229.12945

Damialis, A., Mohammad, A. B., Halley, J. M., & Gange, A. C. (2015). Fungi in a changing world: Growth rates will be elevated, but spore production may decrease in future climates. *International Journal of Biometeorology, 59*(9), 1157–1167. https://doi.org/10.1007/s00484-014-0927-0

Delgado-Baquerizo, M., Guerra, C. A., Cano-Díaz, C., Egidi, E., Wang, J.-T., Eisenhauer, N., Singh, B. K., & Maestre, F. T. (2020). The proportion of soil-borne pathogens increases with warming at the global scale. *Nature Climate Change, 10*(6), 550–554. https://doi.org/10.1038/s41558-020-0759-3

Delmas, C. E. L., Bancal, M., Leyronas, C., Robin, M., Vidal, T., & Launay, M. (2024). Monitoring the phenology of plant pathogenic fungi: Why and how? *Biological Reviews, 99*(3), 1075–1084. https://doi.org/10.1111/brv.13058

Di Nuzzo, L., Benesperi, R., Nascimbene, J., Papini, A., Malaspina, P., Incerti, G., & Giordani, P. (2022). Little time left. Microrefuges may fail in mitigating the effects of climate change on epiphytic lichens. *Science of the Total Environment, 825*, 153943. https://doi.org/10.1016/j.scitotenv.2022.153943

Doweld, A. B. (2001). Prosyllabus tracheophytorum. In: *Tentamen systematis plantarum vascularium (Trachephyta)*. Geos, Moscow.

Duarte, A. G., & Maherali, H. (2022). A meta-analysis of the effects of climate change on the mutualism between plants and arbuscular mycorrhizal fungi. *Ecology and Evolution, 12*(1), e8518. https://doi.org/10.1002/ece3.8518

Dusenge, M. E., Duarte, A. G., & Way, D. A. (2019). Plant carbon metabolism and climate change: Elevated CO2 and temperature impacts on photosynthesis, photorespiration and respiration. *New Phytologist, 221*(1), 32–49. https://doi.org/10.1111/nph.15283

Ellis, C. J., & Eaton, S. (2021). Microclimates hold the key to spatial forest planning under climate change: Cyanolichens in temperate rainforest. *Global Change Biology, 27*(9), 1915–1926. https://doi.org/10.1111/gcb.15514

Erwin, D. H. (2008). Extinction as the loss of evolutionary history. *Proceedings of the National Academy of Sciences, 105*(supplement_1), 11520–11527. https://doi.org/10.1073/pnas.0801913105.

Feng, Y., Zhang, J., Berdugo, M., Guirado, E., Guerra, C. A., Egidi, E., Wang, J., Singh, B. K., & Delgado-Baquerizo, M. (2022). Temperature thresholds drive the global distribution of soil fungal decomposers. *Global Change Biology, 28*(8), 2779–2789. https://doi.org/10.1111/gcb.16096

Franco-Duarte, R., Černáková, L., Kadam, S., Kaushik, K. S., Salehi, B., Bevilacqua, A., et al. (2019). Advances in chemical and biological methods to identify microorganisms—From past to present. *Microorganisms, 7*(5), 130. https://doi.org/10.3390/microorganisms7050130

Galindo, l. j., López-García, p., Torruella, g., Karpov, S., & Moreira, D. (2021). Phylogenomics of a new fungal phylum reveals multiple waves of reductive evolution across Holomycota. *Nature Communications 12*(1), 4973. https://doi.org/10.1038/s41467-021-25308-w

Gornish, E. S., & Tylianakis, J. M. (2013). Community shifts under climate change: Mechanisms at multiple scales. *American Journal of Botany, 100*(7), 1422–1434. https://doi.org/10.3732/ajb.1300046

Grassi, A., Pagliarani, I., Avio, L., Cristani, C., Rossi, F., Turrini, A., Giovannetti, M., & Agnolucci, M. (2024). Bioprospecting for plant resilience to climate change: Mycorrhizal symbionts of European and American beachgrass (Ammophila arenaria and Ammophila breviligulata) from maritime sand dunes. *Mycorrhiza, 34*(3), 159–171. https://doi.org/10.1007/s00572-024-01144-w

Guarro, J., Gené, J., & Stchigel, A. M. (1999). Developments in fungal taxonomy. *Clinical Microbiology Reviews, 12*(3), 454–500. https://doi.org/10.1128/cmr.12.3.454

Gullino, M. L., Gilardi, G., & Garibaldi, A. (2017). Effect of a climate change scenario on Fusarium equiseti leaf spot on wild rocket and radish under phytotron simulation. *Phytoparasitica, 45*(3), 293–298. https://doi.org/10.1007/s12600-017-0590-1

Gusa, A., Williams, J. D., Cho, J.-E., Averette, A. F., Sun, S., Shouse, E. M., Heitman, J., Alspaugh, J. A., & Jinks-Robertson, S. (2020). Transposon mobilization in the human fungal pathogen Cryptococcus is mutagenic during infection and promotes drug resistance in vitro. *Proceedings of the National Academy of Sciences, 117*(18), 9973–9980. https://doi.org/10.1073/pnas.2001451117

Gusa, A., Yadav, V., Roth, C., Williams, J. D., Shouse, E. M., Magwene, P., Heitman, J., & Jinks-Robertson, S. (2023). Genome-wide analysis of heat stress-stimulated transposon mobility in the human fungal pathogen Cryptococcus deneoformans. *Proceedings of the National Academy of Sciences, 120*(4), e2209831120. https://doi.org/10.1073/pnas.2209831120

Haelewaters, D., Gafforov, Y., & Zhou, L.-W. (2022) Editorial: Biodiversity and conservation of fungi and fungus-like organisms. *Frontiers in Fungal Biology, 3*, 973249. https://doi.org/10.3389/ffunb.2022.973249

Hansen, J., Sato, M., & Ruedy, R. (1997). Radiative forcing and climate response. *Journal of Geophysical Research: Atmospheres, 102*(D6), 6831–6864. https://doi.org/10.1029/96JD03436

Harde, H. (2017). Radiation transfer calculations and assessment of global warming by CO2. *International Journal of Atmospheric Sciences, 2017*, 1–30. https://doi.org/10.1155/2017/9251034

Hawksworth, D. L., & Lücking, R. (2017). Fungal diversity revisited: 2.2 to 3.8 million species. *Microbiology Spectrum, 5*(4), 10–1128.

Head, J. R., Sondermeyer-Cooksey, G., Heaney, A. K., Yu, A. T., Jones, I., Bhattachan, A., Campo, S. K., Wagner, R., Mgbara, W., Phillips, S., Keeney, N., Taylor, J., Eisen, E., Lettenmaier, D. P., Hubbard, A., Okin, G. S., Vugia, D. J., Jain, S., & Remais, J. V. (2022). Effects of precipitation, heat, and drought on incidence and expansion of coccidioidomycosis in western USA: A longitudinal surveillance study. *The Lancet Planetary Health, 6*(10), e793–e803. https://doi.org/10.1016/S2542-5196(22)00202-9

Hibbett, D. S., & Donoghue, M. J. (1998). Integrating phylogenetic analysis and classification in fungi. *Mycologia, 90*, 347–356.

Hibbett, D. S., Binder, M., Bischoff, J. F., Blackwell, M., Cannon, P. F., Eriksson, O. E., Huhndorf, S., James, T., Kirk, P. M., Lücking, R., Lumbsch, H. T., Lutzoni, F., Matheny, P. B., McLaughlin, D. J., Powell, M. J., Redhead, S., Schoch, C. L., Spatafora, J. W., Stalpers, J. A., Vilgalys, R., Aime, M. C., Aptroot, A., Bauer, R., Begerow, D., Benny, G., Castlebury, L. A., Crous, P. W., Dai, Y. C., Gams, W., Geiser, D. M., Griffith, G. W., Gueidan, C., Hawksworth, D. L., Hestmark, G., Hosaka, K., Humber, R. A., Hyde, K. D., Ironside, J. E., Kõljalg, U., Kurtzman, C. P., Larsson, K. H., Lichtwardt, R., Longcore, J., Miądlikowska, J., Miller, A., Moncalvo, J. M., Mozley-Standridge, S., Oberwinkler, F., Parmasto, E., Reeb, V., Rogers, J. D., Roux, C., Ryvarden, L., Ampaio, J. P., Schüßler, A., Sugiyama, J., Thorn, R. G., Tibell, L., Untereiner, W. A., Walker, C., Wang, Z., Weir, A., Weiss, M., White, M. M., Winka, K., Yao, Y. J., & Zhang, N. (2007). A higher level phylogenetic classification of the fungi. *Mycological Research, 111*(5), 509–547. https://doi.org/10.1016/j.mycres.2007.03.004

Hull, P. (2015). Life in the aftermath of mass extinctions. *Current Biology, 25*(19), R941–R952. https://doi.org/10.1016/j.cub.2015.08.053

Humber, R. A. (2012). Entomophthoromycota: a new phylum and reclassification for entomophthoroid fungi. *Mycotaxon, 120*(1), 477–492. https://doi.org/10.5248/120.477

Inouye, D. W. (2022). Climate change and phenology. *WIREs Climate Change, 13*(3), e764. https://doi.org/10.1002/wcc.764

Intergovernmental Panel on Climate Change (Ed.). (2014). *Climate change 2013—The physical science basis: Working group I contribution to the fifth assessment report of the intergovernmental Panel on climate change* (1st ed.). Cambridge University Press. https://doi.org/10.1017/CBO9781107415324.

Jackson, L. C., Biastoch, A., Buckley, M. W., Desbruyères, D. G., Frajka-Williams, E., Moat, B., & Robson, J. (2022). The evolution of the North Atlantic meridional overturning circulation since 1980. *Nature Reviews Earth & Environment, 3*(4), 241–254. https://doi.org/10.1038/s43017-022-00263-2

James, T. Y., Kauff, Schoch, C. L., Matheny, P. B., Hofstetter, V., Cox, C. J., Celio, G., Gueidan, C., Fraker, E., Miadlikowska, J., Lumbsch, H. T., Rauhut, A., Reeb, V., Arnold, A. E., Amtoft, A., Stajich, J. E., Hosaka, K., Sung, G.-H., Johnson, D., ... Vilgalys, R. (2006) Reconstructing the early evolution of Fungi using a six-gene phylogeny. *Nature, 443*(7113), 818–822. https://doi.org/10.1038/nature05110

Jouzel, J. (2012). *Evolution of climate in response to human activity*. World Meteorological Organization. https://wmo.int/media/magazine-article/evolution-of-climate-response-human-activity

Karpov, S. A., Mamkaeva, M. A., Aleoshin, V. V., Nassonova, E., Lilje, O., & Gleason, F. H. (2014). Morphology, phylogeny, and ecology of the aphelids (Aphelidea, Opisthokonta) and proposal for the new superphylum Opisthosporidia. *Frontiers in Microbiology, 5*. https://doi.org/10.3389/fmicb.2014.00112

Kauserud, H., Heegaard, E., Büntgen, U., Halvorsen, R., Egli, S., Senn-Irlet, B., Krisai-Greilhuber, I., Dämon, W., Sparks, T., Nordén, J., Høiland, K., Kirk, P., Semenov, M., Boddy, L., & Stenseth, N. C. (2012). Warming-induced shift in European mushroom fruiting phenology. *Proceedings of the National Academy of Sciences, 109*(36), 14488–14493. https://doi.org/10.1073/pnas.1200789109

Kim, M. J., Park, P. W., Ahn, J. Y., Kim, K. H., Seo, J. Y., Jeong, J. H., Park, M. J., Jung, J. W., & Seo, Y. H. (2015). *Rhizopus microsporus sporangium containing sporangiospores [image].* Wikimedia Commons. https://commons.wikimedia.org/wiki/File:Rhizopus_microsporus.png. Licensed under CC BY-SA 3.0.

Koven, C. D., Ringeval, B., Friedlingstein, P., Ciais, P., Cadule, P., Khvorostyanov, D., Krinner, G., & Tarnocai, C. (2011). Permafrost carbon-climate feedbacks accelerate global warming. *Proceedings of the National Academy of Sciences, 108*(36), 14769–14774. https://doi.org/10.1073/pnas.1103910108

Krah, F., Büntgen, U., & Bässler, C. (2023). Temperature affects the timing and duration of fungal fruiting patterns across major terrestrial biomes. *Ecology Letters, 26*(9), 1572–1583. https://doi.org/10.1111/ele.14275

Kryvomaz, T., & Stephenson, S. L. (2017). Preliminary evalution of the possible impact of climate change on myxomycetes. *Nova Hedwigia, 104*(1–3), 5–30. https://doi.org/10.1127/nova_hedwigia/2016/0379

Kryvomaz, T., Michaud, A., & Stephenson, S. L. (2020). Myxomycete biodiversity on five islands of the Seychelles. *Zootaxa, 4851*(2). https://doi.org/10.11646/zootaxa.4851.2.1

Ladrón De Guevara, M., Gozalo, B., Raggio, J., Lafuente, A., Prieto, M., & Maestre, F. T. (2018). Warming reduces the cover, richness and evenness of lichen-dominated biocrusts but promotes moss growth: Insights from an 8 yr experiment. *New Phytologist, 220*(3), 811–823. https://doi.org/10.1111/nph.15000

Lange, O. L., Allan Green, T. G., Meyer, A., & Zellner, H. (2007). Water relations and carbon dioxide exchange of epiphytic lichens in the Namib fog desert. *Flora – Morphology, Distribution, Functional Ecology of Plants, 202*(6), 479–487. https://doi.org/10.1016/j.flora.2006.09.006

Lazarus, K. L., & James, T. Y. (2015). Surveying the biodiversity of the Cryptomycota using a targeted PCR approach. *Fungal Ecology, 14*, 62–70. https://doi.org/10.1016/j.funeco.2014.11.004

Li, Y., Steenwyk, J. L., Chang, Y., Wang, Y., James, T. Y., Stajich, J. E., et al. (2021). A genome-scale phylogeny of the kingdom fungi. *Current Biology, 31*(8), 1653–1665. https://doi.org/10.1016/j.cub.2021.01.074

Li, X., Sun, J., Zhang, H., Tan, H., Hui, R., Qi, J., Zhang, P., & Ward, N. D. (2023). Warming decreases desert ecosystem functioning by altering biocrusts in drylands. *Journal of Applied Ecology, 60*(12), 2676–2687. https://doi.org/10.1111/1365-2664.14528

Lüthi, D., Le Floch, M., Bereiter, B., Blunier, T., Barnola, J.-M., Siegenthaler, U., Raynaud, D., Jouzel, J., Fischer, H., Kawamura, K., & Stocker, T. F. (2008). High-resolution carbon dioxide concentration record 650,000–800,000 years before present. *Nature, 453*(7193), 379–382. https://doi.org/10.1038/nature06949

Mariz, J., Nawaz, A., Bösch, Y., & Wurzbacher, C. (2025). Exploring environmental microfungal diversity through serial single cell screening. *Molecular Ecology Resources, 25*(3), e14055. https://doi.org/10.1111/1755-0998.14055

Martín-Robles, N., Lehmann, A., Seco, E., Aroca, R., Rillig, M. C., & Milla, R. (2018). Impacts of domestication on the arbuscular mycorrhizal symbiosis of 27 crop species. *New Phytologist, 218*(1), 322–334. https://doi.org/10.1111/nph.14962

Masson-Delmotte, V., Pörtner, H.-O., Skea, J., Zhai, P., Roberts, D., Shukla, P. R., Pirani, A., Pidcock, R., Chen, Y., Lonnoy, E., Moufouma-Okia, W., Péan, C., Connors, S., Matthews, J. B. R., Zhou, X., Gomis, M. I., Maycock, T., Tignor, M., & Waterfield, T. (2018). An IPCC Special Report on the impacts of global warming of 1.5°C above pre-industrial levels and related global greenhouse gas emission pathways, in the context of strengthening the global response to the threat of climate change, sustainable development, and efforts to eradicate poverty.

Matvienko, A. I., Evgrafova, S. Y., Kovaleva, N. M., Sideleva, E. V., Sitnikova, M. V., Menyailo, O. V., & Masyagina, O. V. (2024). Greenhouse gas fluxes from the epiphytic lichens: Incubation experiments. *Forests, 15*(1), 107. https://doi.org/10.3390/f15010107

Mazi, P. B., Sahrmann, J. M., Olsen, M. A., Coler-Reilly, A., Rauseo, A. M., Pullen, M., Zuniga-Moya, J. C., Powderly, W. G., & Spec, A. (2023). The geographic distribution of dimorphic mycoses in the United States for the modern era. *Clinical Infectious Diseases, 76*(7), 1295–1301. https://doi.org/10.1093/cid/ciac882

Medmyco. (2015). *Photomicrograph of the sporangiophores and sporangiospores of Mucor plumbeus showing the finger-like projections on the columellae [image].* Wikimedia Commons. https://commons.wikimedia.org/wiki/File:Mucor_plumbeus.jpg. Licensed under CC BY-SA 4.0.

Morera, A., Martínez De Aragón, J., De Cáceres, M., Bonet, J. A., & de-Miguel, S. (2022). Historical and future spatially-explicit climate change impacts on mycorrhizal and saprotrophic macrofungal productivity in Mediterranean pine forests. *Agricultural and Forest Meteorology, 319*, 108918. https://doi.org/10.1016/j.agrformet.2022.108918

Moore, R. T. (1980). Taxonomic proposals for the classification of marine yeasts and other yeast-like fungi including the smuts. *Bot Mar, 23*(6), 361–373.

Naranjo-Ortiz, M. A., & Gabaldón, T. (2019). Fungal evolution: Diversity, taxonomy and phylogeny of the fungi. *Biological Reviews, 94*(6), 2101–2137. https://doi.org/10.1111/brv.12550

Nelsen, M. P., & Lumbsch, H. T. (2020). A data-driven evaluation of lichen climate change indicators in Central Europe. *Biodiversity and Conservation, 29*(14), 3959–3971. https://doi.org/10.1007/s10531-020-02057-8

Neupane, B., Malla, K., Gautam, A., Chaudhary, D., Paudel, S., Timsina, S., & Jamarkattel, N. (2017). Elevational trends in Usnic acid concentration of lichen Parmelia flexilis in relation to temperature and precipitation. *Climate, 5*(2), 40. https://doi.org/10.3390/cli5020040

Norros, V., Halme, P., Norberg, A., & Ovaskainen, O. (2023). Spore production monitoring reveals contrasting seasonal strategies and a trade-off between spore size and number in wood-inhabiting fungi. *Functional Ecology, 37*(3), 551–563. https://doi.org/10.1111/1365-2435.14254

Park, E., & Poulin, R. (2021). Revisiting the phylogeny of microsporidia. *International Journal for Parasitology, 51*(10), 855–864. https://doi.org/10.1016/j.ijpara.2021.02.005

Prank, M., Kenaley, S. C., Bergstrom, G. C., Acevedo, M., & Mahowald, N. M. (2019). Climate change impacts the spread potential of wheat stem rust, a significant crop disease. *Environmental Research Letters, 14*(12), 124053. https://doi.org/10.1088/1748-9326/ab57de

Rodríguez, M. C. H., Evans, H. C., de Abreu, L. M., de Macedo, D. M., Ndacnou, M. K., Bekele, K. B., & Barreto, R. W. (2021). New species and records of Trichoderma isolated as mycoparasites and endophytes from cultivated and wild coffee in Africa. *Scientific Reports, 11*, 5671. https://doi.org/10.1038/s41598-021-84111-1. Licensed under CC BY 4.0.

Ronikier, A., & Lado, C. (2015). Nivicolous Stemonitales from the Austral Andes: Analysis of morphological variability, distribution and phenology as a first step toward testing the large-scale coherence of species and biogeographical properties. *Mycologia, 107*(2), 258–283. https://doi.org/10.3852/14-164

Saccardo, P. A. (1894). *Chromotaxia seu Nomenclator Colorum: Tentamen synopticum triplici lingua, latino, gallico, anglico conscriptum.* Typis Seminarii.

Sahu, N., Singh, S. N., Singh, P., Mishra, S., Karakoti, N., Bajpai, R., Behera, S. K., Nayaka, S., & Upreti, D. K. (2019). Microclimatic variations and their effects on photosynthetic efficiencies and lichen species distribution along elevational gradients in Garhwal Himalayas. *Biodiversity and Conservation, 28*(8–9), 1953–1976. https://doi.org/10.1007/s10531-019-01782-z

Salinger, M. J. (2005). Climate variability and change: Past, present and future—An overview. *Climatic Change, 70*(1–2), 9–29. https://doi.org/10.1007/s10584-005-5936-x

Schneider Von Deimling, T., Meinshausen, M., Levermann, A., Huber, V., Frieler, K., Lawrence, D. M., & Brovkin, V. (2012). Estimating the near-surface permafrost-carbon feedback on global warming. *Biogeosciences, 9*(2), 649–665. https://doi.org/10.5194/bg-9-649-2012

Schüβler, A., Schwarzott, D., & Walker, C. (2001) A new fungal phylum the Glomeromycota: phylogeny and evolution. *Mycological Research, 105*(12), 1413–1421. https://doi.org/10.1017/S0953756201005196

Sirami, C., Caplat, P., Popy, S., Clamens, A., Arlettaz, R., Jiguet, F., Brotons, L., & Martin, J. (2017). Impacts of global change on species distributions: Obstacles and solutions to integrate climate and land use. *Global Ecology and Biogeography, 26*(4), 385–394. https://doi.org/10.1111/geb.12555

Spatafora, J. W., Aime, M. C., Grigoriev, I. V., Martin, F., Stajich, J. E., & Blackwell, M. (2017). The fungal tree of life: From molecular systematics to genome-scale phylogenies. *Microbiology Spectrum, 5*(5), 5.5.03. https://doi.org/10.1128/microbiolspec.FUNK-0053-2016

Steenkamp, E. T., Wright, J., & Baldauf, S. L. (2006). The protistan origins of animals and fungi. *Molecular Biology and Evolution, 23*(1), 93–106. https://doi.org/10.1093/molbev/msj011

Suryanarayanan, T. S., & Shaanker, R. U. (2021). Can fungal endophytes fast-track plant adaptations to climate change? *Fungal Ecology, 50*, 101039. https://doi.org/10.1016/j.funeco.2021.101039

Tedersoo, L., Sánchez-Ramírez, S., Kõljalg, U., Bahram, M., Döring, M., Schigel, D., May, T., Ryberg, M., & Abarenkov, K. (2018). High-level classification of the fungi and a tool for evolutionary ecological analyses. *Fungal Diversity, 90*(1), 135–159. https://doi.org/10.1007/s13225-018-0401-0

Thongbai, B., Miller, S. L., Stadler, M., Wittstein, K., Hyde, K. D., Lumyong, S., & Raspé, O. (2017). Study of three interesting amanita species from Thailand: Morphology, multiple-gene phylogeny and toxin analysis. *PLoS One, 12*(8), e0182131. https://doi.org/10.1371/journal.pone.0182131

Toda, M., Williams, S., Jackson, B. R., Wurster, S., Serpa, J. A., Nigo, M., Grimes, C. Z., Atmar, R. L., Chiller, T. M., Ostrosky-Zeichner, L., & Kontoyiannis, D. P. (2023). Invasive Mold infections following hurricane Harvey—Houston, Texas. *Open Forum Infectious Diseases, 10*(3), ofad093. https://doi.org/10.1093/ofid/ofad093

Tong, D. Q., Wang, J. X. L., Gill, T. E., Lei, H., & Wang, B. (2017). Intensified dust storm activity and valley fever infection in the southwestern United States. *Geophysical Research Letters, 44*(9), 4304–4312. https://doi.org/10.1002/2017GL073524

Trenberth, K. (2011). Changes in precipitation with climate change. *Climate Research, 47*(1), 123–138. https://doi.org/10.3354/cr00953

Trenberth, K. E., & Fasullo, J. T. (2009). Global warming due to increasing absorbed solar radiation. *Geophysical Research Letters, 36*(7), 2009GL037527. https://doi.org/10.1029/2009GL037527

Uhrlaß, S., Verma, S. B., Gräser, Y., Rezaei-Matehkolaei, A., Hatami, M., Schaller, M., & Nenoff, P. (2022). Trichophyton indotineae—An emerging pathogen causing recalcitrant Dermatophytoses in India and worldwide—A multidimensional perspective. *Journal of Fungi, 8*(7), 757. https://doi.org/10.3390/jof8070757

van Rooij P. (2015). *Batrachochytrium species, morphology in culture [photograph]*. Wikimedia Commons. https://commons.wikimedia.org/wiki/File:Batrachochytrium_species,_morphology_in_culture.jpg. Licensed under CC BY 4.0.

Váry, Z., Mullins, E., McElwain, J. C., & Doohan, F. M. (2015). The severity of wheat diseases increases when plants and pathogens are acclimatized to elevated carbon dioxide. *Global Change Biology, 21*(7), 2661–2669. https://doi.org/10.1111/gcb.12899

Větrovský, T., Kohout, P., Kopecký, M., Machac, A., Man, M., Bahnmann, B. D., Brabcová, V., Choi, J., Meszárošová, L., Human, Z. R., Lepinay, C., Lladó, S., López-Mondéjar, R., Martinović, T., Mašínová, T., Morais, D., Navrátilová, D., Odriozola, I., Štursová, M., et al. (2019). A meta-analysis of global fungal distribution reveals climate-driven patterns. *Nature Communications, 10*(1), 5142. https://doi.org/10.1038/s41467-019-13164-8

Vitasse, Y., Ursenbacher, S., Klein, G., Bohnenstengel, T., Chittaro, Y., Delestrade, A., Monnerat, C., Rebetez, M., Rixen, C., Strebel, N., Schmidt, B. R., Wipf, S., Wohlgemuth, T., Yoccoz, N. G., & Lenoir, J. (2021). Phenological and elevational shifts of plants, animals and fungi under climate change in the European Alps. *Biological Reviews, 96*(5), 1816–1835. https://doi.org/10.1111/brv.12727

Voigt, K., James, T. Y., Kirk, P. M., Santiago, A. L. C. M. D. A., Waldman, B., Griffith, G. W., Fu, M., Radek, R., Strassert, J. F. H., Wurzbacher, C., Jerônimo, G. H., Simmons, D. R., Seto, K., Gentekaki, E., Hurdeal, V. G., Hyde, K. D., Nguyen, T. T. T., & Lee, H. B. (2021). Early-

diverging fungal phyla: taxonomy species concept ecology distribution anthropogenic impact and novel phylogenetic proposals Abstract. *Fungal Diversity, 109*(1), 59–98. https://doi.org/10.1007/s13225-021-00480-y
Vogt-Schilb, H., Richard, F., Malaval, J.-C., Rapior, S., Fons, F., Bourgade, V., Schatz, B., Buentgen, U., & Moreau, P.-A. (2022). Climate-induced long-term changes in the phenology of Mediterranean fungi. *Fungal Ecology, 60*, 101166. https://doi.org/10.1016/j.funeco.2022.101166
Walker, L. M., & Stephenson, S. L. (2016). The species problem in Myxomycetes revisited. *Protist, 167*(4), 319–338. https://doi.org/10.1016/j.protis.2016.05.003
Wang, C., Wang, X., Jin, Z., Müller, C., Pugh, T. A. M., Chen, A., Wang, T., Huang, L., Zhang, Y., Li, L. X. Z., & Piao, S. (2021). Occurrence of crop pests and diseases has largely increased in China since 1970. *Nature Food, 3*(1), 57–65. https://doi.org/10.1038/s43016-021-00428-0
Wijayawardene, N., Hyde, K., Dai, D., Sánchez-García, M., Goto, B., Saxena, R., Erdoğdu, M., Selçuk, F., Rajeshkumar, K., Aptroot, A., Błaszkowski, J., Boonyuen, N., da Silva, G., de Souza, F., Dong, W., Ertz, D., Haelewaters, D., Jones, E., Karunarathna, S., ... Thines, M. (2022). Outline of Fungi and fungus-like taxa – 2021. *Mycosphere, 13*(1), 53–453. https://doi.org/10.5943/mycosphere/13/1/2
Wijayawardene, N. N., Hyde, K. D., Mikhailov, K. V., Péter, G., Aptroot, A., Pires-Zottarelli, C. L. A., Goto, B. T., Tokarev, Y. S., Haelewaters, D., Karunarathna, S. C., Kirk, P. M., de A. Santiago A. L. C. M., Saxena, R. K., Schoutteten, N., Wimalasena, M. K., Aleoshin, V. V., Al-Hatmi, A. M., Ariyawansa, K. G. S. U., Assunção, A. R., ... Karpov, S. A. (2024). Classes and phyla of the kingdom Fungi. *Fungal Diversity, 128*(1), 1–165. https://doi.org/10.1007/s13225-024-00540-z
Williams, J. W., Ordonez, A., & Svenning, J.-C. (2020). A unifying framework for studying and managing climate-driven rates of ecological change. *Nature Ecology & Evolution, 5*(1), 17–26. https://doi.org/10.1038/s41559-020-01344-5
Williams, S. L., Toda, M., Chiller, T., Brunkard, J. M., & Litvintseva, A. P. (2024). Effects of climate change on fungal infections. *PLoS Pathogens, 20*(5), e1012219. https://doi.org/10.1371/journal.ppat.1012219
Xu, Z., Tang, Y., Connor, T., Li, D., Li, Y., & Liu, J. (2017). Climate variability and trends at a national scale. *Scientific Reports, 7*(1), 3258. https://doi.org/10.1038/s41598-017-03297-5
Yang, Z. L. (2011). Molecular techniques revolutionize knowledge of basidiomycete evolution. *Fungal Diversity, 50*(1), 47–58. https://doi.org/10.1007/s13225-011-0121-1
Zuo, Y.-B., Han, D.-Y., Wang, Y.-Y., Yang, Q.-X., Ren, Q., Liu, X.-Z., & Wei, X.-L. (2023). Fungal–algal association drives Lichens' mutualistic symbiosis: a case study with Trebouxia-related lichens. *Plants, 12*(17), 3172. https://doi.org/10.3390/plants12173172

Chapter 16
Fungal Diversity in Rural and Urban Plant Landscapes

John Mark De Jesus and Mark Angelo O. Balendres

Abstract Fungi are among Earth's most intriguing and diverse life forms; they are an important food source and support biodiversity and ecological balance. This chapter discusses fungal diversity in rural and urban plant landscapes, methods of assessment, and their roles in various ecological facets. Fungi play diverse roles in plant landscapes, from supporting plant growth to causing plant diseases. Fungal diversity in rural plant landscapes is affected by both plant diversity and landscape context. Fungi in urban environments play an important role in maintaining ecological balance by promoting nutrient cycling, soil health, and plant growth through symbiotic partnerships with urban plants. Urbanization substantially impacts microbial populations, affecting plant health and ecological processes. Although some urban microbes have evolved to digest toxins or improve stress tolerance, microbial diversity is generally lost. Fungi play critical roles in nutrient cycling, soil health, and plant resilience; thus, protecting them necessitates deliberate measures and collaborative efforts. Future research in plant-fungal diversity holds immense potential to address global challenges and unlock ecological and agricultural benefits. The future of fungi research holds incredible potential for innovation, offering solutions to pressing issues like food security, climate change, and biodiversity loss. Safeguarding fungal diversity is key to building a sustainable and resilient ecosystem in the future.

Keywords Endophytic fungi · Mycorrhiza · Plant pathogenic fungi · Urbanization

J. M. De Jesus · M. A. O. Balendres (✉)
Department of Biology, College of Science, De La Salle University, Manila, Philippines
e-mail: mark.angelo.balendres@dlsu.edu.ph

J. J. G. Guerrero et al. (eds.), *Mycology in a Changing Planet: Applications and Perspectives from Southeast Asia*, https://doi.org/10.1007/978-981-95-2640-6_16

Introduction

Among Earth's most intriguing and diverse life forms are fungi, which are essential to preserving ecosystems' functionality and balance (Naranjo-Ortiz & Gabaldón, 2019). They are incredibly diverse, with many species that can flourish in nearly any habitat (Frąc et al., 2018). In biological taxonomy, these species are so distinct from other microbes, plants, and animals that they require their kingdom (Paterson et al., 2023). Fungi are special because they are eukaryotic, meaning their cells have distinct nuclei and are organized like other complex life forms (De Paz et al., 2023; Naranjo-Ortiz & Gabaldón, 2019). Despite this commonality, fungi are remarkably diverse in both form and function. On one end of the spectrum are minute organisms essential to processes like fermentation, like yeast, yet undetectable to human sight. Conversely, they include more recognizable and bigger species, such as mushrooms, which have come to represent the fungal kingdom (Buckley, 2008).

Fungi are ecological powerhouses that are more than just organisms. They play a crucial role in nature, from breaking down organic matter to developing symbiotic connections with plants. Because of their diversity and complexity, they are a fundamental component of life on Earth and a never-ending source of fascination for researchers. Unlike plants, fungi are heterotrophic, which means they cannot produce food. Instead, they take up nutrients from their environment by frequently breaking down organic materials. Unlike the cellulose in plant cell walls, their cell walls are composed of chitin. Fungi are essential microorganisms that influence global ecosystems (De Paz et al., 2023; Van Bael, 2020; Schmit & Mueller, 2007; Buckley, 2008). Ecosystem functions include decomposition, nutrient cycling, symbiosis with plants, food, medicine, bioremediation, industrial applications, ecological balance, and biodiversity (Iqbal et al., 2023; De Paz et al., 2023; Zanne et al., 2019; Frąc et al., 2018; Buckley, 2008).

Furthermore, fungi are an important food source (Buckley, 2008), providing humans with nutritionally dense substances such as mushrooms, which are known for their culinary and health benefits. Fungi have changed medicine (Buckley, 2008) by discovering life-saving medications such as antibiotics, immunosuppressants, and cholesterol-lowering compounds. The symbiotic interactions of fungi and plants (Buckley, 2008), notably mycorrhizal associations, demonstrate their ecological significance (Van Bael, 2020). These mutualistic connections allow plants to absorb critical nutrients and water from the soil efficiently. In contrast, fungi get their energy from the carbohydrates their host plants produce. This dynamic not only promotes plant growth but also strengthens entire ecosystems. Their involvement in bioremediation (Iqbal et al., 2023; De Paz et al., 2023; Buckley, 2008) emphasizes their environmental importance. Particular fungal species such as *Aspergillus niger* and *Pleurotus ostreatus* (Dinakarkumar et al., 2024) can break down and detoxify toxic contaminants such as heavy metals and hydrocarbons, helping to restore the environment. Beyond these applications, fungi have become vital in industrial processes such as manufacturing enzymes, biofuels, and fermented products, demonstrating their versatility in biotechnological breakthroughs.

The number of fungal species has long been debated and remains so today. This is most likely owing to the non-standardized criteria established by academics and mycologists and the estimating methodologies and technologies (Van Bael, 2020). Many publications have now continued to refer to David Hawksworth's accounts. In 1991, Hawksworth estimated that 1.5 million fungal species existed in nature. This is even though this figure has already increased since the introduction of modern DNA sequencing methods, which dramatically revolutionized studies on fungal taxonomy and diversity (Mikryukov et al., 2023; De Paz et al., 2023; Van Bael, 2020; Wu et al., 2019; Money, 2015). These strategies have resulted in revised estimates of fungal species richness, which range from 2.2 to 3.8 million species (Hawksworth & Lücking, 2017) based on host associations and have increased to 11.7–13.2 million species using high-throughput sequencing approaches (Hyde, 2022; Wu et al., 2019). Furthermore, while this technique has improved the number of fungal species estimations, it has introduced new challenges to fungal taxonomy. Obscure molecular phylogenies have been identified (Wu et al., 2019) since using this new approach. As a result, non-culture-dependent investigations in fungus yield more diversified findings. This led to a recent proposal for formally classifying environmental sequences and using DNA sequence data as the basis for fungal nomenclature. However, the mycological community strongly opposed this endeavor (Wu et al., 2019). With the continuing advancement of technology and methodologies for understanding fungal variety, one thing is sure: while numbers change, the diversity of fungi can be substantially greater than previously documented (Hyde, 2022; Van Bael, 2020; Wu et al., 2019).

Fungi support biodiversity and ecological balance. By supporting intricate food webs and offering habitats for microorganisms, they guarantee ecosystem resilience in the face of environmental change. These roles demonstrate the remarkable significance of fungi in supporting life on Earth and advancing scientific understanding in a wide range of disciplines (Zanne et al., 2019). Life on Earth is shaped by human ingenuity and natural ecosystems, which offer countless chances for research and proper development. This chapter discusses fungal diversity in rural and urban plant landscapes, methods of assessment, and their roles in various ecological facets. Specifically, this chapter compares fungi found in plant and rural landscapes, factors influencing fungal diversity, and the evolving approach to fungal identification and detection.

Types of Fungi in Plant Landscapes

Of all the varieties of fungi, plant-associated fungi are the most common and widely explored, followed by human-associated fungi (Konopka et al., 2019). Fungi play diverse roles in plant landscapes, from supporting plant growth to causing plant diseases. Fungi in plant landscapes can be broadly categorized based on their roles and interactions with plants. This includes mycorrhizal fungi, endophytic fungi, pathogenic fungi, and saprophytic fungi (Runnel et al., 2024; Yiallouris et al., 2024).

Mycorrhizal fungi are a fascinating group that forms symbiotic relationships with plant roots, playing a crucial role in plant health and ecosystem functioning. These fungi enhance nutrient and water absorption for plants while receiving carbohydrates produced by the plants through photosynthesis (Liu et al., 2023). They are classified into two main types: ectomycorrhizal fungi and endomycorrhizal fungi (also known as arbuscular mycorrhizal fungi) (Gamalero et al., 2023). The former grows on the surface of plant roots, forming a sheath around them. They are commonly associated with trees and help in nutrient exchange through the Hartig net (network of inward-growing hyphae) (Watkinson, 2015). The latter penetrates the root cells of plants, forming structures like arbuscules within the cells. They are widespread and found in most plant species, including crops (Liu et al., 2023). Mycorrhizal fungi are essential for soil health, plant growth, and resilience against environmental stressors. They also contribute to carbon cycling and biodiversity in ecosystems (Runnel et al., 2024).

Endophytic fungi are a unique group of fungi that live within the tissues of plants without causing any apparent harm. These fungi are found in nearly all plant species and are ubiquitous in terrestrial ecosystems. Their presence is often beneficial to the host plant, as they can enhance growth, improve stress tolerance, and provide protection against pests and pathogens (Afzal et al., 2019). Endophytic fungi share common characteristics such as (1) host specificity: many endophytes are specific to their plant host while others can thrive in a broader host range; (2) mode of transmission: endophytic fungi are commonly passed down from parent plants to their offspring through seeds or by vegetative propagation methods; (3) metabolic product diversity: these fungi are capable of producing a diverse range of secondary metabolites, possessing natural defense mechanism for the plant and for purposes such as antifungal, antibacterial, and even anticancer properties; lastly (4) colonization: proliferation and expansion of endophyte populations within the tissues of the host plant (Aamir et al., 2020).

Pathogenic fungi cause diseases in plants, animals, or humans. These fungi, mainly from the phyla Ascomycota and Basidiomycota, utilize various strategies to infect and colonize plant tissues, causing various symptoms (Rossman et al., 2016). These organisms can range from microscopic yeasts to larger molds and are responsible for various infections and diseases. Some key characteristics of pathogenic fungi are their opportunistic nature, primarily they infect hosts with weakened immune systems or compromised defenses; host diversity: some of them can affect a wide range of hosts, including humans, animals, and plants; and adaptability: they thrive in various environments and can adapt to different conditions, making them challenging to control (Konopka et al., 2019).

Pathogenic fungi pose significant challenges in agriculture, healthcare, and ecosystems. In humans, they can lead to life-threatening infections, especially in individuals with weakened immunity. In plants, fungal pathogens can devastate crops, leading to economic losses and food insecurity (Strange & Scott, 2005; Savary et al., 2019). Plant-pathogenic fungi can be classified into three categories depending on their mode of nutrient acquisition. They can be classified as biotrophs, necrotrophs, and hemi-biotrophs (Priyashantha et al., 2023; Glazebrook, 2005). Biotrophs derive their nutrients and energy from living plant cells and survive within the interstitial space of the cells (Fei & Liu, 2022). Necrotrophic fungi are plant pathogens that kill host cells and then feed on the nutrients from the dead tissue, using enzymes, proteins, or toxins to achieve this (De Cal et al., 2022; Mapuranga et al., 2022; Dickman, 2007). Lastly, hemi-biotroph fungi begin by forming a biotrophic relationship with their host and eventually cause the host cells to die as the infection progresses (De Silva, 2016; Latijnhouwers et al., 2003).

Plant-associated saprophytic fungi are a group of fungi that thrive by decomposing dead organic matter, particularly plant material. These fungi play a vital role in nutrient cycling and soil health by breaking down complex organic compounds like cellulose and lignin into simpler forms that plants and other organisms can utilize (Niego et al., 2022). Some examples of plant-associated saprophytic fungi include mushrooms such as *Agaricus* species, which grow on decaying wood and leaf litter, contributing to forest ecosystem health (Tuomela & Hatakka, 2011). Molds like *Penicillium* and *Aspergillus* species are common saprophytes on rotting fruits, vegetables, and other plant matter (Pitt, 2014). Further, some yeast species are saprophytic, too, feeding on sugars in decaying plant material (Konopka et al., 2019; Volk, 2013). Finally, *Mucor* species, which are often found in decomposing organic matter, are known for their role in breaking down carbohydrates (Konopka et al., 2019). They are primary decomposers, breaking down dead plant material and recycling nutrients into the soil. Decomposing organic matter enriches the soil with essential nutrients, promoting plant growth. Saprophytic fungi contribute to the global carbon cycle by converting organic carbon into forms that plants can reabsorb (Reverchon et al., 2010). These fungi are indispensable for maintaining the balance of ecosystems, particularly in plant landscapes.

Plant-associated fungi are a vital component of ecosystems, playing a range of roles that directly influence plant landscapes and broader environmental health. These fungi act as allies to plants by promoting growth through symbiotic relationships. At the same time, some fungi take on a more destructive role as pathogens, causing plant illnesses and contributing to agricultural losses. Meanwhile, saprophytic fungi work behind the scenes, breaking down dead organic material to recycle nutrients into the soil, enriching it for future plant growth. These diverse fungi create a dynamic balance, shaping the health and functionality of both natural ecosystems and agricultural systems. Their interconnected roles highlight their indispensable contribution to maintaining ecological stability and supporting life on Earth.

Fungal Diversity in Rural Plant Landscapes

Fungal diversity in rural plant landscapes is affected by both plant diversity and landscape context. Fungi are highly diverse, with over a million species, including mycorrhiza, endophytes, pathogenic, and saprophytes. These fungal communities are important in soil health, nitrogen cycling, and plant disease suppression. Many serve as natural biocontrol agents, successfully inhibiting or eradicating fungal infections. This biocontrol function exists in individual fungal species and complex fungal communities linked with plants. Furthermore, a greater variety of plant species results in a more diversified fungal population belowground, which improves rhizosphere soil health and suppresses dangerous diseases (Bollmann-Giolai et al., 2022).

Shen et al. (2021) demonstrated the important relationship between plant variety and soil fungi diversity in maintaining ecological stability. In tests with variable plant diversity and invasive species presence, more plant diversity was observed to enhance soil fungal diversity and microbial resilience. Non-invasive conditions resulted in a strong positive correlation between plant and fungal diversity. In contrast, invasive conditions strengthened some relationships while weakening others. Processes such as species turnover (in bacteria) and species loss (in fungi) influenced these dynamics. Overall, the findings highlight the relevance of plant diversity in maintaining soil microbial health and ecological stability, particularly in the face of difficulties such as plant invasions.

In soil, several fungi create tight symbiotic interactions with plant roots, resulting in a durable mycorrhizal network. Fungal symbionts in legumes are important in nitrogen absorption and nutrient cycling within the plant-soil-fungal system. On the Loess Plateau, these ecological processes demonstrate the close relationship between fungus diversity, soil characteristics, and plant health (Yang et al., 2017). In addition, higher nutrient cycling caused by plant litter and productivity has resulted in more fungal diversity. This comprises two primary mechanisms: direct mutualistic connections between plants and root-associated fungi and indirect effects caused by changed nutrient delivery by free-living fungi (Yang et al., 2017).

Soil fungi are crucial in forest ecosystems' nitrogen cycle and carbon storage. However, the impact of elevation on fungal communities has not been well studied. Research in a subtropical forest in the Qinling Mountains (elevation 1500–1900 m) indicated that soil pH and organic carbon rose with elevation, whereas accessible phosphorus dropped. The most numerous fungal taxa were Ascomycota and Basidiomycota. However, overall fungal diversity, richness, and community structure dropped drastically with elevation. Soil pH, organic carbon, moisture, and phosphorus levels were important factors in determining fungal diversity. These findings demonstrate how elevation-related changes in soil conditions influence fungal diversity in these woodlands (Zhou et al., 2021).

In another study, Reininger et al. (2015) examined the various fungal communities and their symbiotic connections with the soil biota of abundant and relatively rare plants, both native and foreign. The results revealed that native plant species with higher abundance had higher fungal richness than comparatively rare native species. However, this pattern was not seen in foreign species of varied abundance. Among exotic plants, the most abundant species had the lowest levels of mycorrhizal interaction. Plants, other organisms, and non-living environmental elements such as soil pH, moisture, salinity, structure, and temperature influence fungal diversity and activity. These interactions influence fungal dynamics (Frąc et al., 2018; López-Bucio et al., 2015; Rouphael et al., 2015).

The study by Liu et al. (2023) sheds light on how geological and environmental factors influence the ecological diversity of fungal communities. Fungal communities play a crucial role in energy metabolism and material cycling within soil-plant ecosystems, and they uniquely adapt to different climatic conditions. Tedersoo et al. (2020) concluded a thorough regional investigation identifying pH and plant species as significant determinants of soil fungus diversity in Northern Europe. Soil pH is the primary driver of fungal variety, with direct and indirect influences from tree species, nutrients, and mold abundance. Tree species such as *Pinus sylvestris* and ectomycorrhizal plants had a greater impact on biotrophic fungi than saprotrophic fungi. Fungal diversity was highest in woodlands and near ruins, lowest in bogs, and surprisingly high on forest islands due to edge effects. Management measures such as thinning forests lowered fungal variety, but tree diversity increased soil characteristics and ecological interactions.

Rural locations frequently have a greater diversity of fungal species than metropolitan settings. This mismatch originates from the largely unaltered nature of rural ecosystems, where human activity is less intense and natural equilibrium is retained. The abundance of varied habitats in rural environments, such as forests, fields, wetlands, and undisturbed soil, gives perfect circumstances for various fungus species and communities. The complex interplay of resources, such as organic matter from plant decay, diverse vegetation, and soil nutrients, fosters an environment where fungi can thrive in various ecological roles. For example, saprophytic fungi efficiently decompose organic material, recycling nutrients into the soil. In contrast, mycorrhizal fungi form symbiotic relationships with plant roots, enhancing plant growth and soil health. These fungi contribute significantly to nutrient cycling, carbon sequestration, and the overall health of ecosystems.

Furthermore, rural areas are less likely to face disruptions such as pollution, land modification, and deforestation than metropolitan areas. These disruptions can destabilize fungal populations and diminish variety. Thus, rural habitats provide stability and continuity, allowing fungi to form and maintain flourishing populations. The greater fungal diversity found in rural regions emphasizes their relevance in sustaining ecological processes and biodiversity. Fungi in these areas perform important roles in the food chain, contributing to the resilience and sustainability of natural systems. Understanding the dynamics of fungal diversity in rural areas is critical for conservation efforts and encouraging sustainable land-use practices. Plant diversity and habitat fragmentation increase fungi diversity (Raimbault et al., 2024; Su et al., 2022). The pH and plant species are also significant determinants of soil fungus diversity (Tedersoo et al., 2020.

Fungal Diversity in Urban Plant Landscapes

Fungi in urban environments play an important role in maintaining ecological balance by promoting nutrient cycling, soil health, and plant growth through symbiotic partnerships with urban plants. Despite these benefits, urbanization frequently alters fungal diversity and community structures through pollution, habitat fragmentation, changed vegetation, and other consequences (Gómez-Hernández et al., 2021; Newbound et al., 2010). Many mycologists believe habitat destruction and alteration threaten fungal preservation (Abrego et al., 2020; Newbound et al., 2010).

Urban parks and green areas provide habitat for various fungal communities, including ectomycorrhizal fungi that promote tree health and nutrient uptake (Stewart et al., 2023). In certain situations, fungal communities in urban parks mirror those in natural forests, influenced by park age, vegetation, and soil conditions (Hui et al., 2017). Green roofs can serve as important repositories of fungal diversity, increasing urban biodiversity and ecological services (Stewart et al., 2023; Authier et al., 2022).

Green roofs have a significant influence on fungal communities, increasing diversity as compared to ground-level environments. Fungal composition varies depending on plant cover, roof age, and design, while adjoining roofs and ground spaces share several characteristics (Droz et al., 2022). Green roofs are important for urban ecological planning because they serve as biodiversity reservoirs (Cristiano et al., 2020). However, urbanization might damage fungal networks by diminishing their specialization and interconnection (Zheng et al., 2025; Christel et al., 2023). Sustainable urban planning and conservation activities are critical for preserving fungal variety and its contributions to urban ecosystems (Christel et al., 2023). Gómez-Hernández et al. (2021) found that urbanization reduces macromycetes' diversity and functional roles by modifying microclimatic conditions, lowering niche occupation, and leaving resources underutilized. While less urbanized areas have better functional performance, rising urbanization causes a loss of species and functional diversity, thereby jeopardizing ecological stability.

Conversely, Yiallouris et al. (2024) categorically identified climate change and rising temperatures caused by urbanization as a significant threat to fungal diversity. De Oliveira et al. (2025) found that urban pollution harms fungal species richness, but diversity has remained relatively stable. This could be because the area has been investigated recently and may have changed insignificantly by the time research was undertaken. Disrupted fungal communities can impact plant growth (Purahong et al., 2022), increase susceptibility to pathogens (Frąc et al., 2018), degrade soil health (affecting mycorrhiza symbionts), and reduce environmental resilience (De Oliveira et al., 2025; Yiallouris et al., 2024; Abrego et al., 2020). A fall in fungal variety disturbs fungal activities and processes that support plant growth, such as reducing the availability of key nutrients that plants require to flourish (Bonfante & Venice, 2020). Mycorrhizal fungi, which create symbiotic associations with plant roots, help plants absorb nutrients such as nitrogen and phosphorus. The loss of these fungi impairs this collaboration, reducing plant growth and decreasing productivity (Bonfante & Venice, 2020).

Furthermore, fungi such as *Trichoderma* spp., mycorrhizas, endophytes, and yeasts frequently serve as natural defenses against dangerous pathogens (Ghorbanpour et al., 2018). These fungi can also create antibacterial chemicals, which help plants resist infections (Jakubczyk & Dussart, 2020). When fungal populations decline, these protective processes weaken. Without beneficial fungi, plants lose a vital layer of defense, making them vulnerable to disease attacks (Ghorbanpour et al., 2018). Fungal decomposers break down organic debris and supply the soil with nutrients (Bonfante & Venice, 2020; Niego et al., 2022; Frąc et al., 2018). These fungi play an important role in the nutrient cycle, ensuring that essential elements such as nitrogen, phosphorus, and carbon are returned to the soil in forms that plants can absorb. They also aid in developing soil structure by retaining moisture and preventing erosion (Niego et al., 2022). When fungal decomposer populations decrease, the natural recycling process is disrupted. Organic matter accumulates without being properly broken down, resulting in a lack of readily available nutrients. As a result, soil fertility diminishes, negatively impacting plant growth and ecosystem productivity. Over time, the inability to sustain healthy nutrient cycles can lead to soil degradation characterized by poor structure, reduced water retention, and biodiversity loss. Additionally, degraded soils are more vulnerable to erosion and contamination, further exacerbating environmental issues and harming agricultural productivity (Hossain et al., 2020).

Fungi also help to keep urban ecosystems healthy and stable. They help with nutrient cycling, soil health, plant growth, and even air quality through their ecological roles. However, when fungal communities are disrupted, their ability to sustain these processes is dramatically reduced, making urban ecosystems more sensitive to stressors such as pollution and climate change (De Oliveira et al., 2025; Mishra et al., 2024; Dinakarkumar et al., 2024). For example, certain fungi, such as *Aspergillus niger* and *Pleurotus ostreatus* (Dinakarkumar et al., 2024), are excellent for bioremediation that help break down pollutants, decreasing their negative impacts on the environment. Others create symbiotic partnerships with plants, which improve stress tolerance by facilitating nutrient uptake and increasing resistance to adverse weather conditions. These advantages are lost when these fungal communities are disrupted—whether by habitat destruction, soil contamination, or other urban impacts. As a result, plants may struggle to thrive, soil may degrade, and pollutants may accumulate, further compromising the resilience of urban ecosystems (Iqbal et al., 2023).

Healthy fungal communities can protect urban areas from climate change by storing carbon in the soil, improving water retention, and promoting vegetation that reduces urban heat islands (Pandey & Ghosh, 2023). Disruptions to these fungal networks jeopardize these vital functions, rendering cities more vulnerable to environmental stress and incapable of adjusting to future difficulties. Therefore, preserving and restoring fungal variety is critical for developing resilient, sustainable urban settings (Da Gama, 2023; Iqbal et al., 2023; Authier et al., 2022).

Together, urbanization often disrupts fungal diversity through habitat loss, pollution, and climate change, weakening the ecological functions fungi provide. However, green spaces (urban parks and roofs) can help conserve fungal diversity. However, their impact depends on several factors (e.g., proper design, vegetation, and soil management).

Comparative Analysis: Rural Versus Urban Plant Landscapes

Urbanization substantially impacts microbial populations, affecting plant health and ecological processes. Although some urban microbes have evolved to digest toxins or improve stress tolerance, microbial diversity is generally lost. Rusterholz and Baur (2023) found that urbanization has a major impact on forest soil fungus populations, affecting their ecological equilibrium. They used DNA metabarcoding to investigate 20 beech woodlands in Basel, Switzerland. They discovered that urbanization influences fungal richness, diversity, and evenness in various ways. While fungal richness was highest in low-urbanized areas, diversity increased with increasing urbanization. Symbiotic fungi dropped in urban environments, but saprotrophic and pathogenic fungi increased. These alterations illustrate the impact of urbanization on soil fungus dynamics and, perhaps, forest ecosystems.

Similarly, Lin et al. (2021) investigated how urbanization affects the relationship between plants and arbuscular mycorrhizal (AM) fungi, which are essential for plant health. Researchers studied root samples from eight common herbaceous plants in urban Beijing (Fifth Ring) and rural Jiufeng National Forest Park. They discovered 81 AM fungus species and found different trends in urban and rural locations. Urban roots had greater fungal variety and evenness. However, their fungal networks were less specialized and lacked the ordered "nested" structure in rural settings. Urban environments favored generalist fungi (e.g., *Glomus* species and *Claroideoglomus* species) over specialists (e.g., *Paraglomus* species), most likely due to diminished host plant choice. These alterations in fungal connections could impact the balance of urban ecosystems, highlighting how urbanization alters above- and below-ground interactions (Lin et al., 2021).

Gibson and Zimmerman (2023) conducted a related study on the fungal populations found on the leaves of *Metrosideros excelsa*, a common urban tree. They used advanced algorithms to identify 854 fungal sequences, 97% of which formed separate clusters, indicating a high level of diversity. The study concludes that urban elements such as heat islands, vegetation changes, and traffic density will likely impact these fungal populations. Urban plant microbiomes, like those in nature, are complex and dynamic, and understanding their interactions becomes increasingly crucial as urbanization increases.

Furthermore, Tatsumi et al. (2023) reported in a separate study that urbanization and forest edge formation are predicted to absorb more than 20% of US Forest land, posing a hazard to temperate forests. They discovered a breakdown in the mutualism between trees and ectomycorrhizal (ECM) fungi in urban forests, with fewer ECM fungi colonizing fewer tree roots and reduced connectivity in soil microbiome networks. However, urbanization did not lower the relative abundance of ECM fungus in forest soils; forest borders alone did. Urbanization and forest borders formed new "suites" of microorganisms, but urban interior woods had highly homogenized microbiomes. Edge forest microbiomes, on the other hand, were more diverse and unstable, with increased susceptibility to low soil moisture. According to the study, temperate forest areas that have been urbanized may have soil microbiome dysfunction, including increased greenhouse gas emissions.

Li et al. (2023) found that fast urbanization significantly impacts the urban-rural landscape, soil environment, and ecosystem processes. Considering this, microbes are critical to the functioning of the soil ecosystem. The researchers investigated how land use patterns, human activities, and soil conditions affected soil bacterial diversity, composition, and network structure in urban, peri-urban, and rural settings. Peri-urban areas were identified as hotspots for soil microbial diversity. The amount of forestland and the level of human activity on the land surface were the primary factors influencing the soil bacterial community. Human activity and land use may indirectly affect the variety and composition of the soil bacterial population. This research can help us better understand how urbanization patterns affect the microbial soil community and the potential implications for the soil ecosystem and human health.

Finally, Wang et al. (2020) identified urbanization as a crucial factor in determining plant diversity. They examined 134 sites along Shanghai's urban-rural gradient, identifying 526 plant species, 57.8% foreign. They discovered that urbanization had a complex impact on plant variety: whereas overall plant richness, woody plants, and perennial herbs grew, annual herbs and indices of evenness and diversity declined. Surprisingly, native flora exhibited no significant relationship to urbanization levels. These findings demonstrate how urbanization affects plant variety and can help guide conservation efforts in developing metropolitan areas.

Urbanization has a significant impact on microbial populations, which therefore influences plant health and ecological processes. According to studies, urbanization alters soil and plant-associated fungal communities, highlighting complex changes across different environments: soil fungi and forests, mycorrhizal fungi and urban plants, fungal communities on urban trees, forest soil microbiomes, mangrove microbial shifts, and plant diversity patterns. These findings highlight how urbanization affects microbial diversity and complex connections between plants, fungi, and ecosystems. These insights are critical for creating conservation efforts in urban areas.

Methods of Studying Fungal Diversity

Identifying fungi requires a grasp of their physical characteristics, genetic makeup, and ecological roles, making it a critical component of mycology (Haque, 2018). Researchers can recognize various species by looking at physical traits such as the form, size, and structure of fungal spores and fruiting bodies. Phylogenetic analysis goes further, using genetic data to trace evolutionary links and classify fungi better. Meanwhile, ecological research illuminates how fungi interact with their surroundings, showing their roles in the nutrient cycle, symbiosis, and pathogenicity (Naranjo-Ortiz & Gabaldón, 2019). These approaches work together to provide a well-rounded perspective, ensuring that fungal identification is accurate and significant in understanding their variety and function within ecosystems (Corbu et al., 2023).

Researchers employ traditional and molecular approaches to understand the complexities of fungal communities, each providing unique viewpoints. Classical methods entail examining physical attributes and cultivating fungi based on conventional methodologies that reveals morphological and ecological properties. Molecular approaches, on the other hand, take advantage of genetic analysis to provide in-depth insights into fungal taxonomy, phylogenetic relationships, and community dynamics.

Fungal identification using traditional methods often entails crucial procedures that include both environmental and cultured observations. Incubating substrate in wet chambers is one typical strategy, as it produces a controlled environment that promotes fungal development and makes it easier to examine their properties. This method is beneficial for studying myxomycetes and other fungi since incubation promotes the production and quantity of fruiting bodies (Krug, 2004). Another common method is direct sampling of fungal fruiting bodies, in which researchers collect and evaluate visible structures like mushrooms or other reproductive forms observed in nature (Sakamoto, 2018). Culturing endophytes is an important tool for investigating fungi that live within plant tissues, providing insights into their intricate roles and interactions (Mwamburi, 2022). Another essential technique is particle plating, which involves depositing soil, plant debris, or other substrate particles onto a nutrient-rich medium to promote fungal growth and identify species (Kirby et al., 1990).

Molecular approaches have radically altered the study of fungal diversity by allowing researchers to analyze fungi at the genetic level. This is especially useful because many fungi cannot be easily cultivated in the laboratory. Traditional techniques of studying fungi focus on examining physical characteristics or growing them in artificial conditions. However, these methods have severe limitations—many fungal species are either too slow-growing, have complex or specialized nutrient requirements, or are completely unculturable. Researchers can circumvent these barriers by focusing on molecular approaches and gaining direct access to fungi's genetic material. These techniques reveal the identities of well-studied fungi and those previously unknown, such as cryptic species with little or no obvious distinction from others. These techniques include DNA barcoding, which is based on sequencing several barcoding gene regions that include internal transcribed spacer (ITS) sequencing. Sequencing approaches include high-throughput amplicon sequencing and other advanced sequencing methods such as Oxford nanopore technology (ONT) sequencing, multi-locus sequence typing (MLST), and whole genome sequencing (WGS).

The Internal Transcribed Spacer (ITS) region is a genetic "barcode" for identifying fungal species. It varies greatly between species but has conserved areas, making it useful for distinction. Researchers amplify and sequence the ITS region of fungal DNA, then compare the sequence to reference databases to precisely classify fungi, including species that are difficult to culture. This method has proven valuable for cataloging fungi and contributes to investigating some of their potential ecological roles.

Multi-locus sequence typing (MLST) is another technique for identifying and distinguishing fungal strains based on sequences from multiple housekeeping genes, for example, the ITS region, translation elongation factor 1-alpha, β-tubulin, and calmodulin. By comparing allele variants, MLST assigns a distinct profile to each strain, aiding in studying fungal diversity, population genetics, and evolutionary relationships. It is very valuable for identifying pathogenic strains and studying fungal populations.

High-throughput sequencing (HTS) or targeted amplicon sequencing is an essential tool for investigating fungal diversity. It allows researchers to analyze complete fungal communities from environmental samples, such as soil, plants, or air, without cultivating them. The HTS is highly detailed, finding rare species and uncovering hidden diversity. It also explains how fungi interact in ecosystems and highlights their roles (Frąc et al., 2018), such as nutrient cycling or disease transmission.

Whole genome sequencing (WGS) allows accurate species identification, uncovers phylogenetic relationships, and helps in assessing genetic and functional diversity in fungi (Baldrian et al., 2022; Cuadros-Orellana et al., 2013). It identifies functional genes, researches population dynamics, and investigates nutritional cycling and pathogenicity features (Amezrou et al., 2024). This comprehensive approach has improved our understanding of fungal diversity and ecological roles. Oxford Nanopore Technology (ONT) sequencing is a WGS approach for researching fungal diversity. It can sequence lengthy DNA segments, allowing for accurate species identification and classification. It operates in real-time, allowing for quick analysis, and is portable, making it perfect for fieldwork. ONT is also utilized to examine fungal communities by metabarcoding and for cost-effective genome assembly, which provides insights into their genetics and evolution. This method makes studying fungi more efficient and adaptable to many conditions.

Omics-Based Approaches

Omics-based techniques are changing the study of fungal diversity by giving detailed information about fungal genetics, functions, and interactions (Lofgren & Stajich, 2021). Fungal genome sequencing aims to identify species, explain evolutionary relationships, and investigate genetic features such as pathogenicity and symbiosis (Verma et al., 2023; Naranjo-Ortiz & Gabaldón, 2019). Transcriptomics analyzes RNA to identify gene expression patterns, which helps researchers understand how fungi adapt to environmental changes or interact with hosts (Kumar et al., 2023; Verma et al., 2023; Buckley, 2008). Proteomics focuses on fungal proteins, providing insights into their functional significance to the nutrition cycle and stress tolerance (Verma et al., 2023). Metabolomics examines the chemical compounds produced by fungi, such as secondary metabolites, which can have ecological or pharmaceutical significance (Kumar et al., 2023; Verma et al., 2023). This technique analyzes fungal DNA directly from environmental samples, uncovering entire fungal communities and their ecological roles (Verma et al., 2023; Lofgren &

Stajich, 2021; Buckley, 2008). By studying RNA from environmental samples, metatranscriptomics provides a dynamic view of fungal activity and interactions within ecosystems (Kumar et al., 2023; Verma et al., 2023). It also integrates multiple omics approaches, enabling a more comprehensive and holistic understanding of fungal diversity and function (Kumar et al., 2023; Verma et al., 2023). In general, molecular approaches provide new insight into fungal diversity, evolution, and ecological relevance. They are critical for comprehending the entire range of fungal life, including species that would otherwise remain unknown. Nevertheless, the classical approach (cultural, morphological, and bioassays) should still be used with molecular approaches in fungal identification.

Ecological and Functional Implications

Fungi are essential in plant landscapes, providing diverse ecosystem services that support plant health and ecological balance. Saprotrophic fungi are nature's recyclers, decomposing complex organic materials, such as dead plants, animals, and wood. This decomposition releases important nutrients (Niego et al., 2022; Crowther et al., 2012a, b). Saprotrophic fungi work to decompose organic items such as fallen leaves, dead animals, and decaying timber. They break down this waste into simpler components, releasing vital nutrients like nitrogen and phosphorus into the soil. These nutrients are natural fertilizers, increasing soil fertility and promoting healthy plant growth (Crowther et al., 2012a). However, their activity does not end there; throughout decomposition, saprotrophic fungi help to generate humus, a rich, black organic matter that is essential for soil health. Humus enhances soil structure by binding particles together, making it less susceptible to erosion. It also improves water retention, allowing plants to access moisture even during dry periods (Kües, 2015; Crowther et al., 2012a). These mushrooms play an important role in nutrient recycling and soil enrichment, laying the groundwork for thriving ecosystems.

Mycorrhizal fungi act as natural companions for plants, producing beneficial associations. These fungi adhere to plant roots and form enormous networks of thread-like structures known as hyphae, which extend well beyond the roots' reach (Wahab et al., 2023; Volk, 2001). Fungi effectively act as root system extensions, greatly expanding their surface area and ability to absorb water and vital nutrients from the soil, particularly phosphorus (Guo et al., 2024). In exchange for this vital function, plants provide fungi with carbohydrates created during photosynthesis. These sugars provide energy for the fungi, allowing them to grow while supporting the plant. It is a nicely balanced transaction; neither participant takes without returning something (Wahab et al., 2023; Khaliq et al., 2022).

This symbiotic association does more than only enhance growth; it also assists plants in dealing with adverse environmental situations (Khaliq et al., 2022). Mycorrhizal fungi help plants withstand droughts by boosting water availability in the soil. They also help plants to survive high salt levels and detoxify heavy metals in polluted soils (Trocio & Paguntalan, 2023), providing some environmental resilience (Muhammad et al., 2023; Begum et al., 2019). These fungi act as plant lifelines, providing crucial resources and protection that enable plants to grow and thrive under challenging conditions.

Fungi are important soil health and structure builders, working under the surface to maintain stable and productive ecosystems. Their hyphae are tiny thread-like structures that spread throughout the soil, connecting particles to form solid aggregates. These aggregates increase soil texture, making it more suitable for plants and microbes (Lehmann et al., 2016). Fungi allow air to travel freely by binding soil particles together and creating gaps. This ensures that plant roots and soil creatures receive the oxygen they require to survive (Myrold et al., 2011). This construction also improves water infiltration, allowing rain or irrigation to penetrate deeply into the soil rather than draining away. This ensures that plants have continuous access to water and avoids the risk of waterlogging or drought stress. Fungi also serve an important role in controlling erosion by securing soil particles and keeping valuable topsoil from being washed or blown away. This is vital for maintaining rich soil (Ali et al., 2023), which is required for sustainable farming and ecological equilibrium (Rodriguez-Sanchez et al., 2021). Furthermore, the contribution of fungi to soil structure fosters a nutrient-rich environment for other soil organisms by assisting in critical activities like nutrient recycling, decomposition, and disease control (Krause et al., 2022). Fungi are the backbone of soil ecosystems, providing the foundation for healthy plants and microbial life to flourish.

Fungi are like plant bodyguards, discreetly protecting them from deadly infections. Some fungi form physical barriers around plant roots, shielding invaders from reaching the plant's susceptible tissues (Bhardwaj et al., 2023). Others create antimicrobial chemicals and natural antibiotics that target and inhibit disease-causing organisms in soil or on plant surfaces. However, fungi do more than only defend plants; they also help them grow their immunity. Certain fungi can stimulate plants' immune systems, triggering defense systems throughout the plant. Systemic resistance makes the plant stronger and less susceptible to infections, even where fungi are not directly present (Nie et al., 2024). In addition to these protective roles, fungi, like mycorrhizal interaction, often form symbiotic relationships with plants. These fungi enhance nutrients and water absorption and boost the plant's overall health, making it more resilient to stressors and plant diseases (Guo et al., 2024). Fungi are nature's multitaskers, guarding plants, boosting their immunity, and supporting their growth while playing a vital role in maintaining healthy ecosystems.

Fungi are vital contributors to the Earth's carbon cycle. It performs a role that benefits both the environment and plant ecosystems. They act as natural carbon movers, transferring carbon absorbed by plants during photosynthesis into the soil. Once in the soil, this carbon becomes part of the organic matter, enriching the soil with nutrients and improving its health and structure. However, fungi's work goes beyond soil enrichment. It also plays a crucial role in addressing climate change by transferring carbon from the atmosphere, where it exists as carbon dioxide, into the soil; hence, fungi help reduce the level of overall greenhouse gases (Ali et al., 2024). Carbon sequestration in the soil is a crucial step in mitigating the effects of climate change, as it slows the accumulation of heat-trapping gases in the atmosphere

(Malhi et al., 2020). Overall, fungi are not just soil protectors but also environmental champions, bridging the gap between ecosystems and the broader climate system. Their capacity to store carbon in the soil shows their often-overlooked yet essential contribution to combating global climate challenges while supporting healthy plant landscapes (Ali et al., 2024; Branco et al., 2022).

Mycorrhizal networks connect multiple plants, facilitate resource sharing, and enhance plant diversity. This interconnectedness strengthens ecosystem resilience, which allows plant communities to withstand environmental changes and disturbances better (Bortolot et al., 2024; Muhammad et al., 2023). In sustainable agriculture, fungi play a vital role by reducing the need and dependence on chemical fertilizers and pesticides. Mycorrhizal fungi, for example, improve nutrient uptake and water absorption in crops, boosting yields while promoting resilience against environmental stressors like drought and salinity (Nie et al., 2024; Branco et al., 2022). This makes fungi a cornerstone of ecofriendly farming practices (Begum et al., 2019).

Urbanization causes significant modifications in how fungi interact in ecosystems (Newbound et al., 2010). Cities frequently have more diverse fungal communities because of the variety of habitats and human activity, but this comes with a cost (Yiallouris et al., 2024). Fungi that create beneficial associations with plants, such as mycorrhizal fungi, tend to diminish, but decomposer and disease-causing fungi can thrive. Urban soils, which are generally compacted or contaminated, are less favorable to fungi, whereas rural soils support natural fungal communities. The types of plants in cities also influence fungal dynamics, with native species supporting more balanced fungal roles than non-native or decorative plants in urban contexts (Gómez-Hernández et al., 2021). These changes significantly impact ecosystems, including nutrient cycling, soil health, and resilience to environmental change. Rural fungus frequently maintains ecosystems via natural processes. In contrast, urban fungi adapt fast but can cause disruptions such as increased plant diseases or altered soil functions (Rodriguez-Sanchez et al., 2021). Fungi are the hidden builders of ecological balance, maintaining ecosystem health and production while addressing global issues such as food security, climate change, and biodiversity loss (Branco et al., 2022). By executing these roles, fungi help individual plants and contribute to the stability and productivity of whole ecosystems. Their responsibilities are critical for sustainable agriculture, climate mitigation, and biodiversity protection.

The plant-fungal interaction must be conserved and managed to keep ecosystems healthy and make agriculture sustainable. Fungi play critical roles in nutrient cycling, soil health, and plant resilience; thus, protecting them necessitates deliberate measures and collaborative efforts (Tahat et al., 2020). One helpful technique is maintaining fungi habitats, such as forests, grasslands, and wetlands (Pretorius et al., 2023; Baird & Pope, 2022). These niches provide stable habitats for various fungal populations to grow and thrive. Another important strategy is to promote native plants, which help local fungi create symbiotic partnerships, improving both plant growth and soil health. Sustainable farming strategies, including crop rotation, agroforestry, and organic farming, are critical for reducing damage to fungal networks (Zou et al., 2024; Vikas & Ranjan, 2024; Sharma et al., 2024; Çakmakçı et al., 2023). These strategies avoid soil disturbance, limit the use of toxic chemicals, and create settings in which fungi can thrive. Adding organic matter, such as compost or mulch, is especially advantageous since it nourishes fungi and improves soil structure and fertility (Rossi et al., 2024).

Pollution control is also significant. Preventing industrial pollutants from entering soil ecosystems and limiting excessive use of pesticides and synthetic fertilizers contribute to the delicate balance required for fungal life. Addressing soil erosion is critical, as fungi require stable soil to form their networks. These ecosystems can be preserved through cover crop planting, mulch application, and terrace construction (Da Gama, 2023).

Educating communities on the value of fungi is an effective method to boost conservation efforts. Public awareness campaigns, workshops, and instructional materials can encourage sustainable behaviors and emphasize fungi's critical ecological roles. Finally, investing in studying and monitoring fungal diversity enable us to understand their functions better and effectively manage threats, ensuring their survival for future generations. We may develop a supportive framework that protects fungal diversity and improves the ecosystem's health by integrating these measures.

Future Research Directions

Future research in plant-fungal diversity holds immense potential to address global challenges and unlock ecological and agricultural benefits. A significant focus will be on uncovering the vast number of fungal species yet to be documented, especially in underexplored habitats like tropical forests and deep soils, which house unparalleled biodiversity. Understanding the complex interactions between fungi and plants grown in urban settings, particularly through symbiotic relationships like mycorrhizae, will provide insights into how fungi support plant growth, enhance nutrient uptake, and boost stress resilience. As climate change reshapes ecosystems, studying how temperature shifts, altered precipitation, and soil degradation affect fungal diversity and functionality in both rural and urban areas will be crucial to fostering ecosystem adaptability.

Innovative applications of fungi in biotechnology are also an interesting study area. Fungi could play an important part in sustainable agriculture by functioning as biofertilizers, biopesticides, and soil health improvers. Their bioremediation and carbon sequestration capabilities are similarly impressive, providing answers for pollution control and climate change mitigation. Advances in genomics will enable researchers to investigate fungal genetic pathways, revealing their ecological roles and identifying novel uses in health, industry, and environmental control.

Other areas of interest include the study of fungal endophytes, fungi that live within plants without causing apparent damage, which have the potential to improve plant immunity and adapt to environmental challenges. Conservation efforts will also gain traction, focusing on practical techniques to conserve fungal variety from challenges such as habitat loss, pollution, and soil erosion. Establishing fungal databases and monitoring systems could help in tracking changes and protecting beneficial fungi.

Conclusion

Fungi are indispensable to Earth's ecosystems, serving as ecological powerhouses that support nutrient cycling, soil health, plant growth, and biodiversity. Their diverse roles, from symbiotic relationships with plants to decomposition and environmental restoration, highlight their critical importance in sustaining natural systems and addressing global challenges. However, urbanization, habitat loss, and environmental changes threaten fungal diversity, underscoring the need for intentional conservation efforts. By preserving habitats, promoting sustainable practices, and advancing research, we can protect fungi's essential contributions to agriculture, climate resilience, and ecosystem stability. The future of fungi research holds incredible potential for innovation, offering solutions to pressing issues like food security, climate change, and biodiversity loss. Safeguarding fungal diversity is key to building a sustainable and resilient ecosystem in the future.

Acknowledgments John Mark De Jesus is supported by the Department of Science and Technology (DOST)—Science Education Institute through the DOST-Accelerated Science and Technology Human Resource Development Program.

References

Aamir, M., Rai, K. K., Zehra, A., Kumar, S., Yadav, M., Shukla, V., & Upadhyay, R. S. (2020). Fungal endophytes: Classification, diversity, ecological role, and their relevance in sustainable agriculture. In *Elsevier eBooks* (pp. 291–323). https://doi.org/10.1016/b978-0-12-818734-0.00012-7

Abrego, N., Crosier, B., Somervuo, P., Ivanova, N., Abrahamyan, A., Abdi, A., Hämäläinen, K., Junninen, K., Maunula, M., Purhonen, J., & Ovaskainen, O. (2020). Fungal communities decline with urbanization—More in air than in soil. *The ISME Journal, 14*(11), 2806–2815. https://doi.org/10.1038/s41396-020-0732-1

Afzal, I., Shinwari, Z. K., Sikandar, S., & Shahzad, S. (2019). Plant beneficial endophytic bacteria: Mechanisms, diversity, host range and genetic determinants. *Microbiological Research, 221*, 36–49

Ali, N., Naqvi, S. A. A. A. A., Sharif, L., Shafiq, M., Kamran, Z., Masah, N., Haseeb, M., & Shahid, M. A. (2023). Microbial extracellular polymeric substance and impacts on soil aggregation. In *Elsevier eBooks* (pp. 221–237). https://doi.org/10.1016/b978-0-323-95251-4.00021-1

Ali, I., Qaiser, H., Abdullah, R., Kaleem, A., Iqtedar, M., Iqbal, I., & Chen, X. (2024). Prospective roles of extremophilic fungi in climate change mitigation strategies. *Journal of Fungi, 10*(6), 385. https://doi.org/10.3390/jof10060385

Amezrou, R., Ducasse, A., Compain, J., Lapalu, N., Pitarch, A., Dupont, L., et al. (2024). Quantitative pathogenicity and host adaptation in a fungal plant pathogen revealed by whole-genome sequencing. *Nature Communications, 15*(1), 1933.

Authier, L., Violle, C., & Richard, F. (2022). Ectomycorrhizal networks in the Anthropocene: From natural ecosystems to urban planning. *Frontiers in Plant Science, 13*. https://doi.org/10.3389/fpls.2022.900231

Baird, A., & Pope, F. (2022). 'Can't see the forest for the trees': The importance of fungi in the context of UK tree planting. *Food and Energy Security, 11*(4). https://doi.org/10.1002/fes3.371

Baldrian, P., Větrovský, T., Lepinay, C., & Kohout, P. (2022). High-throughput sequencing view on the magnitude of global fungal diversity. *Fungal Diversity, 114*(1), 539–547.

Begum, N., Qin, C., Ahanger, M. A., Raza, S., Khan, M. I., Ashraf, M., Ahmed, N., & Zhang, L. (2019). Role of arbuscular mycorrhizal fungi in plant growth regulation: Implications in abiotic stress tolerance. *Frontiers in Plant Science, 10*. https://doi.org/10.3389/fpls.2019.01068

Bhardwaj, M., Kailoo, S., Khan, R. T., Khan, S. S., & Rasool, S. (2023). Harnessing fungal endophytes for natural management: A biocontrol perspective. *Frontiers in Microbiology, 14*. https://doi.org/10.3389/fmicb.2023.1280258

Bollmann-Giolai, A., Malone, J. G., & Arora, S. (2022). Diversity, detection and exploitation: Linking soil fungi and plant disease. *Current Opinion in Microbiology, 70*, 102199. https://doi.org/10.1016/j.mib.2022.102199

Bonfante, P., & Venice, F. (2020). Mucoromycota: Going to the roots of plant-interacting fungi. *Fungal Biology Reviews, 34*(2), 100–113. https://doi.org/10.1016/j.fbr.2019.12.003

Bortolot, M., Buffoni, B., Mazzarino, S., Hoff, G., Martino, E., Fiorilli, V., & Di Fossalunga, A. S. (2024). The importance of mycorrhizal fungi and their associated bacteria in promoting crops' performance: An applicative perspective. *Horticulturae, 10*(12), 1326. https://doi.org/10.3390/horticulturae10121326

Branco, S., Schauster, A., Liao, H., & Ruytinx, J. (2022). Mechanisms of stress tolerance and their effects on the ecology and evolution of mycorrhizal fungi. *New Phytologist, 235*(6), 2158–2175. https://doi.org/10.1111/nph.18308

Buckley, M. (2008). The Fungal Kingdom: diverse and essential roles in earth's ecosystem. https://doi.org/10.1128/aamcol.2nov.2007

Çakmakçı, R., Salık, M. A., & Çakmakçı, S. (2023). Assessment and principles of environmentally sustainable food and agriculture systems. *Agriculture, 13*(5), 1073. https://doi.org/10.3390/agriculture13051073

Christel, A., Dequiedt, S., Chemidlin-Prevost-Bouré, N., Mercier, F., Tripied, J., Comment, G., Djemiel, C., Bargeot, L., Matagne, E., Fougeron, A., Passi, J. M., Ranjard, L., & Maron, P. (2023). Urban land uses shape soil microbial abundance and diversity. *The Science of the Total Environment, 883*, 163455. https://doi.org/10.1016/j.scitotenv.2023.163455

Corbu, V. M., Gheorghe-Barbu, I., Dumbravă, A. Ş., Vrâncianu, C. O., & Şesan, T. E. (2023). Current insights in fungal importance—A comprehensive review. *Microorganisms, 11*(6), 1384. https://doi.org/10.3390/microorganisms11061384

Cristiano, E., Deidda, R., & Viola, F. (2020). The role of green roofs in urban water-energy-food-ecosystem nexus: A review. *The Science of the Total Environment, 756*, 143876. https://doi.org/10.1016/j.scitotenv.2020.143876

Crowther, T. W., Boddy, L., & Jones, T. H. (2012a). Functional and ecological consequences of saprotrophic fungus–grazer interactions. *The ISME Journal, 6*(11), 1992–2001. https://doi.org/10.1038/ismej.2012.53

Crowther, T. W., Jones, T. H., & Boddy, L. (2012b). Interactions between saprotrophic basidiomycete mycelia and mycophagous soil fauna. *Mycology: An International Journal on Fungal Biology, 3*(1), 77–86. https://doi.org/10.1080/21501203.2012.656723

Cuadros-Orellana, S., Leite, L. R., Smith, A., Medeiros, J. D., Badotti, F., Fonseca, P. L., et al. (2013). Assessment of fungal diversity in the environment using metagenomics: A decade in review. *Fungal Genomics & Biology, 3*(2), 1.

Da Gama, J. T. (2023). The role of soils in sustainability, climate change, and ecosystem services: Challenges and opportunities. *Ecologies, 4*(3), 552–567. https://doi.org/10.3390/ecologies4030036

De Cal, A., Melgarejo, P., & Del Mar Jimenez-Gasco, M. (2022). Editorial: Necrotrophic fungal plant pathogens. *Frontiers in Plant Science, 13*. https://doi.org/10.3389/fpls.2022.839674

De Oliveira, J. B. M., Corrêa, D. J., Parente, C. E. T., & Frases, S. (2025). Fungi in mangrove: Ecological importance, climate change impacts, and the role in environmental remediation. *Microorganisms, 13*(4), 878. https://doi.org/10.3390/microorganisms13040878

De Paz, G. Á., Saucedo, U. C., Morcillo, R. L., & Aranda, E. (2023). *World of fungi and fungal ecosystems* (pp. 1–29). *Elsevier eBooks*. https://doi.org/10.1016/b978-0-323-91872-5.00010-7

De Silva, N. (2016). Mycosphere essays 9: Defining biotrophs and hemibiotrophs. *Mycosphere, 7*(5), 545–559. https://doi.org/10.5943/mycosphere/7/5/2

Dickman, M. B. (2007). Subversion or coersion? Pathogenic deteminants in fungal phytopathogens. *Fungal Biology Reviews, 21*(4), 125–129. https://doi.org/10.1016/j.fbr.2007.06.003

Dinakarkumar, Y., Ramakrishnan, G., Gujjula, K. R., Vasu, V., Balamurugan, P., & Murali, G. (2024). Fungal bioremediation: An overview of the mechanisms, applications and future perspectives. *Environmental Chemistry and Ecotoxicology, 6*, 293–302. https://doi.org/10.1016/j.enceco.2024.07.002

Droz, A. G., Coffman, R. R., Eagar, A. C., & Blackwood, C. B. (2022). Drivers of fungal diversity and community biogeography differ between green roofs and adjacent ground-level green space. *Environmental Microbiology, 24*(12), 5809–5824. https://doi.org/10.1111/1462-2920.16190

Fei, W., & Liu, Y. (2022). Biotrophic fungal pathogens: A critical overview. *Applied Biochemistry and Biotechnology, 195*(1), 1–16. https://doi.org/10.1007/s12010-022-04087-0

Frąc, M., Hannula, S. E., Bełka, M., & Jędryczka, M. (2018). Fungal biodiversity and their role in soil health. *Frontiers in Microbiology, 9*. https://doi.org/10.3389/fmicb.2018.00707

Gamalero, E., Santoyo, G., & Glick, B. R. (2023). Plant growth promoters. In *Elsevier eBooks* (pp. 418–427). https://doi.org/10.1016/b978-0-12-822974-3.00177-4

Ghorbanpour, M., Omidvari, M., Abbaszadeh-Dahaji, P., Omidvar, R., & Kariman, K. (2018). Mechanismsunderlying the protective effects of beneficial fungi against plant diseases. *Biological control, 117*, 147–157.

Gibson, E., & Zimmerman, N. B. (2023). Urban biogeography of fungal endophytes across San Francisco. *Peer J, 11*, e15454. https://doi.org/10.7717/peerj.15454

Glazebrook, J. (2005). Contrasting mechanisms of defense against biotrophic and necrotrophic pathogens. *Annual Review of Phytopathology, 43*, 205–227. https://doi.org/10.1146/annurev.phyto.43.040204.135923

Gómez-Hernández, M., Avendaño-Villegas, E., Toledo-Garibaldi, M., & Gándara, E. (2021). Impact of urbanization on functional diversity in macromycete communities along an urban ecosystem in Southwest Mexico. *Peer J, 9*, e12191. https://doi.org/10.7717/peerj.12191

Guo, S., Xia, L., Xia, D., Li, M., Xu, W., & Liu, L. (2024). Enhancing plant resilience: Arbuscular mycorrhizal fungi's role in alleviating drought stress in vegetation concrete. *Frontiers in Plant Science, 15*. https://doi.org/10.3389/fpls.2024.1401050

Haque, A. K. (2018). Fungal diseases. In *Elsevier eBooks* (pp. 217–243). https://doi.org/10.1016/b978-0-323-39308-9.00012-1

Hawksworth, D. L., & Lücking, R. (2017). Fungal diversity revisited: 2.2 to 3.8 million species. In *The fungal kingdom* (pp. 79–95). ASM Press.

Hossain, A., Krupnik, T. J., Timsina, J., Mahboob, M. G., Chaki, A. K., Farooq, M., ... & Hasanuzzaman, M. (2020). Agricultural land degradation: Processes and problems undermining future food security. *In Environment, climate, plant and vegetation growth* (pp. 17–61). Cham: Springer International Publishing.

Hui, N., Liu, X., Kotze, D. J., Jumpponen, A., Francini, G., & Setälä, H. (2017). Ectomycorrhizal fungal communities in urban parks are similar to those in natural forests but shaped by vegetation and park age. *Applied and Environmental Microbiology, 83*(23). https://doi.org/10.1128/aem.01797-17

Hyde, K. D. (2022). The numbers of fungi. *Fungal Diversity, 114*, 1. https://doi.org/10.1007/s13225-022-00507-y

Iqbal, B., Li, G., Alabbosh, K. F., Hussain, H., Khan, I., Tariq, M., Javed, Q., Naeem, M., & Ahmad, N. (2023). Advancing environmental sustainability through microbial reprogramming in growth improvement, stress alleviation, and phytoremediation. *Plant Stress, 10*, 100283. https://doi.org/10.1016/j.stress.2023.100283

Jakubczyk, D., & Dussart, F. (2020). Selected fungal natural products with antimicrobial properties. *Molecules, 25*(4), 911. https://doi.org/10.3390/molecules25040911

Khaliq, A., Perveen, S., Alamer, K. H., Haq, M. Z. U., Rafique, Z., Alsudays, I. M., Althobaiti, A. T., Saleh, M. A., Hussain, S., & Attia, H. (2022). Arbuscular mycorrhizal fungi symbiosis to enhance plant–soil interaction. *Sustainability, 14*(13), 7840. https://doi.org/10.3390/su14137840

Kirby, J., Webster, J., & Baker, J. (1990). A particle plating method for analysis of fungal community composition and structure. *Mycological Research, 94*(5), 621–626. https://doi.org/10.1016/s0953-7562(09)80663-1

Konopka, J. B., Casadevall, A., Taylor, J. W., Heitman, J., & Cowen, L. (2019). *One health: Fungal pathogens of humans, animals, and plants*. https://doi.org/10.1128/aamcol.18oct.2017

Krause, S. M., Bertilsson, S., Grossart, H., Bodelier, P. L., Van Bodegom, P. M., Lennon, J. T., Philippot, L., & Roux, X. L. (2022). Microbial trait-based approaches for agroecosystems. In *Advances in agronomy* (pp. 259–299). https://doi.org/10.1016/bs.agron.2022.04.002

Krug, J. C. (2004). Moist chambers for the development of fungi. In *Elsevier eBooks* (pp. 589–593). https://doi.org/10.1016/b978-012509551-8/50030-1

Kües, U. (2015). Fungal enzymes for environmental management. *Current Opinion in Biotechnology, 33*, 268–278. https://doi.org/10.1016/j.copbio.2015.03.006

Kumar, U., Raj, S., Sreenikethanam, A., Maddheshiya, R., Kumari, S., Han, S., Kapoor, K. K., Bhaskar, R., Bajhaiya, A. K., & Gahlot, D. K. (2023). Multi-omics approaches in plant–microbe interactions hold enormous promise for sustainable agriculture. *Agronomy, 13*(7), 1804. https://doi.org/10.3390/agronomy13071804

Latijnhouwers, M., de Wit, P. J. G. M., & Govers, F. (2003). Oomycetes and fungi: Similar weaponry to attack plants. *Trends in Microbiology, 11*, 462–469.

Lehmann, A., Leifheit, E., & Rillig, M. (2016). Mycorrhizas and soil aggregation. In *Elsevier eBooks* (pp. 241–262). https://doi.org/10.1016/b978-0-12-804312-7.00014-0

Li, M., Chen, L., Zhao, F., Tang, J., Bu, Q., Wang, X., & Yang, L. (2023). Effects of urban–rural environmentalgradient on soil microbial community in rapidly urbanizing area. *Ecosystem Health and Sustainability, 9*, 0118

Lin, L., Chen, Y., Xu, G., Zhang, Y., Zhang, S., & Ma, K. (2021). Impacts of urbanization undermine nestedness of the plant–Arbuscular mycorrhizal fungal network. *Frontiers in Microbiology, 12*. https://doi.org/10.3389/fmicb.2021.626671

Liu, S., Xiong, C., Lin, L., Keyhani, N. O., Zhu, M., Zhao, Z., Zhang, W., Yang, C., Su, H., Liu, P., Guan, X., & Qiu, J. (2023). Assessing the structure and diversity of fungal community in plant soil under different climatic and vegetation conditions. *Frontiers in Microbiology, 14*. https://doi.org/10.3389/fmicb.2023.1288066

Lofgren, L. A., & Stajich, J. E. (2021). Fungal biodiversity and conservation mycology in light of new technology, big data, and changing attitudes. *Current Biology, 31*(19), R1312–R1325. https://doi.org/10.1016/j.cub.2021.06.083

López-Bucio, J., Pelagio-Flores, R., & Herrera-Estrell, A. (2015). *Trichoderma* as biostimulant: Exploiting the multilevel properties of a plant beneficial fungus. *Scientia Horticulturae, 196*, 109–123. https://doi.org/10.1016/j.scienta.2015.08.043

Malhi, G. S., Kaur, M., Kaushik, P., Alyemeni, M. N., Alsahli, A. A., & Ahmad, P. (2020). Arbuscular mycorrhiza in combating abiotic stresses in vegetables: An eco-friendly approach. *Saudi Journal of Biological Sciences, 28*(2), 1465–1476. https://doi.org/10.1016/j.sjbs.2020.12.001

Mapuranga, J., Zhang, N., Zhang, L., Chang, J., & Yang, W. (2022). Infection strategies and pathogenicity of biotrophic plant fungal pathogens. *Frontiers in Microbiology, 13*. https://doi.org/10.3389/fmicb.2022.799396

Mikryukov, V., Dulya, O., Zizka, A., Bahram, M., Hagh-Doust, N., Anslan, S., Prylutskyi, O., Delgado-Baquerizo, M., Maestre, F. T., Nilsson, H., Pärn, J., Öpik, M., Moora, M., Zobel, M., Espenberg, M., Mander, Ü., Khalid, A. N., Corrales, A., Agan, A., et al. (2023). Connecting the multiple dimensions of global soil fungal diversity. *Science Advances, 9*(48). https://doi.org/10.1126/sciadv.adj8016

Mishra, S., Srivastava, A., Singh, A., Pandey, G. C., & Srivastava, G. (2024). An overview of symbiotic and pathogenic interactions at the fungi-plant interface under environmental constraints. *Frontiers in Fungal Biology, 5*. https://doi.org/10.3389/ffunb.2024.1363460

Money, N. P. (2015). Fungal diversity. In *Elsevier eBooks* (pp. 1–36). https://doi.org/10.1016/b978-0-12-382034-1.00001-3

Schmit, J. P., & Mueller, G. M. (2007). An estimate of the lower limit of global fungal diversity. *Biodiversity and Conservation, 16*, 99–111.

Muhammad, M., Waheed, A., Wahab, A., Majeed, M., Nazim, M., Liu, Y., Li, L., & Li, W. (2023). Soil salinity and drought tolerance: An evaluation of plant growth, productivity, microbial diversity, and amelioration strategies. *Plant Stress, 11*, 100319. https://doi.org/10.1016/j.stress.2023.100319

Mwamburi, L. A. (2022). Cultivation-based studies of endophytic fungi in plants. In *Springer protocols handbooks/springer protocols* (pp. 145–151). https://doi.org/10.1007/978-1-0716-2827-0_15

Myrold, D. D., Pett-Ridge, J., & Bottomley, P. J. (2011). Nitrogen mineralization and assimilation at millimeter scales. In *Methods in enzymology on CD-ROM/methods in enzymology* (pp. 91–114). https://doi.org/10.1016/b978-0-12-386489-5.00004-x

Naranjo-Ortiz, M. A., & Gabaldón, T. (2019). Fungal evolution: Major ecological adaptations and evolutionary transitions. *Biological Reviews/Biological Reviews of the Cambridge Philosophical Society, 94*(4), 1443–1476. https://doi.org/10.1111/brv.12510

Newbound, M., Mccarthy, M. A., & Lebel, T. (2010). Fungi and the urban environment: A review. *Landscape and Urban Planning, 96*(3), 138–145. https://doi.org/10.1016/j.landurbplan.2010.04.005

Nie, W., He, Q., Guo, H., Zhang, W., Ma, L., Li, J., & Wen, D. (2024). Arbuscular mycorrhizal fungi: boosting cropresilience to environmental stresses. *Microorganisms, 12*(12), 2448

Niego, A. G. T., Rapior, S., Thongklang, N., Raspé, O., Hyde, K. D., & Mortimer, P. (2022). Reviewing the contributions of macrofungi to forest ecosystem processes and services. *Fungal Biology Reviews, 44*, 100294. https://doi.org/10.1016/j.fbr.2022.11.002

Pandey, B., & Ghosh, A. (2023). Urban ecosystem services and climate change: A dynamic interplay. *Frontiers in Sustainable Cities, 5*. https://doi.org/10.3389/frsc.2023.1281430

Paterson, R. R. M., Solaiman, Z., & Santamaria, O. (2023). Guest edited collection: Fungal evolution and diversity. *Scientific Reports, 13*, 21438. https://doi.org/10.1038/s41598-023-48471-0

Pitt, J. (2014). PENICILLIUM | Penicillium and talaromyces. In *Elsevier eBooks* (pp. 6–13). https://doi.org/10.1016/b978-0-12-384730-0.00248-2

Pretorius, M., Wimberger, K., & O'Riain, M. J. (2023). Preserving large tracts of natural grassland promotes mammal species richness and occurrence in afforested areas. *Forest Ecology and Management, 542*, 121078. https://doi.org/10.1016/j.foreco.2023.121078

Priyashantha, A. K. H., Dai, D.-Q., Bhat, D. J., Stephenson, S. L., Promputtha, I., Kaushik, P., Tibpromma, S., & Karunarathna, S. C. (2023). Plant–fungi interactions: where it goes? *Biology, 12*(6), 809. https://doi.org/10.3390/biology12060809

Purahong, W., Günther, A., Gminder, A., Tanunchai, B., Gossner, M. M., Buscot, F., & Schulze, E. (2022). City life of mycorrhizal and wood-inhabiting macrofungi: Importance of urban areas for maintaining fungal biodiversity. *Landscape and Urban Planning, 221*, 104360. https://doi.org/10.1016/j.landurbplan.2022.104360

Raimbault, A., Brin, A., Manzi, S., Savoie, J. M., Gandois, L., Oliva, P., Fogliani, O., Roy-Camille, C., Gratacap, L., & Roy, M. (2024). Influence of habitat fragmentation and habitat amount on soil fungi communities in ancient forests. *Landscape Ecology, 39*(2). https://doi.org/10.1007/s10980-024-01821-3

Reininger, V., Martinez-Garcia, L. B., Sanderson, L., & Antunes, P. M. (2015). Composition of fungal soil communities varies with plant abundance and geographic origin. *AoB Plants, plv110*. https://doi.org/10.1093/aobpla/plv110

Reverchon, F., Del Ortega-Larrocea, P. M., & Pérez-Moreno, J. (2010). Saprophytic fungal communities change in diversity and species composition across a volcanic soil chronosequence at Sierra del Chichinautzin, Mexico. *Annals of Microbiology, 60*(2), 217–226. https://doi.org/10.1007/s13213-010-0030-7

Rodriguez-Sanchez, A., Tomasek, A., McMillan, S., Yufra, S., Yupanqui, M., Rondon, R., & Hoagland, L. (2021). Composition and potential functional roles of soil fungal communities on arid farms in Arequipa (Southern Peru) characterized using SMRT sequencing. *Applied Soil Ecology, 169*, 104228. https://doi.org/10.1016/j.apsoil.2021.104228

Rossi, G., Beni, C., & Neri, U. (2024). Organic mulching: A sustainable technique to improve soil quality. *Sustainability, 16*(23), 10261. https://doi.org/10.3390/su162310261

Rossman, A. Y., Allen, W. C., & Castlebury, L. A. (2016). New combinations of plant-associated fungi resulting from the change to one name for fungi. *IMA Fungus, 7*, 1–7. https://doi.org/10.5598/imafungus.2016.07.01.01

Rouphael, Y., Franken, P., Schneider, C., Schwarz, D., Giovannetti, M., Agnolucci, M., et al. (2015). Arbuscular mycorrhizal fungi act as biostimulants in horticultural crops. *Scientia Horticulturae, 196*, 91–108. https://doi.org/10.1016/j.scienta.2015.09.002

Runnel, K., Tedersoo, L., Krah, F., Piepenbring, M., Scheepens, J. F., Hollert, H., Johann, S., Meyer, N., & Bässler, C. (2024). Toward harnessing biodiversity–ecosystem function relationships in fungi. *Trends in Ecology & Evolution*. https://doi.org/10.1016/j.tree.2024.10.004

Rusterholz, H., & Baur, B. (2023). Changes in soil fungal diversity and composition along a rural–urban gradient. *Forests, 14*(11), 2226. https://doi.org/10.3390/f14112226

Sakamoto, Y. (2018). Influences of environmental factors on fruiting body induction, development and maturation in mushroom-forming fungi. *Fungal Biology Reviews, 32*(4), 236–248. https://doi.org/10.1016/j.fbr.2018.02.003

Savary, S., Willocquet, L., Pethybridge, S. J., Esker, P., McRoberts, N., & Nelson, A. (2019). The global burden of pathogens and pests on major food crops. *Nature Ecology & Evolution, 3*, 430–439.

Sharma, P., Sharma, P., & Thakur, N. (2024). Sustainable farming practices and soil health: A pathway to achieving SDGs and future prospects. *Discover Sustainability, 5*(1). https://doi.org/10.1007/s43621-024-00447-4

Shen, C., Wang, J., He, J. Z., Yu, F. H., & Ge, Y. (2021). Plant diversity enhances soil fungal diversity and microbialresistance to plant invasion. *Applied and environmental microbiology, 87*(11), e00251–21

Stewart, J. D., Kiers, E. T., Anthony, M. A., & Kiers, A. H. (2023). Supporting urban greenspace with microbial symbiosis. *Plants, People, Planet, 6*(1), 3–17. https://doi.org/10.1002/ppp3.10403

Strange, R. N., & Scott, P. R. (2005). Plant disease: A threat to global food security. *Annual Review of Phytopathology, 43*, 83–116.

Su, N., Jarvie, S., Yan, Y., Gong, X., Li, F., Han, P., & Zhang, Q. (2022). Landscape context determines soil fungal diversity in a fragmented habitat. *Catena, 213*, 106163. https://doi.org/10.1016/j.catena.2022.106163

Tahat, M. M., Alananbeh, K. M., Othman, Y. A., & Leskovar, D. I. (2020). Soil health and sustainable agriculture. *Sustainability, 12*(12), 4859. https://doi.org/10.3390/su12124859

Tatsumi, C., Atherton, K. F., Garvey, S. M., Conrad-Rooney, E., Morreale, L. L., Hutyra, L. R., Templer, P. H., & Bhatnagar, J. M. (2023). Urbanization and edge effects interact to drive mutualism breakdown and the rise of unstable pathogenic communities in forest soil. *Proceedings of the National Academy of Sciences, 120*(36). https://doi.org/10.1073/pnas.2307519120

Tedersoo, L., Anslan, S., Bahram, M., Drenkhan, R., Pritsch, K., Buegger, F., Padari, A., Hagh-Doust, N., Mikryukov, V., Gohar, D., Amiri, R., Hiiesalu, I., Lutter, R., Rosenvald, R., Rähn, E., Adamson, K., Drenkhan, T., Tullus, H., Jürimaa, K., et al. (2020). Regional-scale in-depth analysis of soil fungal diversity reveals strong pH and plant species effects in northern Europe. *Frontiers in Microbiology, 11*. https://doi.org/10.3389/fmicb.2020.01953

Trocio, D. Y., & Paguntalan, D. (2023). Review on the use of arbuscular mycorrhizal fungi in bioremediation of heavy metal contaminated soils in The Philippines. *The Philippine Journal of Science, 152*(3). https://doi.org/10.56899/152.03.31

Tuomela, M., & Hatakka, A. (2011). Oxidative fungal enzymes for bioremediation. In *Elsevier eBooks* (pp. 183–196). https://doi.org/10.1016/b978-0-08-088504-9.00370-6

Van Bael, S. A. (2020). Fungal diversity. *Diversity, 12*(11), 437. https://doi.org/10.3390/d12110437

Verma, V., Srivastava, A., Garg, S. K., Singh, V. P., & Arora, P. K. (2023). Incorporating omics-based tools into endophytic fungal research. *Biotechnology Notes, 5*, 1–7. https://doi.org/10.1016/j.biotno.2023.12.006

Vikas, N., & Ranjan, R. (2024). Agroecological approaches to sustainable development. *Frontiers in Sustainable Food Systems, 8*. https://doi.org/10.3389/fsufs.2024.1405409

Volk, T. J. (2001). Fungi. In *Elsevier eBooks* (pp. 141–163). https://doi.org/10.1016/b0-12-226865-2/00134-6

Volk, T. J. (2013). Fungi. In *Elsevier eBooks* (pp. 624–640). https://doi.org/10.1016/b978-0-12-384719-5.00062-9

Wahab, A., Muhammad, M., Munir, A., Abdi, G., Zaman, W., Ayaz, A., Khizar, C., & Reddy, S. P. P. (2023). Role of Arbuscular Mycorrhizal fungi in regulating growth, enhancing productivity, and potentially influencing ecosystems under abiotic and biotic stresses. *Plants, 12*(17), 3102. https://doi.org/10.3390/plants12173102

Wang, M., Li, J., Kuang, S., He, Y., Chen, G., Huang, Y., Song, C., Anderson, P., & Łowicki, D. (2020). Plant Diversity Along the Urban–Rural Gradient and Its Relationship with Urbanization Degree in Shanghai, China. *Forests, 11*(2), 171. https://doi.org/10.3390/f11020171

Watkinson, S. C. (2015). Mutualistic symbiosis between fungi and autotrophs. In *Elsevier eBooks* (pp. 205–243). https://doi.org/10.1016/b978-0-12-382034-1.00007-4

Wu, B., Hussain, M., Zhang, W., Stadler, M., Liu, X., & Xiang, M. (2019). Current insights into fungal species diversity and perspective on naming the environmental DNA sequences of fungi. *Mycology, 10*(3), 127–140. https://doi.org/10.1080/21501203.2019.1614106. PMID: 31448147; PMCID: PMC6691916.

Yang, Y., Dou, Y., Huang, Y., & An, S. (2017). Links between soil fungal diversity and plant and soil properties on the loess plateau. *Frontiers in Microbiology, 8*. https://doi.org/10.3389/fmicb.2017.02198

Yiallouris, A., Pana, Z. D., Marangos, G., Tzyrka, I., Karanasios, S., Georgiou, I., Kontopyrgia, K., Triantafyllou, E., Seidel, D., Cornely, O. A., Johnson, E. O., Panagiotou, S., & Filippou, C. (2024). Fungal diversity in the soil Mycobiome: Implications for ONE health. *One Health, 18*, 100720. https://doi.org/10.1016/j.onehlt.2024.100720

Zanne, A. E., Abarenkov, K., Afkhami, M. E., Aguilar-Trigueros, C. A., Bates, S., Bhatnagar, J. M., Busby, P. E., Christian, N., Cornwell, W. K., Crowther, T. W., Flores-Moreno, H., Floudas, D., Gazis, R., Hibbett, D., Kennedy, P., Lindner, D. L., Maynard, D. S., Milo, A. M., Nilsson, R. H., et al. (2019). Fungal functional ecology: Bringing a trait-based approach to plant-associated fungi. *Biological Reviews/Biological Reviews of the Cambridge Philosophical Society, 95*(2), 409–433. https://doi.org/10.1111/brv.12570

Zheng, B., Hui, N., Jumpponen, A., Lu, C., Pouyat, R., Szlavecz, K., Wardle, D. A., Yesilonis, I., Setälä, H., & Kotze, D. J. (2025). Urbanization leads to asynchronous homogenization of soil microbial communities across biomes. *Environmental Science and Ecotechnology, 100547*. https://doi.org/10.1016/j.ese.2025.100547

Zhou, Y., Jia, X., Han, L., Liu, Z., Kang, S., & Zhao, Y. (2021). Fungal community diversity in soils along an elevation gradient in a Quercus aliena var. acuteserrata forest in Qinling Mountains, China. *Applied Soil Ecology, 167*, 104104. https://doi.org/10.1016/j.apsoil.2021.104104

Zou, Y., Liu, Z., Chen, Y., Wang, Y., & Feng, S. (2024). Crop rotation and diversification in China: Enhancing sustainable agriculture and resilience. *Agriculture, 14*(9), 1465. https://doi.org/10.3390/agriculture14091465

Chapter 17
Biodiversity of Forest Fungi and Potential Applications Toward a Sustainable Ecosystem in Southeast Asia Region

Nelson M. Pampolina, Edwin R. Tadiosa, Jessa P. Ata, Janine Kaysee R. Soriano, Jason A. Parlucha, Nik Shawn C. Tabao, Nguyen Manh Ha, and Sri Rahayu

Abstract Southeast Asia harbors an exceptional diversity of forest fungi, accounting for 23.86% of the world's recorded fungal species, playing vital roles in ecosystem resilience, nutrient cycling, and forest productivity. However, the region faces accelerating deforestation, climate change, and land-use conversion, threatening fungal biodiversity and ecological functions. This chapter explores the diversity, ecological significance, and potential applications of forest fungi in ecosystem restoration, alternative food production, biotechnology, and climate change mitigation.

N. M. Pampolina (✉)
Department of Forest Biological Sciences, College of Forestry and Natural Resources, University of the Philippines Los Baños, College, Laguna, Philippines

UPLB Museum of Natural History, University of the Philippines Los Baños, College, Laguna, Philippines
e-mail: nmpampolina@up.edu.ph

E. R. Tadiosa
Science Department, College of Sciences, Bulacan State University, Malolos, Bulacan, Philippines

J. P. Ata · J. K. R. Soriano
Department of Forest Biological Sciences, College of Forestry and Natural Resources, University of the Philippines Los Baños, College, Laguna, Philippines

J. A. Parlucha
Department of Wood Science and Technology, College of Forestry and Environmental Science, Central Mindanao University, Bukidnon, Philippines

N. S. C. Tabao
UPLB Museum of Natural History, University of the Philippines Los Baños, College, Laguna, Philippines

N. M. Ha
Forest Protection Research Centre, Vietnamese Academy of Forest Sciences, Hanoi, Vietnam

S. Rahayu
Faculty of Forestry, Universitas Gadjah Mada, Yogyakarta, Indonesia

J. J. G. Guerrero et al. (eds.), *Mycology in a Changing Planet: Applications and Perspectives from Southeast Asia*, https://doi.org/10.1007/978-981-95-2640-6_17

Despite their critical contributions to forest health, fungi remain understudied and underrepresented in conservation policies, requiring harmonized databases and urgent scientific attention. The discussion highlights mycorrhizal symbiosis in forest regeneration, fungal-based bioremediation strategies, and emerging threats from forest diseases. Additionally, it underscores the need for strengthened conservation efforts, enhanced regional collaboration, and policy integration to harness the full potential of fungal resources for a sustainable ASEAN forest ecosystem.

Keywords Fungal diversity · Functional roles · Mycorrhizas · Conservation

Introduction

The region of Southeast Asia falls under the Indo Malayan floristic realm, with extraordinarily rich species diversity and endemism driven by geographic isolation and variations in climate history (Liu et al., 2023). Species migration and diversification across geologic time scales further contributed to declaring the megadiversity of the region (de Bruyn et al., 2014). The tropical forest, among the most productive ecosystems on planet earth, is a favorable habitat to diverse lifeforms in both terrestrial and marine. However, the forest in this region covers 47% of the 4.5 million square kilometers of land area continuously threatened by anthropogenic and climate-related factors (ASEAN Center for Biodiversity, 2023).

Among the 25 global biodiversity hotspots, four from Southeast Asia, including the Philippines and Sundaland (Indonesia, Malaysia, and Brunei), were considered the "hottest hotspots" in terms of species endemism per area and severity of habitat loss (Myers et al., 2000). Forest degradation, driven mainly by unsustainable land uses, remains one of the key causes of biodiversity loss in Southeast Asia (Tan et al., 2022; Chen et al., 2024). Based on a spatiotemporal modeling and remote sensing approach, an estimated 80 million hectares of Southeast Asian forest lands were lost from 2005 to 2015 (Estoque et al., 2019), where Indonesia and Malaysia have the highest rates of forest cover loss. An average of 70% of protected forests in Sundaland were subject to anthropogenic pressures, restricting geographic ranges of important wildlife (Verma et al., 2020). In the Philippines, almost 15,000 hectares of forest cover were lost. Unfortunately, most of these forests are declared key biodiversity areas (Daipan, 2021). As forests are a major fungal habitat, deforestation can extremely affect the population of fungi that are important in decomposition, nutrient cycling, and productivity in the forest ecosystem (Pampolina et al., 2023).

Fungal diversity is interconnected to forest loss, being sensitive to carbon-rich substrate as an energy source, providing moisture, and balancing pH to function in an eco-physiological way, which are important forest health indicators. For instance, the impact of landscape shifts by monocultures, specifically forest destruction to give way for plantations, has greatly reduced fungal species richness and abundance in the last decades (Botterill-James et al., 2024). Oil palm cultivation across

Southeast Asia has replaced natural forests at an unprecedented rate, significantly reducing ecosystem functions (Dislich et al., 2017). Kerfahi et al. (2014) studied the impacts of oil palm cultivation in Borneo, using soil fungi sampled across various forest land use types. Results revealed that intact forests have the most Basidiomycetes while the Ascomycetes dominate the oil palm plantations. In Peninsular Malaysia, a separate study by McGuire et al. (2015) showed that the oil palm plantations have the least number of recovered ectomycorrhizal (ECM) taxa in contrast with forest-lands having 15–20% more ECM relative abundance, suggesting that the long-term effect of continuous forest clearing prevents the establishment of higher fungal groups dependent on hardwoods. In addition, monoculture forests encouraged the recruitment of plant pathogenic Ascomycetes as observed in rubber and oil palm plantations (Kerfahi et al., 2014; Monkai et al., 2017).

The drastic change in floral diversity in most Southeast Asian forests has adversely affected the distribution of the ectomycorrhizal fungi associations (Brearley, 2012). Mycorrhizas are obligate root symbionts on restricted host taxa from the Dipterocarpaceae, Pinaceae, Myrtaceae, Fagaceae, and Juglandaceae (Corrales et al., 2022). The ECM are consistently more diverse and abundant in natural forest settings, as reported in the lowland Dipterocarp forests of Sumatra, Indonesia (Noorchat et al., 2012), Dipterocarp forests of Thailand (Ruankaew Disyatat et al., 2016), Malaysia (Lee et al., 2003), Philippines (Parlucha et al., 2021), and Vietnam (Ingleby et al., 2000). The prominent ECM in Dipterocarp forests were from the families Amanitaceae, Boletaceae, and Russulaceae (Brearley, 2011). Mycorrhizas have inevitable roles in germinating and establishing seedlings, especially in highly disturbed sites like the Pine forests of Vietnam and the Philippines (Sims et al., 1997; Giao, 1996). Host trees benefit from mycorrhizal symbiosis by increasing soil phosphorus uptake via the extraradical hyphae colonizing the root tips (Plassard & Dell, 2010). Despite the numerous benefits of mycorrhizas in forest ecosystems, more research is needed to harness the potential of fungal resources in biodiversity conservation and forest restoration in Southeast Asia.

This chapter explores the biodiversity of forest fungi from selected Southeast Asian regions, particularly the Philippines, Thailand, Indonesia, Malaysia, Lao-PDR, and Vietnam, though mentioning other ASEAN members whenever information on publications are available. It highlights the significant values of forest fungi in their application (e.g., mycorrhizae) in restoring denuded forest ecosystems in the region. In addition, issues on climate change, emerging and re-emerging diseases, alternative food, and conservation challenges are discussed to enlighten the critical role of forest fungi amid the changing planet.

Status of Forest and Fungal Resources in the ASEAN Region

Southeast Asia harbors nearly 15% of the world's tropical forests (Stibig et al., 2014; Estoque et al., 2019). The region also contains at least four global biodiversity hotspots (Sodhi et al., 2010). Despite its ecological importance, the region is most affected by deforestation, particularly in tropical, humid, and lowland forest areas (Global Forest Watch.org, 2025). According to comprehensive estimates by Estoque et al. (2019), Southeast Asia lost approximately 80 million hectares of forest between 2005 and 2015, translating to an annual deforestation rate of about eight million hectares. Indonesia accounted for nearly two-thirds of this loss, contributing 62.0%. Malaysia followed with 17.4%, while Myanmar and Cambodia experienced deforestation rates of 4.6% and 5.3%, respectively.

The Global Biodiversity Information Facility (GBIF) maintains extensive records of fungal species compiled worldwide from international research networks and scientific publications. Recent GBIF reports provide valuable insights into the current state of fungal diversity across Southeast Asia, as discussed in the following sections. The data from GBIF.org (2025) includes macrofungi and microfungal species, reflecting diverse habitats and ecological niches beyond forest ecosystems alone.

The total global occurrences of fungi stand at an impressive 45,318,602, encompassing approximately 173,660 recognized species worldwide. Southeast Asia contributes significantly to this global biodiversity, with 41,433 fungal species recorded across ASEAN countries. This regional diversity constitutes a substantial proportion of global fungal biodiversity, accounting for about 23.86% of recorded species occurrences (GBIF.org, 2025). Table 17.1 presents the recorded species by country, while Fig. 17.1 highlights the representative genera of tropical forest fungi.

Philippines

The Philippines is known for its rich biological diversity but faces severe ecological degradation, putting its biodiversity at risk of collapse. Ongoing environmental neglect has severely strained ecosystems, leading to serious consequences for the human population. The Philippine forestry statistics (2023) report that the country has a total forest cover of 7,226,394 ha, about 24% of the total land area. This comprises closed, open, and mangrove forests (FMB-DENR, 2023).

The Philippines has identified 7,248 fungal species, with 51,166 recorded occurrences, predominantly Ascomycota (31,841 records) followed by Basidiomycota (18,947 records) (GBIF.org, 2025). One species, *Gymnoderma insulare* Yoshim. & Sharp is categorized as endangered by the IUCN; four species—*Amauroderma renidens* (Bres.) Torrend, *Ganoderma renidens* Bres., *Trichoglossum walteri* (Berk.) E.J. Durand and *Geoglossum walteri* Berk—are vulnerable, and one species, *Sarcosoma globosum* (Schmidel) Casp., is near threatened.

The Philippine forest is home to many plant and animal species and other organisms like forest fungi. The Philippines presently possesses 376 validated nomenclatures of macrofungi, categorized into 66 families and 130 genera (Jacob et al., 2022). The macrofungal taxa documented within the nation are primarily affiliated with the classes Ascomycota and Basidiomycota, exhibiting considerable species diversity, particularly within the families Agaricaceae, Lycoperdaceae,

Table 17.1 Distribution of fungal resources across the Southeast Asian region. (Source: GBIF.org, 2025; IUCN, 2025)

Country	No. of recorded species	IUCN red list species	Conservation status
Thailand	11,826	*Calostoma insigne* (Berk.) Massee	Endangered
		Boletinellus merulioides (Schwein.) Murrill	Vulnerable
Malaysia	7,928	*Spongiforma squarepantsii* Desjardin, Peay & T.D.Bruns	Endangered
		Calostoma insigne (Berk.) Massee	Endangered
		Clavaria zollingeri Lév.	Vulnerable
		Boletinellus merulioides (Schwein.) Murrill	Vulnerable
		Amanita aporema Boedijn	Vulnerable
		Cordyceps sinensis (Berk.) Sacc.	Vulnerable
		Hapalopilus croceus (Pers.) Donk	Vulnerable
Philippines	7,248	*Gymnoderma insulare* Yoshim. & Sharp	Endangered
		Amauroderma renidens (Bres.) Torrend	Vulnerable
		Ganoderma renidens Bres.	Vulnerable
		Trichoglossum walteri (Berk.) E.J.Durand	Vulnerable
		Geoglossum walteri Berk.	Vulnerable
		Sarcosoma globosum (Schmidel) Casp.	Near threatened
Indonesia	6,435	*Concamerella pachyderma* (Hue) W.L.Culb. & C.F.Culb.	Critically endangered
		Cetreliopsis papuae Randlane & Saag	Endangered
		Amanita aporema Boedijn	Vulnerable
		Clavaria zollingeri Lév.	Vulnerable
		Aurantiporus croceus (Pers.) Murrill	Vulnerable
		Geoglossum difforme Fr.	Near threatened
Vietnam	3,230	*Bondarcevomyces taxi* (Bondartsev) Parmasto	Vulnerable
Singapore	1,697	*Boletinellus merulioides* (Schwein.) Murrill	Vulnerable
		Amanita aporema Boedijn	Vulnerable
		Perenniporia medulla-panis (Jacq.) Donk	Near threatened
Myanmar	1,016	Data deficient	Data deficient
Cambodia	692	*Amanita aporema* Boedijn	Vulnerable
Brunei Darussalam	683	*Clavaria zollingeri* Lév.	Vulnerable
Laos P.D.R	641	*Buchwaldoboletus lignicola* (Kallenb.) Pilát	Vulnerable
		Tricholoma matsutake (S. Ito & S. Imai) singer	Vulnerable
Timor Leste	37	Data deficient	Data deficient

Fig. 17.1 Tropical fungi represented by genera of ectomycorrhizal fungi: (**a**). *Pisolithus*, (**b**). *Laccaria*, (**c**). *Cantharellus*, (**d**). *Thelephora*, (**e**). *Lactarius*), saprophytic (**f**). *Stereum*, (**g**). *Xylaria*, (**h**). *Hypholoma*, (**i**). *Dacrymyces*, (**j**). *Ganoderma*), and edible (**k**). *Scleroderma*, (**l**). *Termitomyces*, (**m**). *Pleurotus*, (**n**). *Auricularia*, (**o**). *Schizophyllum*)

Marasmiaceae, and Polyporaceae (Jacob et al., 2022). This heterogeneous assemblage of fungi facilitates biodiversity, enhances ecosystem resilience, establishes mutualistic associations with other life forms, and executes vital ecological roles, all imperative for preserving thriving forest ecosystems.

Among the most studied forest ecosystems in the country are Mount Makiling Forest Reserve (MMFR), Mount Banahaw-San Cristobal Protected Landscape (MBSCPL), and Quezon Protected Landscape (QPL) (Pampolina et al., 2023). Based on the research of Parlucha et al. (2021), these forest ecosystems showed a balanced ratio of saprophytic and mycorrhizal fungi and some parasitic species of economic and cultural significance. A separate investigation carried out in the Mt. Arayat Protected Landscape (MAPL) documented a cumulative total of 224 macrofungi, which are categorized into 2 phyla, 4 classes, 12 orders, 36 families, 53 genera, and 108 species (Bustillos et al., 2024).

Other popularly known forest formations with fungal species assessed include Philippine Teak forest in Lobo, Batangas, Bazal-Baubo Watershed, Aurora, Mt. Maculot, Mt. Iraya Batanes Protected Landscape and Seascape, Mt. Palaypalay-Mataas na Gulod Protected Landscape, Taal Volcano Protected Landscape, Hinulugang Taktak Protected Landscape, Catanduanes Watershed Forest Reserve, Mt. Timpoong Hibok-hibok Natural Monument, Rajah Sikatuna Protected Landscape, San Fernando City Forest, La Union Province, Nueva Vizcaya Watershed, Sierra Madre Mountain Range Cagayan Side, Initao-Buenavista Protected Landscape and Seascape, Subic Watershed Forest Reserve, Malagos Watershed Reservation, Mt. Hamiguitan Range Wildlife Sanctuary, Dinagat Island Forest, Cavinti Underground River and Cave Complex, Mt. Data, Benguet side, Mt. Province, Mt. Isarog, Camarines, Mt. Pangasugan, Leyte, Mt. Pulag, Kabayan, Benguet, Mt. Cleopatra Needle Forest Reserve, Angat Watershed Forest Reserve, Norzagaray Bulacan, Isabela State Univ. Forest Reserve, Mt. Umubi Situated at Upper Casecnan Area of Alfonso Castañeda, Nueva Vizcaya, Kalikasan Forest Park Bicol University, Legazpi City, Bicol, Eastern Philippines. These diverse natural reserves and protected landscapes serve as crucial habitats for various fungal species and play a vital role in preserving the ecological balance of their respective regions.

Lao, PDR

The People's Democratic Republic of Laos is a landlocked country bordered by the other four Southeast Asian countries (Thailand, Cambodia, Vietnam, and Myanmar). Almost 80% of Laos is mountainous with rich species diversity, yet it is considered one of the biodiversity hotspots in Indochina (Myers et al., 2000; Chouangthavy et al., 2020). Documentation of fungal diversity in Laos remains minimal, with studies primarily conducted in national parks and protected areas with ecological significance. Laos has identified 641 fungal species with 2,558 recorded occurrences, composed of Basidiomycota (1,957 records), followed by Ascomycota (561 records) and Mucoromycota (16 records) (GBIF.org, 2025). Two species, *Tricholoma matsutake* (S.Ito & S.Imai) Singer and *Buchwaldoboletus lignicola* (Kallenb.) Pilát are listed as vulnerable under the IUCN category.

A mushroom survey in the Phou Khao Khuoay National Protected Area (PKKNPA) revealed 149 species from 55 families. Almost 40% are saprophytes from Polyporaceae, Ganodermataceae, and Marasmiaceae. Ectomycorrhizal families were also recorded, representing the Russulaceae, Boletaceae, and Amanitaceae (Lee et al., 2018).

Above the PKKNPA lies another conservation site, the Phousbous National Protected Area (PNPA), with 284 fungal species, as reported by Hong et al. (2019). Forty-seven (47%) percent of the total species recorded in PNPA were ectomycorrhizal fungi dominated by the Boletaceae, which are usually found in conifer forests. In Dong Hua Sao National Protected Area (DHSNPA), reports of Hypocreales entomopathogenic fungi were observed, namely *Metarhizium granulomatis*, *Akanthomyces novoguineensis*, *Cordyceps ninchukispora*, *Ophiocordyceps acicularis*, and *O. nutans* (Lee et al., 2021). Documentation of fungal diversity in Laos has significantly improved through molecular advances in fungal taxonomy. New species reports include two agarics, *Lepiota citrophylla* and *L. aureofulvella*, in a deciduous forest (Sysouphanthong et al., 2020), the edible mushroom *Pleurotus djamor* (Phonemany et al., 2021), and wood decay fungi *Trametes cubensis* (Hapuarachchi et al., 2021). The ectomycorrhizal *Amanita princepoides* and *A. chepangianoides* from the pine and deciduous forest (Pedersen & Læssøe, 2024). *Hebeloma parvisporum*, only reported in Laos, is an edible mushroom "wai khom" sold in local markets (Eberhardt et al., 2020).

Indonesia

Indonesia hosts an extraordinary diversity of fungi, thriving across ecosystems ranging from mangrove forests to primary and secondary forests. The country has identified 6,435 fungal species, with 25,503 recorded occurrences, predominantly Ascomycota (14,307 records), followed by Basidiomycota (9,353 records) (GBIF.org, 2025). One species, *Parmotrema pachydermum* (Hue) O. Blanco, A. Crespo, Divakar, Elix & Lumbsch, is critically endangered. At the same time, *Nephromopsis papuae* (Randlane & Saag) Divakar, A. Crespo & Lumbsch is classified as endangered. Three species—*Clavaria zollingeri* Lév., *Hapalopilus croceus* (Pers.) Donk, *Amanita aporema* Boedijn—are vulnerable, and one species, *Geoglossum difforme* Fr., is near threatened, according to IUCN.

Recent research has shed light on their biodiversity and potential applications, emphasizing their ecological and economic significance. For example, high macrofungal diversity was observed in the Kragilan Pine Forest in Magelang, revealing eight species from different genera and seven additional Basidiomycota species (Noviyanti et al., 2022). Additional studies in other Indonesian forest ecosystems have documented diverse macrofungal species, including *Ganoderma applanatum*,

Earliella scabrosa, *Polyporus tenuiculus*, *Polyporus squamosus*, *Microporus xanthopus*, *Favolus grammocephalus*, *Pycnoporus cinnabarinus*, *Ganoderma lucidum*, and *Laccaria vinaceoavellanea* (Pardosi et al., 2020). Another study reported 10 species across six orders and nine families, including *Auricularia* cf. *nigricans*, *Tremella* cf. *fuciformis*, *Gymnopus* sp., and *Coprinellus* sect. *Domestici*, *Macrocybe* sp., *Lycoperdon* cf. *curtisii*, *Ganoderma* sp., *Lentinus* cf. *sajor-caju*, *Mutinus* cf. *elegans*, and *Geastrum* sp. (Augustinus & Putra, 2021).

Similarly, many mycorrhizal species were found in secondary forests dominated by *Tristaniopsis* tree species in Indonesia. Researchers identified 56 ectomycorrhizal fungal species, some of which are also present in primary dipterocarp forests in Malaysia (Helbert et al., 2019). However, no species were shared between *Tristaniopsis* forests in different regions, suggesting that these secondary forests serve as refuges for ectomycorrhizal fungi, which is critical for dipterocarp forest restoration amid deforestation (Helbert et al., 2019).

Some studies have also explored the endophytic fungal communities in forest ecosystems. For example, a study identified 288 fungal species from the mangrove ecosystem (Trianto et al., 2021). Another study found that roots of *Paraserianthes falcataria* and *Sorghum bicolor* from tropical forests in Indonesia were also home to 12 and 21 endophytic fungi, respectively, with species representing several genera such as *Acrocalymma*, *Fusarium*, *Tolypocladium*, *Penicillium*, *Talaromyces*, *Exophiala*, *Dictyosporium*, *Pseudochaetosphaeronema*, *Mariannaea*, *Trichoderma*, and *Mycoleptodiscus* (Maulana et al., 2018). The study further found that plant species significantly influence endophytic fungal communities, with forests containing *Gmelina* sp., *Artocarpus champeden*, mixed dipterocarp forests, and secondary *Macaranga* sp. forests exhibiting higher diversity (Maulana et al., 2018).

The economic potential of fungal species in Indonesia has also been explored. For example, *Fusarium equisetii* isolated from leaves of *Avicennia* sp. exhibited the highest protease activity, highlighting the potential of mangrove-associated fungi to produce commercially important compounds (Trianto et al., 2021). Fungal species documented in Kalipoh Village Forest show promise for food and medicinal applications (Armadhan et al., 2023). Twenty-four out of 34 macrofungal species from 22 genera and 18 families, primarily found on decayed wood and leaf litter, had the potential as a source of food and medicine (Armadhan et al., 2023). Similarly, a study in Kamojang Natural Reserves and Nature Park, West Java, recorded 35 species of edible macroscopic fungi, 10 of which are traditionally used by local villagers for food and medicine (Arko et al., 2017).

Thailand

Thailand has a total of 11,826 identified species in the country, with recorded 100,482 fungal occurrences, predominantly Ascomycota (53,125 records) and Basidiomycota (37,961 records) (GBIF.org, 2025). Only two species have recognized IUCN conservation status: *Calostoma insigne* (Berk.) Massee (endangered) and *Boletinellus merulioides* (Schwein.) Murrill (vulnerable).

Thailand had 16.9 million hectares of natural forest, extending over 33% of its land area as recorded in 2020 (Global Forest Watch, 2025). The extensive forest cover in the western forest complex of Thailand is the main concentration of forest resources as it encompasses national parks such as Huai Kha Khaeng Wildlife Sanctuary, Thung Yai Naresuan Wildlife Sanctuary, Khao Yai National Park, Kaeng Krachan National Park, and other minor national parks.

This extensive forest resource is home to diverse fungi species. Studies conducted in northern Thailand have recorded more than 500 species of macro- and microfungi (Hyde et al., 2018). Dipterocarp forest in Thailand with diverse species was also found to have a greater diversity of ectomycorrhizal fungi compared to monoculture plantations (Ruankaew Disyatat et al., 2016). Compared to other land uses, studies on soil fungi in the country's secondary dry evergreen forest showed the highest diversity (Puangsombat et al., 2010). There are hundreds of studies conducted related to forest fungi and their applications (Siripong et al., 2009; Kumla et al., 2014; Hyde et al., 2018).

Malaysia

Malaysia has identified 7,928 fungal species, with 50,075 recorded occurrences, primarily Ascomycota (27,974 records), followed by Basidiomycota (16,858 records) (GBIF.org, 2025). Two species, *Spongiforma squarepantsii* Desjardin, Peay & T.D.Bruns and *Calostoma insigne* (Berk.) Massee is listed as endangered, while five species—*Clavaria zollingeri* Lév., *Boletinellus merulioides* (Schwein.) Murrill, *Amanita aporema* Boedijn, *Cordyceps sinensis* (Berk.) Sacc., and *Hapalopilus croceus* (Pers.) Donk—are classified as vulnerable by the IUCN.

Some of these fungal species are found in Malaysia's vast forest resource. In 2020, Malaysia had 18.0 million ha of natural forest, extending over 55% of its land area (Global Forest Watch, 2025). Malaysia encompasses a total land area of around 330,345 km^2, featuring roughly 8,840 km of coastline and more than 879 islands. Its forests, characterized by diverse ecosystems and abundant flora and fauna, position Malaysia among the world's megadiverse nations (FAO, 2020). The country's terrestrial biodiversity is found mainly in tropical rainforests, stretching from coastal plains to mountainous regions, including inland water bodies like rivers and lakes. The primary forest classifications present in Malaysia include lowland dipterocarp forests, hill dipterocarp forests, upper hill dipterocarp forests, montane ericaceous forests, peat swamp forests, and mangrove forests. Lowland and hill dipterocarp forests are commonly called mixed dipterocarp forests (FAO, 2020), particularly in Sabah and Sarawak.

Vietnam

Vietnam's diverse forest landscapes support a rich fungal ecosystem with significant medicinal, ecological, and economic potential. The Ministry of Agriculture and Rural Development of Vietnam estimated in 2020 an approximately 14.6 million hectares of forest, accounting for 41.89% of its total country's land area. About 10.29 million hectares consist of primary or naturally regenerated forests. In comparison, roughly 4.3 million hectares comprise plantation forests (MARD, 2020). Vietnam has identified 3,230 fungal species, with 15,114 recorded occurrences, primarily Ascomycota (9.481 records) followed by Basidiomycota (4,349 records) (GBIF.org, 2025). One species, *Bondarcevomyces taxi* (Bondartsev) Parmasto, is vulnerable under the IUCN category.

Conservation studies in nature reserves such as the Phong Dien Nature Reserve, Son Tra Nature Reserve, and Lung Ngoc Hoang Wetland Nature Reserve have identified high macrofungal diversity. Newly recorded species and endangered fungi, like *Lentinus sajor-caju* and *Cookeina tricholoma*, underscore the need for preservation efforts in Phong Dien Nature Reserve (Ngo & Nguyen, 2015). In Son Tra Nature Reserve, Tran et al. (2023) identified 110 fungal taxa belonging to 42 genera, 23 families, 10 orders, and 3 classes of the phylum Basidiomycota. *Microporus xanthopus* and *Stereum hirsutum* are encountered more frequently during the sampling process. In Lung Ngoc Hoang Wetland Nature Reserve, 87 species of mushrooms were found, of which 29 are edible, 16 are medicinal, 13 are wood-destroying, 23 are poisonous, and 17 are of other groups (Duong et al., 2022). Taxonomic research continues to expand knowledge, with new species recorded in Tra Giac Mountain, such as *Stemonitis pallid*, *Thelephora penicillata*, *Coprinellus hiascens*, and *Psathyrella pertinax* (Tran et al., 2019).

Furthermore, medicinal mushrooms were also recorded in some regions in Vietnam. For example, Thua Thien Hue Province, a hotspot for medicinal mushrooms, hosts 111 species, with key families, including Auricalariaceae, Boletaeeae, Coriolaceae, Ganodermataceae, Hymenochaetaceae, Lentinaceae, Lycoperdaceae, Polyporaceae, Sclerodermataceae, Schizophyllaceae, Stereaceae, Tremellaceae, Tricholomataceae, and Xylariaceae. Notably, cultivation efforts have successfully domesticated 35 species in 4 families: Coriolaceae, Ganodermataceae, Hymenochaetaceae, and Lentinaceae, the 6 precious Lingzhi, using local agricultural by-products (Ngo, 2013).

Singapore

Singapore has identified 1,697 fungal species, with 4,712 recorded occurrences, primarily Basidiomycota (2,482 records), followed by Ascomycota (2,164 records) (GBIF.org, 2025). Two species, *Boletinellus merulioides* (Schwein.) Murrill and *Amanita aporema* Boedijn are categorized as vulnerable by IUCN, while one species, *Perenniporia medulla-panis* (Jacq.) Donk is listed as near threatened. A major stake of this diverse fungal species composition is inhabiting the 11,300 ha of natural forest, extending over 16% of the land area of Singapore (Global Forest Watch. Org, 2025). However, Lee (2019) noted that there remains a great many specimens from Singapore nature reserves, particularly Bukit Timah, that had only been identified to the generic level or only to family from the Agaricaceae, Boletaceae, Entolomataceae, Lyophyllaceae, Marasmiaceae, Polyporaceae, Xylariaceae, and many others. Some of the common species in Bukit Timah Nature Reserve include *Pycnoporus sanguineus* (L.) Murrill, *Schizophyllum commune* Fr., *Microporus xanthopus* (Fr.) Kuntze, *Favolaschia manipularis* (Berk.) Teng, and *Scleroderma sinnamariense* Mont. (Lee, 2019). New fungal collections in the nature reserve not listed in published checklists indicate that the fungal diversity in Bukit Timah is higher than previously thought (Lee, 2019). This fungal diversity was also observed even in urban trees in Singapore based on a metagenomic survey. Of the 10,646 annotated operational taxonomic units (OTUs), the majority were identified as either Ascomycota (63.4%) or Basidiomycota (22.5%). Notably, fourteen Basidiomycete OTUs—comprising nine Polyporales, four Hymenochaetales, and one Boletales—along with three ascomycete OTUs belonging to the genus *Scytalidium*, exhibited strong associations with diseased trees (Hong et al., 2023).

Myanmar (Burma)

Myanmar has identified 1,016 fungal species, with 2,612 recorded occurrences, primarily from Ascomycota (1,997 records) and Basidiomycota (482 records). No species are currently listed under the IUCN threatened categories. The country recorded, in 2020, 41.8 million hectares of natural forests, extending over 62% of its land area (Global Forest watch.org, 2025). While mushroom inventory in Myanmar is still in its infancy and many new records and species remain to be described, an effort to characterize mushrooms through molecular methods has recorded 18 new species from nature reserves and botanic gardens in Myanmar (Hosaka et al., 2021). Among these included *Lactarius austrotorminosus*, *Craterellus atratus* (Corner) Yomyart, Watling, Phosri, Piap. & Sihan., *Favolaschia calocera* R. Heim, *Fistulina hepatica* (Schaeff.) With., *Geastrum courtecuissei* P.-A.Moreau & C. Lécuru, *Laccaria vinaceoavellanea* Hongo, *Lactifluus piperatus* (L.) Roussel, *Lactifluus volemus* (Fr.) Kuntze, *Leucocoprinus birnbaumii* (Corda) Singer, *Phallus merulinus* (Berk.) Cooke, *Pisolithus albus* (Cooke & Massee) Priest, *Psathyrella candolleana* (Fr.) Maire, *Pycnoporus sanguineus* (L.) Murrill, *Russula virescens* (Schaeff.) Fr., *Trogia infundibuliformis* Berk. & Broome, *Turbinellus floccosus* (Schwein.) Earle ex Giachini & Castellano (Hosaka et al., 2021). Additionally, a survey of *Ganoderma* spp. in mixed deciduous and tropical rainforests of Myanmar, located within the Greater Mekong Subregion, recorded two species—*G. hoehnelianum* and *G. williamsianum*—as new distributional records for Myanmar (Luangharn et al., 2021). Moreover, *G. myanmarense* was described as a new species from the region (Luangharn et al., 2021).

Cambodia

Cambodia has identified 692 fungal species, with 1,408 recorded occurrences dominated by Ascomycota (986 records), followed by Basidiomycota (386 records). One species, *Amanita aporema* Boedijn, is classified as vulnerable by the IUCN. Recorded in 2020, the country had 6.60 million hectares of natural forest, extending over 36% of its land area (Global Forest watch.org, 2025). A comprehensive survey of fungal diversity in Cambodia from 2009 to 2014 yielded 3,950 mushroom specimens collected from various forested regions, including the Central Cardamom Protected Forest, Kirirom National Park, Seima Biodiversity Conservation Area, Mondulkiri Protected Forest, Phnom Bokor National Park, Western Siem Pang Protected Forest, and Ream National Park (Kim et al., 2017). These specimens were identified through morphological and genetic analyses, resulting in the classification of 302 species across 165 genera and 64 families. Kim et al. (2017) noted that most mushrooms belonged to the phylum Basidiomycota (276 species), with the remainder classified under Ascomycota (26 species). Species-rich families included Polyporaceae (26.2%), Marasmiaceae (9.1%), Ganodermataceae (7.7%), Hymenochaetaceae (7.5%), and Mycenaceae (7.1%).

In the western and eastern forest areas in Cambodia, about 1,383 specimens were further analyzed and grouped into 238 species, 160 genera, 52 families, and 15 orders within three fungal phyla. Dominant genera included *Mycena*, *Ganoderma*, *Microporus*, *Marasmius*, *Marasmiellus*, *Phellinus*, and *Trametes* (Kim et al., 2016). Interestingly, the study showed elevation as a driver species distribution, with the highest diversity observed between 251 and 500 m above sea level. Several species, such as *Daldinia concentrica*, *Microporus vernicipes*, and *Stereum ostrea*, were found across all elevation ranges, while others were restricted to elevations below 500 m (Kim et al., 2016). Regional endemism was also evident, with 141 species unique to the Cardamom region and 20 species exclusive to Seima. Diversity indices for eastern and western regions were 1.83 and 1.77, respectively, with a species similarity index of 0.42, indicating moderate overlap and high regional diversity in Cambodia's fungal communities (Kim et al., 2016).

Brunei Darussalam

Brunei Darussalam has identified 683 fungal species, with 2,114 recorded occurrences, predominantly Ascomycota (1,608 records), followed by Basidiomycota (318 records). One species, *Clavaria zollingeri* Lév, is listed as vulnerable by the IUCN. Brunei maintains extensive forest cover with approximately 488,000 hectares of natural forest, representing 84% of the country's total land area (Global Forest Watch, 2025). The unique and understudied forest ecosystems of Brunei, including mixed dipterocarp and mangrove forests in northwest Borneo, harbor diverse fungal communities with biotechnological potential. In the dipterocarp forests, 19 fungal isolates were obtained using potato dextrose agar and identified through ITS rRNA barcoding (Taha et al., 2020). *Trichoderma*, *Penicillium*, and *Aspergillus* were common to both forest types, while Saksenaea was uniquely isolated from mangroves. In a separate survey on mangrove forests, 28 endophytic Ascomycota strains were isolated from *Rhizophora apiculata*, *Nypa fruticans*, and *Xylocarpus granatum*, with *R. apiculata* exhibiting the highest diversity, particularly in leaf tissues (Mohamad et al., 2023). Studies on manglicolous fungal consortia associated with *N. fruticans* further revealed three distinct assemblages: a commensalistic group dominated by *Astrosphaeriella striatispora*, *Linocarpon nypae*, and *Oxydothis nypae*; a mutualistic group composed of *L. appendiculatum* and *L. bipolaris*; and a low-frequency, possibly antagonistic group including *Anthostomella* spp. and *Trichocladium* sp. (Sarma & Hyde, 2018).

Timor-Leste

Timor-Leste has significantly fewer documented fungal records, with only 37 identified fungal species resulting in 46 occurrences. These occurrences primarily belong to Basidiomycota (36 occurrences), followed by Ascomycota (10 occurrences) (GBIF.org, 2025). No fungal species in Timor-Leste are recognized under the IUCN Red List. Timor-Leste reported approximately 1.01 million hectares of natural forest in 2020, accounting for 67% of its territory, underscoring the critical need for further biodiversity research and documentation efforts in the country (Global Forest Watch, 2025). Research on fungal diversity in the forests of Timor-Leste remains limited; however, notable pathogenic threats to native and agricultural ecosystems have been documented. A significant outbreak of gall rust disease, primarily attributed to *Uromycladium tepperianum*, has severely impacted *Falcataria falcata*, the principal shade tree species in coffee plantations (Old & Cristovao, 2003). By 2001, approximately 67% of coffee-growing areas across all eight districts were affected, with the Ermera district—the country's largest coffee-producing region—experiencing the most extensive damage (Old & Cristovao, 2003). This disease was first recorded in the 1980s and has since spread throughout Southeast Asia (Lelana et al., 2022). Initially identified as *U. tepperianum*, subsequent taxonomic revision recognized the pathogen affecting *F. falcata* as *Uromycladium falcatariae* to reflect its host specificity (Doungsa-ard et al., 2014). This fungus, which belongs within the order Pucciniales (Basidiomycota), induces gall formation on aerial plant parts, particularly shoots and branches, resulting in defoliation, crown dieback, and ultimately, tree mortality (Old & Cristovao, 2003). Given the ecological and economic importance of forest resources, there is a pressing need to expand research on fungal diversity in Timor-Leste to support sustainable forest management and disease mitigation strategies.

Challenges in Sustaining Forest Ecosystems

Climate Change and Emerging Forest Diseases

Global climate change significantly reshapes plant host-pathogen dynamics, influencing forest health and stability. While fungal pathogens play a crucial role in regulating forest ecosystems, shifts in host susceptibility and environmental conditions—primarily driven by anthropogenic factors—can disrupt these ecological balances (Ghelardini et al., 2016; Guégan et al., 2020). Such disruptions may facilitate the emergence of novel fungal phytopathogens or increase the virulence of existing ones. In Southeast Asia, some fungal pathogens have been documented causing diseases in forest stands across newly affected geographic regions (e.g., Mastrili et al., 2024; Pham et al., 2021). Moreover, fungal pathogens that previously had minimal impact on native tree species are now driving widespread tree mortality (Welsh et al., 2009; Worrall et al., 2012).

With its unprecedented magnitude and pace, climate change is expected to drive significant shifts in forest disease dynamics (Desprez-Loustau et al., 2007). Although empirical research on climate change and emerging forest diseases in Southeast Asia remains limited, studies from other regions have identified climate change as a key driver of disease emergence. Abiotic stressors such as drought and flooding can exacerbate tree fungal diseases by weakening host defenses, making them more susceptible to infection (La Porta et al., 2008; Oliva et al., 2014). For instance, certain *Armillaria* species known as secondary agents exploit host vulnerability following drought, temperature extremes, insect infestations, or site degradation (Sturrock et al., 2011). Similarly, physiological or environmental stress can disrupt plant-endophyte relationships, potentially triggering pathogenicity. This was observed in *Discula quercina*, a fungal endophyte that caused a new disease in *Quercus* trees under drought stress (Rabiey et al., 2019). Climate-driven pathogen migration may also expand disease incidence and geographic range as pathogens encounter new hosts or vectors (La Porta et al., 2008).

Due to international trade, many emerging diseases result from the introduction of alien pathogens to new geographic areas (Guégan et al., 2020). The widespread establishment of exotic tree plantations has facilitated the worldwide movement of exotic forest fungi, leading to the emergence of novel fungal diseases in new regions (Wingfield et al., 2001). Newly identified Mycosphaerella species exemplified this, initially reported as the cause of leaf spot disease in eucalypt plantations across Vietnam, Indonesia, and China (Burgess et al., 2007). Alien fungal pathogens may have been introduced alongside their host species and subsequently proliferated due to the plantations' limited genetic and age diversity, which weakens natural resistance and facilitates rapid disease spread (Burgess & Wingfield, 2004). Moreover, the spillover of these pathogens into native ecosystems poses a significant threat to natural forests. For instance, the fungal pathogen *Ceratocystis manginecans* in Vietnam has broadened its host range, transitioning from exotic *Acacia* and *Eucalyptus* species to infect native *Dalbergia tonkinensis* and *Chukrasia tabularis*

(Thu et al., 2021). *Acacia* and *eucalyptus* are Vietnam's two most widely planted tree species, and the area dedicated to their plantations is expanding significantly. However, with the growth of these plantations, various pathogenic fungi have emerged as challenges. Among these, the most prominent issues are the acacia wilt caused by *Ceratocystis manginecans* and the eucalyptus leaf spot caused by *Cryptosporiopsis eucalypti*. *C. manginecans* has been recorded in Vietnam since 2012, damaging *Acacia mangium*, *Acacia auriculiformis,* and *Acacia* hybrid. Symptoms of infected trees are wood discoloration, bleeding, wilting, and dieback. All infected trees could die quickly, affecting the productivity and quality of the forest plantations.

Forest diseases caused by fungi depends on abiotic variables such as humidity, rainfall, and temperature (Singh et al., 2023). As mentioned previously, the effects of climate change have significantly altered atmospheric factors promoting the virulence of fungal pathogens in forests and tree plantations (Boisvenue & Running, 2006). Deforestation has amplified the effects of weather extremes, potentially leading to longer months of dry spells, increased maximum temperature, and more frequent and stronger tropical cyclones (Tölle, 2020). These conditions favor inoculum dissemination and transmission of pathogens like the *Uromycladium* gall rust affecting *Falcataria falcata* (Helfer, 2014; Rahayu et al., 2010). In Indonesia, Lelana et al. (2024) reported that the teliospores of *Uromycladium falcatariae* could spread over 500 m from the infected host trees, which could extend further in the presence of weather extremes. Spores of foliar rust are dispersed by wind and water splashes. In contrast, the germination of infective spores is influenced by prolonged leaf wetness at temperature ranges of 5–30 degrees Celsius (Fraser et al., 2017). *Cryptosporiopsis eucalypti* has been documented in Vietnam since 1990, causing leaf spot disease in *Eucalyptus camaldulensis*, *Eucalyptus urophylla*, and *Eucalyptus hybrids*. *C. eucalypti* has long-lasting fungal fruiting bodies on the tops and small branches, often causing symptoms of prolonged re-infection. This typical symptom appears in most eucalyptus species in all regions of the country, with very different rates and levels of disease. The fungus *C. eucalypti* can develop in areas with an average annual rainfall of over 700 mm to 2596 mm and temperatures from 11 to 35 °C. The most obvious impact of this fungus on eucalyptus in Vietnam is in humid areas (during and after the rainy season), with an average monthly temperature above 20 °C and an average annual rainfall greater than 2000 mm.

Climate change-induced shifts in mean temperature and moisture availability also threaten susceptible host trees across higher altitudinal gradients, increasing disease incidence (Dudney et al., 2021; Fan et al., 2022). In the Philippines, *Falcataria* plantations affected by *Uromycladium* rust were usually less severe at lower elevations. However, changes in climatic conditions would more likely increase disease severity by 400–600 masl (Tering et al., 2025). Physiological disruptions linked to the increasing surface temperatures were also reported in *Hevea brasiliensis*, another important crop in Southeast Asia (Azizan et al., 2023). Leaf fall disease outbreaks in several plantations caused severe defoliation of the tree canopy, ultimately decreasing latex production. The affected host trees undergo disruptions in the phenological cycle, delaying new leaf onset.

While disease severity worsens due to climate change, variable responses were observed, with trees benefitting from mycorrhizal interactions (Duarte & Maherali, 2022). Bakray et al. (2024) studied the effects of elevated CO_2 on the diversity of arbuscular mycorrhizal fungi (AMF) communities associated with Fagaceae tree species in Malaysia. The AM isolates from rhizosphere samples of normal and elevated CO_2 conditions (100–300 ppm) showed no significant difference in the richness of taxa. Interestingly, *Glomus* sp. was the most dominant species of the Glomerales from both sites. Elevated CO_2 and temperature rise stimulate the symbiosis of plants and AMF as an adaptation to stress. As the host plant tries to offset its internal temperatures through an increased transpiration rate, the mycorrhizal roots then expand their area of coverage to accommodate the water and nutrient demands of its host (Milović et al., 2021; Sosa-Rodriguez et al., 2013). On the contrary, climate change impacts the ectomycorrhizal (ECM) fungi unfavorably due to their obligate relationship with trees and possible resource competition with opportunistic soil microbes (Chakraborty et al., 2012).

Gaps in Scientific Knowledge and Conservation Policies

Fungi in forest ecosystems, mainly in tropical Asia, remain largely underexplored despite their critical ecological roles. They significantly influence forest dynamics but are poorly represented in biological collections, environmental policies, and scientific research. This knowledge gap is especially pronounced in emerging economies across tropical Asia. For instance, only 17% of known fungal species are preserved in culture collections, strongly biased toward temperate taxa, leaving regions such as Asia, Central and North Africa, and the Amazon underrepresented (Paton et al., 2020). Tropical Asia harbors the highest number of reported ectomycorrhizal (ECM) host plants, including key species from Dipterocarpaceae, Fagaceae, and Juglandaceae, yet many ECM fungal lineages in the region are likely still undiscovered (Corrales et al., 2022). This has been demonstrated by over a decade of fungal research in Thailand, which has significantly expanded knowledge of fungal biodiversity in the region and revealed remarkably high levels of fungal novelty (Hyde et al., 2018).

As forests—key habitats for fungi—continue to decline across Southeast Asia, conservation policies must be strengthened to mitigate the risk of fungal species extinction. Globally, fungi remain vastly underrepresented in conservation assessments compared to plants, resulting in inadequate protection measures. Between 2017 and 2019, only 285 of the 148,000 described fungal species (0.2%) were assessed for the IUCN Red List (Antonelli et al., 2020). In the UK alone, merely three fungal species are included in the global Red List of threatened species, which comprises 45,000 species (Baird & Pope, 2022). Given the high fungal diversity in Southeast Asia, the number of rare and threatened fungal species in the region is likely much higher. Southeast Asia remains a global hotspot of biodiversity and deforestation, with vast forest loss primarily driven by logging and the conversion of natural forests into plantations (Hoang & Kanemoto, 2021; Stibig et al., 2014). This ongoing deforestation threatens countless fungal species, many of which may already be at risk of extinction. Expanding fungal red-listing efforts, particularly in Southeast Asia, Africa, and Latin America, is crucial for identifying critical habitats and prioritizing conservation efforts for these often-overlooked yet ecologically essential organisms.

With the rising threat of invasive fungal pathogens, strengthening forest and plantation biosecurity in Southeast Asia is critical. Since fungal pathogens and other pests cross borders freely, a biosecurity failure in one country can have cascading effects on its neighbors, especially in Southeast Asia, where many countries share porous land and riverine boundaries and lack natural biophysical barriers (Healey et al., 2022). A key component of effective biosecurity is regular, systematic surveillance; however, many ASEAN countries, including Malaysia and the Philippines, face challenges in implementation due to limited funding and workforce (Lee & Krishnapillay, 2005). To address these gaps, comprehensive forest biosecurity laws must be developed to guide policy coordination, direct on-ground response efforts, and empower relevant agencies to act swiftly against incursions (Healey et al., 2021). Strengthening regional collaboration and resource-sharing will mitigate the spread of invasive fungal pathogens and safeguard Southeast Asia's forests and plantations.

Despite the vital role of fungi in forest health and productivity, further research is needed to fully harness their potential for reforestation. For example, ECM inocula persisting on dipterocarp seedlings in nursery and field conditions have enhanced early seedling growth and survival (Imanuddin et al., 2020; Kettle, 2010). However, significant knowledge gaps remain regarding the adaptability of ECM inocula to specific host species and site conditions and effective inoculation methods for mycorrhiza-deficient soils (Hardwick et al., 2000; Kettle, 2010). Moreover, identifying fungal species remains a major challenge, hindering efforts to optimize their applications in reforestation. Advances in molecular techniques are essential for refining fungal taxonomy and understanding inter- and intraspecific genetic variation within beneficial fungi (Kettle, 2010). Addressing these research gaps is crucial for improving fungal biodiversity conservation and strengthening forest ecosystem resilience in the face of environmental change.

Application and Perspectives of Forest Fungi in the ASEAN Region

Emerging Biotechnologies

The utilization of forest fungi for biotechnologies in the tropics is geared toward tree plantation development, sustainable agroforest production, bioremediation, ecosystem restoration, and discovery of new products. Emerging fungal biotechnologies across SEA include products like biofertilizers, bioactive compounds, growth promoters, and heavy metal accumulators, among others. In the Philippines, the application of ectomycorrhizal inoculation of *Pisolithus tinctorius* and *Scleroderma bovista* isolated from natural pine forest and eucalypt plantations was successful for the reforestation program of the country (Turnbull & de la Cruz, 1990; Brundrett et al., 1996). Commercial production of this biofertilizer was complemented with the development of MycoVAM, a soil inoculant that contains spores and hyphal biomass of *Glomus, Acaulospora, Gigaspora* grown in pot-cultured grass, which are widely applied to most trees for forest rehabilitation, agroforestry, landscaping, and bioremediation in the country. In Indonesia, a mycorrhizal inoculant called Mycosilvi was developed to supplement the nutritional needs of native tree species in degraded soils (Jayani et al., 2018). The *Glomus* spores were produced with the mycorhizzal helper bacteria (MHB) grown on a nutrient agar media. Mycosilvi provided better growth responses and chlorophyll concentration to *Falcataria moluccana*, *Samanea saman*, and *Cassia siamea* (Jayani et al., 2018). Marked increase in soil phosphorus availability (48.3 ppm) was also observed in *Albizia chinensis* treated with Mycosilvi, compost, and post-planting soil tillage compared to controlled treatments (Sahara et al., 2022). In Thailand, AMF inoculant infused with organic fertilizer was produced in alginate pellet form, which significantly improved hyphal spore viability in arid conditions (Pitaktamrong et al., 2018). Recognizing the potential of mycorrhizas in plantation establishment, optimizing spore production for mass inoculation, and extending inoculant shelf life are ongoing research opportunities.

Role of Fungi in Carbon Sequestration

As heterotrophic organisms, fungi source energy through the utilization of organic matter from living or dead bodies using carbon to grow and complete their life cycle. This manner of nutrition exhibited by fungi in most ecosystems would show their ecological function in the degradation of organic matter and cycling of nutrients that sustain productivity in terrestrial and aquatic ecosystems. In the process, fungi and other soil microbial communities like mycorrhizal associations, microalgae, oligotrophs, and copiotrophs are linked to carbon conservation on the forest floor (Buragohain et al., 2019). The diversity of forest fungi was reported in 12 forest formations in the Philippines, where more than a hundred kinds of macrofungi that include common groups of Basidiomycetes and Ascomycetes were recorded (Pampolina et al., 2023). Similar reports on fungal populations were shown in various forest ecosystems from other ASEAN regions in the tropics, as reported elsewhere in this paper.

Generally, the decomposition of organic matter by groups of fungi can store carbon on the forest floor in litter and soil. It was estimated that fungi can store about a third of carbon from fossil emissions. Considering the extensive network of fungal mycelia on belowground biomass along the soil profile, new fungal research estimated that over 13 gigatons of carbon on planet Earth is roughly equivalent to 36% of yearly global fossil fuel emissions (Case et al., 2025; Hawkins et al., 2023).

Mycorrhizas are among the fungi that contribute to plant growth and development through hyphal structures that improve the soil nutrient uptake of available phosphorus, essential elements, and micronutrients. These soil nutrients are necessary for plant growth and development, enhancing carbon sequestration through photosynthetic activities. The emergence of mycorrhizal fungi plays an important role in developing a forest ecosystem and restoring degraded landscapes that consequently follow the sequestration of carbon in soil under elevated CO_2 and nitrogen deposition (Hawkins et al., 2023). These groups of symbiotic mycorrhizal fungi colonize around 80–95% of most fine roots of higher plants that grow in diverse terrestrial habitats on planet earth (Smith & Read, 1997). Based on the evidence, the estimated amount of carbon fixed by terrestrial plants allocated to mycorrhizas was 13.12 Gt of CO_2e, which is stored in underground mycelium per year (Hawkins et al., 2023). The fungi comprise a vast underground network all over the planet underneath the different ecosystems, including all forest formations on Earth. Fungi are crucial for storing carbon and keeping the planet cooler to sustain global biodiversity.

Alternative Food and Medicinal Values

Some species of fungi, including mushrooms, are cultivated to extract polysaccharides, beta-glucans, terpenoids, and secondary metabolites that are nutrient and fiber-rich alternative foods with antioxidant and medicinal properties (Wasser, 2014). Good examples are *Schizophyllum commune*, *Termitomyces eurhizus*, *Scleroderma bovista*, *Pleurotus ostreatus*, *Ganoderma lucidum*, *Ganoderma applanatum, Auricularia* spp., and *Trametes elegans*, among others. Mushroom pharming was on purpose to imply mushroom production intended for pharmaceutical purposes (Reyes, 2024; Wasser, 2014). These tropical mushrooms are extensively found in the ASEAN region, including the Philippines (Dulay et al., 2023; Pampolina et al., 2023).

In the mountainous Cordillera province of the Philippines, where a group of indigenous Ifugao resides, the community believed that polypore *Trametes elegans* has medicinal properties that can cure cancer. This traditional practice supports evidence that species of *Trametes* contain bioactive and phenolic compounds known also for their anticancer and immunomodulatory effects (Wasser, 2014; Zhao et al., 2020). Similarly, *Trametes versicolor* demonstrated the ability to inhibit tumor growth and enhance the immune system, supporting the ethnopharmacological practices of the Ifugao (Cui & Chisti, 2003).

In addition, *Scleroderma bovista,* locally named "buo," an ectomycorrhizal fungi associated with *Pinus kesiya*, is considered precious because of the taste of the Mankayan tribe in the Mountain Province. The presence of the fruiting bodies of *S. bovista* in pine forests is important in conserving the mountain ecosystem since the community protects trees to ensure the production of these fungi. However, there is some caution with edible *Scleroderma* because some species are reportedly poisonous.

In the Northern Sierra Madre Mountain Range, the Bugkalot people use traditional medicine from various species of mushrooms to treat specific ailments (Torres et al., 2020). Examples of fungi are represented by *Formitopsis* sp., which has potential anti-inflammatory and hepatoprotective properties used to address arthritis and hepatitis. *Ganoderma applanatum* was reported to contain gastroprotective and antioxidant values employed to treat gastric ulcers (Boh et al., 2007). On the other hand, *Ganoderma lucidum* is known to cure headaches and skin diseases because of its analgesic and antimicrobial properties (Paterson, 2006). *Polyporus picipes* is traditionally used to improve digestion and alleviate gastrointestinal discomfort, while other kinds of *Polyporus* species are utilized for ulcers and wound healing with anti-inflammatory therapies (Torres et al., 2020). Further, species of *Pleurotus* (P. *dryinus*, *P. citrinopileatus, P. pulmonarius P ostreatus*) are rich in vitamin A, which is essential for good eyesight and prevents blindness.

The edible inky cap mushrooms (e.g., *Coprinus atramentaria*, *C. lagopus*, *C. cinereus*) are commonly utilized as food by the Bugkalot indigenous community (Torres et al., 2020). Coprinus species possess antimicrobial and anticancer properties, with bioactive compounds inhibiting bacterial growth and polysaccharides showing promise for cancer prevention and treatment (Zenkova et al., 2003; Das et al., 2017). Ergothionine, an antioxidant derived from *C. comatus*, could inhibit androgen receptor function in tumor cells associated with reproductive cancers (Ray et al., 2024). In vitro studies also demonstrated the cytotoxic properties of *Coprinus* extract, reducing the viability of brain cancer cells by stopping glioblastoma tumor growth and cell necrosis (Nowakowski et al., 2021).

Forest Restoration Through Mycoremediation of Degraded Ecosystems

The Southeast Asian region has one of the highest deforestation rates, losing around 1.2% of its forest annually, which is exacerbated by environmental pollution (Lai, 2022). Restoring degraded and fragmented forest ecosystems in these tropical areas is crucial to mitigate climate, protect biodiversity, clean the environment, and secure food and livelihood. Sustainable forestry through reforestation, greening programs, agroforestry, and other nature-based solutions are being practiced in the region. Using genetically superior planting materials is important to ensure the growth and survival of out planted trees during rehabilitation and reforestation. The utilization of forest fungi in forest restoration, like the application of mycorrhizal fungi, mycoremediation of mining areas, and the use of compost, biofertilizers, and biopesticides, are environment-friendly and economically efficient. This section highlights the experiences of the ASEAN region, particularly in the Philippines, Thailand, Indonesia, and Vietnam.

Philippines

The restoration of degraded ecosystems brought by mining operations requires the investigation of endomycorrhizal plants adaptive to contaminated heavy metal soil. In a study by Aggangan et al. (2015), an abandoned mining site in Cebu, Philippines, showed a succession of plants that comprised 69 species belonging to 66 genera and 35 families. The kind of plant communities such as ferns (*Lycopodium*, *Nephrolepis*), grass-shrub (*Saccharum spontaneum, Stachytarpheta jamaicensis*), and trees (*Acacia mangium*, *Pithecellobium dulce*, *Muntingia calabura*) suggested a stressed ecosystem due to the presence of heavy metals and infertile soil conditions (Aggangan et al., 2015). The results further revealed the presence of *Glomus* sp., *Acaulospora*, and *Entrophospora,* comprising 24–42% of their spores colonizing 10–100% of the fine roots of identified host plants. Further, were fruiting bodies of ectomycorrhizal fungi (e.g., *Pisolithus*, *Scleroderma*, *Thelephora*, and *Boletellus*) found under *Acacia auriculiformis, A. mangium*, *Eucalyptus urophylla*, and *E. camaldulensis.* The mycorrhizas presented here are potential inoculants that can be purified and mass-produced to fertilize plants during progressive mining site rehabilitation. However, they generally found that the AMF improved plant tolerance to heavy metals, increased bioaccumulation of heavy metals, improved root colonization despite the presence of heavy metals, and improved overall plant growth (Adiova et al., 2013).

A greenhouse experiment in Laguna demonstrated that *Glomus etunicatum*, *Glomus macrocapum*, and *Gigaspora margarita* increased the plant biomass, copper uptake in roots and stem, nutrient phosphorus uptake, promoted symbiotic microbial growth in the root system, and improved overall plant growth of *Desmodium cinereum*. Aggangan et al. (2017), in a field experiment in Marinduque, demonstrated that *Gigaspora margarita*, *Glomus etunicatum*, and *Glomus marocarpum* improved overall plant growth, increased copper, zinc, and lead uptake in roots and reduced their shoot translocation, and increased *Jatropha curcas* survival in the presence of the heavy metals. In a greenhouse (Aggangan et al., 2018) and field experiments (Aggangan et al., 2019) in Marinduque, Anggangan and company demonstrated that *Glomus*, *Gigaspora*, *Acaulospora*, *Scutellospora*, and *Entrophospora* increased the total dry biomass, root colonization, nutrient accumulation, copper uptake, survival percentage of seedlings, wood volume, and improved overall growth of *Pterocarpus indicus* (Narra) despite the presence of heavy metal.

Thailand

A few studies in Thailand demonstrated how fungi can be used for bioremediation of mining areas. These examples, however, do not extend to field testing and actual application. A study reports fungi isolated from natural community forest preservation in Phitsanulok and Pichit provinces, Thailand, that were able to degrade 2,8-DCDD (2,8-dichlorodibenzo-p-dioxin) and DDT

(Dichloro-Diphenyl-Trichloroethane). Two isolates were unidentified, while three were identified as *Trametes* sp., *Polyporus* sp., and *Nigroporus* sp. (Siripong et al., 2009). A different study screened *Pycnoporus coccineus* and *P. sanguineus* basidiocarps collected from Kanchanaburi, Phetchaburi, and Mae Hong Son provinces for laccase activity. While most strains tested positive for laccase, *Pycnoporus coccineus* strain Thongkred 013 BCU had the highest activity. The strain's crude laccase (1 U/ml) was demonstrated to have oxidized polycyclic aromatic hydrocarbons, pyrene and anthracene, fluoranthene, benzoapyrene, and phenanthrene (Thongkred et al., 2011). Another study showed *Astraeus odoratus*, *Phlebopus portentosus*, *Pisolithus albus,* and *Scleroderma sinnamariense* could solubilize toxic metal minerals, specifically, $CdCO_3$, $CoCO_3$, $CuCO_3{\cdot}Cu(OH)_2$, $2Pb(CO_3){\cdot}Pb(OH)_2$, and $ZnCO_3$ (Kumla et al., 2014).

Indonesia

In Indonesia, several studies were conducted concerning bioremediation. The following examples, however, have not reached application in the field. In a study, an *Aspergillus flavus* strain KRP1 isolated from tropical forest soil at Malang, East Java was shown to remove 97.50% and 98.73% from 10 mg/L mercury (II) in potato dextrose broth in shaken and static systems, respectively. The strain was considered to have potential for use in bioremediation (Kurniati et al., 2014). In another study in Bangka post-tin mining, Bangka Island, where Pb was most often found, six fungal isolates were resistant to Pb up to the concentration of 5000 ppm. These isolates were identified as *Rhizoctonia* (I1), *Penicilllium* sp1 (I2), *Papulaspora* (I5), *Penicillium* sp2 (I6), *Penicillium* sp3 (I7), and *Sclerotium* (I9) (Darlingga et al., 2022).

One study determined the combined effects of arbuscular mycorrhizal fungi (AMF) inoculation on the revegetation of post coal-mining lands with *Pongamia pinnata.* Results suggest that adding AMF increased the amount of Fe absorbed by the plant and concluded that the application of AMF drastically improved some chemical soil properties, making this combination suitable for rehabilitation programs in tropical post-mining areas (Agus et al., 2019). A study in a post-nickel mining area of Southeast Sulawesi Province demonstrated how native AMF, *Glomus* and *Acaulospora*, colonized the roots of 16 pioneer plants. Host responses have varied widely, ranging between 1% and 50% of AMF colonization rates (Prayudyaningsih et al., 2024). This work aligns with Agus et al. (2019), on the potential of AMF to rehabilitate the ex-nickel mining area by improving the survival of plants and the absorption of nickel.

A pot experiment by Putra et al. (2022) showed that *Glomus manihotis* increased heavy metal uptake and reduced heavy metal content in the soil, particularly Al, Co, Cr, Fe, and Pb. It also increased the biomass and morphological characteristics of *Pennisetum purpureum* despite the presence of heavy metals. Another study demonstrated white rot fungi from the forests of East Kalimantan to biodegrade polycyclic aromatic hydrocarbons (PAHs). *Trametes polyzona* PBURU 12 was selected for optimization studies. The laccase from this fungus was purified and characterized. The study concluded that live mycelia and crude laccase of PBURU 12 could degrade PAHs up to 50% within 24 h. The fungus was considered a promising candidate for bioremediation of recalcitrant pollutants (Wulandari, 2020).

In the study by Wijaya et al. (2019), fungi isolated from soil taken from Batang Toru, a gold mining site in South Tapanuli, were screened for Pb tolerance. Those that were tolerant were identified as *Trichoderma* spp. These were further assayed for Pb accumulation. Of these, three isolates were shown to accumulate Pb ranging from 36 to 92 mg/g. Hastuty and Hidayat (2019) isolated fungi from Cr(VI)-contaminated soil in the BatuBelubang village, Bangka Belitung province. They successfully screened Cr(VI)-resistant fungi that were molecularly identified as *Trichoderma* spp., specifically *T. crassum* (2 isolates), *T. virescentiflavum* (3 isolates), and *T.* aff. *Tomentosum* (2 isolates). These have the potential to bioremediate Cr(VI) in the area.

Vietnam

Information about the application of fungi in mined-out areas is scarce in Vietnam. However, there is a study by Thi Nguyen et al. (2023) wherein twenty-five yeast strains were isolated from soil samples collected in the gold mining ore in Gia Lai, Vietnam, and screened for Cu^{2+} tolerance. Of these, one isolate with the code GL1T had the highest tolerance for Cu^{2+} and was able to remove Cu^{2+} effectively up to 90.49%. The strain was identified as a new species named *Rhodotorula aurum* sp. nov. based on molecular analysis. Another study by Tang and Bui (2011) evaluated the role of arbuscular mycorrhizal fungi (AMF) in enhancing maize tolerance and lead (Pb) uptake in Pb-contaminated soil. They found a significant increase in dry biomass and Pb accumulation in AM-inoculated maize compared to non-inoculated plants. These findings demonstrate that AMF significantly improved maize growth and Pb absorption. However, high Pb concentrations (2000 ppm) reduce AMF colonization in maize roots and decrease microbial populations in the soil (Tang & Bui, 2011).

Way Forward Toward Fungal Conservation for a Sustainable Forest Ecosystem

Fungal diversity remains critically underrepresented in conservation efforts across the ASEAN region despite its ecological importance, including fungi in the regional and global IUCN Red List is vital for recognizing and protecting threatened species (Baird & Pope, 2022), particularly in biodiversity hotspots like Southeast Asia. While ASEAN countries harbor thousands of fungal species, only a fraction have been assessed for conservation status. Expanding fungal red-listing efforts would help identify species at risk due to deforestation, habitat degradation, and climate change (Mueller et al., 2022). This also helps strengthen the pursuit of emphasizing fungi in national biodiversity action plans and protected area management strategies for a stronger framework for conservation and policymaking (Mueller et al., 2022). Moreover, preserving fungal habitats must be a priority, as habitat loss threatens entire fungal communities. Protecting natural habitats of fungi, such as forests, wetlands, and agroecosystems, will safeguard their ecological functions and unique diversity (Blaalid & Davey, 2022). In support of this, there is a need for a collaborative and integrated long-term ecological monitoring and biodiversity assessment in the ASEAN region where fungal resources are equally considered together with all other organisms.

Comprehensive fungal studies across ASEAN ecosystems are essential to bridge knowledge gaps, linking traditional taxonomic methods with advanced molecular techniques. While morphological identification has long been used in fungal taxonomy, DNA-based methods have greatly improved species classification and ecological research. Molecular approaches enable the discovery of cryptic species, improve fungal phylogenetics, and enhance understanding of their ecological roles (Lin et al., 2023; Mehta et al., 2022). A robust meta-database on ASEAN fungal diversity must centralize and standardize fungal research data. Establishing an open-access regional database would facilitate information-sharing on taxonomy, distribution, and ecological roles while supporting conservation planning and biotechnology applications (Gautier et al., 2022) that are essential inputs in conservation policy making.

Fungal research in ASEAN can be advanced through international collaborations and capacity-building initiatives (Pearce et al., 2020; Ryan et al., 2019). Engagement with non-academic groups, including citizen science initiatives and amateur mycologists, enhances public awareness and expands the scope of fungal surveys, leveraging social media for broader participation toward conservation (Irga et al., 2020). Valuable products generated from fungi that are proven in science to address health issues and environmental concerns should be maximized for commercialization to benefit communities in the ASEAN countryside. Research and development, hence, should encourage more partnerships within and outside the regions that will continue to protect and conserve fungal resources to sustain the forest as a major fungal habitat.

References

Adiova, J. M., Pampolina, N. M., & Aggangan, N. S. (2013). Effect of arbuscular mycorrhizal fungi inoculation on growth and cu uptake and toxicity of Desmodium cinereum (Kunth) D.C. *Philippines Journal of Science, 142*(1), 87–96.

Aggangan, N. S. (2006). Assessment of mycorrhiza diversity in abandoned mine sites in Toledo, Cebu for bioremediation. In *Food and agriculture organization of the United Nations*. FORESPI symposium 2006.

Aggangan, N. S., & Cortes, A. (2018). Screening mined out indigenous mycorrhizal fungi for the rehabilitation of mine tailing areas in The Philippines. *REFORESTA, 6*, 71–85.
Aggangan, N. S., Pampolina, N., Cadiz, N., & Raymundo, A. (2015). Assessment of plant diversity and associated Mycorrhizal fungi in the Minedout sites of atlas mines in Toledo City, Cebu for bioremediation. *Journal of Environmental Science and Management, 18*(1), 71–86.
Aggangan, N. S., Cadiz, N., Llamado, A., & Raymundo, A. (2017). Jatropha curcas for bioenergy and bioremediation in mine tailing area in Mogpog, Marinduque, Philippines. *Energy Procedia, 110*, 471–478.
Aggangan, N. S., Anarna, J. A., & Cadiz, N. M. (2019). Tree Legume–Microbial Symbiosis and other soil amendments as rehabilitation strategies in mine tailings in the Philippines. *Philippine Journal of Science, 148*(3), 481–491.
Agus, C., Primananda, E., Faridah, E., Wulandari, D., & Lestari, T. (2019). Role of arbuscular mycorrhizal fungi and Pongamia pinnata for revegetation of tropical open-pit coal mining soils. *International Journal of Environmental Science and Technology, 16*, 3365–3374.
Antonelli, A., Smith, R. J., Fry, C., Simmonds, M. S. J., Kersey, P. J., Pritchard, H. W., Abbo, M. S., Acedo, J., Adams, J., Ainsworth, A. M., & et al. (2020). *State of the world's plants and fungi*. [Research Report]. Royal Botanic Gardens (Kew); Sfumato Foundation. hal-02957519.
Arko, P. F., Marzuki, B. M., & Kusmoro, J. (2017). The inventory of edible mushroom in Kamojang nature reserve and Nature Park, West Java, Indonesia. *Biodiversitas Journal of Biological Diversity, 18*(2), 530–540.
Armadhan, W. S., Sari, S. P., Aji, M. Y. M. B., Permatasari, D. P., Amalia, B. W., Berlin, G. E., Aszar, A. S., Indrawan, M., Pradhan, P., & Setyawan, A. D. (2023). The macrofungal diversity and its potential from the karst forest of Kalipoh Village, Kebumen District, Indonesia. *Asian Journal of Forestry, 7*(2), 98–105.
ASEAN Center for Biodiversity. (2023). *ASEAN biodiversity outlook 3* (392).
Augustinus, F., & Putra, I. P. (2021). Termitomyces di kawasan urban kecamatan Gunungpati, Kota Semarang, Jawa Tengah. *Buletin Plasma Nutfah, 27*, 101–112.
Azizan, F. A., Astuti, I. S., Young, A., & Aziz, A. A. (2023). Rubber leaf fall phenomenon linked to increased temperature. *Agriculture, Ecosystems & Environment, 352*, 108531.
Baird, A., & Pope, F. (2022). 'Can't see the forest for the trees': The importance of fungi in the context of UK tree planting. *Food and Energy Security, 11*(4), e371. https://doi.org/10.1002/fes3.371
Bakray, N. A. M., Azmi, N. A., Jamar, N. H., Kin, T. B., Mansor, P., Mohti, A., Zohari, A. F., & Said, M. N. M. (2024). The association of non-dipterocarps (Fagaceae) with mycorrhizae in elevated CO2 condition in Tekam Forest reserve, Pahang. In *E3S web of conferences* (Vol. 599, p. 03005). EDP Sciences.
Blaalid, R., & Davey, M. L. (2022). Habitat protection approaches facilitate conservation of overlooked fungal diversity—A Case study from the Norwegian coastal heathland system. *Frontiers in Fungal Biology, 3*, 886685. https://doi.org/10.3389/ffunb.2022.886685
Boisvenue, C., & Running, S. W. (2006). Impacts of climate change on natural forest productivity–evidence since the middle of the twentieth century. *Global Change Biology, 12*(5), 862–882.
Botterill-James, T., Yates, L. A., Buettel, J. C., Aandahl, Z., & Brook, B. W. (2024). Southeast Asian biodiversity is a fifth lower in deforested versus intact forests. *Environmental Research Letters, 19*(11), 113007.
Brearley, F. Q. (2011). The importance of ectomycorrhizas for the growth of dipterocarps and the efficacy of ectomycorrhizal inoculation schemes. In M. Rai & A. Varma (Eds.), *Diversity and biotechnology of ectomycorrhizae* (pp. 3–17). Springer.
Brearley, F. Q. (2012). Ectomycorrhizal associations of the Dipterocarpaceae. *Biotropica, 44*(5), 637–648.
Brundrett, M., Bougher, N., Dell, B., Grove, T., & Malajczuk, N. (1996). *Working with mycorrhizas in forestry and agriculture* (Vol. 32, p. 374). Australian Centre for International Agricultural Research.

Buragohain, P. (2019). Role of microbes on carbon sequestration. *International Journal of Microbiology Research*. ISSN, 0975-5276.

Burgess, T., & Wingfield, M. J. (2004). Impact of fungal pathogens in natural Forest ecosystems: A focus on eucalypts. In K. Sivasithamparama, K. W. Dixon, & R. L. Barrett (Eds.), *Microorganisms in plant conservation and biodiversity* (pp. 285–306). Kluwer Academic Publishers. https://doi.org/10.1007/0-306-48099-9_11

Burgess, T. I., Barber, P. A., Sufaati, S., Xu, D., Hardy, G. E. S. J., & Dell, B. (2007). Mycosphaerella spp. On Eucalyptus in Asia; new species, new hosts and new records. *Fungal Diversity, 24*, 135–157.

Bustillos, R. G., Dulay, R. M. R., Kalaw, S. P., & Reyes, R. G. (2024). Diversity of macrofungi along elevation gradients in Mt. Arayat protected landscape, Arayat, Pampanga, The Philippines. *Studies in Fungi, 9*(1), e017.

Case, N. T., Gurr, S. J., Fisher, M. C., Blehert, D. S., Boone, C., Casadevall, A., Chowdhary, A., Cuomo, C. A., Currie, C. R., Denning, D. W., & Ene, I. V. (2025). Fungal impacts on earth's ecosystems. *Nature, 638*(8049), 49–57.

Chakraborty, S., Pangga, I. B., & Roper, M. M. (2012). Climate change and multitrophic interactions in soil: The primacy of plants and functional domains. *Global Change Biology, 18*(7), 2111–2125.

Chen, S., Woodcock, C., Dong, L., Tarrio, K., Mohammadi, D., & Olofsson, P. (2024). Review of drivers of forest degradation and deforestation in Southeast Asia. *Remote Sensing Applications: Society and Environment, 33*, 101129.

Chouangthavy, B., Bouttavong, K., Louangphan, J., Phewphanh, P., Sibounnavong, P., Souksavat, S., Pinkaew, N., & Babendreier, D. (2020). Beetle biodiversity in forest habitats in Laos depends on the level of human exploitation. *Journal of Insect Conservation, 24*, 833–840.

Corrales, A., Koch, R. A., Vasco-Palacios, A. M., Smith, M. E., Ge, Z.-W., & Henkel, T. W. (2022). Diversity and distribution of tropical ectomycorrhizal fungi. *Mycologia, 114*(6), 919–933. https://doi.org/10.1080/00275514.2022.2115284

Cui, J., & Chisti, Y. (2003). Polysaccharopeptides of Coriolus versicolor: Physiological activity, uses, and production. *Biotechnology Advances, 21*(2), 109–122.

Daipan, B. P. O. (2021). Patterns of forest cover loss in the terrestrial key biodiversity areas in The Philippines: Critical habitat conservation priorities. *Journal of Threatened Taxa, 13*(13), 20019–20032. https://doi.org/10.11609/jott.6904.13.13.20019-20032

Darlingga, M., Lingga, R., & Kurniawan, A. (2022). Isolation, identification, and metal resistance testing of soil fungi (rhizosphere) from Bangka tin post-mining. *BIOLINK (Jurnal Biologi Lingkungan Industri Kesehatan), 8*(2), 180–194.

Das, S., Kundu, A., Ghosh, D., & Bhattacharya, S. (2017). Anticancer activities of *Coprinus cinereus*. *Phytotherapy Research, 31*(5), 865–873.

de Bruyn, M., Stelbrink, B., Morley, R. J., Hall, R., Carvalho, G. R., Cannon, C. H., van den Bergh, G., Meijaard, E., Metcalfe, I., Boitani, L., Maiorano, L., Shoup, R., & von Rintelen, T. (2014). Borneo and Indochina are major evolutionary hotspots for southeast Asian biodiversity. *Systematic Biology, 63*(6), 879–901. https://doi.org/10.1093/sysbio/syu047

Desprez-Loustau, M.-L., Robin, C., Reynaud, G., Déqué, M., Badeau, V., Piou, D., Husson, C., & Marçais, B. (2007). Simulating the effects of a climate-change scenario on the geographical range and activity of forest-pathogenic fungi. *Canadian Journal of Plant Pathology, 29*(2), 101–120. https://doi.org/10.1080/07060660709507447

Dislich, C., Keyel, A. C., Salecker, J., Kisel, Y., Meyer, K. M., Auliya, M., et al. (2017). A review of the ecosystem functions in oil palm plantations, using forests as a reference system. *Biological Reviews, 92*(3), 1539–1569.

Doungsa-ard, C., McTaggart, A. R., Geering, A. D. W., Dalisay, T. U., Ray, J., & Shivas, R. G. (2015). Uromycladium falcatarium sp. Nov., the cause of gall rust on Paraserianthes falcataria in south-east Asia. *Australasian Plant Pathology, 44*(1), 25–30. https://doi.org/10.1007/s13313-014-0301-z

Duarte, A. G., & Maherali, H. (2022). A meta-analysis of the effects of climate change on the mutualism between plants and arbuscular mycorrhizal fungi. *Ecology and Evolution, 12*(1), e8518.

Dudney, J., Willing, C. E., Das, A. J., Latimer, A. M., Nesmith, J. C., & Battles, J. J. (2021). Nonlinear shifts in infectious rust disease due to climate change. *Nature Communications, 12*(1), 5102.

Dulay, R. M. R. (2023). Oudemansiella (Physalacriaceae) mushrooms: A status review on the distribution, cultivation, composition and bioactivity profile. *Studies in Fungi, 8*(1), 13.

Duong, M. T., Truong, H. D., & Ly, V. L. (2022). Determining the values of and the distribution of large mushrooms in lung Ngoc Hoang wetland nature reserve. *Nông Nghiệp & Phát Triển Nông Thôn, 3 and 4*, 144–151.

Eberhardt, U., Beker, H. J., Schütz, N., Pedersen, O. S., Sysouphanthong, P., & Læssøe, T. (2020). Adventurous cuisine in Laos: Hebeloma parvisporum, a new species in Hebeloma section Porphyrospora. *Mycologia, 112*(1), 172–184.

Estoque, R. C., Ooba, M., Avitabile, V., Hijioka, Y., DasGupta, R., Togawa, T., & Murayama, Y. (2019). The future of Southeast Asia's forests. *Nature Communications, 10*(1), 1829.

Fan, Y., Li, J., Zhu, S., Li, H., & Zhou, B. (2022). Trends and variabilities of precipitation and temperature extremes over Southeast Asia during 1981–2017. *Meteorology and Atmospheric Physics, 134*(4), 78.

Food and Agriculture Organization (FAO). (2020). *Global Forest Resources Assessment 2020*. FAO United Nations.

Forest Management Bureau – Department of Environment and Natural Resources (FMB-DENR). (2023). *Philippine forestry statistics 2023*. DENR.

Fraser, S., McTaggart, A. R., Wingfield, M. J., & Roux, J. (2017). Effect of temperature, leaf wetness and the developmental stage of host tissue on infection of Acacia mearnsii by Uromycladium acaciae (Pucciniales). *Australasian Plant Pathology, 46*, 407–419.

Gautier, M., Moreau, P.-A., Boury, B., & Richard, F. (2022). Unravelling the French National Fungal Database: Geography, temporality, taxonomy and ecology of the recorded diversity. *Journal of Fungi, 8*(9), 926. https://doi.org/10.3390/jof8090926

GBIF.org. (2025, March 16). *GBIF occurrence download*. https://doi.org/10.15468/dl.knhp94

Ghelardini, L., Pepori, A. L., Luchi, N., Capretti, P., & Santini, A. (2016). Drivers of emerging fungal diseases of forest trees. *Forest Ecology and Management, 381*, 235–246. https://doi.org/10.1016/j.foreco.2016.09.032

Giao, N. S. (1996). Remarks on the mycorrhizae of some tree species in Vietnam. In *Plantation technology in tropical forest science* (pp. 167–173). Springer.

Global Forest Watch.org. (2025). Global forest watch accessed. Accessed 16 Mar 2025 https://www.globalforestwatch.org/dashboards/country

Guégan, J.-F., Ayouba, A., Cappelle, J., & De Thoisy, B. (2020). Forests and emerging infectious diseases: Unleashing the beast within. *Environmental Research Letters, 15*(8), 083007. https://doi.org/10.1088/1748-9326/ab8dd7

Hapuarachchi, K. K., Karunarathna, S. C., Xu, X. H., Dutta, A. K., Phengsintham, P., Hyde, K. D., & Wen, T. C. (2021). A review on bioactive compounds, beneficial properties and biotechnological approaches of Trametes (Polyporaceae, Polyporales) and a new record from Laos. *Chiang Mai Journal of Science, 48*(3), 674–698.

Hardwick, K. A., Healey, J. R., & Blakesley, D. (2000). Research needs for the ecology of natural regeneration of seasonally dry tropical forests in Southeast Asia. In S. Elliott, J. Kerby, D. Blakesley, K. Hardwick, K. Woods, & V. Anusarnsunthorn (Eds.), *Forest restoration for wildlife conservation* (pp. 165–179). International Tropical Timber Organization and The Forest Restoration Research Unit, Chiang Mai University. Available at: https://www.formad-environnement.org/RNA_harwick.pdf

Hastuty, A., & Hidayat, I. (2019, August). Screening and molecular identification of Cr (VI)-resistant Trichoderma isolated from ex-tin mining soil in Bangka Belitung Province, Indonesia. In *Earth and environmental science* (IOP conference series) (Vol. 308, No. 1, p. 012011). IOP Publishing.

Hawkins, H. J., Cargill, I. M., Van Nuland, M. E., Hagen, J. C., Field, K. J., Sheldrake, M., Soudzilovskaia, N. A., & Kiers, E. T. (2023). Mycorrhizal mycelium as a global carbon Pool. *Current Biology, 33*, R560–R573.

Healey, M. A., Lawson, S. A., Somany, S., Tasen, W., & Ngoc, Q. D. (2021). Forest biosecurity laws in South-East Asia: A review. *International Forestry Review, 23*(4), 418–436. https://doi.org/10.1505/146554821834777215

Healey, M. A., Warman, R., Mohammed, C., & Lawson, S. A. (2022). Acacia and eucalypt plantation biosecurity in Southeast Asia—A history, and directions for future research and engagement. *Australian Forestry, 85*(4), 146–160. https://doi.org/10.1080/00049158.2022.2145721

Helbert, Turjaman, M., & Nara, K. (2019). Ectomycorrhizal fungal communities of secondary tropical forests dominated by Tristaniopsis in Bangka Island, Indonesia. *Plos one, 14*(9), e0221998.

Helfer, S. (2014). Rust fungi and global change. *New Phytologist, 201*, 770–780. https://doi.org/10.1111/nph.12570

Hoang, N. T., & Kanemoto, K. (2021). Mapping the deforestation footprint of nations reveals growing threat to tropical forests. *Nature Ecology & Evolution, 5*(6), 845–853. https://doi.org/10.1038/s41559-021-01417-z

Hong, M. Y., Park, S. W., Kim, D. H., Saysavanh, V., & Lee, J. K. (2019). A checklist of mushrooms of Phousabous National Protected Area (PNPA) of Lao PDR. *Journal of forest and environmental science, 35*(4), 268–271.

Hong, Y., Tan, J. Y., Xue, H., Chow, M. L., Ali, M., Ng, A., Leong, A., Yeo, J., Koh, S. M., Tang, M. S. Y., Lee, Y. Y., Choong, A. M. F., Lee, S. M. L., Delli Ponti, R., Chan, P. M., Lee, D., Wong, J. Y., Mutwil, M., & Fong, Y. K. (2023). A metagenomic survey of wood decay fungi in the urban trees of Singapore. *Journal of Fungi, 9*(4), 460. https://doi.org/10.3390/jof9040460

Hosaka, K., Nam, K.-O., Linn, W., & Aung, M. (2021). Species identification based on DNA of selected mushrooms from Myanmar (1) Lactarius austrotorminosus and 17 other taxa newly reported from Myanmar. *Bulletin of the National Science Museum, Series B, 47*(2), 59–69.

Hyde, K. D., Norphanphoun, C., Chen, J., Dissanayake, A. J., Doilom, M., Hongsanan, S., Jayawardena, R. S., Jeewon, R., Perera, R. H., Thongbai, B., Wanasinghe, D. N., Wisitrassameewong, K., Tibpromma, S., & Stadler, M. (2018). Thailand's amazing diversity: Up to 96% of fungi in northern Thailand may be novel. *Fungal Diversity, 93*(1), 215–239. https://doi.org/10.1007/s13225-018-0415-7

Imanuddin, R., Hidayat, A., Rachmat, H. H., Turjaman, M., Pratiwi, Nurfatriani, F., Indrajaya, Y., & Susilowati, A. (2020). Reforestation and sustainable management of Pinus merkusii Forest plantation in Indonesia: A review. *Forests, 11*(12), 1235. https://doi.org/10.3390/f11121235

Ingleby, K., Thuy, L. T. T., Phong, N. T., & Mason, P. A. (2000). Ectomycorrhizal inoculum potential of soils from forest restoration sites in South Vietnam. *Journal of Tropical Forest Science, 12*(2), 418–422.

Irga, P. J., Dominici, L., & Torpy, F. R. (2020). The mycological social network a way forward for conservation of fungal biodiversity. *Environmental Conservation, 47*(4), 243–250. https://doi.org/10.1017/S0376892920000363

IUCN. (2025). The IUCN Red List of Threatened Species. Version 2025-1. https://www.iucnredlist.org. Accessed on 1 June, 2025.

Jacob, J. K. S., Romorosa, E. S., Arguelles, E. D., & Paguirigan, J. A. G. (2022). Checklist of reported macrofungi in The Philippines. *Philippine Journal of Science, 151*(5), 1657–1676.

Jayani, F. M., Budi, S. W., & Pamoengkas, P. (2018). Response of forest tree species inoculated with MycoSilvi and soil ameliorant addition grown in silica sand. *Asian Journal of Agriculture and Biology, 6*(4), 556–565.

Kerfahi, D., Tripathi, B. M., Lee, J., Edwards, D. P., & Adams, J. M. (2014). The impact of selective-logging and forest clearance for oil palm on fungal communities in Borneo. *PLoS One, 9*(11), e111525. https://doi.org/10.1371/journal.pone.0111525

Kettle, C. J. (2010). Ecological considerations for using dipterocarps for restoration of lowland rainforest in Southeast Asia. *Biodiversity and Conservation, 19*(4), 1137–1151. https://doi.org/10.1007/s10531-009-9772-6

Kim, N. K., Lee, J. H., Jo, J. W., Bunthoeun, R., Ngeth, C., & Lee, J. K. (2016). Fungal biodiversity in cardamom protected forests and Seima biodiversity conservation area of Cambodia. *Journal of Forest and Environmental Science, 32*(2), 158–163. https://doi.org/10.7747/JFES.2016.32.2.158

Kim, N. K., Lee, J. H., Jo, J. W., & Lee, J. K. (2017). A checklist of mushrooms of Cambodia. *Journal of Forest and Environmental Science, 33*(1), 49–65. https://doi.org/10.7747/JFES.2017.33.1.49

Kumla, J., Suwannarach, N., Bussaban, B., Matsui, K., & Lumyong, S. (2014). Indole-3-acetic acid production, solubilization of insoluble metal minerals and metal tolerance of some sclerodermatoid fungi collected from northern Thailand. *Annals of Microbiology, 64*, 707–720.

Kurniati, E., Arfarita, N., Imai, T., Higuchi, T., Kanno, A., Yamamoto, K., & Sekine, M. (2014). Potential bioremediation of mercury-contaminated substrate using filamentous fungi isolated from forest soil. *Journal of Environmental Sciences, 26*(6), 1223–1231.

La Porta, N., Capretti, P., Thomsen, I. M., Kasanen, R., Hietala, A. M., & Von Weissenberg, K. (2008). Forest pathogens with higher damage potential due to climate change in Europe. *Canadian Journal of Plant Pathology, 30*(2), 177–195. https://doi.org/10.1080/07060661.2008.10540534

Lai, J., Cheah, W., Palaniveloo, K., Suwa, R., & Sharma, S. (2022). A systematic review of the physicochemical and microbial diversity of well-preserved, restored, and disturbed mangrove forests: What is known and what is the way forward? *Forests, 13*(12), 2160.

Lee, S. M. L. (2019). Macro-fungal diversity of Bukit Timah nature reserve, Singapore. *Gardens' Bulletin Singapore, 71*(Suppl. 1), 135–144. https://doi.org/10.26492/gbs71(suppl.1).2019-05

Lee, S. S., & Krishnapillay, B. (2005). Biosecurity concerns of the forestry sector in Malaysia (2005/ATCWG/WKSP/011, APEC workshop on building biosecurity planning and surveillance capacity for APEC member economies).

Lee, S. S., Watling, R., & Turnbull, E. (2003). Diversity of putative ectomycorrhizal fungi in Pasoh Forest Reserve. In *Pasoh: ecology of a lowland rain forest in Southeast Asia* (pp. 149–159).

Lee, J. H., Kim, D. H., Yun, J. H., Hong, M. Y., & Lee, J. K. (2018). A checklist of mushrooms of Phou Khao Khuoay National Protected Area (PKKNPA) of Lao-PDR. *Journal of Forest and Environmental Science, 34*(6), 457–460.

Lee, J. K., Kim, D. H., Lee, J. H., Hong, M. Y., Yoon, J. H., Bae, Y. J., Park, S. W., Kim, D. H., Nguyen, M. H., Saysavanh, V., & Keovilai, S. (2021). *Biodiversity of Lao PDR mushrooms*. Geobook Publishing.

Lelana, N. E., Wiyono, S., Giyanto, S., & I. Z., & Anggraeni, I. (2022). Phylogenetic and morphological characteristics of Uromycladium falcatariae, the fungal pathogen that causes gall rust epidemics of Falcataria moluccana in Indonesia. *Journal of Phytopathology, 170*(9), 598–604. https://doi.org/10.1111/jph.13123

Lelana, N. E., Darmawan, U. W., Nuroniah, H. S., Puspitaningtyas, D. M., & Irianto, R. S. B. (2024). Rapid spread and high prevalence of the gall rust pathogen Uromycladium falcatariae in stands of Falcataria falcata. *Forest Science and Technology, 20*(4), 309–315.

Lin, P., Kook, M., Yi, T.-H., & Yan, Z.-F. (2023). Current fungal taxonomy and developments in the identification system. *Current Microbiology, 80*(12), 375. https://doi.org/10.1007/s00284-023-03514-7

Liu, Y., Xu, X., Dimitrov, D., Pellissier, L., Borregaard, M. K., Shrestha, N., Su, X., Luo, A., Zimmermann, N. E., Rahbek, C., & Wang, Z. (2023). An updated floristic map of the world. *Nature Communications, 14*(1), 2990.

Luangharn, T., Karunarathna, S. C., Dutta, A. K., Paloi, S., Promputtha, I., Hyde, K. D., Xu, J., & Mortimer, P. E. (2021). Ganoderma (Ganodermataceae, Basidiomycota) species from the greater Mekong subregion. *Journal of Fungi, 7*(10), 819. https://doi.org/10.3390/jof7100819

Mastrili, R. A. D., Ibarra Caballero, J. R., Malabrigo, P. L., Stewart, J. E., & Ata, J. P. (2024). First report of leaf spot caused by Neopestalotiopsis and Calonectria species on Areca ipot seedlings in Luzon, Philippines. *Forest Pathology, 54*(4), e12883. https://doi.org/10.1111/efp.12883

Maulana, A. F., Turjaman, M., Sato, T., Hashimoto, Y., Cheng, W., & Tawaraya, K. (2018). Isolation of endophytic fungi from tropical forest in Indonesia. *Symbiosis, 76*, 151–162.

McGuire, K. L., D'angelo, H., Brearley, F. Q., Gedallovich, S. M., Babar, N., Yang, N., Gillikin, C. M., Gradoville, R., Bateman, C., Turner, B. L., & Mansor, P. (2015). Responses of soil fungi to logging and oil palm agriculture in southeast Asian tropical forests. *Microbial Ecology, 69*(4), 733–747.

Mehta, N., Jadhav, R., & Baghela, A. (2022). Molecular taxonomy and multigene phylogeny of filamentous fungi. In V. K. Gupta & M. Tuohy (Eds.), *Laboratory protocols in fungal biology* (pp. 175–207). Springer. https://doi.org/10.1007/978-3-030-83749-5_10

Milović, M., Kebert, M., & Orlović, S. (2021). How mycorrhizas can help forests to cope with ongoing climate change? *Šumarski list, 145*(5–6), 279–286.

Ministry of Agriculture and Rural Development (MARD). (2020). *Forest Cover*. Ministry of Agriculture and Rural Development.

Mohamad, N. A. H., Taha, H., Shivanand, P., Ahmad, N., Abdillah, A. N., Alias, M. A., Mohaimin, A. Z. A., & Krishnamoorthy, S. (2023). Endophytic fungal species isolated from mangrove trees Rhizophora apiculata, Nypa fruticans, and Xylocarpus granatum from Brunei Darussalam. *Philippine Journal of Science, 152*(5), 1931–1938.

Monkai, J., Hyde, K. D., Xu, J., & Mortimer, P. E. (2017). Diversity and ecology of soil fungal communities in rubber plantations. *Fungal Biology Reviews, 31*(1), 1–11.

Mueller, G. M., Cunha, K. M., May, T. W., Allen, J. L., Westrip, J. R. S., Canteiro, C., Costa-Rezende, D. H., Drechsler-Santos, E. R., Vasco-Palacios, A. M., Ainsworth, A. M., Alves-Silva, G., Bungartz, F., Chandler, A., Gonçalves, S. C., Krisai-Greilhuber, I., Iršėnaitė, R., Jordal, J. B., Kosmann, T., Lendemer, J., et al. (2022). What Do the first 597 global fungal red list assessments tell us about the threat status of fungi? *Diversity, 14*(9), 736. https://doi.org/10.3390/d14090736

Myers, N., Mittermeier, R. A., Mittermeier, C. G., da Fonseca, G. A., & Kent, J. (2000). Biodiversity hotspots for conservation priorities. *Nature, 403*(6772), 853–858. https://doi.org/10.1038/35002501

Ngo, A. (2013). The diversity of medicinal muhsrooms in Thua Thien Jue Province and cultivation of Lingzhi mushrooms. In *Hoi Nghi Khoa Hoc Cong Nghe Sinh Hoc Toan Quoc* (pp. 679–684). Hanoi.

Ngo, A., & Nguyen, T. P. (2015). Diversity of macromycoflora in the Phong Dien district, Thua Thien-Hue province, Vietnam. In K. D. Long (Ed.), *Proceedings of the sixth national scientific conference on ecology and biological resources* (pp. 454–460).

Noorchat, M. U., Galiana, A., Lee, S. S., Le Roux, C., Patahayah, M., Eny, F., Prin, Y., Duponnois, R., & Ducousso, M. (2012, October 11). Diversity of ectomycorrhizas in lowland and mountain forests of South Sumatra. In *Proceedings of the bio Asia project workshop*. Yogyakarta.

Noviyanti, N., Mahardhika, W. A., Lunggani, A. T., & Putra, I. P. (2022). Macrofungi Inventaritation at the pine Forest of Kragilan, Magelang. *Biovalentia: Biological Research Journal, 8*(2), 138–144.

Nowakowski, P., Markiewicz-Żukowska, R., Gromkowska-Kępka, K., Naliwajko, S. K., Moskwa, J., Bielecka, J., Grabia, M., Borawska, M., & Socha, K. (2021). Mushrooms as potential therapeutic agents in the treatment of cancer: Evaluation of anti-glioma effects of Coprinus comatus, Cantharellus cibarius, Lycoperdon perlatum and Lactarius deliciosus extracts. *Biomedicine & Pharmacotherapy, 133*, 1–14.

Old, K., & Cristovao, C. D. S. (2003). A rust epidemic of the coffee shade tree (Paraserianthes falcataria) in East Timor. In H. Da Costa, C. Piggin, C. J. Da Cruz, & J. J. Fox (Eds.), *Agriculture: New directions for a new nation East Timor (Timor-Leste)*. Australian Centre for International Agricultural Research.

Oliva, J., Stenlid, J., & Martínez-Vilalta, J. (2014). The effect of fungal pathogens on the water and carbon economy of trees: Implications for drought-induced mortality. *New Phytologist, 203*(4), 1028–1035.

Pampolina, N. M., Tadiosa, E. R., Ata, J. P., Soriano, J. K. R., Parlucha, J. A., & Niem, J. M. (2023). Species and functional diversity of forest fungi for conservation and sustainable landscape in The Philippines. In *Mycology in the tropics* (pp. 89–135). Academic.

Pampolina, N. M., Daipan, P. O., Racelis, D., & A. (2024). Unveiling the vulnerability of tropical montane forests in the face of the climate. In *Book of abstracts. Forests and society towards 2050. 28th IUFRO world congress.*

Pardosi, L., Makin, F. M. P., & Wiguna, I. G. A. (2020). The exploration of macroscopic fungus in Oeluan Forest in Timor Tengah Utara District. *Jurnal Saintek Lahan Kering, 3*(1), 4–6.

Parlucha, J. A., Soriano, J. K. R., Yabes, M. D., Pampolina, N. M., & Tadiosa, E. R. (2021). Species and functional diversity of macrofungi from protected areas in mountain forest ecosystems of Southern Luzon, Philippines. *Tropical Ecology, 62*, 359–367.

Paton, A., Antonelli, A., Carine, M., Forzza, R. C., Davies, N., Demissew, S., Dröge, G., Fulcher, T., Grall, A., Holstein, N., Jones, M., Liu, U., Miller, J., Moat, J., Nicolson, N., Ryan, M., Sharrock, S., Smith, D., Thiers, B., et al. (2020). Plant and fungal collections: Current status, future perspectives. *Plants, People, Planet, 2*(5), 499–514. https://doi.org/10.1002/ppp3.10141

Pearce, T. R., Antonelli, A., Brearley, F. Q., Couch, C., Campostrini Forzza, R., Gonçalves, S. C., Magassouba, S., Morim, M. P., Mueller, G. M., Nic Lughadha, E., Obreza, M., Sharrock, S., Simmonds, M. S. J., Tambam, B. B., Utteridge, T. M. A., & Breman, E. (2020). International collaboration between collections-based institutes for halting biodiversity loss and unlocking the useful properties of plants and fungi. *Plants, People, Planet, 2*(5), 515–534. https://doi.org/10.1002/ppp3.10149

Pedersen, O. S., & Læssøe, T. (2024). Amanita species within section Caesareae from the highlands of Laos. *Mycological Progress, 23*(1), 63.

Pham, N. Q., Marincowitz, S., Solís, M., Duong, T. A., Wingfield, B. D., Barnes, I., Slippers, B., Muro Abad, J. I., Durán, A., & Wingfield, M. J. (2021). Eucalyptus scab and shoot malformation: A new and serious foliar disease of Eucalyptus caused by Elsinoe necatrix sp. nov. *Plant Pathology, 70*(5), 1230–1242. https://doi.org/10.1111/ppa.13348

Phanthavong, S., Daly, A., Weir, B., Lee, D., Park, D., Balmas, V., & Burgess, L. (2023). First report of Calonectria pseudoreteaudii in Lao PDR associated with a leaf spot disease of Macadamia integrifolia. *Australasian Plant Pathology, 52*(1), 23–26.

Phengsintham P, Chukeatirote E, McKenzie EHC, Hyde K.D., Braun U. (2013). Monograph of Cercosporoid fungi from Laos. Current Research in Environmental & Applied Mycology 3(1), 34–158, doi https://doi.org/10.5943/cream/3/1/2.

Phonemany, M., Raspé, O., Sysouphanthong, P., Niego, A. G., Niego, T., Thongklang, N., Chuankid, B., & Hyde, K. (2021). Two Pleurotus species (Pleurotaceae) from Lao People's Democratic Republic. *Chiang Mai Journal of Science, 48*(5), 1224–1235.

Pitaktamrong, P., Kingkaew, J., Yooyongwech, S., Cha-Um, S., & Phisalaphong, M. (2018). Development of arbuscular mycorrhizal fungi-organic fertilizer pellets encapsulated with alginate film. *Engineering Journal, 22*(6), 65–79.

Plassard, C., & Dell, B. (2010). Phosphorus nutrition of mycorrhizal trees. *Tree Physiology, 30*(9), 1129–1139.

Prayudyaningsih, R., Mangopang, A. D., Sari, R., & Akib, M. A. (2024). Indigenous symbiotic soil microbes and native tree species for revegetation of nickel postmining area in Indonesia. In *Microbiome drivers of ecosystem function* (pp. 229–247). Academic.

Puangsombat, P., Sangwanit, U., & Marod, D. (2010). Diversity of soil fungi in different land use types in Tha Kum-Huai Raeng forest reserve, Trat Province. *Agriculture and Natural Resources, 44*(6), 1162–1175.

Putra, B., Warly, L., Evitayan, E., & Utama, B. P. (2022). The role of arbuscular mycorrhizal fungi in phytoremediation of heavy metals and their effect on the growth of Pennisetum purpureum cv. Mott on gold mine tailings in Muara Bungo, Jambi, Indonesia. *Biodiversitas, 23*(1), 478–485.

Rabiey, M., Hailey, L. E., Roy, S. R., Grenz, K., Al-Zadjali, M. A. S., Barrett, G. A., & Jackson, R. W. (2019). Endophytes vs tree pathogens and pests: Can they be used as biological control agents to improve tree health? *European Journal of Plant Pathology, 155*(3), 711–729. https://doi.org/10.1007/s10658-019-01814-y

Rahayu, S., Lee, S. S., & Shukor, N. A. A. (2010). Uromycladium tepperianum, the gall rust fungus from Falcataria moluccana in Malaysia and Indonesia. *Mycoscience, 51*(2), 149–153.

Ray, P., Kundu, S., & Paul, D. (2024). Exploring the therapeutic properties of Chinese mushrooms with a focus on their anti-cancer effects: A systemic review. *Pharmacological Research-Modern Chinese Medicine, 100433*, 100433.

Reyes, R. G. (2024). Optimization of culture conditions and forced wood decay assessment of wood-rotting mushrooms of Central Luzon, Philippines. *Asian Journal of Mycology, 7*(2), 119–131.

Ruankaew Disyatat, N., Yomyart, S., Sihanonth, P., & Piapukiew, J. (2016). Community structure and dynamics of ectomycorrhizal fungi in a dipterocarp forest fragment and plantation in Thailand. *Plant Ecology & Diversity, 9*(5–6), 577–588. https://doi.org/10.1080/17550874.2016.1264018

Ryan, M. J., McCluskey, K., Verkleij, G., Robert, V., & Smith, D. (2019). Fungal biological resources to support international development: Challenges and opportunities. *World Journal of Microbiology and Biotechnology, 35*(9), 139. https://doi.org/10.1007/s11274-019-2709-7

Sahara, N., Budi, S. W., & Wibowo, C. (2022). The role of mycoSilvi and silvicultural treatment towards the growth of Albizia chinensis in post-mining area. In *IOP conference series: Earth and environmental science*. (Vol. 959, No. 1 (p. 012040). IOP Publishing.

Sarma, V. V., & Hyde, K. D. (2018). Fungal species consortia on Nypa fruticans at Brunei. *Studies in Fungi, 3*(1), 19–26.

Sims, K., Watling, R., De La Cruz, R., & Jeffries, P. (1997). Ectomycorrhizal fungi of The Philippines: A preliminary survey and notes on the geographic biodiversity of the Sclerodermatales. *Biodiversity and Conservation, 6*, 43–58.

Singh, B. K., Delgado-Baquerizo, M., Egidi, E., Guirado, E., Leach, J. E., Liu, H., & Trivedi, P. (2023). Climate change impacts on plant pathogens, food security and paths forward. *Nature Reviews Microbiology, 21*(10), 640–656.

Siripong, P., Oraphin, B., Sanro, T., & Duanporn, P. (2009). Screening of fungi from natural sources in Thailand for degradation of polychlorinated hydrocarbons. *Eurasian Journal of Agricultural & Environmental Sciences, 5*, 466–472.

Smith, S. E., & Read, D. J. (1997). *Mycorrhizal symbiosis*. Academic.

Sodhi, N. S., et al. (2010). (2010) the state and conservation of southeast Asian biodiversity. *Biodiversity and Conservation, 19*, 317–328.

Sosa-Rodriguez, T., Dupré de Boulois, H., Granet, F., Gaurel, S., Melgarejo, L. M., Carron, M. P., & Declerck, S. (2013). In vitro mycorrhization of the rubber tree Hevea brasiliensis Müll Arg. *In Vitro Cellular & Developmental Biology-Plant, 49*, 207–215.

Stibig, H.-J., Achard, F., Carboni, S., Raši, R., & Miettinen, J. (2014). Change in tropical forest cover of Southeast Asia from 1990 to 2010. *Biogeosciences, 11*(2), 247–258. https://doi.org/10.5194/bg-11-247-2014

Sturrock, R. N., Frankel, S. J., Brown, A. V., Hennon, P. E., Kliejunas, J. T., Lewis, K. J., Worrall, J. J., & Woods, A. J. (2011). Climate change and forest diseases: Climate change and forest diseases. *Plant Pathology, 60*(1), 133–149. https://doi.org/10.1111/j.1365-3059.2010.02406.x

Sysouphanthong, P., Thongklang, N., Suwannapoom, C., Nuangmek, W., & Hyde, K. D. (2020). Lepiota section Stenosporae (Agaricaceae) two new records to Lao People's Democratic Republic. *Chiang Mai Journal of Science, 47*, 49–56.

Taha, H., Shivanand, P., Shahminan, N. I. N., Osman, M., Abdul-Halim, A. M. A.-A., & Abdullah, M. (2020). Isolation and identification of Culturable bacteria and fungi from mixed dipterocarp and mangrove forests of Brunei Darussalam. *Proceedings of the National Academy of*

Sciences, India Section B: Biological Sciences, 90(3), 523–530. https://doi.org/10.1007/s40011-019-01119-4

Tan, Y. L., Chen, J. E., Yiew, T. H., & Habibullah, M. S. (2022). Habitat change and biodiversity loss in south and southeast Asian countries. *Environmental Science and Pollution Research, 29*(42), 63260–63276.

Tang, T. C., & Bui, V. C. (2011). NGHIÊN CỨU SỬ DỤNG NẤM CỘNG SINH ABUSCULAR MYCORHIZA ĐỂ NÂNG CAO HIỆU QUẢ XỬ LÍ ĐẤT NHIỄM CHÌ CỦA CÂY NGÔ. *Vietnam Journal of Science and Technology, 49*(3), 65–71.

Tering, R. R. G., Tulod, A. M., Branzuela, N. E., Batiancela, M. A., Tutor, R. P., Casas, J. V., Rojo, M. J. A., & Gilbero, D. M. (2025). Gall rust disease dynamics in Falcataria falcata L. Across Elevations in Agusan del Norte, Philippines. *Jurnal Sylva Lestari, 13*(1), 332–343. https://doi.org/10.23960/jsl.v13i1.1142

Thi Nguyen, K. C., Truong, P. H., Ho, C. T., Le, C. T., Tran, K. D., Nguyen, T. L., Nguyen, M. T., & Nguyen, P. V. (2023). Copper tolerance of novel Rhodotorula sp. Yeast Isolated from Gold Mining Ore in Gia Lai, Vietnam. *Mycobiology, 51*(6), 379–387. https://doi.org/10.1080/12298093.2023.2274648

Thongkred, P., Lotrakul, P., Prasongsuk, S., Imai, T., & Punnapayak, H. (2011). Oxidation of polycyclic aromatic hydrocarbons by a tropical isolate of Pycnoporus coccineus and its laccase. *Science Asia, 37*, 225–233.

Thu, P. Q., Quang, D. N., Chi, N. M., Hung, T. X., Binh, L. V., & Dell, B. (2021). New and emerging insect Pest and disease threats to Forest plantations in Vietnam. *Forests, 12*(10), 1301. https://doi.org/10.3390/f12101301

Tölle, M. H. (2020). Impact of deforestation on land–atmosphere coupling strength and climate in Southeast Asia. *Sustainability, 12*(15), 6140.

Torres, M. L. S., Ontengco, D. C., Tadiosa, E. R., & Reyes, R. G. (2020). Ethnomycological studies on the tanBugkalots indigenous community in Alfonso Castaneda, Nueva Vizcaya. *International Journal of Pharmaceutical Research and Allied Sciences, 9*(4), 43–54.

Tran, T. P., Vilaivanh, P., Nguyen, H. L. A., & Truong, T. C. V. (2019). Some new collected macro fungi in Tra Giac mountain, northern Tra my district, Quang Nam province. *Tạp Chí Khoa Học – Đại Học Quảng Nam, 14*, 92–104.

Tran, T. P., Do, H. H., & Nguyen, T. T. (2023). Study on species diversity of macrofungi belonging to the phylum Basidiomycota in son Tra nature reserve, Da Nang. *Vietnam Journal of Agricultural Sciences, 21*(12), 1635–1646.

Trianto, A., Radjasa, O. K., Subagiyo, S., Purnaweni, H., Bahry, M. S., Djamaludin, R., Tjoa, A., Singleton, I., Diele, K., & Evan, D. (2021). Potential of fungi isolated from a mangrove ecosystem in northern Sulawesi, Indonesia: Protease, cellulase and anti-microbial capabilities. *Biodiversitas Journal of Biological Diversity, 22*, 1717–1724.

Turnbull, J. W., & de la Cruz, R. E. (1990). Tree technologies with potential to contribute to sustainability in marginal uplands of Southeast Asia. In Technologies for Sustainable Agriculture on marginal uplands in Southeast Asia, ACIAR proceedings(no. 33, pp. 107–113).

Verma, M., Symes, W. S., Watson, J. E., Jones, K. R., Allan, J. R., Venter, O., Rheindt, F. E., Edwards, D. P., & Carrasco, L. R. (2020). Severe human pressures in the Sundaland biodiversity hotspot. *Conservation Science and Practice, 2*(3), e169.

Wasser, S. (2014). Medicinal mushroom science: Current perspectives, advances, evidences, and challenges. *Biomedical Journal, 37*(6), 345.

Welsh, C., Lewis, K., & Woods, A. (2009). The outbreak history of Dothistroma needle blight: An emerging forest disease in northwestern British Columbia, Canada. *Canadian Journal of Forest Research, 39*(12), 2505–2519. https://doi.org/10.1139/X09-159

Wijaya, N., Priyani, N., & Munir, E. (2019, July). Lead accumulation activity of fungi isolated from Batang Toru, south Tapanuli, North Sumatra. In *IOP conference series: Earth and environmental science* (Vol. 305, No. 1, p. 012064). IOP Publishing.

Wingfield, M. J., Slippers, B., Roux, J., & Wingfield, B. D. (2001). Worldwide movement of exotic forest fungi, especially in the tropics and the southern hemisphere. *Bioscience*, 51(2), 134. https://doi.org/10.1641/0006-3568(2001)051[0134:WMOEFF]2.0.CO;2.

Woodruff, D. S. (2010). Biogeography and conservation in Southeast Asia: How 2.7 million years of repeated environmental fluctuations affect today's patterns and the future of the remaining refugial-phase biodiversity. *Biodiversity and Conservation, 19*, 919–941.

Worrall, J. J., Marchetti, S. B., & Mask, R. A. (2012). *An epidemic of needle cast on lodgepole pine in colorado (R2-12–01; Biological evaluation)*. USDA Forest Service Rocky Mountain Region Forest Health Protection.

Wulandari, R. (2020). Biodegradation of polycyclic aromatic hydrocarbons commonly found in coal slurry using white rot fungi isolated from East Kalimantan. Chulalongkorn University theses and dissertations (Chula ETD). 360. https://digital.car.chula.ac.th/chulaetd/360

Zenkova, V. A., Efremenkova, O. V., Ershova, E. Y., Tolstych, I. V., & Dudnik, Y. V. (2003). Antimicrobial activity of medicinal mushrooms from the genus Coprinus (Fr.) SF gray (Agaricomycetideae). *International Journal of Medicinal Mushrooms, 5*(1), 1–6.

Zhao, S., Gao, Q., Rong, C., Wang, S., Zhao, Z., Liu, Y., & Xu, J. (2020). Immunomodulatory effects of edible and medicinal mushrooms and their bioactive immunoregulatory products. *Journal of Fungi, 6*(4), 269.

Part V
Mycology in Shaping Societies

Chapter 18
Mycological Nanotechnology and Applications in Medicine and Agriculture

Jonathan Jaime G. Guerrero, Ric Ryan H. Regalado, John Patrick Cuenca, Martin Joseph Q. Paet, Charmaine A. Malonzo, Maria Czet A. Fulleros, and Marianne Ysabelle Sabdao-Lunas

Abstract Myconanotechnology is an interdisciplinary field that harnesses fungi for the green synthesis of nanomaterials with applications across medicine, agriculture, and environmental sustainability. Southeast Asia, a biodiversity hotspot, offers a rich repository of fungal species with the potential to serve as natural biofactories for nanoparticle production. This chapter provides a synthesis of current research on fungal-mediated nanoparticle synthesis in the region, focusing on mechanisms, biomedical applications, agricultural innovations, and industrial scalability. Fungal systems produce bioactive nanomaterials such as silver, gold, and zinc oxide nanoparticles through eco-friendly and cost-effective processes mediated by fungal enzymes, secretomes, and cellular structures. These nanomaterials have demonstrated potent antimicrobial, anticancer, antioxidant, and antidiabetic activities, with several studies across the region highlighting promising therapeutic effects. In agriculture, fungal-derived nanomaterials enhance crop protection, food preservation,

J. J. G. Guerrero (✉)
College of Medicine, University of the Philippines Manila, Manila, Philippines
e-mail: jgguerrero1@up.edu.ph

R. R. H. Regalado
University of the Philippines Diliman, Quezon City, Philippines

J. P. Cuenca
Department of Otolaryngology-Head and Neck Surgery, College of Medicine, Dankook University, Dandae-ro, South Korea

M. J. Q. Paet · C. A. Malonzo
Department of Biology, College of Science, Bicol University, Legazpi, Philippines

M. C. A. Fulleros
Philippine Coconut Authority – Albay Research Center, Banao, Albay, Philippines

M. Y. Sabdao-Lunas
Department of Biology, College of Science, Bicol University, Legazpi, Philippines

Gyeongsang National University, Jinju, South Korea

J. J. G. Guerrero et al. (eds.), *Mycology in a Changing Planet: Applications and Perspectives from Southeast Asia*, https://doi.org/10.1007/978-981-95-2640-6_18

and environmental remediation, offering sustainable alternatives to conventional agrochemicals and fertilizers. Despite these advancements, challenges such as strain variability, lack of standardization, regulatory gaps, and limited infrastructure hinder wider adoption and commercialization. Strengthening regional research collaborations, investment in biosafety infrastructure, and public-private partnerships are vital in unlocking the full potential of fungal nanotechnology in Southeast Asia.

Keywords Green nanotechnology · Mycoparticles · Nanoparticles · Nanotechnology

Introduction

Fungal nanotechnology is an emerging field that explores the ability of fungi to synthesize nanoparticles (NPs) for applications in medicine, agriculture, and environmental sustainability (Abd-Elsalam, 2021; Hassan et al., 2024). Unlike conventional chemical and physical methods, fungal-mediated nanoparticle synthesis offers an eco-friendly, cost-effective, and scalable alternative that aligns with the principles of green nanotechnology (Abdel-Aziz et al., 2018; Dhanjal et al., 2022; Yadav et al., 2015). In Southeast Asia, where biodiversity is among the richest in the world, indigenous fungal species present an untapped resource for the sustainable production of nanomaterials. Given the region's growing demand for advanced yet accessible technologies, fungal nanotechnology may hold significant potential for addressing local challenges in health, industry, and environmental management.

One of the most promising applications of fungal nanotechnology in Southeast Asia is in healthcare, particularly in combating infectious diseases and antimicrobial resistance (Bueno, 2017; El Enshasy et al., 2018; Varma et al., 2022). Biosynthesized nanoparticles, such as silver (AgNPs), gold (AuNPs), and zinc oxide (ZnO NPs), have demonstrated potent antibacterial, antifungal, and anticancer properties (Baskar et al., 2013; Krishnamoorthi et al., 2021; Seetharaman et al., 2021). With Southeast Asia facing a high burden of drug-resistant infections, vector-borne diseases, and emerging pathogens, fungal-derived nanomaterials could provide alternative therapeutic solutions. Additionally, their biocompatibility and lower cytotoxicity make them attractive candidates for drug delivery, bioimaging, and wound healing applications.

Beyond medicine, fungal nanotechnology has implications for environmental sustainability, particularly in pollution control and agricultural innovation. Industrialization and rapid urbanization have led to increased pollution and heavy metal contamination in Southeast Asia's water and soil systems. Fungal-derived nanoparticles have shown potential for wastewater treatment, pollutant degradation, and soil remediation—offering sustainable strategies for mitigating environmental damage. In agriculture, nanomaterials synthesized by fungi could be integrated into biofertilizers and pesticides to enhance crop yield and resilience while reducing reliance on harmful agrochemicals. These applications align with regional efforts to promote green technology and sustainable development (Anjum et al., 2023; Elsakhawy et al., 2022; Sonawane et al., 2022).

Despite its potential, the large-scale adoption of fungal nanotechnology in Southeast Asia remains limited due to challenges in standardization, scalability, and regulatory frameworks. Variability in fungal strains, biosynthesis conditions, and nanoparticle stability complicates reproducibility and industrial applications. Additionally, insufficient funding, limited access to advanced research infrastructure, and fragmented regulatory policies hinder the commercialization of fungal-derived nanomaterials in the region. However, increasing regional collaborations, particularly through ASEAN science and technology initiatives, provide opportunities for capacity building and interdisciplinary research. Strengthening local research networks and fostering public-private partnerships could accelerate the development and integration of fungal nanotechnology into key industries.

Given these advancements and challenges, a comprehensive review of fungal nanotechnology research in Southeast Asia is necessary to assess its current state, identify gaps, and explore future directions. Thus, this review aims to synthesize recent findings on fungal-mediated nanoparticle synthesis, highlight key applications relevant to the region, and discuss existing barriers to its development. By providing a critical overview of the field, this chapter seeks to support scientists, policymakers, and industry leaders in harnessing fungal nanotechnology for sustainable and impactful innovations in Southeast Asia.

Mechanism of Mycogenic-Synthesized Nanoparticles

Fungal-based NP synthesis belongs to the bottom-up approach of NP production together with chemical-based methods. In this approach, small precursor materials are used as building blocks that lead to NP production. In comparison, the top-down approach involves the breakdown of massive materials through physical destruction to achieve the nanoscale particle size. Between the two approaches, bottom-up synthesis produces more uniform particles with fewer defects, higher yield at lower cost, and more biocompatible NP products (Dhillon et al., 2012; Dorcheh & Vahabi, 2016a, 2016b). Green synthesis of NP has been demonstrated across different fungal systems using different precursor materials. Some of the well-studied systems are silver- (Ag), gold- (Au), and other metal oxides-based NP synthesis with numerous works on process optimization considering important variables such as culture media formulation, pH, temperature, time of reaction, agitation, and concentration of precursors, among others (Dorcheh & Vahabi, 2016a, b; Guilger-Casagrande & Lima, 2019). Some of the commonly used fungal genera for mycogenic NP production include *Agaricus*, *Aspergillus*, *Colletotrichum*, *Fusarium*, *Neurospora*, *Penicillium*, *Rhizopus*, *Saccharomyces*, *Sclerotinia*, *Trichoderma*, and *Verticillium* (Dhillon et al., 2012; Loshchinina et al., 2022).

Fungi have been shown to have a natural resistance against metal ion toxicity whenever present in the environment. Myconanotechnology leverages these various mechanisms and has been utilized in nanomaterial generation since. At present, despite much literature and reviews describing important mycogenic mechanisms of metal NP synthesis, specific pathways remain to be elucidated (Guilger-Casagrande & Lima, 2019). The generally accepted mechanism involves the reaction of the precursor metal ions with any reducing agents produced by the fungi, followed by the nucleation of the elemental or neutral metals, and their stabilization and functionalization via interaction with various other fungi-derived biomolecules (Fig. 18.1). Furthermore, mycogenic NP synthesis can be categorically divided into either extracellularly or intracellularly localized events (or both), depending on the fungi-metal system.

The initial electron exchange between metal ions and reducing agents is the key step to any mycogenic NP synthesis. This electron transfer can be facilitated by various molecules originating from the fungi, including but not limited to proteins, peptides, polysaccharides, vitamins, and many secondary metabolites. This primarily includes the use of reducing enzymes or reductases, which are considered to be the most important players in extracellular metal ion processing. For instance,

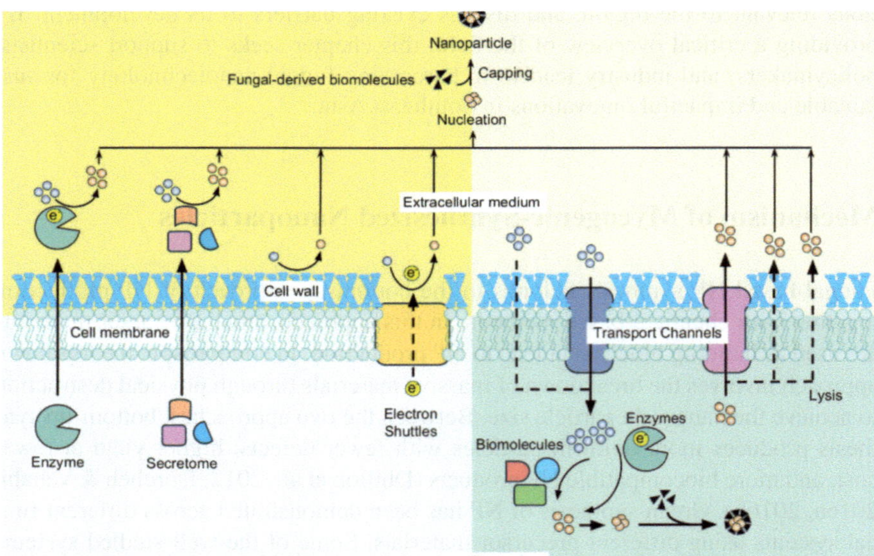

Fig. 18.1 Overview of the mycogenic synthesis of nanoparticles occurring either extracellularly or intracellularly, depending on the fungal-metal system used. Extracellular reduction of metal ions is commonly facilitated by secreted enzymes or other biomolecules (secretome). Adsorption in the cell wall and reduction via cell membrane-associated electron shuttles are also possible. Meanwhile, intracellular processing requires the initial transport of the metal ions into the cytoplasm. Once inside the cell, enzymes and other nonenzyme proteins, peptides, and other biomolecules or organelles will work on neutralizing the ions. Neutralized metal ions are then released extracellularly through channels or lysis of the cell, where they will undergo nucleation and capping to produce the NP. Synthesis of NP can also happen intracellularly, and NP can be recovered via lysis of the cell

Kumar et al. (2007) were able to generate 10–25 nm AgNPs in vitro using an α-NADPH-dependent nitrate reductase purified from *Fusarium oxysporum*. Likewise, various *Aspergillus* sp. with high nitrate reductase activity, as in *A. fumigatus*, demonstrated higher biosynthesis of AgNPs compared to *A. flavus* with low enzyme activity (Zomorodian et al., 2016). Nitrate and sulfite reductases have also been used in AuNP synthesis, as well as other reducing enzymes with similar mechanisms, including sulfate reductase for CdS NP synthesis and FAD-dependent glutathione reductase (Adebayo et al., 2021; Ahmad et al., 2002).

Fungi are also known to secrete numerous metabolites—from enzyme to nonenzyme proteins, peptides, secondary metabolites, and other organic compounds—collectively referred to as the secretome (Dorcheh & Vahabi, 2016a, b; Geetha et al., 2021). Components of the secretome usually contain functional groups that serve as electron donors, helping neutralize metal ions from the environment. For nonenzyme proteins or peptides, the thiol groups of cysteine or the imidazole rings of histidine can participate in electron transfer. Secondary metabolites, including phenols (hydroxyl groups), organic acids (carboxyl groups), alkaloids (nitrogen rings), and tannins (multiple hydroxyl groups), have various sources of electrons to give off. With fungal secretome content, spent media from growing fungal biomass can be readily used for metal neutralization and NP synthesis. For example, *T. atroviride*, *F. oxysporum*, *P. chrysogenum*, *P. expansum*, and *T. harzianum* culture liquids were used for Ag, Au, Pt, Se, and Te NP synthesis, respectively (Loshchinina et al., 2022).

Structurally, the fungal cell wall, composed of polysaccharides like chitin, glucans, and mannans, along with proteins and lipids, provides the cell with mechanical support via the adsorption of metal ions from the surroundings (Li et al., 2022). Their extensive networks of specialized structures, such as hyphae and mycelium, are also essential in providing a large surface area that enhances the fungi-environment interface needed for the uptake or neutralization of metal ions and other precursors. Biosorption of metal ions in the cell wall is possible due to the presence of numerous sorption sites that aid in the electrostatic attraction of the metal ions to the surface. The hydroxyl groups of polysaccharides and the amino groups of chitosan in *Aspergillus* and *Penicillium*'s cell walls are the main binding sites for Cu, Cr, Ni, and Zn (Šebesta et al., 2022). Melanin in the cell walls of *Aureobasidium pullulans* and *Cryptococcus neoformans* also possesses metal-chelating properties against Cu and Ag, respectively, owing to its numerous functional groups (Gadd & de Rome, 1988; Garcia-Rivera & Casadevall, 2001).

Another important extracellular mechanism of metal ion reduction is via the use of electron shuttles, including quinones and cytochrome-c, both of which use reducing agents such as NADH, NADPH, and other fungal metabolites to serve as electron donors. Quinone and its derivatives (anthraquinone, naphthoquinones, and thymoquinone) cycle between reduced and oxidized states by accepting electrons from reducing agents and transferring them to metal ions to convert them to their elemental forms. *F. oxysporum* has been previously described as containing quinones with excellent redox properties that make AgNP synthesis more efficient when coupled with reductases (Durán et al., 2005). Meanwhile, cytochrome-c is a heme-containing protein that utilizes the reduced and oxidized states of its iron-containing moiety to shuttle electrons. In *Aspergillus* sp., cytochromes have been implicated in the synthesis of PbNP together with reductases present in the cell wall (Pavani et al., 2012).

In intracellular synthesis, nanoparticles are formed inside the fungal cell. Fungal uptake of metal ions from their environment can be a passive process through concentration gradients and channels to facilitate their diffusion or via active pumps that would help the influx of metal ions into the cell (Adebayo et al., 2021). In some cases, fungi produce organic molecules to chelate and help trap metal ions, as in the case of *Aspergillus* siderophores for iron uptake (Pecoraro et al., 2021). Upon transport, metal ions are immediately reduced by intracellular enzymes similar to the reductases previously described in the extracellular NP pathway. Enzymes such as hydrogenases produced by *F. oxysporum* also showed reductase activity, producing PtNP (Loshchinina et al., 2022). Moreover, *F. oxysporum* has been reported to produce gold (Au) nanoparticles intracellularly within cytoplasmic vacuoles. This process involves enzymes like plasma membrane-ATPase, 3-glucan binding enzyme, and glyceraldehyde-3-phosphate dehydrogenase (Adebayo et al., 2021). Other fungi also showed intracellular NP synthesis, including *V. luteoalbum*, *A. flavus*, and *N. crassa*, producing AuNP and AgNP (Dhillon et al., 2012). Although they can also be secreted, metallothioneins, peroxidases, phenol-oxidizing enzymes, and other fungal polymers also aid in the detoxification of metals inside the cell with mechanisms similar to how the secretome works (Šebesta et al., 2022). Detoxified metals can then be excreted back into the environment or media to continue the NP synthesis process via similar efflux channels or through autolysis of matured fungal mycelia. For fungal systems that lack the necessary transport systems, NP synthesis is completed inside the cell. They, therefore, require additional fungal lysis steps to release the NP product for downstream use. Mycelial extraction is observed in many fungi-metal NP productions, e.g., *Agaricus* spp. for Ag, Au, Pd, Cu, and Se NP synthesis (Loshchinina et al., 2022). However, due to this added step, extracellular NP synthesis is preferred for easier scaling of production without added cost. Hence, among the studied fungi, *Aspergillus*, *Fusarium*, and *Penicillium* were considered important genera for the fabrication of various metal NPs owing to their extracellular mode of synthesis (Dhillon et al., 2012).

Whether extracellularly or intracellularly reduced, elemental metals immediately interact with various fungi-derived biomolecules. This leads the metal atoms to nucleate and form initial nanoparticle seeds, which grow into larger nanocrystals through further reduction and aggregation. At this stage, the metal nucleus is stabilized by the diverse fungal biomolecules in the reaction mixture, known as capping or functionalization. The attachment of these organic molecules to the inorganic mineral surface contributes to the termination of nanocrystal growth and therefore, their composition determines the final size, shape, and function of the NP products. Although chemical surfactants or phytochelatins from plants are used to stabilize NPs, different fungal-derived biomolecules are enough to stabilize the growing nanocrystal and prevent further aggregation or agglomeration. Chemical characterization and profiling of capping agents in different fungi-NP systems showed the predominance of proteins, peptides, and amino acids. AgNPs produced by *Coriolus versicolor*, *A. niger*, *F. solani*, and *F. semitectum* all showed proteins or peptide enveloping the metal nucleus core (Li et al., 2022). Protein electrophoresis, electron microscopy, and UV-vis spectroscopy showed around 85 kDa band of the protein cap, indicating that the nanocrystal is coated with organic materials rich in oxygen and carbon and that residues of tryptophan and tyrosine are detected, similar to what was present in the fungal filtrate (Guilger-Casagrande & Lima, 2019). Since the final cover of the mycogenic-synthesized NPs is mostly proteins or peptides, NP produced are less toxic than when chemically synthesized. NP caps can also be designed to anchor other molecules or compounds for synergistic treatments, and its biocompatibility ensures relatively easier NP internalization and retention to its target cells (Table 18.1).

Table 18.1 Common fungal models used in the study of mycogenic NP synthesis

Fungi-NP system	Biomolecule involved	References
A. flavus—AgNP	Nitrate reductase	Manjunath and Thiagarajan (2014) and Zomorodian et al. (2016)
A. niger—AgNP	Nitrate reductase	Zomorodian et al. (2016)
	Nitrate reductase and electron shuttle (quinone)	Gade et al. (2008)
A. fumigatus—AgNP	Nitrate reductase	Zomorodian et al. (2016)
	Secretome (proteins)	Bhainsa and D'souza (2006)
A. tubigensis—AgNP	Secretome (proteins)	Ballottin et al. (2016)
Agaricus spp.—AgNP/AuNP	Phenol-oxidizing enzymes	Vetchinkina et al. (2017)
F. oxysporum—AuNP	NADPH-dependent reductase	Thakker et al. (2013)
F. oxysporum—AgNP	NADPH-dependent nitrate reductase and electron shuttle (quinone)	Ishida et al. (2013) and Kumar et al. (2007)
	Nitrate reductase enzyme and electron shuttle (anthraquinones)	Durán et al. (2005)
F. oxysporum—CdS/PbS/ZnS/MoS_2	Sulfate reductase	Ahmad et al. (2002)
N. crassa—PtNP	Hydrogenase	Castro-Longoria et al. (2012)
Penicillium sp.—AgNP	Secretome (proteins)	Shareef et al. (2017) and Pareek et al. (2020)
R. oryzae—AuNP	Cytoplasmic reductases	Das et al. (2012)
T. hamatum—AuNP	Secretome (proteins)	Abdel-Kareem and Zohri (2018)
Trichoderma sp.—gNP	Nitrate reductase	Devi et al., 2013
Verticillium sp.—AuNP	Reductase and electron shuttle (quinone)	Mukherjee et al. (2001)

Applications in Medicine

In Southeast Asia, myconanotechnology represents a new paradigm in the development of nanomedicine. Currently, research in this field is primarily concentrated in Malaysia (50%), followed by Vietnam (17.9%), Indonesia (14.3%), Thailand (10.7%), and the Philippines (7.1%), with papers steadily growing in number from 2013 to 2024. The research trends in the region largely focus on the green synthesis of biometal nanoparticles (BM-NPs), with silver nanoparticles being the most frequently used. Both micro- and macrofungal species are explored for nanoparticle synthesis, *Aspergillus* and *Ganoderma* being the popular sources. Interestingly, lichens from the genera *Parmotrema*, *Ramalina*, and *Usnea* have also been reported (Mie et al., 2014; Din et al., 2015; Nguyen et al., 2024). Most studies examining the applications of these BM-NPs have focused on their antimicrobial efficacy, followed by anticancer, antioxidant, and antidiabetic bioactivities. Moreover, 69% of the studies reported spherical NPs, while 10.3% are equally attributed to other morphologies (i.e., hexagonal, cubic, and rod-shaped).

Nanoparticles as Antimicrobials

The synthesis of biometallic nanoparticles (e.g., gold, silver, iron, zinc, and metal oxides) is most promising as they have shown structural and physicochemical intrigue in terms of biomedical applications, stemming from their high surface area-to-volume ratio and diverse biomedical activities (Shah et al., 2015; Jahangirian et al., 2017; Barabadi et al., 2019). Various studies have demonstrated the bioactive properties of metal nanoparticles, viz., the application of silver nanoparticles (AgNPs) as excellent antibacterial and antifungal agents. Although the precise mechanisms underlying the antimicrobial action of AgNPs are not fully understood, current research indicates that they disrupt cell membrane permeability, interfere with cellular respiration, and induce free radical production. These multipronged actions collectively contribute to their antibacterial efficacy (Prabhu & Poulose, 2012; Lemire et al., 2013). Additionally, it has been proposed that AgNPs may release silver ions, which can bind to thiol groups found in essential bacterial enzymes, rendering them inactive (Prabhu & Poulose, 2012). Silver, which does not react with weak acids, has an innate propensity to react with soft bases like sulfur and phosphorus, which constitute key components within most cells (Li et al., 2010; Seil & Webster, 2012; Agnihotri et al., 2013; Ivask et al., 2014; Bruna et al., 2021). When silver nanoparticles interact with these cellular components, they can trigger reactions that impact cellular viability.

Rangdist et al. (2020) studied the biosynthesis of AgNPs in *Aspergillus* species using live cell filtrate. In their study, they produced exceptionally small particles that effectively inhibited clinically relevant Gram-positive and Gram-negative bacteria, showing bactericidal activity at very low concentrations. Hieu et al. (2022)

examined the effects of AgNPs derived from *A. niger* specifically on *E. coli*. Recently, green synthesis of Ag-doped ZnO nanorods produced from *A. niger* also exhibited antifungal properties and demonstrated superior effectiveness in inhibiting the growth of *Candida albicans* compared to commercial ketoconazole antibiotics (Rilda et al., 2024). Biogenic synthesis of AgNPs produced by a species of *Phomopsis* isolated from zinc mineral rocks in Thailand also showed potential activity against *S. aureus*, *B. cereus*, *P. aeruginosa*, and *E. coli* (Sutjaritvorakul & Chutipaijit, 2018). Additionally, endophytic fungi have also been reported to mediate the synthesis of AgNPs through their extracellular metabolites, exhibiting promising antibacterial activity against several pathogenic bacteria (Choong et al., 2018).

Apparently, AgNPs can also be successfully synthesized from various macrofungal species. Their use in Southeast Asia can be attributed to the region's rich ethnomycological practices, which include a strong tradition of wild mushroom gathering and cultivation practices (Jones & Lim, 1990; Fui et al., 2018; Łuczaj et al., 2021; Khastini et al., 2022; Anh et al., 2023; Guerrero et al., 2025). For example, crude extracts of the bracket fungus *Ganoderma lucidum*, a popular medicinal mushroom in East and Southeast Asia, have been incorporated into AgNPs, displaying strong antimicrobial potential against both bacterial and fungal pathogens with an environmentally benign, straightforward, and easily scalable procedure (Do Dat et al., 2021; Nguyen et al., 2023). Other extracts of macrofungal species, such as *Schizophyllum commune* (San & Don, 2013), *Agaricus bisporus* (Jameel et al., 2022), *Pleurotus sajor-caju* (Musa et al., 2018), and *Hypsizygus ulmarius* (Manimaran et al., 2023), were also reported to be capable of being biogenically synthesized into AgNPs. Meanwhile, sclerotial extract of *Lignosus rhinocerotis* (Katas et al., 2019) has been incorporated into gold nanoparticles (AuNPs). All these exhibit significant antimicrobial properties.

Aside from silver as the primary nano-metallic component, the use of metal oxide and metal oxide-doped particles, specifically zinc oxide (ZnO), also provides literature precedence. Moghaddam et al. (2017) reported, for the first time, the synthesis of ZnO nanoparticles using a green method with a new yeast strain *Pichia kudriavzevii*. Their study demonstrates that these ZnO nanoparticles exhibit a hexagonal wurtzite structure and possess antibacterial activity against both Gram-positive (*S. aureus, B. subtilis, S. epidermidis*) and Gram-negative (*E. coli* and *S. marcescens*) bacteria. Notably, the ZnO nanoparticles showed significant effectiveness against the *S. epidermidis* pathogen. Moreover, biosynthetic metal-doped ZnO nanorods have been explored in two studies by Rilda et al., (2023a, b), aiming to enhance antibacterial activity through morphological modifications achieved by doping with Mg^{2+} and Ti. Their data suggest that both doping strategies improve ZnO's potential as a potent antimicrobial agent. Specifically, Mg^{2+}-doped ZnO exhibited increased antibacterial activity, particularly against Gram-negative bacteria, with a maximum inhibitory zone of 24 mm (Rilda et al., 2023a). Similarly, Ti-doped ZnO nanorods, synthesized through a biosynthetic route, demonstrated improved antimicrobial activity under UV irradiation, with the effectiveness influenced by morphological and synthetic factors (Rilda et al., 2023b).

Biogenic Metal Nanoparticles in Cancer and Diabetes

Remarkably, the biogenic fabrication of BM-NPs to target known cancer cell models has also shown great promise. Gurunathan et al. (2013) demonstrated the synthesis of AgNPs with a mean size of 5 nm using pharmaceutically important *Ganoderma neo-japonicum* mycelia extract. Toxicity studies revealed that biologically synthesized AgNPs exhibited potent cytotoxic effects in MDA-MB-231 breast cancer cells (IC_{50} 6.0 µg/mL), characterized by reduced cell viability and increased membrane leakage in a dose-dependent manner. Furthermore, the cytotoxic effects were corroborated by increased reactive oxygen species and hydroxyl radical production, as well as activation of the caspase 3 apoptotic pathway. Within the same genus, *G. lucidum* AgNPs also exhibit antiproliferative activities against HepG2 (IC_{50} 21.85 µg/mL) and MCF-7 (IC_{50} 67.77 µg/mL) cells (Nguyen et al., 2021), as well as human epidermoid carcinoma KB cells (IC_{50} 190.06 µg/mL) (Do Dat et al., 2021). *Flammulina velutipes*, widely known as enoki mushroom, was used to biosynthesize highly stable AuNPs via a rapid and eco-friendly sonochemical technique (Dheyab et al., 2023). These biosynthesized nano-gold particles showed significant anticancer activity, achieving 98.2% inhibition of MDA-MB-231 cell viability at an 80 µg/mL concentration. Similarly, the edible mushroom *H. ulmarius*, in addition to its previously discussed antibacterial activity, demonstrated dose-dependent cytotoxicity against human lung cancer cell lines (A549; IC_{50} of 50.78 µg/mL), inhibiting cell viability by up to 64.31% post-treatment (Manimaran et al., 2023). In another study, Banu et al. (2021) demonstrated that AgNPs biologically synthesized using *Rhizopus stolonifer* effectively inhibit the growth of esophageal squamous-cell carcinoma (EAC) and colon cancer (HT-29) cell lines in a dose-dependent manner. At the highest concentration tested (8 µg/mL), AgNPs reduced cell viability by up to 68% in EAC cells and 68.46% in HT-29 cells, with IC_{50} values indicating significant cytotoxicity at concentrations as low as 2 µg/mL. Majeed et al. (2020) found that AgNPs synthesized from *Aspergillus clavatus* showed a dose-dependent cytotoxic effect on RAW macrophage cells, with an IC_{50} value at the highest concentration tested (125 mg/mL), indicating their potential for selective anticancer applications.

While the anticancer properties of BM-NPs are well-documented, their therapeutic potential extends beyond cancer treatment. Notably, the antioxidant activities of BM-NPs are also intricately linked to their anticancer effects. By mitigating oxidative stress and neutralizing reactive oxygen species, these nanoparticles can enhance their efficacy in cancer treatment and contribute to overall cellular protection. Two studies in the Philippines, in particular, demonstrated the enhanced antioxidant properties of AgNPs compared to their respective aqueous extracts. Javier et al. (2023) demonstrated that *G. lucidum*-AgNPs showed significantly higher antioxidant activity (% inhibition of 43.87%) than raw *G. lucidum* aqueous extract (35.95%), likely due to the increased surface area-to-volume ratio of the AgNPs. Similarly, Sytu and Camacho (2018) found that *Lenzites betulina*-capped AgNPs exhibited significantly higher antioxidant activity compared to the raw extract, attributed to the enhanced reactivity of AgNPs. These findings suggest that the incorporation of AgNPs enhances the radical scavenging abilities of the natural fungal extracts. Strong DPPH free radical scavenging activity was also observed in the biosynthesized ZnO-NPs from the new yeast strain *P. kudriavzevii* (Moghaddam et al., 2017). Additionally, biogenic production of ZnO nanoparticles from the shiitake mushroom, *Lentinula edodes*, yielded not just antioxidant properties but also demonstrated significant antimicrobial, antipyretic, antidiabetic, and anti-inflammatory activities (Amin et al., 2023). At a concentration of 300 μg/mL, these nanoparticles exhibited notable antioxidant (65.7 ± 1.09%) and anti-inflammatory (86.45 ± 0.60%) effects. They also effectively reduced paw inflammation (1.1 ± 0.06%) and yeast-induced pyrexia (97.4 ± 0.51%) at 10 mg/mL, in a dose-dependent manner. These findings highlight their potential application in the food and nutraceutical industries for managing various health conditions.

To some extent, AgNPs studies in Southeast Asia also explore their potential for antidiabetic effects. In one study, Sutjaritvorakul et al. (2023) found that *Xylaria* AgNPs exhibited significant α-glucosidase inhibitory activities, achieving 63.21% and 54.42% inhibition for maltase and sucrase, respectively. Similarly, the antidiabetic potential of *L. edodes* ZnO-NPs exhibited significant antidiabetic potential through α-amylase inhibition, with a maximum inhibition of 85.18% at 300 μg/mL (Amin et al., 2023). This inhibition is close to that of standard metformin (87.69% at the same concentration), with IC_{50} values of 15.10 μg/mL for the ZnO-NPs compared to 12.48 μg/mL for metformin, representing a promising approach for managing type 2 diabetes mellitus.

Myconanotechnology in Biomaterial Production

The demand for biomaterials continues to rise to replace standard procedures that could suffice structural support like autologous grafts. These materials are optimized to mimic the biological composition of a specific anatomical configuration and elicit regenerative properties. Benefits from such components arise from the reduction of transplant morbidity, to eliminate pain from the site of collection, and to augment a near-identical morphology from the required tissue (Velnar et al., 2016). These biomaterials are widely used from simple wound healing devices to implantable cartilage replacements and even transplanted neural conduits.

Fungal-based biomaterials may provide an alternative source of materials that can be introduced into biological systems. Fungal-based biomaterials have been explored in industrial and agricultural uses and currently focus on the practical uses of fungi in medical biomaterial engineering, particularly in the development in Southeast Asia (SEA). Interestingly, chitosan is a polysaccharide that is produced from the alkalization of chitin and is mainly found in the exoskeleton of crustaceans as well as fungal cell walls. Multiple materials have been correlated with the use of chitosan in different forms, such as electrospun wound patches, thermosensitive or pH-sensitive hydrogels, and nanoparticle delivery systems.

Electrospinning involves the use of high voltage output to induce a charged stream of liquid fiber with the involvement of an electric field and the surface tension of the liquid. This process allows the collection of fibers through micrometer and nanometer scales that can be uniformly produced by controlling the solution, electric output, and collecting device parameters. It has been reported that the use of chitosan has been developed to provide nanometer fibers for wound healing (Charernsriwilaiwat et al., 2013; Manotham et al., 2023; Putra et al., 2020), drug delivery (Chuah et al., 2023; Noor et al., 2020), and cardiac tissue engineering (Mukherjee et al., 2010). Aside from woven fiber nets, chitosan has been used in hydrogels for the regulation of swelling properties and charge-responsive materials. Such material is allowed to retain moisture and form a soft consistency that is beneficial for soft tissue repair. Nanocomposite structures made from hydrogels allow controlled responsive mechanisms to a specific stimulation (Schexnailder & Schmidt, 2009). Chitosan-based hydrogels were already developed (Abdan et al., 2020; Hoang et al., 2022; Wunnoo et al., 2023).

Chitin is a naturally occurring polysaccharide found in fungal cell walls, which can be further processed into chitosan, a nontoxic component majorly used in biomedical applications. The processed material possesses high biocompatibility with biodegradable profiles without immunogen responses. Developments in chitosan fibers have been extensively reported elsewhere, wherein pure chitosan and compounded materials were made on the nanometer scale. A study from Wardhani (Wardhani et al., 2022) showed the capability of chitosan nanofibers to be incorporated in alpha-mangosteen, which may be suited for biomedical applications. The capability of the material to be incorporated with hydrogen bonding shows the stability of the material through chemical analysis and functional group assessments. An extensive study evaluating the effects of chitosan concentrations showed the capability to manipulate fiber diameters as well as fiber texture (Boonpratum et al., 2022). Manipulation of fibers allowed better control release of potential drugs with around 10% release profile changes. The study was focused on a baseline dye for a measurable release profile, which can be further manipulated with different kinds of drugs or proteins. Aside from electrospun materials, chitosan has been reported to be developed into cationic nanoparticle carriers for drug delivery (Nazurah et al., 2024). Nanoparticle preparation of chitosan with silver nanoparticles allowed the effective elimination of *C. albicans* from the disruption of potassium channels at the surface level (Kulatunga et al., 2017). The disruption of ionic cascades leads to the disruption of enzyme activity, leading to apoptotic cascades brought about by the accumulation of reactive oxygen species (ROS).

Aside from nanoscale hydrogels from chitosan, multiple reports of nanoparticle-based products are produced by certain species of fungi, specifically silver nanoparticles. These metal nanoparticles are known for their antibacterial properties, biological sensing and probing, and control of molecular adhesion through zeta potential changes (Mehta, 2017; Ravindran et al., 2013). It has been reported that such nanoparticles were extracted from *Amylomyces rouxii* (Musarrat et al., 2010), *Cochliobolus lunatus* (Salunkhe et al., 2011), and *Pycnoporus sanguineus* (San Chan & Don, 2013), to mention a few. Such components may be functionalized in either electrospun mats or hydrogel composites to increase bioactive responses or anti-bacterial properties that aid in repair. The combination of multiple materials poses one of the benefits of optimizing biomaterials to maximize their efficiency for tissue regenerative tendencies. Silver nanoparticles isolated from white rot fungi showed antibacterial properties that are dependent on particle size. The capability of silver to enter bacterial membranes and interact with sulfur-containing proteins and attack the respiratory chain leads to cell death (Gudikandula et al., 2017). Silver nanoparticles were also reported to have an inhibitory release of alpha-amylase in comparison to plant extract-only samples (Chinnasamy et al., 2019). The increase in inhibitory enzyme was paralleled with increase in glucosidase enzyme upon the addition of nanoparticles. Relative to wound healing studies, silver nanoparticles were used on skin irritations, diabetic wounds, and other components for surface-level epidermal healing applications. An extensive review of nanoparticle interactions with wound therapy showed that nanoparticles have lower toxicity and increase triphosphate and osmotic transport of silver and hydrogen ion antiporters (Nor Azlan et al., 2021). Additionally, the use of silver nanoparticles is efficient as they are capable of destroying bacterial colonies without disruption of tissue cell populations. Another study on silver nanoparticles that was chemically synthesized using vitamin B6 showed better healing and moisture retention properties and reduced inflammatory reactions (Rangasamy et al., 2016).

Applications in Agriculture

In 2008, Mukherjee et al. reported the green biosynthesis of silver nanoparticles by *Trichoderma asperellum*. Vahabi et al. (2011) introduced a novel method for silver nanoparticle synthesis using *T. reesei* for large-scale production. The process involves the exposure of the fungus mycelium to the silver nitrate solution to trigger the production of enzymes and metabolites, followed by the absorption of UV-visible light. This innovative approach has paved the way for eco-friendly nanoparticle production techniques.

Fungal nanotechnology is gaining recognition for its role in food storage, particularly through the use of wrapping materials or food additives via nanoencapsulation and nanoemulsion (Adebayo et al., 2021). Silver nanoparticles produced by fungi like *F. oxysporum*, *G. sessiliforme*, and *Aspergillus clavatus* (Kalshyap et al., 2012; Rai et al., 2021; Mohanta et al., 2018; Alghuthaymi et al., 2015) are known for their antimicrobial properties. These nanoparticles have the potential to be used as alternatives to traditional pesticides or to control food-borne pathogens, offering a promising solution to food safety challenges. Additionally, selenium nanoparticles synthesized by the fungus *Marianneae* sp. act as natural stabilizing agents (Zhang et al., 2019), further demonstrating the versatility and potential of fungal nanoparticles in food applications. Mycosynthesized nanomaterials are also being considered for food packaging, processing, shelf-life improvement, and management of food pathogens.

There are myconano-based coatings that prolong the shelf-life of food products, such as the fungal chitosan nanoparticle studied by Melo et al. (2018) or the pullulan/chitosan-based edible film that utilizes the zinc oxide nanoparticle (Ingle et al., 2023). These innovations are crucial for reducing food waste and ensuring food quality. Chitosan-AgNp composite was also shown to significantly inhibit *Colletotrichum gloeosporioides*, and in vivo assay using detached mango Alphonso had shown that it can be utilized against anthracnose (Chowdappa et al., 2014).

Cerium oxide nanoparticles produced using *Aspergillus niger* have shown significant antibacterial properties (Gopinath et al., 2025). Silver nanoparticles from fungi such as *Aspergillus versicolor*, *Epicoccum nigrum*, and *Guignardia mangiferae* have demonstrated antimicrobial effects against pathogens in strawberry plants and other pathogenic fungi (Elgorban et al., 2016; Qian et al., 2013; Balakumaran et al., 2015). Furthermore, biogenic silver nanoparticles (AgNPs) synthesized from *Alternaria alternata*, *Fusarium semitectum*, and *Penicillium herbarum* have proven effective (Gajbhiye et al., 2009; Gherbawy et al., 2013). These biogenic nanoparticles, when combined with conventional biocides, can help overcome resistance developed by pathogens or parasites against traditional drugs or biocides (Guilger-Casagrande & de Lima, 2019).

In conventional agriculture, pesticides are routinely used to protect crops; however, most pesticides are volatile and are not absorbed by the crop. Instead, they spread or leach into the surrounding environment/ecosystems. Because of this, the utilization of nanoparticles is being studied and considered to improve the delivery and regulate their release. Economically important pests in crops and stored grains, such as Coleoptera and Lepidoptera, can be potentially controlled by selenium and iron nanoparticles (Arunthirumeni et al., 2022; Xu et al., 2021). Fungal pathogens are also culprits of crop productivity loss, and to counter the losses that may occur, a nano-fungicide from *Zataria multiflora* can act as a protectant or noncarriers of a different fungicide (Kutawa et al., 2021).

Extracts from *Fusarium solani* have been used to biosynthesize six types of nanoparticles, with selenium (SeNPs) and titanium dioxide (TiO_2NPs) demonstrating the ability to control larvae of *Callosobruchus chinensis* and *Callosobruchus maculatus* in beans (Helmy et al., 2023). Selenium nanoparticles synthesized from *Trichoderma* fungus have shown efficacy in controlling insect pests. Arunthirumeni et al. (2022) found that these nanoparticles are highly effective as larvicidal and antifeedant agents against *Spodoptera litura* larvae.

One of the first fungal species that was studied for the synthesis of silver nanoparticles (AuNPs) is *Verticillium* sp. Currently, several fungal species are being studied and utilized for the biosynthesis of AuNPs. Some of the fungi that produce silver nanoparticles are *Collectotrichum* sp., *Penicillium* sp., *Harmoconis resinae*, *Trichothecium* sp., and many more (Bahrulolum et al., 2021). Unlike *F. oxysporum* and *A. fumigatus*, which are pathogenic to humans and plants, *Trichoderma* sp. is being studied as an alternative for the synthesis of nanosilver. It also serves as a potent biocontrol agent against soil-borne fungi (Zhao et al., 2021). Goyal et al. (2023) cited that xylanases from *A. niger* and *T. longibrachiatum* can be used to produce biosynthetic metal nanoparticles such as AgNPs to synthesize nanoparticles that are resistant to oxidation and therefore could have broad range applications.

Al-Askar et al. (2013) demonstrated that silver nanoparticles biosynthesized using *F. solani* effectively control phytopathogenic fungi in wheat, barley, and maize seeds. Additionally, nanoparticles such as titanium dioxide (TiO_2) and zinc oxide (ZnO) have intrinsic toxicity that can protect plants from pathogens (Rai et al., 2021). Nanoparticles and nanosensors can also be applied in plant disease diagnosis. Disease management could be done through direct application of the nanoparticle or by encapsulating the nanoparticle with antimicrobial properties. While disease diagnosis could be done by the development of enzyme-based biosensors coated with metals like Au, Ag, Cu, or Ti-nanoparticles (Kahn et al., 2019).

Although research on nanofertilizers is still in its early stages, Aldesuquy et al. (2022) have documented that iron nanoparticles biosynthesized using *F. oxysporum* serve as an excellent biofactory for nano-iron synthesis. It offers a promising solution to the challenges of traditional fertilizers, which often lead to nutrient runoff and environmental pollution. By improving nutrient use efficiency, nanofertilizers can reduce the amount of fertilizer needed, thereby minimizing environmental impact and promoting sustainable agricultural practices.

Way Forward

Fungal nanotechnology in Southeast Asia holds transformative potential for medicine, agriculture, and environmental sustainability. However, its full realization is hindered by significant challenges, particularly in the standardization and scalability of fungal-mediated nanoparticle synthesis. Variability in fungal strains, biosynthesis conditions, and nanoparticle stability remains as key obstacles to industrial applications. Addressing these issues requires the establishment of standardized protocols, quality control measures, and biosafety regulations to ensure consistency and reproducibility. Moreover, investments in nanotechnology research infrastructure, such as advanced characterization facilities and biosafety laboratories, are crucial to bridging the gap between laboratory research and large-scale production.

Despite growing interest in fungal nanotechnology, research efforts across Southeast Asia remain fragmented, with limited cross-border collaborations and knowledge exchange. Strengthening regional networks through ASEAN science and technology initiatives can enhance interdisciplinary collaboration and accelerate technological advancements. In addition, government support in the form of funding, policy development, and intellectual property protections will be critical to fostering innovation and attracting private sector investment.

Beyond research, the commercialization of fungal-derived nanomaterials remains a key challenge due to regulatory uncertainties and limited industry adoption. Establishing public-private partnerships can drive technology transfer, facilitate product development, and create market-ready applications in healthcare, agriculture, and environmental remediation. Encouraging start-ups and local biotech companies to invest in fungal nanotechnology will also help build a robust innovation ecosystem. Furthermore, increasing public awareness and engagement with key stakeholders will be essential to driving acceptance and integration of these novel technologies into mainstream applications.

Acknowledgments The authors are grateful to the authors of cited publications and the numerous researchers on fungal nanotechnology.

References

Abdan, K. B., Yong, S. C., Chiang, E. C. W., Talib, R. A., Hui, T. C., & Hao, L. C. (2020). Barrier properties, antimicrobial and antifungal activities of chitin and chitosan-based IPNs, gels, blends, composites, and nanocomposites. In *Handbook of chitin and chitosan* (pp. 175–227). Elsevier.

Abdel-Aziz, S. M., Prasad, R., Hamed, A. A., & Abdelraof, M. (2018). Fungal nanoparticles: A novel tool for a green biotechnology? In *Fungal nanobionics: Principles and applications* (pp. 61–87).

Abdel-Kareem, M. M., & Zohri, A. A. (2018). Extracellular mycosynthesis of gold nanoparticles using Trichoderma hamatum: Optimization, characterization and antimicrobial activity. *Letters in Applied Microbiology, 67*(5), 465–475.

Abd-Elsalam, K. A. (2021). Fungal nanotechnology. *Journal of Fungi, 7*(8), 583.

Adebayo, E. A., Azeezz, M. A., Alao, M. B., Oke, A. M., & Aina, D. A. (2021). Fungi as veritable tool in current advances in nanotechnology. *Heliyon, 7*(11), e08480. https://doi.org/10.1016/j.heliyon.2021.e08480

Agnihotri, S., Mukherji, S., & Mukherji, S. (2013). Immobilized silver nanoparticles enhance contact killing and show highest efficacy: Elucidation of the mechanism of bactericidal action of silver. *Nanoscale, 5*(16), 7328–7340.

Ahmad, A., Mukherjee, P., Mandal, D., Senapati, S., Khan, M. I., Kumar, R., & Sastry, M. (2002). Enzyme mediated extracellular synthesis of CdS nanoparticles by the fungus, Fusarium oxysporum. *Journal of the American Chemical Society, 124*(41), 12108–12109.

Al-Askar, A. A., Hafez, E. E., Kabeil, S. A., & Meghad, A. (2013). Bioproduction of silver-nano particles by Fusarium oxyporum and their antimicrobial activity against some plant pathogenic bacteria and fungi. *Life Science Journal, 10*(3), 2470–2475.

Aldesuquy, H. S., Abdel-Fattah, G., & Eldeghiedy, A. (2022). *Potentiality of nanoparticle synthesized by fungi as Nanobiofertilizer*. LAB Lambert Academic Publishing. ISBN: 978-620-5-51263-0.

Alghuthaymi, M. A., Almoammar, H., Rai, M., Said-Galiev, & Abd-Alsamal, K. A. (2015). Myconanoparticles: Synthesis and their role in phytopathogens management. *Biotechnology and Biotechnological Equipment, 29*(2), 221–236. https://doi.org/10.1080/13102818.2015.1008194

Amin, Z. S., Afzal, M., Ahmad, J., Ahmed, N., Zeshan, B., Hashim, N. H. H. N., & Yean, C. Y. (2023). Synthesis, characterization and biological activities of zinc oxide nanoparticles derived from secondary metabolites of Lentinula edodes. *Molecules, 28*(8), 3532.

Anh, C. N., Chi, N. M., Kiet, T. T., Long, P. D., Thuy, P. T. T., Van Loi, V., & Dell, B. (2023). Morphological and molecular identification of an edible Russula mushroom in Northeast Vietnam. *Journal of Forestry Science and Technology, 15*, 050–059.

Anjum, S., Vyas, A., & Sofi, T. (2023). Fungi-mediated synthesis of nanoparticles: Characterization process and agricultural applications. *Journal of the Science of Food and Agriculture, 103*(10), 4727–4741.

Arunthirumeni, M., Veerammal, V., & Shivakumar, M. S. (2022). Biocontrol efficacy of Mycosynthesized selenium nanoparticle using Trichoderma sp. on insect Pest Spodoptera litura. *Journal of Cluster Science, 33*, 1645–1653. https://doi.org/10.1007/s10876-021-02095-4

Bahrulolum, H., Nooraei, A., Javanshir, N., Tarrahimofrad, H., Mirbagheri, V. S., Easton, A. J., & Ahmadia, G. (2021). Green synthesis of metal nanoparticles using microorganisms and their application in the agrifood sector. *Journal of Nanobiotechnology, 19*(86). https://doi.org/10.1186/s12951-021-00834-3

Balakumaran, M. D., Ramachandran, R., & Kalaicheilvan, P. T. (2015). Exploitation of endophytic fungus, Guignardia mangiferae for extracellular synthesis of silver nanoparticles and their in vitro biological activities. *Microbiological Research, 178*, 9–17. https://doi.org/10.1016/j.micres.2015.05.009

Ballottin, D., Fulaz, S., Souza, M. L., Corio, P., Rodrigues, A. G., Souza, A. O., et al. (2016). Elucidating protein involvement in the stabilization of the biogenic silver nanoparticles. *Nanoscale Research Letters, 11*, 1–9.

Banu, A., Gousuddin, M., & Yahya, E. B. (2021). Green synthesized monodispersed silver nanoparticles' characterization and their efficacy against cancer cells. *Biomedical Research and Therapy, 8*(8), 4476–4482.

Barabadi, H., Tajani, B., Moradi, M., Damavandi Kamali, K., Meena, R., Honary, S., Mahjoub, M. A., & Saravanan, M. (2019). Penicillium family as emerging nanofactory for biosynthesis of green nanomaterials: A journey into the world of microorganisms. *Journal of Cluster Science, 30*, 843–856.

Baskar, G., Chandhuru, J., Fahad, K. S., & Praveen, A. S. (2013). Mycological synthesis, characterization and antifungal activity of zinc oxide nanoparticles. *Asian Journal of Pharmacy and Technology, 3*(4), 142–146.

Bhainsa, K. C., & D'souza, S. F. (2006). Extracellular biosynthesis of silver nanoparticles using the fungus aspergillus fumigatus. *Colloids and Surfaces B: Biointerfaces, 47*(2), 160–164.

Boonpratum, C., Naemchanthara, P., Limsuwan, P., & Naemchanthara, K. (2022). Effects of chitosan and tween 80 addition on the properties of nanofiber mat through the electrospinning. *E-Polymers, 22*(1), 234–248.

Bruna, T., Maldonado-Bravo, F., Jara, P., & Caro, N. (2021). Silver nanoparticles and their antibacterial applications. *International Journal of Molecular Sciences, 22*(13), 7202.

Bueno, J. (2017). Fungal bionanotechnology, when knowledge merge into a new discipline to combat antimicrobial resistance. In *Fungal Nanotechnology: Applications in Agriculture, Industry, and Medicine* (pp. 189–206).

Castro-Longoria, E., Moreno-Velasquez, S. D., Vilchis-Nestor, A. R., Arenas-Berumen, E., & Avalos-Borja, M. (2012). Production of platinum nanoparticles and nanoaggregates using Neurospora crassa. *Journal of Microbiology and Biotechnology, 22*(7), 1000–1004.

Charernsriwilaiwat, N., Rojanarata, T., Ngawhirunpat, T., Sukma, M., & Opanasopit, P. (2013). Electrospun chitosan-based nanofiber mats loaded with Garcinia mangostana extracts. *International Journal of Pharmaceutics, 452*(1–2), 333–343.

Chinnasamy, G., Chandrasekharan, S., & Bhatnagar, S. (2019). Biosynthesis of silver nanoparticles from Melia azedarach: Enhancement of antibacterial, wound healing, antidiabetic and antioxidant activities. *International Journal of Nanomedicine, Volume 14*, 9823–9836.

Choong, V., Ai, L. O. G., & Suan, T. K. (2018). Synthesis of silver nanoparticles mediated by endophytic fungi associated with orchids and its antibacterial activity. *Materials Today: Proceedings, 5*(10), 22093–22100.

Chowdappa, P., Gowda, S., Chethana, C. S., & Madhura, S. (2014). Antifungal activity of chitosan-silver nanoparticle composite against Collectotrichum gloeosporiodes associated with mango anthracnose. *Academic Journals, 8*(17), 1803–1812. https://doi.org/10.5897/AJMR2013.6584

Chuah, L.-H., Loo, H.-L., Goh, C. F., Fu, J.-Y., & Ng, S.-F. (2023). Chitosan-based drug delivery systems for skin atopic dermatitis: Recent advancements and patent trends. *Drug Delivery and Translational Research, 13*(5), 1436–1455.

Das, S. K., Liang, J., Schmidt, M., Laffir, F., & Marsili, E. (2012). Biomineralization mechanism of gold by zygomycete fungi Rhizopous oryzae. *ACS Nano, 6*(7), 6165–6173.

Devi, T. P., Kulanthaivel, S., Kamil, D., Borah, J. L., Prabhakaran, N., & Srinivasa, N. (2013). Biosynthesis of silver nanoparticles from Trichoderma species. *Indian J Exp Biol, 51*(7), 543–547.

Dhanjal, D. S., Mehra, P., Bhardwaj, S., Singh, R., Sharma, P., Nepovimova, E., et al. (2022). Mycology-nanotechnology interface: Applications in medicine and cosmetology. *International Journal of Nanomedicine, 17*, 2505–2533.

Dheyab, M. A., Aziz, A. A., Al-Mafarjy, S. S., Suardi, N., Ab Razak, N. N. A. N., Ramizy, A., & Jameel, M. S. (2023). Exploring the anticancer potential of biogenic inorganic gold nanoparticles synthesized via mushroom-assisted green route. *Inorganic Chemistry Communications, 157*, 111363.

Dhillon, G. S., Brar, S. K., Kaur, S., & Verma, M. (2012). Green approach for nanoparticle biosynthesis by fungi: Current trends and applications. *Critical Reviews in Biotechnology, 32*(1), 49–73.

Din, L. B., Mie, R., Samsudin, M. W., Ahmad, A., & Ibrahim, N. (2015). Biomimetic synthesis of silver nanoparticles using the lichen Ramalina dumeticola and the antibacterial activity. *Malaysian Journal of Analytical Sciences, 19*, 369–376.

Do Dat, T., Viet, N. D., Dat, N. M., My, P. L. T., Thinh, D. B., Thy, L. T. M., Huong, L. M., Khang, P. T., Hai, N. D., Nam, H. M., Phong, M. T., & Hieu, N. H. (2021). Characterization and bioactivities of silver nanoparticles green synthesized from Vietnamese Ganoderma lucidum. *Surfaces and Interfaces, 27*, 101453.

Dorcheh, S. K., & Vahabi, K. (2016a). In J. M. Mérillon & K. Ramawat (Eds.), Fungal metabolites. Reference series in phytochemistry. *Biosynthesis of nanoparticles by fungi: Large-scale production*. Springer.

Dorcheh, S. K., & Vahabi, K. (2016b). Biosynthesis of nanoparticles by fungi: Large-scale production. *Fungal Metabolites, 5*, 1–20.

Durán, N., Marcato, P. D., Alves, O. L., De Souza, G. I., & Esposito, E. (2005). Mechanistic aspects of biosynthesis of silver nanoparticles by several Fusarium oxysporum strains. *Journal of Nanobiotechnology, 3*, 1–7.

El Enshasy, H. A., El Marzugi, N. A., Elsayed, E. A., Ling, O. M., Malek, R. A., Kepli, A. N., et al. (2018). Medical and cosmetic applications of fungal nanotechnology: Production, characterization, and bioactivity. In *Fungal nanobionics: Principles and applications* (pp. 21–59).

Elgorban, A. M., Aref, S., Seham, S. M., Elhindi, K. M., Bahkali, A. H., Sayed, S. R., et al. (2016). Extracellular synthesis of silver nanoparticles using aspergillus versicolor and evaluation of their activity on plant pathogenic fungi. *Mycosphere, 7*, 844–852. https://doi.org/10.5943/mycosphere/7/6/15

Elsakhawy, T., Omara, A. E. D., Abowaly, M., El-Ramady, H., Badgar, K., Llanaj, X., et al. (2022). Green synthesis of nanoparticles by mushrooms: A crucial dimension for sustainable soil management. *Sustainability, 14*(7), 4328.

Fui, F. S., Saikim, F. H., Kulip, J., & Seelan, J. S. S. (2018). Distribution and ethnomycological knowledge of wild edible mushrooms in Sabah (Northern Borneo), Malaysia. *Journal of Tropical Biology & Conservation (JTBC), 15*, 203â-222.

Gadd, G. M., & de Rome, L. (1988). Biosorption of copper by fungal melanin. *Applied Microbiology and Biotechnology, 29*, 610–617.

Gade, A. K., Bonde, P., Ingle, A. P., Marcato, P. D., Duran, N., & Rai, M. K. (2008). Exploitation of aspergillus Niger for synthesis of silver nanoparticles. *Journal of Biobased Materials and Bioenergy, 2*(3), 243–247.

Gajbhiye, M., Kesharwani, J., Ingle, A., Gade, A., & Rai, M. (2009). Fungus-mediated synthesis of silver nanoparticles and their activity against pathogenic fungi in combination with fluconazole. *Nanomedicine, 5*, 382–386. https://doi.org/10.1016/j.nano.2009.06.005

Garcia-Rivera, J., & Casadevall, A. (2001). Melanization of Cryptococcus neoformans reduces its susceptibility to the antimicrobial effects of silver nitrate. *Sabouraudia, 39*(4), 353–357.

Geetha, N., Bhavya, G., Abhijith, P., Shekhar, R., Dayananda, K., & Jogaiah, S. (2021). Insights into nanomycoremediation: Secretomics and mycogenic biopolymer nanocomposites for heavy metal detoxification. *Journal of Hazardous Materials, 409*, 124541.

Gherbawy, Y. A., Shalaby, I. M., El-Sadek, M. S. A., Elhariry, H. M., & Banaja, A. A. (2013). The anti-fasciolasis properties of silver nanoparticles produced by Trichoderma harzianum and their improvement of the anti-fasciolasis drug triclabendazole. *International Journal of Molecular Sciences, 14*, 21887–21898. https://doi.org/10.3390/ijms141121887

Gopinath, K., Karthika, V., Sundaravadivelan, C., Gowri, A., & Arumugam, A. A. (2025). Mycogenesis of cerium oxide nanoparticles using Apergillus Niger culture filtrate and their applications for antimicrobial and larvicidal activities. *Journal of Nanostructure in Chemistry, 5*, 295–303. https://doi.org/10.1007/s40097-015-0161-2

Goyal, V., Rani, D., Ritika, M. S., Deng, C., & Wang, Y. (2023). Unlocking the potential of nano-enabled precision agriculture for efficient and sustainable farming. *Plants, 12*(21), 3744. https://doi.org/10.3390/plants12213744

Gudikandula, K., Vadapally, P., & Charya, M. S. (2017). Biogenic synthesis of silver nanoparticles from white rot fungi: Their characterization and antibacterial studies. *OpenNano, 2*, 64–78.

Guerrero, J. J. G., Regalado, R. R. H., Penecilla, G. L. M., Malonzo, C. A., General, M. A., & Junsay, A. T. (2025). Mycological biosafety and biosecurity in The Philippines. In J. J. G. Guerrero, R. Cena-Navarro, R. V. Destura, M. P. De Leon, K. I. R. Notarte, & M. A. O. Balendres (Eds.), *Biosafety and biosecurity: Practical insights and applications for low and middle-income countries* (pp. 206–234). CRC Press.

Guilger-Casagrande, M., & de Lima, R. (2019). Synthesis of silver nanoparticle mediated by fungi: A review. Frontiers in bioengineering and biotechnology sec. *NanoBiotechnology, 7*. https://doi.org/10.3389/fbioe.2019.00287

Guilger-Casagrande, M., & Lima, R. D. (2019). Synthesis of silver nanoparticles mediated by fungi: A review. *Frontiers in Bioengineering and Biotechnology, 7*, 287.

Gurunathan, S., Raman, J., Malek, S. N. A., John, P. A., & Vikineswary, S. (2013). Green synthesis of silver nanoparticles using Ganoderma neo-japonicum Imazeki: A potential cytotoxic agent against breast cancer cells. *International Journal of Nanomedicine, 8*, 4399–4413.

Hassan, S., Karaila, G. K., Singh, P., Meenatchi, R., Venkateswaran, A. S., Ahmed, T., et al. (2024). Implications of fungal nanotechnology for sustainable agriculture-applications and future perspectives. Biocatalysis and agricultural. *Biotechnology, 103110*.

Helmy, E. A. M., San, P. P., Zhang, Y. Z., Adarkwah, C., & Tuda, M. (2023). Entomotoxic efficacy of fungus-synthesized nanoparticles against immature stages of stored bean pests. *Scientific Reports, 13*(8508). https://doi.org/10.1038/s41598-023-35697-1

Hieu, H. N., Trang, D. T. H., Hien, V. T. T., Nghia, N. V., Lam, N. T., & Nguyen, T. M. D. (2022). Microorganism-mediated green synthesis of silver nanoparticles using aspergillus Niger and bacillus megaterium. *Digest Journal of Nanomaterials and Biostructures, 17*(1), 359–367.

Hoang, N. H., Le Thanh, T., Sangpueak, R., Treekoon, J., Saengchan, C., Thepbandit, W., et al. (2022). Chitosan nanoparticles-based ionic gelation method: A promising candidate for plant disease management. *Polymers, 14*(4), 662.

Ingle, A. P., Ingle, P., Wypij, M., Golinski, P., & Rai, M. (2023). Chapter 10- Myconanotechnology in food preservation and enhancement of shelf-life of Agri-food and fruits. In *Myconanotechnology emerging trends and application*. CRC Press.

Ishida, K., Cipriano, T. F., Rocha, G. M., Weissmüller, G., Gomes, F., Miranda, K., & Rozental, S. (2013). Silver nanoparticle production by the fungus Fusarium oxysporum: Nanoparticle characterisation and analysis of antifungal activity against pathogenic yeasts. *Memórias do Instituto Oswaldo Cruz, 109*(2), 220–228.

Ivask, A., ElBadawy, A., Kaweeteerawat, C., Boren, D., Fischer, H., Ji, Z., Chang, C. H., Liu, R., Tolaymat, T., Telesca, D., Zink, J. I., Cohen, Y., Holden, P. A., & Godwin, H. A. (2014). Toxicity mechanisms in Escherichia coli vary for silver nanoparticles and differ from ionic silver. *ACS Nano, 8*(1), 374–386.

Jahangirian, H., Lemraski, E. G., Webster, T. J., Rafiee-Moghaddam, R., & Abdollahi, Y. (2017). A review of drug delivery systems based on nanotechnology and green chemistry: Green nanomedicine. *International Journal of Nanomedicine*, 2957–2978.

Jameel, M. S., Aziz, A. A., Dheyab, M. A., Khaniabadi, P. M., Kareem, A. A., Alrosan, M., Ali, A. T., Rabeea, M. A., & Mehrdel, B. (2022). Mycosynthesis of ultrasonically-assisted uniform cubic silver nanoparticles by isolated phenols from Agaricus bisporus and its antibacterial activity. *Surfaces and Interfaces, 29*, 101774.

Javier, K. R. A., Laserna, A. K. C., & Camacho, D. H. (2023, July). Mycosynthesis of silver nanoparticles from Ganoderma lucidum: Optimization, antioxidant activity, and identification of capping agents. In *In 2023 IEEE 23rd international conference on nanotechnology (NANO)* (pp. 415–420). IEEE.

Jones, E. G., & Lim, G. (1990). Edible mushrooms in Singapore and other southeast Asian countries. *Mycologist, 4*(3), 119–124.

Kahn, M., Rizvi, T. F., & Ahamad, F. (2019). Application of nanomaterials in plant disease diagnosis and management. In *Nanotechnology in the Life Science*. Springer. https://doi.org/10.1007/978-3-030-13296-5_2

Kalshyap, P. L., Kumar, S., Srivastava, A. K., & Sharma, A. K. (2012). Myconanotechnology in agriculture: a perspective. *World Journal of Microbiology and Biotechnology, 29*(2), 191–207. https://doi.org/10.1007/s11274-012-1171-6

Katas, H., Lim, C. S., Azlan, A. Y. H. N., Buang, F., & Busra, M. F. M. (2019). Antibacterial activity of biosynthesized gold nanoparticles using biomolecules from Lignosus rhinocerotis and chitosan. *Saudi Pharmaceutical Journal, 27*(2), 283–292.

Khastini, R. O., Maryani, N., Sari, I. J., Lestari, I. D., & Putri, C. F. (2022). Diversity of wild edible mushroom in Gunung Tukung Gede nature reserves, Indonesia. *Bangladesh Journal of Botany, 51*(3), 625–629.

Krishnamoorthi, R., Bharathakumar, S., Malaikozhundan, B., & Mahalingam, P. U. (2021). Mycofabrication of gold nanoparticles: Optimization, characterization, stabilization and evaluation of its antimicrobial potential on selected human pathogens. *Biocatalysis and Agricultural Biotechnology, 35*, 102107.

Kulatunga, D., Dananjaya, S., Godahewa, G., Lee, J., & De Zoysa, M. (2017). Chitosan silver nanocomposite (CAgNC) as an antifungal agent against Candida albicans. *Medical Mycology, 55*(2), 213–222.

Kumar, S. A., Abyaneh, M. K., Gosavi, S. W., Kulkarni, S. K., Pasricha, R., Ahmad, A., Khan, M. I. (2007). Nitrate reductase-mediated synthesis of silver nanoparticles from $AgNO_3$. *Biotechnol Lett, 29*(3), 439–445. https://doi.org/10.1007/s10529-006-9256-7. Epub Jan 20. PMID: 17237973.

Kutawa, A. B., Ahmasd, K., Ali, A., Hussein, M. Z., Wahab, M. A. A., Adamu, A., Ismalia, A., Gunasena, M. T., Rahman, M. Z., & Hossain, M. I. (2021). Trends in nanotechnology and its potential to control plant pathogenic fungi: A review. *Biology, 10*(9), 881. https://doi.org/10.3390/biology10090881

Lemire, J. A., Harrison, J. J., & Turner, R. J. (2013). Antimicrobial activity of metals: Mechanisms, molecular targets and applications. *Nature Reviews Microbiology, 11*(6), 371–384.

Li, W. R., Xie, X. B., Shi, Q. S., Zeng, H. Y., Ou-Yang, Y. S., & Chen, Y. B. (2010). Antibacterial activity and mechanism of silver nanoparticles on Escherichia coli. *Applied Microbiology and Biotechnology, 85*, 1115–1122.

Li, Q., Liu, F., Li, M., Chen, C., & Gadd, G. M. (2022). Nanoparticel and nanomineral production by fungi. *Fungal Biology Reviewa., 41*, 31–44. https://doi.org/10.1016/j.fbr.2021.07.003

Loshchinina, E. A., Vetchinkina, E. P., & Kupryashina, M. A. (2022). Diversity of biogenic nanoparticles obtained by the fungi-mediated synthesis: A review. *Biomimetics, 8*(1). https://doi.org/10.3390/biomimetics8010001

Łuczaj, Ł., Lamxay, V., Tongchan, K., Xayphakatsa, K., Phimmakong, K., Radavanh, S., Kanyasone, V., Pietras, M., & Karbarz, M. (2021). Wild food plants and fungi sold in the markets of Luang Prabang, Lao PDR. *Journal of Ethnobiology and Ethnomedicine, 17*, 1–27.

Majeed, S., Binti Sekeri, S. H., & Tahir Ansari, M. (2020). In vitro cytotoxic effect of aspergillus clavatus generated silver nanoparticles on RAW 264.7 cells. Karbala international journal of modern. *Science, 6*(4), 2.

Manimaran, K., Yanto, D. H. Y., Anita, S. H., Nurhayat, O. D., Selvaraj, K., Basavarajappa, S., Hashem, M. I., Palanisamy, G., Lin, M. C., & Kumarasamy, K. (2023). Synthesis and characterization of Hypsizygus ulmarius extract mediated silver nanoparticles (AgNPs) and test their potentiality on antimicrobial and anticancer effects. *Environmental Research, 235*, 116671.

Manjunath, S., & Thiagarajan, P. (2014). Exploring the mechanism for mycobiosynthesis of silver nanoparticles from aspergillus spp. and optimisation of synthesis parameters. *Micro & Nano Letters, 9*(9), 600–604.

Manotham, S., Khieokae, M., & Butnoi, P. (2023). Electrospun biopolymer polyvinyl alcohol/ Centella asiatica extract nanofibers for antibacterial activity. In *Materials Today: Proceedings*.

Mehta, R. (2017). Synthesis of magnetic nanoparticles and their dispersions with special reference to applications in biomedicine and biotechnology. *Materials Science and Engineering: C, 79*, 901–916.

Melo, N. F. C. B., Mendonça Soares, B. L. D., Diniz, K. M., Leal, C. F., Canto, D., Flores, M. A. P., Tavares-Filho, J. H. D. C., Galembeck, A., Stamford, T. L. M., Stamford-Arnaud, T. M., & Stamford, T. C. M. (2018). Effects of fungal chitosan nanoparticles as eco-friendly edible coatings on the quality of postharvest table grapes. *Postharvest Biolog and Technology, 139*, 56–66. https://doi.org/10.1016/j.postharvbio.2018.01.014

Mie, R., Samsudin, M. W., Din, L. B., Ahmad, A., Ibrahim, N., & Adnan, S. N. A. (2014). Synthesis of silver nanoparticles with antibacterial activity using the lichen Parmotrema praesorediosum. *International Journal of Nanomedicine, 9*, 121–127.

Moghaddam, A. B., Moniri, M., Azizi, S., Abdul Rahim, R., Bin Ariff, A., Zuhainis Saad, W., Namvar, F., Navaderi, M., & Mohamad, R. (2017). Biosynthesis of ZnO nanoparticles by a new Pichia kudriavzevii yeast strain and evaluation of their antimicrobial and antioxidant activities. *Molecules, 22*(6), 872.

Mohanta, Y. K., Nayak, D., Biswas, K., Singdevsachan, S. K., Abd Allah, E. F., Hashem, A., Alqarawi, A., Yadav, D., & Mohanta, T. K. (2018). Silver nanoparticles synthesized using wild mushrooms show potential antimicrobial activities against food borne pathogens. *Molecule, 23*(3), 655. https://doi.org/10.3390/molecules23030655

Mukherjee, P., Ahmad, A., Mandal, D., Senapati, S., Sainkar, S. R., Khan, M. I., et al. (2001). Bioreduction of AuCl4-ions by the fungus, Verticillium sp. and surface trapping of the gold nanoparticles formed. *Angewandte Chemie-International Edition, 40*(19), 3585–3588.

Mukherjee, P., Roy, M., Mandal, B. P., Dey, G. K., Mukherjee, P. K., Ghatak, J., ... & Kale, S. P. (2008). Green synthesis of highly stabilized nanocrystalline silver particles by a non-pathogenic andagriculturally important fungus T. asperellum. *Nanotechnology, 19*(7), 075103.

Mukherjee, S., Venugopal, J. R., Ravichandran, R., Ramakrishna, S., & Raghunath, M. (2010). Multimodal biomaterial strategies for regeneration of infarcted myocardium. *Journal of Materials Chemistry, 20*(40), 8819–8831.

Musa, S. F., Yeat, T. S., Kamal, L. Z. M., Tabana, Y. M., Ahmed, M. A., El Ouweini, A., Lim, V., Keong, L. C., & Sandai, D. (2018). Pleurotus sajor-caju can be used to synthesize silver nanoparticles with antifungal activity against Candida albicans. *Journal of the Science of Food and Agriculture, 98*(3), 1197–1207.

Musarrat, J., Dwivedi, S., Singh, B. R., Al-Khedhairy, A. A., Azam, A., & Naqvi, A. (2010). Production of antimicrobial silver nanoparticles in water extracts of the fungus Amylomyces rouxii strain KSU-09. *Bioresource Technology, 101*(22), 8772–8776.

Nazurah, R. N. F., Noranizan, M., Nor-Khaizura, M., & Hanani, Z. N. (2024). Chitosan nanoparticles incorporate with curry leaf essential oil: Physicochemical characterization and in vitro release properties. *International Journal of Biological Macromolecules, 273*, 132972.

Nguyen, V. P., Le Trung, H., Nguyen, T. H., Hoang, D., Tran, T. H. (2001). Synthesis of biogenic silver nanoparticles with eco-friendly processes using ganoderma lucidum extract and evaluation of their theranostic applications. *Journal of Nanomaterials, 2021*(1), 6135920.

Nguyen, T. H., Nguyen, V. P., Le, T. H., & Tran, T. H. (2023). Green synthesis of silver nanomaterials using Ganoderma lucidum extract as reducing agent and stabilizer with ultrasonic assistance and application as an antibacterial agent. *Hue University Journal of Science: Natural Science, 132*(1D), 15–23.

Nguyen, T. T. T., Nguyen, M. T. L., Nguyen, T. K. N., Huynh, H. T., Le, M. N., & Le, P. H. (2024). Antibacterial and photoprotective activities of silver nanoparticles of lichen extract loaded in activated carbon. *Letters in Applied NanoBioScience, 13*(1), 11.

Noor, N. M., Hashim, Y. Z. H.-Y., Sani, M. S. A., & Salim, W. W. A. W. (2020). Characterization, antibacterial and anti-inflammatory activities of electrospun poly (vinyl alcohol)(PVA) containing Aquilaria Malaccencis leaf extract (ALEX) nanofibers. *International Journal of Recent Technology and Engineering (IJRTE), 9*, 514–549.

Nor Azlan, A. Y. H., Katas, H., Mh Busra, M. F., Salleh, N. A. M., & Smandri, A. (2021). Metal nanoparticles and biomaterials: The multipronged approach for potential diabetic wound therapy. *Nanotechnology Reviews, 10*(1), 653–670.

Pareek, V., Bhargava, A., & Panwar, J. (2020). Biomimetic approach for multifarious synthesis of nanoparticles using metal tolerant fungi: A mechanistic perspective. *Materials Science and Engineering: B, 262*, 114771.

Pavani, K. V., Kumar, N. S., & Sangameswaran, B. B. (2012). Synthesis of lead nanoparticles by aspergillus species. *Polish Journal of Microbiology, 61*(1), 61–63.

Pecoraro, L., Wang, X., Shah, D., Song, X., Kumar, V., Shakoor, A., et al. (2021). Biosynthesis pathways, transport mechanisms and biotechnological applications of fungal siderophores. *Journal of Fungi, 8*(1), 21.

Prabhu, S., & Poulose, E. K. (2012). Silver nanoparticles: Mechanism of antimicrobial action, synthesis, medical applications, and toxicity effects. *International Nano Letters, 2*, 1–10.

Putra, A. P., Fahmadiyah, E., Hikmawati, D., & Aminatun, A. (2020). Chitosan-coated collagen-hyaluronic acid-poly ethylene oxide-based electrospun membrane for corneal ulcers wound dressing candidate. *Walailak Journal of Science and Technology (WJST), 17*(11), 1230–1240.

Qian, Y., Yu, H., He, D., Yang, H., Wang, W., Wan, X., et al. (2013). Biosynthesis of silver nanoparticles by the endophytic fungus Epicoccum nigrum and their activity against pathogenic fungi. *Bioprocess and Biosystems. Engineeering, 36*, 1613–1619. https://doi.org/10.1007/s00449-013-0937-z

Rai, A., Bonde, S., Golinski, P., Trzcinska-Wenzel, J., Gade, A., Abd-Elsalam, K. A., Shende, S., Gaikwad, S., & Ingle, A. (2021). Fusarium as a novel fungus for the synthesis of nanoparticles: Mechanism and applications. *Journal of Fungi, 7*(1), 139. https://doi.org/10.3390/jof7020139

Rangasamy, S., Tak, Y. K., Kim, S., Paul, A., & Song, J. M. (2016). Bifunctional therapeutic high-valence silver-pyridoxine nanoparticles with proliferative and antibacterial wound-healing activities. *Journal of Biomedical Nanotechnology, 12*(1), 182–196.

Rangdist, S., Pakdeesiriwong, N., Chumkiew, S., Musika, J., Pombejra, S. N., Nantapong, N., & Jamklang, M. (2020). Green synthesis and characterization of silver nanoparticles from filamentous fungi isolated from the environment and their anti-microbial activity against healthcare-associated bacterial pathogens. In *Proceedings of the SUT international virtual conference on science and technology*, Nakhon-Ratchasima (vol. 28).

Ravindran, A., Chandran, P., & Khan, S. S. (2013). Biofunctionalized silver nanoparticles: Advances and prospects. *Colloids and Surfaces B: Biointerfaces, 105*, 342–352.

Rilda, Y., Putra, E. S., Arief, S., Agustien, A., Pardi, H., & Sofyan, N. (2023a). Mucor sp.(fungal philospheric) of gambir (Uncaria) leaf surface as a biosynthetic Mg doped ZnO nanorods media for antibacterial applications. *Journal of Dispersion Science and Technology*, 1–11.

Rilda, Y., Valeri, A., Syukri, S., Agustien, A., Pardi, H., & Sofyan, N. (2023b). Biosynthesis, characterization, and antibacterial activity of Ti-doped ZnO (Ti/ZnO) using mediated aspergillus Niger. *South African Journal of Chemical Engineering, 45*(1), 10–19.

Rilda, Y., Syafitri, D. N., Septiani, U., Armaini, A., Refinel, R., Agustien, A., Pardi, H., & Sofyan, N. (2024). A comparative analysis of capping and reducing agents of microbial cell Aspergillus

Niger and Bacillus subtilis for biosynthesis of Ag doped ZnO nanoparticles. *Ceramics International, 50*(16), 28150–28158.
Salunkhe, R. B., Patil, S. V., Patil, C. D., & Salunke, B. K. (2011). Larvicidal potential of silver nanoparticles synthesized using fungus Cochliobolus lunatus against Aedes aegypti (Linnaeus, 1762) and Anopheles stephensi Liston (Diptera; Culicidae). *Parasitology Research, 109*, 823–831.
San Chan, Y., & Don, M. M. (2013). Biosynthesis and structural characterization of Ag nanoparticles from white rot fungi. *Materials Science and Engineering: C, 33*(1), 282–288.
San, C. Y., & Don, M. M. (2013). Biosynthesis of silver nanoparticles from Schizophyllum commune and in-vitro antibacterial and antifungal activity studies. *Journal of Physical Science, 24*(2), 83–96.
Schexnailder, P., & Schmidt, G. (2009). Nanocomposite polymer hydrogels. *Colloid and Polymer Science, 287*(1), 1–11. https://doi.org/10.1007/s00396-008-1949-0
Šebesta, M., Vojtková, H., Cyprichová, V., Ingle, A. P., Urík, M., & Kolenčík, M. (2022). Mycosynthesis of metal-containing nanoparticles—Fungal metal resistance and mechanisms of synthesis. *International Journal of Molecular Sciences, 23*(22), 14084.
Seetharaman, P. K., Chandrasekaran, R., Periakaruppan, R., Gnanasekar, S., Sivaperumal, S., Abd-Elsalam, K. A., et al. (2021). Functional attributes of myco-synthesized silver nanoparticles from endophytic fungi: A new implication in biomedical applications. *Biology, 10*(6), 473.
Seil, J. T., & Webster, T. J. (2012). Antimicrobial applications of nanotechnology: Methods and literature. *International Journal of Nanomedicine, 7*, 2767–2781.
Shah, M., Fawcett, D., Sharma, S., Tripathy, S. K., & Poinern, G. E. J. (2015). Green synthesis of metallic nanoparticles via biological entities. *Materials, 8*(11), 7278–7308.
Shareef, J. U., Rani, M. N., Anand, S., & Rangappa, D. (2017). Synthesis and characterization of silver nanoparticles from Penicillium sps. *Materials Today: Proceedings, 4*(11), 11923–11932.
Sonawane, H., Shelke, D., Chambhare, M., Dixit, N., Math, S., Sen, S., et al. (2022). Fungi-derived agriculturally important nanoparticles and their application in crop stress management–prospects and environmental risks. *Environmental Research, 212*, 113543.
Sutjaritvorakul, T., & Chutipaijit, S. (2018). Biogenic synthesis and antibacterial activity of silver nanoparticles (AgNPs) produced by Phomopsis sp. strain GFPA2. *Journal Of Metals, Materials and Minerals, 28*(2), 41–46.
Sutjaritvorakul, T., Imsuwan, P., Damsud, T., Meksiriporn, B., & Chutipaijit, S. (2023). Myco-mediated synthesis and α-glucosidase inhibitory activity of silver nanoparticles produced by xylariaceous fungi. *Applied Ecology & Environmental Research, 21*(5), 4069.
Sytu, M. R. C., & Camacho, D. H. (2018). Green synthesis of silver nanoparticles (AgNPs) from Lenzites betulina and the potential synergistic effect of AgNP and capping biomolecules in enhancing antioxidant activity. *BioNanoScience, 8*, 835–844.
Thakker, J. N., Dalwadi, P., & Dhandhukia, P. C. (2013). Biosynthesis of gold nanoparticles using Fusarium oxysporum f. sp. cubense JT1, a plant pathogenic fungus. *International Scholarly Research Notices, 2013*(1), 515091.
Vahabi, K., Mansoori, G., & karimi S. (2011). Biosynthesis of silver nanoparticles by fungus Trichoderma reeseiI (A route for large scale production of AgNPs). In *Environmental Science, Materials Science, Biology, Chemistry*. https://doi.org/10.5640/INSC.010165
Varma, S., Jaiswal, N., Shende, R. A., & Chaudhari, B. P. (2022). Insight into fungi-mediated Nano-synthesis for healthcare applications: An Indian perspective. In *Progress in mycology: Biology and biotechnological applications* (pp. 205–243). Springer.
Velnar, T., Bunc, G., Klobucar, R., & Gradisnik, L. (2016). Biomaterials and host versus graft response: A short review. *Bosnian Journal of Basic Medical Sciences, 16*(2), 82.
Vetchinkina, E. P., Loshchinina, E. A., Vodolazov, I. R., Kursky, V. F., Dykman, L. A., & Nikitina, V. E. (2017). Biosynthesis of nanoparticles of metals and metalloids by basidiomycetes. Preparation of gold nanoparticles by using purified fungal phenol oxidases. *Applied Microbiology and Biotechnology, 101*(3), 1047–1062.

Wardhani, R. A. K., Primadona, I., & Hardiansyah, A. (2022). Electrospun α-mangosteen–chitosan–poly (ethylene oxide) nanofibers. *Materials Research Express, 9*(11), 115005.
Wunnoo, S., Lorenzo-Leal, A. C., Voravuthikunchai, S. P., & Bach, H. (2023). Advanced biomaterial agent from chitosan/poloxamer 407-based thermosensitive hydrogen containing biosynthesized silver nanoparticles using Eucalyptus camaldulensis leaf extract. *PLoS One, 18*(10), e0291505.
Xu, J., Huang, Y., Zhu, S., Abbes, N., Jing, X., & Zhang, L. (2021). A review of the green synthesis of ZnO nanoparticles using plant extracts and their prospects for application in antibacterial textiles. *Journal of Engineered Fibers and Fabrics., 16*(1). https://doi.org/10.1177/15589250211046242
Yadav, A., Kon, K., Kratosova, G., Duran, N., Ingle, A. P., & Rai, M. (2015). Fungi as an efficient mycosystem for the synthesis of metal nanoparticles: Progress and key aspects of research. *Biotechnology Letters, 37*, 2099–2120.
Zhang, H., Zhou, H., Bai, J., Li, Y., Yang, J., Ma, Q., & Qu, Y. (2019). Biosynthesis of selenium nanoparticles mediated by fungus Mariannaea sp. and their characterization. *Colloids and Surfaces A: Physicochemical and Engineering Aspects, 57*(10), 9–16. https://doi.org/10.1016/j.colsurfa.2019.02.070
Zhao, L., Gong, K., & Kong, S. (2021). Rapid and environment-friendly preparation of silver nanoparticles and their inhibition against phytopathogenic fungi. *Micro and Nano Letters, 16*(3). https://doi.org/10.1049/MNA2.12027
Zomorodian, K., Pourshahid, S., Sadatsharifi, A., Mehryar, P., Pakshir, K., Rahimi, M. J., & Arabi Monfared, A. (2016). Biosynthesis and characterization of silver nanoparticles by aspergillus species. *BioMed Research International, 2016*(1), 5435397.

Chapter 19
Philippine Fungal Biotechnology: History, Application, and Future Directions

Marian P. De Leon, Nik Shawn C. Tabao, Rose May Anne D. Capanzana, and Ronilo Jose D. Flores

Abstract Fungal biotechnology is a rapidly advancing field that addresses critical global challenges in food security, energy, health, and environmental sustainability. This chapter provides a comprehensive overview of fungal biotechnology in the Philippines, spanning its historical roots, contemporary applications, and future directions. Indigenous practices such as the fermentation of rice wine (*tapuy*) and traditional food products highlight the country's long-standing engagement with fungal processes. Modern research initiatives by different institutions in the Philippines have yielded notable innovations, including *Trichoderma*-based biofertilizers (e.g., *BioQuick* and *BioGreen*), mycorrhizal inoculants (e.g., Mykovam and VAMRI), and biocontrol agents targeting plant pathogens and nematodes. Advances in mushroom cultivation, production of single-cell proteins, and applications of fungal enzymes (amylases, lipases, proteases, cellulases, and pectinases) in the food industry underscore fungi's economic value. Yeast biotechnology also holds significant promise, with yeasts such as *Pseudozyma antarctica* demonstrating potential in producing glycoproteins, recombinant proteins, and mannosylerythritol lipids with industrial and pharmaceutical applications. Environmental applications—including mycoremediation of heavy metals, dye decolorization, and bioenergy production through oleaginous yeasts and bioethanol—further illustrate the breadth of fungal

M. P. De Leon (✉) · N. S. C. Tabao
Museum of Natural History, University of the Philippines Los Baños, Los Baños, Laguna, Philippines
e-mail: mpdeleon1@up.edu.ph

R. M. A. D. Capanzana
National Institute of Molecular Biology and Biotechnology (BIOTECH), University of the Philippines Los Baños, Los Baños, Laguna, Philippines

R. J. D. Flores
Museum of Natural History, University of the Philippines Los Baños, Los Baños, Laguna, Philippines

Institute of Biological Sciences, College of Arts and Sciences, University of the Philippines Los Baños, Los Baños, Laguna, Philippines

J. J. G. Guerrero et al. (eds.), *Mycology in a Changing Planet: Applications and Perspectives from Southeast Asia*, https://doi.org/10.1007/978-981-95-2640-6_19

biotechnology in the country. Despite these achievements, challenges persist in terms of funding, infrastructure, regulatory frameworks, and public awareness. There is a pressing need for expanded bioprospecting, strengthened biological resource centers, and interdisciplinary collaborations to harness the country's rich fungal diversity for sustainable development. Strategic alignment with national priorities and the UN Sustainable Development Goals will ensure that Philippine fungal biotechnology continues to grow as a driver of innovation in agriculture, health, and environmental stewardship in the ASEAN region.

Keywords Philippine fungi · Fungal biotechnology · Mycology · Natural products discovery · Fungal diversity · Applied mycology · Fungi-based innovation

Introduction

The term biotechnology was coined in 1919 by a Hungarian agricultural engineer named Károly (Karl) Ereky. It is in the same year that Pfizer, through the use of biotechnology, produced citric acid from *Aspergillus*, a mold, for various applications (Cairns et al., 2018) and since then revolutionized the European market with multi-billion-dollar industries investing in fungal by-products, leveraging the powers of fungi.

Biotechnology is the industrial use of biological organisms and processes to benefit human endeavors (Roth et al., 2023). The Organisation for Economic Co-operation and Development (Friedrichs & van Beuzekom, 2018) comprehensively defines it as *the application of science and technology to living organisms as well as parts, products, and models thereof, to alter living or non-living materials for the production of knowledge, goods, and services*. Martin et al. (2021) reported that the biotechnology-based industries had an annual growth rate of 1.3% from 2015 to 2020, leading to a substantial increase in the industry's value that amounts to $295 billion (Bolaños et al., 2023). Bolaños et al. (2023) further emphasized that in Southeast Asia, the Philippines has emerged as a pioneer in biotechnology breakthroughs within the region (Dorockis & Boguś, 2014) but still lags behind other ASEAN nations' research and development investments (UIS, 2023).

Fungal biotechnology is a specialized utilization of parts or the whole of the fungus for the fermentation of useful by-products (Meyer et al., 2020). It is also a design and application of fungal biomass, metabolism, or genetics that addresses societal or environmental problems (Roth et al., 2023). Through fungal biotechnology, organic materials are transformed into rich, diverse, and beneficial products, including sources of food, feed, chemicals, enzymes, life-saving drugs and antibiotics, fuels, textiles, and materials for construction, automotive, and transportation industries, for furniture and beyond (Meyer et al., 2020).

Fungal biotechnology has played a very significant role in addressing the world's most pressing issues, such as food security, energy insecurity, human medicine, and environmental sustainability (Roth et al., 2023). The advancements in fungal biotechnology transform the global economy into a sustainable food supply for the increasing human population and are innovative for scaling up industries such as petroleum and environmental clean-up and restoration, such as bioremediation and climate change mitigation. It became promising and exciting as more and more fungi are discovered and explored for high-value products. Not only are these organisms used for breaking down plant materials to prevent too much accumulation of wastes and to recycle their nutrients and metabolic by-products back into their environment (Roth et al., 2023), but through biotechnology, they ferment plant materials into something beneficial and useful that contributes to local and global markets.

This chapter focuses on the history, applications, and future directions of fungal biotechnology in the Philippines. It covers the country's beginnings, significant milestones, key contributors and institutions, applications in agriculture and forestry, food and beverage, medicine and pharmaceuticals, the environment, and the challenges and future directions of fungal biotechnology as it addresses the United Nations Sustainable Development Goals.

History of Fungal Biotechnology in the Philippines

While Biotechnology was coined in 1919, elements of fungal biotechnology started and were found early in the country. Fermentation of rice wines in the Philippines started even before the Spaniards arrived in the country, and Cordillerans have already drunk rice wine, *Tapuy,* for more than 2000 years (Hipol & Almain, 2018). There are various fermented products in the Philippines, and these include fermented derivatives of rice, fish, coconut, palm, mustard leaves, and a plethora of others. Wines are known to be fermented by various microorganisms, including fungi (Sanchez, 1999). Unwittingly, Filipinos have used fungi to ferment different beverages during ancient times and have started fungal biotechnology. This is considered to be under the first level of biotechnological techniques involving the manipulation of microorganisms (Halos, 2000).

The use and production of soy sauce or *toyo*, on the other hand, is considered to be tied to the history of Adobo making during the Spanish occupation of the Philippines. However, soy sauce in the Philippines is thought to have been introduced and produced by the Chinese, possibly through trade and immigration, even before the Spaniards arrived. In 1925, Leopoldo Salazar, a Bachelor of Agriculture student from the University of the Philippines College of Agriculture (now UP Los Baños), developed ways of improving the production of soy sauce in his thesis. It is through this Filipino ingenuity that various raw materials were reported to be used to make *toyo*, although none of these were said to be commercialized. In the 1960s, Japanese-style *toyo* products were introduced. In the 1970s, Filipino manufacturers produced a huge amount of *toyo* in the Philippines (Shurtleff & Aoyagi, 2004). Because the Philippine *toyo* is largely patterned after the Chinese and Japanese way of production or fermentation, it is expected that the fungi found in Philippine *toyo* are species of *Aspergillus* (Abalunan et al., 2013) or derivatives/strains of *A. oryzae*

(=*Aspergillus flavus* var. *oryzae* (Ahlb.) Kurtzman) (Wicklow et al., 2007), although there are more types of fungi involved in the fermentation process (Devanthi & Gkatzionis, 2019).

The Philippines is a pioneer in biotech breakthroughs in the region (Dorockis & Boguś, 2014), including fungal biotechnology, which started as early as the 1920s with the improvement of soy sauce. Table 19.1 provides a brief overview of some developments in fungal biotechnology in food and feed, agriculture, forestry, and the environment.

Developments in Fungal Biotechnology in the Philippines went quiet in the 1980s. It could be attributed to the political and economic atmosphere during the period of civil and economic uncertainty in the country.

Table 19.1 History of fungal biotechnology for food and feeds, agriculture, forestry, and the environment

Year	Milestones	References[a]
1937	The earliest report and mention of the culture of edible mushrooms in the Philippines	Clara (1937)
1957	Experimental cultivation of *Volvariella volvacea* reported	Go (1959)
1960	Technical bulletin on mushroom cultivation released	Alicbusan and Ela (1960)
1961	The use of fodder yeast (*Torulopsis utilis*) as a supplement for man and livestock feed was studied	Hipolito et al. (1961)
	The value of some yeasts has already been evaluated for agricultural purposes, and the yeast *Torulopsis utilis* was reported to have potential as an aquatic fertilizer[b]	Agor and Villadolid (1961)
1966	The cultivation methods for *Auricularia polytricha* were published	Borromeo and Reynolds (1966)
	The cultivation methods for *Agaricus bisporus* were published	Rubio and Alicbusan (1966)
	Tumor-inhibitors were found in Philippine basidiomycetes	Santos and Bernardo (1966)
1967	Submerged culture production of banana mushroom, *Volvariella volvacea* mycelium as a source of protein, B vitamins, and food flavor	Ramos et al. (1967)
1968	The use of Philippine fungi in trapping nematodes was reported as a method of nematode control[a]	Cortado and Davide (1968)
1975–1976	The rennet-forming molds were reported to be useful in cheese production	Santos et al. (1975, 1976)
1977	Yeast strains that excrete amino acids were reported	Dela Cruz-Halos (1977)
	Artificial cultivation of *Auricularia polytricha* was published	Quimio (1977)
	Semisolid cultures of *Trichoderma viridis* were reported to produce cellulases[a]	Torillo et al. (1977)
1978	A method for the indoor cultivation of *Pleurotus ostreatus*, growing it on mostly agricultural waste, was reported	Quimio (1978)

[a]Cited by Quimio and Capilit (1981)
[b]Agriculture, forestry, and environment

Most modern contributions to the field of fungal biotechnology in the country arise from the research and development efforts and body of work of the following institutions, namely, (1) UPLB National Institute of Molecular Biology and Biotechnology (BIOTECH™) formerly the National Institutes of Biotechnology and Applied Microbiology; (2) UPLB Institute of Food Science and Technology (IFST); (3) UPLB Institute of Weed Science, Entomology, and Plant Pathology (IWEP); (4) Center for Tropical Mushroom Research and Development (CTMRD), Central Luzon State University (CLSU) and (5) Benguet State University (BSU).

Applications of Fungal Biotechnology in the Philippines

Fungal biotechnology has different applications in different industries globally. This section will discuss the different applications of fungal biotechnology in agriculture and forestry, the food and beverage industry, medicine and pharmaceuticals, and the environment in the Philippines. It will highlight the technologies applied more recently.

Agriculture and Forestry

Bioprotectants Bioprotection is a sustainable approach to pest management. It plays an important role in regenerative agriculture, which aims to increase resilience in crops, protect the environment, and restore soil health, particularly since soil systems are often degraded and polluted due to the prolonged use of chemical pesticides and fertilizers. Bioprotection products or bioprotectants make use of naturally occurring and non-toxic components that can easily degrade. Bioprotectants have the capacity to outcompete or eliminate specific pests, thereby enhancing biodiversity and promoting the growth of beneficial microorganisms in the soil (Hollis, 2023).

In the Philippines, a bioprotectant composed of endomycorrhizal fungi was developed and branded *Brown Magic* and named after its inventor, Dr. Marilyn B. Brown. *Brown Magic* protects orchids from root diseases and helps reduce the transplant shock of in-vitro cultured orchid seedlings. The product also serves as a biofertilizer, which stimulates flowering, prolongs the blossoms, and promotes more vigorous growth of orchids (BIOTECH, 2023).

Most orchids are photosynthetic. They benefit from the mycorrhizal fungi only during the seed germination stage. Although mycorrhizal associations are present in all photosynthetic orchids throughout their life cycle, the highest level of colonization occurs during their rapid vegetative growth and flowering phases (Xing et al., 2019).

The most commonly isolated orchid mycorrhizal fungi (OMF) are the *Rhizoctonia*-like OMF. Some examples are *Mycena anoectochila* L. Fan and S.X. Guo, *M. dendrobii* L. Fan and S.X. Guo, and *M. orchidicola* L. Fan and S.X. Guo. He et al. (2010) and Xing et al. (2019) showed that the recovered OMF were mostly from the Tulasnellaceae family, along with a few from Ceratobasidiaceae and Sebacinales. Other fungal species found in orchids include members of Cortinariaceae, Marasmiaceae, Russulaceae, Thelephoraceae, and unidentified species from Cantharellales and Atractiellales.

Biopesticides and Biofertilizers Intensive agriculture and related industries generate significant amounts of waste, including crop residues, animal manure, and organic by-products. These agro-industrial wastes contain a variety of biological components, such as plant growth-promoting microorganisms. In the Philippines, inoculants containing *Trichoderma* species have been created as biofertilizers and compost ripeners to help break down household and agro-industrial waste. Several inoculants reported are *Bio Quick* and *Bio Green* that contain *Trichoderma* and a combination of *Trichoderma* with nitrogen-fixing bacteria (BIOTECH, 2023; Office of the Vice Chancellor for Research and Extension (OVCRE), 2018).

An inoculant containing *Trichoderma koningii* (Launio et al., 2020) has been developed by Benguet State University; the product is a compost enhancer and a component of compost fertilizer. Additionally, BioSpark Corporation in the Philippines has created another *Trichoderma*-based technology (Cumagun, 2012).

Trichoderma species play a crucial role in biological decomposition and are known for producing cellulolytic and hemicellulolytic enzymes. These filamentous soil fungi release a well-balanced cellulolytic complex that effectively breaks down cellulosic materials into glucose monomers. *Trichoderma* has been shown to accelerate the decomposition of straw and other organic waste without emitting unpleasant odors (Organo et al., 2022). Cuevas and Banaay (2022b) also highlighted the bioremediation and growth-promoting effects of *Trichoderma* microbial inoculant (TMI) when combined with rice straw compost (RSC). This combination was found to reduce copper and lead contamination in rice paddies while boosting rice yield. Furthermore, TMI in combination with salt has been seen to aid in the fast recovery of mature coconut trees infested with scale insects of coconut (Cuevas et al., 2023). Phytoremediation is a means of rehabilitation of polluted areas such as mining sites by using plants to absorb, detoxify, or immobilize pollutants. An enhanced form of phytoremediation involves the use of Ectomycorrhizal (ECM) fungi to improve the growth and survival of heavy metal-tolerant trees like *Eucalyptus* and *Acacia* in abandoned mine sites. ECM fungi enhance the ability of the plants to tolerate and absorb contaminants, which helps in soil restoration and reforestation in mining-affected areas (Aggangan & Aggangan, 2012).

ECM fungi, including *Pisolithus*, *Astraeus*, *Scleroderma*, and *Suillus*, produce fruiting bodies such as mushrooms and puffballs. Their mycelia form a protective sheath around plant roots, extending into the soil to increase nutrient and water uptake (Dela Cruz et al., 1998). These fungi also release growth-promoting substances, enhancing plant resilience in nutrient-poor soils (Aggangan et al., 2012).

Research on ECM fungi has led to the development of mycorrhizal inoculants like Mycogroe. It is a soil-based biofertilizer that enhances seedling establishment, nutrient absorption, and plant resilience. It improves water and phosphorus uptake, protects against root infections, and reduces the need for chemical fertilizers by 60–80% (BIOTECH, 2023).

Arbuscular Mycorrhizal Fungi (AMF) form symbiotic relationship with plant roots. These fungi improve water and nutrient absorption by extending their hyphal networks into the soil and plant roots. This mutualistic association is critical in environments such as nutrient-poor soils. AMF also offers protection against soil-borne pathogens, reduces the impact of abiotic stresses like drought and temperature extremes, and improves soil structure. In the Philippines, a soil-based biofertilizer called Mykovam contains several AMF strains from genera such as *Glomus*, *Gigaspora*, *Entrophospora*, and *Acaulospora* (Aguilar et al., 2018; BIOTECH, 2023). Mykovam is beneficial to all kinds of fruit trees, ornamental plants, and vegetables except those belonging to the crucifers or cabbage family. It can also be applied to rubber trees, oil palm, coconut, banana, upland rice, sugarcane, and many reforestation tree species such as narra, acacia, gmelina, mahogany, except pines, eucalyptus, agoho, and dipterocarps (BIOTECH, 2008). It has also been documented to help overcome *Fusarium* wilt in banana plantations in Bohol and in Mindanao (Arana, 2022).

Recent developments in the technology include the addition of biochar as a soil amendment in combination with Mykovam. Biochar is a carbon-rich material produced by heating organic materials in an oxygen-depleted environment (pyrolysis). It increases bulk density, water holding capacity, soil pH, and cation exchange capacity, making the soil favorable to mycorrhizal colonization. Biochar further strengthened the efficacy by addressing not only nutrient deficiencies but also improving soil conditions for better plant growth (Jomao-as et al., 2023).

Two vesicular-arbuscular mycorrhizal (VAM) species, *Glomus mosseae* (=*Funneliformis mosseae* (T.H. Nicolson and Gerd.) C. Walker and A. Schüßler) and *G. fasciculatum* (=*Rhizoglomus fasciculatum* (Thaxt.) Sieverd., G.A. Silva and Oehl) were used as inoculants for root crops, vegetables, fruit trees, legumes, sugarcane, bananas, and other crops. The VAM-based product developed in the Philippines is called Vesicular-Arbuscular Mycorrhizal Root Inoculant (VAMRI) (Brown & Monsalud, 2014). It enhances plant growth by improving water and nutrient absorption, particularly phosphorus, and strengthening soil structure. It also provides plants with protection against root pathogens and enhances disease resistance, particularly against soil-borne phytopathogens. It has also been noted for its ability to support beneficial bacteria, like *Pseudochrobactrum asaccharolyticum Kämpfer et al*, which further aids in plant protection against pathogens (Nepomuceno et al., 2019).

Zinc-Solubilizing Fungi Zinc is a vital micronutrient for the physiological functions of living organisms. In plants, zinc is involved in protein synthesis, enzyme activation, chlorophyll production, seed development, and overall growth. Zinc-deficient soils may result in stunted growth, low yields, and reduced plant resilience to different stresses. Ensuring adequate zinc availability in soils is essential for sustaining crop productivity and maintaining ecosystem health (Chaudhary et al., 2022). Some fungi belonging to the genera *Penicillium, Aspergillus, Fusarium, Mucor,* and *Trichoderma* have shown promising results related to zinc-solubilization. Specifically, studies in other Asian countries (India and Iran) showed that *Penicillium* spp., *Aspergillus awamori Nakaz.*, *Fusarium circinatum* Nirenberg and O'Donnell, *Fusarium longifundum* Xia, Lombard, Sand.-Den., Zhang and Crous, and *Mucor circinelloides* Tiegh. showed excellent zinc-solubilization efficiency, making them viable candidates for biofertilizer applications. These organisms convert zinc from the environment to bioavailable forms that can be utilized by plants (Batool et al., 2021; Devi et al., 2022; Islas-Valdez et al., 2024).

Biofertilizers making use of zinc-solubilizing fungi have not yet been developed in the Philippines. The existing technologies for addressing zinc deficiency in soil include two biofertilizers named *Oryzinc* and *Maizinc,* which contain the bacteria *Burkholderia* sp. and *Acinetobacter* sp., respectively (BIOTECH, 2023).

Biocontrol Postharvest losses are a major concern in agricultural countries like the Philippines. The common causes of such types of losses are usually microbial pathogens. Controlling these pathogens at the postharvest stage is paramount so as not to deal with huge losses, and this has been the topic of global research in the academe and industry for almost 40 years (Droby et al., 2016).

Epiphytic microorganisms have been reported to considerably control postharvest pathogens, and such antagonists were first reported in the 1980s (Wilson & Pusey, 1985). The early epiphytic fungal biocontrol agents produced commercially were *Candida oleophila* (=*Yarrowia lipolytica* (Wick., Kurtzman and Herman) Van der Walt and Arx) marketed as Aspire biofungicide (Blachinsky et al., 2007), *Cryptococcus albidus* (=*Naganisha albida* (Saito) Xin Zhan Liu, F.Y. Bai, M. Groenew, and Boekhout) marketed as YieldPlus, and *Candida sake* (Saito and M. Ota) Uden and H.R. Buckley marketed as Candifruit (Teixido et al., 2011). *Aureobasidium pullulans* (de Bary) G. Arnaud, marketed as BoniProtect, was suggested to be used before harvesting pome fruits to control pathogens that enter through wounds in the fruit (Kowalska et al., 2022). *Metschnikowia fructicola* Pitt and M. W. Mill, on the other hand, marketed as Shemer, is known to have applications on fruits and vegetables, including but not limited to apricots, citrus fruit, grapes, peaches, pepper, strawberries, and sweet potatoes (Spadaro & Droby, 2016).

In the Philippines, biocontrol of pathogenic fungi, particularly *Aspergillus parasiticus* Speare, and the detoxification of its aflatoxin in peanuts, copra, and copra meal had been researched as early as the 1990s. One strain of *Cladosporium* sp. had shown antagonistic effects on *A. parasiticus* Speare, which scientists suspect was due to its production of a certain pigment (not disclosed and elaborated) (Beardall et al., 1994).

The epiphytic fungi *Debaryomyces hansenii* Zopf. and *Saccharomyces cerevisiae* (Desm.) Meyen developed and branded HiYEAST Dh and HiYEAST Sc. (Domingo, 2016).

HiYEAST Dh was reported to be effective against *Colletotrichum musae* (Berk. & M.A. Curtis) Arx, a postharvest pathogen found in bananas, and *Aspergillus flavus* Link, which is known to produce aflatoxins in garlic and nuts. HiYEAST Sc, on the other hand, was reported to control anthracnose in mangoes caused by *Colletotrichum gloeosporioides* (Penz.) Penz. & Sacc., a fungus that produces aflatoxins in garlic, and *Penicillium digitatum* (Pers.) Sacc., a fungus that causes green mold disease in calamansi and produces aflatoxin in garlic (Domingo, 2016).

Trichoderma was used to manage the sheath blight of rice caused by *Rhizoctonia solani* J.G. Kuhn (Cumagun & Lapis, 1993). Of the many strains molecularly characterized (Cumagun et al., 2000) and screened for biocontrol activities, one strain was used for a commercialized product (Cumagun, 2012). *Paecilomyces lilacinus* (=*Purpureocillium lilacinum* (Thom) Luangsa-ard, Houbraken, Hywel-Jones and Samson) strain UP1 was used as a biological control agent of the root-knot nematode *Meloidogyne incognita* (Kofoid & White) Chitwood. This fungus can maximally reduce gall number in tomatoes with 7.92×10^6 spores/ml (Oclarit & Cumagun, 2009). AMF also reduced gall by 33% in *Meloidogyne incognita*-infected tissue-cultured banana var. Lakatan (Aggangan et al., 2013).

Food and Beverage Industry

Fungi Cultivation and Spawn Production Research on mycelial growth and fruiting body production of edible mushrooms or macrofungi has grown significantly in recent years from when it started in 1937. In 1993, Quimio described a technology for the indoor cultivation of the straw mushroom, *Volvariella volvacea* (Bull.) Singer. In 1998, Reyes et al. reported the physiological considerations for efficient mycelial colonization of Philippine strains of the fungus. In 2001, Quimio published the potential of low-cost technologies for tropical cultivation of edible mushrooms, after which, Reyes et al. (2009) domesticated a wild nutraceutical mushroom identified as *Coprinus comatus* (O.F. Müll.) Gray (Dulay et al., 2012b). From 2012 onwards, and for over a decade, various optimization techniques and technologies were developed by researchers from the Center for Tropical Mushroom Research and Development and their collaborators, which dominated the R&D landscape for mycelial growth and fruiting body production of edible fungi in the Philippines. *Lentinus tigrinus* (Bull.) Fr., a newly recorded and domesticated wild edible mushroom, was cultivated, and its culture conditions for mycelial growth were optimized (Dulay et al., 2012a; Kalaw et al., 2021a) along with *Coprinus comatus* (O.F. Müll.) Gray (Dulay et al., 2012b) which was cultivated on various pulp and paper wastes. The culture conditions of wild edible mushrooms (Kalaw et al., 2016), *Ganoderma lucidum* (Fr.) P. Karst. (Magday et al., 2014), *Panaeolus antillarum* (Fr.) Dennis, and *Panaeolus cyanescens* Sacc. (Bustillos et al., 2014) as sources of bioactive lipids (Dulay et al., 2015) were also optimized. The nutrient composition and functional activity of different stages in the fruiting body development of Philippine paddy straw mushroom, *Volvariella volvacea* (Bull.) Singer, were described and discussed by Eguchi et al. (2015). Tolentino et al. (2016), on the other hand, described the mycelial growth performance of three *Pleurotus* spp. on corn varieties in the Philippines. De Leon et al. (2017) optimized the mycelial growth and screened the mycochemicals of *Lentinus sajor-caju* (Fr.) Fr., while Dulay et al. (2017) described the mycelial growth and fruiting body production of *Polyporus grammocephalus* (=*Favolus grammocephalus* (Berk.) Imazeki), and Dulay and Rivera (2017) and Dulay et al. (2017) described the same for wild hairy sawgill fungi, *Lentinus strigosus* (=*Panus lecomtei* (Fr.) Corner). Julian et al. (2018) reported on their understanding of the mycelial growth response of *Hericium* on different culture media. During the pandemic, various optimization studies were published by CTMRD researchers. Optimal culture and/or cultivation conditions were described for *Fomitopsis feei* (Fr.) Kreisel (De Leon et al., 2020), *Chlorophyllum molybdites* (G. Mey) Massee (Garcia et al., 2020), *Cyclocybe cylindracea* (D.C.) Vizzini and Angelini (Landingin et al., 2020), *Trametes elegans* (Spreng.) Fr. (Dulay et al., 2021a), *Lentinus* spp. (Dulay et al., 2021b; Kalaw et al., 2021b), *Lentinus swartzii* Berk. (Dulay et al., 2021c), *Ganoderma* spp. (del Rosario et al., 2022), *Xylaria papulis* Lloyd (Lopez et al., 2022). After the pandemic, the optimal culture conditions of wild agaric mushrooms, *Coprinopsis cinerea* (Schaeff.) Redhead (MP11), *Coprinopsis verticillata* (Schulz-Wedd) Redhead, Vilgalys and Moncalvo (DQS7), *Leucoagaricus americanus* (=*Macropsalliota americana* (Peck) Kun L. Yang, Jia Y. Lin and Zhu L. Yang) (CAG10), *Leucoagaricus meleagris* (=*Macropsalliota meleagris* (Gray) Kun L. Yang, Jia Y. Lin and Zhu L.) (CAG9), and *Leucocoprinus cretaceus* (Bull.) Locq. (MP20) were described (Fabros et al., 2023). The various optimization studies have provided various techniques and strategies for the cultivation of macrofungi that may help address food security in the country or contribute to meeting UN Sustainable Development Goal 2 (Zero Hunger). Other than macrofungi, microfungi can also be used as food or food supplements. In the case of yeasts, they can be used as sources of single-cell protein. In the Philippines, Valentino et al. (2015) screened the mycota of yeast sludge from a distillery for single-cell proteins (SCP) and found seven taxa with potential, and these were: *Saccharomyces cerevisiae* (Desm.) Meyen, *Candida parapsilosis* (Ashfod) Langeron and Talice, and *Candida guilliermondii* (=*Meyerozyma guilliermondii* (Wick.) Kurtzman and M. Suzuki), and four species of molds (*Aspergillus flavus* Link, *A. niger* Tiegh, *Rhizopus* sp., and *Aspergillus japonicus* var. *japonicus*).

Production of Traditional Foods and Beverages The Philippines has various fermented products that indigenous Filipinos have regularly consumed. In one stage or another in the fermentation process of these fermented products, a fungus or two may be present or important (Sanchez, 1999; Elegado et al., 2016). Of the various products consumed by Filipinos, some have been characterized mycologically, including Tapuy (Rice Wine) (Hipol & Almain, 2018), Suka (Vinegar), Basi (Sugarcane Wine), Muscovado-based Kefir (Elegado et al., 2016), Kesong Puti, Tuba, Puto, Tinabal, and Toyo (Soy Sauce) (Sanchez, 1999).

Fermentation of wines like *Tapuy* traditionally involves "bubod" or "binokbok," which is a dried disc of old "bubod" or "binokbok," ground rice, *Bidens pilosa* L. (grass) roots, and ginger (Sanchez, 1999). The "bubod" contains various microorganisms as starters of the fermentation process. The starter used for fermentation contains fungi such as *Rhizopus oryzae* (=*Rhizopus arrhizus* A. Fisch.), *Mucor* sp. Fresen. (Kozaki, 1976; Sakai & Caldo, 1985; Sanchez, 1999; Ablang et al., 2017), *Saccharomyces cerevisiae* (Desm.) Meyen (Lim et al., 2006; Ablang et al., 2017), *Pichia kudriavzevii* Boidin, Pignal and Besson (Ablang et al., 2017), *Hansenula anomala* (=*Wickerhamomyces anomalus* (E.C. Hansen) Kurtzman, Robnett and Basehoar-Powers), and *Saccharomycopsis fibuligera* (Lindner) Klöcker (Sakai & Caldo, 1985; Sanchez, 1999; Lim et al., 2006). *Aspergillus oryzae* (=*Aspergillus flavus* var. *oryzae* (Ahlb.) Kurtzman), *Mucor rouxii* (=*Rhizopus arrhizus* A. Fisch.), *Sm. capsularis* Schiønning (Sanchez, 1999), *Sm. burtonii*, and *Saccharomyces uvarumare* are also known to be present in the rice wine fermentation process (Sakai & Caldo, 1985).

Suka (Vinegar) is a food product that all Filipinos use in their dishes or sauces. According to literature, yeasts and acetic acid bacteria are known to ferment it. Unlike the lactic acid and acetic acid bacteria present in the fermentation process, the identities of the yeasts are not yet known (Elegado et al., 2016).

Basi (Sugarcane Wine) is another beverage in which *Saccharomyces cerevisiae* (Desm.) Meyen (Sanchez, 1999) and *Saccharomycopsis* sp. are known to be involved in the fermentation process (Elegado et al., 2016). *Saccharomyces bayanus* Sacc., *S. rosei* (=*Torulaspora delbrueckii* (Lindner) Lindner), *S. capensis* (=*S. cerevisiae var. cerevisiae*), *Saccharomycopsis fibuligera* (Linder) Klöcker, *Sm. vini* (Kreger-van Rij) Van der Walt and D.B. Scott, and *Candida* spp. and *Torulopsis* spp. (=*Metsnikowia* spp.) help maintain the stability and flavor of Basi (Sanchez, 1999).

Muscovado-based Kefir, a beverage from Leyte, is characterized by having *Lachancea fermentati* (H. Nagan) Kurtzman, *Saccharomyces cerevisiae* (Desm.) Meyen, *Candida inconspicua* (=*Pichia inconspicua* (Lodder and Kreger-van Rij) H.Y. Zhu, L.C. Guo and F.Y. Bai), and *Pichia manshurica* Saito as part of its microbial load (Elegado et al., 2016).

Kesong Puti or White Soft Cheese is fermented by a mixed flora of mainly lactic acid bacteria and yeasts of the *Torula* spp. Tuba's fermentation is through its dominant yeast *Saccharomyces chevalieri* Guillierm. Other yeasts present in the fermentation process are *S. cerevisiae* (Desm.) Meyen, *S. capensis* (=*S. cerevisiae* var. *cerevisiae*), *S. rosei* (=*Torulaspora delbrueckii* (Lindner) Lindner), *Sm. vini* (Kreger-van Rij) Van der Walt and D.B. Scott, *S. bailii* (=*Zygosaccharomyces bailii* var. *bailii*), *Candida vartriovaarai* (Capr.) Uden and H.R. Buckley, *C. langeronii* (=*C. albicans* (C.P. Robin) Berkhout), *C. diddensii (Phaff, Mrak and O.B. Williams) Fell and S.A. Mey,* and *Torulopsis haemulonii* (=*Candidozyma haemuli* (Uden and Kolip) Q.M. Wang, A. Yurkovm, Boekhout, F.Y. Bai).

Puto or fermented rice cake is fermented by a mixed flora of lactic acid bacteria and *Saccharomyces cerevisiae* (Desm.) Meyen (Sanchez, 1999). Tinabal or fermented fish is also fermented by a mixed flora of organisms, including yeasts of the genus *Debaryomyces* Lodder and Kreger-van Rij (Sanchez, 1999). *Toyo* fermentation is dominated by *Aspergillus oryzae* (=*Aspergillus flavus* var. *oryzae* (Ahlb.) Kurtzman) and is also accompanied by some yeasts such as *Saccharomycopsis rouxii* and *Zygosaccharomyces sojae* (Saito) Guilliermin. after brining (Sanchez, 1999). Microbial rennet is produced from *Rhizopus chinensis* and is used to coagulate milk in cheese production (Sanchez, 1999).

Fungal Enzymes

Fungal enzymes have various applications in the food and beverage industry. Here, we discuss and give Philippine examples of five enzymes from fungi, which are amylase, lipase, protease, cellulase, and pectinase. Amylases have applications in precooked baby foods and are used for the degradation and removal of starch and control of the volume and crumb structure of bread, and many others. Lipases are used for cheese ripening, improvement of stability of gas cells in dough, cocoa butter equivalents or substitutes, enrichment of ω-3 PUFA, and isolation of oil and fats from seafood products. Proteases are used to degrade protein and enable the aging of cheese, the addition of flavor in chocolates, increase soluble protein and free amino-nitrogen (FAN), and malt improvement. Cellulases have applications in coffee production, specifically, in coffee bean fermentation and coffee concentrates. Pectinases are used to degrade pectins, which are structured polysaccharides present in the cell walls of plants (Chandasekaran et al., 2015).

Amylase/Lipase/Protease Tan Gana et al. (2014) screened yeast strains from commercial banana fruits in the Philippines and found a few with high amylolytic, lipolytic, and proteolytic activities. Of these, a strain of *Hyphopichia burtonii* (Boidin, Pignal, Lehodey, Vey and Abadie) Arx and Van der Walt exhibited amylolytic activity only; a strain of *Pseudozyma hubeiensis* F.Y. Bai and Q.M. Wang exhibited lipolytic activity only; a strain each of *Pichia anomala* (=*Wickerhamomyces anomalus* (E.C. Hansen) Kurtzman), *Candida wangnamkhiaoensis* (=*Hyphopichia wangnamkhiaoensis* Limtong ex L.R. Ribeiro, M. Groenew., N. Jacques, S. Casaregola, Lachance and C.A. Rosa), and *Brandoniozyma complexa* have amylolytic and lipolytic activities; a strain of *Debaryomyces nepalensis* Goto and Sugiy. have amylolytic and proteolytic activities; strains of *Pseudozyma pruni G.Y. Liou, Y.H. Wei and F.L. Lee*, *P. prolifica* (=*Ustilago maydis* (DC.) Corda) (4), *P. hubeiensis* F.Y. Bai and Q.M. Wang have lipolytic and proteolytic activities, and a strain each of *Pichia anomala* (=*Wickerhamomyces anomalus* (E.C. Hansen) Kurtzman) and *P. prolifica* (=*Ustilago maydis* (DC.) Corda) have all three enzymatic activities.

Cellulase Malubag et al. (2021) screened endophytic fungal strains from bananas and found four different strains that have cellulase activity. These fungal strains were identified as *F. solani* (=*Neocosmospora solani* (Mart.) L. Lombard and Crous), *G. candidum* Link, *N. oryzae* (Berk. and Broome) Petch, and *S. commune* Fr.

Pectinase Tapay et al. (1990) screened 371 fungal strains and found eight (8) that have high clarifying values in mango juice. These have high pectinase activity and were identified as *Aspergillus fiicum* (Reichardt) Thom and Currie.

Fungal Colorants

The pigment produced by *Monascus* has been tested for application in different food products globally. In the Philippine setting, UPLB BIOTECH researchers were awarded a large-scale production patent for the *Monascus* pigment. The team of Tambalo et al. (2022) determined the safety of the *Monascus* Red Pigment produced by *Monascus purpureus* Went MTCC 25436, through a battery of tests, which included the AMES test, mammalian micronucleus test, the LD50, dermal sensitization test, and detection of citrinin. They concluded that the pigment was safe for human consumption. This endeavor is the first of its kind in the country.

Medicine and Pharmaceuticals

The discovery of Penicillin from a species of *Penicillium*, a fungal contaminant in the Petri dish of Alexander Fleming, led to the isolation of fungal strains for discovery of therapeutically active molecules for production of medicine for mild to severe infection, to life-threatening chronic diseases (Yocum et al., 1980). Fleming's discovery of penicillin marked the beginning of the age of antibiotic chemotherapy (Ayoub & Mehta, 2024). Houbraken et al. (2011) revealed that the fungal contaminant has been re-identified as *Penicillium rubens* Biourge and produced penicillin F (2-pentenylpenicillin) discovered at Oxford, UK, and not penicillin G or penicillin V, which are the penicillins used in medicine today (Bentley, 2004; Prescott et al., 2023). Ayoub and Mehta (2024) presented the top 10 countries where new fungal species were described in 2019. Among these countries, China led with the most number of new species (377), followed by Thailand (129) and the United States (105). According to Ayoub and Mehta (2024), there is inadequate information from other Asian countries, including Malaysia, Indonesia, and Myanmar.

Novel antibiotics for the therapy of protozoal, bacterial, and fungal infections and several physiological disorders, including cholesterol lowering, are in great demand (Berdy, 2012). The extensive use and misuse of available antibiotics has alarmingly increased the incidence of multiple drug resistance (MDR) and biofilm-forming infections, making treatment very challenging (Correia et al., 2023). As such, an effective measure to prevent MDRs is to ensure that there is a stable arsenal of antimicrobial drugs with a broad spectrum of activity and a repository of novel fungal strains. Using innovative techniques in fungal biotechnology, metabolic and genetic engineering can help boost antibiotic production (Correia et al., 2023). Several species of microfungi, including *Acremonium*, Aspergilli, *Coprinopsis*,

Diaporthe, *Emericella*, *Fusarium*, *Monascus*, Penicilli, *Rhizoctonia*, *Sclerotium*, and *Trichoderma,* produce antibiotics against bacteria, viruses, and fungi (Correia et al., 2023; Prescott et al., 2023; Ayoub & Mehta, 2024; Gemechu, 2024).

Antifungal, antibacterial, and antiviral compounds were also isolated from different stages of fungal growth, fruiting bodies, and cell wall components of mushrooms, including species of *Coprinus*, *Ganoderma*, *Grifola*, *Lentinula,* and *Pleurotus* (Gemechu, 2024).

In the Philippines, Papa et al. (2006) discovered an antibiotic-producing microorganism from a type of mushroom found to be effective in treating diseases of livestock, particularly swine. The mushroom, *Clitopilus passeckerianus* (Pilát) Singer produces the antibiotic called pleuromutilin that prevents the bacteria from producing protein, an essential component of its diet. Without protein, bacteria stop reproducing and consequently die. Pleuromutilin also acts as the building block for the production of tiamulin, a biological compound effective in treating common hog diseases such as mycoplasmas, arthritis, enzootic pneumonia, and dysentery (Papa et al., 2006; Ong & Flores, 2008). Dulay et al. (2023) provided a comprehensive checklist of naturally occurring mushrooms in the Philippines based on the available reports from 2001 to 2021. A total of 2371 identified mushrooms from 64 available reports were taxonomically classified into 447 species, 193 genera, and 72 families. He reported the use of edible mushrooms since olden times as a natural alternative remedy for various diseases. The therapeutic values and medicinal properties of edible mushroom species under *Ganoderma*, *Lentinus*, *Volvariella*, *Collybia*, *Coprinus*, *Panaeolus*, *Termitomyces*, *Schizophyllum*, *Auricularia*, and *Pleurotus* have been found to stem from numerous biologically active compounds or metabolites (Ho et al., 2020; Lu et al., 2020; Painuli et al., 2020, as reported by Dulay et al., 2023). Dulay et al. (2023) reported that the Philippine mushrooms have been revealed to exhibit important bioactivities, namely, (1) antioxidants, (2) antibacterial, (3) hypoglycemic, (4) antihypertensive, (5) anticancer, (6) anticoagulative, (7) anti-inflammatory, (8) teratogenic, (9) aphrodisiac, (10) diuretic, (11) thrombolytic, and (12) potential anti-obesity activities. The TMRD in Central Luzon State University, Philippines, has successfully rescued the cell lines of various wild useful mushrooms and generated their production technologies (Dulay et al., 2023). This will provide biological resources for further research and development of bioactive compounds from mushrooms with pharmacological activities.

Environmental Applications

Mycoremediation Bioremediation technology is a strategy known for restoring degraded or contaminated environments using microbial and fungal organisms. AMF are known biofertilizers and are extensively probed for their capacity to protect plants from the effects of heavy metal toxicity. It is in this regard that AMFs are considered agents for addressing heavy metal contamination in the country (Trocio

& Paguntalan, 2023). Among the reported AMFs that were reported to have applications in heavy metal remediation are: *Glomus* sp., *Acaulospora* spp., *Entresphopora* spp. (Aggangan, 2006; Aggangan et al., 2015; Aggangan & Cortes, 2018; Aggangan et al., 2019), *Scutellospora* spp. (Aggangan et al., 2015; Aggangan & Cortes, 2018; Aggangan et al., 2019), *Glomus etunicatum* (=*Entrophospora etunicata* (W.N. Becker and Gerd.) Błaszk., Niezgoda, B.T. Goto and Magurno), *Glomus macrocarpum* Tul. and C. Tul., *Gigaspora margarita* W.N. Becker and I.R. Hall (Adiova et al., 2013; Aggangan et al., 2017) for copper remediation; *Glomus* sp., *Acaulospora* spp., *Entrophospora* spp. (Aggangan, 2006; Aggangan et al., 2015; Aggangan & Cortes, 2018; Aggangan et al., 2019), *Scutellospora* spp. (Aggangan et al., 2015; Aggangan & Cortes, 2018; Aggangan et al., 2019) for gold remediation; *Glomus etunicatum* (=*Entrophospora etunicata* (W.N. Becker and Gerd.) Błaszk., Niezgoda, B.T. Goto and Magurno), *Glomus macrocarpum* Tul. and C. Tul., *Gigaspora margarita* W.N. Becker and I.R. Hall (Adiova et al., 2013; Aggangan et al., 2017) for zinc and lead remediation as tested in greenhouses and in the field. These AMFs are reported to have one or a combination of the following properties: improve root colonization despite the presence of heavy metals, increase survival of host, improve overall plant growth, increase bioaccumulation of heavy metals, and increase phosphorus uptake, which makes them efficient agents of bioremediation of heavy metal-contaminated soil (Trocio & Paguntalan, 2023). Examples of biofertilizers with AMF as their primary fungal agent are MYKOVAM and MYKORICH, both developed at the UPLB BIOTECH (Aggangan et al., 2022).

Trichoderma Microbial Inoculant (TMI) in combination with rice straw compost was found to reduce total soil Cu and Pb concentrations in contaminated soil in an impoverished farming community in Mogpog, Marinduque, Philippines (Cuevas & Banaay, 2022a). Mechanisms by which *Trichoderma* can reduce soil heavy metals (HM) were provided by Siddiquee et al. (2015) as follows: biosorption of the HM on the cell wall of fungi; metal uptake with the use of a specific carrier molecular through the active transport of H^+ gradient transport wherein once inside the cell, the HM may be immobilized by complexation with SH peptides or by vacuolar compartmentalization, or transformed by various chemical reactions, or volatilized or transferred to other parts of the fungal hyphae; and chelated with fungal products like citric or oxalic acids to form insoluble products.

Decolorization of Synthetic Dyes Synthetic dyes become environmental pollutants when disposed of without treatment. Some researchers in the Philippines have screened fungi for this property to address growing problems of synthetic dye contamination in water systems from textile industries, for example. The following text highlights examples of work on dyes like Congo Red, Remazol Blue RGB, Congo Red, Malachite Green, Methylene Blue, and Rhodamine B.

Marine-derived fungi isolated from Manila Bay and Calatagan Bay were screened for dye decolorization activities. Of these, *Cladosporium* sp. (EMF 14) and *Penicillium* sp. (MF 49) decolorized Congo Red and Crystal Violet, while *Phialophora* sp. (MF 6) decolorized only Congo Red (Torres et al., 2011).

The crude laccase extracted from the strain *Pleurotus florida* (=*Pleuorotus pulmonarius* (Fr.) Quél.) ITDI 6003 grown on wheat grains was shown to significantly decolorize Remazol Blue RGB and four textile wastewaters from dyeing facilities of the Philippine Textile Research Institute. The laccase enzyme activity of ITDI 6003 was found to be 5.36 U/ml based on the oxidation of 2,2′-azino-bis(3-ethylbe nzothiazoline-6-sulfonic acid) diammonium salt (ABTS) (Montiague et al., 2020).

More recently, Obanan et al. (2022) isolated fungal endophytes from leaves of Philippine mangrove (*Avicennia marina (Forssk.) Vierh.*) and screened these for lignolytic enzymes (laccase, lignin peroxidase, and manganese peroxidase). These were also tested for decolorization of Congo Red, Malachite Green, Methylene Blue, and Rhodamine B. All isolates (10) tested positive for laccase; while only *Syncephalastrum racemosum* Cohn, *B. rubra*, *Penicillium citrinum* Thom, and mycelia sterilia showed the ability to produce manganese peroxidase, whereas only *C. lindemuthianum* (Sacc. and Magnus) Briosi and Cavara, *A. niger* Tiegh., *Aspergillus* sp., and *Colletrotrichum* sp. could produce lignin peroxidase. The fungal endophytes *A. niger*, *S. racemosum*, and *Trametes versicolor* (L.) Lloyd were shown to decolorize all dyes, while *A. flavus* Link, *P. citrinum* Thom were able to decolorize all except Malachite green, *Aspergillus* sp., and mycelia sterilia were able to decolorize Congo Red and Malachite Green only, while *B. rubra*, *C. lindemuthianum*, and *Colletotrichum* sp. decolorized Congo Red only.

Bioenergy Production

Researchers have pursued alternative energy sources in the hopes of providing cleaner energy compared to conventional fossil fuels. Bioenergy is an energy source from biomass and is a sustainable alternative to fossil fuels, especially for hard-to-decarbonize sectors such as aviation, heavy transport, and manufacturing (Reid et al., 2020). Examples of alternative bioenergy sources include biodiesel and bioethanol.

Biodiesel Oleaginous yeasts produce microbial oils that are potential sources of sustainable biodiesel production. Twenty-two (22) yeasts from barks, roots, canopy leaves, and associated epiphytic plant material of various forest trees in the Mt. Makiling Forest Reserve were reported by Pajares et al. (2024) to be oleaginous. Out of these, the top strain BUB8 isolated from the bark of a Bagtikan (*Parashorea malaanonan* (Blanco) Merr.) tree, produces the highest biomass and lipid production levels. This was identified as *Rhodotorula* sp. with the closest relative being *R. paludigena* (Fell and Tallman) Q.M. Wang, F.Y. Bai, M. Groenew. and Boekhout. So far, no other work on fungal biodiesel has been published and has been made commercially available.

Bioethanol Fungal pretreatment and treatment of substrates to yield bioethanol have long been researched globally to address energy and waste problems. In the

Philippines, the top agricultural wastes—rice straw and banana pseudostem—are being tested or used to produce bioethanol. Generally, bioethanol is produced using enzymes or common yeasts like *Saccharomyces cerevisiae* (Desm.) Meyen. The study by Gatdula et al. (2021) proved that the thermotolerant yeast *Kluyveromyces marxianus* (E.C. Hansen) Van der Walt is a viable alternative to *S. cerevisiae* that produces bioethanol. It was shown that *K. marxianus* can be used in the simultaneous Saccharification and Fermentation (SSF) of the aforementioned agricultural wastes.

Yeast Biotechnology Yeasts and fungal systems are commonly used to produce recombinant proteins and biopharmaceuticals. Glycoengineered yeasts capable of producing humanized glycoproteins demonstrated tremendous potential and promise as alternatives to mammalian cell cultures (Gerngross 2004). Yeasts are phylogenetically classified into two phyla, the Ascomycota and the Basidiomycota. The ascomycetous yeasts are taxonomically grouped under subphylum Saccharomycotina and Taphrinomycotina (Hibbett et al., 2007). Subphylum Saccharomycotina includes certain genera like *Saccharomyces* and *Candida,* while subphylum Taphrinomycotina includes the fission yeast *Schizosaccharomyces.* On the other hand, the basidiomycetous yeasts are taxonomically grouped under the subphylum Ustilaginomycotina, Agaricomycotina, and Pucciniomycotina (also previously known as classes Ustilaginomycetes, Hymenomycetes, and Urediniomycetes, respectively) (Hibbett et al., 2007; Boekhout, 1991; Bauer et al., 1997; Boekhout et al., 2011). *Mrakia frigida* (Fell, Statzell, I.L. Hunter and Phaff) Y. Yamada and Komag. and *Cystofilobasidium capitatum* (Fell, I.L. Hunter and Tallman) Oberw. and Bandoni are both psychrophilic yeasts that have been commonly isolated from a variety of cold environments, such as marine sediments in Antarctica, soil, frozen fish, and even snow (Yamada & Komagata, 1987; Hua et al., 2010; Nakagawa et al., 2005). It is said that these cold-adapted yeast species can be used in cold-fermentation and may produce some enzymes active in the cold environment (Hamid et al., 2014). The importance of protein glycosylation for the biotech industry is highlighted by the fact that approximately 70% of therapeutic proteins approved or in pre-clinical studies are glycoproteins. Explaining the glycan structures of these yeasts is significant for comprehending the biological roles that these glycans perform in their normal yeast physiology and in infecting host cells, as is the case with *Malassezia furfur* (C.P. Robin) Baill and a few other non-pathogenic and pathogenic yeasts. It is also important in recognizing the tremendous biotechnological applications of these yeasts as being suitable alternatives to mammalian cell cultures in production of heterologous protein (Gerngrosss, 2004). *Pseudozyma antarctica* (Goto, Sugiy. and Iizuka) Boekhout observed to be an excellent producer of recombinant proteins and mannosylerythritol lipids (MELs), which may have a great biotechnological application as a biosurfactant, antitumor, cell-differentiation inducer, moisturizer, and hair-repairer (Avis et al., 2005; Kitamoto et al., 2002; Kitamoto et al., 2009). In addition, enzymes produced by *P. antarctica* have been utilized to degrade biodegradable polymers (Shinozaki et al., 2013). *P. antarctica* itself had been utilized as a host for high-level recombinant protein production of biodegradable plastic-degrading enzyme, which plainly underscores the biotechnological potential of this yeast (Watanabe et al., 2016).

Challenges and Future Directions of Philippine Fungal Biotechnology

The Philippines is one of the world's mega hotspots of biodiversity with its abundant and rich flora, fauna, and fungal resources (De Leon & Siringan, 2023). Endowed with diverse habitat types from lowland rainforests and montane forests to long coastlines including the six world's most colorful coral reefs, the Philippines is probably the most biologically diverse country in the world, in terms of unique terrestrial and marine plant and animal species per unit area (https://www.cepf.net/). Quimio in 2002 (as reported by Dela Cruz et al. in 2023) reported that about 4698 fungal species belonging to 1031 genera are described in the Philippines. Dulay et al. in 2023 reported that from 2001 to 2021, a total of 2371 identified mushrooms from 64 available reports were taxonomically classified into 447 species, 193 genera, and 72 families. For myxomycetes in the country, there are about 162 species identified, about 50 new additions to the 107 species in the annotated list of Reynolds (Reynolds, 1981; Dagamac & dela Cruz, 2015; Macabago et al., 2020). Significant increases in the species list were also reported by dela Cruz (2023), which include 13 species of dictyostelids in the country, two more records from the previous list of 11 species, 1262 published taxa with 1234 validated species names for Philippine lichens, and 68 species of Philippine oomycetes. Indeed, there is a need for new committed mycologists, not only from the institutions strongly working on mycology in the country that aim to discover new species and update the current list for the Philippines but also for collaboration among mycologists, bioinformaticians, and biotechnologists to strengthen local research endeavors on fungal diversity in food, rare environments, clinical samples, animals, and plants and to generate a considerable number of characterized potential resources with biotechnological potential as well as references for future discoveries on novel species (De Leon & Siringan, 2023). Numerous professional organizations in the Philippines (dela Cruz et al., 2023; De Leon & Siringan, 2023) are tasked to promote mycology, microbiology, and/or biotechnology, lead in coming up with a roadmap from fungal diversity to biotechnology, and present to funding agencies for appreciation of the plethora of applications and contributions to Philippine education, research, and economy.

Bioprospecting is another consideration for fungal diversity and biotechnology studies. The Philippine Department of Environment and Natural Resources (DENR) is in charge of the permitting system for collecting and bioprospecting biodiversity in the country. The Philippine Wildlife Resources Conservation and Protection Act of 2001 (RA 9147) does not explicitly mention "fungal bioprospecting." However, the law does address bioprospecting in general, requiring prior informed consent from stakeholders before any bioprospecting activity, including those potentially involving fungi. The law also outlines the requirements for a Bioprospecting Undertaking (BU), which is a permit-like document that resource users need to execute before accessing biological resources for bioprospecting purposes. A more resolute support for relevant research from the department and other agencies and units will make a progressive, productive, and timely research. We need new sources of information; thus, extensive bioprospecting is required for the more than 7000 islands in the archipelago.

Funding opportunities for fungal biodiversity studies and maintaining a repository of cultivated cultures are scarce, though biotechnological applications are promising. Improvement of facilities, curation of fungal strains and bioinformatics, and capacity development and training on fungal biotechnology rely mostly on external funds and less on institutional support. There is a need for sustainable and reliable mycological herbaria (MH) and microbial culture collections (MCC), which will serve as repositories of these resources for future use and applications (De Leon & Siringan, 2023). The involvement of these biological resource centers (BRCs) in such an undertaking will result in the improvement of the operation, increase visibility, and recognition of their relevance in the field of science and technology. Ultimately, MCCs and MH will be appreciated as major infrastructures for the advancement of microbiology and biotechnology in the Philippines and Southeast Asia (De Leon & Siringan, 2023). Commercialization potential and startup opportunities are available but limited. Industry partners, technology takers, and technocrats should be involved early on in the conceptualization of a research and development and a permitting or regulatory agency in the country for compliance. Identification of priority areas for fungal biotechnology should be aligned with national and international priorities, including the UN's SDGs. This alignment will address gaps and sustainability of targets in the development of the country.

The country was the first in the region to establish a biotechnology regulatory committee (Bolaños et al., 2023). At present, most biotech regulations are for plants and animals. Although the Scientific Community at large is receptive to advancements in biotechnology, recently, a Writ of Kalikasan (a Philippine legal remedy protecting one's right to a healthy environment) was enforced to halt further experimentation of the Bt Eggplant, a genetically modified plant with *Bacillus thuringiensis*-derived genes inserted in its genome, developed by the University of the Philippines Los Baños. This means that there may still be a little resistance within the community concerning genetically modified or gene-edited organisms. It is highly likely that when a fungal organism is modified and developed by a group in the Philippines, it may encounter the same challenges and scrutiny faced by the researchers and developers of Bt Eggplant.

Lack of awareness of the value of fungal diversity and biotechnology research can be addressed by science communication and partnership among educators, researchers, science communicators, policymakers, and industry partners. Digital tools like genomic databases, artificial intelligence (AI)-powered applications for fungal identification through image analysis software, citizen watch, and virtual reality can significantly enhance the interactive and engaging learning process and spark interests in mycology and its biotechnology. Improvement of the curricula in biology in academic institutions from primary to tertiary levels will provide more interesting and accurate facts about fungal diversity, metabolic processes, role in the ecosystem, and potential applications. Training for biology teachers through various academic and professional organizations will standardize modules and learning experiences. Internship and volunteering opportunities for teachers, students, and enthusiasts in MCCs, MH, and research centers in mycology, and engaging them in experiential field work activities are training grounds for our young mycologists.

The Philippines is endowed with rich natural resources. Policymakers and researchers should work hand in hand in the beneficial and equitable use of our resources without depriving the next generation. Policy recommendations and implementation for fungal diversity conservation and safe and secure research on fostering fungal biotechnology are necessary to maximize the innovative technologies' potential. As more and more Filipino researchers aspire for a better Philippines, fungal diversity and biotechnology will continue to prosper, given immense institutional, government, and community support.

References

Abalunan, A. J., Teves, F. G., & Madamba, M. R. S. B. (2013). Isolation of fungal species and aflatoxin detection in fermented products. *International Research Journal of Biological Sciences, 2*, 51–54.

Ablang, C. A., Alvar, K. N., Samson, K. A. T., & Santiago, L. A. (2017). Molecular identification of a native wine yeast from bubod used in tapuy fermentation. *Acta Manilana, 65*, 47–54.

Adiova, J. M., Pampolina, N. M., & Aggangan, N. S. (2013). Effect of arbuscular mycorrhizal fungi inoculation on growth and Cu uptake and toxicity of *Desmodium cinereum* (Kunth) D.C. *Philippines Journal of Science, 142*(1), 87–96.

Aggangan, N. S. (2006). Assessment of mycorrhiza diversity in abandoned mine sites in Toledo, Cebu for bioremediation. In *Food and Agriculture Organization of the United Nations. FORESPI symposium 2006*.

Aggangan, N. S., & Aggangan, B. J. S. (2012). Selection of ectomycorrhizal fungi and tree species for rehabilitation of Cu mine tailings in the Philippines. *Journal of Environmental Science and Management, 15*(1).

Aggangan, N. S., & Cortes, A. (2018). Screening mined-out indigenous mycorrhizal fungi for the rehabilitation of mine tailing areas in the Philippines. *REFORESTA, 6*, 71–85.

Aggangan, N. S., Aggangan, J. S., Bulan, J. C. O., & Limos, C. A. S. (2012). Inoculation of dipterocarps *Anisoptera thurifera* and *Shorea guiso* with ectomycorrhizal fungi in Philippine red soil. *Philippine Journal of Science, 141*(2), 229–241.

Aggangan, N. S., Tamayao, P. J. S., Aguilar, E. A., Anarna, J. A., & Dizon, T. O. (2013). Arbuscular mycorrhizal fungi and nitrogen fixing bacteria as growth promoters and as biological control agents against nematodes in tissue-cultured banana var. Lakatan. *Philippine Journal of Science, 142*(2), 153–165.

Aggangan, N. S., Pampolina, N., Cadiz, N. M., & Raymundo, A. K. (2015). Assessment of plant diversity and associated mycorrhizal fungi in the mined out sites of Atlas Mines in Toledo City, Cebu for bioremediation. *Journal of Environmental Science and Management, 18*(1), 71–86.

Aggangan, N. S., Cadiz, N. M., Llamado, A., & Raymundo, A. K. (2017). *Jatropha curcas* for bioenergy and bioremediation in mine tailing area in Mogpog, Marinduque, Philippines. *Energy Procedia, 110*, 471–478.

Aggangan, N. S., Anarna, J. A., & Cadiz, N. M. (2019). Tree legume–microbial symbiosis and other soil amendments as rehabilitation strategies in mine tailings in the Philippines. *Philippine Journal of Science, 148*(3), 481–491.

Aggangan, N. S., Anarna, J. A., Gestiada, E. G., Gilbero, D. M., Magsayo, B. T., Tapi-on, M. C., & Podiotan, V. J. A. (2022). Finding biological solutions on greening nickel areas in Claver, Surigao del Norte, Mindanao, Philippines. In M. Christopher, L. R. J. Cuesta, & F. A. Ruben (Eds.), *19th International Conference on Sustainable Energy Technologies* (pp. 367–376). Department of Architecture and the Built Environment, Faculty of Engineering, University of Nottingham.

Agor, A. B., & Villadolid, D. V. (1961). Report on the laboratory test on the value of yeast (*Torulopsis utilis*) as an aquatic fertilizer. *Araneta Journal of Agriculture, 8*, 132–135.

Aguilar, E. A., Elleva, L. I. F., Fabro, D. M. A., Garcia, G. R., Divina, F. I., & Aggangan, N. (2018). Arbuscular Mycorrhizal Fungi increased root-mycorrhizal association and enhanced seedling growth of abaca, papaya, and sugarcane. *Journal of ISSAAS, 24*(2), 22–29.

Alicbusan, R. V., & Ela, V. M. (1960). Mushroom culture. *UPCA Technical Bulletin, 5*, 1–16.

Arana, J. (2022). It's alive! Mykovam, the hardworking biofertilizer. *The UPLB Horizon*. https://horizon.uplb.edu.ph/horizon-magazine-2022/its-alive-mykovam-the-hardworking-biofertilizer/?fbclid=IwAR2X4mB8YbqAxbpS16s2Km4FURrx1fgCVNbLyw-sQW-iVU4yFqdj1QTw2zI

Avis, T. J., Cheng, Y. L., Zhao, Y. Y., Bolduc, S., Neveu, B., Anguenot, R., & Bélanger, R. R. (2005). The potential of *Pseudozyma* yeast-like epiphytes for the production of heterologous recombinant proteins. *Applied microbiology and biotechnology, 69*(3), 304–311.

Ayoub, Z., & Mehta, A. (2024). Use of fungi in pharmaceuticals and production of antibiotics. In A. C. Shukla (Ed.), *Entrepreneurship with microorganisms*. Academic. https://doi.org/10.1016/B978-0-443-19049-0.00017-7

Batool, S., Asghar, H. N., Shehzad, M. A., Yasin, S., Sohaib, M., Nawaz, F., et al. (2021). Zinc-solubilizing bacteria-mediated enzymatic and physiological regulations confer zinc biofortification in chickpea (*Cicer arietinum* L.). *Journal of Soil Science and Plant Nutrition, 21*(3), 2456–2471.

Bauer, R., Oberwinkler, F., & Vánky, K. (1997). Ultrastructural markers and systematics in smut fungi and allied taxa. *Canadian Journal of Botany, 75*(8), 1273–1314.

Beardall, J., Miller, J. D., & ApSimon, J. W. (1994). *Report on ASEAN fungal and mycotoxin research activities* (p. 41). International Development Research Centre. https://idl-bnc-idrc.dspacedirect.org/server/api/core/bitstreams/896cfdc4-d174-4904-a212-bcf7121a242e/content

Bentley, R. (2004). The molecular structure of penicillin. *Journal of Chemical Education, 81*, 1462–1470. https://doi.org/10.1021/ed081p1462

Bérdy, J. (2012). Thoughts and facts about antibiotics: Where we are now and where we are heading. *The Journal of Antibiotics, 65*, 385–395. https://doi.org/10.1038/ja.2012.27

Blachinsky, D., Antonov, J., Bercovitz, A., Elad, B., Feldman, K., Husid, A., Lazare, M., Marcov, N., Shamai, I., Keren-Zur, M., & Droby, S. (2007). Commercial applications of Shemer for the control of pre- and postharvest diseases. *IOBCWPRS Bulletins, 30*, 75–78.

Boekhout, T. (1991). A revision of ballistoconidia-forming yeasts and fungi. Centraalbureau voor Schimmelcultures.

Boekhout T, Bai FY, Nakashi N (2011). Bullera Derx (1930). In: The yeasts: a taxonomic study (Kurtzman CP, Fell JW, Boekhout T, eds). Elsevier, Amsterdam: 1635–1648.

Bolaños, J. L., Jamias, G. N. A., Merlin, M. L. T., & Alam, Z. F. (2023). The current scenario, challenges and opportunities of biotechnology in the Philippines. In *DLSU Research Congress, Fostering a Humane and Green Future: Pathways to Inclusive Societies and Sustainable Development* (p. 11). DLSU.

Borromeo, J. D., & Reynolds, D. R. (1966). Cultivation of *Auricularia polytricha*. *Philippine Phytopathology, 2*, 49–52.

Brown, M., & Monsalud, R. (2014). Research undertakings of the biotechnology for agriculture and forestry program of BIOTECH. ASEAN Network on Microbial Utilization (AnMicro). Retrieved February 3, 2025, from http://www.anmicro.org/document/meeting/1/Philippines_BIOTECH.pdf

Bustillos, R. G., Dulay, R. M. R., Kalaw, S. P., & Reyes, R. G. (2014). Optimization of culture conditions for mycelial growth and basidiocarp production of Philippine strains of *Panaeolus antillarum* and *Panaeolus cyanescens*. *Mycosphere, 5*(3), 398–404.

Cairns, T. C., Nai, C., & Meyer, V. (2018). How a fungus shapes biotechnology: 100 years of *Aspergillus niger* research. *Fungal Biology and Biotechnology, 2018*(5), 13.

Chandrasekaran, M., Basheer, S. M., Chellappan, S., Krishna, J. G., & Beena, P. S. (2015). Enzymes in food and beverage production: An overview. *Enzyme Food Beverage Process CRC Press, 25*, 133–154.

Chaudhary, P., Singh, S., Chaudhary, A., Sharma, A., & Kumar, G. (2022). Overview of biofertilizers in crop production and stress management for sustainable agriculture. *Frontiers in Plant Science, 13*, 930340.

CLARA, R. P. (1937). Culture of edible mushrooms in the Philippines. *Philippine Agriculturist, 8*, 225–231.

Correia, J., Borges, A., Simões, M., & Simões, L. C. (2023). Beyond Penicillin: The Potential of filamentous fungi for drug discovery in the age of antibiotic resistance. *Antibiotics (Basel), 12*(8), 1250. https://doi.org/10.3390/antibiotics12081250. Erratum in: Antibiotics (Basel). 2024;13(10):944. https://doi.org/10.3390/antibiotics13100944. PMID: 37627670; PMCID: PMC10451904.

Cortado, R. V., & Davide, R. G. (1968). Nematode trapping fungi in the Philippines. *Philippine Phytopathology., 4*, 4.

Cuevas, V. C., & Banaay, C. G. B. (2022a). *In situ* bioremediation and crop growth promotion using *Trichoderma* microbial inoculant (TMI) ameliorate the effects of Cu contamination in lowland rice paddies. *Philippine Journal of Science, 151*(3), 1255–1265.

Cuevas, V., & Banaay, C. G. (2022b). *Trichoderma* inoculants and straw compost improved resilience and yield in Cu-contaminated rice paddies. *Nusantara Bioscience, 14*(1).

Cuevas, V., Caasi-Lit, M., Lit, I., Jr., Alviola, M., Banaay, C. G., & Salazar, A. (2023). Recovery of coconut trees from scale insect infestation with salt and *Trichoderma* microbial inoculant (TMI). *Journal of Environmental Science and Management, 26*(2), 71–82.

Cumagun, C. J. R. (2012). Managing plant disease and promoting sustainability and productivity with *Trichoderma*: The Philippine experience. *Journal of Agricultural Science and Technology, 14*, 699–714.

Cumagun, C. J. R., & Lapis, D. B. (1993). Note: Practical approach in mass production of *Trichoderma* spp. as a means of biological control against sheath blight of rice. *Philippine Agriculturist, 76*, 251–257.

Cumagun, C. J. R., Hockenhull, J., & Lubeck, M. (2000). Characterization of *Trichoderma* isolates from Philippine rice fields by UP-PCR and rDNA-ITS1 analysis: Identification of UP-PCR markers. *Journal of Phytopathology, 143*, 109–115.

Dagamac, N. H. A., & dela Cruz, T. E. E. (2015). Myxomycetes research in the Philippines: Updates and opportunities. *Mycosphere, 6*(6), 784e795.

De Leon, M. P., & Siringan, M. A. T. (2023). Chapter 12 – Culture collections and herbaria: Diverse roles in mycological research in the Philippines. In J. J. G. Guerrero, T. U. Dalisay, M. P. De Leon, M. A. O. Balendres, K. I. R. Notarte, & T. E. E. Dela Cruz (Eds.), *Mycology in the Tropics, Updates on Philippine Fungi, in Developments in Microbiology* (pp. 293–324). Academic. https://doi.org/10.1016/B978-0-323-99489-7.00001-9

De Leon, A. M., Orpilla, J. O. V., Cruz, K. V., Dulay, R. M. R., Kalaw, S. P., & Cruz, T. D. (2017). Optimization of mycelial growth and mycochemical screening of *Lentinus sajor-caju* (fr.) from Banaue, Ifugao province, Philippines. *International Journal of Agricultural Technology, 13*(7.3), 2549–2567.

De Leon, A. M., Dulay, A. R., Villanueva, A. L., & Kalaw, S. P. (2020). Optimal culture conditions and toxicity assessment of *Fomitopsis fee*i (Fr.): A newly documented macro fungus from Philippines. *Studies in Fungi, 5*(1), 491–507.

del Rosario, M. A. G., R, D.,. R. M., Kalaw, S. P., Damaso, E. J., Jr., Ramos, J. P., Abon, M. A., De Leon, A. M., & Reyes, R. G. (2022). Optimization of mycelial culture conditions and fructification of Ganoderma species in rice straw-based substrates. *The International Society of Southeast Asian Agricultural Science, 42*.

Dela Cruz, R. E., Aggangan, N. S., Lorilla, E. B., & Zarate, J. T. (1998). Mycorrhizal inoculation technology. *Application of Biotechnology on Nursery Management.*

Dela Cruz, T. E. E., Guerrero, J. J. G., & De Leon, A. M. (2023). Chapter 1. A brief introduction to Philippine Mycology. In J. J. G. Guerrero, T. U. Dalisay, M. P. De Leon, M. A. O. Balendres, K. I. R. Notarte, & T. E. E. Dela Cruz (Eds.), *Mycology in the Tropics: Updates on Philippine Fungi*. Academic Press.

Dela Cruz-Halos, S. (1977). Yeast strains excreting amino acids. *Kalikasan, Philippine Journal of Biology, 41*, 268–272.

Devanthi, P. V. P., & Gkatzionis, K. (2019). Soy sauce fermentation: Microorganisms, aroma formation, and process modification. *Food Research International, 120*, 364–374.

Devi, D., Gupta, S. B., Mishra, B. K., & Verma, N. P. (2022). Isolation and identification of zinc solubilizing fungal isolates from cumin of semi-arid region of Rajasthan. *Pharma Innovation Journal, 11*, 1036–1040.

Domingo, O. F. (2016). BioCon agent HiYeast controls postharvest diseases in crops. Philippine Council for Agriculture, Aquatic and Natural Resources Research and Development (PCAARRD). Retrieved January 27, 2025, from https://www.pcaarrd.dost.gov.ph/index.php/quick-information-dispatch-qid-articles/biocon-agent-hiyeast-controls-postharvest-diseases-in-crops

Dorockis, S., & Boguś, M. (2014). Regional variety of biotechnology development in Asia. *Procedia-Social and Behavioral Sciences, 120*, 197–212.

Droby, S., Wisniewski, M., Teixidó, N., Spadaro, D., & Jijakli, M. H. (2016). The science, development, and commercialization of postharvest biocontrol products. *Postharvest Biology and Technology, 122*, 22–29.

Dulay, R. M. R., & Rivera, A. G. C. (2017). Mycelial growth and fruiting body production of Philippine (CLSU) strain of *Polyporus grammocephalus* (BIL7749). *Biocatalysis and Agricultural Biotechnology, 11*, 161–165.

Dulay, R. M. R., Kalaw, S. P., Reyes, R. G., Cabrera, E. C., & Alfonso, N. F. (2012a). Optimization of culture conditions for mycelial growth and basidiocarp production of *Lentinus tigrinus* (Bull.) Fr., a new record of domesticated wild edible mushroom in The Philippines. *Philippine Agricultural Scientist, 95*(3), 278–285.

Dulay, R. M. R., Parungao, A. G., Kalaw, S. P., & Reyes, R. G. (2012b). Aseptic cultivation of *Coprinus comatus* (OF Mull.) Gray on various pulp and paper wastes. *Mycosphere, 3*(3), 392–397.

Dulay, R. M. R., Ray, K., & Hou, C. T. (2015). Optimization of liquid culture conditions of Philippine wild edible mushrooms as potential source of bioactive lipids. *Biocatalysis and Agricultural Biotechnology, 4*(3), 409–415.

Dulay, R. M. R., Rivera, A. G. C., & Garcia, E. J. B. (2017). Mycelial growth and basidiocarp production of wild hairy sawgill *Lentinus strigosus*, a new record of naturally occurring mushroom in The Philippines. *Biocatalysis and Agricultural Biotechnology, 10*, 242–246.

Dulay, R. M. R., Alcazar, A. A., Kalaw, S. P., Reyes, R. G., & Cabrera, E. C. (2021a). Nutritional and physical requirements for mycelial growth and basidiocarp production of *Trametes elegans* from The Philippines. *Asian Journal of Agriculture and Biology, 9*(1), 1–9.

Dulay, R. M. R., Cabrera, E. C., Kalaw, S. P., & Reyes, R. G. (2021b). Optimization of submerged culture conditions for mycelial biomass production of fourteen *Lentinus* isolates from Luzon Island, Philippines. *Biocatalysis and Agricultural Biotechnology, 38*, 102226.

Dulay, R. M. R., Cabrera, E. C., Kalaw, S. P., & Reyes, R. G. (2021c). Optimization of culture conditions for mycelial growth and fruiting body production of naturally-occurring Philippine mushroom *Lentinus swartzii* Berk. *Journal of Applied Biology and Biotechnology, 9*(3), 17–25.

Dulay, R. M. R., Batangan, J. N., Kalaw, S. P., De Leon, A. M., Cabrera, E. C., Kimura, K., Eguchi, F., & Reyes, R. G. (2023). Records of wild mushrooms in the Philippines: A review. *Journal of Applied Biology and Biotechnology, 11*(2), 11–32. https://doi.org/10.7324/JABB.2023.110202

Eguchi, F., Kalaw, S. P., Dulay, R. M. R., Miyasawa, N., Yoshimoto, H., Seyama, T., & Reyes, R. G. (2015). Nutrient composition and functional activity of different stages in the fruiting body development of Philippine paddy straw mushroom, *Volvariella volvacea* (Bull.: Fr.) Sing. *Advances in Environmental Biology, 9*(22 S3), 54–66.

Elegado, F. B., Colegio, S. M. T., Lim, V. M. T., Gervasio, A. T. R., Perez, M. T. M., Balolong, M. P., Banaay, C. G. B., & Mendoza, B. C. (2016). Ethnic fermented foods of the Philippines with reference to lactic acid bacteria and yeasts. *Ethnic Fermented Foods and Alcoholic Beverages of Asia*, 323–340.

Ereky, K. (1919). *Biotechnologie der Fleisch-, Fett-, und Milcherzeugung im landwirtschaftlichen Grossbetriebe: für naturwissenschaftlich gebildete Landwirte verfasst*. Paul Parey.

Fabros, J. A., Dulay, R. M. R., Ganareal, K. C. O., Kalaw, S. P., del Rosario, M. A. G., & Reyes, R. G. (2023). Optimization of mycelial culture condition and biomass production of selected wild agaric mushrooms from Luzon Island, Philippines. *Asian Journal of Agriculture and Biology, 2023*(3), 1–10. https://doi.org/10.35495/ajab.2023.021

Friedrichs, S., & van Beuzekom, B. (2018). *OECD Science, Technology and Industry Working Papers: Revised proposal for the revision of the statistical definitions of biotechnology and nanotechnology*. https://doi.org/10.1787/085e0151-en

Garcia, B. L., Undan, J. R., Dulay, R. M. R., Kalaw, S. P., & Reyes, R. G. (2020). Molecular identification and optimization of cultural conditions for mycelial biomass production of wild strain of *Chlorophyllum molybdites* (G. Mey) Massee from the Philippines. *Journal of Applied Biology and Biotechnology, 8*(6), 1–6.

Gatdula, K. M., Blaquera, M. S. F. N., Jimena, C. M. V., Elegado, F. B., Alcantara, J. Z., & Guerrero, G. A. M. (2021). A comparative study on bioethanol production from rice straw and banana pseudostem through simultaneous saccharification and fermentation using *Kluyveromyces marxianus*. In *IOP conference series: Earth and environmental science,* 765 (1). IOP Publishing.

Gerngross, T. U. (2004). Advances in the production of human therapeutic proteins in yeasts and filamentous fungi. *Nature biotechnology, 22*(11), 1409–1414.

Gemechu, G. W. (2024). Antibiotics production from fungi. *Advances in Biotechnology and Microbiology, 18*(2), 555985. https://doi.org/10.19080/AIBM.2024.18.555985

Go, L. K. (1959). Experimental cultivation of *Volvariella volvacea* (Builliard) Quelet. *Philippine Agriculturist, 43*, 44–467.

Halos, S. C. (2000). Agricultural biotechnology research and development in the Philippines: The need for a strategic approach (p. 58). Econstor. PIDS. https://hdl.handle.net/10419/127723

Hamid, B., Ravinder, S. R., Deepak, C., Poonam, S., Fayaz, A. M., Sanjay, S., & Abidi I. (2014). Psychrophilic yeasts and their biotechnological applications-a review. *African Journal of Biotechnology, 13*(22).

He, X., Duan, Y., Chen, Y., & Xu, M. (2010). A 60-year journey of mycorrhizal research in China: Past, present and future directions. *Science China Life Sciences, 53*, 1374–1398.

Hibbett, D. S., Binder, M., Bischoff, J. F., Blackwell, M., Cannon, P. F., Eriksson, O. E., & Huhndorf, S. et al. (2007). A higher-level phylogenetic classification of the Fungi. *Mycological Research, 111*(5), 509–547.

Hipol, R. L. B., & Alma-in, A. B. (2018). Antioxidant potentials of indigenously produced Benguet *tapuy* (rice wine). *International Food Research Journal, 25*(5), 1968–1976.

Hipolito, D. G., Valladolid, D. V., & Gerardo, J. A. (1961). A study in the supplementation of fodder yeast (*Torulopsis utilis*) for man and live-stock. *Araneta Journal for Agriculture, 8*, 92–97.

Ho, L. H., Zulkifli, N. A., & Tan, T. C. (2020). Edible mushroom: Nutritional properties, potential nutraceutical values, and its utilisation in food product development. In: An Introduction to Mushroom. London: IntechOpen; p. 19–38.

Hollis, L. (2023). What are bioprotection products, and how do they work? Center for Agriculture and Bioscience International (CABI) PlantwisePlus Blog. https://blog.plantwise.org/2023/08/16/what-are-bioprotection-products/#:~:text=Bioprotection%20products%20are%20nature%2Dbased,solutions%2C%20such%20as%20bioprotection%20products

Houbraken, J., Frisvad, J. C., & Samson, R. A. (2011). (2011). Fleming's penicillin producing strain is not *Penicillium chrysogenum* but *P. rubens*. *International Mycological Association Fungus, 2*, 87–95. https://doi.org/10.5598/imafungus.2011.02.01.12

Hua, M. X., Chi, Z., Liu, G. L., Buzdar, M. A., & Chi, Z. M. (2010). Production of a novel and cold-active killer toxin by Mrakia frigida 2E00797 isolated from sea sediment in Antarctica. *Extremophiles, 14*(6), 515–521.

Islas-Valdez, S., Afkairin, A., Rovner, B., & Vivanco, J. M. (2024). Isolation of diverse phosphate- and zinc-solubilizing microorganisms from different environments. *Applied Microbiology, 4*(3), 1042–1056.

Jomao-as, J. G., Sabino, N. S., Mendoza, B. C., & Aggangan, N. S. (2023). Influence of bamboo biochar, Arbuscular Mycorrhizal fungi, and nitrogen-fixing bacteria as soil amendments on cacao (*Theobroma cacao* L.) planted in acidic soil. *Philippine Journal of Science, 152*(1).

Julian, A. V., Wright, C. A., & Reyes, R. G. (2018). Prelude to successful cultivation of *Hericium* in the Philippines: Understanding its mycelial growth response on different culture media and its antibacterial activity. *International Journal of Pharmaceutical Research and Allied Sciences, 7*(2), 1–7.

Kalaw, S. P., Alfonso, D. O., Dulay, R. M. R., De Leon, A. M., Undan, J. Q., Undan, J. R., & Reyes, R. G. (2016). Optimization of culture conditions for secondary mycelial growth of wild macrofungi from selected areas in Central Luzon, Philippines. *Current Research in Environmental and Applied Mycology, 6*(4), 277–287.

Kalaw, S. P., De Leon, A. M., Damaso, E. J., Jr., Ramos, J. C., del Rosario, M. A. G., Abon, M. D., Undan, J. Q., Undan, J. R., & Reyes, R. G. (2021a). Cultivation of different strains of *Lentinus tigrinus* from selected areas of Luzon Island, Philippines. *Studies in Fungi, 6*(1), 299–306.

Kalaw, S. P., Dulay, R. M. R., Damaso, E. J., Ramos, J. C., del Rosario, M. A. G., Abon, M. D., De Leon, A. M., Undan, J. R., & Reyes, R. G. (2021b). Optimization of mycelial culture conditions and fructification of *Lentinus* species using rice straw and sawdust based substrates. *Studies in Fungi, 6*(1), 519–530.

Kitamoto, D., Isoda, H., & Nakahara, T. (1990). Functions and potential applications of glycolipid biosurfactants from energy-saving materials to gene delivery carriers. *Journal of bioscience and bioengineering, 94*(3), 187–201.

Kitamoto, D., Isoda, H., & Nakahara, T. (2002). Functions and potential applications of glycolipid biosurfactants from energy-saving materials to gene delivery carriers. *Journal of bioscience and bioengineering, 94*(3), 187–201.

Kitamoto, D., Morita, T., Fukuoka, T., Konishi, M. A., & Imura, T. (2009). Self-assembling properties of glycolipid biosurfactants and their potential applications. *Current Opinion in Colloid & Interface Science, 14*(5), 315–328.

Kowalska, J., Krzymińska, J., & Tyburski, J. (2022). Yeasts as a potential biological agent in plant disease protection and yield improvement – A short review. *Agriculture, 12*(9), 1404.

Kozaki, M. (1976). Fermented foods and related microorganisms Southeast Asia. *Proc Jap Assoc Mycotoxicol, 2*,1–9.

Landingin, H. R. R., Francisco, B. E., Dulay, R. M. R., Kalaw, S., & Reyes, R. (2020). Optimization of culture conditions for mycelial growth and basidiocarp production of *Cyclocybe cylindracea* (Maire). *CLSU International Journal of Science & Technology, 4*(1), 1–17.

Launio, C. C., Labon, K. O., Bañez, A. A., & Batani, R. S. (2020). Adoption and economic analysis of using biological control in Philippine highland farms: Case of *Trichoderma koningii* strain KA. *Crop Protection, 136*, 105177.

Lim, E. A. Y., Panes, V. A., & Romero, G. O. (2006). Species identification and genetic diversity analysis by DNA fingerprinting of yeast isolates from Philippine rice wine starters. *Philippine Agricultural Scientist, 89*(4), 326.

Lopez, M. K. S., Kalaw, S. P., Dulay, R. M. R., De Leon, A. M., & Reyes, R. G. (2022). Optimization of mycelial growth of *Xylaria papulis* Lloyd (Xylariaceae) in indigenous liquid culture conditions, science city of Muñoz, Nueva Ecija, Philippines. *Studies in Fungi, 7*(1), 1–7.

Lu, J., He, R., Sun, P., Zhang, F., Linhardt, R. J., & Zhang A. (2020). Molecular mechanisms of bioactive polysaccharides from *Ganoderma lucidum* (Lingzhi), a review. *International Journal of Biological Macromolecules, 150*, 765–774.

Macabago, S. A. B., Dagamac, N. H. A., dela Cruz, T. E. E., & Stephenson, S. L. (2020). Myxomycetes of the Caramoan Islands and update on the myxomycetes of the Bicol Peninsula in the Philippines. *The Philippine Journal of Systematic Biology, 14*, 1.

Malubag, A. G., Parayao, A. M., & Waing, K. G. D. (2021). Identification of endophytic fungi associated in banana (*Musa paradisiaca* L.) and evaluation of its enzymatic abilities. *International Journal of Agricultural Technology, 17*(3), 941–958.

Martin, D. K., Vicente, O., Beccari, T., Kellermayer, M., Koller, M., Lal, R., Marks, R. S., Marova, I., Mechler, A., Tapaloaga, D., Žnidaršič-Plazl, P., & Dundar, M. (2021). A brief overview of global biotechnology. *Biotechnology & Biotechnological Equipment, 35*(sup1), S5–S14. https://doi.org/10.1080/13102818.2021.1878933

Meyer, V., Basenko, E. Y., Benz, J. P., Braus, G. H., Caddick, M. X., Csukai, M., de Vries, R. P., Endy, D., Frisvad, J. C., Gunde-Cimerman, N., Haarmann, T., Hadar, Y., Hansen, K., Johnson, R. I., Keller, N. P., Kraševec, N., Mortensen, U. H., Perez, R., Ram, A. F. J., Record, E., Ross, P., Shapaval, V., Steiniger, C., van den Brink, H., van Munster, J., & Wösten, H. A. (2020). Growing a circular economy with fungal biotechnology: A white paper. *Fungal Biology and Biotechnology, 7*, 5. https://doi.org/10.1186/s40694-020-00095-z

Montiague, E. N., Guzman, J. P. M. D., Unciano, N. M., Panerio, E. G., Bigol, U. G., Castro, I. J. L., Jose, J. P. G., & Mantaring, S. D. A. (2020). Biodecolourization of textile dye and wastewaters by crude laccase from *Pleurotus florida* ITDI 6003 cultivated in wheat grains. *Current Research in Environmental and Applied Mycology (Journal of Fungal Biology), 10*(1), 167–175. https://doi.org/10.5943/cream/10/1/17

Nakagawa, T., Nagaoka, T., Miyaji, T., & Tomizuka, N. (2005). Cold-active polygalacturonase from psychrophilic-basidiomycetous yeast *Cystofilobasidium capitatum* strain PPY-1. *Bioscience, biotechnology, and biochemistry, 69*(2), 419–421.

National Institute of Molecular Biology and Biotechnology (BIOTECH). (2008). Mykovam for a Greener and Safer Environment [Brochure]. Retrieved February 3, 2025, from https://ati2.da.gov.ph/ati-4b/content/sites/default/files/2022-11/Bio-N-and-MykoVam-1.pdf.

National Institute of Molecular Biology and Biotechnology (BIOTECH). (2023). *Products*. https://biotech.uplb.edu.ph/

Nepomuceno, R. A., Brown, C. M. B., Mojica, P. N., & Brown, M. B. (2019). Biological control potential of vesicular arbuscular mycorrhizal root inoculant (VAMRI) and associated phosphate solubilizing bacteria, *Pseudochrobactrum asaccharolyticum* against soilborne phytopathogens of onion (*Allium cepa* L. var. red creole). *Archives of Phytopathology and Plant Protection, 52*(7–8), 714–732.

Obanan, S. P., Sadaba, R. B., Alva-Gatchalian, P., Cadiz, A. V., Basilio, E. M., & Bela-ong, D. B. (2022). Decolorization of synthetic textile dyes by fungal endophytes isolated from the leaves of Philippine mangrove (*Avicennia marina*). *EnvironmentAsia, 15*(1), 21–32. https://doi.org/10.14456/ea.2022.3

Oclarit, E., & Cumagun, C. (2009). Evaluation of efficacy of *Paecilomyces lilacinus* as biological control agent of *Meloidogyne incognita* attacking tomato. *Journal of Plant Protection Research, 49*(4), 337–340.

Office of the Vice Chancellor for Research and Extension (OVCRE). (2018). Our technologies. https://ovcre.uplb.edu.ph/research/our-technologies

Ong, G., & Flores, H. (2008). Antibiotic-producing mushroom discovered. https://www.philstar.com/headlines/2008/07/18/73616/antibiotic-produng-mushroom-discovered

Organo, N. D., Granada, S. M. J. M., Pineda, H. G. S., Sandro, J. M., Nguyen, V. H., & Gummert, M. (2022). Assessing the potential of a *Trichoderma*-based compost activator to hasten the decomposition of incorporated rice straw. *Scientific Reports, 12*(1), 448.

Painuli, S., Semwal, P., Egbuna, C. (2020). Mushroom: Nutraceutical, mineral, proximate constituents and bioactive component. In: Functional Foods and Nutraceuticals. Berlin: Springer; p. 307–36.

Pajares, I. G., Requiso, P. J., Fabro, L. M., Jr., Ramos, K. R. M., & Raymundo, A. K. (2024). Biodiesel production of oleaginous yeast isolated from the Mount Makiling Forest Reserve. *Journal of Applied Biology and Biotechnology, 12*, 67–75.

Papa, I. A., Zulaybar, T. O., & Raymundo, A. K. (2006). Increasing pleuromutilin activity of *Clitopilus passeckerianus* by chemical mutagenesis and improvement of production medium. *Philippine Agricultural Scientist, 89*, 20–33.

Prescott, T. A. K., Hill, R., Mas-Claret, E., Gaya, E., & Burns, E. (2023). Fungal drug discovery for chronic disease: History, new discoveries and new approaches. *Biomolecules, 13*(6), 986. https://doi.org/10.3390/biom13060986

Quimio, T. H. (1977). Artificial cultivation of *Auricularia polytricha* (Mont. Sacc.). Kalikasan, Philippine. *Journal of Biology, 6*, 75.

Quimio, T. H. (1978). Indoor cultivation of *Pleurotus ostreatus. Philippine Agriculturist, 61*(7/8), 253–262.

Quimio, T. H. (1993). Indoor cultivation of the straw mushroom *Volvariella volvacea. Mushroom Research, 2*(2), 87–90.

Quimio, T. H. (2001). Edible mushrooms: Potential and low cost technologies for tropical cultivation. In *Trichomycetes and other fungal groups* (pp. 335–366). CRC Press.

Quimio, T. H. (2002). Checklist and database of Philippine fungi (1806e2001). ASEAN Regional Center for Biodiversity Conservation.

Quimio, T. H., & Capilit, A. (1981). Enumeration and bibliography of Philippine fungi (1936–1977) (p. 79). BIOTECH, University of the Philippines at Los Baños.

Ramos, M. A., Palo, M. A., Villadolid, D. V., & Cruz, D. S. (1967). A study on submerged culture production of banana mushroom (*Volvariella volvacea* (Bull. ex. Fr.) Sing.) mycelium as a source of protein, B-vitamins and food flavor. *Philippine Journal of Science, 96*, 191–213.

Reid, W. V., Ali, M. K., & Field, C. B. (2020). The future of bioenergy. *Global Change Biology, 26*(1), 274–286.

Reyes, R. G., Eguchi, F., Iijima, T., & Higaki, M. (1998). Physiological considerations for efficient mycelial colonization of Philippine strains of *Volvariella volvacea. Journal of Wood Science, 44*, 408–413.

Reyes, R. G., Lopez, L. L. M. A., Kumakura, K., Kalaw, S. P., Kikukawa, T., & Eguchi, F. (2009). *Coprinus comatus*, a newly domesticated wild nutriceutical mushroom in the Philippines. *Journal of Agricultural Technology, 5*(2), 299–316.

Reynolds, D. R. (1981). Southeast Asian myxomycetes: II. Philippines. *Kalikasan. Philippine Journal of Biology, 10*(2e3), 127e150.

Roth, M. G., Westrick, N. M., & Baldwin, T. T. (2023). Fungal biotechnology: From yesterday to tomorrow. *Frontiers in Fungal Biology, 4*. https://doi.org/10.3389/ffunb.2023.1135263

Rubio, P., & Alicbusan, R. V. (1966). Synthetic compost for spawn production of *Agaricus bisporus. Philippine Phytopathology., 2*, 38–48.

Sakai, H., & Caldo, A. G. (1985). Microbiological and chemical changes in tapuy fermentation. *Journal of Fermentation Technology, 63*(1), 11–16.

Sanchez, P. (1999). Microorganisms and technology of Philippine fermented foods. *Japanese Journal of Lactic Acid Bacteria, 10*(1), 19–28.

Santos, P. S., & Bernardo, Z. P. (1966). Tumor inhibitors in Philippine basidiomycetes. *Philippine Journal of Science, 95*, 189–197.

Santos, S. P., Palo, M. A., Felix, L. S., Palo, N. D., & Yares, A. Y. (1975). The rennet-forming molds in cheese production from defatted coconut milk or gata. *NRCP Research Bulletin, 30*, 170–177.

Santos, S. P., Palo, M. A., Felix, L. S., Palo, N. D., & Yares, A. Y. (1976). Fungal rennet in cheese production. *NRCP Research Bulletin, 31*, 14–31.

Shinozaki, Y., Morita, T., Cao, X. H., Yoshida, S., Koitabashi, M., Watanabe, T., & Kitamoto, H. K. (2013). Biodegradable plastic-degrading enzyme from *Pseudozyma antarctica*: cloning, sequencing, and characterization. *Applied Microbiology and Biotechnology, 97*(7), 2951–2959.

Shurtleff, W., & Aoyagi, A. (2004). History of soy sauce, shoyu, and tamari. In *History of soybeans and soyfoods, 1100 B.C. to the 1980s* (p. 8). Soyfoods Center. https://www.soyinfocenter.com/HSS/soy_sauce2.php

Siddiquee, S., Rovina, K., Azad, S. A., Naher, L., & Suryani, S. (2015). Heavy metal contaminants removal from wastewater using the potential filamentous fungi biomass: A review. *Journal of Microbial and Biochemical Technology, 7*(6), 384–393.

Spadaro, D., & Droby, S. (2016). Development of biocontrol products for postharvest diseases of fruit: The importance of elucidating the mechanisms of action of yeast antagonists. *Trends in Food Science & Technology, 47*, 39–49.

Tambalo, F. M. Z., Garcia, J. F., & Estrellana, C. D. (2022). Safety assessment of a fungal-based red colorant produced by *Monascus purpureus* MTCC 25436. *Philippine Journal of Science, 151*(4), 1507–1516.

Tan Gana, N. H., Mendoza, B. C., & Monsalud, R. G. (2014). Isolation, screening and characterization of yeasts with amylolytic, lipolytic, and proteolytic activities from the surface of Philippine bananas (*Musa* spp.). *Philippine Journal of Science, 143*(1), 81–87.

Tapay, L., Espino, T., Mercado, M., Mercado, S., & Espino, T. (1990). Isolation and identification of fungi for pectinase production. *Philippine Journal of Biotechnology, 1*(1), 25–42.

Teixidó, N., Torres, R., Viñas, I., Abadias, M., & Usall, J. (2011). Biological control of postharvest diseases in fruit and vegetables. In C. Lacroix (Ed.), *Protective cultures, antimicrobial metabolites and bacteriophages for food and beverage biopreservation* (Woodhead Publishing Series in Food Science, Technology and Nutrition No. 201).

Tolentino, J. J. V., Kalaw, S. P., Reyes, R. G., & Undan, J. R. (2016). Mycelial growth performance of three *Pleurotus* species on corn varieties in the Philippines. *Advances in Environmental Biology, 10*(7), 155–161.

Torillo, A. R., De Ocampo, A., Vilela, L. C., & Del Rosario, E. J. (1977). Cellulase production in semisolid cultures of *Trichoderma viridis*. *Kalikasan, Philippine Journal of Biology, 6*, 76.

Torres, J. M. O., Cardenas, C. V., Moron, L. S., Guzman, A. P. A., & dela Cruz, T. E. E. (2011). Dye decolorization activities of marine-derived fungi isolated from Manila Bay and Calatagan Bay, Philippines. *Philippine Journal of Science, 140*(2), 133–143.

Trocio, D. Y. C., & Paguntalan, D. P. (2023). Review on the use of arbuscular mycorrhizal fungi in bioremediation of heavy metal contaminated soils in the Philippines. *Philippine Journal of Science, 152*(3), 1139–1159.

UNESCO Institute for Statistics (UIS). (2023). http://data.uis.unesco.org/Index.aspx?DataSetCode=SCN_DS&lang=en#

Watanabe, T., Morita, T., Koike, H., Yarimizu, T., Shinozaki, Y., Sameshima-Yamashita, Y., & Kitamoto, H. (2016). High-level recombinant protein production by the basidiomycetous yeast *Pseudozyma antarctica* under a xylose-inducible xylanase promoter. *Applied microbiology and biotechnology, 100*(7), 3207–3217.

Wicklow, D. T., McAlpin, C. E., & Yeoh, Q. L. (2007). Diversity of *Aspergillus oryzae* genotypes (RFLP) isolated from traditional soy sauce production within Malaysia and Southeast Asia. *Mycoscience, 48*(6), 373–380.

Wilson, C. L., & Pusey, P. L. (1985). Potential for biological control of postharvest plant diseases. *Plant Disease, 69*(1985), 375–378.

Xing, X., Jacquemyn, H., Gai, X., Gao, Y., Liu, Q., Zhao, Z., & Guo, S. (2019). The impact of life form on the architecture of orchid mycorrhizal networks in tropical forests. *Oikos, 128*(9), 1254–1264.

Yamada, Y., & Komagata, K. (1987). Mrakia gen. nov., a heterobasidiomycetous yeast genus for the q8-equipped, self-sporulating organisms which produce a unicellular metabasidium, formerly classified in the genus Leucosporidium. *The Journal of General and Applied Microbiology, 33*(5), 455–457.

Yocum, R. R., Rasmussen, J. R., & Strominger, J. L. (1980). The mechanism of action of penicillin. Penicillin acylates the active site of *Bacillus stearothermophilus* D-alanine carboxypeptidase. *Journal of Biological Chemistry, 255*, 3977–3986. https://doi.org/10.1016/S0021-9258(19)85621-1

Chapter 20
Wood Fungal Decay and Strategies in Heritage Conservation Across Southeast Asia

Marian P. De Leon, Cheek S. Fadriquela, and Mark Kevin S. Lopez

Abstract Cultural heritage embodies the true significance of cultural development processes. It serves as a positive representation of historical traditions that form a national identity passed down through generations. A considerable number of these cultural and natural artifacts are crafted from wood or similar cellulosic materials, which are relatively durable, yet susceptible to concurrent deterioration, presenting challenges for their preservation and upkeep. Biodeterioration, such as decay caused by fungi, threatens these invaluable treasures, particularly affecting wooden cultural or natural monuments and religious artifacts. Wood rot in wooden natural and cultural monuments and artifacts across Southeast Asia is caused by fungal species generally under three divisions, namely, Ascomycota, Basidiomycota, and Mucoromycota. The diverse lifestyle and extensive enzymatic systems of fungi enable them to thrive even in often extremophilic conditions associated with cultural heritage and the built environments. In Southeast Asia, traditional knowledge has significantly contributed to the preservation of historic wooden structures by offering insights into the timing and types of trees designated for use, and also provides guidance on which organic materials can be employed as fungicides. A variety of repair methods and chemicals, including carbon disulfide and thymol, as well as a 10% thymol solution in alcohol or ether for fungicidal purposes, are utilized for wood-based paper conservation. The architectural marvels of wooden-built heritage exhibit intricate details and are rich in historical milieu; hence, it necessitates more extensive studies on microbial communities contributing to the complexities of bio-

M. P. De Leon
Museum of Natural History, University of the Philippines Los Baños, Los Baños, Laguna, Philippines
e-mail: mpdeleon1@up.edu.ph

C. S. Fadriquela
University of Santo Tomas, Manila, Philippines

M. K. S. Lopez (✉)
Don Mariano Marcos Memorial State University, Bacnotan, La Union, Philippines

J. J. G. Guerrero et al. (eds.), *Mycology in a Changing Planet: Applications and Perspectives from Southeast Asia*, https://doi.org/10.1007/978-981-95-2640-6_20

deterioration to gain a comprehensive understanding of the biodegradative processes and effective conservation methods for these valuable cultural assets.

Keywords Biodeterioration · Cultural heritage · Preservation strategies · Southeast Asia · Wood rot fungi

Introduction

Bioeconomy, or the use of biological sciences in the sustainability of innovations and economy, has positively pushed the transition of the use of wood in construction materials for modern structures (OECD, 2009). The Food and Agriculture Organization (FAO) of the United Nations, through its Advisory Committee on Sustainable Forest-based Industries (ACSFI), listed wood encouragement policies (WEP) adopted by several countries with an economically important forest sector, including Australia, Canada, Finland, France, Germany, Japan, New Zealand, Sweden, Switzerland, and the United Kingdom (ACSFI, FAO, 2020). Tesoro, in 1987, reported in his review that the present and future use of wood for construction in several ASEAN countries is devoted to increasing the amount of forest resources to meet domestic demand for rising housing needs and growing populations. He suggested that future requirements will be met through the onset of unconventional resources and increased use of new technologies in wood processing. In 2003, de la Roche, O'Connor, and Têtu emphasized that wood-built structures score very high in about a dozen attributes in building homes, which include *appearance and style, flexible design, healthy, durable, safe, affordable, low maintenance, adaptable, multi-functional, comfortable, energy efficient, and environmentally friendly*. They noted that despite these attributes, there were notable concerns about the *strength, fire safety, durability, and environmental merit* of wood construction. For instance, the durability of wood materials threatened by wood decay or rot caused by species of fungi, excessive moisture, and damage from termites has been considered a perennial issue. De la Roche et al. (2003) emphasized that *there is a need for further research around the moisture management of wood structures, as well as more aggressive technology transfer and training programs for architects and the construction industry*.

Wooden artifacts define the complexity of the past, provide insights into the materiality of the wood and skills of the craftsmen, and reflect the culture and tradition passed on from one generation to another. Fadriquela (2013) reported that since ancient times, Filipino ancestors used wood in transporting and housing the living with *barotos* (dugout boats), *balangays* (plank-built boats), *bale* (Ifugao house) and *torogans* (Maranao house), and the dead with *kabaong* (coffins).

Tesoro, in 1987, envisioned wood as a major construction material in Southeast Asia in the coming decades. However, technology and innovations will produce more nontraditional materials with superior durability and strength. The conservation of cultural heritage across the globe is an important endeavor of countries.

Southeast Asia is home to many wooden cultural properties threatened by fungi. Among them are wood rots and fungal decay.

Cultural heritage reflects the true value of cultural formation processes and the positive expression of historical tradition and constitutes one's national characteristics. It is a legacy from our ancestors, a window of humanity's creation throughout history, and a unique and irreplaceable bridge between the past and the future. Cultural heritage plays a significant role in people's lives as it reflects one's social values, religions, beliefs, and traditions (Savković et al., 2022; Oliveira et al., 2023; Dán et al., 2024). There is also a natural heritage that pertains to natural monuments, including natural parks distinguished by their natural beauty and distinct and robust biodiversity, like the *Kinabalu National Park* in Borneo, Malaysia, and the *Tubbataha Reef Natural Park* in Palawan, Philippines, in Southeast Asia. Cultural or historical monuments, on the other hand, pertain to artifacts, monuments, buildings, and museums that hold artistic, symbolic, historical, ethnological, anthropological, or social values like *dluwang manuscripts* (made from the bark of mulberry where old Javanese scripts were written) in Indonesia, *Balla Lompoa Ri Galesong* (traditional house of Takalar Regency) in South Sulawesi, Indonesia, and the historic *Baker Memorial Hall* in the University of the Philippines Los Baños, Laguna, Philippines (Oetari et al., 2016; Fadriquela & Rivarez, 2019; Oliveira et al., 2023; Arif et al., 2024).

The beauty and architectural wonders that make up these natural, historical, or cultural monuments, which are intricately detailed and rich in historical milieu, must be acknowledged as living archives of our shared past rather than silent structures. Truly, their rather echoing historical, cultural, and artistic values resonate, making them invaluable treasures for humanity. However, these priceless treasures are threatened by biodeterioration, among others, such as fungal decay, especially on wooden natural or cultural monuments and religious items. This chapter explores the initiatives employed to preserve these socio-culturally important items and the future of heritage conservation in mycology.

Wood Rot and Its Diversity

Most of the cultural heritage items globally were constructed using wood. Based on the United Nations Educational, Scientific and Cultural Organization (UNESCO), wooden cultural heritage (WCH) can be generally grouped into three types, namely, (i) movable WCH like furniture, sculptures, instruments, frames, and the like, (ii) immovable WCH such as churches, monasteries, temples, bridges, windmills, and the like, and (iii) underwater WCH like shipwrecks, wooden cargo, foundation piles, and similar structures that were periodically or continuously submerged (partially or entirely) for over hundred years (Corbu et al., 2023). An interesting example of underwater WCH can be found in Southeast Asia, particularly in Indonesia's territorial waters known as the "Maritime Silk and Spice Route", harboring over 700 shipwrecks dating between the tenth and twentieth centuries. Significantly, the underwater WCH of Indonesia is abundant and holds a remarkable history of global maritime trade between Japan, China, India, the western Indian Ocean, European countries, and Southeast Asia (Polkinghorne et al., 2024).

A significant portion of these natural and cultural monuments is made up of wood or related materials, which are considerably durable cellulosic materials. However, they are prone to simultaneous degradation by biotic and abiotic agents, which pose challenges to preserving and conserving these wooden cultural items (Goodell et al., 2020). This section will present the nature and composition of wood, biotic agents of wood degradation, focusing mainly on wood rot fungi along with abiotic factors, and the mechanisms of different wood fungal decay. It will also highlight the diversity of varying wood rot fungi found in several natural and cultural heritage sites across Southeast Asia.

Wood Structure and Composition

Since prehistoric times, wood has been regarded as a primary construction material due to its outstanding mechanical qualities, portability, lightweightness, and ease of shaping. Wood is also among the most durable polymeric materials. It is composed mainly of cellulose microfibrils (40–44%), lignin (18–35%), hemicellulose (15–32%), and other components (2–15%) such as starch, sugars, and other polysaccharides as well as pectin, proteins, and fatty acids (Goodell et al., 2020; Isola et al., 2024). The discrete units of cellulose that make up elementary fibrils at the nanoscale are composed of a densely packed crystalline substance at their core. At the same time, the outer layers are amorphous cellulose that is more loosely packed. At the same scale, wood's remarkable tensile strength is comparable to steel and aramid (Kevlar) due to the crystalline nature of the inner core of the primary cellulose fibrils. In addition, compared to the other amorphous carbohydrate polymers found in the wood cell wall, the inner part of the cellulose primary fibril is more resistant to degradation, again, due to its crystallinity (Mittal et al., 2018; Goodell et al., 2020). The lignin component is a heteropolymer composed of repeating phenylpropane units with three different forms of the monomer joined by a diverse array of bonds. Lignin's diverse bonding patterns and the integration of cellulosic components make wood resistant to degradation, of course, with few exceptions. Some microorganisms, like fungi, can degrade wood with their extensive enzymatic systems on top of abiotic factors involved in wood degradation (Goodell et al., 2020). The linking cellulose and lignin components of the wood matrix are hemicellulose. Hemicellulose comprises branching heteropolymers of sugar monomers such as pentose and hexose. Hemicellulose is an integral part of the wood structure. However, it is prone to various abiotic and biotic degradation agents. It is a weak link initially targeted by fungi in the primary stages of wood fungal decay. Also, abiotic agents like excessive heat or acidic environments profoundly affect the rigidity of the wood structural system. Other wood components include organic compounds (which may confer additional protective functions) such as pectins, tannins, resins, oils, waxes, and various secondary metabolites (Goodell et al., 2020; Isola et al., 2024).

Wood is the xylem tissue of woody vascular plants, mainly located across the root, trunk, and branches of woody plants, such as the gymnosperms (naked-seeded trees) and angiosperms (flowering trees). Interestingly, the wood composition and characteristics of these two types of woody trees vary considerably. Gymnosperms (like coniferous trees such as spruce and pines) are anatomically characterized by a homogenous structure known as *homoxyl wood* (softwood). In contrast, angiosperms are anatomically characterized by a heterogeneous structure known as *heteroxyl wood* (hardwood) (Isola et al., 2024). Hardwoods tend to be denser and harder than softwoods. However, some hardwoods are relatively softer than certain softwoods. The key distinction is not in the wood's hardness but in the botanical classification, such that angiosperms produce hardwood while gymnosperms produce softwood. In terms of wood fungal decay, hardwoods are generally more decay-resistant than softwoods. However, the extent of resistance among species within each group can vary significantly. Because hardwood has more extractives—natural antifungal compounds—than softwood, especially in the heartwood (dense inner part), it is often more resistant to fungal rot. By preventing fungal development and enzymatic activity, these extractives—which include tannins, flavonoids, and terpenes—function as organic preservatives. Furthermore, hardwoods frequently have higher densities and lignin content, which can restrict fungal access and growth (Broda, 2020; Kirker et al., 2024). Examples of hardwoods are Oak (*Quercus* sp.) and Teak (*Tectona grandis*), which are used for carvings, flooring, furniture, decorative elements, and the like. Softwood also constitutes natural resistance to insects and rot, and there are considered "hard" softwoods such as Cedar (*Cedrus* sp.), Hemlock (*Conium* sp.), and Yew (*Taxus* sp.). Other examples of softwoods are Juniper (*Juniperus* sp.) and Pine (*Pinus* sp.), used in monuments as frames, sheathes, structural elements, or decorative details (Oliveira et al., 2023). In general, however, the biodegradation path and mechanism of wood vary depending on the diversity of fungi and tree species, as fungi possess selectivity in the degradation of versatile wood components (Bari et al., 2020; Qi et al., 2023), which will be discussed in the next section.

The plant cell wall is composed of several layers at the cellular level. Starting externally is the thinnest layer of (i) *middle lamella* (mainly composed of lignin and pectin) holding adjacent cells together, followed by a thin and elastic (ii) *primary cell wall* (mainly composed of cellulose microfibrils embedded in a matrix of lignin, hemicellulose, and pectin) and then followed by the layers of (iii) *secondary cell walls* (usually three layers—S1, S2, and S3) with different chemical composition, thickness, and orientation of microfibrils (Pornou, 2020). Knowing the structure and composition of wood is important in understanding how certain microorganisms, like fungi, unravel the wood structural systems with their key enzymes during wood fungal decay.

Wood Fungal Decay

Most cultural heritage properties in Southeast Asia were built using wood and are prone to degradation inevitably over time if neglected. Degradation of WCH can be attributed to abiotic (such as temperature, humidity, and weathering) and biotic agents (such as insects and microorganisms like bacteria and fungi). Because of these agents of degradation, the structure and quality of WCH, like traditional houses or paper manuscripts and sculptures, are altered and affected, showing undesirable changes resulting in structural or aesthetic damage (Corbu et al., 2023; Avdanina & Zhgun, 2024). Microorganisms are the most important agents in the deterioration of WCH due to their ability to utilize them as substrates to support their growth and reproduction. The growth of microorganisms internally or on the surface of WCH can cause mechanical and chemical transformation, leading to long-term deterioration and affecting the durability of wooden monumental artifacts (El-sayed et al., 2024).

Fungi are well-established wood decomposers since they can break down wood's complex high molecular weight components—cellulose, hemicellulose, and lignin. The ability of fungi to degrade a wide variety of organic and inorganic materials makes them ubiquitous saprotrophs (Ding et al., 2020). If we are to consider, the biodeterioration of wooden cultural heritage is primarily caused by wood rot fungi. The penetrative and explorative lifestyle, sporulation, and morphological adaptations of fungi enable efficient dispersal and colonization of various substrates. Aside from that, many fungal species are also resistant to environmental stresses such as low water activity, high salinity, UV radiation, and toxic substances. Interestingly, many can survive even in nutrient-limited environments and produce resistant cells for long periods of adverse conditions (Gadd et al., 2024).

Several detrimental factors, weather, moisture, fire, pests, and biological degradation, endanger heritage wooden structures and artifacts. The latter is of particular concern because its effects take longer to manifest but are expected to cause greater damage. Termites and wood-degrading microorganisms are some of the pests causing serious damage (Unger et al., 2001; Fadriquela 2013). *Wood-rotting microorganisms include many species of fungi that secrete enzymes such as cellulase, hemicellulase, lignin-peroxidase, and manganese-peroxidase, which degrade a wood structure down to its monomeric units* (Deacon, 2005; Schmidt, 2006, 2007). Depending on its mode of action and physical appearance, wood rot may be classified as brown, white, or soft rot. Wood rot requires high relative humidity (RH) and moderately low temperature. It is caused by several Basidiomycota and some Ascomycota species. Soft rot fungi attack wood with low lignin content, producing unique holes in the secondary cell wall layers (Schmidt, 2006). White rot is caused by fungi that attack lignin, cellulose, and hemicellulose components, while brown rot is caused by fungi that only attack cellulose (Ochoa, 1978; Schmidt, 2006).

Wood fungal decay can be categorized into three types: brown rot, white rot, and soft rot, depending on the patterns of wood decay observed, such as color, texture, and chemical changes. Brown rot fungi are known due to crumbly and brown discoloration of the decaying wood because of the degradation of all wood carbohydrates, including the crystalline cellulose, except lignin. Degradation of wood involves both enzymatic and nonenzymatic systems (Goodell et al., 2008). Enzymes such as cellulase, mannanase, and xylanase, along with nonenzyme compounds such as oxalic acid, hydrogen peroxide, and Fenton reagent, are released by penetrating fungal hyphae, and the degradation happens far from the point of hyphal growth (Goodell et al., 2020; Isola et al., 2024). Brown rot fungi often target softwood, although they can also degrade hardwood in some environments. Another mechanism exhibited by brown rot fungi is the chelator-mediated Fenton (CMF) system, allowing brown rot fungi to generate powerful hydroxyl radicals, causing lignin to depolymerize (to access cellulosic components) and then re-polymerize into smaller, discrete, irregular masses and separate from cellulosic components. Interestingly, CMF chemistry is being studied in biorefineries for an efficient mechanism of lignin processing—a potential biotechnological application (Goodell et al., 2020).

White rot fungi are named after the bleaching of wood caused by the degradation of all wood components, including lignin. Due to lignin oxidation, other colorations in the wood, such as yellow to violet or red, can be observed. Degradation of wood is mainly enzymatic, i.e., through carbohydrate-active enzymes (CAZymes) and lignin-degrading enzymes targeting all wood components simultaneously (simultaneous rot) or by preference (preferential rot). Simultaneous rot happens when all cell wall components are degraded enzymatically from the cell lumen to the middle lamella, resulting in a uniform and gradual loss of the cell wall. Conversely, preferential rot selectively degrades lignin before the degradation of cellulose and hemicellulose results in a granular appearance (Goodell et al., 2020; Isola et al., 2024).

Soft rot fungi derive its name from the spongy soft texture developed on damaged wood with higher moisture content (waterlogged) and lower lignin composition. Soft rot fungi exhibit two decay pattern types: Type I (cavity formation) and Type II (cell wall erosion). The Type I decay pattern is characterized by chains of conical-shaped cavities within the secondary cell walls. In contrast, the Type II decay pattern generally erodes the cell wall layers but not the middle lamella. Degradation by soft rot fungi tends to have large but localized effects on wood. Soft rot fungi show enzymatic activity against cellulose and hemicellulose and also show the ability to degrade lignin, as evidenced by the cavity formation (Goodell et al., 2008, 2020; Isola et al., 2024). There are also pigmented fungi responsible for wood alteration, leading to permanent discoloration of wooden materials, and are known as bluestain or sapstain, producing black, grey, brown, blue, yellow, green, pink, or reddish coloration. Discoloration is caused by staining fungi that act on free sugars, wood extractives, and other nutrients present in wood (Isola et al., 2024) (Fig. 20.1).

All brown rot fungi belong to the division Basidiomycota and include *Coniophora puteana* (Schumach.) P. Karst., *Schizophyllum commune* Fr., and *Gloeophyllum sepiarium* (Wulfen) P. Karst. White rot decay is mainly caused by Basidiomycota, particularly members of the Order Polyporales. Examples are *Trametes versicolor* (L.) Lloyd, *Fomes fomentarius* (L.) Fr., and *Phlebia radiata* Fr. Phylogenetically speaking, white rot fungi are derived from ancestral soft rot fungi conserved across

Fig. 20.1 Various types of wood rot: brown rot (top left) taken from *molave* (*Vitex parviflora*) post and brace of Teus Mansion in Manila, Philippines; soft rot (top right) taken from the *yakal* (*Shorea astylosa*) panel of Malate Church (Philippines) left main facade door; white rot (bottom left) taken from CC by 4.0 and bluestain (bottom right) caused by pigmented fungi taken from CC by 4.0

Ascomycota and Basidiomycota. Soft rot decay pattern is associated with Ascomycota; examples are *Chaetomium globosum* Kunze and *Cadophora malorum* (Kidd and Beaumont) W. Gams (Isola et al., 2024).

While the mechanisms of wood rot and fungal decay in terrestrial environments have been thoroughly discussed previously, the mechanisms of wood rot and fungal decay in aquatic or underwater wooden cultural heritage are slightly different compared to the terrestrial environment. Marine wooden cultural relics, such as wooden sunken ships, are classical examples with large volumes and severe corrosion due to long-term immersion in the aquatic environment. Owing to long-term immersion, in this case, in the marine aquatic environment, WCHs are characterized by having (i) extremely high moisture content, (ii) high degree of decay, serious corrosion and degradation with relatively lower wood cellulose and hemicellulose contents compared to fresh wood, and (iii) high concentration of soluble salts (NaCl) and insoluble salts (sulfur iron compounds) which promote cellulose degradation and fracture in crystallization and dissolution exhibited by fragmentation of cellulose aggregates and a significant loss of crystallinity (Han et al., 2021; Wang et al., 2024; Chu et al., 2025).

Underwater shipwrecks are often equipped with iron tools such as nails and pans. In the presence of sulfate-reducing microorganisms and the iron tools, sulfur-iron compounds can be formed, which, when oxidized, generate sulfates and sulfuric acid, leading to wood penetration and acidification (Wang et al., 2024). *Fusarium oxysporum* Schltdl. is a fungus capable of reducing elemental sulfur to hydrogen sulfide even in an anoxic environment. This is attributed to a novel dissimilation mechanism of *F. oxysporum* by adapting to anaerobic conditions through replacement of the energy-producing mechanism of oxygen respiration with sulfur reduction (Abe et al., 2007). Interestingly, Han et al. (2021) showed that the genus *Fusarium* has the highest relative abundance in sea mud samples along with other genera such as *Emericellopsis, Volutella,* and other fungal genera associated with the hull of the Nanhai No. 1 shipwreck.

Meanwhile, in stream freshwater habitats, Jones et al. (2019) found that mass loss occurred 9% faster in streams than on land; however, the loss of cellulose, hemicellulose, and lignin did not differ in these two habitats. Wood decomposition in stream habitats is affected by fragmentation and leaching of organic materials from wood on top of biotic decomposition. Factors such as flowing streams limit fungal hyphal connections and access to nutrients from sediments, thus growth is limited by substrate and stream nutrient content. Generally, the primary type of biodegradation observed in waterlogged wooden cultural heritage is soft rot based on morphological characteristics (Chu et al., 2025), as previously discussed.

Diversity of Wood Rot Fungi in Natural and Cultural Monuments Across Southeast Asia

Generally, fungi are highly diverse monophyletic and saprotrophic eukaryotes with profound ecological and economic impacts. They can proliferate in various substrates such as trees, wood, soil, and leaves. Many are microscopic and have cryptic life cycles, with around 140 thousand described fungal species. However, their biodiversity is believed to reach a staggering millions of species (He & Zhao, 2022). Interestingly, wood rot fungi form well-defined groups within the Ascomycota and Basidiomycota. Most of these wood rot fungi produce overt fruiting bodies, and they belong to the class Agaricomycetes under the Basidiomycota. Other wood rot fungi are Polypore fungi (a polyphyletic group) with representatives under the order Polyporales, Hymenochaetales, and Gloeophyllales, and some others belong to Ascomycota (Park et al., 2020; Isola et al., 2024).

Based on reports on the different wood rot fungal species found in wooden natural and cultural monuments and artifacts across Southeast Asia, wood rot fungal species are generally under three divisions, namely, Ascomycota, Basidiomycota, and Mucoromycota. The majority of these belong to Ascomycota (84%), followed by Basidiomycota (15%), and the least is Mucoromycota (1%), as shown in Fig. 20.2a. Wood rot fungal species are further classified into seven classes, namely, Eurotiomycetes (32%), Sordariomycetes (25%), Dothideomycetes (21%), Agaricomycetes (15%), Leotiomycetes (3%), Saccharomycetes (3%), and Mucoromycetes (1%). Furthermore, they are classified into 16 orders, namely,

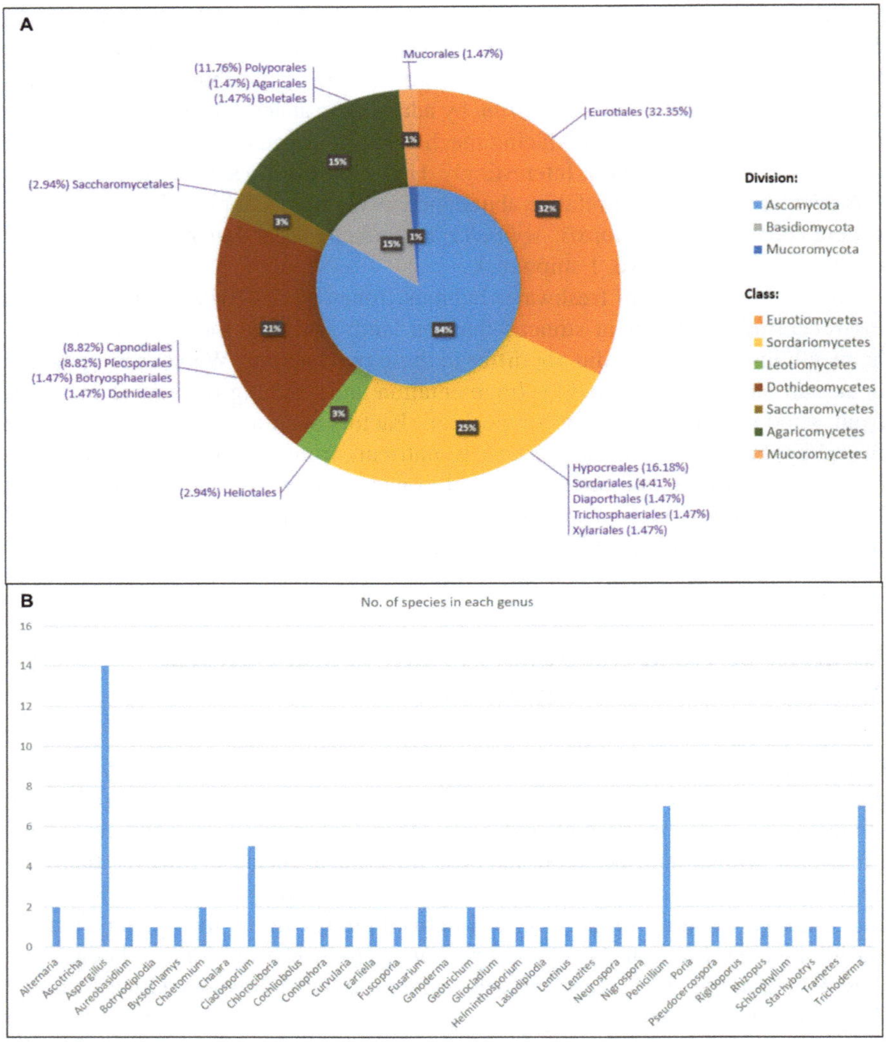

Fig. 20.2 Taxonomic rank distribution of the 64 fungal species retrieved from available records on wooden cultural heritage across Southeast Asia: The inner ring shows Division, the outer ring shows Class, and the Orders (**a**) are labeled in purple. The frequency of fungal species in every genus is based on available records (**b**)

Eurotiales (32.35%), Hypocreales (16.18%), Polyporales (11.76%), Capnodiales (8.82%), Pleosporales (8.82%), Sordariales (4.41%), Heliotales (2.94%), Saccharomycetales (2.94%), Diaporthales (1.47%), Agaricales (1.47%), Boletales (1.47%), Trichosphaeriales (1.47%), Xylariales (1.47%), Botryosphaeriales (1.47%), Dothideales (1.47%), and Mucorales (1.47%). All in all, there are 34 fungal genera and 64 fungal species. Regarding the frequency of fungal species, most wood rot fungal species are under the genus *Aspergillus* (14 species), followed by *Trichoderma* and *Penicillium* (each having seven species), as shown in Fig. 20.2b. Table 20.1, on the other hand, shows the list of the different wood-rotting fungal species in natural and cultural heritage across Southeast Asia.

By nature, fungi are ubiquitous and play an important role in the biosphere as decomposers of a massive range of organic and inorganic substances, underpinning their biodeteriogenic capabilities in cultural artifacts and built environments. The diversity of wood rot fungi is associated with their fungal lifestyle and chemoorganotrophic metabolism, enabling them to adapt and thrive in the often extremophilic conditions associated with indoor and outdoor cultural heritage. These conditions include aridity, salinity, potentially harmful organic and inorganic substances, and ultraviolet radiation. It can also be correlated with other environmental factors such as substrate composition, temperature, pH, and humidity. However, it is interesting to note that much of the wood rot fungal taxa and the microbial community attacking cultural heritage items are still unknown. The culture-dependent method only represents a small account of the diversity of wood rot fungi. Advancement in molecular research—the culture-independent methods, like metagenomic approaches, reveal that culturable fungal taxa are only a small portion of the microbial community responsible for deterioration, as various unculturable fungal taxa and other microorganisms are also present (Montanari et al., 2012; Migliore et al., 2019; Leplat et al., 2020; Gadd et al., 2024).

Factors Affecting Fungal Decay of Wooden Cultural Heritage

Many factors affect the fungal decay of wooden cultural heritage. Gadd et al. (2024) summarized this into three main categories, namely, (i) the nature and chemical composition of the artifacts or built environments, (ii) the macro- and microclimate of the environment and exogenous or endogenous exposure to environmental influences, and (iii) housekeeping management. Moreover, wooden cultural heritage (WCH) comprises organic materials. Accordingly, the climatic conditions outdoors (arid climates, rain or snow, wind blow, UV radiation, air pollution, temperature fluctuation, etc.) and indoors (exposure to light, temperature and humidity fluctuation, dust, flood waters, etc.) exert exogenous or endogenous environmental influences affecting fungal biodeterioration of WCH. Furthermore, housekeeping management (a technique for microenvironmental control—automated or manual, regular monitoring of microenvironmental parameters and the level of biocontamination, the manner and frequency of household management like dusting or surface cleaning, and appropriate ventilation) cannot be underestimated. These factors determine the initiation and progress of the biodeterioration process, triggering fungal growth and colonization.

In addition to the nature of the wood substrate itself, several requirements must be met for colonization to proceed. Generally, when "bound" water saturates the wood cell wall (fiber saturation point or FSP), fungal spores or hyphal fragments

Table 20.1 List of wood-rotting fungal species in natural and cultural heritage across Southeast Asia retrieved from reviewed papers

Fungal species	Natural or cultural heritage	Southeast Asian country	References
Alternaria alternata (Fr.) Keissl.	Books in libraries	Thailand	Chingduang et al. (1995)
Alternaria sp.	Wooden temple	Thailand	Luksamanapha et al. (1995)
Ascotricha guamensis L.M. Ames	Books in libraries	Thailand	Chingduang et al. (1995)
Aspergillus awamori Nakaz.	Wooden artifact	Indonesia	Oetari et al. (2016)
Aspergillus clavatus Desm.	Wooden artifact	Indonesia	Oetari et al. (2016)
Aspergillus flavus Link	Wooden artifact	Indonesia	Oetari et al. (2016)
	Paper and wooden objects	Thailand	Chaisrisook et al. (1995)
	Books in libraries	Thailand	Chingduang et al. (1995)
Aspergillus fumigatus Fresen.	Wooden artifact	Indonesia	Oetari et al. (2016)
	Historical temples	Thailand	Senbua and Wichitwechkara (2019)
	Church doors	Philippines	Fadriquela and De Leon (unpublished)
	Paper and wooden objects	Thailand	Chaisrisook et al. (1995)
	Books in libraries	Thailand	Chingduang et al. (1995)
Aspergillus jensenii Jurjević, S.W. Peterson & B.W. Horn	Wooden artifact	Indonesia	Oetari et al. (2016)
Aspergillus nidulans (Eidam) G. Winter	Paper and wooden objects	Thailand	Chaisrisook et al. (1995)
Aspergillus niger Tiegh.	Timber woods	Malaysia	Teoh et al. (2011)
	Historical museum/ building	Indonesia	Arif et al. (2019)
	Wooden artifact	Indonesia	Oetari et al. (2016)
	Paper and wooden objects	Thailand	Chaisrisook et al. (1995)
	Books in libraries	Thailand	Chingduang et al. (1995)
Aspergillus pulvericola Visagie, Seifert, Frisvad & Samson	Wooden artifact	Indonesia	Oetari et al. (2016)
Aspergillus ruber (Jos. König, Spieck. & W. Bremer) Thom & Church	Wooden artifact	Indonesia	Oetari et al. (2016)
Aspergillus sydowii (Bainier & Sartory) Thom & Church	Books in libraries	Thailand	Chingduang et al. (1995)

(continued)

Table 20.1 (continued)

Fungal species	Natural or cultural heritage	Southeast Asian country	References
Aspergillus terreus Thom	Books in libraries	Thailand	Chingduang et al. (1995)
Aspergillus unguis (Émile-Weill & L. Gaudin) Thom & Raper	Historical temples	Thailand	Senbua and Wichitwechkara (2019)
Aspergillus versicolor (Vuill.) Tirab.	Wooden artifact	Indonesia	Oetari et al. (2016)
Aspergillus sp.	Historical temples	Thailand	Senbua and Wichitwechkara (2019)
	Wooden temple	Thailand	Luksamanapha et al. (1995)
	Paper and wooden objects	Thailand	Chaisrisook et al. (1995)
Aureobasidium pullulans (de Bary & Löwenthal) G. Arnaud	Church doors	Philippines	Fadriquela and De Leon (unpublished)
Botryodiplodia theobromae Pat.	Timber woods	Malaysia	Teoh et al. (2011)
Byssochlamys fulva Oliver & G. Sm.	Church doors	Philippines	Fadriquela and De Leon (unpublished)
Chaetomium globosum Kunze	Books in libraries	Thailand	Chingduang et al. (1995)
Chaetomium sp.	Paper and Wooden Objects	Thailand	Chaisrisook et al. (1995)
Chalara sp.	Wooden temple	Thailand	Luksamanapha et al. (1995)
Chlorociboria sp.	Historical house	Indonesia	Arif et al. (2024)
Cladosporium cladosporioides (Fresen.) G.A. de Vries	Church doors	Philippines	Fadriquela and De Leon (unpublished)
Cladosporium macrocarpum Preuss	Church doors	Philippines	Fadriquela and De Leon (unpublished)
Cladosporium oxysporum Berk. & M.A. Curtis	Books in libraries	Thailand	Chingduang et al. (1995)
Cladosporium sphaerospermum Penz.	Church doors	Philippines	Fadriquela and De Leon (unpublished)
Cladosporium sp.	Wooden temple	Thailand	Luksamanapha et al. (1995)
	Paper and wooden objects	Thailand	Chaisrisook et al. (1995)
Cochliobolus geniculatus R.R. Nelson	Books in libraries	Thailand	Chingduang et al. (1995)
Coniophora sp.	Historical building	Philippines	Fadriquela and Rivarez (2019)
Curvularia lunata (Wakker) Boedijn	Books in libraries	Thailand	Chingduang et al. (1995)

(continued)

Table 20.1 (continued)

Fungal species	Natural or cultural heritage	Southeast Asian country	References
Curvularia sp.	Historical temples	Thailand	Senbua and Wichitwechkara (2019)
	Wooden temple	Thailand	Luksamanapha et al. (1995)
Earliella scabrosa (Pers.) Gilb. & Ryvarden	Timber woods	Malaysia	Teoh et al. (2011)
Fusarium sambucinum Fuckel, Jb. nassau. Ver. Naturk.	Paper and wooden objects	Thailand	Chaisrisook et al. (1995)
Fusarium sp.	Historical temples	Thailand	Senbua and Wichitwechkara (2019)
	Wooden temple	Thailand	Luksamanapha et al. (1995)
Fuscoporia senex (Nees & Mont.) Ghob.-Nejh.	Timber woods	Malaysia	Teoh et al. (2011)
Ganoderma applanatum (Pers.) Pat.	Timber woods	Malaysia	Teoh et al. (2011)
Geotrichum candidum Link	Church doors	Philippines	Fadriquela and De Leon (unpublished)
Geotrichum sp.	Paper and wooden objects	Thailand	Chaisrisook et al. (1995)
Gliocladium sp.	Historical house	Indonesia	Arif et al. (2024)
Helminthosporium sp.	Paper and wooden objects	Thailand	Chaisrisook et al. (1995)
Lasiodiplodia sp.	Wooden temple	Thailand	Luksamanapha et al. (1995)
Lentinus blepharodes Berk. & M.A. Curtis	Timber woods	Malaysia	Teoh et al. (2011)
Lenzites palisotii (Fr.) Fr.	Timber woods	Malaysia	Teoh et al. (2011)
Neurospora sp.	Historical temples	Thailand	Senbua and Wichitwechkara (2019)
Nigrospora sp.	Paper and wooden objects	Thailand	Chaisrisook et al. (1995)
Penicillium citrinum Thom	Wooden artifact	Indonesia	Oetari et al. (2016)
	Books in libraries	Thailand	Chingduang et al. (1995)
Penicillium expansum Link	Church doors	Philippines	Fadriquela and De Leon (unpublished)
Penicillium glabrum (Wehmer) Westling	Church doors	Philippines	Fadriquela and De Leon (unpublished)
Penicillium nalgiovense Laxa	Church doors	Philippines	Fadriquela and De Leon (unpublished)

(continued)

Table 20.1 (continued)

Fungal species	Natural or cultural heritage	Southeast Asian country	References
Penicillium oxalicum Currie & Thom	Wooden artifact	Indonesia	Oetari et al. (2016)
	Books in libraries	Thailand	Chingduang et al. (1995)
Penicillium rubens Biourge	Wooden artifact	Indonesia	Oetari et al. (2016)
Penicillium sp.	Historical temples	Thailand	Senbua and Wichitwechkara (2019)
	Historical house	Indonesia	Arif et al. (2024)
	Paper and wooden objects	Thailand	Chaisrisook et al. (1995)
Poria sp.	Timber woods	Malaysia	Teoh et al. (2011)
Pseudocercospora chiangmaiensis Cheew., K.D. Hyde & Crous	Wooden artifact	Indonesia	Oetari et al. (2016)
Rhizopus sp.	Historical building	Philippines	Fadriquela and Rivarez (2019)
Rigidoporus lineatus (Pers.) Ryvarden	Timber woods	Malaysia	Teoh et al. (2011)
Schizophyllum commune Fr.	Timber woods	Malaysia	Teoh et al. (2011)
Stachybotrys echinatus (Rivolta) G. Sm.	Books in libraries	Thailand	Chingduang et al. (1995)
Trametes versicolor (L.) Lloyd	Timber woods	Malaysia	Teoh et al. (2011)
Trichoderma crassum Bissett	Historical building	Philippines	Fadriquela and Rivarez (2019)
Trichoderma harzianum Rifai	Books in libraries	Thailand	Chingduang et al. (1995)
Trichoderma longibrachiatum Rifai	Wooden artifact	Indonesia	Oetari et al. (2016)
Trichoderma pseudokoningii Rifai	Books in libraries	Thailand	Chingduang et al. (1995)
Trichoderma reesei E.G. Simmons	Historical building	Philippines	Fadriquela and Rivarez (2019)
Trichoderma viride Pers.	Historical building	Philippines	Fadriquela and Rivarez 2019)
Trichoderma sp.	Historical house	Indonesia	Arif et al. (2024)
	Paper and wooden objects	Thailand	Chaisrisook et al. (1995)

Fungal species names are reported in this book based on Index Fungorum

start germinating and growing. However, Brischke and Alfredsen (2021) reviewed that several fungi can start to germinate and grow at around 15% moisture content below the FSP of 30%. The range of optimum moisture content for colonization is 40% to 80%, and then declining fungal growth as it reaches 100% moisture content due to oxygen-limiting conditions (Goodell et al., 2020; Jakes et al., 2020). Water is often the limiting factor in wood fungal decay, as are other factors such as oxygen and temperature. Although oxygen is a rarely limiting factor for fungal growth in wood, complete saturation of logs or very high moisture content can preclude suitable oxygen levels. This is why ponding (submersion of logs in bodies of water) is often employed for logs to limit decay for several months, leaving its pristine condition due to lack of oxygen (Goodell et al., 2020). Temperature is undeniably critical in most physiological reactions. In most fungi, metabolic reaction increases as the temperature increases, with their growth temperature optima between 24 °C and 32 °C. However, exceeding 50 °C, fungal growth is no longer favorable in most, except for thermophilic fungi. Conversely, wood fungal decay stops at freezing temperatures, although some fungi may continue growing even at near-freezing levels (Flyen et al., 2020; Goodell et al., 2020).

Notably, factors such as climate change and the highly "anthropogenic" characteristics that give "cities" a unique environmental condition can favor a new ecological niche jump (unnatural ecological role), i.e., the spread of certain fungal species and the inhabitants of synthetic materials made by humans. Global climate change significantly lowers the accuracy of weather forecasting models locally and worldwide. Some of the effects of the ecosystem-level changes are unpredictable, affecting specific microbial communities in the built and natural environments (Ma et al., 2020; Gadd et al., 2024). Gadd et al. (2024) presented one example—the fungus *Eurotium halophilicum* C.M. Chr., Papav. and C.R. Benj, an obligate xerophilic ascomycete, was first described to be associated with foods with low moisture content. However, *E. halophilicum* is increasingly linked to monospecific infections in libraries, archives, and museums. Understanding the different factors governing the biodeterioration of wooden cultural heritage may give insight into a possible change in certain taxa's ecological niche and biogeography.

Conservation Strategies on Cultural Heritage Across Southeast Asia Against Wood Rot and Fungal Decay

Limited scholarly research exists on biodeterioration due to wood rot and fungal decay of cultural properties in Southeast Asia, and even less on the measures of conserving them. However, one cannot discount the possibility that there could be a wealth of related publications in the vernacular language that this study could not access. Notwithstanding, this section attempts to give an overview of the conservation approaches of some of the countries in the region.

Wooden archaeological and anthropological artifacts, and architectural and structural works, are finite and irreplaceable resources that must be conserved for the meanings, identities, and inspiration they provide for present and future generations. Their conservation is a multidisciplinary endeavor that involves all operations of preventing, retarding, and consolidating physical loss or deterioration, as well as all processes of protecting, keeping, and respecting the aesthetic, historic, scientific, social, or spiritual significance ascribed to them. Conservation works entail keeping their authenticity and integrity throughout their existence.

Traditional Knowledge in Wood Conservation

However, as historic wooden artifacts and structures are imbued with rich traditions, conservation is not only a process after their fabrication or construction but also an undertaking that commenced from their inception with the selection of wood species and construction techniques that are best suited to withstand the agents of deterioration through time. Moreover, traditional knowledge has also played an important role in giving insights into when timber materials will be sourced. It also provides information on what organic materials can be utilized as fungicides.

Felling Season

There is a Filipino belief that trees should be cut in the moonlight to prevent trunk infestation and during the dry season when they are strongest and most resistant to termites (Fadriquela, 2013). A 1735 manual on architecture by Fray Juan Albarrán, who built the Sto. Niño church in Cebu prescribes similar measures in felling trees based on the teachings of Vitruvius (José & Fadriquela, 2022).

Likewise, the traditional wisdom of the Minangkabau society in Indonesia for preparing good-quality wood for the traditional house *rumah gadang* necessitated that the selected tree should not be in a blossoming or flowering state nor should it be felled in the rainy season (Puteri, 2018). In Cambodia, cutting of wood was also preferably done during the dry season; like monks or species of the vegetable kingdom, trees "follow" the annual "retreat of the rainy season," the vassā, and cutting during this period was considered inauspicious (Tainturier, 2006).

The aforementioned traditional knowledge of the season, when trees ought to be felled, can be correlated to the seasonal contents of moisture and starch in wood, which are nutrients for fungi and xylophagous insects to thrive.

Wood Species

Choosing the most appropriate timber is also paramount in conserving wooden structures. *Molave* (*Vitex parviflora* Juss.) was the most utilized wood in constructing houses for dwellings, places of worship, and afterlife vessels in the Philippines from the pre-Hispanic period until the end of the nineteenth century (Fadriquela, 2013). Apart from its availability due to its wide distribution in the country, this preference is primarily due to the natural durability of *molave* to moisture incursion because of tyloses that occlude its cell cavities.

Kòkir (*Hopea odorata* Roxb.) is a precious timber in Cambodia that can only be utilized for monastery buildings and royal residences, such that the *kòkir* columns from a deteriorated *vihāra* could not be repurposed, especially as house posts (Tainturier, 2006). *Kòkir* wood is close and even-grained, hard and heavy, strong, durable, and resistant to termites and other wood-eating organisms (Fern, 2024).

In Vietnam, archaeological waterlogged *lim xanh* (*Erythrophleum fordii* Oliver) wood artifacts that were commonly excavated from the Thang Long Imperial Citadel site were found highly resistant to degradation, and this is due to the abundance of condensed type lignin and thick wall wood fibers with closed fiber lumina of *lim xanh* (Nguyen, 2018). *Lim xanh* is an indigenous tree in Vietnam, and it is also known there as 'ironwood' (Sein & Mitlöhner, 2011).

Most of the Buddhist monasteries, also called *kyaung* or *pongyi kyaung* in Myanmar, were built using teak (*Tectona grandis* L. f.) (Daw Soe Soe Win, 2008). *Joglo,* which is a traditional Javanese wooden house, also employs teak (Mida, 2014). Teak is the most valuable and important material for construction of wooden shops and residential houses in the Trok Baan Jeen community in Thailand (Khwansuwan, 2012). Valued for its durability, beauty, and mechanical resistance, teak is a timber species with high resistance to biological deterioration (Cárcamo-Ibarra et al., 2022). Its resistance to fungal decay can be attributed to the synergistic effect of various extractive compounds such as anthraquinones, naphthoquinones, and tectoquinones (Broda, 2020).

Natural Fungicides

In Malaysia, a study (Barkesli & Halim, 2014) was conducted on the efficacy of garlic (*Allium sativum* L.) and vinegar, which are traditionally applied in timber buildings as a fungicide for wood staining and mold fungi. *Aspergillus niger* Tiegh. is used as a subject of an experiment, which is very commonly found on deteriorated wooden surfaces and on artifacts displayed where temperature and humidity are difficult to control, especially in a country like Malaysia, where rain is present all year round. Traditional methods for making garlic extract showed that after 14 days, the color of the garlic extract changed from whitish yellow to light brown, and it imparted a terrible odor. However, vinegar did not show any change in color or smell. *A. niger* rapidly grows on potato dextrose agar (PDA) (control set) within 2 weeks, while there is no sign of growth on the medium of PDA + vinegar and PDA + garlic extract (treated set). The study recommends exploring how to improve the quality of these natural fungicides to minimize the use of inorganic and synthetic organic fungicides used in conservation treatment, especially for wooden cultural heritage.

Researchers from Thailand and Japan explored the possibility of using medicinal plants and spices to inhibit fungal growth, which can be utilized as a biological control measure (Chinguduang et al., 1995). Seven medicinal plants and spices, i.e., *Kaempferia galanga* L., *Amomum cardamomum* L., *Carum carvil* L. (caraway seed), *Cinnamomum iners* Bl. (cinnamon), *Myristica fragrans* Houtt. (nutmeg),

Eugenia carayphyllus Bullock and Harrison (clove), and *Cariandrum sativum* L. (coriander) were evaluated against cultures of *Aspergillus flavus*, *A. versicolor (Vuill.) Tirab* (Vuill.) Tirab, *A. niger*, *Penicillium citrinum* Thom, *P. viridicatum* Westling and *Cladosporium cladosporioides* (Fresen.) G.A. de Vries grown on agar media in Petri dishes. The results showed that clove was the most promising species, which could completely inhibit all tested fungi, while *Kaempferia galanga* was the second tentative species. *P. viridicatum* was the most sensitive fungus to all medicinal plant species, with some medicinal plants, particularly coriander, enhancing the growth of some fungi.

The traditional water soaking formula consisting of cloves (*Syzgium aromaticum* (L.) Merr. and Perry), tobacco (*Nicotiana tabacum* L.), and banana (*Musa paradisiaca* L.) stems are still being used for cleaning the *joglos* in Indonesia (Mida, 2014; Mahirta, 2012). Clove contains flavonoids, eugenol, eugenol acetate, and B-varyophyllene, terpenoids, and phenolic compounds that are anti-microbial and insecticidal, while alkaloids in tobacco can inhibit insect development (Mida, 2014; Mahirta, 2012). However, although a study showed that banana pseudo-stem extract showed no impact on many fungal species, banana leaves showed considerable antifungal activity, with the widest inhibition zone diameters against species such as *Penicillium*, *Candida*, and *Alternaria* sp. (Mostafa, 2021).

Teak buildings in Myanmar were coated with "earth oil," which is crude petroleum oil, as protection from the sun, as well as fungal and insect attacks (Christian, 2020). This coating made the teak buildings exhibit a dark brown color. Despite serving its purpose as a sunblock and inhibiting fungal activity, termites can attack wood surfaces that were not treated, such as the end grain of the columns.

Traditional lacquer found in wax tree (*Rhus succcedanea* L.) is applied to wooden buildings in Vietnam (Do, 2008). This lacquer is not only for decoration but also for preservation as it has the ability to resist water, moss, and fungi. For the background lacquer, raw lacquer is mixed with calcium carbonate in Ha Tay (northern part of Vietnam) or clay in Hue. Preservation lacquer is prepared with Trau oil (*Vernicia montana* Lour.) in 20% of the mass in Ha Tay or 10% Trau oil and 10% colophony in Hue.

Structural Repair Methods As the mechanical properties of rotten wood decline, structural wooden components may no longer be viable to withstand the load and warrant replacement, but repair must be endeavored. Rotten parts are removed in a repair and consolidated using epoxy resin to impart strength when required. Prior to consolidation, the components are fumigated and applied with wood preservatives. Use of the same wood species is preferred for new wooden components for compatibility with physical and mechanical properties and to preserve the historic structure's authenticity.

Puteri (2018) reported that weathered and rotten timbers in a *rumah gadang* are usually repaired and replaced in Indonesia. The rotten and termite-infested *molave* timber posts at the portico of the nineteenth-century Teus Mansion in Manila had to be repaired by removing the rotten part and replacing it with new timber of the same *molave* wood species (Manalo, 2025).

The Vietnamese standard prescribes the use of *kiền kiền* (*Hopea pierrei* Hance) and *lim xanh* for preserving and restoring the timber structures of the Hue royal architectural monuments (Duong, 2018). Do (2008) reports that the traditional method for filling cracked and rotten timber surfaces in Hue utilizes a composite of natural lacquer mixed with wood powder, which requires time, money, and good skill. As such, epoxy resin is also being used, which is easier to use than traditional lacquer. However, he cautions that cracks may appear in some parts of the filling epoxy, and the joinery becomes weak after 10 years of restoration.

The use of epoxy was also reported in the repair of wooden structural members in Thai roofing (Ketkul, 2003). The mixture of epoxy with methanol, a hardener called "10% cobalt naphthenate" and a filler (diatomaceous earth) is used, and this can be diluted to facilitate the ease of work. For consolidating the original wood to the repair wood with epoxy, it must be fixed with stainless steel rods or glass fiber rods.

The National Historical Institute (now referred to as the National Historical Commission of the Philippines) recommends the use of an array of chemicals as fumigants, which include carbon disulfide and thymol, and 10% thymol in alcohol or ether as fungicide solution for wood conservation (Fadriquela, 2013). The same institution proposes the mixture of 1 part linseed oil and 1 part mineral spirit with 1 part 7% acetic acid and ¼ part ethanol or methanol as wood preservatives (Fadriquela, 2013).

A report on the preservation of a traditional folk house built using "ironwood" in Bac Ninh province in Vietnam indicated the use of wood preservatives for structural members attacked by agents of biodeterioration (Nguyen, 2004). For such components attacked by worms, moths, and fungi, multiple applications by brushing of PBB, LN2, LN5, or SOLTOX from Poland and TERMIDOC produced by ROHONE POULENC in France were prescribed.

The painted main façade doors of Malate Church in the Philippines were restored by dismantling them from the façade (Escuela Taller, n.d.). Before dismantling, the fungal species causing wood deterioration were investigated. Wood species were then determined after the disassembly of the timber components of each panel. Upon documentation, disassembled mostly *yakal* (*Shorea astylosa* Foxw.) timber components were stripped of paint, fumigated using the food-grade aluminum phosphide, and applied with a commercially available wood preservative. The paint was removed to reveal the natural color of the wooden components, as desired by the Columban fathers. Components that could no longer be consolidated were replaced with timbers of the same or compatible wood species. The consolidated components were reassembled and reintegrated into the door panels, brushed with three coats of paraffin wax. Paraffin wax has the highest moisture exclusion property among finishes tested by the US National Park Service.

Wood Derived Materials

Apart from timber components of historic buildings and artifacts, wood and other plant-derived materials, such as paper and books, are susceptible to fungal attack. It is a common concern for all institutions that care for them, such as libraries, archives, and museums.

The National Archives of Malaysia has employed a solution of 100 g of thymol in one liter of methylated spirit that is either sprayed or brushed on archival materials (Talib, 1989). Thymol vapor in a gas-tight chamber, which was slightly warmed with a heat bulb, was utilized to address fungal attacks in ancient Thai books or *Samut Thai* (Aranyanak, 1988). Aranyanak (1988) also reported that in some libraries, strips of thymol-impregnated paper are used for interleaving between infected papers and are kept inside the storage cabinet to act as a preventive measure against fungal growth.

Conclusion

Cultural heritage, whether a natural monument, church, furniture, sculpture, or instrument, reflects the true value of cultural formation processes and the positive expression of historical tradition and constitutes one's national characteristics passed from generation to generation. Cultural heritage plays a significant role in people's lives as it reflects their social values, religions, beliefs, and traditions.

A significant portion of these natural and cultural monuments and items is made up of wood or related materials, which are considerably durable cellulosic materials, making them prone to simultaneous degradation, posing challenges to their preservation and conservation. These priceless treasures are threatened by biodeterioration, such as fungal decay, especially on wooden natural or cultural monuments and religious items.

Fungal decay or wood rot in wooden natural and cultural monuments and artifacts across Southeast Asia is caused by fungal species generally under three divisions, namely, Ascomycota, Basidiomycota, and Mucoromycota. Their biodeteriogenic capabilities in cultural artifacts and built environments are attributed to the diverse fungal species' lifestyle and chemoorganotrophic metabolism, enabling them to adapt and thrive in the often extremophilic conditions associated with indoor and outdoor cultural heritage.

Though there are few published studies on the biodeterioration due to wood rot and fungal decay of cultural properties in Southeast Asia and even fewer on the measures of conserving them, one cannot discount the possibility that there could be a wealth of related publications in the vernacular language that this study was not able to access. Traditional knowledge has played an important role in conserving historic wooden structures, giving insights into when and what types of trees will be felled. It also provides information on what organic materials can be utilized as fungicides. Repairs and use of an array of chemicals as fumigants, which include carbon disulfide and thymol, and 10% thymol in alcohol or ether as fungicide solution, are also being applied for wood conservation.

Natural, historical, or cultural monuments and their architectural wonders are intricately detailed and rich in historical milieu; thus, a need for more extensive research on the microbial community and its complexity to fully understand the biodegradative action and applicable and efficient conservation of these rich cultural treasures for appreciation of the next generation.

References

Abe, T., Hoshino, T., Nakamura, A., & Takaya, N. (2007). Anaerobic elemental sulfur reduction by fungus *Fusarium oxysporum. Bioscience, Biotechnology, and Biochemistry, 10*, 2402–2407.

Aranyanak, C. (1988). Conservation of ancient Thai books. *Studies in Conservation, 33*, 22–24.

Arif, A., Syahidah, & Ilyas, M. M. (2019). Wood museum biodeterioration of karaeng patingalloang benteng somba opu, Gowa regency. *IOP Conference Series: Earth and Environmental Science*, 343.

Arif, A., Putri, G., Soma, A. S., Syahidah, & Syahribulan. (2024). Biological deterioration of wooden components of Balla Lompoa Ri Galesong in South Sulawesi, Indonesia. *Jurnal Sylva Lestari, 12*, 580–594.

Avdanina, D. A., & Zhgun, A. A. (2024). Rainbow code of biodeterioration to cultural heritage objects. *Heritage Science, 12*, 1–24.

Bari, E., Daniel, G., Yilgor, N., Kim, J. S., Tajick-Ghanbary, M. A., Singh, A. P., & Ribera, J. (2020). Comparison of the decay behavior of two white-rot fungi in relation to wood type and exposure conditions. *Microorganisms, 12*, 1–20.

Barkesli, M., & Halim, N. A. (2014). Scientific analysis in traditional preventive measure using garlic and vinegar as a wood fungicide in Malaysia. In *Proceedings of the 7th international conference on the biodeterioration of cultural property*. International Council for Biodeterioration of Cultural Property.

Brischke, C., & Alfredsen, G. (2021). Wood-water relationships and their role for wood susceptibility to fungal decay. *Applied Microbiology and Biotechnology, 105*, 3787–3795.

Broda, M. (2020). Natural compounds for wood protection against fungi – A review. *Molecules, 15*, 1–24.

Cárcamo-Ibarra, E., Martínez-Pacheco, M. M., Munro-Rojas, A., Ambriz-Parra, J. E., & Velázquez-Becerra, C. (2022). Antifungal activity of crude extracts of *Tectona grandis* L.f. against wood decay fungi. *Phyton-International Journal of Experimental Botany, 91(8)*, 1795–1808.

Chaisrisook, C., Suwanarit, P., & Aranyanak, C. (1995). Fungal deterioration of paper in the national libraries and of wooden objects in the National Museum Bangkok, Thailand. In *3rd international conference on biodeterioration of cultural property. Preprints* (pp. 461–471). The organizing Committee of ICBCP-3.

Chingduan, S., Siriacha, P., & Saito, M. (1995). Effect of some medicinal plants and spices on growth of fungi. In *3rd international conference on biodeterioration of cultural property. Preprints*. The organizing Committee of ICBCP-3.

Christian, R. (2020). *Challenges of teak architecture conservation in Myanmar* (Architectural Conservation Working Paper Series, No. 01).

Chu, S., Li, Y., Wang, X., Li, N., Song, J., & Lanying, L. (2025). Degradation condition and microbial analysis of waterlogged archaeological wood from the second shipwreck site on the northwestern continental slope of the South China Sea. *npj Heritage Science, 13*, 1–11.

Corbu, V. M., Gheorghe-Barbu, I., Dumbravă, A. Ş., Vrâncianu, C. O., & Şesan, T. E. (2023). Current insights in fungal importance – A comprehensive review. *Microorganisms, 11*, 1–52.

Dán, K., Kocsubé, S., Tóth, L., Farkas, A., Rákhely, G., & Galgóczy, L. (2024). Isolation and identification of fungal biodeteriogens from the wall of a cultural heritage church and potential applicability of antifungal proteins in protection. *Journal of Cultural Heritage, 67*, 194–202.

Daw Soe Soe Win. (2008). Problems and needs for cultural heritage protection and restoration activities in Myanmar. In *Training Course on Cultural Heritage Protection in the Asia Pacific Region 2007: Preservation and restoration of wooden structures* (pp. 80–91). Cultural Heritage Protection Cooperation Office, Asia/Pacific Cultural Centre for UNESCO.

de la Roche, I. A., O'Connor, J., & Têtu, P. (2003). *Wood products and sustainable construction*. XII World Forestry Congress.

Deacon, J. (2005). Wood decay and wood rooting fungi. University of Edinburgh Printing Press.

Ding, S., Hu, H., & Gu, J.-D. (2020). Diversity, abundance, and distribution of wood-decay fungi in major parks of Hong Kong. *Forests, 11*, 1–16.

Do, H. T. (2008). Some thoughts on the consolidation and preservation of wooden structures in Hue. In *Training course on Cultural Heritage Protection in the Asia Pacific Region 2007: Preservation and restoration of wooden structures* (pp. 211–220). Cultural Heritage Protection Cooperation Office, Asia/Pacific Cultural Centre for UNESCO.

Duong, N. A. (2018). Some difficulties of preservation and restoration of wooden structures in the Hue's royal architectural monuments. In *Training course on Cultural Heritage Protection in the Asia-Pacific Region 2017: Preservation and restoration of wooden structures*. Cultural Heritage Protection Cooperation Office, Asia-Pacific Cultural Center for UNESCO (ACCU).

El-Sayed, M. H., AlHarbi, M., Elsehemy, I. A., Haggag, W. M., Refaat, B. M., Ali, S. M., & Elkelish, A. (2024). Natural inhibitory treatment of fungi-induced deterioration of carbonate and cellulosic ancient monuments: Isolation, identification and simulation of biogenic deterioration. *Journal of Microbiology and Biotechnology, 34*, 2049–2069.

Escuela Taller de Filipinas Foundation Inc. *Restoration of Malate Church Doors* (unpublished).

Fadriquela, C. (2013). *Kahoy: Wood in the Philippines, wood and its uses from Pre-Hispanic to Spanish colonial Philippines* (p. 400). UST Publishing House.

Fadriquela, C. S., & De Leon, M. P. *Identification of fungal (Molds) isolates from the Malate Church Main Façade Doors* (unpublished).

Fadriquela, C. S., & Rivarez, M. P. (2019). Morpho-cultural identification of wood-damaging fungi in the historic baker memorial hall at the University of the Philippines, Los Baños. *Phillipp Agric Scientist, 102*, 340–349.

Fern, K. (2024). *Tropical plants database*. tropical.theferns.info. 2025-03-03. tropical.theferns.info/viewtropical.php?id=Hopea%20

Flyen, A. C., Flyen, C., & Mattson, J. (2020). Climate change impacts and fungal decay in vulnerable historic structures at Svalbard. *E3S Web of Conference, 172*, 1–6.

Gadd, G. M., Fomina, M., & Pinzari, F. (2024). Fungal biodeterioration and preservation of cultural heritage, artwork, and historical artifacts: Extremophily and adaptation. *Microbiology and Molecular Biology Reviews, 88*, 1–62.

Goodell, B., Qian, Y., & Jellison, J. (2008). Fungal decay of wood: Soft rot – Brown rot – White rot. In T. P. Schultz et al. (Eds.), *Development of commercial wood preservatives* (pp. 9–31). ACS Publications.

Goodell, B., Winandy, J. E., & Morrell, J. J. (2020). Fungal degradation of wood: Emerging data, new insights and changing perceptions. *Coatings, 10*, 2–19.

Han, Y., Huang, X., Wang, Y., Du, J., Ma, K., Chen, Y., Li, N., Zhang, Z., & Pan, J. (2021). Fungal community and biodeterioration analysis of hull wood and its storage environment of the Nanhai No. 1 Shipwreck. *Frontiers in Microbiology, 11*, 1–14.

He, X., & Zhao, C.-L. (2022). Diversity of wood-decaying fungi in Wuliangshan Area, Yunnan Province, P.R. China. *Diversity, 14*, 1–29.

Isola, D., Lee, H.-J., Chung, Y.-J., Zucconi, L., & Pelosi, C. (2024). Once upon a time, there was a piece of wood: Present knowledge and future perspectives in fungal deterioration of wooden cultural heritage in terrestrial ecosystems and diagnostic tools. *Journal of Fungi, 10*, 1–63.

Jakes, J. E., Zelinka, S. L., Hunt, C. G., Ciesielski, P., Frihart, C. R., Yelle, D., Passarini, L., Gleber, S. C., Vine, D., Vogt, S. (2020). Measurement of moisture-dependent ion diffusion constants in wood cell wall layers using time-lapse micro X-ray fluorescence microscopy. *Scientific Reports, 10*(1):9919. https://doi.org/10.1038/s41598-020-66916-8. PMID: 32555373; PMCID: PMC7303177.

Jones, J. M., Heath, K. D., Ferrer, A., Brown, S. P., Canam, T., & Walling, J. W. (2019). Wood decomposition in aquatic and terrestrial ecosystems in the tropics: Contrasting biotic and abiotic processes. *FEMS Microbiology Ecology, 95*, 1–13.

José, R. T., & Fadriquela, C. (2022). An 18th century manual on architecture: Fray Juan Albarrán's *Barias Reglas de Arquitectura*, 1735 (Part Two). *Philippiniana Sacra, 57*, 529–556.

Ketkul, V. (2003). Characteristics of Thai houses and problems in cultural heritage protection activities in Thailand. In *Training course on Cultural Heritage Protection in Asia and the Pacific 2002: Preservation and restoration of wooden structures* (pp. 90–95). Cultural Heritage Protection Cooperation Office, Asia/Pacific Cultural Centre for UNESCO.

Khwansuwan, P. (2012). Gaps in the conservation of traditional communities in Thailand. In *Training course on Cultural Heritage Protection in the Asia Pacific Region 2011: Preservation and restoration of wooden structures* (pp. 200–213). Cultural Heritage Protection Cooperation Office, Asia/Pacific Cultural Centre for UNESCO.

Kirker, G. T., Hassan, B., Mankowski, M. E., & Eller, F. J. (2024). Critical review on the use of extractives of naturally durable woods as natural wood protectants. *Insects., 15*, 1–16.

Leplat, J., François, A., & Bousta, F. (2020). *Parengyodontium album*, a frequently reported fungal species in the cultural heritage environment. *Fungal Biology Reviews, 34*, 126–135.

Luksamanapha, P., Sardsud, U., & Sukchotiratana, M. (1995). Isolation of molds from Ton Gwen Wooden Temple. In *3rd international conference on biodeterioration of cultural property. Preprints* (pp. 369–374). The organizing Committee of ICBCP-3.

Ma, J., Zhou, L., Foltz, G. R., Qu, X., Ying, J., Tokinaga, H., Mechoso, C. R., Li, J., & Gu, X. (2020). Hydrological cycle changes under global warming and their effects on multiscale climate variability. *Annals of the New York Academy of Sciences, 1472*, 21–48.

Mahirta. (2012). Protection of Rumah Pencu in Kudus (Northern Central Java, Indonesia). In *Training course on Cultural Heritage Protection in the Asia Pacific Region 2011: Preservation and restoration of wooden structures* (pp. 85–99). Cultural Heritage Protection Cooperation Office, Asia/Pacific Cultural Centre for UNESCO.

Manalo, M. (2025). *As informed through private communication*.

Maravilla, N. (1995). Biodeterioration of Paper. In *3rd International conference on Biodeterioration of cultural property. Preprints*. The Organizing Committee of ICBCP-3.

Mida, A. (2014). Conservation of traditional wooden houses (Joglo) at Kotagede, Yogyakarta, Indonesia. In *Revisiting the philosophy of preserving wooden structures: Restoration method for wooden structures and its philosophy* (pp. 69–77). Cultural Heritage Protection Cooperation Office, Asia/Pacific Cultural Centre for UNESCO.

Migliore, L., Perini, N., Mercuri, F., Orlanducci, S., Rubechini, A., & Thaller, M. C. (2019). Three ancient documents solve the jigsaw of the parchment purple spot deterioration and validate the microbial succession model. *Scientific Reports, 9*, 1–13.

Mittal, N., Ansari, F., Gowda, V. K., Brouzet, C., Chen, P., Larsson, P. T., & Söderberg, L. D. (2018). Multiscale control of Nanocellulose Assembly: Transferring remarkable nanoscale fibril mechanics to macroscale fibers. *ACS Nano, 12*, 6378–6388.

Montanari, M., Melloni, V., Pinzari, F., & Innocenti, G. (2012). Fungal biodeterioration of historical library materials stored in compactus movable shelves. *International Biodeterioration and Biodegradation, 75*, 83–88.

Mostafa, H. S. (2021). Banana plant as a source of valuable antimicrobial compounds and its current applications in the food sector. *Journal of Food Science, 86*(9), 3778–3797.

Nguyen, A. D. (2004). Pilot investigation, research, preservation and conservation of traditional folk houses of Vietnam. In *Training course on Cultural Heritage Protection in the Asia-Pacific Region 2004: Preservation and restoration of wooden structures* (pp. 132–138). Cultural Heritage Protection Cooperation Office, Asia/Pacific Cultural Centre for UNESCO.

Nguyen, D. T. (2018). Study on conservation of archaeological waterlogged wood in Vietnam. *Sustainable Humanosphere, 14*, 21.

Ochoa, A. K. (1978). Wood-rotting by lignin- degrading fungi. A thesis manuscript submitted for publication. University of the Philippines, Los Banos, Laguna.

Oetari, A., Susetyo-Salim, T., Sjamsuridzal, W., Suherman, E. A., Monica, M., Wongso, R., Fitri, R., Nurlaili, D. G., Ayu, D. C., & Teja, T. P. (2016). Occurrence of fungi on deteriorated old Dluwang manuscripts from Indonesia. *International Biodeterioration and Biodegradation, 114*, 94–103.

Oliveira, M., Reis, J., & Almaguer, M. (2023). Fungal diversity and heritage. *Preprint, 2023100824.*

Organisation for Economic Co-Operation And Development. (2009). *The Bioeconomy for 2030.* Available online at www.oecd.org/publishing/corrigenda

Park, J.-H., Pavlov, I. N., Kim, M.-J., Park, M. S., Oh, S.-Y., Park, K. H., Fong, J. J., & Lim, Y. W. (2020). Investigating wood decaying fungi diversity in Central Siberia, Russia using ITS sequence analysis and interaction with host trees. *Sustainability, 12*, 1–13.

Polkinghorne, M., Pearson, N., van Duivenvoorde, W., Nayati, W., Tahir, Z., Ridwan, N. N. H., Forrest, C., Tan, N. H., Popelka-Filcoff, R., Morton, C., Kowlessar, J., & Staniforth, M. (2024). Reuniting orphaned cargoes: Recovering cultural knowledge from salvaged and dispersed underwater cultural heritage in Southeast Asia. *Marine Policy, 163*, 106074.

Pournou, A. (2020). Biodeterioration of wooden cultural heritage. In *Organisms and decay mechanisms in aquatic and terrestrial ecosystems* (p. 538). Springer.

Puteri, T. N. H. (2018). Problems and needs for the preservation of Rumah Dadan in West Sumatra Province, Indonesia. In *Training course on Cultural Heritage Protection in the Asia-Pacific Region 2017: Preservation and restoration of wooden structures* (pp. 68–73). Nara.

Qi, J., Li, F., Jia, L., Zhang, X., Deng, S., Luo, B., Zhou, Y., Fan, M., & Xia, Y. (2023). Fungal selectivity and biodegradation effects by white and brown rot fungi for wood biomass pretreatment. *Polymers, 15*, 1–15.

Savković, Ž., Stupar, M., Unković, N., Knežević, A., Vukojević, J., & Ljaljević Grbić, M. (2022). *Fungal deterioration of cultural heritage objects*. IntechOpen.

Schmidt, O. (2006). Wood and tree fungi. Biology, damage, protection, and use. Springer, Berlin, pp. 343.

Schmidt, O. (2007). Indoor wood-decay basidiomycetes: damage, causal fungi, physiology, identification and characterization, prevention and control. *Mycol. Progress, 6*, 261–279.

Secretariat of the Advisory Committee on Sustainable Forest-based Industries, Food and Agriculture Organization. (2020). *Status of public policies encouraging wood use in construction – An overview.*

Sein, C. C., & Mitlöhner, R. (2011). *Erythrophloeum fordii Oliver: Ecology and silviculture in Vietnam. Bogor, Indonesia* (pp. 1–8). CIFOR.

Senbua, W., & Wichitwechkarn, J. (2019). Molecular identification of fungi colonizing art objects in Thailand and their growth inhibition by local plant extracts. *3 Biotech, 9*, 1–10.

Talib, N. (1989). Conservation of National Archives of Malaysia. In *SPAFA-ICCROM seminar on Conservation standards in Southeast Asia, National Museum* (Final Report) (pp. 41–50). SPAFA Regional Centre.

Tanturier, F. (2006). Building in wood: Notes on a vanishing age-old tradition. In *Wooden architecture of Cambodia: A disappearing heritage* (pp. 12–37). Center for Khmer Studies.

Teoh, Y. P., Don, M. M., & Ujang, S. (2011). Assessment of the properties, utilization, and preservation of rubberwood (*Hevea brasiliensis*): A case study in Malaysia. *Journal of Wood Science, 57*, 255–266.

Tesoro, F. (1987). *Tropical limbers for construction in the ASEAN countries* (Unasylva – No. 156 – Tropical Rain Forest Management a Status Report) (Vol. 39).

Unger, A., Schniewind, A. P., & Unger, W. (2001). Biological Deterioration of Wood. In: Conservation of Wood Artifacts. Natural Science in Archaeology. Springer, Berlin, Heidelberg. https://doi.org/10.1007/978-3-662-06398-9_6

Wang, Y., Han, Y., Li, N., Wang, C., Ma, K., Huang, X., Du, J., Guo, H., & Pan, J. (2024). Study on biodegradation mechanism of *Fusarium solani* NK-NH1 on the hull wood of the Nanhai No. 1 shipwreck. *Frontiers in Microbiology, 15*, 1–13.

MIX
Papier aus verantwortungsvollen Quellen
Paper from responsible sources
FSC® C105338

If you have any concerns about our products,
you can contact us on
ProductSafety@springernature.com

In case Publisher is established outside the EU,
the EU authorized representative is:
Springer Nature Customer Service Center GmbH
Europaplatz 3, 69115 Heidelberg, Germany

Printed by Libri Plureos GmbH
in Hamburg, Germany